VISCOUS FLOW

GW00597427

McGraw–Hill Series in Mechanical Engineering

Jack P. Holman, *Southern Methodist University*
Consulting Editor

VISCOUS FLOW

Frederick S. Sherman

University of California at Berkeley

McGraw-Hill Publishing Company

New York St. Louis San Francisco Auckland Bogotá Caracas
Hamburg Lisbon London Madrid Mexico Milan Montreal
New Delhi Oklahoma City Paris San Juan São Paulo
Singapore Sydney Tokyo Toronto

VISCOUS FLOW
INTERNATIONAL EDITION 1990

Exclusive rights by McGraw-Hill Book Co. – Singapore
for manufacture and export. This book cannot be re-exported
from the country to which it is consigned by McGraw-Hill.

3 4 5 6 7 8 9 0 CMO PMP 9 5 4 3 2

Copyright © 1990 by McGraw-Hill, Inc. All rights reserved.
Except as permitted under the United States Copyright Act of
1976, no part of this publication may be reproduced or
distributed in any form or by any means, or stored in a data
base or retrieval system, without the prior written
permission of the publisher.

This book was set in Times Roman.
The editors were John Corrigan and John M. Morriss;
the production supervisor was Denise L. Puryear.
The cover was designed by Karen Quigley.
Project supervision was done by the Universities Press.

Jacket photo credit: photo is a "coating flow" encountered in the
making of color film, made available by the kindness of Peter M.
Schweizer, ILFORD AG., Fribourg, Switzerland.

Library of Congress Cataloging-in-Publication Data

Sherman, Frederick S., (date).
    Viscous flow / Frederick S. Sherman.
    p.   cm.—(McGraw-Hill series in mechanical engineering)
    Includes index.
    ISBN 0-07-056579-1
    1. Fluid dynamics.   2. Viscous flow.   I. Title.   II. Series.
TA357.S453   1990
620.1'064—dc20   89-12106

When ordering this title use ISBN 0-07-100928-0

Printed in Singapore

ABOUT THE AUTHOR

Frederick Sherman was an undergraduate at Harvard, where Howard Emmons introduced him to the mysteries of fluid mechanics in 1949. He returned to his home state for graduate study, and in 1954 received the PhD. in Mechanical Engineering at the University of California at Berkeley. He has been on the faculty at Berkeley ever since, except for $2\frac{1}{2}$ years with the Mechanics Branch of the Office of Naval Research in the late 1950s. He is currently Professor of Mechanical Engineering.

Until the mid 1950s, his major research interests concerned high-speed rarefied gas flows and flight in the upper atmosphere. He then studied meteorology and oceanography, with a special interest in the mixing of stably stratified flows, while retaining an interest in transonic throat flows and hypersonic free jets. In recent years, he has been particularly interested in hydrodynamic instability, free convection, unsteady boundary layers, and the computational simultation of viscous flows by the Random Vortex Method.

Teaching, especially of fluid dynamics of all sorts, is his professional true love. He has received awards for distinguished teaching from the Academic Senate of U.C. at Berkeley, and from the U.S. and California Societies of Professional Engineers. The teaching of viscous flow has long been a favorite challenge.

CONTENTS

7 Flows with Nearly Constant Density and Transport Properties

8 Vorticity

9 Analytical Solutions of the Full Navier–Stokes Equations

10 Numerical Solutions of the Full Navier–Stokes Equations

11 Creeping Flows

12 Laminar Boundary Layers

13 Instability of Viscous Flows

14 Turbulent Flow

Appendix A Mathematical Aids

Appendix B FORTRAN Programs

Index

PREFACE

This text has grown out of a series of courses, given at the University of California, Berkeley, to seniors and graduate students who had adopted fluid mechanics or heat transfer as major fields of study. The courses were presented in the Department of Mechanical Engineering, but also served considerable numbers of students from Naval Architecture, Civil Engineering, and Chemical engineering.

Our courses in fluid mechanics dealt, more or less separately, with classical hydrodynamics, compressible inviscid flow, and viscous flow. Hydrodynamic instability and turbulent flow were topics in the course in viscous flow. Hydrodynamic instability and turbulent flow were topics in the course in viscous flow, and in recent years the engineering treatment of turbulent flow has been studied in a separate course. For a number of years, a whole quarter was given to the numerical analysis of laminar boundary layers, while the most advanced course in viscous flow has provided an introduction to more complicated numerical analysis, and to systematic techniques of approximation based on the smallness of some relevant physical parameter.

This book is intended to support the first year of study of viscous flow. The reader will need to have had an introductory course in fluid mechanics, and a good background in multivariable calculus and ordinary differential equations. The text is aimed at the development of analytical skills along with conceptual understanding, and the reader will find it helpful to pursue some study of partial differential equations, and of classical hydrodynamics, in parallel with the study of viscous flow. Readers who are anxious about the adequacy of their preparation in mathematics will find, I hope, that this book will help them gain confidence as well as skill. In early chapters, mathematical derivations are given in great detail; in later chapters, the reader is expected to fill in more of the details, but the logic of any analysis is fully presented.

During thirty years of teaching many branches of fluid mechanics, I have found that students regard viscous flow as the hardest to learn. I do not dispute this judgment, even though viscous flow is my personal favorite. I believe that the difficulty arises from the following sources.

1. The motion of a deforming body is hard to visualize, but we need to analyze it in great detail, until concepts such as vorticity and rate of deformation become comfortably familiar.

2. The typical phenomena associated with different ranges of the Reynolds number are greatly different, and the mathematical approximations that are effective in each range are correspondingly different. When a real-life flow problem appears, in engineering practice or in an examination, there may be a moment of panic while one wonders which approximate analysis is appropriate, and another panic when one tries to remember the details of any of them.

3. Some of the quantities that play crucial roles in theoretical analyses are terribly hard to measure, and make little impression on one's ordinary sensory perceptions. This is particularly true of vorticity, and of stresses on imaginary surfaces within the fluid.

4. The standard mathematical form of the theory of viscous flow, with its coupled nonlinear partial differential equations adjoined to initial and boundary conditions, has a formidable aspect that conveys little sense of cause and effect.

5. It is difficult to find exercises that develop analytical skill and strengthen one's grasp of concepts but that are not discouragingly difficult or time-consuming.

I do not believe that any author can make these difficulties go away, but I have tried hard to help the reader by providing carefully worded explanations, logically complete derivations, references, sample calculations, and well-tested exercises. I particularly urge the reader to take the exercises seriously. They are designed to help you come to grips with each topic to lead you toward a more concrete understanding, or toward the mastery of some technique. These objectives can only be reached if you work hard, and mostly by yourself. Many of the exercises bring out specialized information which is not in the text itself, and which is an extra reward for readers who are willing to dig for it.

Any modern text on fluid mechanics must embody a point of view about the importance of numerical analysis, which forms such a large component of our technical literature. I have personally found it rewarding to try to understand the logic of flow-simulation algorithms, and to see, by application of the algorithm, the implications of some mathematical theorem that had previously seemed hopelessly abstract. I have also discovered that some algorithms, in particular the randomvortex scheme which is introduced in Chapter 7, have a logical structure that reveals a particularly vivid scenario of cause and effect. Finally, I believe that some classic algorithms are so generally useful that every student of fluid mechanics should learn how they work, and that numerical analysis is such an important research tool that every student should learn at least what kinds of flows are particularly easy or hard to describe with this tool. The effects of this point of view will be found sprinkled throughout the book from Chapter 7 to the end.

The first nine chapters present those introductory topics that need to be thoroughly understood before individual flow fields are analyzed. The first three chapters and much of Chapter 5 could apply to any fluid; the constitutive equation for a Newtonian fluid is introduced in Chapter 4 and is used, along with the Fourier law of thermal conduction, from the latter part of Chapter 5 to the end.

In Chapter 7, the scope is narrowed again, to exclude any further consideration of fluid elasticity, or of temperature-dependent viscosity or thermal conductivity. This does not reflect any lack of interest in important topics such as shockwave-boundary layer interaction, but expresses the judgment that the essential effects of viscosity can all be illustrated in less complicated flows. We do retain, under conditions carefully delineated in Chapter 7, an interest in temperature as a variable quantity, so we can discuss phenomena such as free and forced convective heat transfer.

The discussion of vorticity in Chapter 8 is extended beyond the usual scope of a text at this level to include a general description of the random-vortex method for the numerical simulation of viscous flows. I believe that the algorithmic structure of this method allows one to dissect the complexities of viscous flows into a sequence of processes, each of which is easily visualized and understood. The chapter is, in a sense, the heart of the book, to which frequent reference is made in later chapters.

Chapters 9 and 10 analyze sample exact solutions of the Navier–Stokes equations. The examples are selected to display important phenomena singly or in simple combinations, but are presented in an order that gradually increases the sophistication of the necessary mathematical and/or numerical analysis. Many similar examples are held in reserve, as exercises. When it seems safe to begin skipping over some of the analytical details, the reader is asked to work them out as exercises.

Chapter 11 deals with very viscous flows, first in the simple but important context of hydrodynamic lubrication. This is followed by an analysis of creeping flows in thin films with free surfaces, and of recirculating flows in corners. External creeping flows are introduced through the classic example of flow around a sphere. After illustration of the effects of body shape and orientation, a brief introduction to the analysis of topics such as the swimming of microorganisms is provided. Systematic attempts to account for small effects of inertia by use of regular and singular perturbation expansions are analyzed to close the chapter.

Chapter 12 analyzes laminar boundary layers, first by Thwaites' integral method, which allows a quick survey of many important trends, and then by accurate numerical methods. A number of previously unpublished numerical results are shown, to illustrate the development of velocity profiles, and to confirm predictions from the integral method. Some exercises involving programming and computation are included for readers who wish to develop those skills. Complications due to unsteadiness and three-dimensional flow are mentioned just sufficiently to identify key analytical concepts and necessary numerical procedures. The concept of boundary-layer separation is discussed in some detail. The chapter ends with an introduction to second-order boundary-layer theory in the context of flow over a parabolic cylinder. The last sections of this chapter deal with relatively advanced topics that may not fit in a first-year graduate course but may stimulate the interest of ambitious students.

Chapter 13 introduces the vast subject of hydrodynamic instability, showing the main ideas and a few standard techniques of the linear theory, largely in the context of external flows such as boundary layers, jets, and slipstreams. Centrifugal instabilities and the instabilities of displacement flows in porous media are also analyzed, to give some sense of phenomenological variety. The transition from laminar to turbulent flow, which has been the subject of intense and interesting research in recent years, is described in the light of this research. This final section again includes some material that is rather advanced, but I hope provocative.

Chapter 14 deals with turbulent flow by a careful and up-to-date sketch of experimental methods and findings, by analysis of some theoretical cartoons that are thought to illuminate some important aspects of turbulence, and by a brief discussion of the prospects and problems inherent in various approaches to the task of numerical simulation.

References are used sparingly in chapters that contain mostly classical material, but are cited frequently in the last few chapters, where an important aim of study is to prepare oneself to read current technical literature. Appendices give analytical data that makes it easier to work problems in convenient coordinate systems, and a number of typical FORTRAN programs for numerical work.

I have been greatly influenced, in my studies and teaching of viscous flow, by the books and monographs of G. K. Batchelor, S. Goldstein, P. Lagerstrom, and H. Schlichting, and by the authors who contributed to *Laminar Boundary Layers,* edited by L. Rosenhead. It becomes obvious in the last few chapters that I find much of value in the *Journal of Fluid Mechanics,* and in the *Annual Reviews of Fluid Mechanics.* My interest in numerical analysis of boundary layers was awakened by Professor Harry Dwyer of the University of California at Davis.

Two persons require special mention, because my present understanding of viscous flow embodies so many of their ideas. The first is Professor Alexandre Chorin of our Department of Mathematics, whose development of the random-vortex representation of slightly viscous flows has allowed me to understand, for the first time, precisely why, where, and when vorticity enters a viscous fluid; and whose lectures on turbulent flow have clarified many issues about which I had previously entertained only hazy ideas. The second is my office mate of thirty years, Professor Gilles Corcos, who is in every way, excepting responsibility for error, a co-author of this book.

I would also like to thank the following reviewers for their many helpful comments and suggestions: Christina Amon, Carnegie Mellon University; Edward Boguca, Syracuse University; Fred R. DeJarnette, North Carolina State University; Alan T. McDonald, Purdue University; Safwat Moustafa, California Polytechnic State University; Ted H. Okiishi, Iowa State University; and Stephen Pope, Cornell University.

Finally, I am deeply grateful to Michael Dooley, whose help in transforming numerical files into computer-drawn figures was freely given and indispensable.

Frederick S. Sherman

CHAPTER

1

VISCOSITY

1.1 INTRODUCTION

A *fluid*, by definition, will yield to any system of external forces that tends to deform it without changing its volume. That is, it will move, or *flow*, as long as such forces are present. However, in almost all fluids, a finite deforming force will produce only a finite *rate* of deformation. In many common and practically important fluids, such as air, water, oil, or mercury, the stress associated with a given rate of deformation is a *linear* function of that rate. We call such fluids *Newtonian*, and name the coefficients of proportionality in the linear function *viscosity* coefficients. The value of the viscosity depends on the chemical identity and thermodynamic state of the fluid. We call the stresses that depend on rates of deformation *viscous stresses*.

To develop a theory that shows how viscosity affects fluid dynamics, one must first give precise meaning to terms such as stress and rate of deformation, and must show what it means, algebraically and geometrically, for such things to be linearly related. Constraints must be applied to the most general possible linear relationship, to exclude phenomena that are not observed in ordinary fluids. All this will be the business of Chapters 2, 3, and 4.

Next, one may consider, as in Chapter 5, how viscous stresses enter mathematical formulations of the principles of conservation of linear momentum and energy. The resulting equations are quite formidable, but a review of the ways in which viscosity, thermal conductivity, and surface tension depend on the chemical identity and thermodynamic state of the fluid (Chapter 6), and a careful analysis of some helpful thermodynamic simplifications (Chapter 7), allows one to disregard some of the complexities without losing sight of the phenomena of essential interest. One of these phenomena, the rotation of fluid elements, will receive special attention in Chapter 8, after which you will be ready to study a wide variety of specific viscous flows.

Before all that, let us review in this chapter some of the things you already know, or soon will know, about the qualitative effects of viscosity. This should help you to sense both the practical importance and the intellectual fascination of viscous flow. The following sections will also serve to introduce some technical terms, shown in *italics,* which you will see frequently throughout the text.

1.2 TRANSMISSION OF TANGENTIAL FORCES AND SLIDING MOTION

If a flat solid wall, in contact with a fluid, slides tangentially, it drags the adjacent fluid along. According to the *no-slip condition,* which we can take to be a simple fact of observation, the fluid that actually touches the wall moves along with it. Because of viscosity, that layer of fluid exerts a tangential stress on the fluid next closest to the wall, setting it in motion, and so on. The motion, and the viscous stresses which accompany it, appear to *diffuse* through the fluid, propagating in a direction in which there is no motion of the fluid itself. This phenomenon is exploited in the design of *viscosity pumps* and *film lubrication* systems, in which a sliding solid surface drags a viscous fluid from a region of low pressure into a region of high pressure. At the exit of the pump, the high pressure serves to transport the fluid onward, against resistance which may also be due to viscosity; in the bearing, the high pressure serves to keep neighboring solid surfaces away from each other.

1.3 TRANSMISSION OF VORTICITY THROUGH A FLUID

Viscous stresses often vary from one side of a fluid mass to the other so as to apply a net torque to the mass, giving it an angular acceleration. In a fluid of uniform density, there is usually no other way to make a fluid particle spin about its center of mass. The result, which seems at first entirely abstract, but which eventually will help you to develop an intuitive feeling for the behavior of viscous fluids, is that a fluid particle can only acquire a nonzero angular velocity by rubbing up against a neighboring particle that is already spinning. The mathematical description of this process corresponds closely to that of the process of molecular *diffusion,* the relevant coefficient of diffusion being the *kinematic viscosity.* The diffusing quantity is usually chosen to be the *vorticity,* which is twice the angular velocity.

In many flows, the first particles to be set spinning are those next to a solid wall, and one can visualize a scenario in which vorticity is created at the wall, at a rate sufficient to enforce the no-slip condition, and then diffused out into the fluid. After this first crucial step, which depends on viscosity, vorticity may be further redistributed by the motion of the fluid, i.e., by *convection* and by continuing diffusion. All of this is useful, because of a straightforward mathematical procedure that yields the velocity distribution that is *induced* by any given vorticity distribution.

1.4 INERTIA VERSUS VISCOSITY, OR CONVECTION VERSUS DIFFUSION: THE REYNOLDS NUMBER

Inertia allows a moving fluid to coast, after removal of any propulsive force. Thus we can stir our coffee, remove the spoon, and be entertained by motions which persist for a minute or two. Viscosity resists the deformations associated with these motions, and eventually brings the fluid to rest.

What is seen in this simple experiment depends very much on the value of the *Reynolds number,* $Re = UD/v$, where U is a typical velocity of the spoon, D is the diameter of the cup, and v is the kinematic viscosity of the coffee. Imagine that we retain the same cup, and the same trajectory of the spoon, but replace the coffee with honey—need I say more?

The Reynolds number is used to assess the importance of inertia, relative to that of viscosity; or the importance of convection, relative to that of diffusion. For the stirring of coffee, it has a moderately high value: say $Re = 1500$ if $U = 1$ cm/sec, $D = 6$ cm, and $v = 4 \times 10^{-3}$ cm^2/sec. The corresponding value for honey is probably less than 1.

1.5 DRAG AND PROPULSION

Fluid through which a foreign body moves is both deformed and accelerated. If the body is small, the motion slow, and the fluid viscous, so that the Reynolds number is small, the force required to propel the body through a viscous fluid is due mostly to the deformation, rather than the acceleration. We say that, in such circumstances, the drag is due to viscosity. Such drag is responsible, for example, for the fact that a relatively weak updraft can keep a normal cloud up in the sky where it belongs, even though water is about 800 times more dense than air. There are many technical processes in which the key element is the *creeping motion* of tiny droplets, bubbles, or solid particles relative to a carrier fluid, or relative to one another.

Viscosity is also crucial to the lifestyle of many microorganisms that operate at tiny Reynolds numbers. Water's resistance to deformation gives them something against which to push, so that they can swim. We larger creatures, on the other hand, swim by pushing against the inertia of the water.

At higher values of the Reynolds numbers, viscosity contributes to drag in two ways: (1) through *skin friction,* which is the tangential viscous stress that the fluid exerts on the body and (2) by affecting the pattern of fluid acceleration, and consequently the distribution of normal stress acting on the body, leading to *form drag*. The second of these effects is important also for lift and propulsion, and deserves separate mention.

1.6 FLOW SEPARATION

The most distinctive way in which viscosity affects the pattern of fluid acceleration is through the phenomenon of *flow separation*. We recognize this phenomenon when we see flow that had been closely following the contour of a wall suddenly turn away from the wall at a *separation locus*. Downstream of this locus, the fluid near the wall approaches the locus and then forms part of the flow that erupts away from the wall. Separation is responsible for the formation of jets and other flow features so familiar that we can hardly imagine a world without them. Most of the prominent acoustic effects generated by flow, both noisy and musical, are associated with separation. It may greatly affect the distribution of forces on immersed bodies, and the pressure recovery possible in diffusers.

Separation can sometimes be understood as a result of diffusion alone, as is the case in creeping corner flows, and sometimes of a subtle interplay between convection and diffusion of vorticity, as in separating boundary layers.

1.7 HYDRODYNAMIC INSTABILITY

Flows are said to be *hydrodynamically unstable* if they can be modified by a measurable amount as the result of an *immeasurably small perturbation*. Two such flows, developing from nominally identical initial states, and driven by nominally identical motions of bounding walls, betray their instability when measurable differences between them appear.

Instability is often associated with large values of the Reynolds number, which suggests that viscosity has a stabilizing effect. It often does, but there is another side to the coin. It can be shown mathematically that incompressible flows without vorticity, which we call *irrotational* flows, are stable. Since viscosity can often be blamed for getting the vorticity into a flow,[1] one could say that viscosity is destabilizing. Sometimes, as when a less viscous liquid is used to drive a more viscous one out of a porous medium, there is an instability due to the contrast in values of viscosity.

---

[1] The other principal agency that can set a fluid spinning involves density gradients inclined to pressure gradients. This is called *baroclinicity*. The associated instabilities are important in the atmosphere and oceans.

1.8 DISSIPATION OF MECHANICAL ENERGY AND PRODUCTION OF ENTROPY

An identified mass of viscous fluid may be viewed as a thermodynamic system that stores various forms of energy. Whenever any portion of this fluid is being deformed, there results an irreversible transformation of *mechanical energy* (kinetic energy of macroscopic motion, plus potential energy associated with an external force field such as gravity) into *internal* or *thermal energy*. This transfer increases the *entropy* of the fluid. We pay for this dissipation when we supply electricity or fuel to pumps and propulsion devices.

This dissipation is due to viscosity, and the rate of dissipation per unit volume is proportional to the viscosity coefficient and to the square of the rate of deformation. However, when Re $\gg$ 1, the integral of this local rate of dissipation, over the whole volume of a pump or pipe, may depend only weakly on the value of the viscosity coefficient, if other factors such as the flow rate are held constant. The volume-averaged rate of deformation must go down when the viscosity goes up.

1.9 HISTORICAL EFFECTS OF VISCOSITY VERSUS IMMEDIATE EFFECTS

Viscosity may carry vorticity far away from the wall at which it was created, if there is sufficient time. Something may then happen suddenly to that rotational flow in a process that seems entirely uninfluenced by viscosity. One may see an example whenever a Pitot tube is introduced into a pipe flow, to measure the velocity profile. The profile to be measured was established gradually, throughout the entry length of the pipe, by what we may call an *historical effect* of viscosity. The deceleration of that flow, as it approaches the tip of the probe, is a comparatively rapid process, in which the *immediate effect* of viscosity is quite negligible. Another example involves the instability of the flow at the edge of a jet. Diffusion and convection of vorticity establish a rotational flow that is stable as long as it is bounded on one side by the wall of the nozzle. When it separates, at the lip of the nozzle or exit of the pipe, the flow suddenly becomes unstable. Its future evolution is very swift, and little affected by viscosity. As far as the instability is concerned, viscosity is again historically, but not immediately, important. This sort of situation is likely to occur when a relevant Reynolds number is large.

CHAPTER
2

STRESSES ACTING ON AN ELEMENT OF FLUID

2.1 INTRODUCTION

To start an analysis of the dynamics of viscous fluids, one needs a conceptual model of the mechanical interactions between an *identified mass of fluid*, to be called the *system*, and the matter that surrounds it. We assume at the outset that the fluid may be treated as a *continuum*, so one can imagine an unbroken *material surface*, which always marks the boundary between system and surroundings. Every point of this material boundary moves with the local *macroscopic* velocity of the fluid. Individual molecules, whose velocities scatter somewhat around the local macroscopic velocity, may wander back and forth across the material surface, but that is irrelevant to the present analysis.

If accelerated motion of the system is observed, the system must be subject to some *external forces*. We imagine that these forces act on the system in either of two ways:

1. At a distance, on each element of volume of the system, usually in proportion to the mass within that volume (gravitational force), but occasionally in proportion to other quantities, such as the amount of

electric current flowing in the volume if a magnetic field is present (Lorentz, or electromotive, force). Forces that act like this are called *body forces*.

2. At the boundary, on each element of area in proportion to its size, and in proportion to the local value of a quantity we call *stress*. Forces that act like this are called *surface forces*.

Note that this classification is a logical companion to the concept of matter as a continuum. It has been so useful that it is frequently employed even in the kinetic theory of rarefied gas flows, where the concept of mechanical interactions involves relatively infrequent and distantly spaced collisions of molecules, and where the concept of a material surface has no easily observed reality. The concept of stress in kinetic theory is then adapted to the purposes of continuum mechanics in some discussions of turbulent flow. For the present, we think of the point of application of a surface force as lying precisely on a boundary of zero thickness.

The aim of this chapter is to establish the concept of a local *state of stress*, to become familiar with mathematical representations of that concept by means of a *stress tensor*, and to predict the linear and angular accelerations of an infinitesimally small system, which result from stress that varies from point to point of the system boundary. In the process, we shall also have to analyze the effects of body forces.

2.2 THE STATE OF STRESS

The *stress vector*, **σ**, is defined to be the surface force per unit area, acting upon the system at some point of the material surface that separates the system from its surroundings. The local orientation of the surface is specified by a unit vector **n**, pointing outward from the system into the surroundings. Figure 2.1 will remind you of these definitions.

To measure **σ** at a point inside a flowing fluid, one would have to introduce a tiny, force-detecting *test surface*, orient it to the desired value of **n**,

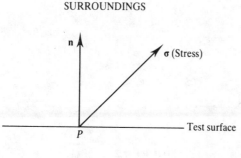

SURROUNDINGS

n

σ (Stress)

Test surface

P

SYSTEM

FIGURE 2.1
Definition of stress vector.

and arrange for it to move so as to cause no disturbance of the fluid motion. There is no practical way to do this; so one must be content to imagine the experiment, waiting a while to see whether theories that embody the concept of stress prove valuable.

Even without the ability to carry out the program of measurement, we suspect that σ depends on $\mathbf{n}$. To know the *state of stress* at a point P in the fluid, we must know σ for any specified value of $\mathbf{n}$. A little analysis will show, however, that knowledge of the values of σ for any three mutually orthogonal test surfaces is sufficient. Given these three values, one can algebraically compute the stress on any other test surface. The proof, and indeed all the analysis of this chapter, is very old.

Consider the small tetrahedron, shown in Fig. 2.2. It has three mutually orthogonal faces, with areas A_x, A_y, and A_z; and a slant face, with area A and unit outward normal vector $\mathbf{n}$. Let n_x, n_y, and n_z denote the x, y, and z components of $\mathbf{n}$, and then notice that

$$A_x = An_x \qquad A_y = An_y \qquad \text{and} \qquad A_z = An_z$$

Now imagine that the four faces of the tetrahedron are material surfaces, and consider the force exerted on the enclosed matter by its surroundings. Let σ_j denote the average stress exerted, on the system, at the surface A_j, where j can be x, y, or z.

Let σ denote the average stress, on the system, at surface A. Finally, let $\mathbf{f}$, V, and ρ denote the average body force per unit volume, the volume of the tetrahedron, and the density of the fluid, respectively. The total external force on the system will be

$$\mathbf{F} = \sigma A + \sigma_x A_x + \sigma_y A_y + \sigma_z A_z + \mathbf{f}V$$

and the average acceleration of the system will be, according to Newton's

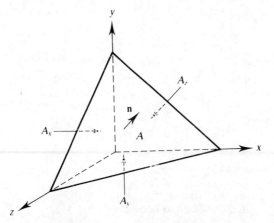

FIGURE 2.2
Tetrahedron for force balance.

second law,

$$\mathbf{a} = \frac{\mathbf{F}}{\rho V} = \frac{A}{\rho V}(\boldsymbol{\sigma} + \boldsymbol{\sigma}_x n_x + \boldsymbol{\sigma}_y n_y + \boldsymbol{\sigma}_z n_z) + \frac{\mathbf{f}}{\rho} \tag{2.1}$$

Now let the size of the tetrahedron approach zero, while its shape remains constant. If L is the length of one of the edges, then A/V will be proportional to $1/L$, and will approach infinity in this limiting process. On the other hand, $\mathbf{a}$ must remain finite. The last term is presumably approaching a finite constant, so the only way to avoid a physical absurdity is to assert that

$$\boldsymbol{\sigma} = -\{\boldsymbol{\sigma}_x n_x + \boldsymbol{\sigma}_y n_y + \boldsymbol{\sigma}_z n_z\} \tag{2.2}$$

2.3 THE STRESS TENSOR

The given stresses, $\boldsymbol{\sigma}_x$, $\boldsymbol{\sigma}_y$, and $\boldsymbol{\sigma}_z$, may now be decomposed into components along the coordinate axes. We write

$$\begin{aligned}
\boldsymbol{\sigma}_x &= -(\sigma_{xx}\mathbf{e}_x + \sigma_{xy}\mathbf{e}_y + \sigma_{xz}\mathbf{e}_z) \\
\boldsymbol{\sigma}_y &= -(\sigma_{yx}\mathbf{e}_x + \sigma_{yy}\mathbf{e}_y + \sigma_{yz}\mathbf{e}_z) \\
\boldsymbol{\sigma}_z &= -(\sigma_{zx}\mathbf{e}_x + \sigma_{zy}\mathbf{e}_y + \sigma_{zz}\mathbf{e}_z)
\end{aligned} \tag{2.3}$$

thereby introducing an arbitrary convention,[1] that

σ_{xy} denotes the y component of the stress that acts on a surface

oriented by the outward unit normal vector $\mathbf{n} = \mathbf{e}_x$

Combining (2.2) and (2.3), we display the result for $\boldsymbol{\sigma}(\mathbf{n})$, separated into x, y, and z components

$$\begin{aligned}
\boldsymbol{\sigma}(\mathbf{n}) = (n_x\sigma_{xx} + n_y\sigma_{yx} + n_z\sigma_{zx})\mathbf{e}_x + (n_x\sigma_{xy} + n_y\sigma_{yy} + n_z\sigma_{zy})\mathbf{e}_y \\
+ (n_x\sigma_{xz} + n_y\sigma_{yz} + n_z\sigma_{zz})\mathbf{e}_z
\end{aligned} \tag{2.4}$$

A compact notation for such an equation will save a lot of writing in the future. We shall use two of these, one which makes reference to an orthogonal triad of coordinate axes while representing (2.4) as

$$\sigma_i = n_j\sigma_{ji} \tag{2.5}$$

Here a *summation convention* is invoked whenever an index, such as j above, is repeated in the same quantity or algebraic term.[2] Each subscript refers to one of the three orthogonal coordinate axes, so (2.5) represents any one of the three scalar components of the vector equation (2.4).

---

[1] Unfortunately, this convention is not universally employed, even in the field of fluid mechanics. Batchelor's *Introduction to Fluid Mechanics,* for example, uses the opposite convention, assigning the first subscript to the component of stress, and the second to the orientation of the test surface.

[2] A detailed description of both compact notations is given in Appendix A.

The alternative shorthand,

$$\boldsymbol{\sigma} = \mathbf{n}\boldsymbol{\sigma}, \tag{2.6}$$

reminds us that vectors, such as $\boldsymbol{\sigma}$ and $\mathbf{n}$, are essentially independent of any coordinate system, and introduces another quantity, the *stress tensor* $\boldsymbol{\sigma}$, which shares that independence. In this book, this *symbolic notation* will be used when mathematics is employed as an aid to logical discussion; the index notation, or the expanded matrix notation

$$(\sigma_x, \sigma_y, \sigma_z) = (n_x, n_y, n_z) \begin{bmatrix} \sigma_{xx} & \sigma_{xy} & \sigma_{xz} \\ \sigma_{yx} & \sigma_{yy} & \sigma_{yz} \\ \sigma_{zx} & \sigma_{zy} & \sigma_{zz} \end{bmatrix} \tag{2.7}$$

will be used for computation.

See Sample Calculation A, at the end of the chapter.

Equation (2.7) represents the stress tensor by a 3×3 matrix, and makes it clear that the stress tensor is, mathematically, a linear operator that multiplies the vector $\mathbf{n}$ in a special way, to produce the vector $\boldsymbol{\sigma}$. It is harder to comprehend the essence of the stress tensor than that of the stress vector, but the following analysis will help.

2.4 SYMMETRY OF THE STRESS TENSOR. PRINCIPAL STRESSES

While studying Sample Calculation A, you may have noticed that the given values of σ_{ji} and σ_{ij} were equal. This equality holds for any continuous medium that is free of *internal couples*.[3] The proof of this, which is given adequately in nearly every introductory text of fluid mechanics, rests on the physical requirement that any angular acceleration resulting from the stresses acting on a tiny cube of fluid must be finite, in the limit as the size of the cube approaches zero. This proof will be expanded in a later section; for the moment, the result is simply accepted. Noting the symmetry of the matrix representation of $\boldsymbol{\sigma}$, we say that the stress tensor is *symmetric*.

In a deforming fluid, the stress on an arbitrarily oriented test surface has components both parallel to $\mathbf{n}$ (*normal stress*) and perpendicular to $\mathbf{n}$ (*tangential stress*). However, because of the symmetry of the stress tensor, there must be three mutually orthogonal orientations, $\mathbf{n}_1$, $\mathbf{n}_2$, and $\mathbf{n}_3$, for which the tangential stress vanishes. The corresponding, purely normal, stresses are designated $\boldsymbol{\sigma}^{(1)}$, $\boldsymbol{\sigma}^{(2)}$, and $\boldsymbol{\sigma}^{(3)}$, and are called *principal stresses*. The proof

---

[3] A fluid possessing internal couples might be something like a slurry containing many tiny iron filings. By introducing a magnetic field, one might be able to apply a torque to each tiny volume element of the slurry, much as gravity applies a volume-distributed force. Most fluids do not have this property.

involves nothing but algebra, and starts with the assertion to be proved:

$$\boldsymbol{\sigma}^{(\alpha)} = \mathbf{n}^{(\alpha)}\boldsymbol{\sigma} = \mathbf{n}^{(\alpha)}\sigma^{(\alpha)} \qquad \text{for } \alpha = 1, 2, \text{ or } 3 \qquad (2.8)$$

Note that $\sigma^{(\alpha)}$ is a scalar, so we have asserted that $\boldsymbol{\sigma}^{(\alpha)}$ and $\mathbf{n}^{(\alpha)}$ are parallel.

We now expand the second equation of (2.8) into three scalar equations:

$$n_x^{(\alpha)}\sigma_{xx} + n_y^{(\alpha)}\sigma_{yx} + n_z^{(\alpha)}\sigma_{zx} = n_x^{(\alpha)}\sigma^{(\alpha)}$$
$$n_x^{(\alpha)}\sigma_{xy} + n_y^{(\alpha)}\sigma_{yy} + n_z^{(\alpha)}\sigma_{zy} = n_y^{(\alpha)}\sigma^{(\alpha)} \qquad (2.9)$$
$$n_x^{(\alpha)}\sigma_{xz} + n_y^{(\alpha)}\sigma_{yz} + n_z^{(\alpha)}\sigma_{zz} = n_z^{(\alpha)}\sigma^{(\alpha)}$$

and prepare to solve these for the three scalar components of $\mathbf{n}^{(\alpha)}$, treating $\sigma^{(\alpha)}$ as a parameter. Since the equations are homogeneous, there can be nonzero solutions only for special values of the parameter, such that

$$\begin{vmatrix} \sigma_{xx} - \sigma^{(\alpha)} & \sigma_{yx} & \sigma_{zx} \\ \sigma_{xy} & \sigma_{yy} - \sigma^{(\alpha)} & \sigma_{zy} \\ \sigma_{xz} & \sigma_{yz} & \sigma_{zz} - \sigma^{(\alpha)} \end{vmatrix} = 0 \qquad (2.10)$$

This is a cubic equation, the roots of which are the *eigenvalues* $\sigma^{(1)}$, $\sigma^{(2)}$, and $\sigma^{(3)}$. Because $\boldsymbol{\sigma}$ is symmetric, the roots are all real.[4] Each root corresponds to one of the three *principal directions*, which can be computed from the equations

$$\frac{n_y^{(\alpha)}}{n_x^{(\alpha)}} = \frac{(\sigma_{xx} - \sigma^{(\alpha)})\sigma_{zy} - \sigma_{xy}\sigma_{zx}}{(\sigma_{yy} - \sigma^{(\alpha)})\sigma_{zx} - \sigma_{yx}\sigma_{zy}}$$

$$\frac{n_z^{(\alpha)}}{n_x^{(\alpha)}} = \frac{(\sigma_{xx} - \sigma^{(\alpha)})\sigma_{yz} - \sigma_{xz}\sigma_{yx}}{(\sigma_{zz} - \sigma^{(\alpha)})\sigma_{yx} - \sigma_{zx}\sigma_{yz}} \qquad (2.11)$$

and

$$(n_x^{(\alpha)})^2 + (n_y^{(\alpha)})^2 + (n_z^{(\alpha)})^2 = 1 \qquad (2.12)$$

2.5 THE ESSENCE OF THE STRESS TENSOR

Even after acquiring some skill in matrix multiplication, you may reasonably ask: "What *is* the stress tensor?" It was originally introduced as a linear algebraic operator that produces a specified effect on a given unit normal vector, but now we know enough to state an alternative definition, as follows:

> The *essence* of the stress tensor is the orthogonal triad of principal stress vectors. These three vectors, associated with a given point in the fluid, describe the local state of stress, just as the velocity vector describes the local state of translation.

---

[4] An algorithm for solving a cubic equation may be found, for example, in the *Handbook of Chemistry and Physics*. The analysis of this section may be found in any text on linear algebra, under the heading *diagonalization of a symmetric matrix*.

This statement helps to differentiate the essence of σ from a *representation* of σ as a matrix. The representation is like a photograph; its appearance depends not only on the nature of the object, but on the line of sight chosen by the photographer. Given one stress tensor, one can generate infinitely many different matrix representations, depending on the orientation of the coordinate axes, on which the principal stress vectors are projected.

When a body of fluid is in a state of *uniform translation*, the vector **u** is the same at all points of the body. The scalar components (u_1, u_2, u_3) that we may use to represent **u** may or may not be the same at all points, depending on whether the unit vectors of the chosen coordinate system are the same at all points. In *rectangular* Cartesian coordinates, they are; in *polar*, or other *curvilinear* coordinates, they are not. Similarly, when a body of fluid is in a state of *uniform stress*, the three principal stress vectors are the same at all points of the body. The nine scalars that fill out the matrix representation may or may not be the same, for the reasons outlined in the case of **u**.

2.6 RESULTANT FORCE PER UNIT VOLUME DUE TO NONUNIFORM STRESS

In our first analysis of the resultant force due to surface stresses, it was only proved that the surface forces on a body that is in a state of uniform stress must add up to zero, to prevent infinite linear acceleration of the body when its size approached zero. Let us now analyze the resultant force that appears when the state of stress is nonuniform. Consider a fluid system that momentarily occupies a small cube, bounded by the planes $x = \pm L$, $y = \pm L$, and $z = \pm L$. Figure 2.3 shows the cube, with the stress components acting at typical points on the surfaces $x = +L$, and $y = +L$. The x-component of the

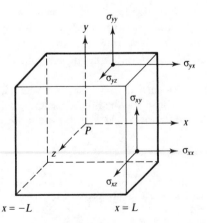

$x = -L$ $\qquad\qquad$ $x = L$

FIGURE 2.3
Stresses on a cube.

total surface force is found by integrating over all six faces of the cube, with the initial result.

$$F_x = \int_{-L}^{L} dy \int_{-L}^{L} \{\sigma_{xx}(+L, y, z) - \sigma_{xx}(-L, y, z)\} \, dz$$

$$+ \int_{-L}^{L} dz \int_{-L}^{L} \{\sigma_{yx}(x, +L, z) - \sigma_{yx}(x, -L, z)\} \, dx$$

$$+ \int_{-L}^{L} dx \int_{-L}^{L} \{\sigma_{zx}(x, y, +L) - \sigma_{zx}(x, y, -L)\} \, dy \qquad (2.13)$$

If the cube is very small, the stress components on the faces can be evaluated by use of a *Taylor series* expansion, given their values at the center of the cube. This requires a *notation for partial differentiation*. We shall let $f_{,x}$ denotes the partial derivative of $f(x, y, z, t)$ with respect to x, taken while y, z, and t are held constant. The second partial derivative with respect to x and y, will be written $f_{,xy}$. With this notation, the first few terms of the series for σ_{xx} are

$$\sigma_{xx}(x, y, z) = \sigma_{xx}(P) + x\sigma_{xx,x}(P) + y\sigma_{xx,y}(P) + z\sigma_{xx,z}(P)$$

$$+ \tfrac{1}{2}\{x^2\sigma_{xx,xx}(P) + y^2\sigma_{xx,yy}(P) + z^2\sigma_{xx,zz}(P)\}$$

$$+ xy\sigma_{xx,xy}(P) + yz\sigma_{xx,yz}(P) + zx\sigma_{xx,zx}(P) + \cdots$$

The notation $\sigma_{xx}(P)$ indicates the value of σ_{xx} at P, the center of the cube. All such values will be constants in the integrations that follow.

Inserting this and corresponding expansions for the other components of σ into the expression for F_x, and setting x, y or z equal to $+L$ or $-L$ as appropriate for each face of the cube, we note that many of the remaining integrals involve odd-function integrands, and hence equal zero. All that is left is

$$F_x = 8L^3\{\sigma_{xx,x}(P) + \sigma_{yx,y}(P) + \sigma_{zx,z}(P)\} + \text{terms of order } L^5$$

Dividing by the volume of the cube, $8L^3$, and letting $L \to 0$, we find the x component of the *resultant surface force per unit volume*. The derivation is easily modified to give the y and z components, and the result for any one of these can be written compactly as

$$\frac{F_i}{V} \to \sigma_{ji,j} \qquad (2.14)$$

in which we employ the summation convention. In the coordinate-free symbolic notation, this result is written as

$$\frac{\mathbf{F}}{V} \to \lim_{V \to 0} \left\{ \frac{1}{V} \int \sigma \, dA \right\} = \lim_{V \to 0} \left\{ \frac{1}{V} \int \mathbf{n}\sigma \, dA \right\} \equiv \text{div } \sigma \qquad (2.15)$$

You should remember that the *divergence* of a vector is a scalar; here you see that the divergence of the stress tensor, div σ, is a vector. The coordinate-free definition of div σ is very convenient when the correct representation of this quantity in a curvilinear coordinate system is to be derived. Such derivations are given in Sample Calculation B, and Appendix A.

2.7 TORQUE DUE TO NONUNIFORM STRESS

As was promised in Section 2.4, we now return to the evaluation of torque, to show the consequences of a nonuniform state of stress. We shall calculate the torque around the z axis, T_z, for the cube shown in Fig. 2.3. Look again at the figure, noting the distance between the line of action of each stress component and the z axis. Note also that nonuniform stresses on the faces $z = \pm L$ can contribute to T_z, although they were left off Fig. 2.3, to avoid cluttering the picture. The first result is

$$
T_z = \int_{-L}^{L} dy \int_{-L}^{L} [(-y\sigma_{xx} + x\sigma_{xy})_{x=L} - (-y\sigma_{xx} + x\sigma_{xy})_{x=-L}] \, dz
$$

$$
+ \int_{-L}^{L} dz \int_{-L}^{L} [(-y\sigma_{yx} + x\sigma_{yy})_{x=L} - (-y\sigma_{yx} + x\sigma_{yy})_{x=-L}] \, dx
$$

$$
+ \int_{-L}^{L} dx \int_{-L}^{L} [(-y\sigma_{zx} + x\sigma_{zy})_{x=L} - (-y\sigma_{zx} + x\sigma_{zy})_{x=-L}] \, dy \quad (2.16)
$$

To evaluate the integrals, one again employs Taylor series for the stress components on the faces. This time, all the terms involving first derivatives drop out; what is left is

$$
T_z = 8L^3(\sigma_{xy} - \sigma_{yx}) + \tfrac{4}{3}L^5[(3\sigma_{xy} - \sigma_{yx})_{,xx} - (3\sigma_{xy} - \sigma_{yx})_{,yy}
$$

$$
+ (\sigma_{xy} - \sigma_{yx})_{,zz} + 2(\sigma_{yy} - \sigma_{xx})_{,xy} + 2(\sigma_{zy,xz} - \sigma_{zx,yz})]
$$

$$
+ \text{terms of order } L^7
$$

A nonzero torque will change the angular momentum of a fluid particle in accordance with the vector equation

$$
\mathbf{T} = \frac{d(\omega \mathsf{I})}{dt} = \alpha \mathsf{I} + \omega \frac{d\mathsf{I}}{dt}. \quad (2.17)
$$

The *angular inertia* of the particle, I, is a second-rank tensor, each component of which is proportional to L^5. If *angular velocity*, ω, and *angular acceleration*, α, are to remain finite, but typically nonzero, in the limit as $L \to 0$, the first group of terms in our expression for T_z must be identically zero, and the contribution of torque to angular acceleration will be given by the second group of terms. Because $\sigma_{xy} = \sigma_{yx}$, the expression for T_z is greatly simplified.

When the remaining terms are slightly rearranged, we get

$$T_z = \frac{8L^5}{3}[(\sigma_{xy,x} + \sigma_{yy,y} + \sigma_{zy,z})_{,x} - (\sigma_{xx,x} + \sigma_{yx,y} + \sigma_{zx,z})_{,y}]$$

Comparison with (2.14) and with the familiar expression for the z component of the *curl* of a vector shows that T_z is proportional to the z component of the vector curl div $\boldsymbol{\sigma}$. This result can be shown to be independent of the shape of the chosen fluid body, and of the choice of coordinate system. Thus, stresses contribute to angular acceleration of a fluid particle, only if the vector field of resultant surface force per unit volume curls around the particle. We shall reexamine these ideas in the next chapter, and in Chapter 8.

2.8 TORQUE DUE TO A BODY FORCE

A body force may also exert a torque around the center of volume of a small fluid mass. We shall deal here only with the gravitational body force which, of course, acts through the center of mass. If the center of mass is horizontally displaced from the center of volume, there will be a torque about the latter.

For the cube of Fig. 2.3, it is easy to calculate the displacement of the center of mass, relative to the center of volume. It is, to the lowest order in L,

$$\mathbf{r} = \frac{L^2}{3\rho}\nabla\rho \qquad (2.18)$$

in which $\nabla\rho$ is the *density gradient* and ρ is the average density of the cube. The torque exerted by the weight, $\mathbf{W}$, of the cube, around the center of volume, is then $\mathbf{T} = \mathbf{r} \times \mathbf{W}$. Substituting for $\mathbf{r}$, and setting $\mathbf{W} = 8L^3\rho\mathbf{g}$, where $\mathbf{g}$ is the acceleration of gravity, we get

$$\mathbf{T} = \frac{8L^5}{3}\nabla\rho \times \mathbf{g} \qquad (2.19)$$

Note that the same power of L appears here, in equations for the torque due to nonuniform body force, as in the equation for the angular inertia. In Chapter 8, we shall look briefly at some consequences of (2.19).

2.9 SUMMARY

In this chapter, we have defined a *stress vector*, a *stress tensor*, and a *state of stress*. The stress vector acting on an arbitrarily chosen test surface will have both normal and tangential components. However, in ordinary fluids, there will be through any point three mutually orthogonal test surfaces on which the stress is purely normal. The three corresponding *principal stress vectors* completely determine the state of stress at that point, and constitute the essence of the stress tensor.

Uniform stresses cause no acceleration of the fluid; but nonuniform stresses contribute $(1/\rho)\,\mathrm{div}\,\sigma$ to the linear acceleration and $(1/2\rho)\,\mathrm{curl}\,\mathrm{div}\,\sigma$ to the angular acceleration of a fluid particle of infinitesimal size.

2.10 SAMPLE CALCULATIONS

A. Matrix manipulations. Given the matrix representation of a stress tensor:

$$\xrightarrow{} j \text{ increasing}$$

$$\sigma = \begin{Vmatrix} -2 & 4 & 0 \\ 4 & -3 & 1 \\ 0 & 1 & -6 \end{Vmatrix}$$

$$i \text{ increasing}$$

(1) Find the stress on a plane oriented by the unit normal vector $n = (1, 2, 3)/\sqrt{14}$. Using $\sigma_i = n_j\sigma_{ji}$, we get

$$\sigma_1 = n_1\sigma_{11} + n_2\sigma_{21} + n_3\sigma_{31} = \{(1)(-2) + (2)(4) + (3)(0)\}/\sqrt{14} = 6/\sqrt{14}$$

$$\sigma_2 = n_1\sigma_{12} + n_2\sigma_{22} + n_3\sigma_{32} = \{(1)(4) + (2)(-3) + (3)(1)\}/\sqrt{14} = 1/\sqrt{14}$$

$$\sigma_3 = n_1\sigma_{13} + n_2\sigma_{23} + n_3\sigma_{33} = \{(1)(0) + (2)(1) + (3)(-6)\}/\sqrt{14} = -16/\sqrt{14}$$

(2) Find the principal stresses. Using (2.10), we write out the characteristic equation, using the symmetry of σ:

$$D = (\sigma_{11} - \sigma)(\sigma_{22} - \sigma)(\sigma_{33} - \sigma) + 2\sigma_{12}\sigma_{23}\sigma_{31} - (\sigma_{11} - \sigma)\sigma_{23}^2$$
$$- (\sigma_{22} - \sigma)\sigma_{13}^2 - (\sigma_{33} - \sigma)\sigma_{12}^2 = 0$$

Substituting the numbers and multiplying out the factors, we get

$$\sigma^3 + 11\sigma^2 + 19\sigma - 62 = 0$$

To find the roots of this, let us use the Newton–Raphson iterative method, which is a suitable root-finding method for many, more complicated, equations. A quick calculation and sketch shows that $D(\sigma) = 0$ at about $\sigma = -7.2$, -5.5, and $+1.7$. To refine these estimates, we substitute into the recursion formula

$$\sigma_{n+1} = \sigma_n - \frac{D(\sigma_n)}{D'(\sigma_n)}$$

in which D' stands for $dD/d\sigma$. For our equation, this reduces to

$$\sigma_{n+1} = \frac{2\sigma_n^3 + 11\sigma_n^2 + 62}{3\sigma s_n^2 + 22\sigma_n + 19}$$

When we insert our rough approximations, the calculation quickly converges to the values of the *magnitudes* of the three principal stresses:

$$\sigma^{(1)} = -7.07820 \qquad \sigma^{(2)} = -5.51117 \qquad \sigma^{(3)} = 1.58937$$

(two compressive, and one tensile)

To find the *direction* of each principal stress, we use the equations (2.11) and

(2.12). For $\sigma^{(1)}$, we find

$$\frac{n_2^{(1)}}{n_1^{(1)}} = \frac{(-2 + 7.07820)(1) - (4)(0)\}}{(-3 + 7.07820)(0) - (4)(1)}$$

$$= -1.26955$$

$$\frac{n_3^{(1)}}{n_1^{(1)}} = \frac{(-2 + 7.07820)(1) - (4)(0)}{(-6 + 7.07820)(0) - (0)(1)}$$

$$= 1.17747$$

$$n_1^{(1)} = [1 + (-1.26955)^2 + (1.17747)^2]^{-1/2} = 0.50011$$

$$n_2^{(1)} = -0.63492 \qquad n_3^{(1)} = 0.58887$$

Corresponding calculations for $\mathbf{n}^{(2)}$ and $\mathbf{n}^{(3)}$ give the results

$$\mathbf{n}^{(2)} = (0.44743, -0.39275, -0.80346) \quad \text{and} \quad \mathbf{n}^{(3)} = (0.74141, 0.66530, 0.08766).$$

We should now test the *orthogonality* of the three principal stresses. First, check to see that $\mathbf{n}^{(1)} \cdot \mathbf{n}^{(2)} = 0$. We find

$$\mathbf{n}^{(1)} \cdot \mathbf{n}^{(2)} = (0.50011)(0.44743) + (-0.63492)(-0.39275) + (0.58887)(-0.80346)$$

$$= -0.000004$$

which meets the test, within our margin of roundoff error. For variety, let us check $\mathbf{n}^{(3)}$ against a formula which defines a *right-handed orthogonal triad* of unit vectors,

$$\mathbf{n}^{(3)} = \mathbf{n}^{(1)} \times \mathbf{n}^{(2)}$$

This gives

$$n_1^{(3)} = (-0.63492)(-0.80346) - (-0.39275)(0.58887) = 0.74141$$

$$n_2^{(3)} = (0.58887)(0.44743) - (-0.80346)(0.50011) = 0.66530$$

$$n_3^{(3)} = (0.50011)(-0.39275) - (0.44743)(-0.63492) = 0.08766,$$

all of which agree, within roundoff error, with the values found directly from (2.11) and (2.12).

B. Calculus in curvilinear coordinate systems

(1) Evaluate div σ in *cylindrical polar coordinates*. We start by sketching the element of volume that is enclosed between neighboring coordinate surfaces in this system (Fig. 2.4).

There are two special features of this coordinate system, of which we must beware:

1. The unit vectors $\mathbf{e}_r$ and $\mathbf{e}_\theta$, depend on θ. Specifically, $\mathbf{e}_{r,\theta} = \mathbf{e}_\theta$, and $\mathbf{e}_{\theta,\theta} = -\mathbf{e}_r$.

2. The area of the face normal to $\mathbf{e}_r$ is proportional to r.

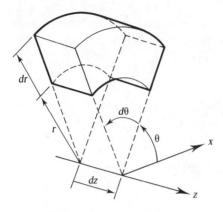

FIGURE 2.4
Cylindrical polar coordinates.

Keeping these facts in mind, we sum the surface forces acting on the six faces, using the stress at the center of each face as an adequate approximation to the average stress on that face. We get

$$
\begin{aligned}
\mathbf{F} = \{&[(\sigma_{rr}\mathbf{e}_r + \sigma_{r\theta}\mathbf{e}_\theta + \sigma_{rz}\mathbf{e}_z)r]_{r+dr,\,\theta+d\theta/2,\,z+dz/2} \\
&- [(\sigma_{rr}\mathbf{e}_r + \sigma_{r\theta}\mathbf{e}_\theta + \sigma_{rz}\mathbf{e}_z)r]_{r,\,\theta+d\theta/2,\,z+dz/2}\} \; d\theta \, dz \\
&+ [(\sigma_{\theta r}\mathbf{e}_r + \sigma_{\theta\theta}\mathbf{e}_\theta + \sigma_{\theta z}\mathbf{e}_z)]_{r+dr/2,\,\theta+d\theta,\,z+dz/2} \\
&- (\sigma_{\theta r}\mathbf{e}_r + \sigma_{\theta\theta}\mathbf{e}_\theta + \sigma_{\theta z}\mathbf{e}_z)_{r+dr/2,\,\theta,\,z+dz/2}] \; dz \, dr \\
&+ \{[(\sigma_{zr}\mathbf{e}_r + \sigma_{z\theta}\mathbf{e}_\theta + \sigma_{zz}\mathbf{e}_z)r]_{r+dr/2,\,\theta+d\theta/2,\,z+dz} \\
&- [(\sigma_{zr}\mathbf{e}_r + \sigma_{z\theta}\mathbf{e}_\theta + \sigma_{zz}\mathbf{e}_z)r]_{r+dr/2,\,\theta+d\theta/2,\,z}\} \; dr \, d\theta
\end{aligned}
$$

To the lowest order in the differential quantities, this becomes

$$
\begin{aligned}
\mathbf{F} = \{&[(\sigma_{rr}\mathbf{e}_r + \sigma_{r\theta}\mathbf{e}_\theta + \sigma_{rz}\mathbf{e}_z)r]_{,r} + (\sigma_{\theta r}\mathbf{e}_r + \sigma_{\theta\theta}\mathbf{e}_\theta + \sigma_{\theta z}\mathbf{e}_z)_{,\theta} \\
&+ [(\sigma_{zr}\mathbf{e}_r + \sigma_{z\theta}\mathbf{e}_\theta + \sigma_{zz}\mathbf{e}_z)r]_{,z}\} \; dr \, d\theta \, dz
\end{aligned}
$$

Expanding the derivatives of the products, and collecting coefficients of the unit vectors, we find

$$
F_r = \{(r\sigma_{rr})_{,r} + \sigma_{\theta r,\theta} - \sigma_{\theta\theta} + (r\sigma_{zr})_{,z}\} \; dr \, d\theta \, dz
$$

$$
F_\theta = \{(r\sigma_{r\theta})_{,r} + \sigma_{\theta r} + \sigma_{\theta\theta,\theta} + (r\sigma_{z\theta})_{,z}\} \; dr \, d\theta \, dz
$$

$$
F_z = \{(r\sigma_{rz})_{,r} + \sigma_{\theta z,\theta} + (r\sigma_{zz})_{,z}\} \; dr \, d\theta \, dz
$$

We have only to divide by the differential volume, $r \, dr \, d\theta \, dz$, to get the force per unit volume.

(2) Given the stress on the surface of a sphere, in *spherical polar coordinates*, integrate to find the net surface force on the sphere. The given expressions are

$$
\sigma_{RR} = -P + \frac{3\mu U}{2a}\cos\theta \qquad \sigma_{R\theta} = -\frac{3\mu U}{2a}\sin\theta \qquad \sigma_{R\phi} = 0
$$

P and U are the pressure and speed of the flow far from the sphere; μ is the viscosity of the fluid, and a is the radius of the sphere. All are constants. The coordinates, R, θ, and ϕ, are shown in Fig. 2.5.

Polar coordinates are convenient for the analysis of flow over a sphere, but they must be handled with care in an integration like this. We must note that in the representation of stress, $\boldsymbol{\sigma} = \sigma_{RR}\mathbf{e}_R + \sigma_{R\theta}\mathbf{e}_\theta$, both the scalar components of the stress tensor and the unit vectors vary with θ. The latter variation can be deduced from the sketch, and is

$$\mathbf{e}_R = \sin\theta\cos\phi\,\mathbf{e}_x + \sin\theta\sin\phi\,\mathbf{e}_y + \cos\theta\,\mathbf{e}_z$$
$$\mathbf{e}_\theta = \cos\theta\cos\phi\,\mathbf{e}_x + \cos\theta\sin\phi\,\mathbf{e}_y - \sin\theta\,\mathbf{e}_z$$
$$\mathbf{e}_\phi = -\sin\phi\,\mathbf{e}_x + \cos\phi\,\mathbf{e}_y$$

These allow us to reformulate our given viscous stress data, in a representation with constant unit vectors,

$$\boldsymbol{\sigma} = -P\mathbf{e}_R + \frac{3\mu U}{2a}\mathbf{e}_z$$

The first term obviously integrates to zero, because the contributions from area elements at opposite ends of any diameter cancel. The second integrand is constant, so its integral is simply

$$F = \frac{3\mu U}{2a}\,4\pi a^2\,\mathbf{e}_z = 6\pi\mu Ua\,\mathbf{e}_z$$

You may recognize this as the famous formula of Stokes, for the drag of a solid sphere in creeping flow.

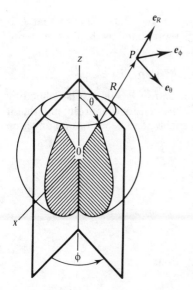

FIGURE 2.5
Spherical polar coordinates.

EXERCISES

2.1. Prove that there will be no tangential stress component on *any* test plane through point P, if the magnitudes of the three principal stresses at P are equal. (*Hint*: Consider a force balance on a small tetrahedron, whose three orthogonal edges are parallel to the principal stress vectors.)

2.2. Given the stress matrix

$$\sigma = \begin{bmatrix} -1 & 4 & 5 \\ 4 & -2 & 6 \\ 5 & 6 & -3 \end{bmatrix}$$

find the magnitudes and directions of the principal stresses, and verify the orthogonality of the three vectors. (*Hint*: The magnitude of one of the stresses is exactly 8, so all work can be done analytically, rather than numerically.)

2.3. A *uniform* states of stress is represented in cylindrical polar coordinates, at a point on the ray $\theta = 45°$, by the matrix given in Problem 2.2. What is the corresponding matrix for a point on the ray $\theta = 135°$?

2.4. Given the magnitudes of the three principal stresses, $\sigma^{(1)}$, $\sigma^{(2)}$, and $\sigma^{(3)}$, and the nine direction cosines

$$l^{(\alpha)} = \mathbf{n}^{(\alpha)} \cdot \mathbf{e}_x \qquad m^{(\alpha)} = \mathbf{n}^{(\alpha)} \cdot \mathbf{e}_y \qquad \text{and} \qquad n^{(\alpha)} = \mathbf{n}^{(\alpha)} \cdot \mathbf{e}_z$$

derive expressions for the matrix representation of σ in rectangular Cartesian coordinates.

2.5. Fill in the steps of analysis, that lead from Eq. (2.16) to the next equation below it.

2.6. From the sketch of the spherical polar coordinate system, deduce expressions for the following partial derivatives of the unit vectors: $\mathbf{e}_{\theta,\theta}$; $\mathbf{e}_{R,\phi}$; $\mathbf{e}_{\phi,\theta}$.

2.7. Imitating Sample Calculation **B.1**, derive an expression for the R-component of div σ, in spherical polar coordinates.

2.8. Find the center of mass of the fluid cube shown in Fig. 2.3, assuming that L is small enough so that the density variation can be represented by the first few terms of the Taylor series

$$\rho(x, y, z) = \rho(P) + x\rho_{,x}(P) + y\rho_{,y}(P) + z\rho_{,z}(P) + \cdots$$

2.9. A *centroidal product of inertia* of the fluid cube shown in Fig. 2.3 is defined by $M_{xy} = \iiint \rho\, xy\, dx\, dy\, dz$. Show that M_{xy} is zero if the density is uniform, and that it is at most proportional to L^7 if the density is variable, and hence has no effect on the angular acceleration of the fluid.

CHAPTER
3

KINEMATICS
OF ROTATING
AND DEFORMING
FLOW

3.1 INTRODUCTION

Having established the concept and mathematical representation of stress in a fluid, and having defined a viscous fluid as one which resists a finite *rate of deformation*, we must now establish the concept and mathematical description of that rate.

It will first be shown that the relative velocity of two neighboring *material points*[1] can be decomposed into a velocity due to rigid-body rotation, plus a velocity due to deformation. This will be done in Section 3.2 for a simple example; the concepts revealed in this example will be generalized in Section 3.3. Special emphasis is given to the concepts of *rate of rotation* and *rate of expansion*.

In Section 3.4 this local analysis of a nonuniform velocity field is followed by an important mathematical result that allows a global synthesis of the

---

[1] A *material point* moves with the macroscopic velocity of the fluid. Material *lines*, *surfaces*, and *volumes* are composed of material points.

velocity field, given the spatial distribution of the rate of rotation and the rate of expansion, and given boundary conditions on the velocity field. An important part of this result is called the *Biot–Savart law*.

You will again encounter second-rank tensors, and should make a special effort to visualize the invariant quantities that are the essence of the rates of rotation and deformation. This preliminary work will make it easier to understand the formulation, in Chapter 4, of a physical postulate concerning the relation between viscous stress and rate of deformation.

3.2 KINEMATICS OF A SIMPLE SHEAR FLOW

Consider a simple *linear shear flow,* with the velocity field $\mathbf{u} = ay\mathbf{e}_x$. The distribution of velocity, relative to a particle that moves at the level $y = y_P$, is sketched in Fig. 3.1(a).

The shearing motion will rotate and stretch any straight material line that passes through P into the first or third quadrants of the x–y plane. A material line passing through P into the second or fourth quadrants will be rotated and shortened. Only the material line coinciding with $y = y_P$ will be unaffected.

We provisionally define the rate of rotation of the fluid at P to be the

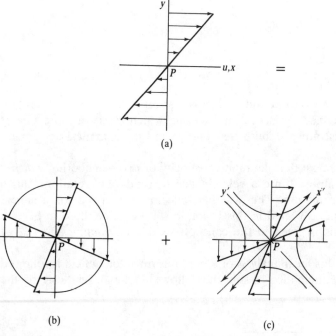

FIGURE 3.1
(a) Shear; (b) rotation; (c) deformation.

average of the angular velocities of two perpendicular material lines through P, say the lines $x = x_P$ and $y = y_P$. Defining counterclockwise angular velocity to be positive, and noting that the second line has zero angular velocity, we calculate

$$\boldsymbol{\omega} = -\frac{1}{2}\frac{u(y)}{y}\,\mathbf{e}_z = -\frac{a}{2}\mathbf{e}_z \tag{3.1}$$

Figure 3.1(b) shows the velocity field that corresponds to a rigid-body motion with this angular velocity, calculated from the equation $\mathbf{u} = \boldsymbol{\omega} \times \mathbf{r}$. Note that its velocity components are

$$u_R = \frac{a}{2}(y - y_P) \qquad v_R = -\frac{a}{2}(x - x_P) \tag{3.2a}$$

The remaining part of the original relative velocity has the components

$$u_D = \frac{a}{2}(y - y_P) \qquad v_D = +\frac{a}{2}(x - x_P) \tag{3.2b}$$

and is sketched in Fig. 3.1(c). The streamlines of this last field are hyperbolas, asymptotic to the orthogonal lines $y - y_p = \pm(x - x_p)$. Its velocity field is particularly easy to describe in a new rectangular coordinate system, with x' (and y') axes lying along these asymptotes, as shown in Fig. 3.1(c). The corresponding components of $\mathbf{u}_D$ are simply

$$u_{D'} = \frac{a}{2}x' \qquad v_{D'} = -\frac{a}{2}y' \tag{3.3}$$

In this example, the relative velocity due to deformation, $\mathbf{u}_D$, is the sum of two orthogonal velocity fields: one that produces a uniform proportional rate of *extension* of material elements, in the x' direction, plus one that produces a uniform proportional rate of *contraction,* in the y' direction. For this example, the *rate of deformation* at any point in the fluid may be specified by giving the three orthogonal vectors

$$\mathbf{d}^{(1)} = \frac{a}{2}\mathbf{e}_x, \qquad \mathbf{d}^{(2)} = -\frac{a}{2}\mathbf{e}_y, \qquad \mathbf{d}^{(3)} = (0)\mathbf{e}_z, \tag{3.4}$$

which should remind you of the orthogonal principal stress vectors, which specify the state of stress.

Since pure extension and contraction cannot rotate a fluid element, our provisional way of subtracting out the relative velocity due to rotation has been successful.[2]

---

[2] Exercise 3.2 amplifies and clarifies this point.

3.3   GENERAL ANALYSIS OF RELATIVE MOTION OF NEIGHBORING POINTS

The example just given was very simple, because the flow was unidirectional, and the rate of deformation was the same at all points. A generalized analysis is now given, and again involves the notion of a second-rank tensor: at first the *velocity gradient tensor*.

Velocity gradient

We need a mathematical operator that describes the rate of change of velocity as a result of spatial displacement. This kind of rate of change, of *any* continuous function of the displacement vector $\mathbf{r}$, say $F(\mathbf{r})$, is called the *gradient* of F. If F is a *scalar*, its gradient will be a *vector*, with three scalar components that describe the rate of change of F for each of three orthogonal components of spatial displacement. If $\mathbf{F}$ is a vector, its gradient will be a second-rank tensor with nine scalar components, three for each scalar component of $\mathbf{F}$. Thus, the velocity gradient tensor has a 3×3 matrix representation. For rectangular cartesian coordinates, in which $\mathbf{u}$ and $\mathbf{r}$ have scalar components (u, v, w) and (x, y, z), the matrix is

$$\text{grad } \mathbf{u} = \begin{bmatrix} u_{,x} & v_{,x} & w_{,x} \\ u_{,y} & v_{,y} & w_{,y} \\ u_{,z} & v_{,z} & w_{,z} \end{bmatrix} \tag{3.5}$$

The arrangement of entries is such that the chain rule for differentiation of a function of several variables can be written as

$$(du, dv, dw) = (dx, dy, dz) \begin{bmatrix} u_{,x} & v_{,x} & w_{,x} \\ u_{,y} & v_{,y} & w_{,y} \\ u_{,z} & v_{,z} & w_{,z} \end{bmatrix}$$

or in index notation, as $du_i = dx_j u_{i,j}$ (3.6)

The coordinate-free symbolic notation is simply

$$d\mathbf{u} = d\mathbf{r} \text{ grad } \mathbf{u} \tag{3.7}$$

An equivalent symbolic definition of grad $\mathbf{u}$, involving the limit of a surface integral, is presented and explained in Appendix A.

Rate of rotation

The relative velocity in an infinitesimal neighborhood of a material point can be decomposed into a velocity field of rigid-body rotation plus a velocity field of pure extension and contraction, by simple algebraic manipulations of grad $\mathbf{u}$. Consider first the *transpose* of grad $\mathbf{u}$. Any second-rank tensor, say T, has a transpose, which we denote by T*.

The relation between T and T* is expressed symbolically by the equation

$$\mathbf{c} \cdot (\mathbf{d}\mathsf{T}) = \mathbf{d} \cdot (\mathbf{c}\mathsf{T}^*), \tag{3.8}$$

which must be satisfied foɪ any two vectors $\mathbf{c}$ and $\mathbf{d}$. When we rewrite this in index notation, as

$$c_i(d_j T_{ji}) = d_i(c_j T_{ji}^*)$$

we see that

$$T_{ji} = T_{ij}^* \tag{3.9}$$

Thus, the matrix representation of T* is obtained from that of T by *transposing* rows and columns.

Consider the tensor $\mathsf{B} = \mathsf{T} - \mathsf{T}^*$. In index notation, $B_{ij} = T_{ij} - T_{ij}^*$. Because of (3.9), this can be written

$$B_{ij} = T_{ij} - T_{ji} \tag{3.10}$$

This tensor is called *skew-symmetric*. Obviously, the terms on the upper-left to lower-right diagonal of its matrix are zero, and the off-diagonal terms are related by $B_{ji} = -B_{ij}$. This means that a skew-symmetric tensor can be defined by just three scalar values, and suggests that its essence is a single vector, say $\mathbf{b}$. This is true, the relation between $\mathbf{b}$ and B being, for an arbitrary vector $\mathbf{c}$,

$$\mathbf{c}\mathsf{B} = \mathbf{b} \times \mathbf{c} \tag{3.11}$$

Now consider a special skew-symmetric tensor, formed from the velocity gradient tensor,

$$\text{rot } \mathbf{u} = \text{grad } \mathbf{u} - (\text{grad } \mathbf{u})^* \tag{3.12}$$

For rectangular cartesian coordinates, the matrix representation of rot $\mathbf{u}$ is

$$\text{rot } \mathbf{u} = \begin{bmatrix} 0 & \Omega_z & -\Omega_y \\ -\Omega_z & 0 & \Omega_x \\ \Omega_y & -\Omega_x & 0 \end{bmatrix} \tag{3.13}$$

In this, $\Omega_k = u_{j,i} - u_{i,j}$ is the kth Cartesian component of the vector curl $\mathbf{u}$, for which we use the symbol $\mathbf{\Omega}$. This suggests that $\mathbf{\Omega}$ is in some sense the essence of rot $\mathbf{u}$, and indeed one can verify that

$$d\mathbf{r} \text{ rot } \mathbf{u} = \mathbf{\Omega} \times d\mathbf{r} \tag{3.14}$$

for any infinitesimal displacement vector, $d\mathbf{r}$. Since the differential velocity due to a rigid-body rotation, with angular velocity $\boldsymbol{\omega}$, is given by $d\mathbf{u}_R = \boldsymbol{\omega} \times d\mathbf{r}$, we are led to the following generalization of Eq. (3.2) of the simple example:

$$d\mathbf{u}_R = \tfrac{1}{2}\mathbf{\Omega} \times d\mathbf{r} = \tfrac{1}{2} d\mathbf{r} \text{ rot } \mathbf{u} \tag{3.15}$$

The reason for the factor $\frac{1}{2}$ is shown by the simple example, and will soon be further demonstrated. The immediate implication is that $\mathbf{\Omega} = 2\boldsymbol{\omega}$. You will learn much more about $\mathbf{\Omega}$, which we name *vorticity*, in Chapter 8.

Rate of deformation

The part of the relative velocity that is attributed to deformation can now be found by subtraction, as was done in the example. We find, with the help of (3.12)

$$d\mathbf{u}_D = d\mathbf{u} - d\mathbf{u}_R = d\mathbf{r}\{grad\ \mathbf{u} - \tfrac{1}{2}[grad\ \mathbf{u} - (grad\ \mathbf{u})^*]\}$$

$$= \tfrac{1}{2}\,d\mathbf{r}[grad\ \mathbf{u} + (grad\ \mathbf{u})^*] = \tfrac{1}{2}\,d\mathbf{r}\ \text{def}\ \mathbf{u} \qquad (3.16)$$

The final step introduces the *rate of deformation tensor*,

$$\text{def}\ \mathbf{u} = \text{grad}\ \mathbf{u} + (\text{grad}\ \mathbf{u})^* \qquad (3.17)$$

The rate of deformation is itself a vector, equal to the ratio of $d\mathbf{u}_D$ to the magnitude of $d\mathbf{r}$. Denoting the direction of $d\mathbf{r}$ by the unit vector $\mathbf{s} = d\mathbf{r}/dr$, we obtain the important result

$$\text{Rate of deformation} = \frac{d\mathbf{u}_D}{dr} = \tfrac{1}{2}\mathbf{s}\ \text{def}\ \mathbf{u} \qquad (3.18)$$

The rate of deformation tensor, like the stress tensor, is symmetric.[3] Its matrix representation for rectangular Cartesian coordinates is

$$\text{def}\ \mathbf{u} = \begin{bmatrix} 2u_{,x} & v_{,x}+u_{,y} & w_{,x}+u_{,z} \\ u_{,y}+v_{,x} & 2v_{,y} & w_{,y}+v_{,z} \\ u_{,z}+w_{,x} & v_{,z}+w_{,y} & 2w_{,z} \end{bmatrix}. \qquad (3.19)$$

For an arbitrary choice of the unit vector $\mathbf{s}$, the rate of deformation has components both parallel to $\mathbf{s}$ (extension or contraction), and perpendicular to $\mathbf{s}$ (*shear*). However, the symmetry of def $\mathbf{u}$ guarantees that there will always be three mutually orthogonal orientations of $\mathbf{s}$, for each of which the corresponding rate of deformation is either pure extension or pure contraction. These are the *principal rates of deformation*, to which we assign the symbols $\mathbf{d}^{(1)}$, $\mathbf{d}^{(2)}$, and $\mathbf{d}^{(3)}$. Each one equals the proportional rate of increase of length, $L^{-1}\,dL/dt$, of an infinitesimal material segment of the corresponding *principal axis of deformation*. In the example, the principal axes are the x'- and y'- axes, shown in Fig. 3.1(c), and the z-axis, along which there happened to be neither extension nor contraction.

The matrix representation of def $\mathbf{u}$ that employs coordinate axes aligned

---

[3] Note that a symmetric tensor equals its transpose, and that an arbitrary second-rank tensor can be decomposed into symmetric and skew-symmetric parts:

$$\mathsf{T} = \tfrac{1}{2}(\mathsf{T}+\mathsf{T}^*) + \tfrac{1}{2}(\mathsf{T}-\mathsf{T}^*)$$
$$\text{symmetric} \quad \text{skew-symmetric}$$

with the principal axes is simply

$$\text{def } \mathbf{u} = 2 \begin{bmatrix} d^{(1)} & 0 & 0 \\ 0 & d^{(2)} & 0 \\ 0 & 0 & d^{(3)} \end{bmatrix} \tag{3.20}$$

$d^{(1)}$, $d^{(2)}$, and $d^{(3)}$ being the magnitudes of the principal rates of deformation.

Proportional rate of increase of volume

The sum of the magnitudes of the three principal rates of deformation gives the proportional rate of increase of an infinitesimal material volume. This quantity,

$$\text{div } \mathbf{u} = d^{(1)} + d^{(2)} + d^{(3)} \tag{3.21}$$

is called the *divergence* of the velocity field. A definition independent of coordinate system is given by the equation

$$\text{div } \mathbf{u} \equiv \lim_{V \to 0} \left(V^{-1} \frac{dV}{dt} \right) = \lim_{V \to 0} \left(V^{-1} \int \mathbf{n} \cdot \mathbf{u} \, dA \right) \tag{3.22}$$

Its representation in rectangular cartesian coordinates is

$$\text{div } \mathbf{u} = u_{,x} + v_{,y} + w_{,z} \tag{3.23}$$

Representations in other orthogonal coordinate systems can easily be derived, either from the definition (3.22), or by calculating the entries in the matrix representation of grad $\mathbf{u}$, and then summing the terms on the main diagonal. The scalar sum is independent of the orientation of coordinate axes.

Proportional rate of change of an infinitesimal material line

One final item may help to integrate these ideas. Consider what happens to an infinitesimal element of a straight material line, as a result of a nonuniform velocity field. We see it in Fig. 3.2.

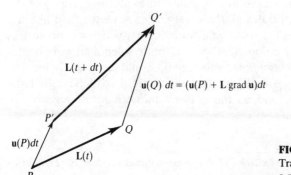

FIGURE 3.2
Translation, rotation, and stretching of a material line element.

From the sketch, it is clear that

$$d\mathbf{L}/dt = \mathbf{L}\,\mathrm{grad}\,\mathbf{u} \qquad (3.24)$$

in the limit as $\mathbf{L} \to 0$. The evident stretching of $\mathbf{L}$ can be due only to the rate of deformation, but for an arbitrary initial orientation of $\mathbf{L}$, the rotation of $\mathbf{L}$ is due partly to deformation, partly to rigid-body rotation. The part due to deformation tends to rotate $\mathbf{L}$ towards the direction of the principal axis along which the rate of extension is greatest.

3.4 SYNTHESIS OF A VELOCITY FIELD

We have seen how the velocity gradient, $\mathrm{grad}\,\mathbf{u}$, can be calculated when the velocity field is known. The required mathematical operations involve spatial differentiation of the known function, $\mathbf{u}(\mathbf{r}, t)$.

It is clear from the relation between $\mathrm{grad}\,\mathbf{u}$ and a directional derivative, that $\mathbf{u}(\mathbf{r}, t)$ can be calculated at any desired point, if it is known at some other point, and if $\mathrm{grad}\,\mathbf{u}$ is known all along a path that connects the two points. The necessary operation is integration, specifically line integration. However, this procedure is seldom useful in fluid mechanics, because the evolution of the field of $\mathrm{grad}\,\mathbf{u}$ is not governed by any simple laws.

However, there is a useful way of synthesizing the field of $\mathbf{u}$, given only the fields of $\mathrm{curl}\,\mathbf{u}$ and $\mathrm{div}\,\mathbf{u}$. You may have seen simple examples in introductory fluid mechanics, or in classical hydrodynamics, these being the two-dimensional flows due to *line vortices* or *line sources*. We shall return to these examples, but first state the relevant mathematical theorem,[4] which applies to any continuous vector field, not just to velocity. The theorem asserts that the field of $\mathbf{u}$ can be synthesized out of three qualitatively different fields,

$$\mathbf{u} = \mathbf{u}_v + \mathbf{u}_d + \mathbf{u}_p \qquad (3.25)$$

such that
$$\mathrm{curl}\,\mathbf{u}_v = \mathrm{curl}\,\mathbf{u} \qquad \mathrm{div}\,\mathbf{u}_v = 0$$
$$\mathrm{div}\,\mathbf{u}_d = \mathrm{div}\,\mathbf{u} \qquad \mathrm{curl}\,\mathbf{u}_v = 0$$
$$\mathrm{curl}\,\mathbf{u}_p = \mathrm{div}\,\mathbf{u}_p = 0$$

If $\mathbf{u}$ does represent a velocity, we see that $\mathbf{u}_v$ contributes all the vorticity of the composite field, while $\mathbf{u}_d$ contributes all the divergence. A vector field like $\mathbf{u}_v$, which has no divergence, is called *solenoidal*; one like $\mathbf{u}_d$, which has no curl, is called *irrotational*. The third component, $\mathbf{u}_p$, is both solenoidal and irrotational; we shall call it the *potential flow*, although $\mathbf{u}_d$ would also qualify for that name. Because of the way we shall use this theorem, $\mathbf{u}_v$ will be called the *velocity induced by vorticity*, and $\mathbf{u}_d$ the *velocity induced by divergence*. In

---

[4] See, for example, A. Sommerfeld, Mechanics of Deformable Bodies, Section 20, *Lectures on Theoretical Physics*, Vol. II, Academic, New York, 1960.

what follows, the usual symbol, Ω, is used for vorticity; and the shorthand notation, $\Delta \equiv \text{div}\,\mathbf{u}$, is used for divergence.

The Biot–Savart Integral

Under certain circumstances, $\mathbf{u}_v$ can be calculated by integration, from the formula

$$\mathbf{u}_v(\mathbf{r}, t) = \frac{1}{4\pi} \iiint_V \xi^{-3}\Omega(r', t) \times \xi \, dV(\mathbf{r}') \tag{3.26}$$

The integration is carried out over all the volume of fluid in which there is any vorticity, and the *displacement vector* extends from the point $\mathbf{r}'$, where the vorticity is located, to the point $\mathbf{r}$, where the velocity is induced. Thus $\xi = \mathbf{r} - \mathbf{r}'$. This integral expression will ordinarily be meaningful only if the field of Ω has *compact support*,[5] although it may suffice that Ω be a periodic function of position.

Equation (3.26) is often called the *Biot–Savart Law*, in honor of the men who discovered it empirically while studying the magnetic field induced by various configurations of electric current.

There is a corresponding integral formula for $\mathbf{u}_d$:

$$\mathbf{u}_d(\mathbf{r}, t) = \frac{1}{4\pi} \iiint_V \xi^{-3}\Delta(\mathbf{r}', t)\xi \, dV(\mathbf{r}') \tag{3.27}$$

Simple sample calculations show that the fields of $\mathbf{u}_v$ and $\mathbf{u}_d$, constructed by these formulas, do not add up to a realistic velocity field in most practical cases, because they do not meet realistic boundary conditions. In particular, the flow represented by $\mathbf{u}_v + \mathbf{u}_d$ would pass right through a solid wall, and would have vanishingly small velocity at points sufficiently far from the region that contains all the vorticity and/or dilatation. The third component, $\mathbf{u}_p$, must then be constructed so that the sum of all three is physically realistic.

You may well wonder, at this point, what good can come of such ominous-looking formulas, which require knowledge of the spatial distribution of Ω and Δ. These are quantities to which you have barely been introduced, and for which you probably have little intuitive feeling. It turns out, however, that Ω and Δ are frequently negligible everywhere except in a very limited part

---

[5] The nicely descriptive mathematical term *support* describes the set of points on which some function has nonzero values. Thus, for example, the area of contact between your feet and the ground is the support for the field of external stress which prevents your falling through the earth. The support of a function is *compact* if every point at which the value of the function is nonzero lies within a finite distance of some one given point. The integration required by the Biot–Savart formula is perhaps analogous to wiping up spilt milk: a finite operation if the support of the milk is compact, an endless task if it is not.

of the flow field. In Chapter 7, it will be shown that there is a large and important class of flows in which $\mathbf{u}_d$ is negligible everywhere; and in Chapter 8 you will see that there are straightforward ways of predicting where Ω will be nonzero at any given time, so that Eq. (3.26) can be of use. Equation (3.24) has recently found good use in approximate theories of the motion of flames, a topic which is far beyond the scope of this text. A very simple example, taken from classical hydrodynamics, is given below in Sample Calculation D.

3.5 SUMMARY

Analysis. The local field of relative velocity in a deforming fluid can be completely and usefully described by specifying the values of four vectors. These are the vorticity, Ω, and the three orthogonal principal rates of deformation, $\mathbf{d}^{(1)}$, $\mathbf{d}^{(2)}$, and $\mathbf{d}^{(3)}$. These four vectors can be constructed whenever the nine scalar components of the velocity-gradient tensor are known.

Synthesis. Given the fields of vorticity and divergence, the velocity field can be partially reconstructed by use of the Biot–Savart integral formulas. Besides the velocity induced by vorticity and divergence, a typical velocity field contains an irrotational, solenoidal component, constructed so that the flow meets realistic kinematic boundary conditions.

3.6 SAMPLE CALCULATIONS

A. *Given* the velocity field $\mathbf{u} = u(x, y)\mathbf{e}_x + v(x, y)\mathbf{e}_y$, *find* the principal rates of deformation. Investigate the special case in which div $\mathbf{u} = 0$. For this simple flow, Eq. (3.19) is reduced to

$$\text{def } \mathbf{u} = \begin{bmatrix} 2u_{,x} & v_{,x} + u_{,y} & 0 \\ u_{,y} + v_{,x} & 2v_{,y} & 0 \\ 0 & 0 & 0 \end{bmatrix}$$

Let $d^{(\alpha)}$ denote the magnitude of the principal rate of deformation that corresponds to the direction vector $\mathbf{s}^{(\alpha)}$. Follow the procedure used to find the principal stresses in Chapter 2. The principal axes will be aligned with the $\mathbf{s}^{(\alpha)}$ if

$$2u_{,x}s_x^{(\alpha)} + (u_{,y} + v_{,x})s_y^{(\alpha)} = 2d^{(\alpha)}s_x^{(\alpha)}$$

and

$$(v_{,x} + u_{,y})s_x^{(\alpha)} + 2v_{,y}s_y^{(\alpha)} = 2d^{(\alpha)}s_y^{(\alpha)}$$

from which we get the characteristic equation

$$4(u_{,x} - d^{(\alpha)})(v_{,y} - d^{(\alpha)}) - (v_{,x} + u_{,y})^2 = 0.$$

The solution is

$$d^{(\alpha)} = \tfrac{1}{2}[(u_{,x} + v_{,y}) \pm [(u_{,x} - v_{,y})^2 + (v_{,x} + u_{,y})^2]^{1/2}]$$

The corresponding slope of the principal axis is

$$\frac{s_y^{(\alpha)}}{s_x^{(\alpha)}} = \frac{(v,_y - u,_x) \pm [(u,_x - v,_y)^2 + (v,_x + u,_y)^2]^{1/2}}{v,_x + u,_y}$$

Special cases: (1) div **u** = 0. Here this reduces to $u,_x + v,_y = 0$, so that the principal rates of deformation are equal and opposite. We find

$$d^{(1)} = -d^{(2)} = \tfrac{1}{2}[4v_{,y}^2 + (v,_x + u,_y)^2]^{1/2}$$

$$\frac{s_y^{(\alpha)}}{s_x^{(\alpha)}} = \frac{2v,_y \pm [4v_{,y}^2 + (v,_x + u,_y)^2]^{1/2}}{v,_x + u,_y}$$

A *special subcase* of this is the *simple shear flow*, in which $v = 0$. Then we find the very simple results

$$d^{(1)} = -d^{(2)} = \tfrac{1}{2}u,_y \qquad \text{and} \qquad \frac{s_y^{(\alpha)}}{s_x^{(\alpha)}} = \pm 1 \tag{3.28}$$

Note that these results, which appeared in our initial example, do not depend on the linearity of the function $u(y)$.

B. Let us, in a sense, work sample calculation (A) backwards. *Given* the same two-dimensional velocity field, now characterized by its *essential* parameters Ω_z, $d^{(1)}$, and $d^{(2)}$: *calculate* the scalar components of grad **u** for an x, y, z coordinate system, in which the principal axis for extension makes an angle θ^* with the positive x axis. (Refer to Fig. 3.1(c), in which $\theta^* = 45°$.)

We start with the general formula for an infinitesimal relative velocity,

$$d\mathbf{u} = \tfrac{1}{2}\{\mathbf{\Omega} \times d\mathbf{r} + d\mathbf{r}\ \text{def } \mathbf{u}\} \tag{3.29}$$

We have two equivalent representations of $d\mathbf{r}$, namely

$$d\mathbf{r} = dx\ \mathbf{e}_x + dy\ \mathbf{e}_y = dx'\ \mathbf{e}_{x'} + dy'\ \mathbf{e}_{y'}$$

and shall use both, to rewrite (3.29) as

$$d\mathbf{u} = \tfrac{1}{2}\Omega_z(-dy\ \mathbf{e}_x + dx\ \mathbf{e}_y) + d^{(1)}\ dx'\ \mathbf{e}_{x'} + d^{(2)}\ dy'\ \mathbf{e}_{y'}$$

Next we do some projective geometry, to get

$$\mathbf{e}_{x'} = \mathbf{e}_x \cos\theta^* + \mathbf{e}_y \sin\theta^* \qquad \mathbf{e}_{y'} = \mathbf{e}_y \cos\theta^* - \mathbf{e}_x \sin\theta^*$$

$$dx' = dx \cos\theta^* + dy \sin\theta^* \qquad \text{and} \qquad dy' = dy \cos\theta^* - dx \sin\theta^*$$

Inserting these into the expression for $d\mathbf{u}$, and separating the scalar components, we find

$$du = [d^{(1)} \cos^2\theta^* + d^{(2)} \sin^2\theta^*]\ dx + [(d^{(1)} - d^{(2)}) \sin\theta^* \cos\theta^* - \tfrac{1}{2}\Omega_z]\ dy$$

$$dv = [(d^{(1)} - d^{(2)}) \sin\theta^* \cos\theta^* + \tfrac{1}{2}\Omega_z]\ dx + [d^{(2)} \cos^2\theta^* + d^{(1)} \sin^2\theta^*]\ dy$$

Comparing these with simple chain-rule expressions, e.g., $du = u,_x\ dx + u,_y\ dy$, we can read out the desired expressions:

$$u_{,x} = d^{(1)} \cos^2 \theta^* + d^{(2)} \sin^2 \theta^* \qquad u_{,y} = \tfrac{1}{2}[(d^{(1)} - d^{(2)}) \sin 2\theta^* - \Omega_z]$$

$$v_{,x} = \tfrac{1}{2}[(d^{(1)} - d^{(2)}) \sin 2\theta^* + \Omega_z] \qquad v_{,y} = d^{(1)} \sin^2 \theta^* + d^{(2)} \cos^2 \theta^*$$

Various combinations of these are of special interest. For example, we see that

$$u_{,x} + v_{,y} = d^{(1)} + d^{(2)} \qquad \text{and that} \qquad v_{,x} - u_{,y} = \Omega_z$$

whatever the value of θ^*. On the other hand,

$$v_{,x} + u_{,y} = (d^{(1)} - d^{(2)}) \sin 2\theta^*$$

which vanishes when the chosen axes coincide with the principal axes.

C. Given the velocity field $\mathbf{u} = v(r)\mathbf{e}_\theta$ in cylindrical polar coordinates, calculate the directional derivatives of $\mathbf{u}$, first in the direction of increasing r, and then in the direction of increasing θ. From these results, and any other necessary calculations, fill in the matrix representation of grad $\mathbf{u}$ for this flow and this coordinate system.

By *directional derivative*, we mean the rate of change with *distance*, in a specified direction. Thus we need $\mathbf{u}_{,r}$ and $r^{-1}\mathbf{u}_{,\theta}$. These are

$$\mathbf{u}_{,r} = v_{,r}\mathbf{e}_\theta \qquad \text{and} \qquad r^{-1}\mathbf{u}_{,\theta} = \frac{v}{r}\mathbf{e}_{\theta,\theta} = -\frac{v}{r}\mathbf{e}_r$$

The ij component of a second-rank tensor T is calculated from the formula

$$T_{ij} = \mathbf{e}_j \cdot (\mathbf{e}_i \mathsf{T}) \tag{3.30}$$

for any orthogonal coordinate system with unit vectors $\mathbf{e}_i$, $\mathbf{e}_j$, and $\mathbf{e}_k$. According to the definition of grad $\mathbf{u}$, the vector $\mathbf{e}_j$ grad $\mathbf{u}$ is the directional derivative of $\mathbf{u}$ in the direction of $\mathbf{e}_j$. Thus, for example,

$$(\text{grad } \mathbf{u})_{\theta\theta} = \frac{\mathbf{e}_\theta}{r} \cdot \mathbf{u}_{,\theta}$$

which vanishes in our example, because $\mathbf{u}_{,\theta}$ has no component in the θ direction.

Since our $\mathbf{u}$ does not depend on z, the matrix for grad $\mathbf{u}$ has only a few nonzero terms, and

$$\text{grad } \mathbf{u} = \begin{bmatrix} 0 & v_{,r} & 0 \\ -v/r & 0 & 0 \\ 0 & 0 & 0 \end{bmatrix} \tag{3.31}$$

D. Construct the velocity field that has the following properties:

1. At $x = 1$, $y = 1$, there is a *line vortex* of circulation Γ.
2. At $x = 2$, $y = 2$, there is a *line source* of strength m.
3. The flow is bounded below by a solid wall in the plane $y = 0$.
4. At $x = \pm\infty$, $\mathbf{u} = U\mathbf{e}_x$.

The situation and the solution are sketched in Fig. 3.3.

The line of the vortex is parallel to the z axis, and we understand that $\mathbf{\Omega}(x', y') = \Omega(x', y')\mathbf{e}_z$. The vorticity distribution of a line vortex is *singular*, in that $\Omega(x', y')$ is infinite at the point $x = x'$, $y = y'$ and zero everywhere else. Furthermore, the circulation of the vortex is defined to be the integral

$$\iint\limits_{-\infty}^{\infty} \Omega(x', y')\, dx'\, dy' = \Gamma$$

Similarly, for the line source the dilatation is infinite along the line $x = x'$, $y = y'$ and zero everywhere else, and the strength of the source is defined by the integral

$$\iint\limits_{-\infty}^{\infty} \Delta(x', y')\, dx'\, dy' = m$$

We now turn to Eqs. (3.23) and (3.24), to evaluate $\mathbf{u}_v$ and $\mathbf{u}_d$. We use cartesian coordinates, so that $dV(\mathbf{r}') = dx'\, dy'\, dz'$. When Ω and Δ are independent of z, the integration over z, from $-\infty$ to ∞, can be done first. The details are left to Exercise 3.6. We next observe that the relevant displacement

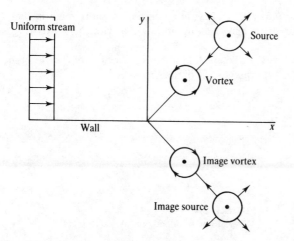

FIGURE 3.3
Images introduced to prevent flow through a wall.

vector has only x and y components, and that it can be taken outside of the integral signs, because the remaining factor in the integrand is zero everywhere except where $x = x'$ and $y = y'$. The results of these considerations are the classic formulas

$$\mathbf{u}_v = \frac{\Gamma}{2\pi r^2}\{-(y-y')\mathbf{e}_x + (x-x')\mathbf{e}_y\}$$

and

$$\mathbf{u}_d = \frac{m}{2\pi r^2}\{(x-x')\mathbf{e}_x + (y-y')\mathbf{e}_y$$

where

$$r^2 = (x-x')^2 + (y-y')^2$$

We see immediately that both of these flows pass right through the wall at $y = 0$, and that they both die away as $r \to \infty$. To correct for these shortcomings, we construct $\mathbf{u}_p$ as the sum of flows due to *images* of the vortex and the source, as shown in the sketch, and a uniform stream, also shown. Note that, in the fluid, the velocity field induced by the images is both solenoidal and irrotational. The image method does not always work so easily, but a wall of quite arbitrary shape can be made approximately leakproof by distributing singularities in an appropriate way throughout the region outside the fluid.

EXERCISES

3.1. From Fig. 3.1(c), derive expressions for the unit vectors $\mathbf{e}_{x'}$ and $\mathbf{e}_{y'}$ in terms of $\mathbf{e}_x$ and $\mathbf{e}_y$, and then the steps leading from Eq. (3.2) to Eq. (3.3). Show every step of the derivation.

3.2. Using rectangular cartesian coordinates, and assuming that

$$\mathbf{u} = u(x, y, z)\mathbf{e}_x + v(x, y, z)\mathbf{e}_y + w(x, y, z)\mathbf{e}_z,$$

derive the expression for the rate of deformation that corresponds to the direction $\mathbf{s} = \mathbf{e}_{y'}$.

3.3. Given the velocity field $\mathbf{u} = u(x, y)\mathbf{e}_x + v(x, y)\mathbf{e}_y$, consider a short material line segment that passes through the origin, making an angle θ with the x axis (Fig. 3.4)

(a) Calculate the angular velocity, $d\theta/dt$, of that line, as a function of θ and of the partial derivatives, $u_{,x}$, etc., evaluated at the origin.

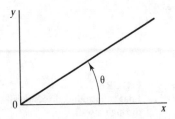

FIGURE 3.4
Material line segment. End 0 is stationary.

(b) Calculate the average angular velocity of all such lines, $0 \le \theta \le 2\pi$.

(c) Show that the answer found in (b) is the same as the average for any two mutually perpendicular material lines, say $\theta = \theta_0$ and $\theta = \theta_0 + \pi/2$.

(d) Verify that there are just two material lines that rotate at the average angular speed, and that these are identified by

$$\theta_1 = \tfrac{1}{2} \tan^{-1} \frac{v_{,x} + u_{,y}}{u_{,x} - v_{,y}} \quad \text{and} \quad \theta_2 = \theta_1 + \frac{\pi}{2}$$

(e) Verify, by use of a trigonometric identity, that the two lines found in (d) are the axes of the principal rates of deformation.

(f) Specify, additionally, that div $\mathbf{u} = 0$. Then show that your result for part (a) can be expressed in the form

$$\frac{d\theta}{dt} = \tfrac{1}{2}\Omega_z - d^{(1)} \sin\left[2(\theta - \theta^*)\right]$$

in which Ω_z, the only nonzero component of the vorticity, equals $v_{,x} - u_{,y}$; $d^{(1)}$ is the principal rate of extension; and $\theta = \theta^*$ locates the principal axis for extension. Check this result against what you see in Fig. 3.1(c). Note the implication that all material lines through 0 rotate in the same sense (though with different speeds) if $\Omega_z \le 2d^{(1)}$.

3.4. Continue the analysis of Sample Calculation C, to evaluate the vorticity Ω and to find the magnitude and direction of the principal rates of deformation. Test your final equations by letting $v = \omega r$, corresponding to a rigid-body rotation around the origin, with angular speed ω. Do your answers make good physical sense?

3.5. Given the velocity field $\mathbf{u} = (\Gamma/2\pi r)[1 - \exp(-r^2/4vt)]\mathbf{e}_\theta$, which describes a *dissolving line vortex*, calculate the principal rate of extension, $d^{(1)}$, as a function of r and t. The parameter Γ (circulation) and v (kinematic viscosity) are constants.

Analyze and discuss your solution in each of two limits: $\eta \ll 1$, and $\eta \gg 1$, where $\eta = r^2/4vt$.

3.6. Given the velocity field of Problem 3.5, and a material line segment that occupies, at time $t = 0$, the positive x axis from $x = a$ to $x = 4a$,

(a) Make an accurate sketch of the locus of that material line segment at time $t = 2a^2/v$, for a value of the circulation $\Gamma = 8\pi v$.

(b) Show that $d(\tan \Psi)/dt = -(\Gamma/\pi r^2)[(1 + \eta) \exp(-\eta) - 1]$, where Ψ is the angle shown in Fig. 3.5. The dimensionless variable η is defined by $\eta = r^2/4vt$.

(c) Using the initial condition, $\Psi(0) = 0$, derive a closed-form expression for $\Psi(r, t)$. Check your result against the sketch made in part (a).

(d) Calculate the rate of extension of this material line, as a function of r and t. Call the rate e.

(e) Find the limiting behavior of Ψ and e, when $\eta \ll 1$. Verify that in this limit, the line becomes an equiangular spiral.

3.7. Show that the Biot–Savart integral can be reduced to an area integral, rather than a volume integral, if $\Omega = \Omega(x, y, t)\mathbf{e}_z$. Specifically, show that in these circumstances

$$u_v(x, y, t) = -\frac{1}{2\pi} \iint \Omega(x', y', t)(y - y')r^{-2} \, dx' \, dy' \tag{3.32}$$

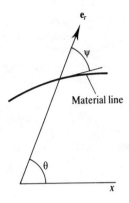

FIGURE 3.5
Segment of a material line, being wound up by a vortex.

and

$$v_v(x, y, t) = \frac{1}{2\pi} \int\int \Omega(x', y', t)(x - x')r^{-2} \, dx' \, dy' \qquad (3.33)$$

where

$$r^2 = (x - x')^2 + (y - y')^2$$

(*Hint:* Substitute the assumed form of Ω into Eq. (3.23), and integrate over z, from $-\infty$ to ∞.)

3.8. Consider a rectangular parallelepiped of fluid, bounded by the planes $x = \pm a$, $y = \pm b$, and $z = \pm c$. Suppose that the coordinate axes coincide with the principal axes of the rate of deformation. Assume that the density is uniform.

(a) Show that the moments of inertia around the principal axes are

$$I_{xx} = \tfrac{1}{3}m(b^2 + c^2) \qquad I_{yy} = \tfrac{1}{3}m(c^2 + a^2) \qquad \text{and} \qquad I_{zz} = \tfrac{1}{3}m(a^2 + b^2)$$

with $m = 8\rho abc$.

(b) Assuming that a, b, and c are infinitesimal, calculate the temporal rate of change of these moments in terms of the principal rates of deformation $d^{(1)}$, $d^{(2)}$, and $d^{(3)}$.

(c) In your answer to part (b), set $a = b = c$, so that the parallelepiped becomes a cube. Show that then $dI_{xx}/dt = I_{xx}(d^{(2)} + d^{(3)})$, with similar expressions for I_{yy} and I_{zz}.

(d) Evaluating the angular momentum, **M**, of the fluid cube by use of the equation

$$\mathbf{M} = \omega_1 I_{xx} + \omega_2 I_{yy} + \omega_3 I_{zz}$$

where ω_1, ω_2, and ω_3 are components of the angular velocity along the x-, y-, and z-axes, (which are here the principal axes), show that

$$\frac{d\mathbf{M}}{dt} = \frac{I_{xx} \, d\omega_1}{dt} + \frac{I_{yy} \, d\omega_2}{dt} + \frac{I_{zz} \, d\omega_3}{dt}$$

$$+ \omega_1 I_{xx}(d^{(2)} + d^{(3)}) + \omega_2 I_{yy}(d^{(3)} + d^{(1)}) + \omega_3 I_{zz}(d^{(1)} + d^{(2)}) \quad (3.34)$$

3.9. A sphere spins steadily around its polar axis, which is vertical, while immersed in a viscous fluid. The fluid is at rest, except for the motion produced by the spinning sphere. The velocity field of a general steady flow, represented in spherical polar coordinates, is written as

$$\mathbf{u} = u(R, \theta, \phi)\mathbf{e}_R + v(R, \theta, \phi)\mathbf{e}_\theta + w(R, \theta, \phi)\mathbf{e}_\phi$$

How can this be simplified for the flow induced by the sphere? (The answer requires some physical common sense, but no formal mathematics.)

CHAPTER
4

CONSERVATION EQUATIONS

4.1 INTRODUCTION

This chapter contains a review of material you learned in an introductory course, but will probably give a more elaborate and detailed treatment of some results that will be particularly important for the following chapters. The system chosen for analysis may be new to you; the results are of course independent of such matters of taste.

The derivations employ coordinate-free vector and tensor notations from the start, on the assumption that Chapters 2 and 3 have shown you the meaning of the various symbols. Appendix A may be helpful if you are troubled by compact notations.

The principles of conservation of mass, momentum, and energy are formulated here for a fluid that consists of a single chemical substance, such as water, and is present in a single phase, such as vapor or liquid. We call such a substance *homogeneous*, and note that we have ruled out the rigorous study of such two-phase mixtures such as wet steam. However, a chemical mixture such as air may be effectively treated as a homogeneous substance, except when the diffusive separation of its constituents is of interest. The relevant difference between air and wet steam is that the components of the former have about the same mass and rarely interact chemically, whereas the components of the latter have enormously different densities and may be exchanging mass by evaporation or condensation.

The *system* on which attention is focused is an infinitesimal portion of the fluid continuum, consisting always of the same matter, so that it is a system of constant mass. In thermodynamic terminology, it is a *closed system*; in this book it is called a *fluid particle*. It is small enough so that the variation of properties throughout it can be well approximated by the first few terms of a Taylor series that is centered on any point of the system. At the same time, it is large enough to consist of an enormous number of molecules. Its boundary is a material surface, every point of which moves with the local fluid velocity **u**. This boundary is often crossed by individual molecules, but that fact plays no part in the analysis.

First, in Sections 4.2–4.5, conservation principles are presented in *differential* form, applicable in the limit as one considers smaller and smaller systems containing the same material point. The initial point of view is *lagrangian,* i.e., attention is fixed on the system, wherever it moves, rather than on what happens at a fixed point when the system goes by. Then the latter, *eulerian,* viewpoint is adopted, and results are presented in two commonly useful forms: one called the *convective* form, the other the *divergence* form.

Section 4.6 exhibits a trick called *splitting the energy equation,* which allows one to think separately about mechanisms that affect the kinetic and potential energy of the system, and those that affect the thermal, or internal energy. This leads to a discussion of temperature and entropy variations, and requires a definition of pressure, all presented in Section 4.7. This work sets the stage for the discussion, in Chapter 5, of thermodynamic constraints on possible values of the viscosity coefficients and thermal conductivity, which are defined there. It also prepares you for the analysis of useful thermodynamic approximations, which is carried out in Chapter 7.

Integral forms of the conservation principles are derived from the differential forms in Section 4.8, for various, frequently useful, kinds of finite control volumes.

Finally, the conservation principles are applied to initially finite, pancake-shaped control volumes, which include a portion of a material interface between fluid and solid, or between immiscible fluids, to deduce rules governing the variations of stress and heat flux across such an interface. This leads to a brief discussion of *surface tension,* which will appear again in Chapter 11.

4.2 CONSERVATION OF MASS—THE CONTINUITY EQUATION

Let the fluid particle have constant mass m, but possibly variable volume V, and correspondingly variable mean mass density ρ^*. Throughout this chapter, an asterisk will denote the average value of some property of a small, but finite mass of fluid. The asterisk is removed when the mass shrinks to that of an infinitesimal particle. Thus

$$m = \rho^* V \tag{4.1}$$

and
$$\frac{dm}{dt} = V \frac{d\rho^*}{dt} + \rho^* \frac{dV}{dt} = 0 \tag{4.2}$$

Divide through by the small, but finite, volume, and then let $V \to 0$.[1] In this limit, $\rho^* \to \rho$, the density at the point to which the particle shrinks. The result is

$$\frac{D\rho}{Dt} = -\rho \lim_{V \to 0} \left(V^{-1} \frac{dV}{dt} \right) = -\rho \operatorname{div} \mathbf{u} \tag{4.3}$$

The symbol $D\rho/Dt$ will be used to denote the *lagrangian time derivative* of the density. It is the temporal rate of change of density, observed by one who moves with the particle.

When we adopt the *eulerian* viewpoint, in which $\rho = \rho(\mathbf{r}, t)$, infinitesimal changes in t and $\mathbf{r}$ produce the density change $d\rho = (\partial\rho/\partial t) \, dt + d\mathbf{r} \cdot \operatorname{grad} \rho$. When $d\mathbf{r}$ is the result of motion of a given fluid particle, it will be related to dt by $d\mathbf{r} = \mathbf{u} \, dt$. The eulerian representation of the rate of change of density of a fluid particle is thus

$$\frac{D\rho}{Dt} = \frac{\partial\rho}{\partial t} + \mathbf{u} \cdot \operatorname{grad} \rho \tag{4.4}$$

The first term on the right is called the *local rate of change*; the second is called the *convective rate of change*. These terms are applied to corresponding rates of change of other properties of the particle. The convective rate of change of density, $\mathbf{u} \cdot \operatorname{grad} \rho$, equals the speed of the fluid times the directional derivative of density in the direction of motion. This is emphasized by the popular compact notation $(\mathbf{u} \cdot \nabla)\rho$, in which ∇ symbolizes the gradient operator. This emphasizes that $(\mathbf{u} \cdot \nabla)$ is a scalar operator, which does not change the tensorial character of the quantity upon which it operates.

The identification of $\operatorname{div} \mathbf{u}$ as the proportional rate of increase of an infinitesimal material volume has been made in Chapter 3. It actually involves the assumption that no holes or voids appear in the continuum, hence the name *continuity equation* for Eq. (4.3).

4.3   CONSERVATION OF MOMENTUM

Suppose that the symbol $\mathbf{u}$ denotes velocity relative to an *inertial frame of reference*. Newton's Second Law of Motion may be applied directly to our

---

[1] The symbol $V \to 0$ refers to an imaginary process, in which we consider ever smaller portions of the continuum, all of which have at least one material point in common. It does *not* imply that we take one particle, and condense its mass to infinite density.

particle, which has linear momentum $m\mathbf{u}^*$. We use an asterisk again, to denote an average property of the small but finite system. Since the particle has constant mass, Newton's law says

$$m\frac{d\mathbf{u}^*}{dt} = \text{net force exerted on the particle by its surroundings.}$$

Dividing again by V, and letting $V \to 0$, we find

$$\rho\frac{D\mathbf{u}}{Dt} = \lim_{V \to 0}\left(\frac{\text{net force}}{V}\right) = \text{div } \sigma + \mathbf{f}, \tag{4.5}$$

where σ is the stress tensor, and $\mathbf{f}$ is the body force per unit volume. The last equality in (4.5) has been established in Chapter 2. The three scalar components of this equation are presented for easy reference in Appendix A, for rectangular-cartesian, cylindrical-polar, and spherical-polar coordinates.

4.4 CONSERVATION OF ENERGY

As noted before, the particle is a closed thermodynamic system. Thus, according to the First Law of Thermodynamics:

> In any process undergone by the system, the increase of energy stored in the system = the *heat* transferred to the system from its surroundings + the *work* done on the system by its surroundings.

The particle stores *kinetic energy*, associated with its *coherent* motion, in the amount $\frac{1}{2}\mathbf{u}^* \cdot \mathbf{u}^*$ per unit mass. Note that this is due entirely to translational motion; although the particle has an average angular velocity, the kinetic energy associated with rotation vanishes in the limit as $V \to 0$, so we shall not bother to include it in the expression for a small finite mass. The particle also stores *internal energy*, associated with the *random* motion of its component microscopic particles and with the potential energy of chemical and electronic bonds, etc., in amount e^* per unit mass. One can choose either to say that the system stores *gravitational potential energy* or to count gravity among the body forces that do work on the system. For the time being, we make the latter choice.

The *heat flux* at any point in the fluid has direction (from hot to cold regions) and magnitude. Hence it is a vector, usually assigned the symbol $\mathbf{q}$. The heat comes in across the bounding surface A, by either conduction or radiation, at a net rate equal to the surface integral

$$-\iint_A \mathbf{n} \cdot \mathbf{q}\, dA \tag{4.6}$$

The minus sign appears because $\mathbf{n}$ points out of the system, so that (4.6) gives the rate of heat transferred into the system.

No mention is made here of heat release due to chemical reactions. That is beyond the scope of this book, but would be treated as a transfer from one type of internal energy to another.

Work is done on the particle by body forces, if the particle moves coherently, at a rate

$$(\mathbf{f}^* \cdot \mathbf{u}^*)V \tag{4.7}$$

Surface forces do work too, if the material surface element on which they act is moving. The rate is given by the surface integral

$$\iint_A \boldsymbol{\sigma} \cdot \mathbf{u} \, dA \tag{4.8}$$

in which $\boldsymbol{\sigma}$ is the stress vector. Note again that the sign is correct, because $\boldsymbol{\sigma}$ acts on the system, and $\mathbf{u}$ is the velocity of the part of the system on which the force acts. Surface forces do work on the particle, in association with both coherent motion of the entire particle, and relative motions associated with deformation of the particle.

The rate of increase of stored energy can be written as

$$d\{m(\mathbf{u}^* \cdot \mathbf{u}^*/2 + e^*)\}/dt.$$

As before, m is constant. For the small but finite volume, the First Law comes down to

$$m\frac{d(\tfrac{1}{2}\mathbf{u}^* \cdot \mathbf{u}^* + e^*)}{dt} = -\iint_A \mathbf{n} \cdot \mathbf{q} \, dA + \mathbf{f}^* \cdot \mathbf{u}^* V + \iint_A \boldsymbol{\sigma} \cdot \mathbf{u} \, dA$$

We divide by V, let $V \to 0$, and get

$$\rho\frac{D(\tfrac{1}{2}\mathbf{u} \cdot \mathbf{u} + e)}{Dt} = -\operatorname{div}\mathbf{q} + \mathbf{f} \cdot \mathbf{u} + \lim_{V \to 0}\left(V^{-1}\iint_A \boldsymbol{\sigma} \cdot \mathbf{u} \, dA\right)$$

The last term can be transformed into the divergence of a vector, as follows. Since $\boldsymbol{\sigma} = \mathbf{n}\sigma$ and σ is symmetric, we can use (3.8) to write

$$\boldsymbol{\sigma} \cdot \mathbf{u} = \mathbf{u} \cdot \boldsymbol{\sigma} = \mathbf{u} \cdot (\mathbf{n}\sigma) = \mathbf{n} \cdot (\mathbf{u}\sigma^*) = \mathbf{n} \cdot (\mathbf{u}\sigma)$$

With this substitution, the stress-work term has the same mathematical form as the term that gave rise to $\operatorname{div}\mathbf{q}$, and we recognize it as $\operatorname{div}(\mathbf{u}\sigma)$. Our final representation of the First Law is then

$$\rho\frac{D(\tfrac{1}{2}\mathbf{u} \cdot \mathbf{u} + e)}{Dt} = \operatorname{div}(\mathbf{u}\sigma - \mathbf{q}) + \mathbf{f} \cdot \mathbf{u} \tag{4.9}$$

4.5   CONSERVATION EQUATIONS IN DIVERGENCE FORM

It is occasionally useful, for either analysis or numerical calculation, to recast equations (4.3), (4.5), and (4.9) into what is called *divergence form*. The reason for the name will soon be obvious.

For variety, and to set forth some equations for future reference, we use rectangular cartesian coordinates for this derivation, going back to a coordinate-free notation to express the final results.

The *continuity* equation is first written out as

$$\rho_{,t} + u\rho_{,x} + v\rho_{,y} + w\rho_{,z} + \rho(u_{,x} + v_{,y} + w_{,z}) = 0$$

Next, terms are combined, into

$$\rho_{,t} + (u\rho)_{,x} + (v\rho)_{,y} + (e\rho)_{,z} = 0 \tag{4.10}$$

which can be compressed into coordinate-free notation, as

$$\rho_{,t} + \text{div}\,(\mathbf{u}\rho) = 0 \tag{4.11}$$

The *momentum* equation has three scalar components; we shall examine only the x component. To the inertia term,

$$\mathbf{e}_x \cdot \rho\,\frac{D\mathbf{u}}{Dt} = \rho(u_{,t} + uu_{,x} + vu_{,y} + wu_{,z})$$

we add $u\{\rho_{,t} + (\rho u)_{,x} + (\rho u)_{,y} + (\rho u)_{,z}\}$, which equals zero by virtue of the continuity equation (4.11). The result can be rearranged, by grouping terms, into

$$\mathbf{e}_x \cdot \rho\,\frac{D\mathbf{u}}{Dt} = (\rho u)_{,t} + (\rho uu)_{,x} + (\rho uv)_{,y} + (\rho uw)_{,z}$$

This can be brought together with corresponding expressions for the y and z components, into the coordinate-free expression

$$\rho\,\frac{D\mathbf{u}}{Dt} = (\rho\mathbf{u})_{,t} + \text{div}\,(\rho\mathbf{u}; \mathbf{u})$$

so that the divergence form of the momentum equation is

$$(\rho\mathbf{u})_{,t} + \text{div}\,(\rho\mathbf{u}; \mathbf{u} - \sigma) = \rho\mathbf{f} \tag{4.12}$$

The unfamiliar quantity $\rho\mathbf{u}; \mathbf{u}$ in this equation is called the *momentum flux tensor*. Its matrix array is, in any orthogonal coordinate system,

$$\rho\mathbf{u}; \mathbf{u} = \rho \begin{bmatrix} u_1u_1 & u_1u_2 & u_1u_3 \\ u_2u_1 & u_2u_2 & u_2u_3 \\ u_3u_1 & u_3u_2 & u_3u_3 \end{bmatrix} \tag{4.13}$$

Its name is understood as follows. Take a stationary reference plane in the fluid, oriented by the unit vector $\mathbf{n}$. Multiply $\mathbf{n}$ by $\rho\mathbf{u}; \mathbf{u}$. Carry this out with index notation, to get

$$\{\mathbf{n}(\rho\mathbf{u}; \mathbf{u})\}_i = n_j\rho u_j u_i = \rho u_i(n_j u_j)$$

The physical significance of this is suggested very clearly in the coordinate-free notation, which is

$$\mathbf{n}(\rho\mathbf{u}; \mathbf{u}) = \rho\mathbf{u}(\mathbf{n} \cdot \mathbf{u}) \tag{4.14}$$

Since $\rho\mathbf{u}$ is the momentum per unit volume of fluid, and $(\mathbf{n} \cdot \mathbf{u})\, dA\, dt$ is the volume of fluid that crosses area dA of the reference plane in the direction of $\mathbf{n}$ in time dt, it follows that $\mathbf{n}(\rho\mathbf{u};\mathbf{u})$ equals the rate at which momentum is *convected* across unit area of the reference plane, in the direction of $\mathbf{n}$. This is called *momentum flux*, just as $\mathbf{n} \cdot \mathbf{q}$ is called heat flux.

The quantity $\operatorname{div}(\rho\mathbf{u};\mathbf{u})$, which appears in the divergence form of the momentum equation, equals the net rate at which momentum is being convected away from a fixed point in the flow field, per unit volume. Note that this concept is intrinsically wedded to the eulerian viewpoint. The original system, a material particle, exchanges nothing with its surroundings by convection, because it is a closed system. With the eulerian viewpoint, attention is focused on a fixed, infinitesimal control volume, the contents of which can be changed by convection.

Finally, we can add $(\frac{1}{2}\mathbf{u} \cdot \mathbf{u} + e)\{\rho_{,t} + \operatorname{div}(\mathbf{u}\rho)\}$, which again equals zero, to the left-hand side of Eq. (4.9), the *energy* equation, and write the result as

$$[\rho(\tfrac{1}{2}\mathbf{u} \cdot \mathbf{u} + e)]_{,t} + \operatorname{div}\{\rho\mathbf{u}(\tfrac{1}{2}\mathbf{u} \cdot \mathbf{u} + e) - \mathbf{u}\sigma + \mathbf{q}\} = \mathbf{f} \cdot \mathbf{u} \qquad (4.15)$$

4.6 SPLITTING THE ENERGY EQUATION

The First Law of Thermodynamics is concerned with the total energy stored in the system, whether it resides in the kinetic energy of coherent motion, or in the various forms of internal energy. Changes in the coherent kinetic energy of our material particle can be caused only by coherent forces, like the body force $\mathbf{f}$, or by the resultant of partially canceling forces, such as the surface forces. The separate accounting for these coherent effects is expressed by the *mechanical energy equation*,

$$\rho\frac{D(\tfrac{1}{2}\mathbf{u} \cdot \mathbf{u})}{Dt} = \mathbf{u} \cdot (\operatorname{div}\sigma + \mathbf{f}) \qquad (4.16)$$

The equation obtained by subtracting this from the full energy equation (4.9) is called the *thermal energy equation*. It is

$$\rho\frac{De}{Dt} = \operatorname{div}(\mathbf{u}\sigma) - \mathbf{u} \cdot \operatorname{div}\sigma - \operatorname{div}\mathbf{q}$$

The first two terms on the right can be combined, by symbolic manipulations shown in Appendix A. We shall establish the result by use of the cartesian index notation, as follows:

$$\operatorname{div}(\mathbf{u}\sigma) = (u_j\sigma_{ji})_{,i} = u_{j,i}\sigma_{ji} + u_j\sigma_{ji,i} = \sigma \cdot \operatorname{grad}\mathbf{u} + \mathbf{u} \cdot \operatorname{div}\sigma$$

This allows us to write the final form of the internal energy equation, as

$$\rho\frac{De}{Dt} = \sigma \cdot \operatorname{grad}\mathbf{u} - \operatorname{div}\mathbf{q} \qquad (4.17)$$

Each term of this is a scalar; when $\sigma \cdot \text{grad } \mathbf{u}$ is represented with an orthogonal coordinate system, it is a sum of nine terms. Such a quantity is called the *scalar product* of two second-rank tensors. Note the following special property of the scalar product: it vanishes if one of the tensors is symmetric, while the other is skew-symmetric. Thus,

$$\sigma \cdot \text{grad } \mathbf{u} = \tfrac{1}{2}\sigma \cdot (\text{def } \mathbf{u} + \text{rot } \mathbf{u}) = \tfrac{1}{2}\sigma \cdot \text{def } \mathbf{u}$$

because σ is symmetric, while rot $\mathbf{u}$ is skew-symmetric.

Using the principal axes for def u, we can reduce this to

$$\sigma \cdot \text{grad } \mathbf{u} = \sigma^{11}d^{(1)} + \sigma^{22}d^{(2)} + \sigma^{33}d^{(3)}$$

4.7 PRESSURE, TEMPERATURE, AND ENTROPY

In order to proceed with the thermodynamic aspects of the subject, we must now introduce the concept of *pressure*. For the moment, it is only asserted that when the fluid is not deforming, the stress on any material test surface is purely normal, and its magnitude is the same for all orientations of the surface. Symbolically, $\sigma = -\mathbf{n}p$. For a more general situation, it is customary to postulate the relationship

$$\sigma = -\mathbf{n}p + \tau \tag{4.18}$$

in which p is called the *pressure*, and τ is a *deviatoric stress* that is in some way associated with the deformation. In Chapter 5, you will encounter a special and widely useful relationship between the deviatoric stress and the rate of deformation. For the present, it is asserted that the p in (4.18) is the same thermodynamic state variable as the p that appears in the equation for the reversible work of a *simple compressible substance*,[2]

$$dW_{\text{rev}} = -p \, dV \tag{4.19}$$

Equation (4.18) implies the tensor equation

$$\sigma = -p\mathsf{I} + \tau \tag{4.20}$$

in which the *unit tensor*, I, leaves a vector unchanged when it multiplies it. Its matrix array is simply

$$\mathsf{I} = \begin{bmatrix} 1 & 0 & 0 \\ 0 & 1 & 0 \\ 0 & 0 & 1 \end{bmatrix}$$

---

[2] The terminology is that of W. C. Reynolds and H. C. Perkins, *Engineering Thermodynamics*, 2d ed., McGraw-Hill, 1977. Some authors prefer to identify the symbol p with the quantity $(\sigma^{(1)} + \sigma^{(2)} + \sigma^{(3)})/3$, even in a deforming fluid. Since the principal stresses are not directly measurable in those circumstances, I prefer the present identification.

A simple calculation shows that $\mathbf{I} \cdot \mathrm{grad}\, \mathbf{u} = \mathrm{div}\, \mathbf{u}$, so we can now insert (4.20) into (4.17) and find $\rho \dfrac{De}{Dt} + p\, \mathrm{div}\, \mathbf{u} = \boldsymbol{\tau} \cdot \mathrm{grad}\, \mathbf{u} - \mathrm{div}\, \mathbf{q}$. Eliminating $\mathrm{div}\, \mathbf{u}$ by use of (4.3), and using v to represent *specific volume*, $(v = \rho^{-1})$, we obtain

$$\rho\left(\frac{De}{Dt} + p\frac{Dv}{Dt}\right) = \boldsymbol{\tau} \cdot \mathrm{grad}\, \mathbf{u} - \mathrm{div}\, \mathbf{q} \tag{4.21}$$

Finally, we introduce *temperature, T,* and *entropy, S,* noting that

$$T\, dS = de + p\, dv$$

during an infinitesimal change of state of a simple compressible substance. This leads to the important equation

$$\rho T \frac{DS}{Dt} = \boldsymbol{\tau} \cdot \mathrm{grad}\, \mathbf{u} - \mathrm{div}\, \mathbf{q} = \tfrac{1}{2}\boldsymbol{\tau} \cdot \mathrm{def}\, \mathbf{u} - \mathrm{div}\, \mathbf{q} \tag{4.22}$$

This equation attributes any entropy increase of the material particle to the action of stresses associated with a finite rate of deformation, and/or to that of heating.

As a final step, we may analyze the effect of heating into a *reversible* component, and an *irreversible* component. To evaluate the former, return to the surface integral that represented the heating effect, and divide the integrand by the local value of T, recalling that $dQ_{\mathrm{rev}} = T\, dS$. This implies that the reversible part of the entropy increase of the particle, per unit mass, is $-\mathrm{div}\,(\mathbf{q}/T)$. But it is easy to show that

$$\mathrm{div}\left(\frac{\mathbf{q}}{T}\right) = T^{-1}\, \mathrm{div}\, \mathbf{q} - T^{-2}\, \mathbf{q} \cdot \mathrm{grad}\, T$$

so that

$$\rho T \frac{DS}{Dt} = -T\, \mathrm{div}\left(\frac{\mathbf{q}}{T}\right) - T^{-1}\mathbf{q} \cdot \mathrm{grad}\, T + \tfrac{1}{2}\boldsymbol{\tau} \cdot \mathrm{def}\, \mathbf{u}, \tag{4.23}$$

$$\text{(reversible)} \qquad \text{(irreversible)}$$

This equation gives mathematical expression to the notion that irreversible increases in entropy are caused by departures from thermal and mechanical equilibrium. Note that the irreversible effects represented in (4.23) are in some sense higher-order effects, which can become negligible compared to the corresponding reversible effects when temperature gradients and rates of deformation (the symptoms of thermal and mechanical disequilibrium), approach zero. However, in naturally evolving fluid flows, there are almost always regions in which the irreversible effects are very prominent. This theme, particularly as it concerns the effects of finite rates of deformation, will echo repeatedly through subsequent chapters.

Equation (4.22) can be combined with standard thermodynamic equations to yield a useful equation for the *temperature* of the particle. We start with a generalized equation of state for a simple compressible substance,

$S = S(T, p)$, and use the definitions

$$c_p = T\left(\frac{\partial S}{\partial T}\right)_p \qquad \text{(specific heat at constant pressure)}$$

$$\beta = v^{-1}\left(\frac{\partial v}{\partial T}\right)_p \qquad \text{(thermal expansion coefficient)}$$

The chain rule of differentiation and the Maxwell relation

$$(\partial S/\partial p)_T = -(\partial v/\partial T)_p$$

are then used, to show that $T\,DS/DT = c_p\,DT/Dt - \beta Tv\,Dp/Dt$. This is used to eliminate $T\,DS/DT$ from (4.22), leaving

$$\rho c_p \frac{DT}{Dt} = \beta T \frac{Dp}{Dt} + \tfrac{1}{2}\tau \cdot \text{def}\,\mathbf{u} - \text{div}\,\mathbf{q} \qquad (4.24)$$

4.8 CONSERVATION EQUATIONS FOR FINITE CONTROL VOLUMES

We shall now derive equations that are frequently formulated at the beginning of a discussion of conservation principles. To start, multiply Eqs. (4.11), (4.12), and (4.15), which state the conservation laws in divergence form, by an increment of volume dV, and integrate the results over the entire interior of a finite control volume V. Having done that, convert some of the volume integrals to surface integrals by use of the *Divergence Theorems*,

$$\iiint_V \text{div}\,\mathbf{w}\,dV = \iint_A \mathbf{n} \cdot \mathbf{w}\,dA \qquad \text{and} \qquad \iiint_V \text{div}\,\mathsf{T}\,dV = \iint_A \mathbf{n}\mathsf{T}\,dA$$

for a vector $\mathbf{w}$, or a second-rank tensor T. The bounding surface, A, is now finite and not necessarily a material surface. As before, $\mathbf{n}$ points outward from V. The surface A may consist of internal and external boundaries, as sketched in Fig. 4.1.

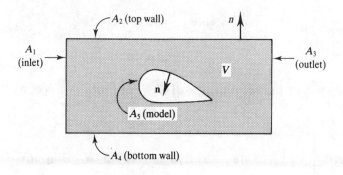

$$A = A_1 + A_2 + A_3 + A_4 + A_5$$

FIGURE 4.1
Control volume with inner and outer bounding surfaces.

The initial results are,

for *mass*

$$\iiint_V \rho_{,t}\, dV + \iint_A \mathbf{n} \cdot \rho\mathbf{u}\, dA = 0$$

for *momentum*

$$\iiint_V (\rho\mathbf{u})_{,t}\, dV + \iint_A \mathbf{n}(\rho\mathbf{u}; \mathbf{u} - \sigma)\, dA - \iiint_V \mathbf{f}\, dV = 0$$

for *energy*

$$\iiint_V [\rho(\tfrac{1}{2}\mathbf{u} \cdot \mathbf{u} + e)]_{,t}\, dV + \iint_A \mathbf{n} \cdot \{\rho\mathbf{u}(\tfrac{1}{2}\mathbf{u} \cdot \mathbf{u} + e) - \mathbf{u}\sigma + \mathbf{q}\}\, dA$$
$$- \iiint_V \mathbf{f} \cdot \mathbf{u}\, dV = 0$$

The first term in each equation is not well suited for practical applications, because it involves the sum of local transient effects, which it is likely to be difficult or impossible to evaluate. One would like to take the time derivative outside of the integral sign, and must be sure to do this properly. The correct procedure is given by the Leibnitz rule,

$$\frac{d}{dt}\left[\iiint_V f(\mathbf{r}, t)\, dV\right] = \iiint_V f_{,t}\, dV + \iint_A \mathbf{n} \cdot \mathbf{u}_A f\, dA = 0$$

In this, $\mathbf{u}_A$ denotes the local velocity of dA, a portion of the boundary of V. Thus, the time derivative may be taken outside the integral with no further ado if the boundaries of V are stationary. If the boundaries are moving, one must account carefully for the capture or loss of the quantity f that occurs when volume is added to, or lost from, the system. In the general case, the integral balances can be rewritten as

for *mass*

$$\frac{d}{dt}\left[\iiint_V \rho\, dV\right] = \iint_A \mathbf{n} \cdot \rho(\mathbf{u}_A - \mathbf{u})\, dA \qquad (4.25)$$

for *momentum*

$$\frac{d}{dt}\left[\iiint_V \rho\mathbf{u}\, dV\right] = \iiint_A [\mathbf{n} \cdot (\mathbf{u}_A - \mathbf{u})\rho\mathbf{u} + \sigma]\, dA + \iiint_V \mathbf{f}\, dV = 0 \qquad (4.26)$$

for *energy*

$$\frac{d}{dt}\left[\iiint_V \rho(\tfrac{1}{2}\mathbf{u} \cdot \mathbf{u} + e)\, dV\right] = \iiint_A \{\mathbf{n} \cdot (\mathbf{u}_A - \mathbf{u})\rho(\tfrac{1}{2}\mathbf{u} \cdot \mathbf{u} + e) + \mathbf{u} \cdot \sigma - \mathbf{n} \cdot \mathbf{q}\}\, dA$$
$$+ \iiint_V \mathbf{f} \cdot \mathbf{u}\, dV = 0 \qquad (4.27)$$

In deriving these, we have used $\boldsymbol{\sigma} = \mathbf{n}\sigma$ and $\mathbf{n} \cdot (\mathbf{u}\sigma) = \mathbf{u} \cdot (\mathbf{n}\sigma) = \mathbf{u} \cdot \boldsymbol{\sigma}$, which were seen before.

It is very important to understand why $\mathbf{u}_A$ appears only in the factor $\mathbf{n} \cdot (\mathbf{u}_A - \mathbf{u})$ that accounts for the convective capture of mass, momentum, and energy from the surroundings, but not in the terms describing work. Work is done only when the matter of the system moves; a moving boundary, such as a piston or pump impeller, does work on the fluid only because the fluid in contact with that boundary moves. Notice that a thermodynamically closed system is one for which $\mathbf{n} \cdot (\mathbf{u}_A - \mathbf{u}) = 0$ at all points of the boundary.

4.9 APPLICATION OF CONSERVATION LAWS AT BOUNDARIES

The conservation principles, as presented in this chapter, do not apply only to fluids, and can be applied to a composite system that includes the interface between dissimilar materials, such as fluid and solid, or two immiscible fluids. Such a system is sketched in Fig. 4.2.

Adopting a simplistic mechanical view of a microscopically complex phenomenon, one can imagine that the interface itself is like an elastic membrane, in which there may exist a force per unit length, parallel to the tangent plane, and normal to a test line in the interface. This force, which we may expect to find in the interface between immiscible fluids, is called *surface tension*. Here it is assigned the symbol Σ.

Consider the limit in which the surfaces A_1 and A_2 approach the interface, while the edges of the pancake remain fixed. In this limit, the mass and the heat capacity of the system vanish, while the areas at which $\boldsymbol{\sigma}$ and $\mathbf{q}$ act remain finite. In order to avoid infinite acceleration or rate of increase of temperature, the resultant force on the system and net heat transfer to the system must also vanish.

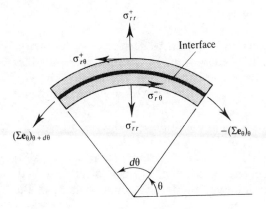

FIGURE 4.2
Force balance on an interface.

Figure 4.2 shows a bit of interface, curved in two dimensions and locally fitted with a cylindrical polar coordinate system. A three-dimensionally curved interface is only a little harder to analyze, but much harder to sketch.

Resolving the net force along $\mathbf{e}_r$ and $\mathbf{e}_\theta$, and then taking the limit $d\theta \to 0$, we find

$$\sigma_{rr}^+ - \sigma_{rr}^- = r^{-1}\Sigma \quad \text{and} \quad \sigma_{r\theta}^+ - \sigma_{r\theta}^- = -r^{-1}\Sigma,_\theta \tag{4.28}$$

The corresponding analysis for the internal energy balance might have to take account of a surface energy source, such as latent heat of phase change; if this is excluded, the result is simply $q_r^+ = q_r^-$.

4.10 SUMMARY

The fundamental principles of conservation of mass, momentum, and energy have been given mathematical expression for three different, and frequently useful, choices of the system of interest. The first was a material particle, the second an infinitesimal fixed control volume, the third a finite control volume bounded by a surface, or surfaces, which might move, but need not move with the fluid.

The energy equation can be split into two: one part accounting for the energy of coherent motion, and one for internal energy. The latter accounting brings our subject into contact with the study of thermodynamics, and allows the derivation of equations governing the history of the entropy and the temperature of a fluid particle. Our discussion goes beyond that of an introductory thermodynamics text, yielding definite expressions for the irreversible production of entropy due to thermal and mechanical disequilibrium.

A short analysis of conservation principles, applied to the interface between fluid and solid, or between immiscible fluids, provides some of the auxiliary data needed for eventual solution of the conservation equations in specified cases.

4.11 SAMPLE CALCULATION

A. Consider the control volume shown in Fig. 4.3. All surfaces are stationary. On A_1 and A_2, $\mathbf{u} = u(x, y)\mathbf{e}_x + v(x, y)\mathbf{e}_y$, $p = p(x)$, and τ_{xx} is negligible. On A_3, $\mathbf{u} = U(x)\mathbf{e}_x + v(x, h)\mathbf{e}_y$, $p = p(x)$, and τ_{yx} is negligible. On A_4, $\mathbf{u} = v_w(x)\mathbf{e}_y$, $p = p(x)$, and τ_{yx} equals τ_w. The density, ρ, is uniform and independent of time.

Suppose dx is infinitesimal, but h is finite and independent of x.

(a) Use the given data, with Eq. (4.24), to find $v(x, h)$ in terms of the x derivative of an appropriate integral. We note that the first term in (4.24) vanishes, because everything is independent of t. In the second term, $\mathbf{u}_A = 0$, because the control surfaces are stationary. The density comes outside of the

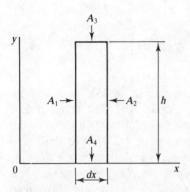

FIGURE 4.3
Stationary control volume.

integral sign in the second term, because it is uniform. There remains only

$$\iint_{A_1} \mathbf{n} \cdot \mathbf{u} \, dA + \iint_{A_2} \mathbf{n} \cdot \mathbf{u} \, dA + \iint_{A_3} \mathbf{n} \cdot \mathbf{u} \, dA + \iint_{A_4} \mathbf{n} \cdot \mathbf{u} \, dA = 0$$

On A_3 and A_4, $dA = b \, dx$, where b is the span of the flow, normal to the paper. Since dx is infinitesimal, the integrals are adequately approximated as follows:

$$\iint_{A_3} \mathbf{n} \cdot \mathbf{u} \, dA = \mathbf{e}_y \cdot \{U(x + dx/2)\mathbf{e}_x + v(x + dx/2, h)\mathbf{e}_y\} b \, dx = v(x + dx/2, h) b \, dx$$

$$\iint_{A_4} \mathbf{n} \cdot \mathbf{u} \, dA = -\mathbf{e}_y \cdot \{v_w(x + dx/2)\mathbf{e}_y\} b \, dx = -v_w(x + dx/2) b \, dx$$

The integrals over A_1 and A_2 must be left in integral form, with the appropriate evaluations of the integrals. Thus,

$$\iint_{A_1} \mathbf{n} \cdot \mathbf{u} \, dA = \int_0^h -\mathbf{e}_x \cdot \{u(x, y)\mathbf{e}_x + v(x, y)\mathbf{e}_y\} \, b \, dy = -\int_0^h u(x, y) b \, dy$$

$$\iint_{A_2} \mathbf{n} \cdot \mathbf{u} \, dA = \int_0^h \mathbf{e}_x \cdot \{u(x + dx, y)\mathbf{e}_x + v(x + dx, y)\mathbf{e}_y\} b \, dy$$

$$= \int_0^h u(x + dx, y) b \, dy$$

Since h is independent of x, the sum of the integrals over A_1 and A_2 can be approximated by

$$\left[\frac{d}{dx} \left\{ \int_0^h u(x, y) b \, dy \right\} \right] dx$$

and the desired expression for $v(x, h)$ becomes

$$v(x, h) = v_w(x) - \frac{d}{dx}\left\{\int_0^h u(x, y)b\, dy\right\}$$

(b) Make a similar evaluation of the integrals in (4.25), and find an expression for $\tau_0(x)$. Ignore the body force $\mathbf{f}$. Again, the first term in Eq. (4.25) vanishes, because of steady conditions. Since τ_0 represents a force in the x direction, we need only consider the x component of (4.25). We shall also have to use (4.18). Thus, on A_1, where $\mathbf{n} = -\mathbf{e}_x$, we find $\mathbf{u} \cdot \mathbf{n} = -u(x, y)$, $\mathbf{e}_x \cdot \mathbf{u} = u(x, y)$, and $\boldsymbol{\sigma} = -\mathbf{n}p(x) + \boldsymbol{\tau}(x, y) = p\mathbf{e}_x - \tau_{xx}(x, y)\mathbf{e}_x - \tau_{xy}(x, y)\mathbf{e}_y$. Formally, the integrand is $-\rho u(x, y)^2 - p(x) + \tau_{xx}(x, y)$, but we are allowed to drop the τ_{xx}.

When we substitute the given data in the other integrals, and then proceed as in (a), we find

$$\tau_0 = -\frac{d}{dx}\left[\int_0^h [\rho u^2(x, y) + p(x)]\, dy\right] - \rho U v(x, h)$$

EXERCISES

4.1. Show, by matrix multiplication, that $\mathbf{I} \cdot \text{grad } \mathbf{u} = \text{div } \mathbf{u}$.

4.2. Verify, using x, y, z coordinates, that $\text{div } (\mathbf{q}/T) = T^{-1} \text{div } \mathbf{q} - T^{-2}\mathbf{q} \cdot \text{grad } T$.

4.3. Consider *unaccelerated* flow in a pipe of circular cross section. Assume that τ depends only upon r (distance from the axis of the pipe), and that $p + \rho gh$ depends only upon z (distance along the axis.) The only body force is due to gravity, g is the acceleration of gravity, h is the elevation of a point in the fluid, and the value of ρ is a constant.

Show, by a straightforward balance of forces on an axial fluid cylinder of radius r, that

$$2\tau_{rz} = r\frac{d(p + \rho gh)}{dz}$$

4.4. Following the techniques of Sections 4.5 and 4.8, derive a *mechanical* energy equation for a finite control volume.

4.5. Assume that the body force is due solely to gravity, and hence the force per unit volume can be represented as

$$\mathbf{f} = -\rho \text{ grad } P$$

where P is a time-independent potential. Show that the rate of work done by this force, per unit volume of fluid, equals $-\rho\, DP/Dt$. Then show that Eq. (4.26) can be written without the volume integral of $\mathbf{u} \cdot \mathbf{f}$, providing that P is added to $\frac{1}{2}\mathbf{u} \cdot \mathbf{u} + e$, wherever the latter appears.

4.6. Apply Eq. (4.26) to a control volume that contains all the fluid inside a centrifugal pump. Thus, let the bounding surface be composed of four parts, as follows: A_1 is a stationary cross section of the inlet pipe, A_2 is a stationary cross section of the outlet pipe, A_3 coincides with the wetted surface of the pump impeller, and A_4 coincides with the wetted surface of the pump casing, including inlet, stationary vanes, and diffuser.

Assume uniform velocity, pressure, and temperature fields at the surfaces A_1 and A_2, and assume that A_3 and A_4 are thermally insulated.

Identify the integral or integrals that can logically be called *shaft work* and *flow work*.

CHAPTER

5

NEWTONIAN FLUIDS AND THE NAVIER–STOKES EQUATIONS

5.1 INTRODUCTION

Much of what has been developed in the last three chapters is quite independent of the properties of any particular substance. It has only been assumed that one may treat matter as a continuum, and think of velocity, stress, and rate of deformation as continuous point functions. One cannot predict the flow of a specified fluid until more is known about its physical properties. This point is emphasized by a count of the dependent variables that appear in the conservation equations; they outnumber the equations by a large margin.

Now it is time to focus attention on specific substances, called *Newtonian fluids,* for which experimental data suggest a particularly simple relationship between stress and rate of deformation.

In Section 5.2, certain qualitative properties of Newtonian fluids are described. The statements tend to be negative; i.e., they help to identify substances that are not Newtonian fluids, by citing various phenomena that a Newtonian fluid does not exhibit.

In Section 5.3, quantitative empirical conclusions are drawn about a class of flows in which the stress exerted by the fluid on a solid boundary can be measured, and in which the state of stress in the interior of the fluid can be deduced analytically from the measurements made at the wall. In such *viscometric* flows, the fluid motion is unaccelerated, and any stress that is caused by the motion is associated with the rate of deformation of the fluid. This study is concluded with a positive statement of the properties of a Newtonian fluid, and gives a formal introduction to the *viscosity coefficients.*

Because viscous flow is often accompanied by heat conduction, the empirical law relating conductive heat flux to temperature gradient in Newtonian fluids is presented in this chapter. This introduces *thermal conductivity,* in Section 5.4.

The dependence of viscosity and thermal conductivity on temperature, pressure, and chemical composition must be determined empirically. However, the possible values of these properties are constrained by the Second Law of Thermodynamics. These constraints are stated in Section 5.5, but the proof makes a lovely exercise in logic and simple calculus, and is left to the reader.

Equations that mathematically relate deviatoric stress to rate of deformation, and conductive heat flux to temperature gradient, are called *constitutive equations.* When the constitutive equations of a Newtonian fluid are used in the conservation equations, the result is called the *Navier–Stokes* equations. These are recorded in coordinate-free notation in Section 5.6.

In Section 5.7, empirical laws governing the variation of velocity and temperature across an interface between a fluid and a solid, or between two chemically disparate fluids, are stated. These accompany the laws for the variation of stress and heat flux across an interface, which were derived from conservation principles in Chapter 4.

Finally, in Section 5.8, some of the common fluids that ordinarily exhibit Newtonian behavior are listed, and some of the most naturally or technologically important exceptions are briefly mentioned.

5.2 QUALITATIVE PROPERTIES OF NEWTONIAN FLUIDS

A substance is called a *Newtonian fluid* if, as far as can be determined by actual measurement, it meets the following criteria.

1. When the fluid is not being deformed, the stress reduces to a normal pressure, the same for all orientations of the test surface. Mathematically, $\sigma = -\mathbf{n}p$.

2. A deviatoric stress, $\tau = \sigma + \mathbf{n}p$, appears when the fluid is being deformed at a finite rate. The local, instantaneous value of this stress depends only upon the local, instantaneous rates of deformation. There is no action at a distance, and no memory of past configurations or deformations. This rules out *viscoelastic* substances, which tend to return to a past configuration when a deforming force system is removed.

    The corresponding statement for heat conduction (not for radiation) is that the local, instantaneous heat flux depends only upon the local, instantaneous value of the temperature gradient. Radiation transport through a partially absorbing medium is a prime example of action at a distance; so is momentum transport in the free-molecule flow of rarefied gases.

3. The fluid is *isotropic*, in the sense that it exhibits no preferred directions, along which it can be stretched with special ease. This means that the principal axes for the deviatoric stress coincide with those of the rate of deformation, and each principal deviatoric stress is related to the corresponding principal rate of deformation in the same way. This might rule out liquids containing small amounts of special long-chain polymers.

   A corresponding statement applies to heat conduction; the heat flux and temperature gradient are parallel, and the relation between them is the same for all directions of these vectors.

4. There is no coupling between deviatoric stress and heat flux; thus there is no stress due to temperature gradient alone, and no heat flux due to rate of deformation alone.

5.3 QUANTITATIVE CONSTITUTIVE RELATIONS FOR NEWTONIAN FLUIDS

Before stating the mathematical forms of the constitutive relations, let us analyze some of the experiments that suggested these forms. The first involves the simple linear shear flow, which was analyzed kinematically in Chapter 3. That idealized flow can be closely approximated in practice by enclosing the fluid between coaxial, differently-rotating cylinders. Another useful special flow is that in a long, very straight capillary tube.

The useful special feature of these flows is that the conservation of momentum equation can be solved for the spatial distribution of the state of stress, without our having to know the constitutive equation. Consider, for example, the flow between coaxial cylinders that have been rotating at constant angular speed for a long time. One can assume, subject to experimental verification, that the material particles move in the simplest imaginable way, i.e., at constant speed along circular paths centered on the axis of the cylinders. You will see in Chapter 13 that this is wishful thinking in general, but that there is a useful range of conditions in which it can be very accurately true.

Consider now the net torque around the axis exerted on an annulus of fluid of infinitesimal radial thickness. It has the value

$$T_z = [(\tau_{r\theta})(2\pi rL)(r)]_{r+dr} - [(\tau_{r\theta})(2\pi rL)(r)]_r$$

The factors set apart by parentheses are the tangential stress, the area on which it acts, and the lever arm.

Since the angular momentum of the annulus around its axis is constant, T_z must be zero. This means that $\tau_{r\theta} = Ar^{-2}$. All that is needed to evaluate A

is a measurement of the torque applied by the fluid to either cylinder, and the radius of that cylinder.

The velocity distribution has the form $\mathbf{u} = v(r)\mathbf{e}_\theta$. The form of the function is unknown, but the values of v can be measured at the inner and outer cylinder. Suppose that the inner cylinder is stationary; then the rates of deformation must surely be proportional to the speed of the outer cylinder, if all other factors that might influence the relation between stress and rate of deformation are held constant. The identity of influential factors is not known a priori, but it was eventually discovered that only temperature was important for a given fluid.

When the torque required to restrain the inner cylinder is measured as a function of the speed of the outer cylinder, the two quantities are found, for many fluids and for a considerable range of speeds, to be linearly proportional. This implies a linear relationship between deviatoric stress and rate of deformation. Similar inferences are drawn from measurements of the pressure drop associated with a given flow rate through a long, straight tube: another flow in which the stress distribution can be determined before the constitutive relation is known.

Viscosity

As a first effort to predict these experimental facts, and to build in the notion of isotropy, one need only postulate a linear proportionality between principal deviatoric stresses, and corresponding principal rates of deformation. Thus, one writes

$$\boldsymbol{\tau}^{(\alpha)} = 2\mu \mathbf{d}^{(\alpha)} \tag{5.1}$$

understanding that the coefficient μ is a scalar property of the fluid, depending on the thermodynamic state, but not the state of motion. One calls μ the *viscosity coefficient*, the *dynamic viscosity*, or usually just the *viscosity*. From now on, the ponderous name, deviatoric stress, is dropped, and $\boldsymbol{\tau}$ is simply called the *viscous stress*. The factor 2 is present for historical reasons, and is absent in the corresponding tensor equation

$$\tau = \mu \operatorname{def} \mathbf{u} \tag{5.2}$$

The second viscosity coefficient

Equation (5.1) implies that the stress required to produce a specified rate of extension or contraction along any one of the principal axes is independent of the values of the other two principal rates of deformation.

This implication is not easily verified or refuted by direct measurement. The analogous statement is not true for elastic solids, and a theory with two moduli of elasticity is needed to explain experimental data. There is also a clue from the kinetic theory of simple gases, from which an equation like (5.1) appears if $Ld^{(\alpha)} \ll c$, where L is the molecular mean free path, and c is the most probable random molecular speed.[1] The specific equation is

$$\tau^{(\alpha)} = 2\mu \mathbf{d}^{(\alpha)} + \lambda \mathbf{n}^{(\alpha)} \operatorname{div} \mathbf{u} \tag{5.3}$$

or its equivalent

$$\tau = \mu \operatorname{def} \mathbf{u} + \lambda I \operatorname{div} \mathbf{u}. \tag{5.4}$$

The new coefficient, λ, is called the *second viscosity*. For a simple monatomic gas, the kinetic theory predicts that $\lambda = -2\mu/3$. A thermodynamic analysis, which we shall soon examine, shows that the ratio of λ to μ cannot be more negative than this.

The implication of a negative value of λ is that the tension required to produce a specified proportional rate of stretching along one principal axis is reduced if the fluid is locally expanding. Unfortunately, there seems to be no way to test this, or any other idea about possible values of λ, by direct mechanical measurements. Such values as are usually reported are deduced indirectly, from observations of phenomena such as the spatial attenuation of plane sound waves, or the spreading of weak shock waves. These measurements confirm the theoretical value for monatomic gases; they also suggest that λ is very nearly zero for air, and that λ/μ is large and positive for gases such as carbon dioxide, which have a more complex molecular structure.

The whole subject of second viscosity remains somewhat open, even after many years of study.[2] In some studies, it has been disclosed that observable macroscopic effects, which may be tolerably well predicted by an analysis that embodies (5.4), may be even better predicted by an analysis that completely ignores viscous stresses but conceives of multiple reservoirs of internal energy, with some resistance to transfer of energy between them.[3] Fortunately, there appear to be very few practical circumstances in which $\lambda \operatorname{div} \mathbf{u}$ is not completely negligible compared to the pressure, so (5.1) will serve as.an adequate basis for most of the analyses of this text.

THE VISCOUS DISSIPATION FUNCTION. Before we forget second viscosity, insert (5.3) into the general expression, (4.23), for the rate of work done by the viscous stresses, in association with the rates of deformation. The result, in terms of the principal rates of deformation, is

$$\Phi \equiv \tfrac{1}{2}\tau \cdot \operatorname{def} \mathbf{u} = 2\mu\{[d^{(1)}]^2 + [d^{(2)}]^2 + [d^{(3)}]^2\} + \lambda\{d^{(1)} + d^{(2)} + d^{(3)}\}^2. \tag{5.5}$$

Because of its direct association with the viscosity coefficients, Φ is called the

---

[1] See, for example, J. O. Hirschfelder, C. F. Curtiss, and R. B. Bird, *Molecular Theory of Gases and Liquids,* Wiley, 1954.

[2] For a thoughtful discussion and references to the literature, see W. G. Vincenti and C. H. Kruger, *Physical Gas Dynamics* Wiley, 1965.

[3] W. G. Griffith and A. Kenny, "On fully-dispersed shock waves in carbon dioxide," *Journal of Fluid Mechanics* **3**: 286–288 (1976).

rate of viscous dissipation, or simply the *dissipation function*. For applications, one often needs to evaluate Φ in various coordinate systems. The necessary equations are given in Appendix A.

In many cases, the sum of nine terms is dominated by a single term; for example, in the rectilinear shear flow, $\mathbf{u} = u(y)\mathbf{e}_x$, the complete expression reduces exactly to $\Phi = \mu u_{,y}^2$.

5.4 HEAT CONDUCTION

In most fluids for which Eq. (5.3) accurately represents the connection between viscous stress and rate of deformation, heat conduction is accurately represented by a companion relationship, the *Fourier* law. This is

$$\mathbf{q} = -k \operatorname{grad} T \tag{5.6}$$

in which k denotes the *thermal conductivity*. The corresponding irreversible contribution to entropy increase, from Eq. (4.25), is

$$-\mathbf{q} \cdot \operatorname{grad} T = k(\operatorname{grad} T)^2. \tag{5.7}$$

We may call this quantity the *thermal dissipation* rate.[4]

5.5 IMPLICATIONS OF THE SECOND LAW OF THERMODYNAMICS

One standard way of stating the Second Law of Thermodynamics asserts, in effect, that the sum of viscous and thermal dissipation rates must be positive; i.e.,

$$\Phi - \mathbf{q} \cdot \operatorname{grad} T = 2\mu\{[d^{(1)}]^2 + [d^{(2)}]^2 + [d^{(3)}]^2\}$$
$$+ \lambda\{d^{(1)} + d^{(2)} + d^{(3)}\}^2 + k(\operatorname{grad} T)^2 \geq 0. \tag{5.8}$$

By considering easily-imagined special circumstances, one can deduce from this that

$$k \geq 0 \qquad \mu \geq 0 \qquad \text{and} \qquad \lambda \geq -\frac{2\mu}{3}. \tag{5.9}$$

The proof is left to you, in Exercise 5.5.

5.6 THE NAVIER–STOKES EQUATIONS

When the constitutive equations for a Newtonian fluid are substituted into the conservation equations, the resulting set of partial differential equations is

---

[4] There is a semantic difficulty with the word *dissipation*, which can refer either to something which scatters away from a center of concentration, or to something which stays home and becomes useless. Clearly, the second meaning is involved here, and we shall always use the words *dispersion* or *diffusion* to convey the first meaning.

ordinarily called the *Navier–Stokes equations*, in honor of C.-L.-M.-H. Navier, and Sir G. G. Stokes, who formulated them independently in 1822 and 1845, respectively.[5] They are recorded here, for future reference, in coordinate-free notation.

Mass:
$$\frac{D\rho}{Dt} + \rho \operatorname{div} \mathbf{u} = 0 \tag{5.10}$$

Momentum:
$$\rho \frac{D\mathbf{u}}{Dt} + \operatorname{grad} p = \operatorname{div}(\mu \operatorname{def} \mathbf{u}) + \operatorname{grad}(\lambda \operatorname{div} \mathbf{u}) + \mathbf{f} \tag{5.11}$$

Temperature:

$$\rho c_p \frac{DT}{Dt} = \beta T \frac{Dp}{Dt} + \frac{\mu}{2} \operatorname{def} \mathbf{u} \cdot \operatorname{def} \mathbf{u} + \lambda(\operatorname{div} \mathbf{u})^2 + \operatorname{div}(k \operatorname{grad} T) \tag{5.12}$$

A count of the unknowns gives one vector ($\mathbf{u}$) and eight scalars (p, T, ρ, μ, λ, c_p, β, and k). The missing information is provided by various *equations of state*, which determine values of the last six scalars in terms of the first two. This property data is often available only in numerical or graphical form. Some examples are given in the following chapter.

5.7   EMPIRICAL CONDITIONS AT A PHYSICAL INTERFACE

In Chapter 4, certain constraints on stress and heat flux at an interface between fluid and solid, or between immiscible fluids, were deduced. We now record two empirical statements that seem to hold true whenever the constitutive equations presented in this chapter are accurate.

1. At a physical interface, the velocity is the same in both media. The media do not separate from each other, penetrate each other, or slip tangentially past each other. Mathematically, this one statement is often separated into two:

$$(\mathbf{u}^+ - \mathbf{u}^-) \cdot \mathbf{n} = 0 \qquad \text{(no relative velocity normal to the interface)} \tag{5.13}$$

$$(\mathbf{u}^+ - \mathbf{u}^-) \times \mathbf{n} = 0 \qquad \text{(no relative velocity along the interface)} \tag{5.14}$$

The first of these is often called the *no-penetration condition*; the second is usually called the *no-slip condition*.

---

[5] G. G. Stokes, "On the theories of the internal friction of fluids in motion, and of the equilibrium and motion of elastic solids," *Trans. Camb. Phil. Soc.*, **8**: 287–305 (1845). C. L. M. H. Navier, "Memoire sur les lois de mouvement des fluides," *Mem. Acad. R. Sci., Paris*, **6**: 389–416 (1823).

2. At a physical interface, the temperature is the same in both media.

$$T^+ - T^- = 0 \quad \text{(\textit{no temperature jump} across the interface)} \quad (5.15)$$

A physical explanation of these conditions cannot come from continuum theory. The kinetic theory of gases predicts that slip and temperature jump actually exist, but that they are approximately equal to the product of the mean free path and the rate of change of tangential velocity or temperature with distance normal to the interface. In most common circumstances, the effects are too small to be observable. In everything which follows in this text, equations (5.13) to (5.15) will be treated as exact.

5.8 SCOPE OF THE NAVIER–STOKES EQUATIONS

The equations set forth in the last two sections provide a satisfactory starting point for the analysis of the flow of most common fluids, such as air, water, gasoline, oils, and mercury, over an astonishingly wide range of thermo-dynamic conditions, rates of deformation, and temperature gradients.

As has been said above, there are only a few, highly special flow fields that serve as a direct testing ground for the basic hypothesis relating stress and rate of deformation, but confidence in the Navier–Stokes equations is bolstered by a long history of successful predictions of observable effects: successes that could hardly be explained if the fundamental equations were wrong. In one field, that of rarefied gas dynamics, one can use the more comprehensive kinetic theory of gases to identify conditions under which the Navier–Stokes equations begin to lose accuracy. The criteria that emerge from that study imply that the local behavior of gases at normal atmospheric densities should be accurately described by the Navier–Stokes equations under all conditions except those that occur within shock waves and sound fields of ultrahigh frequency. Even in turbulent flows, it appears that the highest rates of deformation are naturally constrained to be much smaller than the characteristic rate of the molecular collision processes that maintain the macroscopic relations on which the Navier–Stokes equations are based.

For liquids, there is no comparably reassuring molecular theory, but a large body of data demonstrates dynamical similarity of liquid and gas flows under conditions suggested by the Navier–Stokes equations. There is also a large body of experimental data on film lubrication, which involves enormous rates of deformation and exhibits good agreement with calculations based on the Navier–Stokes equations. There are, however, easily-made dilute solutions of long-chain polymers in water for which fundamentally and practically interesting deviations from Newtonian behavior have been intensively studied. As will be mentioned in Chapter 14, the addition of these polymers to water flowing turbulently along a smooth solid wall has a dramatic effect on important small-scale flow features, and may substantially reduce the mean stress exerted by the fluid on the wall.[6]

5.9 SUMMARY

In Chapter 2, it was shown that the state of stress at a point in a fluid can be represented by three mutually orthogonal principal stress vectors. In Chapter 3, it was shown that the rate of deformation can be similarly represented by three mutually orthogonal vectors, each of which indicates the proportional rate of extension or contraction of an infinitesimal material line segment. In this chapter, it is recognized that an equal part of each of the principal stresses is a thermodynamic pressure, which is independent of the rate of deformation. The remainder of each principal stress is called a viscous stress, and we postulate that it is linearly proportional to the corresponding principal rate of deformation. The coefficient of proportionality has the same value, 2μ, for each principal axis. We call μ the *viscosity,* and assert that its value depends only upon the thermodynamic state of the fluid, not on the state of motion. The mathematical expression of these postulates is called a constitutive equation.

Thermodynamic considerations show that $\mu \geq 0$, so that tension begets extension, while compression begets contraction.

There is room in the postulate for a second contribution to the principal viscous stresses, $\lambda \operatorname{div} \mathbf{u}$, equal for all three directions. Values of λ are not easily measured, because $|\lambda \operatorname{div} \mathbf{u}| \ll p$ in all but the most unusual circumstances. In subsequent chapters, we shall set λ equal to zero.

A companion postulate asserts that the conductive component of heat flux is negatively proportional to the temperature gradient. The coefficient of proportionality, $k,$ is called thermal conductivity; its values, like those of the viscosity, depend only upon the thermodynamic state. Again, thermodynamic arguments prove that $k \geq 0$.

When the constitutive equations are combined with the conservation equations, we get the Navier–Stokes equations. These, together with boundary conditions that prescribe continuity of velocity and temperature across an interface between fluid and solid or between immiscible fluids, and that relate possible discontinuities in stress or heat flux to surface tension or surface heat sources, provide a reliable theoretical base for the study of a wide range of important fluid flows. All that need to be added, in the way of physical

---

[6] Some of the intriguing phenomena observed in non-Newtonian liquids and in rarefied gases have been documented in the following educational films: *Rheological Behavior of Fluids,* by Hershel Markovitz, and *Flow of Rarefield Gases,* by F. C. Hurlbut and F. S. Sherman. Both films, and others recommended later in this text, were made under the sponsorship of the U.S. National Committee for Fluid Mechanics Films and the National Science Foundation, and are available from Encyclopedia Britannica Films, 425 N. Michigan Ave., Chicago, IL 60611.

information, are equations of state, which will be discussed briefly in the next chapter.

5.10 SAMPLE CALCULATIONS

A. The rotating-cylinder viscosimeter. Let us analyze the flow in a viscosimeter made of long coaxial rotating cylinders, to derive a calibration formula for the apparatus. We shall assume that both cylinders are rotating, although in practice the inner cylinder may be stationary.

In Chapter 3, the matrix for grad $\mathbf{u}$ was evaluated for the simple velocity field $\mathbf{u} = v(r)\mathbf{e}_\theta$. From the result, (3.24), the matrix for def $\mathbf{u}$ can readily be formed. We find that

$$(\text{def } \mathbf{u})_{r\theta} = (\text{def } \mathbf{u})_{\theta r} = \frac{dv}{dr} - \frac{v}{r} = r\frac{d(v/r)}{dr} \tag{5.16}$$

From the analysis of Section 5.3, we deduced that in the gap between coaxial, steadily-rotating cylinders $\tau_{r\theta} = Ar^{-2}$. Now, with the constitutive equation $\tau_{r\theta} = \mu(\text{def } \mathbf{u})_{r\theta}$, we can write

$$\frac{d(v/r)}{dr} = \frac{\tau_{r\theta}}{\mu}r^{-1} = \frac{A}{\mu}r^{-3}$$

Integrating this from $r = r_1$, where v/r has the given value ω_1; to $r = r_2$, where v/r has the given value ω_2, we find an equation for the constant A.

$$A = (\omega_2 - \omega_1)\left(\int_{r_1}^{r_2} \mu^{-1}r^{-3}\, dr\right)^{-1} \tag{5.17}$$

If the viscosity is constant,

$$A = 2\mu(\omega_2 - \omega_1)(r_1^{-2} - r_2^{-2})^{-1}$$

Denote by T_z the torque exerted on the inner cylinder, per unit length of cylinders. Then solve for the viscosity, given the torque, the angular speeds, and the radii. The calibration formula for this type of viscosimeter is thus

$$\mu = \frac{T_z}{4\pi}\frac{r_1^{-2} - r_2^{-2}}{\omega_2 - \omega_1} \tag{5.18}$$

B. Viscous dissipation and heat conduction. It is quite conceivable that the viscosity may *not* be constant in the example above, particularly if the fluid is an oil. The internal energy generated by viscous dissipation in oil is not easily removed by heat conduction, and the viscosity of oil may depend quite strongly on temperature. Let us take an initial look at this problem, estimating the temperature distribution in the fluid with an approximate analysis that assumes that both μ and k are constant.

To be specific, assume that the inner cylinder is insulated, while the outer cylinder is maintained at room temperature. At steady state, the temperature should depend only on the radial position, r. If the cylinder axis is vertical, the fluid particles will move in horizontal planes, and each particle will experience a constant pressure. Thus the entropy of each particle remains constant, and Eq. (4.22) reduces to $\Phi = \text{div } \mathbf{q}$.

Since $\text{div } \mathbf{u} = 0$ in this flow, the general expression for the viscous dissipation function can be written directly in terms of the viscous stress tensor:

$$\Phi = \frac{1}{2\mu} \tau \cdot \tau \tag{5.19}$$

In this simple flow, the only nonzero components of τ are $\tau_{r\theta}$ and $\tau_{\theta r}$, which are equal. Using the result $\tau_{r\theta} = Ar^{-2}$, which was established for this simple flow in Section 5.3, we get $\Phi = (A^2/2\mu)r^{-4}$.

If T depends only on r, then $\mathbf{q}$ will have only a radial component,

$$q_r = -k \frac{dT}{dr}$$

Applying the coordinate-free definition of $\text{div } \mathbf{q}$ to the elemental volume shown in Fig. 2.3, we easily show that, in this simple case

$$\text{div } \mathbf{q} = r^{-1} \frac{d(rq_r)}{dr}$$

Combining all these results, we get

$$\Phi = \frac{A^2}{2\mu} r^{-4} = \text{div } \mathbf{q} = -r^{-1} \frac{d(rk \, dT/dr)}{dr}$$

whence we get a differential equation for T alone:

$$\frac{d(rk \, dT/dr)}{dr} = -\frac{A^2}{2\mu} r^{-3} \tag{5.20}$$

The insulation of the inner wall means that $dT/dr = 0$ at $r = r_1$. Integrate (5.20) from $r = r_1$ to some larger value $r \le r_2$, using this boundary condition. The result is

$$\frac{dT}{dr} = -\frac{A^2}{rk} \int_{r_1}^{r} \mu^{-1} r^{-3} \, dr \tag{5.21}$$

Call room temperature T_2, and integrate again, this time inward from r_2, where the wall is at room temperature. The result is

$$T = T_2 - \int_{r_1}^{r_2} \frac{dT}{dr} \, dr = T_2 + A^2 \int_{r}^{r_2} (kx)^{-1} \left\{ \int_{r_1}^{x} \mu^{-1} z^{-3} \, dz \right\} dx \tag{5.22}$$

This looks formidable, but the integrals are easy to evaluate if μ and k are constants. The result is

$$T = T_2 + \frac{A^2}{2\mu k} \left[r_1^{-2} \ln \left(\frac{r_2}{r} \right) - \tfrac{1}{2}(r^{-2} - r_2^{-2}) \right] \tag{5.23}$$

When A is replaced by the expression found in calculation (A), we get our final answer. The highest temperature occurs at the insulated wall; if the inner wall is stationary, the highest temperature is given by

$$T_1 = T_2 + \frac{\mu}{k}(\omega_2 r_2)^2 f(x)$$

where

$$x = \left(\frac{r_2}{r_1}\right)^2 \quad \text{and} \quad f(x) = \frac{x \ln x + 1 - x}{(x - 1)^2}$$

The last function decreases slowly from a value of $1/2$ at $x = 1$ (the case of vanishing clearance between cylinders), to 0.368 when $r_2 = 1.5r_1$. To estimate whether a significant temperature rise occurs, one can use the first limit. Noting that $v_2 = \omega_2 r_2$, we write

$$T_1 \cong T_2 + \frac{\mu}{2k} v_2^2$$

To assess the practical significance of this, one needs some typical values of μ and k, which will be discussed in the next chapter.

C. Given the velocity field, in cylindrical polar coordinates,

$$\mathbf{u} = \frac{\Gamma}{2\pi r}\left[1 - \exp\left(\frac{r^2}{4vt}\right)\right]\mathbf{e}_\theta$$

(a) Calculate all nonzero components of the pressure gradient, assuming $\mathbf{f} = 0$. To do this, one must evaluate $\rho \, D\mathbf{u}/Dt$ and div τ, and then find ∇p by subtraction, using the momentum equation.

The local acceleration is

$$\mathbf{u}_{,t} = \frac{\Gamma}{2\pi r}\frac{r^2}{4vt^2}\exp\left(\frac{r^2}{4vt}\right)\mathbf{e}_\theta$$

and the convective acceleration is

$$(\mathbf{u} \cdot \nabla)\mathbf{u} = \frac{v}{r}\mathbf{u}_{,\theta} = -\frac{v^2}{r}\mathbf{e}_r$$

$$= -\left(\frac{\Gamma}{2\pi}\right)^2 r^{-3}\left[1 - \exp\left(\frac{r^2}{4vt}\right)\right]^2 \mathbf{e}_r \qquad (5.24)$$

Given the *radial symmetry* of the flow, and the lack of any body force, we conclude that the pressure will depend only on r and t; thus $\nabla p = p_{,r}\mathbf{e}_r$.

To evaluate div τ, we can use the fact, discussed in Sample Calculation A, that the only nonzero components of τ are $\tau_{r\theta}$ and $\tau_{\theta r}$. Further, it was shown that these components are, for this flow, independent of θ. A force balance on the elemental volume sketched in Fig. 2.3 immediately shows that div τ has no radial component. Hence the r-component of the momentum

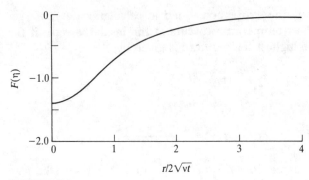

FIGURE 5.1
Dimensionless pressure distribution in a diffusing line vortex.

equation reduces to

$$p_{,r} = \frac{\rho v^2}{r} \tag{5.25}$$

The necessary expression for v^2/r is given in Eq. (5.24).

(b) Calculate the pressure, $p(r, t)$, given $p(\infty, t)$. To prepare for the integration over r, we introduce the shorthand $\eta = r^2/4vt$. Then we have, with a little algebra which you can supply,

$$p(r) - p(0) = \rho \left(\frac{\Gamma}{2\pi}\right)^2 \frac{1}{8vt} \int_0^\eta [1 - \exp(-x)]^2 x^{-2}\, dx$$

This integral looks formidable, but it is easily evaluated numerically. A check on accuracy is given by the known value of the integral for $\eta = \infty$, which is $2\ln 2$. Using this value, we get

$$p(\infty) - p(0) = \frac{\ln 2}{16\pi^2} \frac{\rho\Gamma^2}{vt}$$

The pressure profile is expressed as $p(r, t) - p(\infty) = (\rho\Gamma^2/32\pi^2 vt)F(\eta)$. Figure 5.1 shows $F(\eta)$ versus $\eta^{1/2}$, which more directly reveals the dependence of p upon r.

EXERCISES

5.1. Given the velocity field $\mathbf{u} = U \, \mathrm{erf}\,[y/\delta(t)]\mathbf{e}_x$:
   (a) Calculate all nonzero components of the viscous stress tensor, τ, as functions of y.
   (b) Calculate the viscous force per unit volume, div τ, as a function of y.
   (c) Calculate the dissipation function, Φ, as a function of y.
   (d) Calculate the integral of Φ, from $y = 0$ to $y = \infty$.
5.2. Given the velocity field, in cylindrical polar coordinates, $\mathbf{u} = U\{1 - (r/a)^2\}\mathbf{e}_z$, in the domain $0 \le r \le a$.

(a) Calculate all nonzero components of the viscous stress tensor, τ, as functions of r.

(b) Calculate the dissipation function, Φ, as a function of r.

(c) Calculate the integral of Φ over the circular area $0 \le r \le a$.

5.3. Given the velocity field, in spherical polar coordinates,

$$\mathbf{u} = U\left\{\cos\theta\left[1 - \left(\frac{a}{R}\right)^3\right]\mathbf{e}_R - \sin\theta\left[1 + \frac{1}{2}\left(\frac{a}{R}\right)^3\right]\mathbf{e}_\theta\right\}$$

(a) Calculate all nonzero components of the viscous stress tensor, τ, as functions of R and θ.

(b) Calculate the dissipation function, Φ, as a function of R and θ.

(c) Calculate the integral of Φ over the volume $a \le r \le \infty$.

5.4. Consider the ratio of the viscous normal stress to the pressure, at the interface between a Newtonian fluid and a plane, rigid wall.

(a) Letting the wall be normal to the x axis, derive a formula for τ_{xx}/p, as a function of $D\rho/Dt$ and the appropriate properties of the fluid. (*Hint:* use the continuity equation, the no-slip condition, and the fact that the wall is rigid.)

(b) What would have to be the proportional rate of change of density, to make $\tau_{xx}/p = 0.01$ in air at S.T.P.? (Property data are in Chapter 6.)

5.5. Supply the logical and analytical details necessary to prove the results cited in Eq. (5.9).

CHAPTER
6

PHYSICAL
PROPERTIES
OF FLUIDS

6.1 INTRODUCTION

Many of the most important concepts of viscous flow can be understood without much knowledge of the behavior of physical properties, such as viscosity, as functions of temperature and pressure, because the variability of the properties introduces no qualitatively novel flow features. Water and air, for example, are often used interchangeably in experiments aimed at the elucidation of some major phenomenon, such as the transition from laminar to turbulent flow in a boundary layer, although the viscosity of water decreases with increasing temperature, while that of air increases.

Important phenomena do, however, depend on the variability of properties, and a brief discussion will prepare you to anticipate some of these.

Tabular values of some fluid properties are given here for handy reference. More extensive data collections can be found, for example, in the *Handbook of Chemistry and Physics,* but these will suffice for the purposes of this book. The tables will also serve as a reminder of the basic dimensions and units associated with each property, preparing the way for applications of dimensional analysis which appear in later chapters.

The identification of conditions under which it may be quantitatively acceptable to ignore property variations entirely follows from the information given here, but involves a special analysis, which is given in Chapter 7.

6.2 DENSITY

A convenient and frequently used differential equation of state for density is

$$\frac{d\rho}{\rho} = -\beta \, dT + \kappa \, dp \tag{6.1}$$

in which β is the *coefficient of thermal expansion*, and κ is the *isothermal compressibility*. Both β and κ are typically positive; a famous exception is that of water near the freezing point, for which β is negative, giving rise to fascinating effects in the flow around melting blocks of ice.

For a *thermally perfect gas*, we can differentiate the algebraic equation of state, $p = \rho R T$, to show that $\beta = T^{-1}$ and $\kappa = p^{-1}$.

A second convenient differential form expresses the dependence of density on pressure and *entropy*. This is

$$\frac{d\rho}{\rho} = -\frac{\beta T}{c_p} \, dS + \kappa_s \, dp \tag{6.2}$$

in which c_p is the specific heat at constant pressure, and κ_s is the *isentropic compressibility*. There is an important relationship between κ_s and the *speed of sound*: $\kappa_s = (\rho c^2)^{-1}$. For a thermally and calorically perfect gas, $c_p = \gamma R / (\gamma - 1)$ and $c^2 = \gamma R T$. The ratio of specific heats, γ, varies between 1 and 5/3. A good approximation for air near room temperature is $\gamma = 7/5$. In liquids, κ_s and κ are very nearly equal.

We note that βT, κp, and $\kappa_s p$ are all dimensionless, so they provide convenient measures by which to compare different fluids. A few sample data are given in Table 6.1. Dimensional quantities are evaluated in SI units throughout this text.

Gases are colloquially said to be compressible, and liquids incompressible. This corresponds to the fact that κp has much larger values in gases than

TABLE 6.1
Equilibrium properties of sample fluids at $p = 1$ atm, $T = 293$ K

Fluid	ρ kg m	βT	κp	C_p J kg-K	γ	c m/s
Air	1.205	1	1	1005	1.40	344
Helium	0.167	1	1	5227	1.67	1010
CO_2	1.841	1	1	832	1.30	269
Water	1,000	0.0607	4.91×10^{-5}	4182	1.006	1461
Glycerol	1,250	0.148	2.2×10^{-6}	2333	1.134	2044
Mercury	13,579	0.0533	3.76×10^{-6}	1391	1.014	1409

in liquids. We note from the table that gases are also more subject to thermal expansion.

6.3 SPECIFIC HEAT

Typical values of the specific heats of gases are not much different from those of liquids, and there is little sensitivity to changes in temperature in either case. There are important exceptions to the latter statement in the case of very hot gases, in which quantized internal energy states associated with molecular rotations, vibrations, dissociation, electronic excitation or ionization, become accessible as temperature increases. These phenomena are, however, more important in the study of compressible flow, than in our present study.

6.4 VISCOSITY

Gases. In *gases,* at densities far below the critical-point density, viscosity is proportional to the mass and mean random speed of the molecules, and to the inverse of a molecular collision cross section.[1] None of these quantities depends on the number density of the molecules, so the viscosity depends only on temperature. The mean random speed is proportional to $(RT)^{1/2}$, and the collision cross section ordinarily decreases somewhat as the mean speed increases. The overall effect is well fitted by the Sutherland law,

$$\mu = \mu_0 \left(\frac{T}{273.2} \right)^{3/2} \frac{C + 273.2}{C + T} . \tag{6.3}$$

The constant, C, varies somewhat from gas to gas; some values are given in Table 6.2.

A quantity of interest in discussions of the dynamical effects of viscosity variations is the *dimensionless sensitivity,*

$$S_\mu = \frac{T}{\mu} \frac{d\mu}{dT} \tag{6.4}$$

TABLE 6.2
Viscosity trends for gases

Gas	μ (293 K) kg/m-s	ν (293 K, 1 atm) m²/s	C K	S_μ (293 K)
Air	1.81×10^{-5}	1.50×10^{-5}	111	0.77
Helium	1.97×10^{-5}	1.18×10^{-4}	79.4	0.71
CO$_2$	1.46×10^{-5}	0.79×10^{-5}	222	0.93

[1] A clear exposition of the essential ideas is given in Sir James Jeans, *An Introduction to the Kinetic Theory of Gases,* Cambridge University Press, 1940.

For the Sutherland law, we find $S_\mu = (T + 3C)/2(T + C)$. For air ($C \cong 111$ K), at $T = 288$ K, the sensitivity is about 0.78, so a 10% change in T produces only a 7.8% change in viscosity.

Notice that the SI units for viscosity are kg/m-s. Remembering that the SI name for the unit of pressure is the *pascal*, ($1 \text{ Pa} = 1 \text{ Newton/m}^2$), we see that the units of viscosity can be cited as pascal-seconds.

Liquids. The molecular picture of liquid behavior is much more complex, and less well established, than the kinetic theory of gases. There are at least three qualitatively different kinds of pure liquids, all of which exhibit Newtonian behavior. These are *liquid metals,* such as mercury; *polar liquids,* such as water; and *nonpolar liquids,* such as glycerol. The differences are associated with the degree of ionization and the electrical conductivity; liquid metals are good conductors, nonpolar liquids are good insulators, and polar liquids fall somewhere in between.

Most theories of the viscosity of liquids emphasize the concept of a potential energy barrier which must be surmounted by molecules that are being displaced past each other in a macroscopically deforming fluid. Viscosity is imagined to be roughly analogous to the resistance one would encounter in trying to pull one corrugated sheet of metal past another one, the two being constrained from moving apart laterally. The viscosity is observed to decrease as temperature increases, and there is a substantial correlation between the decrease in viscosity and the corresponding thermal expansion of the liquid. The correlation is of the form $\mu \approx 1/(v - v^*)$, which represents a very rapid decrease of μ as the specific volume increases away from a value at which μ would be infinite. The correlation is not universal, however, failing for water in the notorious range of temperature 0–4°C, where μ and v simultaneously decrease as T increases. However, it reinforces the idea of a potential energy barrier, and suggests that the temperature dependence of μ should contain a factor $\exp(E^*/kT)$.[2]

Tolerably accurate fitting formulas for measured viscosities of liquids can indeed be found, in the form

$$\mu(T) = \mu(T_0)\left(\frac{T}{T_0}\right)^n \exp[B(T^{-1} - T_0^{-1})] \qquad (6.5)$$

to which corresponds the following formula for the sensitivity:

$$S_\mu = n - \frac{B}{T} \qquad (6.6)$$

---

[2] See Chapter 16 of E. A. Moelwyn-Hughes, *Physical Chemistry* Second Revised Edition, Pergamon Press, 1961, for an extensive discussion of the liquid state.

TABLE 6.3
Viscosity trends for liquids

Liquid	n	B K	μ (293 K) kg/m-s	ν (293 K) m²/s	S_μ (293 K)
Mercury	0.32	414	1.554×10^{-3}	1.554×10^{-7}	¬1.09
Water	8.9	4,700	1.005×10^{-3}	1.005×10^{-6}	-7.1
Glycerol	52.4	23,100	1.49×10^{1}	4.19×10^{-2}	-26.4

Values of n, B, μ (293 K), and S_μ (293 K) are presented in Table 6.3, for mercury, water, and glycerol. Note how very sensitive the viscosity of glycerol is to temperature; a 1% change in T produces approximately a 26% change in μ at $T = 293$ K! Large sensitivities are typical of many oils, and have great practical importance for the design of lubrication systems.

Kinematic viscosity. When a fluid's resistance to deformation is compared to its resistance to acceleration, the ratio of viscosity to density appears as the important fluid property. This ratio is the *kinematic viscosity*,

$$\nu = \frac{\mu}{\rho} \tag{6.7}$$

We note that water is more viscous than air when values of μ are compared; but that air has about 14 times the kinematic viscosity of water, at STP. Note also that it is relatively easy to vary the kinematic viscosity of gases by several orders of magnitude, by pressurizing or partially evacuating the test apparatus.

Mixed liquids. For laboratory investigations of viscous flows, it is often desirable to be able to vary the viscosity. This can be done very conveniently with aqueous solutions of glycerin, which are easy to prepare and handle. The first 1% addition of glycerin to pure water increases the viscosity by about 2.5%; the first 1% addition of water to pure glycerin decreases the viscosity by about 20%![3] Over the full range of concentration, the viscosity varies by about a factor of 1500, at room temperature. Fortunately, the diffusivity of glycerin in water is very small, so the mixture, once well prepared, tends to remain homogeneous.

6.5 THERMAL CONDUCTIVITY

For gases the molecular theory of thermal conductivity parallels that of viscosity, but the same cannot be said for liquids.

---

[3] Data for 1% increments of concentration can be found in the *Handbook of Chemistry and Physics*.

Gases. For gases, the kinetic theory shows that the thermal conductivity is directly proportional to the viscosity and the specific heat. A useful formula, due to Euken, is

$$k = \frac{9\gamma - 5}{4\gamma} \mu c_p \qquad (6.8)$$

This shows that the dimensionless *Prandtl number*, $\text{Pr} = \mu c_p / k$, is nearly unity in gases. The formula gives $\text{Pr} = 2/3$ for monatomic gases ($\gamma = 5/3$), and $\text{Pr} = 14/19$ for diatomic gases. The latter value is a good estimate for air.

Liquids. The thermal conductivity of liquids is much less sensitive to temperature changes, than is viscosity. This may be because much of the internal energy of liquids is associated with the vibration of molecules around equilibrium positions in temporarily ordered lattices. Molecules do not have to overcome the same potential energy barrier to transfer energy as they must to achieve macroscopic relative displacements, and there is no obvious correlation between thermal conductivity and specific volume.

There is a large difference, however, between the thermal conductivities of different liquids. Liquid metals, such as mercury or molten sodium, have very large values of k because they contain free electrons, which are effective carriers of internal energy as well as of electric charge. Polar liquids contain some free ions, but these are relatively immobile, compared to electrons, so the thermal conductivities of polar and nonpolar liquids do not differ much. Many lubricating oils have quite low values of k.

6.6 THERMAL DIFFUSIVITY AND PRANDTL NUMBER

The quantity that most directly affects the transient evolution of temperature fields due to heat conduction is the *thermal diffusivity*,

$$\alpha = \frac{k}{\rho c_p}$$

TABLE 6.4
Thermal conductivity

Fluid	k (293 K) W/m-K	α (293 K, 1 atm) m^2/s	Pr (293 K)
Air	0.0256	2.11×10^{-5}	0.71
Helium	0.147	1.68×10^{-4}	0.70
CO_2	0.0158	1.03×10^{-5}	0.77
Water	0.597	1.43×10^{-7}	7.03
Glycerin	0.264	0.91×10^{-7}	4.6×10^5
Mercury	8.69	4.66×10^{-6}	0.0249

Its dimensions are the same as those of kinematic viscosity, or of a molecular diffusion coefficient. The Prandtl number, introduced above, is just the dimensionless ratio v/α. Liquids exhibit a wide range of values Pr; sample data are shown in Table 6.4.

6.7 SURFACE TENSION

Surface tension and its variability interact with viscosity in a number of technically very important processes, sometimes with intuitively unexpected effects. The more familiar phenomena involve droplets or bubbles, which tend towards a spherical shape because of the jump in normal stress associated with surface tension.[4]

Surface tension depends somewhat on temperature, in a pure liquid. More importantly, it depends strongly on the nature of the molecules that congregate at the surface, and hence is easily modified by the addition of relatively tiny amounts of specific impurities, called *surfactants*. This is particularly true of water. In many circumstances, the concentration of surfactants varies over the interface, inducing a gradient of surface tension. Usually, the surface tension is low where the concentration of surfactants is high, and vice versa. The consequence, as shown in Section 4.9, is a discontinuity in tangential stress, which may cause a profound alteration of the flow field. The implications are of interest in fields as diverse as chemical engineering and naval architecture.[5]

The foregoing discussion treats surface tension as something static, more akin to elasticity than to viscosity, but in impure fluids there is often a dynamic aspect that complicates analysis. If a material portion of interface is suddenly stretched, say by the passage of a surface wave, new sites for impurity molecules become available at the interface. They are not instanteously filled, however, because it takes time for the impurities to reach the surface by some combination of convection and diffusion. The surface tension is still a fluid property that, like viscosity, is conceptually independent of the fluid motion, but its value at any given place and time can be strongly affected by the motion. The interactions are frequently so complex that it is difficult to formulate mathematical models and to test their effectiveness. A concept of

---

[4] The phenomenon we have treated by endowing the interface with a mechanical property, surface *tension*, is more fundamentally treated by the concept of an excess *energy* possessed by surface molecules, by virtue of their location. This concept is combined with a postulate that the fluid, say of a droplet, tends towards an equilibrium of minimum energy, including the surface energy. Since the spherical shape minimizes the ratio of surface to volume, it is the equilibrium shape.

[5] A fascinating peek at some of these phenomena is provided by the NCFMF film, *Surface Tension in Fluid Mechanics*, by L. Trefethen. For a very helpful review, which informed much of what is said here about interfacial properties, see J. C. Berg, *Interfacial Phenomena in Fluid Phase Separation Processes*, in Vol. II of Recent Developments in Separation Science, CRC Press, 1972.

TABLE 6.5
**Surface tension of pure liquids,
against air**

Fluid	Σ (293 K) N/m
Water	0.073
Glycerol	0.063
Mercury	0.487

surface viscosity has been advanced to explain some of the dynamic effects, but it seems that the concept of variable surface tension is sufficient to explain the most dramatic ones.

A few values of surface tension, for *pure* liquids against air, are listed in Table 6.5.

CHAPTER
7

FLOWS WITH NEARLY CONSTANT DENSITY AND TRANSPORT PROPERTIES

7.1 INTRODUCTION

In many viscous flows, some of the physical properties that appear in the conservation equations remain so nearly constant that they can be treated mathematically as given parameters, rather than as dependent variables. This is often true of the density, and of the transport properties such as viscosity and thermal conductivity. The complexity of the mathematical analysis is thereby reduced; for example, the coupling between the energy equation and the other two conservation equations is broken if ρ and μ are constant.

There, is however, a risk that qualitatively important phenomena may be excluded from theoretical description if these approximations are made too quickly, without sufficient attention to their consequences. For example, the propagation of sound waves involves only tiny variations in density, as does the mild free convection that redistributes nutrients and wastes in lakes and oceans. These phenomena are inconceivable in a fluid of strictly constant density.

The objective of this chapter is to show how to estimate in advance, by evaluating appropriate dimensionless parameters, whether acceptable accuracy can be obtained with a theory that ignores variations of density and/or transport properties.

Sections 7.2 and 7.3 deal with density variations: first with their dynamic effects, and then with their kinematic implications. Section 7.4 introduces the concept of *anelastic flow*, the theory of which rules out acoustic effects while still allowing significant variability of ρ, μ, and k. Sections 7.5 and 7.6 deal with the variations of transport properties.

7.2 DYNAMIC EFFECTS OF DENSITY VARIATIONS

Density variations can affect the motion of a fluid in a variety of ways, some of which can be associated quite directly with individual terms in the conservation equations. Here are some examples.

1. A local change of density can upset the balance of forces that keeps a fluid in hydrostatic equilibrium, and hence can be the agency that initiates motion. The result is called *free convection* or *natural convection*. To account for it theoretically, ρ must be allowed to vary in the expression for the gravitational body force, $\mathbf{f} = \rho\mathbf{g}$, even if it is treated as a constant parameter wherever else it appears in the conservation equations.
2. If there is already a nonzero force per unit volume, which is uniform throughout a finite body of fluid, spatial variations of density will result in nonuniform acceleration. In an adequate theory, ρ must be allowed to vary in the term $\rho\,D\mathbf{u}/Dt$.

7.3 KINEMATIC EFFECTS OF DENSITY VARIATION

The continuity equation gives a direct connection between the proportional rate of change of density, and the kinematic quantity div $\mathbf{u}$. In Section 3.3, we have seen how a field of div $\mathbf{u}$ makes a contribution $\mathbf{u}_d$ to the velocity at any point in the fluid. It is not easy to calculate $\mathbf{u}_d$ in your head, given some complicated distribution of div $\mathbf{u}$, but it is easy to imagine that in many circumstances $\mathbf{u}_d$ may be negligible compared to the sum of $\mathbf{u}_v$ and $\mathbf{u}_p$. Obviously, $\mathbf{u}_d$ vanishes if ρ is strictly constant; what is now needed is a systematic way of anticipating the conditions in which it will be negligible, although not zero. We shall analyze a pair of examples in some detail, hoping that the method is thus made clear. The first step is to identify, for each case, the specific benefit to be expected if one can safely assume that div $\mathbf{u} = 0$.

EFFECT OF EXPANSION ON ACCELERATION. As will be seen in Chapter 10, an important class of special solutions of the conservation equations describes flow in which the convective acceleration, $(\mathbf{u} \cdot \nabla)\mathbf{u}$, is absent, because of constraints imposed by the bounding walls and by the equation div $\mathbf{u} = 0$. These are flows in pipes, and between parallel walls.

Suppose that $\mathbf{u} = u(x, y)\mathbf{e}_x$. The flow is maintained by a pressure gradient $p_{,x}$ and we wish to be assured that $\rho u u_{,x} \ll p_{,x}$. The first step is to use the continuity equation, in this case simply $\rho u_{,x} = -u\rho_{,x}$, to eliminate $u_{,x}$. This turns attention to the search for conditions that guarantee that $u^2 \rho_{,x} \ll p_{,x}$.

Consider *isothermal* flow of a perfect gas. Then $\rho_{,x} = p_{,x}/RT$, and the criterion becomes $u^2 \ll RT = c^2/\gamma$, where c is the speed of sound. In terms of the *Mach number*, $M = u/c$, the condition becomes $M^2 \ll 1/\gamma$. It is shown in the one-dimensional treatment of this flow in introductory texts that a cross-sectional average volume of M increases steadily with distance down the pipe. Thus, if $M^2 \ll 1/\gamma$ at the pipe exit, we can safely neglect $(\mathbf{u} \cdot \nabla)\mathbf{u}$ everywhere.

EFFECT OF EXPANSION ON SHAPE OF STREAMLINES. In analysis of a less-constrained flow, such as that in a laminar boundary layer, one cannot expect to neglect $(\mathbf{u} \cdot \nabla)\mathbf{u}$, but may want to compute the field of one velocity component, say $v(x, y)$, given the field of the other component, $u(x, y)$. The exact continuity equation gives $v_{,y} = -u_{,x} - \rho^{-1} D\rho/Dt$, and one needs to know when this can accurately be replaced by $v_{,y} = -u_{,x}$. Thus, $|\rho^{-1} D\rho/Dt|$ must be small, not just in some abstract sense, but in comparison to $|u_{,x}|$.

To relate $r^{-1} Dr/Dt$ to quantities that may be more easily estimated, we combine Eqs. (6.2) and (4.22) into

$$\rho^{-1}\frac{D\rho}{Dt} = (\rho c^2)^{-1}\frac{D\rho}{Dt} - \frac{\beta}{\rho c_p}(\Phi - \text{div }\mathbf{q}) \tag{7.1}$$

This equation identifies three ways to change the density of a given fluid particle:

1. Isentropic compression
2. Viscous dissipation, accompanied by thermal expansion
3. Heating by conduction or radiation, accompanied by thermal expansion

As a rule of thumb, (1) is important in high-speed gas flows, and in sound waves propagating through either liquids or gases; (2) is important in high-speed gas bearings and in supersonic boundary layers; and (3) is important in fires and in boundary layers on very hot or cold bodies.

Our present objective is to see how to restrict the values of the flow parameters such as flight speed, wall temperature, bearing clearance, chimney height, etc., so that one may successfully set div $\mathbf{u} = 0$ in a theoretical study of a given flow. If one already has a theoretical description that includes this restriction, it can be used to estimate the size of terms omitted from the more complete theory. In particular, one can estimate typical values of $D\rho/Dt$, Φ, and div $\mathbf{q}$, and then see how each contribution to $|\rho^{-1} D\rho/Dt|$ compares to

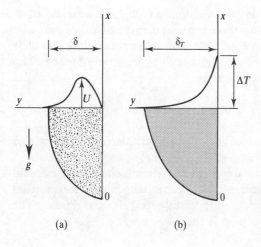

FIGURE 7.1
Definition sketch for free convection on vertical flat plate: (a) velocity; (b) temperature.

$|u_{,x}|$. By seeing how each comparison depends on certain dimensionless flow parameters, one learns how the values of these parameters must be restricted. Let us try this for a sample flow: laminar free convection along a vertical flat plate. Suppose the flow is steady, and that x increases upward from the bottom edge of the plate. A sketch of the flow, to define symbols, is given in Fig. 7.1.

A detailed theoretical description of this flow may be found in introductory texts on heat transfer. All we need for present purposes is the general notion that in the accounting for vertical momentum three groups of terms are typically of equal importance. These terms describe

1. The buoyancy force per unit mass, which is proportional to $\beta \Delta T g$
2. The viscous force per unit mass, which is proportional to $\nu U / \delta^2$
3. The vertical acceleration, which is proportional to U^2 / x

The statement that these effects are equally important then yields the estimates

$$U^2 = \approx \beta \Delta T gx \qquad \text{and} \qquad \delta^2 \approx \frac{\nu x}{U}$$

The corresponding estimate of $u_{,x}$ is simply U/x.

To estimate the effect of isentropic compression, assume that the vertical pressure gradient is little changed from its hydrostatic value, which is $p_{,x} = -\rho g$. Then $D\rho/Dt \approx up_{,x} \approx -\rho g U$. With these estimates, we get $|(\rho c^2)^{-1} Dp/Dt|/|u_{,x}| \approx gx/c^2$, and arrive at the restriction of parameters,

$$\frac{gx}{c^2} \ll 1. \tag{7.2}$$

If the fluid is air, for which $c \approx 350 \text{ m/s}$, this means $x \ll 12$ kilometers, a condition easily satisfied in most applications.

Next we estimate the effect of viscous dissipation. A good estimate of Φ in a boundary layer is $\mu(U/\delta)^2$. Using the former estimates of U, δ, and $u_{,x}$, we find that $\beta\Phi/\rho c_p u_{,x} \approx (\beta^2/c_p)\,\Delta T gx$. We can introduce more familiar thermodynamic quantities by use of the identity[1] $\beta^2/c_p = (\gamma - 1)/Tc^2$. The corresponding restriction of parameters is thus

$$\frac{\Delta T}{T}\frac{gx}{c^2} \ll 1 \qquad (7.3)$$

Since $\Delta T/T$ is not likely to be much larger than unity, this restriction is also easily met in practice.

Finally, we use the estimate $\operatorname{div}\mathbf{q} \approx k\,\Delta T/\delta_T^2$, in which δ_T is the thickness of the thermal boundary layer. The restricted theory shows that $\delta_T^2/\delta^2 \approx k/\mu c_p = 1/\text{Pr}$; so the estimate for $\beta\operatorname{div}\mathbf{q}/\rho c_p$ is $\beta\,\Delta T U/x$. The restriction that makes $|\beta\operatorname{div}\mathbf{q}/\rho c_p| \ll |u_{,x}|$ is thus

$$\beta\,\Delta T \ll 1 \qquad (7.4)$$

This is, by far, the most binding restriction. It shows that $\operatorname{div}\mathbf{u}$ may be safely ignored in a theory of free convection around slightly warm or cool bodies, but must be retained in a theory of fires.

Other sample flows, defined by different parameters, could be analyzed in this same way. For example, if a driving pressure difference Δp is given, the associated restriction is determined by the isentropic compression and is $\Delta p/\rho c^2 \ll 1$. If a flight speed U is given, the restriction may be determined by isentropic compression, in which case it is $(U/c)^2 \ll 1$, or by viscous dissipation, in which case it is $\beta U^2/c_p \ll 1$. For a perfect gas, these last two restrictions amount to the same thing: the flow must be very *subsonic*.

If the fluid is driven by an oscillating boundary, with given frequency f and velocity amplitude U, the restriction again comes from isentropic compression. It is

$$\frac{U}{c}\frac{fL}{c} \ll 1, \qquad (7.5)$$

in which L is a typical diameter of the flow field. Since c/f is the wavelength of a sound wave of frequency f, the restriction involves both the amplitude of the sound waves, U/c, and the phase shift associated with them. If $fL/c \ll 1$, the motion associated with the sound wave is all in phase, and the effects of compressibility will not be noticed.

7.4 TEMPERATURE VARIATIONS

Values of the transport properties, μ and k, depend mostly on temperature. A theory that ignores variations of μ and k will be accurate only if $(\Delta T/T)S_\mu \ll 1$

---

[1] See, for example, Section 8.7 of W. C. Reynolds and H. C. Perkins, *Engineering Thermodynamics*, McGraw-Hill, New York, 1977.

and $(\Delta T/T)S_k \ll 1$. Here we use the dimensionless temperature sensitivities, defined by Eqs. (6.4) and (6.6). The ΔT in these inequalities represents a typical temperature change in the flow, which may be estimated if one knows DT/Dt and a typical time over which DT/Dt acts. For steady flows, characterized by a typical speed U, and a typical particle path length x, one can use the estimate $\Delta T \approx (x/U)DT/Dt$. For the rate of change of T, we use Eq. (4.24),

$$\frac{DT}{Dt} = \frac{\beta T}{\rho c_p}\frac{Dp}{Dt} + \frac{1}{\rho c_p}(\Phi - \text{div } \mathbf{q})$$

Specific classes of flow must again be analyzed one-by-one. For the example of free convection, we already have estimates of Dp/Dt, Φ, and div $\mathbf{q}$, and find the corresponding restrictions if viscosity variations are to be ignored:

1. From isentropic compression: $(\gamma - 1)(\beta T)^{-1}S_\mu(gx/c^2) \ll 1$
2. From viscous dissipation: $(\gamma - 1)S_\mu(\beta \Delta T)(gx/c^2) \ll 1$
3. From heat conduction: $S_\mu(\Delta T/T) \ll 1$

Again, the dominant cause of temperature variation is heat conduction, as it is in many subsonic flows. When this is true, the temperature history of a particle is accurately described by the equation

$$\frac{DT}{Dt} = (\rho c_p)^{-1} \text{ div } (k \text{ grad } T) \tag{7.6}$$

A conservation equation in this simplified form is called a *convection-diffusion* equation.

Before leaving this topic, consider a very different but fairly common example, for which the foregoing estimates are inappropriate. It involves the flow in a journal bearing, or rotating-cylinder viscosimeter, and has been substantially analyzed in Sample Calculation B of Chapter 5. In that steady flow, $DT/Dt = 0$, but T may vary strongly with r. The restricted analysis showed that ΔT, in this case the temperature change across the annulus of fluid, is approximately equal to $(\mu/2k)U^2$ if U designates the peripheral speed of the rotating cylinder. The corresponding restrictions on U, if variations of μ and k are to be negligible, are $(\mu U^2/2kT)S_\mu \ll 1$ and $(\mu U^2/2kT)S_k \ll 1$, respectively.

Some sample physical data are given below.

Engine oil at 40°C	$\mu/kT = 5.3 \times 10^{-2}(\text{s/m})^2$
Water at 40°C	$\mu/kT = 5.8 \times 10^{-6}(\text{s/m})^2$
Air at 40°C	$\mu/kT = 2.5 \times 10^{-6}(\text{s/m})^2$
Mercury at 40°C	$\mu/kT = 6.3 \times 10^{-7}(\text{s/m})^2$

Typical oils have relatively high values of μ/kT because they are very viscous and poor conductors of heat. If $U = 10 \text{ m/s}$, the oil would get intolerably hot.

It must not be allowed to come to the equilibrium represented by $DT/Dt = 0$, but must be recirculated through a cooler. At this same value of U, the other three fluids would hardly warm up at all. On the other hand, they would have very little load-bearing capacity, as we shall see in Chapter 11.

7.5 ANELASTIC FLOW

The term *incompressible flow* is literally equivalent to the equation $D\rho/Dt = 0$. It implies that the density of each fluid particle never changes, although different particles may have different densities. The term *anelastic flow* is newer, and may not have such a settled meaning. It usually implies that the density of a particle varies only as a result of isobaric thermal expansion, so that the differential equation of state, Eq. (6.1), can be approximated by

$$\frac{D\rho}{Dt} = -\rho\beta\frac{DT}{Dt} \tag{7.7}$$

For the temperature, Eq. (4.24) is correspondingly reduced to Eq. (7.6). Together with the full continuity equation, (7.6) and (7.7) yield the result

$$\operatorname{div} \mathbf{u} + \frac{\rho\beta}{c_p}\operatorname{div} \mathbf{q} = 0 \tag{7.8}$$

The effect of the anelastic approximation is to remove *acoustic* phenomena from theoretical consideration, while still allowing an accurate accounting for: (1) the buoyancy and inertial effects of variable density, (2) the modification of viscous forces and heat conduction by the variation of transport properties, and (3) the velocity induced by moderately slow expansion or contraction of the fluid particles. Note that the term anelastic applies to the flow, rather than to the fluid. Air, which is certainly an elastic substance, can flow anelastically.

Equation (7.7) has interesting special implications for a thermally perfect gas. Then $\beta T = 1$, and (7.7) leads to $D(\rho T)/Dt = 0$. If all fluid particles start out with the same value of ρT, the field of ρT will be uniform for all time. *However,* these thermodynamic conclusions do not imply that the pressure is absolutely uniform. Pressure gradients may be large enough to be important in the momentum balance, but simultaneously small enough to have no significant influence on the variations of density or temperature. Equations (7.6) and (7.7) may be very accurate representations of a given state of flow, but they are not exact, and their inexactness allows us to retain $\operatorname{grad} p$ in the momentum equation.[2]

---

[2] The effect of significant viscous dissipation might well be encompassed within a concept of anelastic flow, but the term has evolved in the contexts of combustion and strong free convection, in both of which the thermal effects of viscous dissipation are usually negligible.

We summarize the anelastic approximation in the following equations, assuming that conduction is the only mode of heat transfer, and that the fluid is a *thermally perfect gas*.

$$\text{div} \{P\mathbf{u} + (1 - \gamma^{-1})k \text{ grad } T\} = 0 \tag{7.9}$$

$$\rho \frac{D\mathbf{u}}{Dt} + \text{grad } p = \text{div} (\mu \text{ def } \mathbf{u}) + \rho\mathbf{g} \tag{7.10}$$

$$\rho c_p \frac{DT}{Dt} = \text{div} (k \text{ grad } T) \tag{7.11}$$

$$\rho T = P \quad \text{(a constant)} \tag{7.12}$$

$$\mu = \mu(T) \quad k = k(T) \tag{7.13}$$

The value of P is determined by the initial state of the fluid. If the fluid is rigidly contained in a volume V, and then heated at a rate Q, our analysis must be modified a little, to allow P to increase at the rate

$$\frac{dP}{dt} = (\gamma - 1)\frac{Q}{V} \tag{7.14}$$

If you are dismayed by the reasoning of this section, which asserts that a statement that is adequately true in one context may have false implications in another, be patient and return to it from time to time. A careful, and frequently somewhat subtle, construction of approximate equations suitable for various special kinds of flow lies at the heart of the theory of viscous flow.

7.6 REGULAR PERTURBATION SERIES, TO CORRECT FOR SMALL PROPERTY VARIATIONS

Once we have identified the parameter or parameters that must be small if ρ, μ, and k are to be treated as constants, we can sometimes calculate the quantitative consequences of small property variations by use of regular perturbation series. For example, if $\text{div } \mathbf{u} = 0$ is to represent accurately the steady isentropic flow over a body flying at Mach number M, we must have $M^2 \ll 1$. We can then postulate, for small but finite M^2, series of the form

$$\mathbf{u}(\mathbf{r}, M^2) = \mathbf{u}_0(\mathbf{r}) + \mathbf{u}_1(\mathbf{r})M^2 + \mathbf{u}_2(\mathbf{r})M^4 + \cdots$$

$$p(\mathbf{r}, M^2) = p_0(\mathbf{r}) + p_1(\mathbf{r})M^2 + p_2(\mathbf{r})M^4 + \cdots$$

$$\rho(\mathbf{r}, M^2) = \rho_0 + \rho_1(\mathbf{r})M^2 + \rho_2(\mathbf{r})M^4 + \cdots$$

We substitute these series into the conservation equations and boundary conditions and find the equations for, say, $\mathbf{u}_1(\mathbf{r})$, $p_1(\mathbf{r})$, and $\rho_1(\mathbf{r})$, by gathering together all the terms having M^2 as a factor. The lower-order functions, $\mathbf{u}_0(\mathbf{r})$, $p_0(\mathbf{r})$, and ρ_0, and the spatial derivatives of these functions, will appear in these equations. If the zeroth-order functions are sufficiently simple, we may be able to calculate the first-order functions analytically. Finally, by examining the flow property of greatest interest to us, we can fix a precise limit on the value of M^2, such that the corrections proportional to M^2 are acceptably small. We expect that the contributions of higher-order corrections will, individually and collectively, be negligible if the contribution of the first-order correction is almost negligible.

7.7 THE BOUSSINESQ EQUATIONS

When the thermodynamic effects of isentropic compression and viscous dissipation are negligible, and heat transfer is too weak to induce a significant value of div $\mathbf{u}$ or to cause significant changes in the transport properties, the Navier–Stokes equations may accurately be replaced by the following set:

$$\text{div } \mathbf{u} = 0$$

$$\rho_0 \, D\mathbf{u}/Dt + \text{grad } p' = \mu_0 \text{ div (def } \mathbf{u}) + (\rho - \rho_0)\mathbf{g},$$

$$\frac{DT}{Dt} = \left(\frac{k}{\rho c_p}\right)_0 \text{ div (grad } T)$$

$$\rho - \rho_0 = -\beta_0 \rho_0 (T - T_0)$$

$$p' = p + \rho_0 gh$$

In all of these, subscript zero denotes a constant that serves as an accurate average value of the subscripted property; the symbol h denotes elevation above a stationary horizon. With the further notation: $\mu_0/\rho_0 = \nu$ (kinematic viscosity), $(k/\rho c_p)_0 = \alpha$ (thermal diffusivity), and $\Pi = p'/\rho_0$, these can be written in the form associated with the names of Boussinesq and Oberbeck.[3]

$$\text{div } \mathbf{u} = 0 \tag{7.15}$$

$$\frac{D\mathbf{u}}{Dt} + \text{grad } \Pi = \nu \text{ div (def } \mathbf{u}) - \beta(T - T_0)\mathbf{g} \tag{7.16}$$

$$\frac{DT}{Dt} = \alpha \text{ div (grad } T) \tag{7.17}$$

---

[3] A. Oberbeck, "Ueber die Wärmeleitung der Flussigkeiten bei Berücksichtigung der Strömungen infolge von Temperaturdifferenzen." *Annalen der Physik und Chemie*, Neue Folge, **7**, 271–292, (1879). See, in particular, pp. 272–275. This paper seems very modern, and presents the anelastic treatment of density variations. See also J. Boussinesq, *Theorie Analytique de la Chaleur*, Gauthier–Villars, Paris, (1903). The ideas for the Boussinesq approximation appear in Vol. 2, 34th Lesson, pp. 154–176.

The subscript has been dropped from β_0, just to give a tidy appearance. Note that the momentum and energy equations are still coupled, through the buoyancy force.

7.8 THE NAVIER–STOKES EQUATIONS FOR CONSTANT DENSITY AND VISCOSITY

Finally, if density variations are so small that even buoyancy forces are negligible, we find the form of the Navier–Stokes equations with which most of the rest of this text is concerned:

$$\text{div } \mathbf{u} = 0 \tag{7.18}$$

$$\frac{D\mathbf{u}}{Dt} + \text{grad } \Pi = -\nu \text{ curl curl } \mathbf{u} \tag{7.19}$$

$$\frac{DT}{Dt} = \alpha \text{ div (grad } T) \tag{7.20}$$

Now the equation for T need no longer be solved simultaneously with the continuity and momentum equations; it can be solved later if the temperature field is of separate interest.

Note that the net viscous force per unit mass has been rewritten to show its dependence on the vorticity, by use of the identity

$$\text{div (def } \mathbf{u}) = -\text{curl (curl } \mathbf{u}) + \text{grad (div } \mathbf{u}).$$

This shows that, when ρ and μ are constant, viscosity has no effect on the velocity field, unless vorticity is present. This may seem paradoxical, since viscous stress and viscous dissipation are both present in irrotational flows.

7.9 SUMMARY

Solutions of the unabridged Navier–Stokes equations represent a number of phenomena that may be unimportant in the restricted circumstances of a particular flow. It is mathematically and computationally very helpful if the equations can be abridged so that their solutions no longer exhibit these phenomena.

The phenomena that we wish to set aside, so as to see more clearly the effects of viscosity, are mainly associated with variations of density, viscosity, and thermal conductivity. These variations are traced to three causes: isentropic compression, viscous dissipation, and heat transfer.

Isentropic compression is important when the fluid velocity approaches or exceeds the speed of sound, or when the wavelength of a strong acoustic oscillation is short compared to the diameter of the flow field. When these conditions are avoided, one can use an abridged set of equations, in which only viscous dissipation and heat transfer remain to cause variation of ρ, μ, and k.

Viscous dissipation is likely to be important in the boundary layers of supersonic flows, but can also have serious effects in very subsonic flows, such as in lubricating oil films. Watch out for fluids that are very viscous, but poor conductors of heat.

When both isentropic compression and viscous dissipation are negligible, one can derive an abridged set of equations, called the *anelastic equations*. These introduce the notion that pressure variations that significantly affect the coherent motion of the fluid may negligibly affect its thermodynamic state. Similarly, a rate of viscous dissipation that constitutes a major loss of coherent mechanical energy may not suffice to raise the fluid temperature significantly. This is particularly true in water.

High rates of heat transfer can cause significant variations of transport properties, significant values of div **u**, and significant coherent motions due to buoyancy. As **q** is reduced, the first two effects may become negligible, while buoyancy remains important. If the only significant effect of property variation is associated with buoyancy, the appropriate abridged equations are the *Boussinesq equations*.

An extensive and important class of viscous flows is accurately described by the abridged Navier–Stokes equations, in which ρ, μ, and k are all held strictly constant. In these equations, the term describing net viscous force per unit mass is proportional to the curl of the vorticity, and hence vanishes in an irrotational flow.

7.10 SAMPLE CALCULATIONS

A. Errors in isothermal theory of a viscosimeter. Let us continue the analysis of a viscometer, which was started in Chapter 5, Sample Calculation B, to illustrate the analysis of property variations by perturbation series. To ease the burden of calculation, assume that the geometry involves plane parallel plates instead of circular cylinders. Thus we shall analyze the equations

$$\frac{d(\mu \, du/dy)}{dy} = 0 \tag{7.21}$$

and

$$\frac{d(k \, dT/dy)}{dy} = -\mu \left(\frac{du}{dy}\right)^2 \tag{7.22}$$

subject to the boundary conditions

$$u(0) = 0 \qquad u(h) = V \qquad \frac{dT}{dy}(0) = 0 \qquad \text{and} \qquad T(h) = T_0$$

In Chapter 5, we found that the temperature difference across the flow is proportional to $\mu V^2/k$, according to a theory in which μ and k are held constant. Following the discussion of Section 7.4, we now choose a small parameter ε, which in that section was identified with M^2. Here we make the identification

$$\varepsilon = \left(\frac{\mu V^2}{kT}\right)_0$$

For convenience, we introduce dimensionless variables

$$\eta = \frac{y}{h} \qquad \phi = \frac{u}{V} \qquad \theta = \frac{T - T_0}{T_0}$$

and remind ourselves of the dimensionless sensitivities

$$S_\mu = \left(\frac{T}{\mu}\frac{d\mu}{dT}\right)_0 \qquad \text{and} \qquad S_k = \left(\frac{T}{k}\frac{dk}{dT}\right)_0$$

The constant-property solution to the problem, corresponding to the limit $\varepsilon \to 0$, is simply $u = Vy/h$ and $T = T_0$. Taking advantage of that, we expand ϕ and θ as follows:

$$\phi(\eta, \varepsilon) = \eta + \varepsilon\phi_1(\eta) + \varepsilon^2\phi_2(\eta) + \cdots$$

$$\theta(\eta, \varepsilon) = \varepsilon\theta_1(\eta) + \varepsilon^2\theta_2(\eta) + \cdots$$

and expand μ and k in Taylor series around their values at T_0. These expansions take the form

$$\mu = \mu_0(1 + S_\mu\theta + \cdots) = \mu_0[1 + \varepsilon S_\mu\theta_1(\eta) + \cdots]$$

$$k = k_0(1 + S_k\theta + \cdots) = k_0[1 + \varepsilon S_k\theta_1(\eta) + \cdots]$$

We now substitute the series into the differential equations, and gather terms according to the power of ε. From (7.20), we get

$$\varepsilon^0(0) + \varepsilon(S_\mu\theta_1' + \phi_1'') + \varepsilon^2(\ ) + \cdots = 0$$

while (7.21) yields

$$\varepsilon^0(\theta_1'' + 1) + \varepsilon[S_k(\theta_1\theta_1')' + \theta_2'' + S_\mu\theta_1 + 2\phi_1'] + \varepsilon^2(\ ) + \cdots = 0$$

In these, $\theta_1' \equiv d\theta_1/d\eta$, $\theta_1'' \equiv d^2\theta_1/d\eta^2$, etc.

We solve the expanded equations approximately, remembering that we first seek answers in the limit as $\varepsilon \to 0$. Thus we set $\varepsilon = 0$ in the equations, and find

$$0 = 0 \text{ from the first equation} \qquad \text{and} \qquad \theta_1'' = -1 \text{ from the second.}$$

Integrating the latter, we get $\theta_1' = -\eta + a$, then $\theta_1 = -\eta^2/2 + a\eta + b$. To evaluate the constants a and b, we need to substitute the expansions into the boundary conditions. This gives

$$\phi_n(0) = 0 \qquad \phi_n(1) = 0 \qquad \theta_n'(0) = 0 \qquad \text{and} \qquad \theta_n(1) = 0 \qquad \text{for } n = 1, 2, 3, \ldots$$

Applying the appropriate conditions to θ_1, we find $a = 0$ and $b = \frac{1}{2}$. Thus, our first-order estimate of the temperature perturbation is

$$\theta \approx \varepsilon(1 - \eta^2)/2 \tag{7.23}$$

Now return to the expanded equations, in which the terms proportional to ε^0 have been made to vanish. Divide out a factor ε, and again let $\varepsilon \to 0$. This time we get

$$S_\mu \theta_1' + \phi_1'' = 0 \qquad \text{from the first equation}$$

and $\qquad S_k(\theta_1 \theta_1')' + \theta_2'' + S_\mu \theta_1 + 2\phi_1' = 0 \qquad \text{from the second}$

Knowing that $\theta_1' = -\eta$, we calculate that $\phi_1'' = S_\mu \eta$, and therefore that $\phi_1 = S_\mu \eta^3/6 + c\eta + d$. The boundary conditions on ϕ dictate the values of c and d, namely $d = 0$ and $c = -S_\mu/6$. Thus the velocity profile, corrected to the first order in ε, is

$$\phi \approx \eta - \varepsilon S_\mu(\eta - \eta^3)/6 \tag{7.24}$$

This is enough to show the method of analysis. Before leaving it, let us assess the error in a measurement of μ, which would be committed if the heating by viscous dissipation were ignored. The isothermal calibration formula is $\mu = \tau h/V$, where τ is the measured value of the stress. In terms of the real, temperature-dependent, viscosity,

$$\tau = \mu \frac{du}{dy} \approx \frac{\mu_0 V}{h}[1 + \varepsilon S_\mu(1 - \eta^2)/2 + \cdots][1 - \varepsilon S_\mu(1 - 3\eta^2)/6 + \cdots]$$

$$\approx \frac{\mu_0 V}{h}[1 + \varepsilon S_\mu/3 + \varepsilon^2(\ ) + \cdots]$$

We notice that our estimate of τ is, up through the term proportional to ε, independent of η. An exact answer would be exactly independent of y, as is seen from (7.21). If we attribute the measured value of viscosity, μ_m, to the temperature T_0, we make a proportional error

$$\frac{\mu_m - \mu_0}{\mu_0} \approx \frac{\varepsilon S_\mu}{3}$$

However, if we measure the temperature of both walls, and attribute the measured viscosity to the temperature $T_m = (2T(0) + T_0)/3$, the proportional error becomes proportional to ε^2.

B. Numerical treatment of A. The problem posed in Sample Calculation A can be solved with high numerical accuracy by a simple iterative algorithm. The basic analysis follows the line of Sample Calculation B of Chapter 5, simplified in accordance with the flat, rather than circular, shape of the walls. Integrating (7.21) from $y = 0$, where $u = 0$, to some positive value of y, we find

$$\phi(\eta, \varepsilon) = \frac{u}{V} = \frac{\tau h}{\mu_0 V} v(\eta, \varepsilon) \tag{7.25}$$

in which

$$v(\eta, \varepsilon) = \int_0^\eta \frac{\mu_0}{\mu(\xi)} d\xi \tag{7.26}$$

Remember that τ, the viscous stress, is independent of y in this flow. From the boundary condition $\phi(1, \varepsilon) = 1$, it follows that $\tau h/\mu_0 V = 1/v(1, \varepsilon)$.

Integrating $d(k\, dT/dy)/dy = -\tau^2/\mu$ from $y = 0$, where $dT/dy = 0$, to some positive value of y, we find

$$k\frac{dT}{dy} = -\frac{\tau^2 h}{\mu_0} v(\eta, \varepsilon)$$

Divide by k and integrate again, this time from $y = h$, where $T = T_0$, to a lower value of y. The result, in nondimensional variables, is

$$\theta(\eta, \varepsilon) = \varepsilon[v(1, \varepsilon)]^{-2} \int_\eta^1 v(\xi, \varepsilon) \frac{k_0}{k(\xi)} d\xi \qquad (7.27)$$

in this expression, and in (7.26), ξ is just a dummy variable of integration.

Given empirical equations for μ_0/μ and k_0/k, we solve (7.26) and (7.27) iteratively, first using $\theta = \varepsilon(1 - \eta^2)/2$ to evaluate the integrals. This yields a new distribution of θ, which is put back into the integrals to get a better distribution, and so on, until the answer settles down. A FORTRAN program for such a calculation, called SHEAR, appears in Appendix B. The fluid is specified to be glycerol, for which the empirical equations

$$\frac{\mu_0}{\mu} = \left(\frac{T_0}{T}\right)^{52.4} \exp\left[23,100(T_0^{-1} - T^{-1})\right] \quad \text{and} \quad k_0/k = 1$$

are adequate for a modest range of temperatures around room temperature.

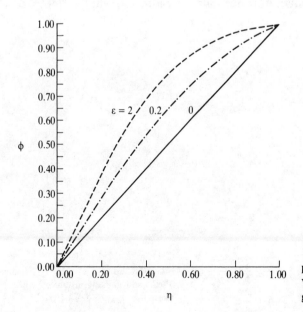

FIGURE 7.2
Velocity profiles in Couette flow of glycerol.

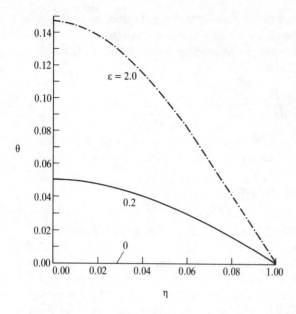

FIGURE 7.3
Temperature profiles.

For $T_0 = 293$ K, Tables 6.3 and 6.4 give $\mu_0 = 14.9$ kg/m-s, and $k_0 = 0.264$ W/m-K. The perturbation parameter takes the value $\varepsilon = 0.193V^2$, when V is measured in m/s.

Program SHEAR executes the quadratures shown in Eq. (7.27) by the Trapezoidal Rule, using 60 equal intervals from $\eta = 0$ to $\eta = 1$. The iterations initially showed a tendency to oscillate, which was cured by *under-relaxation*, i.e., by inserting into the quadrature formula an estimate composed 50% of

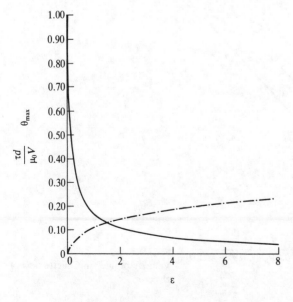

FIGURE 7.4
Shear and maximum temperature rise: —— θ_{max}; ---- $\tau d/\mu_0 V$.

the latest calculated temperature, and 50% of the next-to-latest calculated temperature. Some resulting velocity and temperature profiles are plotted in Figs. 7.2 and 7.3. Figure 7.4 shows the dimensionless shear, $\tau h / \mu_0 V$, and the maximum value of θ, as functions of ε.

We find the linear trends predicted by the first-order theory only when $\varepsilon \leq 0.01$. The effects of increasing ε grow less than linearly thereafter, because $|S_\mu|$ is a rapidly decreasing function of T.

EXERCISES

7.1. Suppose that in the viscosimeter experiment we maintain a constant *stress, τ,* rather than a constant differential velocity, V, while viscous dissipation is warming the fluid. To be specific, assume that one wall is thermally insulated while the other is kept at room temperature. What will happen to V, and the temperature of the insulated wall, as time goes on? Does it matter whether the fluid is a liquid or a gas? Explain your conclusions.

7.2. Analyze the relative contributions of isentropic compression and viscous dissipation to the rate of change of density in a fully-developed laminar flow of air in a round pipe. Assume that the velocity profile has the same shape as in incompressible flow, namely

$$\mathbf{u} = \frac{1}{4\mu} \frac{dp}{dz} (r^2 - a^2) \mathbf{e}_z$$

where a is the radius of the pipe, and cylindrical polar coordinates have been used.

7.3. Carry out a perturbation expansion, on the model of Sample Calculation A, for fully-developed laminar flow in a round pipe. Assume that $\mathbf{f} = 0$, and that

$$\mathbf{u} = w(r)\mathbf{e}_z \qquad T = T(r) \qquad \text{and} \qquad p = p(z)$$

a. Show, by applying these assumptions to the z component of Eq. (5.11), that

$$\frac{d(r\mu \, dw/dr)}{dr} = r \frac{dp}{dz}$$

b. Integrate this, insisting that dw/dr be finite at $r = 0$, to show that

$$\mu \frac{dw}{dr} = \frac{r}{2} dp/dz \tag{7.28}$$

c. Apply the assumptions to Eq. (5.12), and show that

$$\frac{d(rk \, dT/dr)}{dr} = r \left[\beta T w \frac{dp}{dz} - \mu \left(\frac{dw}{dr} \right)^2 \right] \tag{7.29}$$

d. Adopting the perturbation parameter $\varepsilon = \mu_0 V^2 / k_0 T_0$, where V is the mean velocity, defined by

$$a^2 V = 2 \int_0^a rw \, dr \tag{7.30}$$

in which a is the radius of the pipe. Call $\eta = r/a$, and set up the perturbation series as

$$w = V[\phi_0(\eta) + \varepsilon\phi_1(\eta) + \cdots]$$

$$T = T_0[1 + \varepsilon\theta_1(\eta) + \cdots]$$

and

$$dp/dz = -\frac{8\mu_0 V}{a^2}(1 + \varepsilon\pi_1 + \cdots)$$

Calculate $\phi_0(\eta)$ and $\theta_1(\eta)$, imitating the procedure of Sample Calculation A.

e. Continuing the analysis, find a relationship between $\phi_1(\eta)$ and π_1.

f. Use Eq. (7.30) as an additional constraint on ϕ_1, and thus evaluate π_1.

CHAPTER
8

VORTICITY

8.1 INTRODUCTION

As was shown in Chapter 2, the tangential stresses that arise in a deforming viscous fluid can set fluid particles spinning. It will be shown here that, in a barotropic flow that starts from rest under the influence of conservative body forces, only those fluid particles that have been subjected to viscous torques ever acquire any angular velocity. The rest of the fluid moves in accordance with the relatively simple laws of irrotational flow. Consequently, one's understanding of viscous flows, particularly at high Reynolds numbers, may be much enhanced, and interesting and useful schemes for theoretical modeling of these flows may be discovered, by a study of the generation of vorticity at a bounding surface, and the subsequently motion of vorticity by convection and diffusion. Fortunately, this formidable-sounding task can be broken down into a few small and conceptually easy steps. Patience and a good memory, rather than a dazzling burst of intuition, should be all you need to gain a trustworthy general comprehension of quite complicated real situations.

At first, it may seem odd to focus so much attention on the *rotation* of the fluid particles, which is hard to visualize and to measure, when practical interest is more closely associated with their *translation*. However, Section 3.3 has shown how one can compute the velocity field, given the vorticity field and some other information that is usually readily at hand. Thus, there is a very good chance that an early study of vorticity will not be a detour into useless abstraction,[1] but a flanking movement that leads to solid predictions of interesting quantities more effectively than does a frontal attack.

Sections 8.2 to 8.10 are concerned with definitions and with the dynamics of vorticity. They show, in some detail, the implications of the principle of conservation of momentum for the generation, convection, and diffusion of vorticity.[2] Section 8.11 elaborates and exemplifies the material of Section 3.4, concerning the synthesis of the velocity field that corresponds to a specified vorticity field. Section 8.12 analyzes the kinematic consequences of the no-slip condition. Finally, Section 8.13 sketches an algorithm with which numerical simulations of the evolution of a fairly broad class of viscous flows may be made. The intent is not to introduce you to numerical analysis per se, but to redeem the claim that the considerations of this chapter can lead to a sure-footed sense of cause and effect.

This chapter will probably not be easy at a first reading, and you will be referred back to it frequently in later chapters; so be prepared to read it more than once.

8.2 VORTICITY AND CIRCULATION

As was shown in Chapter 3, vorticity is a *point function*, equal to twice the average angular velocity of a fluid particle, as observed from an inertial frame of reference. It is related to the spatial derivatives of the velocity by the formula

$$\mathbf{\Omega} = \operatorname{curl} \mathbf{u} \tag{8.1}$$

A closely related *functional* is the *circulation*, defined as the line integral of velocity around an arbitrary closed path in the fluid.

$$\Gamma = \int_C \mathbf{u} \cdot d\mathbf{r} \tag{8.2}$$

in which $d\mathbf{r}$ is a element of arc length, directed so that the top side of any surface that spans the curve C lies on the left, as one follows $d\mathbf{r}$ around the curve.

---

[1] The classic educational film *Vorticity*, by A. H. Shapiro, makes the concept of vorticity very real, and truly fascinating. Professor Shapiro was the driving force behind the National Committee of Fluid Mechanics Films, and deserves the heartfelt thanks of any ardent student of the field.

[2] The content of these sections is strongly influenced by the admirably clear treatment in G. K. Batchelor, *An Introduction to Fluid Dynamics*, Cambridge University Press, 1967.

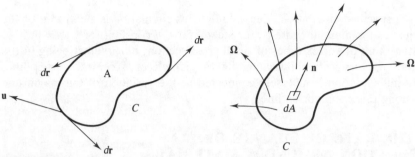

FIGURE 8.1
Circulation and the flux of vorticity.

By Stokes' theorem of vector calculus, we have

$$\int_C \mathbf{u} \cdot d\mathbf{r} = \iint_A \mathbf{n} \cdot \text{curl } \mathbf{u} \, dA = \iint_A \mathbf{n} \cdot \mathbf{\Omega} \, dA \qquad (8.3)$$

in which A is any simply-connected surface spanning the closed curve C. The local unit vector, normal to A and pointing out the top side of A, is **n**. Figure 8.1 shows these geometrical features.

The final integral over A is mathematically identical to expressions called volume flux or heat flux when **u** or **q** takes place of $\mathbf{\Omega}$; so it is called *flux of vorticity*. This term is not ideal, because vorticity does not flow in the direction indicated by the vector $\mathbf{\Omega}$. In a later section of this chapter, the term *diffusive flux of vorticity* will be used to describe actual transfer of vorticity across a material surface by the action of viscous forces. In words, Stokes' theorem tells us that the circulation around the closed curve C equals the flux of vorticity through the spanning surface A.

8.3 VORTEX LINES AND TUBES

Curves that are everywhere tangent to $\mathbf{\Omega}$ are called *vortex lines,* just as curves everywhere tangent to **u** are called streamlines. Because $\mathbf{\Omega} = \text{curl } \mathbf{u}$, and the divergence of the curl of any vector field is zero, we have

$$\text{div } \mathbf{\Omega} = 0 \qquad (8.4)$$

Under ordinary circumstances, the curves tangent to a solenoidal vector field cannot simply end in the region within which the vector field is defined, but must either form closed loops or connect points located on the boundary of the region. Thus, vortex lines have these properties. Since vorticity is a single-valued function, there is ordinarily only one vortex line through any given point; exceptions occur where $\mathbf{\Omega} = 0$, because the direction of a vector of zero length is undefined. Examples will be shown later.

A vortex *tube* is quite analogous to a streamtube. It is bounded by all those vortex lines that pass through points on a closed curve, around which there is a specified circulation. The value of this circulation is a constant of the tube, just as the mass flowrate is, for steady flow, a constant for a streamtube. We often think of vortex tubes of small cross section, our interest being in the variation of cross-sectional area along the length of a central vortex line. Clearly this variation is inversely connected to the variation of the magnitude of Ω, along this same vortex line.

8.4 THE RATE OF CHANGE OF CIRCULATION AROUND A MATERIAL LOOP

Suppose the curve C, used in the definition of circulation, is a material curve, composed always of the same fluid particles. The circulation around such a loop varies in a special way, described by famous theorems that were established by Lord Kelvin and by V. Bjerknes.

To evaluate $d\Gamma/dt$, we first note that in the line integral that defines Γ both $\mathbf{u}$ and $d\mathbf{r}$ may vary with time. If C is composed of material particles, it is clear that the relevant rate of change of $\mathbf{u}$ is $D\mathbf{u}/Dt$, the material or lagrangian rate of change. Also, since $d\mathbf{r}$ is a material line segment, we may refer back to Eq. (3.24) to find that

$$\frac{d(d\mathbf{r})}{dt} = d\mathbf{r} \, \mathrm{grad}\, \mathbf{u}$$

Hence, for a material loop C, the rate of change of circulation is

$$\frac{d\Gamma}{dt} = \int_C \frac{D\mathbf{u}}{Dt} \cdot d\mathbf{r} + \int_C \mathbf{u} \cdot (d\mathbf{r} \, \mathrm{grad}\, \mathbf{u})$$

Recall the definition of the transpose of a second-rank tensor, to see that

$$\mathbf{u} \cdot (d\mathbf{r} \, \mathrm{grad}\, \mathbf{u}) = d\mathbf{r} \cdot (\mathbf{u}(\mathrm{grad}\, \mathbf{u})^*) = d\mathbf{r} \cdot \mathrm{grad}\, (\mathbf{u} \cdot \mathbf{u}/2)$$

The last step of this is easily checked by cartesian index notation, and is left as an exercise.

Now, $d\mathbf{r} \cdot \mathrm{grad}\, (\mathbf{u} \cdot \mathbf{u}/2)$ is the exact differential of a single-valued function of position; thus its line integral around the closed path C is zero. We are left with

$$\frac{d\Gamma}{dt} = \int_C \frac{D\mathbf{u}}{Dt} \cdot d\mathbf{r} \qquad (8.5)$$

Using the momentum equation to evaluate the acceleration, we find

$$\frac{d\Gamma}{dt} = \int_C (-\rho^{-1} \, \mathrm{grad}\, p + \rho^{-1} \, \mathrm{div}\, \tau + \rho^{-1} \mathbf{f}) \cdot d\mathbf{r} \qquad (8.6)$$

If the body force per unit mass can be derived from a scalar potential, as is the case with gravity, then $\rho^{-1}\mathbf{f} \cdot d\mathbf{r}$ is an exact differential and contributes nothing to the integral. Such body forces are called *conservative*.

If the density is either constant or a function only of pressure, then $\rho^{-1}\operatorname{grad} p \cdot d\mathbf{r}$ is an exact differential and contributes nothing to the integral. A flow in which this is true is called *barotropic*.

Finally, if viscous stresses are absent or negligible, $\rho^{-1}\operatorname{div}\tau \cdot d\mathbf{r} = 0$. A flow in which this is true is called *inviscid*. We are left with the following conclusion:

The circulation around a material loop remains constant, *if*

1. The body force is *conservative*

2. The flow is *barotropic*

3. The flow is *inviscid*

This is Kelvin's theorem. Bjerknes's theorem focuses on *baroclinic* flow, in which the contribution of $\rho^{-1}\operatorname{grad} p \cdot d\mathbf{r}$ to the integral is essential.

8.5 THE RATE OF CHANGE OF THE VORTICITY OF A FLUID PARTICLE

The angular acceleration of a fluid particle could be deduced directly by equating the net torque of external forces, which we analyzed in Chapter 2, to the rate of change of angular momentum. This would involve, however, an analysis of the rate of change of the moment of inertia of a deforming body. An easier procedure starts by taking the curl, term by term, of the linear momentum equation. For this purpose, it is helpful to rewrite the convective acceleration, using the vector identity.

$$\mathbf{u} \operatorname{grad} \mathbf{u} = \operatorname{grad}(\mathbf{u} \cdot \mathbf{u}/2) + \mathbf{\Omega} \times \mathbf{u}. \tag{8.7}$$

Then $\quad \operatorname{Curl}\left(\dfrac{D\mathbf{u}}{Dt}\right) = \dfrac{\partial \mathbf{\Omega}}{\partial t} + \operatorname{curl}(\mathbf{\Omega} \times \mathbf{u}) = \dfrac{D\mathbf{\Omega}}{Dt} - \mathbf{\Omega} \operatorname{grad} \mathbf{u} + \mathbf{\Omega} \operatorname{div} \mathbf{u}$

To get this, we have used $\operatorname{div}\mathbf{\Omega} = 0$. Now we take the curl of the fluid acceleration

$$\frac{D\mathbf{u}}{Dt} = -\rho^{-1}\operatorname{grad} p + \rho^{-1}\operatorname{div}\tau + \rho^{-1}\mathbf{f}$$

while assuming that the body force is potential. The result is

$$\frac{D\mathbf{\Omega}}{Dt} = \mathbf{\Omega} \operatorname{grad} \mathbf{u} - \mathbf{\Omega} \operatorname{div} \mathbf{u} + \rho^{-2}\operatorname{grad}\rho \times (\operatorname{grad} p - \operatorname{div}\tau) + \rho^{-1}\operatorname{curl}\operatorname{div}\tau$$

$$\tag{8.8}$$

The last term in this, the torque per unit mass due to viscous stresses, was calculated directly in Chapter 2. Note that $\operatorname{curl}\operatorname{div}\tau = \operatorname{curl}\operatorname{div}\sigma$, because $\operatorname{curl}\operatorname{grad} p = 0$.

Recalling that in barotropic flow $\operatorname{grad}\rho \times \operatorname{grad} p = 0$, and that $\operatorname{div}\tau = 0$ in inviscid flow, we see that the vorticity history that corresponds to the provisos of Kelvin's theorem is governed by

$$\frac{D\Omega}{Dt} = \Omega \, \text{grad } \mathbf{u} - \Omega \, \text{div } \mathbf{u} \qquad (8.9)$$

Before analyzing this, let us record an alternative form of (8.8), useful in calculations in which reference to the pressure is avoided. This is, if $\mathbf{f} = \rho\mathbf{g}$,

$$\frac{D\Omega}{Dt} = \Omega \, \text{grad } \mathbf{u} - \Omega \, \text{div } \mathbf{u} + \rho^{-1} \text{grad } \rho \times \left(\mathbf{g} - \frac{D\mathbf{u}}{Dt} \right) + \rho^{-1} \text{curl div } \tau \quad (8.10)$$

Return now to (8.9), and note that it describes the case in which the external forces apply no net torque to the particle. Thus, the right-hand side must describe the effects of changing moments of inertia. If, as in Chapter 3, $\mathbf{d}^{(1)}$, $\mathbf{d}^{(2)}$, and $\mathbf{d}^{(3)}$ represent the principal rates of deformation, and if Ω is decomposed into components Ω_1, Ω_2, and Ω_3 along the three principal axes, we can rewrite (8.9) as

$$\frac{D\Omega}{Dt} = -\Omega_1(d^{(2)} + d^{(3)}) - \Omega_2(d^{(3)} + d^{(1)}) - \Omega_3(d^{(1)} + d^{(2)}) \qquad (8.11)$$

This shows that each principle component of Ω is amplified by a factor equal to the combined rates of contraction of the fluid in the directions normal to that component. This is qualitatively what we would expect from watching dancers and figure skaters, and it agrees with what was deduced, in Problem 3.8, by direct analysis of the angular momentum of a fluid cube.

An important special case is that of *two-dimensional* flow, in which the vorticity points normal to the plane of motion, in a direction in which $\mathbf{u}$ does not vary. It follows directly that $\Omega \, \text{grad } \mathbf{u} = 0$. If such a flow is also *solenoidal*, it follows that

$$\frac{D\Omega}{Dt} = 0 \qquad (8.12)$$

In words, the vorticity of a two-dimensional, inviscid, solenoidal, barotropic flow with potential body forces is carried along, unchanging, with the fluid.

8.6 ROTATING FRAMES OF REFERENCE

In all of this analysis, the velocity $\mathbf{u}$ is understood to be measured in an inertial frame of reference. If the frame of reference is in turn rotating, with angular velocity ω, we need only replace Ω by $\Omega + 2\omega$ in these equations to have a correct theory.

8.7 DIFFUSION OF VORTICITY

The general vorticity equation (8.8), is greatly simplified if density and viscosity are taken constant. With the warnings of Chapter 7 in mind, we adopt this simplification for the remainder of this chapter. Equation (8.8) is reduced to

$$\frac{D\mathbf{\Omega}}{Dt} = \mathbf{\Omega} \operatorname{grad} \mathbf{u} + v \operatorname{div} \operatorname{grad} \mathbf{\Omega} \tag{8.13}$$

Convection-Diffusion Equations

To focus on the phenomenon of *diffusion of vorticity*, consider the further simplification to two-dimensional flow, in the x–y plane. Then $\mathbf{\Omega} = \Omega(x, y)\mathbf{e}_z$, and Eq. (8.13) reduces further, to

$$\frac{D\mathbf{\Omega}}{Dt} = v \, \nabla^2 \mathbf{\Omega}, \tag{8.14}$$

in which
$$\nabla^2 = \frac{\partial^2}{\partial x^2} + \frac{\partial^2}{\partial y^2}$$

Equation (8.14) invites comparison with Eq. (7.16), rewritten as

$$\frac{DT}{Dt} = \alpha \, \nabla^2 T$$

and the mathematically similar equation for the ordinary diffusion of chemical species,

$$\frac{DC}{Dt} = D \, \nabla^2 C$$

These are all called *convection-diffusion equations*. D is the species diffusion coefficient; α is the thermal diffusivity, and v is called the *diffusivity of vorticity*, or sometimes the diffusivity of momentum The first nickname for v is better, because momentum can be transferred by pressure waves, which have nothing to do with diffusion.

Diffusive flux of vorticity. A diffusive flux of vorticity can now be defined, even for cases in which $\mathbf{\Omega}$ is a vector with three components. Recognizing that $\operatorname{grad} \mathbf{\Omega}$ is a second-rank tensor, one can use the divergence theorem to write

$$\int_V \operatorname{div} \operatorname{grad} \mathbf{\Omega} \, dV = \int_A \mathbf{n} \operatorname{grad} \mathbf{\Omega} \, dA = \int_A \frac{\partial \mathbf{\Omega}}{\partial n} \, dA$$

where $\partial \mathbf{\Omega}/\partial n$ is the directional derivative of $\mathbf{\Omega}$ in the direction of the outward normal, $\mathbf{n}$. By analogy to species diffusion, for which $-D \, \partial C/\partial n$ represents the rate at which concentration of species diffuses across unit area normal to $\mathbf{n}$, the expression $-v \, \partial \mathbf{\Omega}/\partial n$ is now said to represent the rate at which vorticity diffuses across unit area normal to $\mathbf{n}$. By definition, these diffusive fluxes proceed from the region out of which $\mathbf{n}$ points, to the region into which $\mathbf{n}$ points.

Diffusion of vorticity at a solid wall. It is particularly interesting to consider the rate at which vorticity diffuses into a fluid from a solid wall at which the no-slip condition is in effect. To see this for the case of constant density and viscosity, we return to the momentum equation in the form of Eq. (7.18):

$$\frac{D\mathbf{u}}{Dt} + \operatorname{grad} P = -v \operatorname{curl} \mathbf{\Omega}$$

Consider first the case of two-dimensional flow in the x–y plane, and let the wall, at $y = 0$, be stationary. Because of the no-slip condition, $\mathbf{u} = 0$ at the wall and the x component of the momentum equation reduces locally to

$$\Pi_{,x} = -u\mathbf{\Omega}_{,y} \tag{8.15}$$

Thus the diffusive flux of vorticity out of the wall is equal to the pressure gradient along the wall (minus the corresponding component of the pressure gradient necessary to support the weight of the fluid at rest.) This result can be used in two ways.

1. When $\Pi_{,x}$ is known in advance, as in boundary-layer theory, it specifies a source of vorticity which has important effects on the flow.
2. Certain computational schemes, which will be sketched later in this chapter, keep account of the amount of vorticity added to the flow at the wall, during each increment of time. Thus, they effect a sort of calorimetric measurement of the vorticity flux through the wall, which can be used to calculate the pressure gradient. A detailed discussion of this is given in Section 8.13.

The Permanence of Irrotational Flow

A very significant general deduction can be based on Eq. (8.13), concerning a flow that starts from rest, and is then set in motion by a moving solid wall. Any material particle in such a flow will start out with $\mathbf{\Omega} = 0$, and (8.13) says that it will retain that value until it comes up against another particle that has $\mathbf{\Omega} \neq 0$, because it can receive its first bit of vorticity only by diffusion. However, the same proposition applies to the other particle; it must have received its vorticity from still another particle, and so on. Where does this bucket brigade start?

The source of vorticity is found at the moving wall, where vorticity appears as soon as there is relative tangential motion between wall and fluid, in order to satisfy the no-slip condition.

This effect of the no-slip condition will be discussed at length later in this chapter, and illustrated in Chapters 9 and 10. Meanwhile, try to imagine this general scenario:

1. The wall is set in motion, and vorticity immediately appears in the fluid at the wall, wherever it is needed to satisfy the no-slip condition.

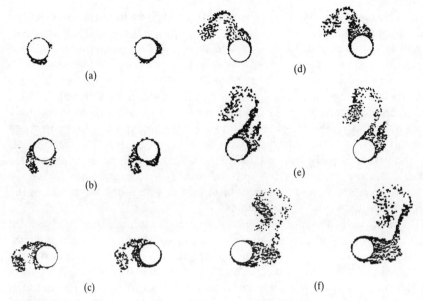

(a)

(b)

(c)

(d)

(e)

(f)

FIGURE 8.2

Vorticity being transported away from a circular cylinder. The flow far from the cylinder varies in speed and direction, so that the vorticity stays close to the cylinder. The flow is started at time $t = 0$; (a) to (f) are taken after equal increments of t. In each pair of pictures, computation elements that carry positive vorticity are shown on the left, elements that carry negative vorticity are shown on the right. (Reproduced, by permission, from E. C. Tiemroth, "The Simulation of the Viscous Flow Around a Cylinder by the Random Vortex Method", Ph.D. Dissertation, Dept. of Naval Arch. and Offshore Engr., Univ. of Calif., Berkeley, May 1986.)

2. This vorticity is diffused out into the fluid farther from the wall, and convected along the wall.

3. These processes continue, with vorticity getting farther from the wall, particularly where a finite body, like a ball or a wing, leaves vorticity in its wake.

The relative importance of convection and diffusion in shaping the distribution of vorticity around a body of a given shape is determined by the value of the Reynolds number. An example, computed by the algorithm to be sketched in Section 8.13, is exhibited in Fig. 8.2. The cylinder is horizontally submerged in water, the motion having been started by the arrival, from the left, of a train of water waves.

8.8 AN EXTREMAL THEOREM FOR VORTICITY IN TWO-DIMENSIONAL FLOW

The convection-diffusion equation (8.14), implies an important limitation on the extreme values which can be attained by Ω within a specified space–time domain.[3] The equation says that the vorticity of a given material particle can only be changed, in such a flow, by diffusion. If $\nu = 0$, the remnant of Eq. (8.14) shows that the value of Ω attached to a given material particle cannot change at all.

Suppose there is one material particle that enters the time–space domain, either by being in the space domain at the initial time or by entering it later, with a larger positive value of Ω than is brought in by any other. It can only acquire a more positive value of Ω by diffusion, and this can only happen when it comes up against another particle with an even larger positive value of Ω. This is, however, equally true for all particles, so it will never meet such a neighbor. Diffusive exchanges can only smooth away the extremes which are carried into the domain, and all values of Ω that can be found anywhere in the domain must lie between these extremes.

This result provides one immediate test of the plausibility of the local values of Ω produced in approximate numerical simulations of viscous flows. Remember, however, that it applies only to two-dimensional flows. In three-dimensional flow, the vorticity amplification due to decreasing angular inertia can overpower diffusion at times, and can lead to spectacular extrema of vorticity. Examples will be shown in Chapters 9 and 14.

8.9 THE MOTION OF VORTEX LINES AND FILAMENTS

It is often helpful, when one tries to rationalize observations of complicated flows, to add a sense of physical reality to the concept of a vortex line. This is repeatedly illustrated in the film *Vorticity*, when the characteristic flow in a channel bend or around the base of a bridge piling in a river is explained. The principal helpful idea is that vortex lines are somehow embedded in the fluid, so that there is, under appropriate conditions, a permanent association between a particular vortex line and a particular material line.

It is fairly easy to imagine what the above-mentioned appropriate conditions must be; they must rule out diffusion, because that passes vorticity from one body of fluid to another. They must also rule out any sources of vorticity, such as baroclinicity or action of a nonpotential body force. Indeed, they are just the conditions for which the circulation around a material fluid loop remains independent of time.

The permanent association that can be demonstrated under these conditions has two parts:

1. The vortex line does not move sideways through the fluid. This is readily proved by considering two material loops, one in each of two intersecting

---

[3] Usually a fixed volume of space, observed between specified initial and final times.

vortex surfaces. There is initially no circulation around either loop, because there is no flux of vorticity through a vortex surface. Kelvin's theorem shows that there will never be any circulation around either loop if the prescribed conditions are met. Thus, the material surfaces remain vortex surfaces, and their intersection remains both a vortex line and a material line. Since no material line can ever leave the material surface of which it was once a part, the intersection of two material surfaces is always the same material line. The same statement can be made about a vortex line, so the vortex line and the material line will forever follow a common space curve.

2. If an infinitesimal segment of the material curve, represented by the vector **L**, is identified, and if Ω and ρ are the vorticity and density at the midpoint of the segment, the magnitude of Ω/ρ remains proportional to that of **L**. Thus, we speak of the stretching of vortex lines as a cause of intensification of vorticity.

A mathematically rigorous proof of these propositions would carry us beyond the intended scope of this text.[4] The proof which follows is not quite rigorous, but adequately suggests the main line of reasoning.

We start with Eq. (8.9) and the continuity equation in the form (4.3). Eliminating div **u** between them, we find

$$\frac{D\Omega}{Dt} = \Omega \operatorname{grad} \mathbf{u} - \frac{\Omega}{\rho}\frac{D\rho}{Dt}$$

and then

$$\frac{D(\Omega/\rho)}{Dt} = \frac{\Omega}{\rho}\operatorname{grad}\mathbf{u} \tag{8.16}$$

Recall Eq. (3.24),

$$\frac{D\mathbf{L}}{Dt} = \mathbf{L}\operatorname{grad}\mathbf{u} \tag{8.17}$$

Evidently, Eqs. (8.16) and (8.17) agree if

$$\mathbf{L} = \frac{A\Omega}{\rho} \tag{8.18}$$

in which A is a scalar, independent of time. Equation (8.18) summarizes both of the verbal propositions. It requires a little work, but you can show that this is just another way of describing the physical content of Eq. (8.10).

---

[4] A very comprehensive treatment appears in Section 4.2 of S. Goldstein, *Lectures on Fluid Mechanics,* Interscience Publishers, London, 1960. See also Sections 5.2 and 5.3 of G. K. Batchelor, *Introduction to Fluid Dynamics,* Cambridge University Press, 1967. The basic ideas are due to Helmholtz.

This discussion, with its attention to the deformation of vortex lines, seems to be peculiarly relevant to three-dimensional flow. In two-dimensional flow there is no such deformation, and we focus on the motion of identifiable concentrations of vorticity. We wish to show that points, actually vortex lines in end view, that mark such concentrations, move with the fluid, in the stipulated kind of flow.

Our analytic tool, the equation

$$\frac{D(\Omega/\rho)}{Dt} = 0 \tag{8.19}$$

certainly suffices for incompressible flow. If an initial vorticity field exhibits easily identified extrema, like the tall persons in a crowd, and if these extrema are marked initially by vortex lines, the equation $\dfrac{D\Omega}{Dt} = 0$ shows that the extrema stay with the fluid. By proposing that the vortex lines move with the fluid, we maintain their association with the extrema, as desired.

8.10 VORTICITY IN AN INVISCID FLOW?

All of this discussion of vorticity in an inviscid, barotropic flow with conservative body forces, following quickly after proofs that a flow that starts from rest can only acquire vorticity by action of viscosity, baroclinicity, or nonpotential body forces, may seem nonsensical. The situation it pictures is not, however, unfamiliar in all sorts of affairs that have a significant history. Thus, it is quite possible for vorticity to have been established in the past, by agencies that are no longer effective. This happens especially when vorticity-producing-and-redistributing agencies have worked slowly, over a long time, to produce a vorticity field that is suddenly perturbed. The response to the perturbation may then be entirely inertial.

A good example is the flow around a Pitot tube, used to measure the velocity profile in a pipe flow or boundary layer. The flow that is being probed is full of vorticity, produced and redistributed throughout a long upstream region. A given vortex line is deflected by the tiny probe only when it gets very close to it, and the effect of the probe on the trajectory of the vortex line lasts only a very short time, during which the viscous diffusion has no appreciable effect. We shall see in Chapters 9 and 11 that viscosity may indeed affect the reading of such a probe, but in a way that has nothing to do with the motion of vortex lines in the oncoming flow.

8.11 EFFECTS OF VISCOSITY ON THE EVOLUTION OF VORTEX LINES

When vorticity can diffuse through the fluid, the configuration of vortex lines can evolve in surprising ways. One of these is illustrated in Fig. 8.3. Vortex

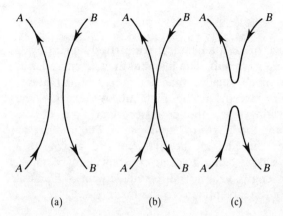

FIGURE 8.3
Reconnection of vortex lines, due to diffusion.

(a)   (b)   (c)

filaments of equal strength AA and BB, roughly antiparallel in (a), approach each other by diffusion, finally touching in (b), and reconnecting in (c), to become AB and BA.[5] This process can sometimes be seen in the vapor trails that mark the wingtip vortices of high-flying airplanes.

The diffusive cancelation of vorticities of opposite sign is a common feature of many viscous flows, particularly in tubes and channels. This phenomenon severely limits the utility of an evolving pattern of vortex lines as a tool for visualizing the evolution of a highly viscous flow field. For two-dimensional flows, a contour map of the single scalar component of vorticity, like a map of isotherms in a temperature field, often conveys useful insights. In particular, it shows the diffusive spreading of local concentrations of vorticity very nicely.

8.12 VELOCITY INDUCED BY IDEALIZED CONCENTRATIONS OF VORTICITY

To make good use of an ability to predict the evolution of the vorticity distribution, one needs to be able to calculate the velocity distributions induced in free space by various idealized concentrations of vorticity. The necessary study was initiated in Section 3.4. The one example presented there was of a line vortex of circulation Γ, which induces a flow with circular streamlines centered on the line, and with speed $v = \Gamma/2\pi r$, where r is distance from the line.

---

[5] For a similar discussion of vortex-line reconnection, see B. R. Morton, "The Generation and Decay of Vorticity", *Geophys, Astrophys, Fluid Dynamics*, **28**, 277–308 (1984). See, in particular, article 5.3 and Fig. 10.

Vortex Filament

Consider the more general case of a filament, characterized again by a circulation Γ, that is constant along its length. The filament may be curved, but we assume that its diameter is much smaller than its radius of curvature. Assuming that $\Omega = 0$ everywhere except in the filament, we execute the volume integral by first integrating over the cross-sectional area of the filament, and then integrating along the length of the filament. The integration over the cross section measures the flux in the filament and assigns it a direction along the filament. The flux, being a scalar constant, can be taken outside the final integration over length. Calling a directed element of filament length $d\mathbf{s}$, we can reduce the Biot–Savart integral to the form

$$\mathbf{u}_v(\mathbf{r}) = -\frac{\Gamma}{4\pi} \int_C \xi^{-3}\boldsymbol{\xi} \times d\mathbf{s}(\mathbf{r}') \tag{8.20}$$

The locus of the filament is the curve C; the integration is carried out over the entire length of the filament. As in Chapter 3, $\boldsymbol{\xi} = \mathbf{r} - \mathbf{r}'$, and ξ is the magnitude of $\boldsymbol{\xi}$. The formula is only expected to be accurate at points outside of the filament. The velocity induced on itself by a curved filament must be analyzed with great care, so that it is not a simple matter, for example, to calculate the forward speed of a circular vortex ring.[6]

Of course, the vorticity in a real flow will not ordinarily collect itself into a single filament, but the Biot–Savart law describes a *linear* relationship between vorticity and induced velocity, so the principle of superposition applies.

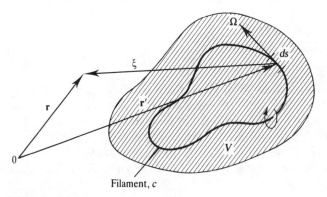

Filament, c

FIGURE 8.4
An isolated vortex filament.

---

[6] Batchelor gives an introductory discussion on pages 522 to 524.

Two-Dimensional Induced Flows

All of the examples to be presented here involve two-dimensional flow, induced by collections of finitely long, straight, parallel filaments. The vectors **u**, **r**, and **r**′ thus lie in a plane normal to the filaments and one can work with the length-integrated form of the Biot–Savart Law

$$\mathbf{u}_v(\mathbf{r}) = \frac{1}{2\pi} \iint_A \xi^{-1}\mathbf{\Omega}(\mathbf{r}') \times \xi \, dA \tag{8.21}$$

Vortex sheets. It is sometimes advantageous to think of vorticity as being concentrated in sheets, rather than in isolated filaments. For the purpose of calculation, the sheet is assembled by laying vortex filaments down side-by-side to cover a specified surface. Calculations of the sort presented here are frequently found in books on electromagnetism, where they describe the magnetic field induced by sheets of electric current. One calculation will be shown here in detail; then some other interesting results will simply be cited, the calculations being left for exercises.

The plane sheet of uniform strength. Suppose vorticity is distributed uniformly along the x axis, from $x = -\infty$ to $x = \infty$. The strength of the sheet, at a given value of x', is defined by the integral

$$U(x') = -\int_{-\infty}^{\infty} \Omega(x', y') \, dy'$$

so that one actually integrates the formula

$$u(x, y) = -\frac{1}{2\pi} \lim_{A \to \infty} \left\{ \int_{-A}^{A} (y - y')[(x - x')^2 + (y - y')^2]^{-1} U(x') \, dx' \right\}$$

The reason for the careful treatment of the limits of integration will soon appear. In this case, U is constant and $y' = 0$. We define $\zeta = (x - x')/y$, in terms of which

$$u(x, y) = \frac{U}{2\pi} \lim_{A \to \infty} \left[\int_{(x+A)/y}^{(x-A)/y} (\zeta^2 + 1)^{-1} \, d\zeta \right]$$

$$= \frac{U}{2\pi} \lim_{A \to \infty} \{ \arctan[(x - A)/y] - \arctan[(x + A)/y] \}$$

Notice that if $y > 0$, the argument of the first arctan approaches $-\infty$, while that of the second arctan approaches $+\infty$. If $y < 0$, just the opposite happens. Thus, the answer is

$$u(x, y) = -\frac{U}{2} \quad \text{if } y > 0 \qquad \text{and} \qquad u(x, y) = +\frac{U}{2} \quad \text{if } y < 0$$

In the calculation for $v(x, y)$, the integrand is an odd function of ζ, so we find $v = 0$ everywhere, without any further calculation. Figure 8.5 summarizes the result.

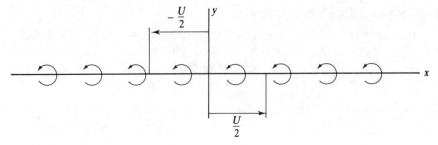

FIGURE 8.5
Plane vortex sheet of uniform strength $d\Gamma/dx = U$.

The appearance of arctangents in calculations of this sort is very typical, because the tangential component of velocity jumps discontinuously across a vortex sheet. The velocity normal to the sheet is continuous. Near a sheet-like distribution of dilatation, instead of vorticity, just the opposite would appear. Note that, by symmetry, this flat sheet of uniform strength induces no velocity on itself. This is a special result, which can be upset by either curvature or variability of strength.

Circular sheet of uniform strength. Consider a sheet of uniform strength, $U = a^{-1} \, d\Gamma/d\theta$, wrapped around the surface of the circular cylinder $r = a$. It is shown in Fig. 8.6. The induced velocity is:

$$u = 0 \quad \text{everywhere} \qquad v = 0, \quad \text{if } r < a \qquad \text{and} \qquad v = \frac{Ua}{r} \quad \text{if } r > a$$

Note that the total circulation of the sheet is $\Gamma = 2\pi a U$, so that the induced field outside the sheet is the same as would be induced by a single line vortex of this strength, placed on the axis of the cylinder.

Circular sheet with sinusoidally varying strength. Change the sheet strength of the last example to be $d\Gamma/d\theta = -2Ua \sin \theta$. The resulting velocity components

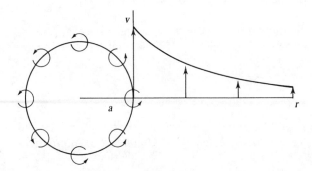

FIGURE 8.6
Cylindrical vortex sheet of uniform strength.

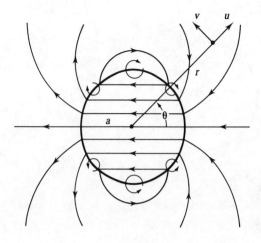

FIGURE 8.7
Cylindrical vortex sheet with
$d\Gamma/d\theta = -2Ua\sin\theta$.

will be

$$u(r,\theta) = -U(a/r)^2\cos\theta \quad \text{if } r > a$$

and
$$u(r,\theta) = -U\cos\theta \quad \text{if } r < a$$

$$v(r,\theta) = -U(a/r)^2\sin\theta \quad \text{if } r > a$$

and
$$v(r,\theta) = U\sin\theta \quad \text{if } r < a.$$

The streamlines of this flow are very interesting, and are shown in Fig. 8.7. Inside the cylinder, there is uniform flow from right to left, with speed U. Outside, the flow is identical to that produced by a volume-flow line doublet on the axis. This is just the potential flow that would be produced by a solid cylinder, moving from right to left, with speed U.

The Effects of Walls. Image Systems

In Section 3.4, a way of accounting for the presence of solid walls by the introduction of *image distributions* of vorticity or dilatation was introduced. The example given there was the simplest imaginable; here are a few others.

Vortex in a rectangular corner. Consider the line vortex of strength Γ, shown with its images in Fig. 8.8(a). It is clear by symmetry that no flow crosses either the x axis or y axis. The details are easily worked out, and are left for an exercise.

Vortex outside a circular cylinder. If a line vortex lies at (r, θ), outside a circular cylinder of radius a, a single counterrotating vortex of equal strength at the *reciprocal point* $(a^2/r, \theta)$ will prevent flow through the surface of the

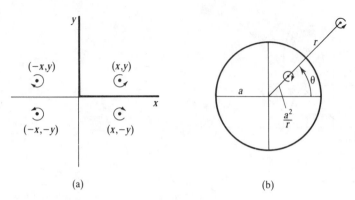

FIGURE 8.8
Simple image systems.

cylinder. This is easily shown with the Law of Cosines; the proof is left for an exercise. The arrangement of vortex an image is shown in Fig. 8.8(b).

This extremely simple example already exhibits some very interesting properties. Suppose the flow started from rest, and the vorticity at P arrived there after being generated at the surface of the cylinder, which is moving. The circulation around a large loop, which encircles both the cylinder and the external vortex, should be zero because none of the circulation-producing mechanisms identified in Kelvin's theorem can yet have acted on the fluid of the loop. A direct calculation of the flux of vorticity through the loop, including both the flux of the external vortex and that of the image, gives the value zero. This means that the flow due to the single image is the physically meaningful $\mathbf{u}_p$ for this problem.

Vortex between two plane, parallel walls. Let the vortex lie at $(a, 0)$, between the walls $x = 0$ and $x = h$. We place primary images at

$$x = -a \qquad \text{and} \qquad x = h + (h - a) = 2h - a,$$

but then observe that the image at $x = -a$ induces some flow across the wall at $x = h$, while that at $x = 2h - a$ violates the boundary condition at $y = 0$. This is avoided by adding secondary images at $x = 2h + a$ and $x = -2h + a$, and so on. Figure 8.9 shows that if the pattern of equal and opposite vortices, seen at $x = a$ and $x = -a$, is repeated periodically around the planes $x = \pm 2nh$, with $n = 1, 2, 3, \ldots$, the resulting flow will not cross any of the planes $x = \pm nh$, with $n = 0, 1, 2, 3, \ldots$.

There are thus two arrays of equally-spaced line vortices, one with $\Gamma > 0$, the other with $\Gamma < 0$. The flow due to each is given by an infinite series, which can be summed. This yields a fairly compact expression for the streamfunction:

$$\Psi(x, y) = \frac{\Gamma}{4\pi} \left\{ \ln \left[\cosh \left(\frac{\pi y}{h} \right) - \cos \left(\frac{\pi(x - a)}{h} \right) \right] \right.$$

$$\left. - \ln \left[\cosh \left(\frac{\pi y}{h} \right) - \cos \left(\frac{\pi(x + a)}{h} \right) \right] \right\}$$

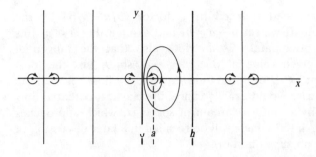

FIGURE 8.9
Line vortex between plane parallel walls.

from which the velocity components are derived, as

$$u = \Psi_{,y} \quad \text{and} \quad v = -\Psi_{,x} \tag{8.22}$$

Vortex sheet next to an infinite plane wall. Consider a vortex sheet, made up of filaments, each of which is parallel to the plane $y = 0$. Each element of the sheet, of strength $d\Gamma$, at point (x, y), can be assigned an image, of strength $-d\Gamma$, at a point $(x, -y)$. In this way, an image sheet is constructed.

The simplest, but very useful, example is that of a flat sheet of uniform strength, parallel to the wall. This induces the simple plug flow shown in Fig. 8.10. Notice that material points exactly on the vortex sheet have the velocity $-U/2$, induced by the image sheet.

Cylindrical sheets, outside a solid cylinder. Suppose that a solid cylinder of radius a is surrounded by a vortex sheet of radius b. For this case, each element of the sheet, of strength $d\Gamma$, at point (b, θ), can be assigned an image, of strength $-d\Gamma$, at its reciprocal point $(a^2/b, \theta)$. The flow field induced by the real sheet plus the image sheet can be found by superposition.

Note the similarity between the case of the cylindrical sheet of uniform strength, and that of a uniform plane sheet. The resultant flow is entirely confined to the region between sheet and image. For the cylindrical sheet with sinusoidally varying strength, there is still a doublet-induced disturbance at

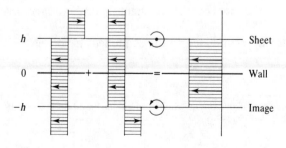

FIGURE 8.10
Vortex sheet and image in a flat wall.

$r > b$, but its strength has been reduced by a factor $(b/a - a/b)$ by the presence of the image sheet. If we set $b - a = h$, and assume that $h \ll a$, this factor of reduction is approximately $2h/a$. Note also that the tangential velocity at the wall at a given value of θ equals $-2U \sin \theta$, just the local velocity jump across the vortex sheet in the fluid. In fact, the flow between the sheet and its image in this case would exactly cancel the classical potential flow over the cylinder, caused by a uniform stream of speed U, which approaches the cylinder from the left. All these facts will be recalled in a later discussion of the kinematic effect of the no-slip condition.

8.13   KINEMATIC CONSEQUENCES OF THE NO-SLIP CONDITION

One can now begin to see just how the no-slip condition affects a fluid motion. Let us imagine that the processes that affect the vorticity distribution, namely convection, generation, and diffusion, occur sequentially, rather than simultaneously, during each tiny increment of time. It can be proved that this is mathematically legitimate,[7] and it facilitates analysis. Suppose that the flow in its initial state satisfies the no-slip condition.

Let convection act for a brief increment of time. Vorticity which is already in the fluid, will be displaced a bit. In its new configuration the vorticity field will induce a slightly different velocity field than it did an instant before. The field of image vorticity will also have been displaced, and its contribution to $\mathbf{u}_p$ will have changed correspondingly. Meanwhile, the other component of $\mathbf{u}_p$, which makes conditions correct at infinity, may have changed a bit to reflect, for example, a change in wind-tunnel speed.

The net effect of these changes is that the new velocity field will probably not satisfy the no-slip condition. Nature corrects this error by generating *new vorticity* wherever a slip velocity appears. We have seen in previous examples how this must happen. A new vortex sheet appears in the fluid right in front of the wall. An image sheet appears right behind the wall, to keep the no-penetration condition in force. Together with its amage, the new vortex sheet produces a plug flow in the infinitesimally wide gap between them, and induces only infinitesimal velocities anywhere else. The strength of the new sheet is everywhere just sufficient so that the plug flow cancels the illegal slip flow.

This picture is expected to be valid everywhere along the wall, except possibly at sharp bends, where the concept of a sheet and its image is problematical.

We think of this process as the second step in the sequence. If vorticity is

---

[7] J. E. Marsden, "A formula for the Solution of the Navier–Stokes equations based on a method of Chorin", *Bull. Amer. Math. Soc.* **80**: 154–158 (1974).

being generated within the bulk of the fluid, by baroclinicity or nonpotential body forces, that can be calculated as part of this step.

The final process to be imagined is *diffusion*. Vorticity that was newly generated at the wall is thereby carried out into the fluid, and the vorticity already in the fluid is spread around a bit. The newly generated wall vorticity, now in a slightly diffuse sheet, is still so close to its image that it still effectively cancels the slip. Everything is now ready for the clock to tick again![8]

8.14 SIMULATION OF VISCOUS DIFFUSION. THE RANDOM-VORTEX METHOD

Before exploring, in Chapter 9, some illustrative special solutions of the Navier–Stokes equations, let us consider briefly an algorithm for the numerical simulation of two-dimensional viscous flows.[9] This is presented so early in the book, not because you are expected to master its details, which are still under development, but because the algorithm conveys a very clear idea of the basic behavior of a viscous fluid. It may help you to see the special analytical solutions in a broad context, and may remove some of the mystery from the procedures that lead to those solutions.

We have, in fact, already collected most of the ideas of the algorithm, but need to decide how to simulate viscous diffusion, and must be a little more specific about the way the vorticity field is to be represented by computational elements.

Representation of the Vorticity Field

At any given time, the vorticity that was in the fluid at an initial time, plus any vorticity that has been subsequently added to satisfy the no-slip condition, is represented by a number of discrete computational elements. Each element is characterized by a value of its circulation, by a diameter or length, and by the velocity field that it induces in free space. Vorticity may be distributed

---

[8] In many textbooks, the no-slip condition is described as being due to viscosity, and inapplicable to inviscid-flow theory. What we see here is a bit more precise: without viscosity to diffuse vorticity away from the wall, no effect of the no-slip condition would ever be observed. The vortex sheet and its image, needed to annihilate slip, would remain at the wall, affecting the velocity in only an infinitely thin region.

[9] The essential concepts and techniques embodied in the algorithm were set forth by A. J. Chorin in three papers: "Numerical Study of Slightly Viscous Flow," *Jour. Fluid Mech.*, **57**: 785 (1973); "Vortex Sheet Approximation of Boundary Layers," *Jour. Comput. Phys.*, **27**: 428–442 (1973); "Vortex Models and Boundary Layer Instability," *SIAM Jour. Sci. Comput.* **1**: 1–21 (1980). See A. F. Ghoniem and F. S. Sherman, "Grid-free Simulation of Diffusion Using Random Walk Methods," *Jour. Comput. Phys.* **61**: 1–37 (1985) for an extensive review of the title subject.

throughout the element in various ways, but always so that the velocity induced by the element at any point is finite.

To represent an initially continuous field of vorticity, the support of that field is covered by a convenient coordinate mesh. The circulation around each mesh cell is calculated, and a discrete vorticity element with that circulation is assigned to a point in the center of the cell. The elements so created ordinarily have a radially-symmetric distribution of vorticity, and are called *blobs*. An accurate representation of the initial field of u_v can be obtained by spreading the vorticity of each blob over an area somewhat larger than the area of the grid cell, so that the blobs overlap.

The vorticity newly created at a wall is assigned to sheet-like elements. The no-slip condition is enforced only at a discrete set of wall points, so each sheet is given a circulation equal to the slip velocity observed at the point of its creation, times the distance between wall points. Usually the circulation is partitioned among a number of weaker, but otherwise identical, sheets; this improves the accuracy of the simulation of diffusion. Each sheet is born with an image sheet, like the ones exemplified in Section 8.11, so that the effect of a sheet and its image is entirely confined to the zone between them.

Sheet elements are used only very close to the wall; when a sheet moves past a certain distance from the wall, it is converted into a blob of equal circulation.

Time-splitting The simultaneous processes of convection, creation, and diffusion of vorticity are dealt with sequentially during each time increment.

Convection is simulated by calculating $u = u_v + u_p$ at the position of each vortex blob or sheet, and displacing the element by an amount $d\mathbf{r} = \mathbf{u}\,dt$. This solves, approximately, the equation $D\Omega/Dt = 0$. The accuracy of the approximation improves as the number of elements used to simulate the vorticity field increases. The necessary values of u_p may be calculated by the method of images, or by any other technique that, with adequate accuracy, prevents flow through solid walls and generates the desired flow at infinity.

Creation is simulated by the scheme described in Section 8.12. At a finite number of test stations along the wall, the velocity induced by the vorticity in its displaced configuration, and the correspondingly updated values of u_p are evaluated. Any resulting slip velocity is canceled by the creation of new vorticity elements at the wall.

Diffusion is simulated by use of an analogy between diffusion and *random walk*. This is the feature of the algorithm that gives the method its name. It is easily demonstrated that solutions to the diffusion equation, $\partial\Omega/\partial t = v\,\nabla^2\Omega$, may be accurately approximated by giving discrete elements of Ω random displacements, drawn from a Gaussian distribution of mean zero and variance $\sigma^2 = 2v\,\Delta t$. Examples are shown in Chapter 9.

In the Random-Vortex Method, sheets are given random displacements in only one direction, *normal* to the wall. The probability of drawing a step of length h falls off with increasing h, according to the law $P(h) = P(0) \exp(-h^2/2\sigma^2)$. Blobs are given a two-dimensional random displacement.

The great advantage of this means of simulation of diffusion is that it requires no evaluation of spatial derivatives, while it retains all the qualitative features on which our discussion of vorticity dynamics depends. In particular it shows very clearly that a given particle can acquire its first vorticity only by being close to another particle that has some already. In the simulation, the vorticity simply hops from one particle to another, the average length of the hop being about σ, which is very small if the time step is small.

This treatment of viscous diffusion has one conceptually disturbing feature, which strongly influences the appearance of computed vorticity distributions. As described, the method has no recipe for the destruction of vorticity, although it is clear that viscosity destroys vorticity whenever counterrotating fluid masses come together. Were it not so, the vorticity created by the flight of birds from time immemorial would overwhelm us!

When opposite-signed vortex elements of the algorithm come very close together, they are not abandoned, but their contributions to $\mathbf{u}_v$ are, to some extent, mutually canceling. This means that a dense cloud of vortex elements, such as is shown in Fig. 8.2, does not necessarily correspond to the presence of a large coherent vortex. One has to examine the computed velocity field to check for such features.

Forces on Immersed Bodies

One of the main objectives of fluid flow theory is to predict the forces on immersed bodies. An unexpected feature of the Random-Vortex Method is the ease with which it allows calculation of the pressure distribution along a stationary solid wall that bounds a two-dimensional flow. We have only to look at Eq. (8.15), which equates the tangential pressure gradient to the density times the diffusive flux of vorticity through the wall, and to realize that the algorithm effects a sort of calorimetric measurement of the flux. The local flux times the time increment Δt simply equals the amount of vorticity locally created during Δt, to keep the no-slip condition in force, because all of that vorticity is diffused out into the fluid before Δt is over. This accounting is summarized by the formula

$$\Pi_{,x} = -\frac{U_s}{\Delta t} \tag{8.23}$$

in which U_s is the slip velocity, taken positive in the direction of increasing x, that had to be annihilated. Remember that $\Pi = p/\rho + gh$. The same result can be reached in another way, which emphasizes the wonders of time-splitting. We can think of the first, convective, process in the algorithm, during which the no-slip condition and viscous diffusion are held in abeyance, as causing the fluid at the wall to accelerate from rest (because it satisfied the no-slip condition at the end of the previous time step) to a speed U_s. The version of the momentum equation that describes this is just

$$U_{s,t} + U_s U_{s,x} + \Pi_{,x} = 0.$$

Since U_s is zero at the beginning of the time step, and presumably small at the end, the term that involves U_s quadratically is negligible. Our best estimate of $U_{s,t}$ is just $U_s/\Delta t$, and Eq. (8.21) follows.

The irony of the Random-Vortex Method is that it deals entirely with vorticity, but does not readily yield a good estimate of the value of Ω at the wall. One can form an estimate, such as $\Omega \approx -u(y^*)/y^*$, where y^* is some small distance from the wall, perhaps equal to σ, the standard deviation of the population of random step lengths. However, the result is likely to exhibit fairly severe statistical fluctuations, because it will change suddenly whenever a vortex sheet crosses the level $y = \sigma$.

Overview

The injection of a random element into the computation scheme means that the answers from a given computation will depend somewhat on the precise sequence of random numbers drawn during the calculation, and that there will be some statistical scatter among repeated calculations, in which nothing was changed but the point of entry to the computer's random number string. This may, however, not be all bad, as we shall see in Chapters 13 and 14.

For the purposes of this chapter, the accuracy and cost of the Random-Vortex Method are not really important. The method is fairly new, and these characteristics are continually being improved. It suffices that the method sets forth, in step-by-step algorithmic form, everything needed in principle to predict two-dimensional barotropic viscous flows. The method embodies no preconceptions about the relative importance of different terms in the equations of motion, other than to rule out three-dimensional and baroclinic effects, which can be included at a large increase in computing cost. It makes no a priori assumptions about the appearance of the flow field. Although other computational methods may be more efficient for selected problems, this one seems to me to convey the clearest mental picture of what the flow must do, as mandated by the Navier–Stokes equations, the initial conditions, and the boundary conditions.

8.15 DEFORMATION AND DISSIPATION ASSOCIATED WITH CONCENTRATED VORTICITY

The notion of a concentrated vortex filament is probably as old as or older than the scientific study of fluid dynamics, because such filaments often engender striking visible evidence of their presence. Sometimes, as in a whirlpool, this is associated with the pressure variations that were analyzed in Sample Calculation C of Chapter 5. Sometimes it is a consequence of the cumulative winding-up of a nearby interface between fluid masses that are identified by smoke or color.

Many of the technical applications of fluid motion depend on rapid and continued stretching of the interfaces between fluid masses that differ in some essential characteristic such as temperature or chemical composition. This stretching of the interface is accompanied by a convective steepening of the gradients of temperature or concentration normal to the interface, so that the intrinscially slow processes of conduction or diffusion are driven hard, on a very wide front.

Regions of rapid deformation tend to be closely associated with regions of concentrated vorticity, so it is wise to learn something about that association. One example has already been considered in Exercises 3.5 and 3.6; others will be shown in Chapters 9 and 14.

8.16 SUMMARY

Vorticity is a uniquely interesting property of a viscous flow field, because it can usually only be acquired by the action of viscous torques. The exceptions involve vorticity production by baroclinity, or nonpotential body forces.

When density and viscosity are constant, the action of viscous torques is mathematically equivalent to a process of diffusion. This has many logical consequences, the most striking being that usually vorticity can be traced back to its point of creation, at a boundary of the flowfield where the no-slip condition must be enforced.

Once the vorticity distribution is known, the velocity distribution can be calculated by straightforward, although computationally expensive, procedures. After that, one can calculate how vorticity is convected with the fluid, allowing for changes in Ω due to changes in the angular inertia of deforming fluid particles. Then one can calculate the amount of fresh vorticity that must be created at any point of the boundary to enforce the no-slip condition. Finally, one can calculate the diffusion of vorticity through the motionless fluid, before time is allowed to increase again.

Nature is able to manage all these processes simultaneously; mathematicians have proved that it is perfectly logical to think of them happening sequentially, as long as every process is duly considered during each tiny increment of time.

Although vorticity is conceptually quite distinct from rates of deformation or of viscous dissipation, intense local concentrations of vorticity are often accompanied by high values of these other quantities.

SAMPLE CALCULATIONS

A. Use of Stokes' theorem to compute vorticity components. It is sometimes difficult to remember, or to find, the correct formula for a particular component of Ω in a curvilinear coordinate system. It is, however, very easy to

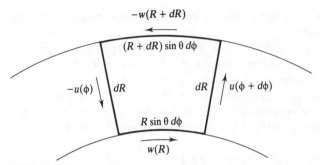

FIGURE 8.11
Circuit for calculation of the θ-component of $\mathbf{\Omega}$.

calculate it directly by applying Stokes' theorem to an elementary area of the coordinate surface normal to the desired component of $\mathbf{\Omega}$. For example, let us derive the formula for Ω_θ in spherical polar coordinates. Look back along the unit vector $\mathbf{e}_\theta$ in this coordinate system, to see the curves of intersection of the planes of constant ϕ, and the spheres of constant R, with a cone of constant θ. Figure 2.5 will remind you of what to expect, which is sketched in Fig. 8.11. The scalar velocity components are symbolized by u, v, and w in $\mathbf{u} = u(R, \theta, \phi)\mathbf{e}_R + v(R, \theta, \phi)\mathbf{e}_\theta + w(R, \theta, \phi)\mathbf{e}_\phi$.

Applied to the elementary area, Eq. (8.3) gives

$$\Omega_\theta R^2 \sin\theta \, dR \, d\phi = [u(\phi + d\phi) - u(\phi)]R \, d\theta$$
$$+ [Rw(R) - (R + dR)w(R + dR)] \sin\theta \, d\phi$$

Dividing by $R^2 \sin\theta \, dR \, d\phi$, and letting dR and $d\phi$ approach zero, we find

$$\Omega_\theta = (R \sin\theta)^{-1} u_{,\phi} - R^{-1}(Rw)_{,R}$$

B. Vorticity equation for axisymmetric, nonswirling flow. If the velocity is $\mathbf{u} = u(r, z)\mathbf{e}_r + w(r, z)\mathbf{e}_z$ in cylindrical polar coordinates, there will be only one scalar component of vorticity, i.e., $\mathbf{\Omega} = \Omega(r, z)\mathbf{e}_\theta$. Show that $D(\Omega/\rho r)/Dt = 0$, if the flow is inviscid, barotropic, and free of nonpotential body forces.

Start by expanding the D/Dt operator and the derivatives of a product in the expression to be validated. First we get

$$\frac{D(\Omega/\rho r)}{Dt} = \left(\frac{\Omega}{\rho r}\right)_{,t} + u\left(\frac{\Omega}{\rho r}\right)_{,r} + w\left(\frac{\Omega}{\rho r}\right)_{,z} = 0$$

Then, since r is independent of t and z,

$$\frac{D(\Omega/\rho r)}{dt} = r^{-1}\left[\left(\frac{\Omega}{\rho}\right)_{,t} + u\left(\frac{\Omega}{\rho}\right)_{,r} + w\left(\frac{\Omega}{\rho}\right)_{,z} - \Omega\frac{u}{\rho r}\right] = 0$$

Recomposing the D/Dt operator, we get

$$\frac{D(\Omega/\rho)}{dT} = \frac{\Omega u}{\rho r}$$

Now compare this with the form of Eq. (8.16), which applies to axisymmetric, nonswirling flow. At first sight, it seems that grad $\mathbf{u}$ is zero, because speed is independent of θ. However, velocity depends on θ, because $\mathbf{e}_{r,\theta} = \mathbf{e}_\theta$. Thus, we get

$$\mathbf{\Omega} \text{ grad } \mathbf{u} = \Omega r^{-1}\mathbf{u}_{,\theta} = \Omega \frac{u}{r}\mathbf{e}_\theta$$

and the one scalar component of Eq. (8.16) becomes $D(\Omega/\rho)/Dt = \Omega u/\rho r$, just what we wanted. Can you think of a simple physical explanation of this result, in terms of the variation of angular inertia of a fluid particle?

C. More exercise with the velocity-gradient tensor. Derive Eq. (8.11) from (8.9). Start by decomposing $\text{grad } \mathbf{u} = \frac{1}{2}(\text{rot } \mathbf{u} + \text{def } \mathbf{u})$. Recall that $\mathbf{b} \text{ rot } \mathbf{u} = \mathbf{\Omega} \times \mathbf{b}$ for any vector $\mathbf{b}$, and that the cross product of any vector with itself is zero. It follows that $\mathbf{\Omega} \text{ rot } \mathbf{u} = 0$. What is left of (8.9) is

$$\frac{D\mathbf{\Omega}}{Dt} = \frac{1}{2}\mathbf{\Omega} \text{ def } \mathbf{u} - \mathbf{\Omega} \text{ div } \mathbf{u}$$

The first term on the right is easily evaluated if $\mathbf{\Omega}$ is projected onto the principal axes of def $\mathbf{u}$. In this coordinate system, the matrix representation of $\frac{1}{2}$ def $\mathbf{u}$ has the simple form (3.20), and the matrix multiplication gives the result

$$\tfrac{1}{2}\mathbf{\Omega} \text{ def } \mathbf{u} = \Omega_1 d^{(1)} + \Omega_2 d^{(2)} + \Omega_3 d^{(3)}$$

By definition of components of a vector, $\mathbf{\Omega} = \mathbf{\Omega}_1 + \mathbf{\Omega}_2 + \mathbf{\Omega}_3$; and we recall that $\text{div } \mathbf{u} = d^{(1)} + d^{(2)} + d^{(3)}$. The rest of the derivation is a simple vector addition.

D. Biot–Savart integrations. Evaluation of the Biot–Savart integral, when it can be done analytically at all, is often fairly tricky. Consider the following example.

Vortex sheet in stagnation-point flow. The two-dimensional potential flow, $\mathbf{u} = Ax\mathbf{e}_x - Ay\mathbf{e}_y$, is called a stagnation-point flow. Imagine it approaching a wall, $y = 0$, and dividing to flow left or right, according to the sign of x. The wall is stationary, and the no-slip condition is enforced. This requires the creation of a vortex sheet at $y = 0^+$, and an image sheet at $y = 0^-$, the strength of the former being $d\Gamma = -Ax$.

Suppose now that this vortex sheet is placed at $y = h$, with the image at $y = -h$. Find the resulting velocity field, as a sum of

1. The velocity induced by the vortex sheet and its image
2. The original potential flow

The scheme for calculation (1) is sketched below.

At point P, the bit of sheet at (x', h) and the bit of image sheet at $(x', -h)$ make the following contributions to the velocity.

$$du = -\frac{Ax'}{2\pi}\{(h-y)[(x'-x)^2+(h-y)^2]^{-1}$$
$$+ (h+y)[(x'-x)^2+(h+y)^2]^{-1}\}\,dx'$$

$$dv = -\frac{Ax'}{2\pi}\{(x'-x)[(x'-x)^2+(h-y)^2]^{-1}$$
$$- (x'-x)[(x'-x)^2+(h+y)^2]^{-1}\}\,dx'$$

Call $x'-x=\xi$, so $dx'=d\xi$. Note that as x' runs from $-\infty$ to $+\infty$, so does ξ. To account for all the sheet, we integrate over ξ from $-\infty$ to $+\infty$. Note that this problem violates the requirement that the vorticity have compact support. The fact that we shall be able to get a finite answer shows that our statement of the requirement was a little conservative. We get, for u:

$$u = -\frac{A}{2\pi}\int_{-\infty}^{\infty}\{(h-y)(x+\xi)[\xi^2+(h-y)^2]^{-1}$$
$$+ (h+y)(x+\xi)[\xi^2+(h+y)^2]^{-1}\}\,d\xi$$

Here, and in the evaluation of v, we use the fact that

$$\int_{-\infty}^{\infty}\xi[\xi^2+(h\pm y)^2]^{-1}\,d\xi = 0$$

because the integrand is an odd function of ξ. This leaves

$$u = -\frac{Ax}{2\pi}\int_{-\infty}^{\infty}\{(h-y)[\xi^2+(h-y)^2]^{-1}+(h+y)[\xi^2+(h+y)^2]^{-1}\}\,d\xi$$

whence $\quad u = -\dfrac{Ax}{2\pi}\{\arctan[\xi/(h-y)]+\arctan[\xi/(h+y)]\}|_{\xi=-\infty}^{\xi=+\infty}$

We are interested in positive values of y, and expect a jump across the plane of the sheet, $y = h$. Indeed, we find

If $y>h$ $\qquad u = -\dfrac{Ax}{2\pi}(-\pi/2-\pi/2+\pi/2+\pi/2)=0$

If $0<y<h$ $\qquad u = -\dfrac{Ax}{2\pi}(+\pi/2+\pi/2+\pi/2+\pi/2)=-Ax$

We see that this exactly cancels the x component of the basic potential flow, in the region between the sheet and the wall, and is zero elsewhere.

To get v, we proceed in the same way, setting up the expression

$$v = \frac{A}{2\pi}\int_{-\infty}^{\infty}\{\xi(x+\xi)[\xi^2+(h-y)^2]^{-1}+\xi(x+\xi)[\xi^2+(h+y)^2]^{-1}\}\,d\xi$$

Dropping out the odd-function integrands, we are left with

$$v = \frac{A}{2\pi} \int_{-\infty}^{\infty} \{\xi^2[\xi^2 + (h-y)^2]^{-1} + \xi^2[\xi^2 + (h+y)^2]^{-1}\} \, d\xi$$

This looks like the difference between two infinite quantities, but the trouble can be avoided by a simple algebraic trick. We write

$$\xi^2 = \xi^2 + (h-y)^2 - (h-y)^2$$

in the first integrand, and $\xi^2 = \xi^2 + (h+y)^2 - (h+y)^2$ in the second. This leads to the form

$$v = -\frac{A}{2\pi} \int_{-\infty}^{\infty} \{(h-y)^2[\xi^2 + (h-y)^2]^{-1} + (h+y)^2[\xi^2 + (h+y)^2]^{-1}\} \, d\xi$$

and to

$$v = -\frac{A}{2\pi} \left\{(h-y) \arctan\left(\frac{\xi}{(h-y)}\right) - (h+y) \arctan\left(\frac{\xi}{(h+y)}\right)\right\}_{\xi=-\infty}^{\xi=+\infty}$$

$$\text{If } y > h \qquad v = -\frac{A}{2\pi}\{-(h-y)\pi - (h+y)\pi\} = Ah$$

Notice that this induced velocity, outside of the region between a sheet and its image, is again proportional to the distance between them.

$$\text{If } 0 < y < h \qquad v = -\frac{A}{2\pi}[+(h-y)\pi - (h+y)\pi] = Ay$$

which again cancels exactly the y component of the basic potential flow. The final, combined velocity field is then

$$\text{If } y > h \qquad \mathbf{u} = Ax\mathbf{e}_x - A(y-h)\mathbf{e}_y$$
$$\text{If } 0 < y < h \qquad \mathbf{u} = 0$$

The effect of the sheet on the flow above it is simply to *displace* the wall from $y = 0$ to $y = h$. Keep this in mind when you get to Chapter 12.

E. Use of random-walk ideas. This thought-calculation illustrates, in a very simple setting, how we impose the no-slip condition while using a random-walk simulation of viscosity. It also introduces us to a different kind of *method of images*, useful in diffusion problems.

Consider an initially stationary fluid, between the infinite, solid planes $y = 0$ and $y = h$. Suddenly, the wall at $y = 0$ starts to move, with constant velocity $\mathbf{u} = U\mathbf{e}_x$. The other wall remains stationary. We want to simulate the simplest possible resulting flow, by random walk of vortex sheets. The specific question we wish to answer is: what should we do when one of the sheets steps out of the fluid?

We start by recognizing that the initial effect of the no-slip condition is to create a uniform vortex sheet of strength U, in the fluid at $y = 0$. We divide the strength of this sheet equally among a large number of microsheets, and disperse them from $y = 0$, giving each one a random displacement in the y-direction, once every time step. Since the probability of a step of a given length is independent of the direction of the step, about half of the microsheets will immediately step out of the fluid. We must take some corrective action, because the boundary conditions require that the same amount of vorticity must remain between the two walls forever. (Prove this to yourself, assuming only that $\mathbf{u} = u(y, t)\mathbf{e}_x$ for all time.)

A successful technique emerges immediately from consideration of the following imaginary experiment. Suppose there were no walls, only an infinite body of fluid in which vortex sheets of strength $2U$ were introduced initially in the planes $y = \pm 2nh$, with $n = 0, 1, 2, 3, \ldots$. What will happen, after each sheet has been divided into microsheets, and each microsheet has taken a few random steps? Consider first the neighborhood of $y = 0$.

In the first step, approximately half the microsheets will acquire positive values of y, so the velocity difference between $y = 0$ and $y = h$ will be approximately what it is in the original problem.

In the second step, some sheets may be displaced from $y > 0$ to $y < 0$. By symmetry, we conclude that each such event is very likely to be matched, at least quite closely, by an event in which just the opposite move was made. Thus, the velocity difference is again approximately maintained.

At some later step, one of the microsheets that started out at $y = 0$ will pass $y = h$. Again we conclude from symmetry that this event is likely to be matched by the passage, in the opposite direction, of a microsheet that started out from $y = 2h$. This logic can be applied over and over, with microsheets from more distant origins coming into consideration at later times. We shall see, in Chapter 9, an analytic solution to this problem, constructed by use of these same symmetry arguments. You can see a sort of analogy to the logic that helped us deduce the image system for a line vortex between two planes.

These considerations are built into the following algorithm.

1. Draw random displacements for only N microsheets, each of strength U/N, and reverse the sign of any negative displacements. This exactly enforces the condition that the velocity difference between $y = 0$ and $y = h$ be equal to U, without significantly altering the statistical distribution of the positive displacements.

2. Whenever a microsheet is subsequently moved from $y > 0$ to $y < 0$, note the final value of y, and reverse its sign before giving the microsheet its next displacement.

3. Whenever a microsheet is subsequently moved from $y < h$ to $y > h$, note the final value of $y - h$, and reverse its sign before giving the microsheet its next displacement.

In short, we make sure that we start out with the right number of microsheets between 0 and h, and we never let one get away. Sample results will be shown in Chapter 9, when we have the analytic solution for comparison.

EXERCISES

8.1. Verify, using cartesian index notation, that $d\mathbf{r} \cdot (\mathbf{u}(\text{grad } \mathbf{u})^*) = d\mathbf{r} \cdot \text{grad} (\mathbf{u} \cdot \mathbf{u}/2)$.

8.2. Verify, using cartesian index notation, that $\mathbf{u} \text{ grad } \mathbf{u} = \text{grad} (\mathbf{u} \cdot \mathbf{u}/2) + \mathbf{\Omega} \times \mathbf{u}$.

8.3. Verify, using cartesian index notation, that

$$\text{curl} (\mathbf{\Omega} \times \mathbf{u}) = \mathbf{u} \text{ grad } \mathbf{\Omega} - \mathbf{\Omega} \text{ grad } \mathbf{u} + \mathbf{\Omega} \text{ div } \mathbf{u}$$

8.4. Show that if $\mathbf{u} \text{ grad } \mathbf{u} = 0$ and $\mathbf{u} \text{ grad } \mathbf{\Omega} = 0$, vortex lines are also lines of constant $\mathbf{u}$. As an example, suppose that $\mathbf{u} = w(x, y, t)\mathbf{e}_z$. Show that in such a rectilinear, fully-developed flow, $\mathbf{\Omega} = -\mathbf{e}_z \times \text{grad } \mathbf{u}$.

8.5. Show, by analysis of the force balance on an element of stationary fluid, in which there is a horizontal density gradient and a gravitational body force, that the expression $\rho^{-2} \nabla\rho \times \nabla p$ has the correct sign, to describe the angular acceleration that results from baroclinicity.

8.6. In free convection flow along an infinitely long vertical wall, the velocity has the form $\mathbf{u} = u(y, t)\mathbf{e}_x$, where u rises from zero at the wall to a maximum, and then decreases again to zero, as distance, y, from the wall increases. Meanwhile, the temperature decreases monotonically from a maximum at the wall.

Describe the vorticity distribution across such a boundary layer. Using the ideas of this chapter, explain how such a distribution can be established in a fluid that was initially at rest and isothermal. (Imagine that the wall temperature jumped suddenly, at time zero.)

8.7. Verify the statement in Section 8.6 about the correct form of the vorticity equation in a rotating frame of reference.

8.8. Derive a generalized form of Eq. (8.15) for the case in which the solid wall is accelerating. Explain your logic in detail.

8.9. Calculate the trajectory followed by the line vortex in a right-angled bend, shown with its images in Fig. 8.8(a).

8.10. Verify, essentially by trigonometry, that the image of a line vortex in a circular cylinder is as shown in Fig. 8.8(b). Derive a formula for the tangential component of velocity, at the surface of the cylinder, induced by the vortex and its image. Let the position of the vortex be $r = b$, $\theta = 0$.

8.11. Show that a line vortex placed parallel to a circular cylinder will move around the cylinder on a circular trajectory. Let the vortex be at $r = b$, and the cylinder surface be at $r = a$. The strength of the vortex is Γ. What is the period of the circular orbit,

(a) If there is no circulation around a concentric circle of radius $>b$?

(b) If the circulation around such a circle equals Γ?

8.12. Show that the velocity induced at the point $(0, 0)$ by the line vortex between parallel planes (see Fig. 8.9) is given by the expression

$$v(0, 0) = -\frac{\Gamma}{2h} \text{ctn} \frac{\pi a}{2h}$$

From this, find the smallest value of h/a, for which the presence of the wall at $x = h$ affects the value of $v(0, 0)$ by less than 1%.

8.13. Derive expressions for radial and circumferential velocity components, $u(r, \theta)$ and $v(r, \theta)$, induced by a cylindrical vortex sheet of uniform strength and radius a. Call the total strength of the vortex sheet Γ. Show, especially, what velocity is induced at a point of the sheet itself.

8.14. Suppose a vortex sheet of uniform strength $d\Gamma/d\theta = Ua$, lies on the cylinder $r = b$, surrounding a coaxial solid cylinder of radius $a = 0.75b$. Calculate and sketch the resulting distribution of circumferential velocity $v(r)$. Assume that there is no velocity at $r = \infty$.

8.15. Consider what will happen to the vortex sheet of Sample Calculation C, if each point of it moves with the velocity induced by the sheet itself, the image sheet, and the given, steady, potential flow.

As a first step, show that A, the strength factor for the vortex sheet, will be a function of t, and derive the form of that function. Let $A = A_0$ when $t = 0$, and the sheet is at $y = h$. In the steady potential flow, $A = A_0$ for all values of t.

CHAPTER
9

ANALYTICAL SOLUTIONS OF THE FULL NAVIER–STOKES EQUATIONS

9.1 INTRODUCTION

In spite of the complexity of the Navier–Stokes equations, a number of special solutions can be described in terms of well-known mathematical functions, and others can be generated by numerical methods that can easily provide any degree of accuracy necessary for a comparison between theory and experiment. These solutions meet idealized initial and boundary conditions, which may be hard to reproduce in practice. When small deviations from the idealized auxiliary conditions produce only small changes in the resultant flow, these solutions may be sufficiently realistic to provide a useful guide for engineering design. However, the solutions are primarily interesting as simple illustrations of the principal qualitative effects of viscosity. By studying these solutions, and the analysis that leads to them, you will begin to gain a feeling for what is possible, or impossible, in a viscous flow.

The usual presentation of this material is organized along mathematical lines. Unaccelerated motions appear first; then come flows without convective acceleration; next come the so-called shape-preserving, or self-similar, flows; and finally, some examples of numerical simulation are presented to illustrate phenomena, such as flow separation, that do not appear in the simpler examples.

This chapter and the next are organized slightly differently, along more phenomenological lines. One aim is to provide immediate reinforcement of the concepts presented in Chapter 8. Thus, the first topic is viscous diffusion, especially diffusion of vorticity without any competition from convection. The generation of vorticity, as required to enforce the no-slip condition, is illustrated in a number of examples. One example involving baroclinic generation of vorticity is included. All examples involve transient response of the fluid to sudden changes, or harmonic oscillations, of the boundary conditions. All these are the subject of Section 9.2.

Having seen how steady flows may be gradually brought about by viscous diffusion, we encounter a detailed study of a few classic examples. Emphasis is on the balance of forces in such flows, and how certain flows can be used for measurement of viscosity. These are analyzed in Section 9.3.

Competition or collaboration between diffusion and convection is the third theme of the chapter; illustrative examples are presented in Section 9.4. Special emphasis is given to convection that limits the spread of vorticity. Qualitative changes associated with changing values of a Reynolds number are emphasized whenever they appear. The principal example used for this purpose is that of an axisymmetric jet produced by a point force.

By virtue of the idealizations inherent in these examples, any discussion of general mathematical properties of the Navier–Stokes equations can be postponed to Chapter 10. The analytical techniques used, and to some extent taught in this chapter, are mostly those associated with heat conduction in solids, and you may find considerable help in a good book on that subject.[1] They include the systematic search for self-similar solutions, separation of variables, Fourier synthesis, and the replacement of boundary-value problems by initial-value problems that satisfy the boundary conditions by virtue of symmetry.

9.2 VISCOUS DIFFUSION

As presented in the discussion of vorticity, the most characteristic effect of viscosity is a diffusive one, by which initial concentrations of vorticity are dispersed. The first examples will exhibit this diffusion, free from any competition with convection.

Diffusing Vortex Sheet

Consider a plane vortex sheet, initially occupying the plane $y = 0$. Let the velocity jump across the sheet be perfectly uniform, say from $\mathbf{u} = -U\mathbf{e}_x$ where $y < 0$, to $\mathbf{u} = +U\mathbf{e}_x$ where $y > 0$. Suppose that the velocity at all times is the $\mathbf{u}_v$

---

[1] H. S. Carslaw and J. C. Jaeger, *Conduction of Heat in Solids,* Oxford University Press, 1948.

induced by the vorticity in the diffusing sheet, and assume that both **u** and Ω will be forever independent of x and z. Finally, assume that both ρ and μ may be treated as constants.

Since the motion is in the x direction, but nothing depends on x, the convective part of D/Dt vanishes, and Eq. (8.14) reduces simply to

$$\Omega_{,t} = v\Omega_{,yy} \tag{9.1}$$

Before any mathematical analysis, consider the quantities on which Ω can depend. First is the overall velocity jump, $2U$. Since U and Ω are linearly related by the Biot–Savart law, it is clear that Ω will be linearly proportional to U. The only other quantities that affect the value of Ω are y, t, and v. Thus, one can write an implicit equation,

$$F\left(\frac{\Omega}{U}, y, t, v\right) = 0 \tag{9.2}$$

Consider the dimensions of each quantity in this equation, using the symbol $[A]$ to denote the dimensions of the quantity A. Thus

$$[\Omega/U] = L^{-1} \qquad [y] = L \qquad [t] = T \qquad [v] = L^2 T^{-1}$$

It follows from the principle of *dimensional homogeneity* that it must be possible to combine the quantities in an equation such as (9.2) into *dimensionless groups*, if the equation represents a quantitative physical relationship. Noting that the problem statement mentions no characteristic length, we eliminate the dimension T between t and v, to form a quantity with the dimension L, namely $(vt)^{1/2}$. We then use $(vt)^{1/2}$ to make both y and Ω/U dimensionless. As a result, Eq. (9.2) can be compressed into

$$F\left(\frac{\Omega(vt)^{1/2}}{U}, \frac{y}{(vt)^{1/2}}\right) = 0$$

or into the equivalent explicit equation

$$\Omega = U(vt)^{-1/2} f(\eta) \qquad \text{in which} \qquad \eta = y(vt)^{-1/2}$$

We shall need the partial derivatives of η, with respect to y and t. They are

$$\eta_{,y} = (vt)^{-1/2} \qquad \text{and} \qquad \eta_{,t} = -\frac{1}{2}\frac{\eta}{t}$$

Thus,

$$\Omega_{,t} = U(vt)^{1/2}\left[-\frac{1}{2t}f(\eta) + f'(\eta)\eta_{,t}\right]$$

$$= -\frac{U}{2t}(vt)^{1/2}[f(\eta) + \eta f'(\eta)]$$

$$\Omega_{,y} = U(vt)^{1/2}f'(\eta)\eta_{,y} = Uf'(\eta)$$

and

$$\Omega_{,yy} = Uf''(\eta)\eta_{,y} = U(vt)^{-1/2}f''(\eta)$$

In all of these, primes denote differentiation with respect to η. Substituting these expressions into (9.3), we obtain the ordinary differential equation,

$$2f'' + \eta f' + f = 2f'' + (\eta f)' = 0$$

integrating once, we get

$$2f' + \eta f = A$$

The vorticity distribution is bound, by symmetry, to be an even function of y, and hence of η. Thus $f'(0) = 0$, and $A = 0$. The remaining equation is *separable*, into

$$\frac{df}{f} = -\eta \frac{d\eta}{2} = -d\left(\frac{\eta^2}{4}\right),$$

so

$$f(\eta) = B \exp\left(-\frac{\eta^2}{4}\right)$$

The constant B is determined by the fact that the velocity jump across the diffused vortex sheet remains always equal to $2U$. Thus,

$$u(\infty, t) - u(-\infty, t) = -\int_{-\infty}^{\infty} \Omega \, dy = 2U$$

Inserting

$$\Omega = U(vt)^{-1/2} B \exp\left(-\frac{\eta^2}{4}\right)$$

and integrating, we find

$$B = -(\pi)^{-1/2}$$

In the original variables, the final answer is

$$\Omega(y, t) = -U(\pi vt)^{-1/2} \exp\left(-\frac{y^2}{4vt}\right), \tag{9.3}$$

and

$$u(y, t) = U \, \text{erf}\left[y(4vt)^{-1/2}\right] \tag{9.4}$$

These functions are plotted versus y, for various values of t, in Fig. 9.1. Note how the maximum value of Ω decreases as the vorticity spreads out.

Characteristic thicknesses of a shear layer. A layer of diffused vorticity, such as this, is often called a *free shear layer*. Its thickness at a given time is subject to somewhat arbitrary definition. Three popular measures of thickness are:

1. Vorticity thickness

$$\delta_\omega = -\frac{2U}{\Omega_{\text{max}}} \tag{9.5}$$

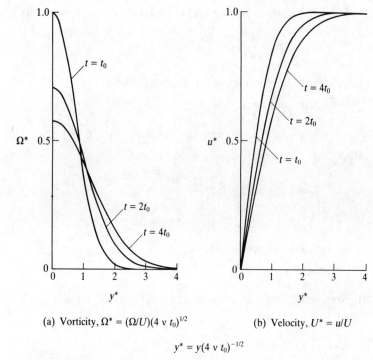

(a) Vorticity, $\Omega^* = (\Omega/U)(4 \nu t_0)^{1/2}$ (b) Velocity, $U^* = u/U$

$$y^* = y(4 \nu t_0)^{-1/2}$$

FIGURE 9.1
Diffusing vortex sheet.

2. Displacement thickness

$$\delta^* = 2U^{-1} \int_0^\infty (U - u)\, dy \tag{9.6}$$

3. Momentum thickness

$$\theta = (2U)^{-2} \int_{-\infty}^\infty (U^2 - u^2)\, dy \tag{9.7}$$

Using Eqs. (9.3) and (9.4), we calculate

$$\delta_\omega = (\pi \nu t)^{1/2} \qquad \delta^* = \left(\frac{\nu t}{\pi}\right)^{1/2} \qquad \text{and} \qquad \theta = \left(\frac{\nu t}{8\pi}\right)^{1/2}$$

These quantities will be encountered again in Chapters 12 and 14.

Diffusing Line Vortex

Consider the radially-symmetric diffusion of a single line vortex, which initially occupies the z axis of a cylindrical polar-coordinate system. The vorticity has only a z component, which depends only on r and t. Equation (8.14) reduces to

$$\Omega_{,t} = \frac{v}{r}(r\Omega_{,r})_{,r} \tag{9.8}$$

and dimensional analysis again leads to a special form of solution,

$$\Omega = \Gamma(vt)^{-1}f(\eta)$$

This time, $\eta = r(vt)^{-1/2}$, and Γ denotes the circulation of the vortex. Thus

$$\Omega_{,t} = \Gamma(2vt^2)^{-1}(2f + \eta f')$$

$$\frac{v}{r}(r\Omega_{,r})_{,r} = \Gamma(vt^2)^{-1}\frac{(\eta f')'}{\eta}$$

and so

$$2(\eta f')' + \eta(2f + \eta f') = 0$$

The last group of terms forms an exact derivative, so the first integration gives

$$2\eta f' + \eta^2 f = A.$$

If $f'(0)$ and $f(0)$ are finite, as is expected, $A = 0$. One again finds $df/f = -\eta \, d\eta/2$, and

$$f(\eta) = B \exp\left(-\frac{\eta^2}{4}\right)$$

This time, the value of B is determined by the equation

$$\Gamma = 2\pi \int_0^\infty r\Omega \, dr$$

which gives $B = (4\pi)^{-1}$, and

$$\Omega(r, t) = \Gamma(4\pi vt)^{-1} \exp\left(-\frac{r^2}{4vt}\right) \tag{9.9}$$

The corresponding distribution of v is obtained from the equation $r\Omega = (rv)_{,r}$, and is

$$v(r, t) = \frac{\Gamma}{2\pi r}\left[1 - \exp\left(-\frac{r^2}{4vt}\right)\right] \tag{9.10}$$

Flow Induced by a Sliding Plane. Stokes' First Problem and Related Self-Similar Flows

Flow induced by an impulsively accelerated wall. Reverse the sign of the vorticity in the diffusing vortex sheet, and add the uniform flow $\mathbf{u} = U\mathbf{e}_x$; you will have a solution which describes, in the half-space $y \geq 0$, the flow induced when an infinite solid wall is impulsively set into sliding motion.

Mathematically,

$$u(y, t) = U\{1 - \text{erf}\,[y(4vt)^{-1/2}]\} = U\,\text{erfc}\,[y(4vt)^{-1/2}] \qquad (9.11)$$

This, like many of the solutions presented in this chapter, was found long ago.[2] At the time, Stokes was interested in the flows created by pendulums. Later, Lord Rayleigh used the solution as a model for a flat-plate boundary layer.

Flow induced by a wall that applies a constant stress. Suppose that the solid wall in Stokes' First Problem, instead of suddenly jumping to a constant velocity, suddenly applies a constant tangential stress τ_0 to the fluid. The corresponding solution is easily found when one realizes that the stress, $\mu u_{,y}$, satisfies the same partial differential equation and boundary conditions in the new problem as does the velocity in Stokes' First Problem. Thus the stress distribution must be

$$\tau_{yx}(y, t) = -\tau_0\,\text{erfc}\,[y(4vt)^{-1/2}]$$

The minus sign appears because $\tau_{xy}(0, t)$ represents the stress exerted on the wall by the fluid, while τ_0 represents the stress exerted on the fluid by the wall. The velocity is now found by integration, as

$$u(y, t) = \int_{\infty}^{y} \left(\frac{\tau_{yx}}{\mu}\right) dy$$

the limits of integration being chosen so that there is no motion at $y = \infty$, for any finite value of t. The calculation involves integration by parts, and gives the result

$$u(y, t) = \frac{\tau_0}{\mu}(4vt)^{1/2}[\pi^{-1/2}\exp(-\eta^2) - \eta\,\text{erfc}\,(\eta)] \qquad (9.12)$$

in which $\eta = y(4vt)^{-1/2}$. Note that the velocity of the wall itself increases as the square root of time. The corresponding distribution of vorticity is

$$\Omega(y, t) = \frac{\tau_0}{\mu}\text{erfc}\,[y(4vt)^{-1/2}] \qquad (9.13)$$

Flow induced by a constantly accelerating wall. Suppose now that the wall has, after time zero, a constant acceleration A. Because of Eq. (9.1), this implies a constant value of $u_{,yy}$ at the wall. The corresponding solution can again be obtained by a simple integration over y; this time Eq. (9.12) provides the

---

[2] G. G. Stokes, "On the effect of the internal friction of fluids on the motion of pendulums," *Trans. Camb. Phil. Soc.* **9**, Pt. II: 8–106 (1851).

integrand. The details are left as an exercise, the answer being

$$u(y, t) = At\left[(1 + 2\eta^2)\operatorname{erfc}(\eta) - \frac{2}{\sqrt{\pi}}\eta\exp(-\eta^2)\right] \tag{9.14}$$

and $\qquad \Omega(y, t) = 2A(t/v)^{1/2}\{\pi^{-1/2}\exp(-\eta^2) - \eta\operatorname{erfc}(\eta)\}.$ (9.15)

Other self-similar flows induced by sliding walls. All the solutions shown so far in this chapter have a special analytic form,

$$u(y, t) = U(t)f(\eta) \qquad \text{and} \qquad \Omega(y, t) = W(t)g(\eta)$$

where $\eta = y(4vt)^{-1/2}$.

This form goes by many names; two of the most popular are *self-similar* and *shape-preserving*. Both refer to the property that the velocity and vorticity profiles have the same shape at all times. This happens only under exceptional circumstances, which can be anticipated intuitively or investigated formally. Let us do the latter, for the general class of flows induced by a sliding motion of a single wall, the velocity of which is $U(t)$.

Insert $u(y, t) = U(t)f(\eta)$ into the partial differential equation $u_{,t} = vu_{,yy}$, getting

$$f'' + 2\eta f' - 4m(t)f = 0 \tag{9.16}$$

in which $\qquad\qquad\qquad m(t) = \frac{t}{U}\frac{dU}{dt}$

Note the following facts about the *similarity variable*, η. When $y = 0$ and $t > 0$, $\eta = 0$. When $y \to \infty$ and t is finite, $\eta \to \infty$. When $t = 0$ and $y > 0$, $\eta = \infty$. The boundary conditions are

1. No slip at the wall; hence $f(0) = 1$
2. No velocity at infinity, for any finite time; hence $f(\infty) = 0$

The initial condition is

3. No velocity at $t = 0$, for any $y > 0$; hence $f(\infty) = 0$, again

This shows that the initial and boundary conditions can be specified entirely in terms of the required behavior of $f(\eta)$. However, solutions of (9.16) will be functions of both η and t, unless the parameter m is independent of t. That will happen only if

$$U(t) = U_0\left(\frac{t}{t_0}\right)^m \tag{9.17}$$

The examples previously worked out in closed form correspond to $m = 0, \frac{1}{2}$, and 1. For other values of m, solutions are easily computed numerically.

A Self-Similar Free-Convection Problem

The following problem exemplifies the simultaneous diffusion of heat and vorticity, and the *baroclinic generation* of vorticity.

Imagine an infinite vertical plane wall, $y = 0$, adjacent to an initially isothermal, motionless fluid. At time zero, the temperature of the wall is suddenly increased by an amount ΔT, and is subsequently held constant. Suppose that the motion and temperature distribution are adequately described by the Boussinesq equations, Eqs. (7.14) to (7.16). Let the x axis point vertically upward, and assume that

$$\mathbf{u} = u(y, t)\mathbf{e}_x \quad \text{while} \quad T = T(y, t)$$

This is reasonable, because the boundary and initial conditions provide no means to distinguish one value of x or z from another. Further, note that $\Pi_{,x} = 0$, as can be proved by consideration of the y component of the momentum equations, and the boundary condition that Π remains uniform at $y = \infty$. Thus, the partial differential equations for u and T reduce to

$$u_{,t} = \nu u_{,yy} + \beta g T \quad \text{and} \quad T_{,t} = \alpha T_{,yy}$$

The initial conditions are $u(y, 0) = T(y, 0) = 0$, for $y > 0$. The boundary conditions are

$u(0, t) = 0,$ and $T(0, t) = \Delta T,$ for $t > 0$ (no-slip and continuity of temperature)

and

$u(\infty, t) = 0,$ and $T(\infty, y) = 0,$ for $t > 0$ (no motion or heating at infinity)

These conditions, plus the lack of any specified scales of time or distance, suggest a search for self-similar solutions, in the form

$$u(y, t) = (\beta \, \Delta T \, gt)F(\eta, \text{Pr}), \quad \text{and} \quad T(y, t) = \Delta T \, G(\eta, \text{Pr}) \quad (9.18)$$

in which $$\eta = y(4\nu t)^{-1/2}$$

A complete study of this system, for arbitrary value of the Prandtl number, has been published;[3] it suffices for our purposes to consider the simplest case, in which $\text{Pr} = 1$. The equations and boundary conditions for F and G are then

$$F'' + 2\eta F' - 4F + 4G = 0$$
$$G'' + 2\eta G = 0$$

with $\quad F(0) = 0, \quad F(\infty) = 0, \quad G(0) = 1, \quad G(\infty) = 0$

---

[3] C. R. Illingworth, "Unsteady Laminar Flow of Gas Near an Infinite Flat Plate", *Proc. Camb. Phil. Soc.* **46**: 603–613 (1950). His equations (37) and (39) describe the velocity profile of interest here.

We have seen the problem for G before; the answer is simply

$$G = \text{erfc}\,(\eta) \tag{9.19}$$

A particular solution for F is obviously $F_p = \text{erfc}\,(\eta)$. The homogeneous equation for F is also familiar; it is Eq. (9.16) with $m = 1$. Its solution appears in Eq. (9.14). The boundary conditions are satisfied by the linear combination of the particular and homogeneous solutions

$$F = \frac{2}{\sqrt{\pi}}\,\eta \exp\,(-\eta^2) - 2\eta^2\,\text{erfc}\,(\eta) \tag{9.20}$$

The functions F and G are plotted in Fig. 9.2(a); in Fig. 9.2(b) we see a simulation by random walk of temperature-jump sheets and vortex sheets.

The baroclinic generation of vorticity is very simply simulated by introucing a new vortex sheet of strength $\delta u = (\beta g\,\delta T)\,\Delta t$, at the location of a temperature-jump sheet of strength δT, at the end of each time step of duration Δt. The random steps of the temperature-jump sheets are drawn from a Gaussian distribution with variance $2\alpha\,\Delta t$, while those for the vortex sheets are drawn from a Gaussian distribution with variance $2\nu\,\Delta t$. For the example shown, the two Gaussians were the same.

Impulsively Started Couette Flow

Suppose the flow of Stokes's First Problem is modified by the presence of a second, stationary, wall in the plane $y = h$. The second wall will constrain the free outward diffusion of vorticity, so that the channel between the parallel planes eventually becomes uniformly full of vorticity. Correspondingly,

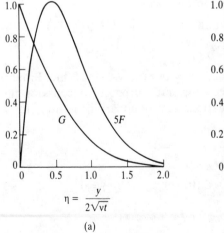

(a)

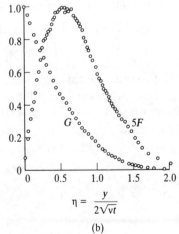

(b)

FIGURE 9.2
Free-convection functions: (a) Eqns. (9.19) and (9.20); (b) random-walk simulation after 5 time steps, with 900 temperature-jump sheets, 7200 vortex sheets.

$u(y, t) \rightarrow Uy/h$ as $t \rightarrow \infty$. This final, steady, flow is called *plane Couette* flow; one of the objectives of this analysis is to predict the duration of the startup phase of the motion.

There are two classical approaches to this analysis; one is especially effective for early times, the other for late times. It is already clear from the examples previously analyzed that the appropriate measure of time, by means of which terms such as early and late can be defined, is the value of the dimensionless quantity vt/h^2.

Analysis for early times. For early times, the superposition of flows due to diffusing vortex sheets, which was presented in Sample Calculation D of Chapter 8, is the most effective. The results are

$$\Omega(y, t) = U(\pi vt)^{-1/2}\left\{\exp\left(\frac{\xi^2}{\tau}\right) + \exp\left(\frac{(2 - \xi)^2}{\tau}\right) + \exp\left(\frac{(2 + \xi)^2}{\tau}\right)\right.$$

$$\left. + \exp\left(\frac{(4 - \xi)^2}{\tau}\right) + \exp\left(\frac{(4 + \xi)^2}{\tau}\right) + \cdots\right\} \tag{9.21}$$

and

$$u(y, t) = U\left\{\operatorname{erfc}\frac{\xi}{\sqrt{\tau}} - \operatorname{erfc}\left[\frac{(2 - \xi)}{\sqrt{\tau}}\right] + \operatorname{erfc}\left[\frac{(2 + \xi)}{\sqrt{\tau}}\right]\right.$$

$$\left. + -\operatorname{erfc}\left[\frac{(4 - \xi)}{\sqrt{\tau}}\right] + \operatorname{erfc}\left[\frac{(4 + \xi)}{\sqrt{\tau}}\right] - \cdots\right\} \tag{9.22}$$

in which

$$\xi = \frac{y}{h} \quad \text{and} \quad \tau = \frac{4vt}{h^2}$$

Each successive term in the series becomes important at a later time. The second term, for example, represents the arrival of vorticity from the sheet initially generated at $y = 2h$. This appears first at $y = h$. The third term represents the arrival of vorticity from the sheet generated at $y = -2h$. This appears first at $y = 0$, and is completely negligible at the time when the second term first becomes important. For $0 \le \tau \le \frac{1}{2}$, only the first three terms of either series is required for five-figure accuracy.

Analysis for late times. For larger values of τ, a more efficient analysis employs the classic method of *separation of variables*. One actually constructs a solution for the dimensionless quantity $\phi(\xi, \tau) = u(y, t)/U - y/h$, which decays to zero at $t \rightarrow \infty$, and which equals zero at both walls.

Inserting $u/U = \phi + \xi$ into the equation $u_{,t} = vu_{,yy}$, we obtain the partial differential equation

$$4\phi_{,\tau} = \phi_{,\xi\xi} \tag{9.23}$$

to be solved subject to the boundary conditions $\phi(0, \tau) = 0$, and $\phi(1, \tau) = 0$, and to the initial condition $\phi(\xi, 0) = \xi - 1$.

With the method of *separation of variables,* one seeks a solution in the form of a sum of functions, each of which is the product of a function of ξ alone and a function of τ alone. Each product function meets the boundary conditions; the final sum is constructed so as to meet the initial condition. The method depends on the principle of superposition, and therefore works only when the differential equation and auxiliary conditions are linear.

Substituting the product form, $\phi = F(\xi)G(\tau)$, into (9.23), we get

$$4F'G = FG''$$

Dividing by FG, which should not be zero, we write

$$\frac{4F'}{F} = \frac{G''}{G} = -\lambda^2$$

Each of the first two terms can be set equal to the third, a constant, because of logical necessity. The first term, which is independent of ξ, equals the second term, which is independent of τ. This can only be true when each term is independent of both ξ and τ. We have made the constant negative so that G will involve trigonometric, rather than exponential, functions, since the latter could never meet the boundary conditions. Thus, the typical product solution is

$$\phi_n = (A_n \cos \lambda_n\xi + B_n \sin \lambda_n\xi) \exp\left(-\lambda_n^2 \frac{\tau}{4}\right)$$

which must vanish at $\xi = 0$, so that $A_n = 0$; and at $\xi = 1$, so that $\lambda_n = n\pi$, with $n = 1, 2, 3, \dots$. Now add solutions of this type, for $n = 1, 2, 3$, etc., adjusting the values of B_n so that the initial condition is met. This is done by requiring that

$$\sum_{n=1}^{\infty} B_n \sin n\pi\xi = \xi - 1$$

The coefficients in this *Fourier series* are calculated from the formula[4]

$$B_n = 2 \int_0^1 (\xi - 1) \sin (n\pi\xi) \, d\xi = -\frac{2}{n\pi}$$

so the final solution is

$$u(y, t) = U\left\{1 - \xi - \frac{2}{\pi}\left[\sin (\pi\xi) \exp\left(-\frac{\pi^2\tau}{4}\right) + \tfrac{1}{2} \sin (2\pi\xi) \exp\left(-\frac{4\pi^2\tau}{4}\right)\right.\right.$$

$$\left.\left. + \tfrac{1}{3} \sin (3\pi\xi) \exp\left(-\frac{9\pi^2\tau}{4}\right) + \cdots\right]\right\}$$

$$(9.24)$$

---

[4] See, for example, R. V. Churchill, *Fourier Series and Boundary Value Problems,* McGraw-Hill, 1941, or C. R. Wylie, Jr., *Advanced Engineering Mathematics,* 2nd ed., McGraw-Hill, 1960.

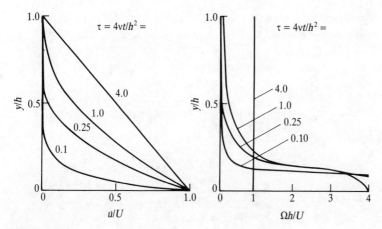

FIGURE 9.3
Impulsively started Couette flow; closed-form solution.

This number of terms suffices for five-figure accuracy, if $\tau \geq \frac{1}{2}$. The corresponding representation of the vorticity is

$$\Omega(y, t) = -\frac{U}{h}\left\{1 - 2\left[\cos(\pi\xi)\exp\left(-\frac{\pi^2\tau}{4}\right) + \cos(2\pi\xi)\exp\left(-\frac{\pi^2\tau}{4}\right)\right.\right.$$
$$\left.\left. + \cos(3\pi\xi)\exp\left(-\frac{9\pi^2\tau}{4}\right) + \cdots\right]\right\} \tag{9.25}$$

The later the time, the fewer terms are needed for accuracy. To decide when the flow is effectively steady, one can demand 1% accuracy of u at $y = h/2$, which gives the value $\tau = 1.9643$, or one can demand 1% accuracy of Ω at $y = h$, which gives $\tau = 2.1473$. Compromising on the value, $\tau = 2$, we find that the startup phase lasts until about $t = h^2/2\nu$.

Figure 9.3 shows these results graphically. For comparison, Fig. 9.4 shows a numerical simulation of the velocity profile, by the random walk of vortex sheets. For the simulation, the total vorticity is shared among 1000 sheets of equal strength.

Flow Over an Oscillating Wall. Stokes' Second Problem

Imagine that the plane $y = 0$ is a solid wall that slides back and forth in a harmonic oscillation, so that $u(0, t) = U\cos\omega t$. We seek a solution of $u_{,t} = \nu u_{,yy}$ that contains harmonic time factors $\cos\omega t$ and $\sin\omega t$. A convenient way to do this is to employ complex exponential notation,

$$u(y, t) = \mathcal{R}\{f(y)\exp(i\omega t)\} \tag{9.26}$$

in which the function $f(y)$ has both real and imaginary parts, and the symbol

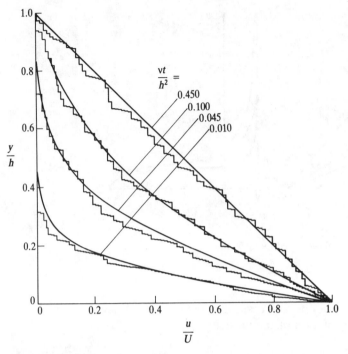

FIGURE 9.4
Impulsively started Couette flow; random-walk simulation with 100 vortex sheets, $\Delta t = 0.001 h^2/v$. (Adapted, by permission, from Ghoniem and Sherman (1985) op. cit.)

$\mathcal{R}$ denotes the real part of a complex expression. The no-slip boundary condition is then equivalent to the requirement, $f(0) = U$.

During the preliminary analysis, we shall treat u itself as a complex quantity, remembering that only its real part represents the physical velocity. The necessary partial derivatives are $u_{,t} = i\omega f \exp(i\omega t)$, and $u_{,yy} = f'' \exp(i\omega t)$, so the partial differential equation is replaced by the ordinary differential equation

$$i\omega f = v f''$$

The general solution of this is

$$f = A \exp\left[\left(\frac{i\omega}{v}\right)^{1/2} y\right] + B \exp\left[-\left(\frac{i\omega}{v}\right)^{1/2} y\right]$$

To be definite about the square root of i, we choose the root with positive real part,

$$\sqrt{i} = (1 + i)/\sqrt{2}$$

We then specify that the velocity should remain finite as $y \to \infty$, which requires that $A = 0$. The no-slip condition then gives $B = U$, and the final solution is

$$u(y, t) = \mathscr{R}\left\{ U \exp\left[-(1+i)\left(\frac{\omega}{2v}\right)^{1/2} y + i\omega t\right]\right\}$$

$$= U \exp\left[-\left(\frac{\omega}{2v}\right)^{1/2} y\right] \cos\left[\omega t - \left(\frac{\omega}{2v}\right)^{1/2} y\right] \qquad (9.27)$$

This solution can be interpreted as an upward-running harmonic wave, the amplitude of which suffers an exponential damping with increasing distance from the wall. The phase of the cosine wave is constant along lines $y = (2\omega v)^{1/2} t$, so one may say that the *speed of propagation* of the wave equals $(2\omega v)^{1/2}$.

Note that above $y = 6.4(v/\omega)^{1/2}$, the value of u never exceeds $0.01U$, so the motion is confined to a *Stokes layer* of thickness

$$\delta = 6.4\left(\frac{\omega}{v}\right)^{1/2} \qquad (9.28)$$

It is interesting to imagine the simulation of this solution by the random walk of discrete vortex microsheets. The momentum equation, in the form $u_{,t} = -v\Omega_{,y}$ shows that the wall is a source of positive vorticity while it is accelerating, and of negative vorticity while it is decelerating. In the simulation, each time step involves the release of a certain number of microsheets at the wall, which subsequently wander out into the fluid just as they do in Stokes' First Problem. As they do so, sometimes stepping forwards and sometimes backwards, they mingle with the microsheets that left the wall at earlier and later times. The mingling of positive and negative vorticity is manifest in the exponential damping factor in the analytical solution.

Phase relationships are interesting in an oscillatory flow. Note, for example, that vorticity leads velocity by 45°. This statement is given graphic meaning, first by the complex equation

$$\Omega = -u_{,y} = (1+i)\left(\frac{\omega}{2v}\right)^{1/2} u \qquad (9.29)$$

and then by the sketch of the product of two complex numbers, shown in Fig. 9.5. In that figure, u is represented by a vector that rotates steadily around the

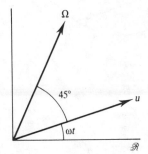

FIGURE 9.5
Phase relationships in an oscillating Stokes layer.

origin, at angular speed ω. The length of the vector depends on y. Equation (3.30) shows that Ω is a second vector, obtained by multiplying u by a complex constant, the argument of which is $+\pi/4$, or 45°. Thus, the argument of Ω exceeds that of u by 45°, and the vector that represents Ω leads the vector that represents u around the circle. In particular, the maximum forward stress exerted by the wall on the fluid leads the maximum forward velocity of the fluid at the wall by 45°.

The Ekman Layer

Because of the linearity of the partial differential equation and boundary conditions for Stokes' Second Problem, the sum of two such solutions is a third solution. Consider, then, the motion in fluid at $z > 0$ when the solid plane $z = 0$ executes the motion $\mathbf{u}(0, t) = U\{\cos \omega t\, \mathbf{e}_x + \sin \omega t\, \mathbf{e}_y\}$. Using the symbol λ as shorthand for $(\omega/2\nu)^{1/2}$, one can immediately write down the following expression for the fluid velocity:

$$\mathbf{u}(z, t) = U \exp(-\lambda z)[\cos(\omega t - \lambda z)\mathbf{e}_x + \sin(\omega t - \lambda z)\mathbf{e}_y] \qquad (9.30)$$

An interesting feature of this flow is that it appears steady in a rotating frame of reference. Define new coordinate axes, as sketched in Fig. 9.6(a). The old unit vectors are related to the new by the equations

$$\mathbf{e}_x = \cos(\omega t)\mathbf{e}_{x'} - \sin(\omega t)\mathbf{e}_{y'} \qquad \text{and} \qquad \mathbf{e}_y = \sin(\omega t)\mathbf{e}_{x'} + \cos(\omega t)\mathbf{e}_{y'}$$

If we expand $\cos(\omega t - \lambda z)$ and $\sin(\omega t - \lambda z)$, and substitute for $\mathbf{e}_x$ and $\mathbf{e}_y$, we can transform Eq. (9.30) into

$$\mathbf{u}(z) = U \exp(-\lambda z)[\cos(\lambda y)\mathbf{e}_{x'} - \sin(\lambda z)\mathbf{e}_{y'}] \qquad (9.31)$$

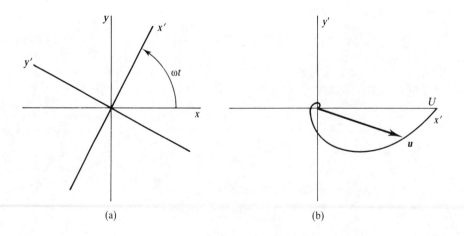

(a)   (b)

FIGURE 9.6
The Ekman layer: (a) rotating axes; (b) Ekman sprial.

from which t has disappeared. This velocity distribution, seen in plan view in Fig. 9.6(b), forms an *Ekman spiral*.

The corresponding viscous stress on a plane of constant z is

$$\tau(z) = -\mu U \left(\frac{\omega}{\nu}\right)^{1/2} \exp{(-\lambda z)}[\cos{(\lambda y)}\mathbf{e}_{z''} - \sin{(\lambda y)}\mathbf{e}_{x''}] \qquad (9.32)$$

in which $\mathbf{e}_{z''}$ and $\mathbf{e}_{x''}$ are unit vectors that lead by $45°$ the rotation of vectors $\mathbf{e}_{z'}$ and $\mathbf{e}_{x'}$. Thus, in the original unsteady problem the stress applied to the fluid by the wall is a rotating vector that leads the rotation of the surface velocity vector by $45°$.

A final interesting feature of the Stokes layer and the Ekman layer is the depth-integrated velocity,

$$\mathbf{Q} = \int_0^\infty \mathbf{u} \, dz \qquad (9.33)$$

A simple calculation shows that this quantity lags $45°$ behind the oscillation or rotation of the surface velocity, and hence $90°$ behind the applied stress.

Although the model we have analyzed is far too idealized to represent phenomena like ocean currents with any great accuracy, many of the insights gained from it, particularly that the depth-integrated flow moves at right angles to the stress applied to the ocean surface by a wind, are invaluable to the physical oceanographer.

Flows Outside a Spinning Cylinder

When a solid cylinder of radius a spins about its axis, with a time-varying angular velocity, some very interesting phenomena may be observed. The mathematical descriptions will be more complicated than those for flow over a plane wall, or for a diffusing line vortex, because of the presence of a new scale of length, the cylinder radius. One might expect intuitively that flow near the cylinder will closely resemble a corresponding plane flow if the vorticity remains close to the cylinder, and that flow very far from the cylinder may closely correspond to some evolution of a line source of vorticity.

The cylinder with harmonically oscillating angular velocity. The analog to Stokes' Second Problem will be analyzed first, because the mathematical treatment is fairly simple. We start with the circumferential momentum equation,

$$v_{,t} = \nu(v_{,rr} + r^{-1}v_{,r} - r^{-2}v) \qquad (9.34)$$

and seek a harmonically oscillating solution,

$$v(r, t) = \mathcal{R}[f(\xi) \exp{(i\omega t)}] \qquad (9.35)$$

in which $\xi = r/a$. The specified velocity at the cylinder surface is

$$v(a, t) = \omega a \cos{(\omega t)}$$

The resulting equation for the complex function $f(x)$ is then:

$$f'' + \xi^{-1}f' - \left[\left(\frac{i\omega a^2}{\nu}\right) + \xi^{-2}\right]f = 0 \tag{9.36}$$

This differential equation is well known; its solutions are the modified Bessel functions of complex argument,[5]

$$f = AI_1\left[\left(\frac{i\omega a^2}{\nu}\right)^{1/2}\xi\right] + BK_1\left[\left(\frac{i\omega a^2}{\nu}\right)^{1/2}\xi\right] \tag{9.37}$$

the boundary conditions on f are

$$f(1) = \omega a \quad \text{(no-slip)} \quad \text{and} \quad f(\infty) \text{ is finite}$$

It is known that $I_1[z] \to \infty$ as $|z| \to \infty$, so we must set $A = 0$. The no-slip condition then determines that

$$B = \frac{\omega a}{K_1\left[\left(\frac{i\omega a^2}{\nu}\right)^{1/2}\right]}$$

Quasisteady oscillations. In Stokes' Second Problem, there is no meaningful way to judge whether the oscillations have high or low frequency, because the given data define no standard of comparison. In the present case there is, on the contrary, a dimensionless parameter, $\omega a^2/\nu$, which is clearly a kind of Reynolds number. When this parameter has very small values, we may expect that the fluid will respond quasisteadily to the oscillations of the cylinder. To see this, we use the first term in a power series expansion of the Bessel function for small values of the argument. This is $K_1(z) \approx 1/z$, so we find the low-frequency limit, $f(\xi) \approx \omega a/\xi$, and

$$v(r, t) \approx \frac{\omega a^2}{r} \cos(\omega t) \tag{9.38}$$

This is, conceptually, quite a startling result! A quick check shows that this flow, which can be brought about only by the viscous diffusion of vorticity, is irrotational! Where did the vorticity go?

High-frequency oscillations. Before trying to answer this provocative question, let us examine the opposite limit, in which $\omega a^2/\nu \to \infty$. For this, we use the

---

[5] See, for example, Sections 9.6 and 9.7 of M. Abramowitz and I. A. Stegun, *Handbook of Mathematical Functions*, Dover Publications, New York, 1965.

asymptotic behavior of the Bessel function,

$$K_1(z) \approx \exp(-z) \qquad \text{for } |z| \gg 1$$

The corresponding limit of $f(\xi)$ is then $f(\xi) \approx \omega a \exp[(i\omega a^2/\nu)^{1/2}(1-\xi)]$. If we adopt the notation, $\omega a = U$, and $r - a = y$, we can rewrite this approximation as

$$f \approx U \exp\left[\left(\frac{\omega}{2\nu}\right)^{1/2}(i-1)y\right]$$

The corresponding approxmation for v is

$$v(r, t) \approx U \exp\left[-\left(\frac{\omega}{2\nu}\right)^{1/2}y\right]\cos\left[\omega t - \left(\frac{\omega}{2\nu}\right)^{1/2}y\right]$$

which is just the solution to Stokes' Second Problem. We would expect this, because $\omega a^2/\nu \gg 1$ guarantees that the Stokes layer thickness is very small compared to the radius of the cylinder. We shall make frequent use, in Chapter 12, of the notion that curvature of a wall has negligible effect on the transport of vorticity in a thin layer near the wall if the thickness of the layer is very small compared to the radii of curvature of the wall. Meanwhile, we note that the high-frequency limit of the present solution exhibits plenty of vorticity, so that there must be something special about a very low-frequency oscillation that lets the vorticity, but not the velocity, disappear.

The cylinder with impulsive angular acceleration. To get a direct analog to Stokes' First Problem, and a graphic answer to what happens to the vorticity in the quasi-steady oscillations described above, let the angular velocity of the cylinder be zero for $t < 0$, and equal to a constant, ω, thereafter. The no-slip condition introduces a vortex sheet of uniform strength, $d\Gamma/d\theta = \omega a^2$, which is accompanied by an image sheet of the same strength, so that there is initially no fluid motion except right at the wall.

The mathematical analysis makes use of the Laplace transform[6] and arrives finally at the results

$$v(r, t) = \omega a\left\{\xi^{-1} + \frac{2}{\pi}\int_0^\infty \exp(-p^2\tau)[J_1(p\xi)Y_1(p) - Y_1(p\xi)J_1(p)]D^{-1}\right\}dp$$

(9.39)

in which $\qquad \xi = \dfrac{r}{a} \qquad \tau = \dfrac{\nu t}{a^2} \qquad$ and $\qquad D = p[J_1^2(p) + Y_1^2(p)]$

---

[6] See, for example, Carslaw and Jaeger, *op. cit.*, Chapter 12. A very similar problem is analyzed in their Section 127.

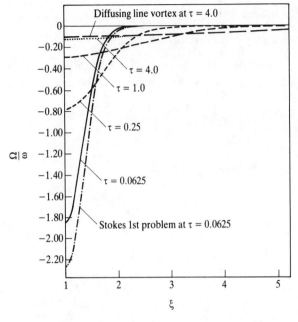

FIGURE 9.7
Vorticity diffusing from an impulsively spun-up cylinder: $\tau = vt/a^2$; $\xi = r/a$.

The functions J_1 and Y_1 are ordinary Bessel functions of the first order. The corresponding representation of the vorticity is

$$\Omega(r, t) = \frac{2\omega}{\pi} \int_0^\infty \exp(-p^2\tau)[J_0(p\xi)Y_1(p) - Y_0(p\xi)J_1(p)]D^{-1}\, dp \quad (9.40)$$

This function has been plotted versus ξ, with τ as a parameter, in Fig. 9.7. Two approximations are shown for comparison, these being

$$\Omega(r, t) = -\omega(\pi\tau)^{-1/2} \exp\left(-\frac{(\xi - 1)^2}{4\tau}\right)$$

from Stokes' First Problem, and

$$\Omega(r, t) = -\omega(2\tau)^{-1} \exp\left(-\frac{\xi^2}{4\tau}\right)$$

from the diffusing line vortex.

In the limit as $\tau \to \infty$, the velocity at any finite value of ξ approaches the limiting distribution $v = \omega a^2/r$, which describes an irrotational flow. This is quite easily understood from our work in Chapter 8. Recall that a cylindrical vortex sheet of uniform strength, which surrounds a circular cylinder, and which has an image sheet of equal strength inside the cylinder, induces a circular potential flow between sheet and image, and no flow elsewhere. Now think of the vortex sheet, which is produced when the cylinder starts rotating, as being partitioned into microvortices. As time advances, each microvortex takes a succession of random steps in the x–y plane. The probability of a step of given length is independent of the direction of the step, except that the microvortices are not allowed to leave the fluid, because that would violate the no-slip condition. On the average, all the microvortices eventually wander farther and farther from the surface of the cylinder, and the image vortex, needed to prevent flow through the cylinder, approaches the point $r = 0$. The velocity field that is left between $r = a$ and any finite value of $r > a$ is just the potential flow induced by all the images, concentrated at $r = 0$. In a quasi-steady oscillation, each fresh charge of vorticity associated with a change in surface velocity has time to diffuse out to infinity, before the next charge of signficant magnitude is generated.

Flows with Uniform Pressure Gradients

In all of the foregoing examples, vorticity has been generated as the result of tangential motion of a wall, in a fluid that would otherwise be stationary. Very similar phenomena arise when there is a potential-flow component, $\mathbf{u}_p(t)$, with which is associated a uniform pressure gradient along a stationary wall. Two examples will be presented now, each involving axial flow in a straight pipe of circular cross section.

Harmonically oscillating pipe flow. Suppose $\mathbf{u}_p(t) = W \cos(\omega t)\mathbf{e}_z$, with which there is associated the pressure gradient $\Pi_{,z} = -w_{,t} = \omega W \sin(\omega t)$. We seek $\mathbf{u}_v(r, t)$, such that $\mathbf{u}_p + \mathbf{u}_v$ satisfies the no-slip condition at the pipe wall, $r = a$. This is another variant of Stokes' Second Problem, in which there is a parameter, again $\omega a^2/v$, that can be used to distinguish high- and low-frequency behavior.

We assume the simplest possible form for $\mathbf{u}_v$, namely $\mathbf{u}_v = w(r, t)\mathbf{e}_z$. The partial differential equation governing w is the simple diffusion equation

$$w_{,t} = \frac{v}{r}(rw_{,r})_{,r} \tag{9.41}$$

We extract a harmonic time factor, setting $w = \mathscr{R}[f(\xi)\exp(i\omega t)]$, with $\xi = r/a$. The differential equation and boundary conditions for $f(\xi)$ are

$$f'' + \xi^{-1}f' - \frac{i\omega a^2}{v}f = 0$$

with $f(1) = -W$ (no-slip); and $f(0)$ finite. The solution again involves Bessel functions, this time of zeroth order. Specifically,

$$f(\xi) = AI_0\left[\left(\frac{i\omega a^2}{v}\right)^{1/2}\xi\right] + BK_0\left[\left(\frac{i\omega a^2}{v}\right)^{1/2}\xi\right] \tag{9.42}$$

Because the condition of finiteness now applies at $\xi = 0$ instead of $\xi = \infty$, and $K_0(0)$ is logarithmically infinite, we must set $B = 0$. Adjusting A to meet the no-slip condition, we get

$$f(\xi) = -\frac{I_0[(i\omega a^2/v)^{1/2}\xi]}{I_0[(i\omega a^2/v)^{1/2}]} W \qquad (9.43)$$

Quasi-steady oscillations. Noting that for $|z| \ll 1$, $I_0(z) \approx 1 + z^2/4$, we find the low-frequency limit,

$$f(\xi) \approx -\frac{[1 + i(\omega a^2/4v)\xi^2 + \cdots]}{[1 + i(\omega a^2/4v) + \cdots]} W$$

$$\approx -W\left[1 + i\left(\frac{\omega a^2}{4v}\right)(\xi^2 - 1) + \cdots\right]$$

Thus, $\qquad w(r, t) \approx -W\left\{\cos(\omega t) - \left(\frac{\omega a^2}{4v}\right)(\xi^2 - 1)\sin(\omega t)\right\}$

When we add on the contribution of $\mathbf{u}_p$, and eliminate $\omega W \sin(\omega t)$ in favor of $\Pi_{,z}$, we get the famous Hagen–Poiseuille formula:

$$w(r, t) \approx -\frac{\Pi_{,z}}{4v}(a^2 - r^2) \qquad (9.44)$$

By use of the asymptotic formula for $I_0(z)$ in the limit as $|z| \to \infty$, we again recover the solution for thin Stokes layers.

Sudden application of a constant pressure gradient. Suppose a uniform, constant, axial pressure gradient is applied suddenly at time zero, and held constant thereafter. We wish to study the approach of the flow to a steady state in which viscous forces eventually arrest the acceleration of the fluid. For this purpose, it is convenient to study the difference between the transient value of $w(r, t)$, and the steady-state limit given by Eq. (9.42). For variety, let us return to a plane-flow geometry, with stationary walls at $y = 0$ and $y = h$. The steady-state flow for a uniform pressure gradient $\Pi_{,x}$ is described by the solution to $\Pi_{,x} = vu_{,yy}$. This is

$$u(y, \infty) = \frac{1}{2v}\Pi_{,x}(y^2 - yh) \qquad (9.45)$$

Let us employ the two approaches used to study the starting process of a plane Couette flow. We again use the variables $\xi = y/h$, and $\tau = 4vt/h^2$.

For early times, $\tau < 1$, we use the method of image flows. At $y = 0$, the pressure gradient would cause a constant acceleration of the flow, if it acted alone. The no-slip condition thus acts to generate clockwise vorticity at a constant rate, starting at time zero. At $y = h$, counterclockwise vorticity is generated at the same rate.

We have a solution for the diffusion of vorticity from such a source, in the absence of a second wall, in Eqs. (9.14) and (9.15). The solution we want

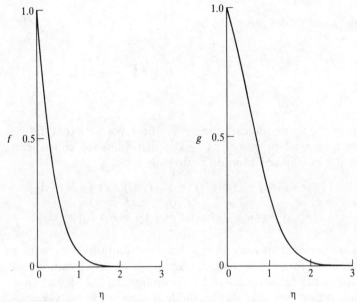

FIGURE 9.8
Similarity functions for constantly accelerated flow a stationary wall.

here describes a uniform, constantly accelerating flow, minus the effects of vorticity diffusing into the channel from the real and image walls. To compute it, we first evaluate the functions

$$f(\eta) = \frac{2}{\sqrt{\pi}} \, \eta \exp\left(-\eta^2\right) - (1 + 2\eta^2) \operatorname{erfc}(\eta)$$

and
$$g(h) = \exp\left(-\eta^2\right) - \sqrt{\pi} \, \eta \operatorname{erfc}(\eta)$$

at a suitable sequence of values of h. The functions of η are plotted in Fig. 9.8, to give an idea of the range of η for which computations will be necessary.

Figure 9.9 indicates a numbering system for a sequence of values of η.

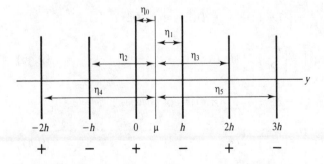

FIGURE 9.9
Enumeration of images for impulsively started plane Poiseuille flow.

Analytically, the required values are given by

$$\eta_0 = \frac{\xi}{\sqrt{\tau}} \qquad \eta_1 = \frac{1 - \xi}{\sqrt{\tau}} \qquad \eta_2 = \frac{1 + \xi}{\sqrt{\tau}}$$

$$\eta_3 = \frac{2 - \xi}{\sqrt{\tau}} \qquad \eta_4 = \frac{2 + \xi}{\sqrt{\tau}}, \qquad \text{etc.}$$

The $+$ and $-$ signs in the figure indicate the sense of the vorticity that diffuses into the fluid from each wall or image wall. The algorithm for constructing solutions that meet the no-slip condition on both walls is

$$u(y, t) = -\Pi_{,x} t \left[1 + f(\eta_0) + f(\eta_1) - f(\eta_2) - f(\eta_3) + f(\eta_4) + f(\eta_5) - \cdots \right]$$

$$\Omega(y, t) = -2\Pi_{,x} \left(\frac{t}{\pi v} \right)^{1/2} \left[g(\eta_0) - g(\eta_1) - g(\eta_2) + g(\eta_3) + g(\eta_4) - g(\eta_5) - \cdots \right]$$

Each added term in these expressins brings the contribution from a more distant image, so at early times only a small number of the contributions will have arrived. Details of the calculation will be shown in Sample Calculation A.

For late times, $\tau > 1$, we use the method of separation of variables, solving again for the difference between $u(y, t)$ and $u(y, \infty)$. This difference, which we may call $W(y, t)$, satisfies the simple diffusion equation, $W_{,t} = vW_{,yy}$. It vanishes at both walls, so the solution again has the form

$$W(y, t) = \sum_{n=1}^{\infty} B_n \exp\left[-\left(\frac{n\pi}{2} \right)^2 \tau \right] \sin(n\pi\xi).$$

The initial condition is $W(y, 0) = -(\Pi_{,x} h^2 / 2v)(\xi^2 - \xi)$, so

$$B_n = -2 \frac{\Pi_{,x} h^2}{2v} \int_0^1 (\xi^2 - \xi) \sin(n\pi\xi) \, d\xi$$

$$= -4(n\pi)^{-3}[\cos(n\pi) - 1]\left(\frac{\Pi_{,x} h^2}{2v} \right)$$

The final solution,

$$u(y, t) = -\frac{\Pi_{,x} h^2}{2v} \left\{ \xi - \xi^2 + 4 \sum_{n=1}^{\infty} (n\pi)^{-3}[\cos(n\pi) - 1] \right.$$

$$\left. \times \exp\left[-\left(\frac{n\pi}{2} \right)^2 t \right] \sin(n\pi\xi) \right\} \tag{9.46}$$

and

$$\Omega(y, t) = \frac{\Pi_{,x} h}{2v} \left\{ 1 - 2\xi + 4 \sum_{n=1}^{\infty} (n\pi)^{-2}[\cos(n\pi) - 1] \right.$$

$$\left. \times \exp\left[-\left(\frac{n\pi}{2} \right)^2 t \right] \cos(n\pi\xi) \right\} \tag{9.47}$$

shows that the mid-channel velocity has attained 99% of its asymptotic value when $t = 0.47 h^2/v$, while the vorticity at the wall has attained 99% of its asymptotic value when $t = 0.45h^2/\nu$. We can compare these figures with the startup time for flow in a round pipe of diameter d, which is about $t = 0.19d^2/\nu$. Presumably, the tube flow comes more quickly to steady state because the vorticity diffuses in from all sides. The values for an impulsively-started plane Couette flow are very nearly the same as for the corresponding Poiseuille flow.

9.3 STEADY VISCOMETRIC FLOWS

The simplest solutions of the Navier–Stokes equations describe steady flow conditions that arise, as we have seen, some time after the establishment of steady motion of a driving boundary, or application of a constant pressure gradient. They may also appear as a response to low-frequency oscillatory driving motions or forces.

Several of these appear in elementary texts, and have been used as examples in earlier chapters of this text. Many of them are called *viscometric flows*, because the solutions closely approximate flows in devices used to measure viscosity.

Couette Flows and Poiseuille Flows

The name of *M. Couette* is given to a variety of flows driven by differential tangential motion of enclosing walls, while that of *J. L. M. Poiseuille* is often associated with rectilinear pressure-driven flows in stationary conduits.[7] We have already used a number of these flows for our discussions of kinematics and the variability of thermodynamic and transport properties. Let us now derive, in detail, one such solution, involving coaxial circular cylinders in differential axial translation, with a imposed axial pressure gradient. We start with the assumption that the flow is already steady, and that the velocity is independent of z and θ in a cylindrical polar, (r, θ, z), coordinate system. We must eventually say something about the accuracy of these assumptions in real cases, but that discussion needs more advanced theoretical tools, and will be postponed to Chapters 10, 12, and 13. If

$$\mathbf{u} = u(r)\mathbf{e}_r + v(r)\mathbf{e}_q + w(r)\mathbf{e}_z$$

the continuity equation,

$$(ru),_r + v,_\theta + (rw),_z = 0 \tag{9.48}$$

reduces to $(ru),_r = 0$. If either cylinder is impervious, we deduce that $u = 0$. If

---

[7] M. Couette, "Etudes sur le Frottement des Liquides", *Ann. Chim. Phys.*, **21**: 433–510 (1890). J. L. M. Poiseuille "Sur le mouvement des liquides dans le tube de trés-petit diametre", *Comptes Rendues*, **9**: 487 (1839).

$v = 0$ at the walls, it may be assumed to be zero everywhere, leaving only $\mathbf{u} = w(r)\mathbf{e}_z$. It follows from this that the viscous stress tensor has only two, equal, nonzero components, τ_{rz} and τ_{zr}, which depend only on r.

From the r- and θ-momentum equations, we find that Π can depend only on z. We then consider a force balance on a small annular segment of fluid, bounded by cylinders of radius r and $r + dr$, and by planes at $z = Z$ and $z = Z + dz$. The pressure force plus weight of fluid has the resultant $-2\pi r \, dr \, \rho \Pi_{,z} \, dz \, \mathbf{e}_z$, to the first order in the small quantity dr. The viscous force has the resultant $2\pi(r\mu w_{,r})_{,r} \, dr \, dz \, \mathbf{e}_z$. Since the flow is unaccelerated, these two forces must add to zero, giving the equation

$$v(rw_{,r})_{,r} = -\Pi_{,z} \tag{9.49}$$

A first integration gives

$$rw_{,r} = -\frac{\Pi_{,z}}{2v} r^2 + A$$

and a second gives

$$w = \frac{\Pi_{,z}}{4v} r^2 + A \ln r + B$$

Let the inner cylinder, of radius a, be stationary; while the outer, of radius b, moves with speed W. The no-slip conditions determine the values of A and B, and the final expression for the velocity profile is

$$w(r) = -\frac{\Pi_{,z}}{4v}(r^2 - a^2) + \left[W + \frac{\Pi_{,z}}{4v}(b^2 - a^2) \right] \left[\frac{\ln (r/a)}{\ln (b/a)} \right] \tag{9.50}$$

This rather awkward expression involves the geometrical factor b/a, and special interest attaches to the limiting cases, $b - a \ll a$, and $b \gg a$. We examine the first, the *narrow gap*, case by setting $r = a(1 + \xi)$ and $b = a(1 + \varepsilon)$, where ξ and ε are both $\ll 1$. We then use the power series,

$$\ln (1 + x) = x - \frac{x^2}{2} + \frac{x^3}{3} - \cdots$$

retaining the indicated number of terms, to show that

$$w(r) = -\frac{\Pi_{,z} a^2}{2v} \xi(\varepsilon - \xi)\left(1 - \frac{2\xi - \varepsilon}{6} + \cdots\right) + W\left(\frac{\xi}{\varepsilon}\right)\left(1 - \frac{\xi - \varepsilon}{2} + \cdots\right) \tag{9.51}$$

The leading terms describe the flow between plane parallel walls; the correction terms show that the effect of curvature is everywhere less than 1% if $\varepsilon < 0.06$ (for the pressure-driven component of flow) or $\varepsilon < 0.02$ (for the component driven by differential wall motion.)

In the opposite limit, $a \ll b$, we visualize a tiny wire running along the axis of a large pipe. Since the wire is stationary, it puts a dent in the velocity profile that would exist without the wire. The corresponding effect on the flow rate is quite remarkable.

Substituting the exact expression for w, given by (9.50), into the integral

$$Q = 2\pi \int_a^b wr\, dr$$

and introducing the symbol $\alpha = a/b$, we calculate that

$$Q = -\frac{\pi b^4 \Pi_{,z}}{8v}\left(1 + \frac{(1-\alpha^2)^2}{\ln \alpha}\right) + \pi b^2 W\left(1 + \frac{(1-\alpha^2)}{2\ln \alpha}\right) \qquad (9.52)$$

Even when $\alpha = 0.001$, the pressure-driven flow rate is reduced by 14%, while the flow rate dragged along with the moving outer pipe is reduced by 7%. Of course, when we set $\alpha = 0$ and $W = 0$, Eq. (9.52) reduces to the famous *Hagen–Poiseuille* formula for the rate of steady, laminar flow in a circular tube of radius b.

Notice one more thing before we leave this example. We have analyzed an unaccelerated flow that is driven by two separate mechanisms, the imposed pressure gradient and the axial motion of the outer cylinder. Because the nonlinear terms of the Navier–Stokes equation vanish for this idealized flow, and because the boundary conditions also involve no nonlinear mathematical operations, we could have treated this as two separate problems, one for each driving mechanism, and then superposed the answers. We could now, by continuation of this procedure, add on a flow driven by differential rotation of the cylinders around their common axis. Although that flow is not strictly unaccelerated, because $a_r = -v^2/r$, the differential equation and boundary conditions for $v(r)$ are linear. This is *not* typical of most fluid flow calculations, but we shall encounter a few more interesting examples in Section 9.4.

Poiseuille-flow solutions are known for some cylindrical conduits with unusual cross sections, and for some simple flows of immiscible fluids of different density and viscosity. We shall give one example of each of these.

Poiseuille flow in a tube of elliptical cross section. It can easily be verified that the velocity distribution

$$\mathbf{u}(x, y) = -\frac{\Pi_{,z}}{2v}\frac{a^2 b^2 - b^2 x^2 - a^2 y^2}{a^2 + b^2}\mathbf{e}_z \qquad (9.53)$$

satisfies the equations of motion and the boundary condition, $\mathbf{u} = 0$, on the elliptical cylinder

$$\frac{x^2}{a^2} + \frac{y^2}{b^2} = 1$$

If $a = b$, we recover the formula for a circular cross section; if $a \to \infty$ with a fixed value of b, we get $w(x, y) \to -(\Pi_{,z}/2v)[(b^2 - y^2 - b^2 x^2)/a^2]$, in which the last term is negligible for any finite value of the ratio x/b. This is just the formula for plane Poiseuille flow.

Equilibrium Flow of Two Immiscible Fluids Down an Inclined Plane

Suppose two superposed layers of immiscible fluid, each of constant, uniform depth, flow down an inclined plane. We suppose that the pressure on the upper surface is uniform. Let the x axis point down the fall line of the plane, as shown in Fig. 9.10.

We assume that the flow is rectilinear, fully-developed, and steady, so that $\mathbf{u} = u(y)\mathbf{e}_x$. The momentum equation reduces to

$$\Pi_{,x} = vu_{,yy} \quad \text{and} \quad \Pi_{,y} = 0$$

At the upper surface of the fluid layer, the pressure is a constant, which we may call P. Thus, at $y = h$, $\Pi = P/\rho + gh$ and $\Pi_{,x} = gh_{,x} = -g \sin a$. Since Π is independent of y, this value of $\Pi_{,x}$ acts uniformly through both liquids. The velocity profile will then consist of two parabolas,

$$u_1 = -\frac{g \sin \alpha}{2v_1} y^2 + A_1 y + B_1 \quad \text{in the lower fluid, } 0 \le y \le \beta h$$

and
$$u_2 = -\frac{g \sin \alpha}{2v_2} y^2 + A_2 y + B_2 \quad \text{in the upper fluid, } \beta h \le y \le h$$

The boundary conditions are applied in the following way:

1. At $y = 0$, $u = 0$ (no slip), so $B_1 = 0$.
2. At $y = h$, $\tau_{yx} = \mu_2 u_{,y} = 0$. This is an idealized condition, representing the idea that any fluid, e.g., air, that lies above the liquids exerts negligible tangential stress on the top liquid surface. It implies that $A_2 = (gh \sin \alpha)/v_2$.
3. At $y = \beta h$, the liquid–liquid interface, τ_{yx} is continuous; to $\mu_2 u_{2,y} = \mu_1 u_{1,y}$, which gives $A_1 = [(g \sin \alpha)/\mu_1]\{\rho_1 \beta h + \rho_2(1 - \beta)h\}$.
4. Also, at $y = \beta h$, u is continuous; so $u_2 = u_1$. This determines the value of B_2.

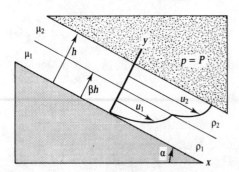

FIGURE 9.10
Flow down an inclined plane.

For the final expressions for the velocity profiles, we introduce the symbols $M = \mu_2/\mu_1$, $R = \rho_2/\rho_1$, $\phi = (2v_1/gh^2 \sin \alpha)u$, and $\eta = y/h$. The results are

$$\phi_1 = 2[\beta + R(1 - \beta)]\eta - \eta^2 \tag{9.54}$$

and
$$\phi_2 = \frac{R}{M}(\eta - \beta)(2 - \eta - \beta) + \beta^2 + 2R\beta(1 - \beta) \tag{9.55}$$

As a check on the algebra, we note that both of these equations reduce to $\phi = 2\eta - \eta^2$, the result for a homogeneous layer, when $M = 1$ and $R = 1$. Note that only the density, but not the viscosity, of the upper liquid affects the flow in the lower layer. This is because the value of τ_{yx} at the interface equals $\rho_2 gh(1 - \beta) \sin \alpha$, as can be shown by integrating the x-momentum equation in its primitive form, $\tau_{yx,y} = -\rho g \sin \alpha$, from the free surface down to the liquid–liquid interface.

Some special cases are of interest. Suppose we have a very viscous oil over water. Typical values might be $M = 1000$, $R = 0.9$, $\beta = 0.1$. These correspond to

$$\phi_1 = 1.82\eta - \eta^2,$$

which is not much affected by the lower density of the oil, and

$$\frac{M}{R}\phi_2 = 2\eta - \eta^2 + 198.21$$

which is enormously affected by the low viscosity of the water! The velocity profile of the oil would extrapolate to a large slip velocity at the wall. It is interesting to compare the speed of the free top surface to that of the interface between the liquids. The general formula is $\phi_2(1) - \phi_2(\beta) = (R/M)(1 - \beta)^2$. For our example, this gives $\phi_2(1)/\phi_2(\beta) = 1.0043$. The speed of the oil hardly varies at all!

Effects of Sidewalls in Plane Couette Flow

The idealization that a channel between plane parallel walls is infinitely wide in the spanwise direction can be criticized with the mathematical tools we have already at hand. Suppose that the boundary conditions on a plane, steady Couette flow include not only $u = 0$ at $y = 0$, and $u = U$ at $y = H$, but the *sidewall conditons* $u = 0$ at $z = \pm W$. Our approach parallels that used to analyze the late transient phase of the flow between infinite plane walls after one has been set impulsively in motion. We set

$$u(y, z) = \frac{Uy}{H} + f(y, z)$$

and solve for $f(x, y)$ by the method of separation of variables. The governing equation comes from the x-momentum equation, under the assumption that

$\mathbf{u} = u(y, z)\mathbf{e}_x$. This time, the continuity equation does not rule out the possibility that v and w might both be nonzero, but shows that if one vanishes everywhere, so does the other. The x-momentum equation reduces to

$$\Pi_{,x} = v(u_{,yy} + u_{,zz})$$

If this is differentiated again, with respect to x, we find that $\Pi_{,xx} = 0$. Then, if Π has the same value at two distinct values of x, $P_{,x}$ will vanish everywhere, and we have only to solve $u_{,yy} + u_{,zz} = 0$. The corresponding problem for $f(y, z)$ is then

$$f_{,yy} + f_{,zz} = 0 \tag{9.56}$$

with $f = 0$ for $y = 0$ and $y = H$, $-W < z < W$; and $f = -Uy/H$ for $z = \pm W$, $0 \le y < H$. Notice that the velocity at the junctions of the moving top wall and the stationary side walls is not specified. We now try for a separation-type solution of (9.56), in the form

$$f_n = (A_n \cosh \lambda_n z + B_n \sinh \lambda_n z)(C_n \cos \lambda_n y + D_n \sin \lambda_n y)$$

Trigonometric functions are chosen for the y direction, so that Fourier synthesis can be used to meet the sidewall boundary conditions. Because of symmetry, f will be an even function of z, so we set $B_n = 0$. Because f must vanish at $y = 0$ and $y = H$, we must set $C_n = 0$ and $\lambda_n = n\pi/H$, for $n = 1, 2, 3$, etc. This leaves

$$f_n = A_n \cosh\left(\frac{n\pi z}{H}\right) \sin\left(\frac{n\pi y}{H}\right)$$

with the values of the A_n to be determined so that

$$-\frac{Uy}{H} = \sum_{n=1}^{\infty} A_n \cosh\left(\frac{n\pi W}{H}\right) \sin\left(\frac{n\pi y}{H}\right)$$

The final result is

$$u(y, z) = U\left\{\frac{y}{H} + z \sum_{n=1}^{\infty} (-1)^n \frac{1}{n\pi}\left[\frac{\cosh(n\pi z/H)}{\cosh(n\pi W/H)}\right] \sin\left(\frac{n\pi y}{H}\right)\right\} \tag{9.57}$$

For practical purposes, we wish to know how the stress exerted on the bottom wall varies with z, for a given value of the *aspect ratio* of the channel, W/H. This is described by the equation

$$\tau_{yx}(0, z) = \frac{\mu U}{H}\left\{1 + z \sum_{n=1}^{\infty} (-1)^n\left[\frac{\cosh(n\pi z/H)}{\cosh(n\pi W/H)}\right]\right\} \tag{9.58}$$

Consider first the stress in the middle of the bottom wall, where $z = 0$. Its value very rapidly approaches $\mu U/H$ as W/H increases, being reduced by less than one part in a million when $W = 5H$. For this same value of W/H, the stress is nearly uniform over the middle one-half of the span of the channel, the reduction from the value $\mu U/H$ being only 0.0078% at $z = \pm W/2$. Thus, if only the central half of the bottom wall were instrumented for a measurement of drag, the measured value would be essentially unaffected by the presence of the sidewalls.

It is interesting to note that the series like (9.58), for the stress on the top, moving, wall, does not converge. This is a mathematical reflection of the physical impossibility of our specification that $u(H, \pm W)$ has two different values.

An even more practically interesting case would involve coaxial rotating cylinders with endwalls, but the simplest possible flow then involves all three scalar components of velocity, and nonlinear equations.

9.4 COMPETITION BETWEEN CONVECTION AND DIFFUSION

In the examples of the last section there has been convection, but no competition between convection and diffusion. Specifically, there has been no velocity component in the one direction in which vorticity was diffusing. We now consider some simple examples in which convection opposes diffusion.

A Diffusing Vortex Sheet With Stretched Vortex Lines

Let us reconsider the diffusion of a plane, uniform vortex sheet, with $\mathbf{\Omega} = \Omega(y, t)\mathbf{e}_z$, and $\mathbf{u}_v = u(y, t)\mathbf{e}_x$; adding a potential flow $\mathbf{u}_p = -Ay\mathbf{e}_y + Az\mathbf{e}_z$. This brings two new terms into the vorticity equation:

$$\mathbf{u} \operatorname{grad} \mathbf{\Omega} = v\Omega_{,y}\mathbf{e}_z = -Ay\Omega_{,y}\mathbf{e}_z \quad \text{and} \quad \mathbf{\Omega} \operatorname{grad} \mathbf{u} = \Omega w_{,z}\mathbf{e}_z = \Omega A\mathbf{e}_z$$

The vorticity equation thus becomes

$$\Omega_{,t} - Ay\Omega_{,y} - \Omega A = v\Omega_{,yy} \tag{9.59}$$

The potential flow has a characteristic time, A^{-1}, but no characteristic length; so we may still hope to find some sort of shape-preserving solution. We try the form

$$\Omega = -\frac{4}{\sqrt{\pi}}\frac{U}{\delta}f(\eta)$$

where $\eta = y/\delta$, and δ is an unknown function of t. U is, as before, the constant strength of the vortex sheet. The factor $-(4/\sqrt{\pi})$ is inserted so that we recover Eq. (9.3) if $\delta^2 = 4vt$, as it does when $A = 0$. We need the partial derivatives,

$$\Omega_{,t} = \frac{4}{\sqrt{\pi}}\frac{U}{\delta}\frac{\delta'}{\delta}(f + \eta f') \qquad \Omega_{,y} = -\frac{4}{\sqrt{\pi}}\frac{U}{\delta^2}f'$$

$$\Omega_{,yy} = -\frac{4}{\sqrt{\pi}}\frac{U}{\delta^3}f''$$

After substituting and collecting terms, we have

$$-\left(\frac{\delta'}{\delta} + A\right)(f + \eta f') = \frac{v}{\delta^2} f''$$

We can again use the logic of separation of variables, rearranging this into

$$-\left(\frac{\delta'}{\delta} + A\right)\bigg/\frac{v}{\delta^2} \quad = \quad f''/(f + \eta f') \quad = \quad -2B^2$$

(independent of η)   (independent of t)   (constant)

The reason for the negative constant is seen when we solve for $f(\eta)$, which again turns out to be a Gaussian function, $f = \exp(-B^2 \eta^2)$.

The equation for δ is easily put into the form $(\delta^2)' + 2A\delta^2 = 4vB^2$, with initial condition $\delta(0) = 0$. The solution is

$$\delta^2 = \frac{4vB^2}{2A}[1 - \exp(-2At)]$$

The value of B must be such that we recover the solution for pure diffusion in the limit as $A \to 0$. Referring back to Eq. (9.3), we see that $B = 1$. Thus, the final result is

$$\Omega = -\frac{4}{\sqrt{\pi}}\frac{U}{\delta}\exp\left(-\frac{y^2}{\delta^2}\right) \tag{9.60}$$

with

$$\delta^2 = \frac{2v}{A}[1 - \exp(-2At)] \tag{9.61}$$

When δ is very small, the diffusion takes place where the competing convection is very weak; thus the value of A has no effect in the limit as $t \to 0$. A standoff between diffusion and convection is reached when $\delta^2 = (2v/A)$. Notice that the steady-state limit of δ is proportional to $v^{1/2}$, but that the time to reach steady state is independent of the value of v.

The stretched axisymmetric vortex. There is an axisymmetric analog to the problem just analysed. The initial conditions are composed of a line vortex of strength Γ, along the z axis, plus the potential flow $\mathbf{u}_p = -Ar\mathbf{e}_r + 2Az\mathbf{e}_z$. The analysis exactly parallels that given for the stretched vortex sheet, and the answer is analyzed in some detail in Section 14.4.[8]

---

[8] J. M. Burgers, "A Mathematical Model Illustrating the Theory of Turbulence", *Advances in Applied Mechanics*, Vol. 1, p. 171–198, Academic, New York, 1948.

The Cross-Stretched Vortex Sheet

Consider now a slight, but interesting modification of the first example of this section, in which we turn the superposed potential flow so that $\mathbf{u}_p = -Ay\mathbf{e}_y + Ax\mathbf{e}_x$. This new potential flow does not stretch the vortex lines of the sheet, but moves them farther apart. Thus the velocity jump U induced by the sheet will decrease with time.

Since the length of any material line segment parallel to the x axis will grow exponentially as $X(t) = X(0) \exp(At)$, the corresponding behavior of the shear will be $U(t) = U(0) \exp(-At)$. We then try again for a solution that has the correct form in the limit $A \to 0$, namely

$$\Omega = -\frac{4}{\sqrt{\pi}} \frac{U(0)}{\delta} \exp(-At) f(\eta)$$

and discover that $f(\eta)$ and $\delta(t)$ are exactly the same functions as when the sheet is stretched in the other direction. However, in the present case the flow never becomes steady; by the time the thickness of the layer has stabilized, the vorticity has been pumped out of it.

Diffusion Against a Uniform Flow

Suppose that fluid can actually be withdrawn through a porous wall that is a source of vorticity for a fluid region adjoining the wall. The usual model of such a wall assumes that there is still no slip *along* the wall, but that there may be a specified, nonzero, velocity *normal* to the wall. Physical surfaces, usually composed of a porous ceramic, sintered metal, or compressed screen, have actually been tested, and it seems that the model is reasonable. There are three solutions, very easily obtained, that show the possibilities of the resulting competition, or collaboration, between convection and diffusion.

The asymptotic suction boundary layer. Suppose the velocity in the half space $y \geq 0$ has the form $\mathbf{u} = u(y)\mathbf{e}_x - V\mathbf{e}_y$, for some positive constant V. The bondary conditions on $u(y)$ are $u(0) = 0$ (no-slip); $u(\infty) = U$. The x component of the momentum equation reduces to $vu_{,y} = vu_{,yy}$, or to $v\Omega = v\Omega_{,y}$. The latter represents an exact balance between the rate at which vorticity diffuses outward, and the rate at which it is convected back toward the wall. The velocity and vorticity profiles are

$$u(y) = U\left[1 - \exp\left(\frac{-Vy}{v}\right)\right] \tag{9.62}$$

and

$$\Omega(y) = \frac{-UV}{v} \exp\left(\frac{-Vy}{v}\right) \tag{9.63}$$

This is called the *asymptotic suction layer,* because it represents the state of flow that appears far from the leading edge of a semi-infinite plane wall, on which the boundary conditions specified here apply only where $x > 0$.

Flow between infinite, plane, parallel, porous walls. If a second porous wall, on which the no-slip condition must be met, occupies the plane $y = h$, and the flow is sustained by the uniform pressure gradient $\Pi_{,x}$, the x-momentum equation becomes $v\Omega - \Pi_{,x} = v\Omega_{,y}$. Let $v = V$, this time a positive constant. The solution for Ω has the form

$$\Omega(y) = \frac{\Pi_{,x}}{V} + A \exp\left(\frac{Vy}{v}\right)$$

and the corresponding equation for u is

$$u(y) = -\frac{\Pi_{,x}}{V} y - \frac{Av}{V} \exp\left(\frac{Vy}{v}\right) + B$$

We adjust A and B so that $u = 0$ at both $y = 0$ and $y = h$, and introduce the symbols

$$\xi = \frac{y}{h} \qquad \text{Re} = \frac{Vh}{v}$$

The final results are

$$u(y) = -\frac{\Pi_{,x}h}{V}\left(\xi - \frac{\exp(\text{Re}\,\xi) - 1}{\exp(\text{Re}) - 1}\right), \tag{9.64}$$

and

$$\Omega(y) = \frac{\Pi_{,x}}{V}\left(1 - \frac{\text{Re}\exp(\text{Re}\,\xi)}{\exp(\text{Re}) - 1}\right) \tag{9.65}$$

These are normalized and plotted, with Re as a parameter, in Fig. 9.11. They have interesting limits as the *cross-flow Reynolds number*, Re, approaches 0 or

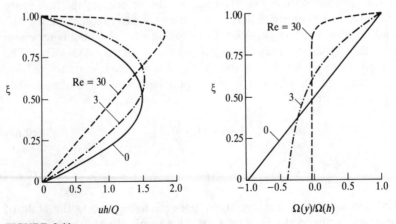

FIGURE 9.11
Flow in a porous channel, with uniform crossflow.

∞. As Re $\to 0$,

$$u(y) \to \frac{-\Pi_{,x} h^2}{2\nu} (\xi - \xi^2) \quad \text{and} \quad \Omega(y) \to \frac{\Pi_{,x} h}{2\nu} (1 - 2\xi)$$

just the results for plane Poiseuille flow. As Re $\to \infty$,

$$u(y) \to \frac{\Pi_{,x} h}{V} \{\exp(-\text{Re}(1 - \xi)] - \xi\}$$

and

$$\Omega(y) \to -\frac{\Pi_{,x}}{V} \{\text{Re} \exp[-\text{Re}(1 - \xi)] - 1\}$$

The last result shows that the velocity profile is just a straight line, while vorticity is uniform, and relatively weak, outside of a thin layer adjacent to the top wall. If we extrapolate the straight part of the velocity profile, to the wall where $\xi = 1$, and call the resulting intercept U, we find that the velocity and vorticity profiles inside the thin layer are very well approximated by the formulas

$$u(y) = U\left[1 - \exp\left(-\frac{V(h - y)}{\nu}\right)\right] \quad \text{and} \quad \Omega(y) = \frac{UV}{\nu} \exp\left(\frac{-V(h - y)}{\nu}\right)$$

These are just the formulas for the asymptotic suction layer.

When Re $\gg 1$, diffusion is unimportant where it would assist convection, but important where convection and diffusion are opposed.

Flow outside a porous, spinning cylinder. Suppose that the surface, $r = a$, of a circular cylinder is porous, and that flow is drawn uniformly through the surface, producing the radial flow $\mathbf{u}_p = -(Ua/r)\mathbf{e}_r$. We shall imagine that this flow is steady, so that U is a constant. The cylinder is spinning around its axis, with angular velocity ω, which has been constant for a long time. We seek a solution for the velocity associated with vorticity, in the form $\mathbf{u}_v = v(r)\mathbf{e}_\theta$, requiring that $v(a) = \omega a$ (no-slip), and that $v(\infty)$ is finite. The θ-component of the momentum equation reduces to

$$-\text{Re}\,\Omega = r\Omega_{,r}$$

in which Re $= Ua/\nu$. The solution may be written as $\Omega(r) = (2 - \text{Re})Br^{-\text{Re}}$, with a corresponding velocity profile $v(r) = Ar^{-1} + Br^{1-\text{Re}}$.

The value of B that will keep v finite at $r = \infty$ depends on the value of Re. If Re ≥ 1, any value of B will do. If Re < 1, we must set $B = 0$.

In the latter case, $A = \omega a^2$ and the field of v is just the same as though there were no suction at all! The suction has been too weak to prevent the diffusion of vorticity all the way out to infinity.

When Re ≥ 1, only A or B can be eliminated, unless something more explicit is said about the flow at infinity. Suppose, for example, there is a second porous cylinder, with surface $r = b \gg a$, which is not spinning. Then we find

$$\Omega(r) = (2 - \mathrm{Re})\omega\left[1 - \left(\frac{b}{a}\right)^{2-\mathrm{Re}}\right]^{-1}\left(\frac{r}{a}\right)^{-\mathrm{Re}} \tag{9.66}$$

When both b/a and Re are very large, we find the same qualitative result as in the last previous example, when its Re was very large; there is some vorticity everywhere between the walls, but it is intensely concentrated in a thin layer next to the wall into which fluid is being sucked.

Axisymmetric Flow Induced By a Point Force

An analytically simple but phenomenologically very rich exact solution was discovered by Slezkin, and developed independently by Landau and by Squire.[9] It represents the simplest motion that could be induced in an infinite fluid by a fixed, constant, point source of momentum.

One might imagine a tiny ducted propellor with straightening vanes as the cause of the momentum flux or force. Such details do not appear in the solution, which is characterized by only three physical parameters: the force **F**, the density ρ, and the kinematic viscosity v. These can be combined into a single dimensionless parameter $F/\rho v^2$, which is, as we shall see, a kind of Reynolds number.

When the force **F** is first turned on, it accelerates fluid along its line of action, and generates a very concentrated vortex ring, which encircles the line of action. This ring moves swiftly away from 0, the point of application of the force, inducing a velocity field that draws fresh fluid toward point 0. The force acts on this fresh fluid, imparting vorticity to it. The new vorticity contributes to the velocity field, and the process continues until the starting vortex has moved off to infinity and a steady state is asymptotically established in any region of finite radius centered on 0.

Viscosity diffuses the vorticity away from the line of action of the force, to an extent that depends on the value of the parameter $F/\rho v^2$. If no solid walls are present, the velocity field will consist exclusively of the component $\mathbf{u}_v$, and the simplest imaginable flow will be axisymmetric, with no swirl. The preferred coordinate system for analysis will be the spherical polar system, with **F** directed along the polar axis, where $\theta = 0$. Then

$$\mathbf{u} = u(R, \theta, t)\mathbf{e}_R + v(R, \theta, t)\mathbf{e}_\theta$$

It will be convenient to introduce a streamfunction $\Psi(R, \theta, t)$ from which

[9] N. A. Slezkin, "On an exact solution of the equations of viscous flow", *Moscow State Univ. Uch. Zap. MGV Sci. Rec.* **2** (1934). L. D. Landau, "A New Exact Solution of Navier–Stokes Equations", *Doklady Akad. Nauk S.S.S.R.* **43**: No. 7, 286–288 (1944). (In English). H. B. Squire, "The Round Laminar Jet", *Quart. Jour. Mech. Appl. Math.* **4**: 321–329 (1951).

the velocity components are calculated. The necessary equations are

$$u = (R^2 \sin \theta)^{-1}\Psi_{,\theta} \quad \text{and} \quad v = -(R \sin \theta)^{-1}\Psi_{,R} \qquad (9.67)$$

The way in which Ψ can depend on R and t is constrained by dimensional considerations. Consider the implicit relationship

$$F(\ \Psi;\ R, \theta, t; F/\rho v^2,\ v\ ) = 0$$
$$ L^3T^{-1}\ L\ 1\ T\ \ 1\ \ L^2T^{-1}$$

The dimensions of each quantity appear below it. Let us use v to remove the dimension of time, wherever it appears. This gives

$$F(\Psi/v; R, \theta, vt; F/\rho v^2) = 0$$
$$ L\ \ L\ 1\ L^2\ \ 1$$

Now we see that only R remains, as a quantity with which the dimension L can be removed from Ψ/v and vt. Using it, we find

$$F(\Psi/vR; \theta, vt/R^2\ F/\rho v^2) = 0$$
$$ 1\ \ 1\ \ 1\ \ \ 1$$

which can be written explicitly as

$$\Psi = vR f\left(\theta, \frac{vt}{R^2}; \frac{F}{\rho v^2}\right)$$

The analysis of the transient case, for which **F** is turned on suddenly at time zero, has been carried out, but is far from simple.[10] We consider here only the limit as $vt/R^2 \to \infty$, for which

$$\Psi(R, \theta) = vR f\left(\theta; \frac{F}{\rho v^2}\right) \qquad (9.68)$$

A parallel dimensional analysis exhibits constraints on the pressure field, and leads to the form

$$\Pi(R, \theta) - \Pi_\infty = \frac{v^2}{R^2} g\left(\theta; \frac{F}{\rho v^2}\right) \qquad (9.69)$$

Here Π_∞ represents the uniform value of Π that exists at $R = \infty$.

For subsequent analysis, we shall need expressions for velocity, vorticity, and viscous stress. For convenience, we replace θ by the variable $\xi = \cos \theta$.

---

[10] B. J. Cantwell (1981) "Transition in the axisymmetric jet." *J. Fluid Mech.* **104**: 369–386.

The necessary equations are

$$u = -vR^{-1}f'(\xi), \qquad v = -vR^{-1}(1-\xi^2)^{-1/2}f(\xi) \tag{9.70}$$

$$\Omega = -vR^{-1}(1-\xi^2)^{1/2}f''(\xi) \tag{9.71}$$

$$\tau_{RR} = 2\rho v^2 R^{-2}f'(\xi) \tag{9.72}$$

$$\tau_{R\theta} = \rho v^2 R^{-2}[2(1-\xi^2)^{-1/2}f(\xi) + (1-\xi^2)^{1/2}f''(\xi)] \tag{9.73}$$

$$\tau_{\theta\theta} = 2\rho v^2 R^{-2}\xi(1-\xi^2)^{-1}f(\xi) \tag{9.74}$$

$$\tau_{\phi\phi} = -2\rho v^2 R^{-2}[f'(\xi) + \xi(1-\xi^2)^{-1}f(\xi)] \tag{9.75}$$

The viscous force per unit volume is

$$\operatorname{div}\tau = -\rho v^2 R^{-3}\{[(1-\xi^2)f''(\xi)]'\mathbf{e}_R - (1-\xi^2)^{1/2}f''(\xi)\mathbf{e}_\theta\} \tag{9.76}$$

the resultant pressure force per unit volume is

$$-\rho\,\nabla\Pi = \rho v^2 R^{-3}[2g\mathbf{e}_R + (1-\xi^2)^{1/2}g'\mathbf{e}_\theta] \tag{9.77}$$

and the inertial reaction per unit volume is

$$\rho\frac{D\mathbf{u}}{Dt} = -\rho v^2 R^{-3}\{[(f(\xi)f'(\xi))' + (1-\xi^2)^{-1}f^2(\xi)]\mathbf{e}_R$$
$$+ (1-\xi^2)^{-1/2}[\xi(1-\xi^2)^{-1}f^2(\xi) + f(\xi)f'(\xi)]\mathbf{e}_\theta\} \tag{9.78}$$

The R-momentum equation gives

$$g = -\tfrac{1}{2}\{(1-\xi^2)^{-1}f^2 + [ff' - (1-\xi^2)f'']'\}$$

The θ-momentum equation gives

$$g' = -f'' - \tfrac{1}{2}[(1-\xi^2)^{-1}f^2]'$$

which can be integrated to give

$$g = -f' - \tfrac{1}{2}[(1-\xi^2)^{-1}f^2] + C_1$$

The pressure function g can now be eliminated, and the resulting equation is readily integrated once more, leaving

$$-f + C_1\xi + C_2 = -\tfrac{1}{2}[ff' - (1-\xi^2)f''] \tag{9.79}$$

Let us now introduce two boundary conditions on the positive polar axis, $\xi = 1$. We expect this, by symmetry, to be a streamline, and we choose it to be the *zero streamline*. Thus, we set

$$f(1) = 0 \tag{9.80}$$

The vorticity and the velocity must be finite on the axis, if $R > 0$. These requirements mean that both $(1-\xi^2)^{1/2}f''$ and f' are finite when $\xi = 1$, so that the right-hand side of (9.79) vanishes there. It follows, from the remnant of (9.79) and (9.80), that $C_1 + C_2 = 0$.

Noting that $2f + ff' - (1 - \xi^2)f'' = [2\xi f + (1 - \xi^2)f']'$, we can integrate (9.79) again, getting $2\xi f + (1 - \xi^2)f' - \frac{1}{2}f^2 = C_1(\xi^2 - 2\xi) + C_3$.

Because $f(1) = 0$ and $f'(1)$ is finite, one can show that $C_3 = C_1$. This leaves

$$4\xi f + 2(1 - \xi^2)f' - f^2 = 2C_1(\xi - 1)^2$$

Because of symmetry, the negative polar axis, $\xi = -1$, is part of the zero streamline. The velocity must be finite there. These considerations show that $f(-1) = 0$, and that $f'(-1)$ is finite, and thus imply that $C_1 = 0$. The remaining, first-order, differential equation for f is easily integrated if we try the substitution $f = (1 - \xi^2)h(\xi)$. Then we find that

$$2h' - h^2 = 0 \qquad \text{so that} \qquad h = 2(C_4 - \xi)^{-1}$$

For later convenience, we substitute $C_4 - 1 = C$, and write the final formula for $f(\xi)$ as

$$f(\xi) = \frac{2(1 - \xi^2)}{C + 1 - \xi} \tag{9.81}$$

The value of C depends on the value of the parameter $F/\rho v^2$. To find the relationship between them, we apply the principle of conservation of momentum to a fixed control volume bounded by a sphere of constant R. We take the z component of all forces and momentum fluxes, and obtain the equation

$$F = \iint\limits_{A} [\rho u(u \cos \theta - v \sin \theta) + \rho \Pi \cos \theta - \tau_{RR} \cos \theta + \tau_{R\theta} \sin \theta] \, dA$$

When we substitute for u, v, Π, τ_{RR} and $\tau_{R\theta}$ from Eqs. (9.69) to (9.73), this becomes

$$F = 2\pi\rho v^2 \int_{-1}^{1} [f'(\xi f' - f) + \xi(g - 2f') + 2f + (1 - \xi^2)f''] \, d\xi$$

Using (9.81) to evaluate f and its derivatives, and evaluating g with the equation

$$g(\xi) = \frac{-4[1 - (C + 1)\xi]}{(C + 1 - \xi)^2} \tag{9.82}$$

we find
$$\frac{F}{2\pi\rho v^2} = \frac{32}{3C}\frac{1 + C}{2 + C} + 4(1 + C)^2 \ln\left(\frac{C}{2 + C}\right) + 8(1 + C) \tag{9.83}$$

This relationship is plotted in Fig. 9.12.

It is easy to show that

$$\frac{F}{2\pi\rho v^2} \to \frac{16}{3C} \qquad \text{when } C \to 0$$

and that
$$\frac{F}{2\pi\rho v^2} \to \frac{8}{C} \qquad \text{when } C \to \infty$$

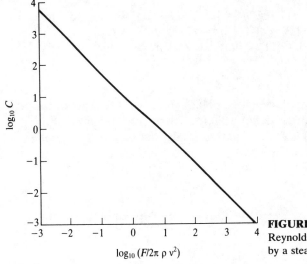

FIGURE 9.12
Reynolds number of the flow produced by a steady point force.

Another interesting interpretation of C appears in the equation

$$\frac{Ru(R, 0)}{v} = \frac{4}{C} \tag{9.84}$$

which shows C to be inversely proportional to a Reynolds number based on the centerline speed of the jet and the distance from the origin.

Variation of the flow field, with variations in the value of C. Figure 9.13(a) shows streamlines, isobars, and equivorticity contours, corresponding to the value $C = 1$. Certain features correspond to characteristic values of the polar angle, θ. For example:

1. At $\theta = \theta_1$, the radial component of velocity changes sign. If $\theta < \theta_1$, there is outflow; otherwise, there is inflow. If $u = 0$, $f'(\xi) = 0$. From (9.81), we calculate

$$f'(\xi) = 2[\xi^2 - 2(C + 1)\xi + 1](C + 1 - \xi)^{-2} \tag{9.85}$$

so that f' vanishes when $\xi = \xi_1 = 1 - [C(C + 2)]^{1/2}$

2. At $\theta = \theta_2$, the streamlines become parallel to the polar axis, and the pressure perturbation vanishes. Assign the symbol q to the velocity component normal to the polar axis. Then

$$q = u \sin \theta + v \cos \theta = -(v/R \sin \theta)[(1 - \xi^2)f' + \xi f]$$

A straightforward calculation shows that q vanishes where $f = 2\xi$, which means that $\xi = \xi_2 = (C + 1)^{-1}$. Equation (9.82) shows that $g(\xi_2) = 0$, so that $\Pi = \Pi_\infty$ along this ray. Positive pressure perturbations are found between this ray and the jet axis. This is shown in Fig. 9.13(b).

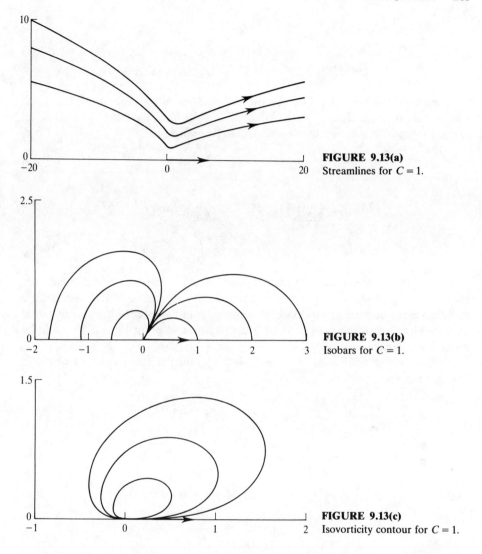

FIGURE 9.13(a)
Streamlines for $C = 1$.

FIGURE 9.13(b)
Isobars for $C = 1$.

FIGURE 9.13(c)
Isovorticity contour for $C = 1$.

3. At $\theta = \theta_3$, there is a ridge in the vorticity distribution; i.e., the angular variation of Ω reaches a maximum, as shown in Fig. 9.13(c).

From the general formula

$$\Omega = \frac{4\nu C(C + 2)}{R^2} \frac{(1 - \xi^2)^{1/2}}{(1 + C - \xi)^3} \tag{9.86}$$

we find that the maximum appears where $(2\xi + 3)(\xi - 1) + C\xi = 0$: thus, where $\xi = \xi_3 = \{[(1 + C)^2 + 24]^{1/2} - (C + 1)\}/4$.

All these special angles approach zero as $C \rightarrow 0$, i.e., as the Reynolds number approaches infinity. The formula, $\xi \approx 1 - \theta^2/2$, may be used to show that, as $C \to 0$,

$$\theta_1 \approx (8C)^{1/4} \qquad \theta_2 \approx (2C)^{1/2} \qquad \theta_3 \approx (2C/5)^{1/2} \qquad (9.87)$$
$$(u = 0) \qquad (q = 0, \Pi = \Pi_\infty) \qquad (\Omega_{,\theta} = 0)$$

Limiting behavior for low Reynolds number. Consider now the limit $C \to \infty$, appropriate for a weak force in a very viscous fluid. The limiting expressions for streamfunction, vorticity and pressure are

$$\Psi = 2\nu R \frac{1 - \xi^2}{C} = \frac{FR}{8\pi\mu} \sin^2 \theta \qquad (9.88)$$

$$\Omega = \frac{4\nu}{R^2 C} (1 - \xi^2)^{1/2} = \frac{F}{4\pi\mu R^2} \sin \theta \qquad (9.89)$$

$$\Pi(R, q) - \Pi_\infty = \frac{(\nu^2/R^2)\xi}{C} = \frac{F}{16\pi\rho R^2} \cos \theta \qquad (9.90)$$

The corresponding streamlines, isovorticity contours, and isobars are shown in Fig. 9.14. This solution is intrinsically interesting, and serves as a building block for the calculation, by superposition, of very viscous flows that may be imagined to be produced by a distribution of point forces. This will be mentioned again in Chapter 11.

Before leaving this limiting case, we compare the three kinds of forces in the momentum equation, finding

$$\text{div } \boldsymbol{\tau} = -\frac{F}{4\pi R^3} (2 \cos \theta \, \mathbf{e}_R + \sin \theta \, \mathbf{e}_\theta)$$

$$-\rho \text{ grad } \Pi = \frac{F}{4\pi R^3} (2 \cos \theta \, \mathbf{e}_R + \sin \theta \, \mathbf{e}_\theta)$$

and

$$\rho \frac{D\mathbf{u}}{Dt} = \frac{F}{4\pi R^3 C} [(1 - 5 \cos^2 \theta)\mathbf{e}_R + \sin \theta \cos \theta \, \mathbf{e}_\theta]$$

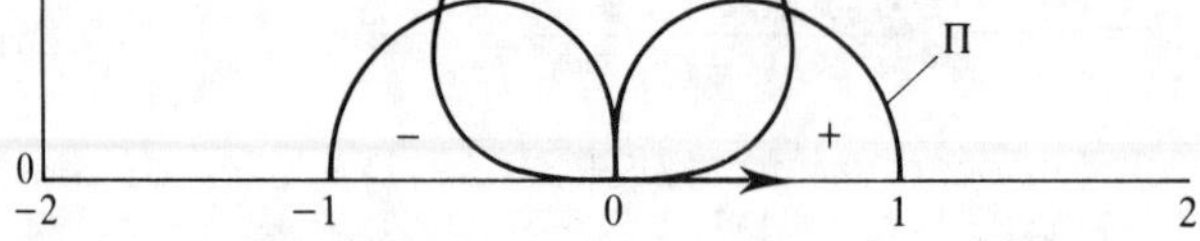

FIGURE 9.14
Characteristic contours in the limit $C \to \infty$.

Note that in this limiting flow, which is called *creeping flow*, the viscous and pressure forces almost exactly cancel each other, while the inertial reaction is smaller by a factor proportional to C^{-1}. Similarly, the rate of convection of vorticity, $u\Omega$, is less than the rate of diffusion, $-v \operatorname{grad} \Omega$, by a factor proportional to C^{-1}.

Notice also that neither the streamlines nor the equivorticity contours reveal the sense (+ or −) of the driving force. This is, however, shown by the pressure field, the higher pressures being on the side towards which the force is driving the fluid.

Limiting behavior for high Reynolds number. Consider now the opposite limit, in which $C \to 0$. We find a considerably more complicated situation, and need to treat the region near $\xi = 1$ with special care. We have seen that the qualitatively interesting features of the flow crowd closely against the axis of the jet when C is very small. If we wish to see these features, we must somehow magnify the angle θ. The way to do this is suggested by Eqs. (9.87).

The outer limit. Let us first ignore the small-θ problem, and see what happens to Ψ, Ω, and Π when $C \to 0$, with a fixed value of ξ, $\xi \neq 1$. We shall call the resulting formulas the *outer limits* of Ψ, Ω, and Π. The term has quite general meaning in the theory of non-uniformly valid approximations, but for this problem we can associate it with the region outside of the jet itself. In the theory of high-Reynolds-number flow, this limit is sometimes called the *Euler* limit. Thus, we adopt the following notation

$$\lim_{\substack{C \to 0 \\ \theta \text{ fixed and} \neq 0}} [f(\theta, C)] = f_E(\theta)$$

Simple calculations show that

$$\Psi_E = 2vR(1 + \xi) \tag{9.91}$$

$$\Omega_E = \frac{8vC}{R^2} (1 + \xi)^{1/2} (1 - \xi)^{-5/2} \tag{9.92}$$

and

$$\Pi_E(R, \theta) - \Pi_\infty = -\frac{v^2}{R^2} (1 - \xi)^{-1} \tag{9.93}$$

Notice the extra factor of C in the vorticity. If ρ and v are held constant, and C is driven toward zero by letting $F \to \infty$, Ψ_E and Π_E approach the nonzero limits shown by (9.88) and (9.90), while Ω_E vanishes. We say that *the outer flow is irrotational*. The streamlines and isobars of the outer flow are shown in Fig. 9.15. The velocity components of the outer flow are

$$u_E = -\frac{2v}{R} \quad \text{and} \quad v_E = -\frac{2v}{R} \operatorname{ctn} \frac{\theta}{2} \tag{9.94}$$

Notice that u_E is always negative, and that v_E has the following interesting

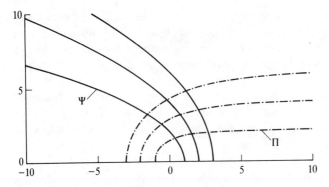

FIGURE 9.15
Streamlines and isobars of the outer flow in the limit $C \to 0$.

limit as $\theta \to 0$:

$$v_E \approx -\frac{4v}{r} \qquad \text{where } r \text{ is distance from the polar axis}$$

This is the expression we would expect to find, if the effect of the jet on the outer flow were the same as a *line sink of uniform strength*, $dQ/dz = 2\pi r v_E = -8\pi v$, laid along the positive z axis. In fact, a calculation which makes a nice exercise shows that this is exactly true. It is also easy to show that the streamlines of the outer flow are parabolas, described by the equation $r^2 = 4z_0(z_0 - z)$, where $z_0 = \Psi/4v$.

The inner limit. To see the detailed structure of the jet itself, we introduce a magnified version of the angle θ,

$$\eta = \frac{\theta}{\theta_2} = \theta(2C)^{-1/2} \qquad (9.95)$$

We use θ_2 instead of θ_1, because the image of the jet, if magnified only by the transformation $\zeta = \theta/\theta_1$, would still disappear into the z axis as $C \to 0$.

For a fixed value of η, and for $C \to 0$, we can approximate $\sin \theta$ and $\cos \theta$ by the small-angle formulas

$$\sin \theta = \sin [\eta(2C)^{1/2}] \approx \eta(2C)^{1/2} \qquad \text{and} \qquad \cos \theta = \cos [\eta(2C)^{1/2}] \approx 1 - 2C\eta^2$$

The limit approached as $C \to 0$, with η fixed and finite, is called an *inner limit*, or in the present context, a *Prandtl limit*. We denote such a limit by the subscript P, and easily calculate that

$$\Psi_P = 4vR\eta^2(1 + \eta^2)^{-1} \qquad (9.96)$$

$$\Omega_P = \frac{32v}{R^2}(2C)^{-3/2}\eta(1 + \eta^2)^{-3} \qquad (9.97)$$

and $\qquad \Pi_P(R, \theta) - P_\infty = -\frac{4v^2}{R^2}C^{-1}(\eta^2 - 1)(1 + \eta^2)^{-2} \qquad (9.98)$

The velocity components are

$$u_P = \frac{4v}{R} C^{-1}(1 + \eta^2)^{-2} \tag{9.99}$$

and

$$v_P = -\frac{4v}{R} (2C)^{-1/2}\eta(1 + \eta^2)^{-1} \tag{9.100}$$

Some of these expressions still contain C, in ways that suggest the need for some new standards of comparison for the corresponding quantities. It is easy to see, for example, that an average value of u in the jet must be enormously larger than the value of u in the outer flow, because all the flow that enters a sphere of radius R in the outer region, which is very nearly equal to $4\pi R^2(-u_E) = 8\pi vR$, must exit through the jet. The area of the jet is approximately

$$A_j = \int_0^{\theta_1} 2\pi R^2 \sin\theta \, d\theta \approx \pi R^2\theta_1^2 = 2\pi R^2 C$$

so the mean value of u_P must approximately equal $8\pi vR/A_j = (4v/R)C^{-1}$. From Eq. (9.82) we see that this is actually the maximum value of u_P, indicating that we have underestimated by some factor which is independent of C. In any event, we are lead to $u_{max} = (4v/R)C^{-1}$ as an appropriate *scale for velocity in the jet*. Clearly, a correspondingly appropriate *scale for the local width of the jet* is $\delta = R\theta_1 = R(2C)^{1/2}$. Using these scales, we rewrite Eqs. (9.93) to (9.97) as

$$\Psi_P = u_{max}\,\delta^2\eta^2(1 + \eta^2)^{-1} \tag{9.101}$$

$$\Omega_P = 4\frac{u_{max}}{\delta}\eta(1 + \eta^2)^{-3} \tag{9.102}$$

$$\Pi_P(R, \theta) - \Pi_\infty = -\frac{C}{4}u_{max}^2(\eta^2 - 1)(1 + \eta^2)^{-2} \tag{9.103}$$

$$u_P = u_{max}(1 + \eta^2)^{-2} \tag{9.104}$$

and

$$v_P = -\left(u_{max}\frac{\delta}{2R}\right)\eta(1 + \eta^2)^{-1} \tag{9.105}$$

With these equations, we forecast results that are quite generally valid for steady viscous flow in slender regions. They will be deduced again in Chapters 11 and 12, by order-of-magnitude arguments, as the basis for lubrication theory and the theory of laminar boundary layers. The velocity and vorticity profiles of the inner flow are shown in Fig. 9.16.

Matching the inner and outer descriptions. Further profitable analysis can be made of the inner limit, but we leave it and ask, "In what sense do the inner and outer limits describe the same flow?" Consider the limit of the outer descriptions, as $\theta \to 0$, and that of the inner descriptions, as $\eta \to \infty$. These both

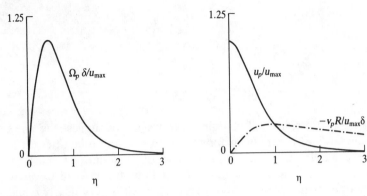

FIGURE 9.16
Velocity and vorticity profiles of the inner flow in the limit $C \to 0$.

describe flow at the edge of the jet, and they should agree. Thus, we find

$$\Psi_E \to 4vR \quad \text{as } \theta \to 0, \qquad \text{while} \qquad \Psi_P \to 4vR \quad \text{as } \eta \to \infty$$

$$\Omega_E \frac{\delta}{u_{max}} \to 4\left(\frac{2C}{\theta^2}\right)^{5/2} \quad \text{as } \theta \to 0 \qquad \text{while} \qquad \Omega_P \frac{\delta}{u_{max}} \to \frac{4}{\eta^5} \quad \text{as } \eta \to \infty$$

(These agree because of the definition of η)

$$\Pi_E - \Pi_\infty \to -\frac{2v^2}{R^2\theta^2} \quad \text{as } \theta \to 0, \qquad \text{while} \qquad \Pi_P - \Pi_\infty \to -\frac{4v^2}{R^2C\eta^2} \quad \text{as } \eta \to \infty$$

(These agree, because of the definition of η)

$$\frac{u_P}{u_{max}} = -\frac{C}{2} \quad \text{for all } \theta > 0, \qquad \text{while} \qquad \frac{u_P}{u_{max}} \to 0 \quad \text{as } \eta \to \infty$$

(These agree, with a discrepancy of order C, which is permissible)

$$v_E \to -\frac{4v}{R\theta} \quad \text{as } \theta \to 0, \qquad \text{while} \qquad v_P \to -\frac{4v}{R\theta} \quad \text{as } \eta \to \infty$$

Thus, we see that either description is adequate near the edge of the jet. The existence of such limiting descriptions, which may be accurate in only a limited portion of the flow field, motivates a search for them, starting with appropriately approximated equations of motion. Thus, in Chapter 12, we shall rediscover the Prandtl limit of the present solution as a solution to the boundary-layer equations.

Bounded jets. Since real jets ordinarily issue from a hole in a wall, there has been much interest in modifications of the above solution to account for the presence of a solid conical boundary, located by $\theta = \theta_w$. Both walls that allow slip and walls on which $\mathbf{u} = 0$ have been considered, particularly in the limit as $C \to 0$. In a very interesting study,[11] it has been shown that the irrotational

---

[11] W. Schneider, "Flow induced by jets and plumes," *Jour. Fluid. Mech.* **108**: 55–65 (1981).

outer flow disappears when a wall with no slip is introduced. The convection in that region is so slow that the weak vorticity generated at the wall diffuses throughout the entire outer region. An exact, self-similar, solution still exists, but must be found by numerical integration.

9.5 SUMMARY

Either the tangential motion of a bounding wall or the imposition of a tangential pressure gradient along a stationary wall creates a localized source of vorticity. The vorticity then diffuses away from the wall, typically moving a distance L in a time L^2/ν. Impulsively started flows in a channel of height h thus attain a steady state in a time of order h^2/ν.

If the source of vorticity oscillates harmonically, with a frequency ω, the effect of the source is limited to a Stokes layer of thickness of order $(\nu/\omega)^{1/2}$. This examplifies the cancelation, by diffusion, of vorticity concentrations of opposite sign. Within the Stokes, or Ekman, layer, oscillations of vorticity lead oscillations of tangential velocity by 45° of phase.

When diffusion competes with convection to redistribute the vorticity, there will be a characteristic Reynolds number of some sort. When this has very small values, the convection is relatively ineffective. When it has large values, diffusion is relatively ineffective, *except* in subregions of the flow, in which the boundary conditions reduce the effectiveness of convection, or in which convection and diffusion tend to drive the vorticity in opposite directions. If the convecting velocity field can be characterized by a speed U and a length L, the width of the region within which diffusion is important is then typically of order $(\nu L/U)^{1/2}$.

Most of our study of the competition of convection and diffusion involved a superposed convection field of external origin, but in the last example (of the Slezkin–Landau–Squire jet) convection and diffusion are both natural responses of the same cause—a point force that threatens to punch a hole through the fluid. The resistance of the fluid is both viscous and inertial, and is characterized by the material parameter $\rho\nu^2$, which has the dimensions of force. If $F \ll \rho\nu^2$, the inertial resistance is unimportant, throughout the entire flowfield. If $F \gg \rho\nu^2$, the viscous resistance is unimportant everywhere, *except very near the forward line of action of the force*, where it is essential. Analysis of the flow in this case involves a separate characterization of an *outer region*, in which there is negligible vorticity and viscous force, and of an *inner region*, which must be suitably magnified to reveal the variations within it. Within this region, the important component of diffusive flux is the one at right angles to the convective flux. Many of these results will reappear in Chapters 11 and 12.

SAMPLE CALCULATION

   A. To illustrate the use of the image method for diffusion problems between plane walls, we calculate the velocity and vorticity profiles for the impulsively started plane Poiseuille flow, using the algorithm described in connection with Figs. 9.8 and 9.9. We pick the dimensionless time $\tau = 1$, and calculate u and Ω for $\xi = 0.25$. The formulas for velocity and vorticity are

$$u(y, t) = -\Pi_{,x}t[1 + f(\eta_0) + f(\eta_1) - f(\eta_2) - f(\eta_3) + f(\eta_4) + f(\eta_5) + \cdots]$$

$$\Omega(y, t) = -2\Pi_{,x}\left(\frac{t}{\pi v}\right)^{1/2}\{g(\eta_0) - g(\eta_1) - g(\eta_2) + g(\eta_3) + g(\eta_4) - g(\eta_5) + \cdots\}$$

where

$$\eta_0 = \frac{\xi}{\sqrt{\tau}} \qquad \eta_1 = \frac{1 - \xi}{\sqrt{\tau}} \qquad \eta_2 = \frac{1 + \xi}{\sqrt{\tau}} \qquad \eta_3 = \frac{2 - \xi}{\sqrt{\tau}} \qquad \eta_4 = \frac{2 + \xi}{\sqrt{\tau}} \qquad \text{etc.}$$

For the given values of ξ and τ, these become

$$\eta_0 = 0.25 \qquad \eta_1 = 0.75 \qquad \eta_2 = 1.25 \qquad \eta_3 = 1.75 \qquad \eta_4 = 2.25$$

$$\eta_5 = 2.75 \qquad \eta_6 = 3.25 \qquad \text{etc.}$$

The functions f and g are deduced from Eqs. (9.14) and (9.15), and are

$$f(\eta) = (1 + 2\eta^2)\,\text{erfc}\,(\eta) - \frac{2}{\sqrt{\pi}}\,\eta\,\exp\,(-\eta^2)$$

and

$$g(\eta) = \exp\,(-\eta^2) - \sqrt{\pi}\eta\,\text{erfc}\,(\eta)$$

Their values for the point and time of interest are shown in the following table

η	0.25	0.75	1.25	1.75	2.25	2.75	3.25
$f(\eta)$	0.54913	0.13159	0.02239	0.00262	0.00017	0.00001	0.00000
$g(\eta)$	0.61874	0.18581	0.03879	0.00542	0.00051	0.00003	0.00000

The resulting values of u and Ω are $u = -0.34411\Pi_{,x}t$, and

$$\Omega = -0.45140\Pi_{,x}(t/v)^{1/2}.$$

It is interesting to compare this with the solution for late times, given by Eqs. (9.60) and (9.61). For the given values of ξ and τ, only the first term of the Fourier series is needed for five-decimal precision, and gives $u = -0.34406\Pi_{,x}t$, and $\Omega = -0.45140\Pi_{,x}(t/v)^{1/2}$. As expected, the two methods of solution agree; for this example, the late-time (Fourier series) method is more economical of calculation.

EXERCISES

9.1. Using Eq. (9.4) in Eqs. (9.5, 6, and 7), verify the values given for δ_ω, δ^*, and θ, for the diffusing vortex sheet.

9.2. Do the dimensional analysis that shows that the vorticity distribution in a diffusing line vortex, of circulation Γ, may be represented in the self-similar form

$$\Omega = \Gamma(vt)^{-1}f(\eta) \qquad \text{where} \qquad \eta = r(vt)^{-1/2}$$

9.3. Show, by dimensional analysis, that the vorticity distribution in an initially stationary fluid outside a circular cylinder of radius a cannot have a self-similar form, even if the angular velocity of the cylinder increases impulsively from zero to a constant value ω at time zero.

9.4. Do the integration by parts that leads to Eq. (9.12).

9.5. Analyze steady plane Couette flow, for which $\mathbf{u} = u(y)\mathbf{e}_x$, for the parallel flow of two immiscible liquids. Liquid with density ρ_1 and viscosity μ_1 fills the channel from $y = 0$ to $y = h/2$; liquid with density ρ_2 and viscosity μ_2 fills the channel from $y = h/2$ to $y = h$. The boundary conditions at the walls are: $u = 0$ at $y = 0$, $u = U$ at $y = h$.

What role does the density ratio ρ_1/ρ_2 play in this analysis? What role might it play in an attempt to produce this flow in the laboratory?

9.6. Suppose that the no-slip condition is replaced by a condition of zero tangential stress at $y = 0$ in the problem of vertical flow induced by a suddenly hot wall. The temperature at $y = 0$ is the same as in the example worked out in the text. Find the velocity and vorticity profiles, and compare them with what we find when the no-slip condition is enforced.

(*Hint*: The differential equation is unchanged, and hence the particular solution and the solution to the homogeneous equations are the same as for the example in the text. You have only to combine them now so that F', rather than F, vanishes at $y = 0$.)

9.7. Compare the velocity profiles given by Eqs. (9.19) and (9.25) at time $\tau = 1/2$.

9.8. Calculate the profile of a steady Ekman layer in a fluid layer of finite depth h. Let the velocity at $z = 0$ be $\mathbf{u} = U\mathbf{e}_y$, in the rotating frame of reference in which the flow is steady, and calculate the angle between this vector and the tangential stress vector at $z = 0$ as a function of the dimensionless parameter λh. The velocity of the plane $z = h$ is zero in the rotating frame, in which the momentum equations reduce to

$$u_{,zz} = -2\lambda^2 v \quad \text{and} \quad v_{,zz} = 2\lambda^2 u$$

These represent a balance between viscous force and Coriolis force. To ease the analysis, define $w = u + iv$, a complex number. The two equations above can then be written as a single complex equation, $w_{,zz} = 2i\lambda^2 w$.

Check to be sure that you can recover Eq. (9.31) in the limit as $\lambda h \to \infty$, and describe the opposite limit, in which $\lambda h \to 0$.

9.9. Describe a possible laboratory setup for investigation of the Ekman spiral produced by an applied surface stress. Suppose water at 20°C is the experimental fluid. Specify, if possible, realistic values of the applied stress, angular velocity of the frame of reference, total water depth, and horizontal diameter D of the test region. It is required that the depth of Ekman layer be very small compared to D, and that the bottom have no direct effect on the spiral.

9.10. Suppose that Stokes' First Problem, in which vorticity diffuses away from a plane wall that is accelerated at time zero to a constant speed U, is modified so that the wall suddenly stops moving at time t_1. Find the velocity profile at a time $t > t_1$. (Recall that the equations and boundary conditions are linear, so that superposition can be used.)

9.11. Find, by superposition of solutions that are properly displaced in time, the tangential stress exerted by fluid in the halfspace $y \geq 0$ on a sliding wall, $y = 0$, when the speed of the wall is the ramp function $U(t) = 0$ if $t < 0$, $U(t) = Vt/t_1$ if $0 \leq t \leq t_1$, and $U(t) = V$ if $t \geq t_1$. Compare the stress history with that which would exist had $U(t)$ jumped suddenly from 0 to V at time $t_1/2$.

9.12. Generalize Stokes' Second Problem, which involves flow over a plane that slides back and forth harmonically, by adding a uniform suction velocity, $v = V$. Identify a dimensionless parameter that can be varied to exhibit high-frequency and quasisteady regimes of fluid response. Describe the flow in these limits.

9.13. Verify, by use of the asymptotic formula $I_0(z) \approx \cos(z + \pi/4)$, that Eq. (9.41) leads us back to Eq. (9.25).

9.14. In both the quasisteady and the high-frequency limits of flow outside a circular cylinder that undergoes harmonic rotational oscillations around its central axis, the flow is found to be irrotational for any value of r that is finitely greater than the cylinder radius, a. Compare these irrotational flows and discuss the comparison.

9.15. Combining Eqs. (9.36), (9.37), and the given expression for the constant B, form an expression for the velocity profile outside an angularly oscillating cylinder, for time $t = 0$, and for the value of the frequency parameter, $\omega a^2/\nu = 1$. Use a table of Bessel functions, and plot the profile, comparing it with the corresponding profile of velocity versus distance from a plane oscillating wall.

9.16. If an elastic cylinder undergoes standing *torsional oscillations* that produce a surface velocity $\mathbf{u}(a, z, t) = V \cos(kz) \cos(\omega t)\mathbf{e}_\theta$, can the motion of a surrounding viscous fluid be described by $\mathbf{u}(r, z, t) = v(r, z, t)\mathbf{e}_\theta$? Base your answer on a detailed examination of the momentum equations.

9.17 Suppose that steady flow is driven through a channel between the infinite planes $y = 0$ and $y = h$ by a tangential motion of the plane $y = h$, where the velocity is $\mathbf{V}$, and by a uniform pressure gradient such that $\Pi_{,y} = 0$. Show that the velocity profile is given by

$$\mathbf{u} = -\frac{\nabla\Pi}{2\nu}y(h - y) + \frac{\mathbf{V}y}{h}$$

and confirm that all terms in the convective acceleration vanish, no matter what the angle between $\mathbf{V}$ and $\nabla\Pi$. (This equation is a building block for lubrication theory.)

9.18. Show that the velocity distribution $\mathbf{u} = A(3\xi^2 - \eta^2)(\eta - 1)\mathbf{e}_z$ can represent steady pressure-driven flow in a stationary pipe, the cross section of which is an equilateral triangle. The corners of the triangle lie at the points ($\xi = x/a = 0$, $\eta = y/a = 0$), ($\xi = 1/\sqrt{3}$, $\eta = 1$), and ($\xi = -1/\sqrt{3}$, $\eta = 1$). Evaluate the constant A in terms of the pressure gradient $\Pi_{,z}$ and the kinematic viscosity. Calculate the volumetric flowrate, and compare it to that in a pipe of circular cross section of equal area for the same values of $\Pi_{,z}$ and ν.

9.19. By separation of variables, compute the velocity profile for pressure-driven flow down a channel bounded by the planes $z = \pm W$, $y = \pm H$. For large values of the aspect ratio, W/H, calculate the reduction of the volumetric flowrate, due to the vanishing of $\mathbf{u}$ at the sidewalls, where $z = \pm W$. To be precise, compare the flowrate through the channel with sidewalls to that through an equal width, $2W$, of a channel without sidewalls, holding $\Pi_{,x}$ and ν constant. Find the value of the aspect ratio for which the effect is 1%. (Answer: $W/H = 62.7$)

(*Hint*: Solve for $f(y, z)$, which equals $u(y, z) - (\Pi_{,x}/2\nu)(y^2 - h^2)$.)

9.20. Verify, by substitution and differentiation, that Eqs. (9.58) and (9.59) provide a solution to Eq. (9.57).

9.21. Equation (9.58) asserts that the vorticity distribution in a plane vortex sheet that is diffusing and stretching is gaussian at all times, but this obviously does not apply to arbitrary, initial distributions of vorticity. An argument can nevertheless be made that the same steady-state limit (given by (9.60) and (9.61) as $At \to \infty$) is approached, whatever the initial distribution, so long as it has the form $\Omega(\mathbf{r}, 0) = f(y)\mathbf{e}_z$. Write down the arguments that seem to you to support this conclusion.

9.22. Reconsidering the problem of pressure-driven flow along a channel with plane, parallel, porous walls, start with the governing differential equation,

$$Vu_{,y} + \Pi_{,x} = vu_{,yy}$$

and try a perturbation analysis for small values of the dimensionless parameter Vh/v. Specifically, call the parameter Re, and substitute the series approximation,

$$u(y, \text{Re}) = u_0(y) + \text{Re}\, u_1(y) + \text{Re}^2 u_2(y) + \cdots$$

into the differential equation and the boundary conditions $u = 0$ at $y = 0$ and at $y = h$. Require equality in the limit as $\text{Re} \to 0$, successively determining $u_0(y)$, $u_1(y)$, etc. Carry the process out through the determination of $u_1(y)$.

9.23. (This exercise is a companion to Exercise 9.22) Suppose we now try to investigate the opposite limit, $\text{Re} \to \infty$, by postulating an expansion in the inverse Reynolds number, $\varepsilon = v/Vh$. Substitute the series approximation,

$$u(y, \varepsilon) = u_0(y) + \varepsilon u_1(y) + \varepsilon^2 u_2(y) + \cdots$$

into the differential equation and boundary conditions, and try to find $u_0(y)$, $u_1(y)$, etc. successively. What happens?

Considering the procedures needed to analyze the high-Reynolds-number behavior of the jet induced by a point force, reconsider the approach you have just tried, and see whether you can determine the inner and outer limits of the flow, as $\varepsilon \to 0$.

9.24. Calculate the velocity field induced by a semi-infinite line sink of uniform intensity, $dQ/dz = -8\pi v$, which occupies the positive z axis. The velocity induced by the sink of strength dQ, placed at $z = \zeta$, is directed towards the sink and has magnitude equal to $dQ/4\pi L^2$, where L is the distance from the sink. Verify that the corresponding stream function is given by $\Psi = 2vR(1 + \cos \theta)$.

9.25. Consider a steady plane flow, produced by a force that acts uniformly at points along the z axis. The force, per unit length of the z axis, is specified by the equation $d\mathbf{F}/dz = F'\mathbf{e}_x$. Show, by dimensional analysis, that the resulting flow will not be self-similar.

Consider, on the other hand, flows driven by a volumetric sink or source, placed at the apex of a wedge or a cone. The plane flow will be characterized by a volumetric flowrate per unit length, say Q', and the axisymmetric flow by a simple flowrate, Q. Show by dimensional analysis whether either of these flows can be self-similar.

9.26. For the jet produced by a point force, evaluate the *inner limit* of the viscous force per unit mass, the pressure force per unit mass, and the fluid acceleration. Show that the θ-components of $\mathbf{u}$, $D\mathbf{u}/Dt$, and div $\boldsymbol{\tau}$, are all small, (of order $C^{1/2}$), compared to the corresponding R-components. Show that the R-component of $\rho \nabla \Pi$ is small compared to the corresponding component of div $\boldsymbol{\tau}$, but that the θ-components of these two quantities are of equal order of smallness.

CHAPTER 10

NUMERICAL SOLUTIONS OF THE FULL NAVIER–STOKES EQUATIONS

10.1 INTRODUCTION

Numerical integration of differential equations has come to play such a major role in the development and illustration of the theory of viscous flow that every serious student of the subject needs a working knowledge of the most common numerical algorithms, and should have a general acquaintance with the issues that arise in this sort of work. This chapter aims, in a very modest way, to respond to these needs. This objective would, however, scarcely justify the inclusion of the chapter except for the existence of a large number of very precisely computed solutions of the Navier–Stokes equations that illustrate important flow phenomena, especially flow separation, which cannot be described by closed-form analytic solutions.

Many of the solutions included here were first worked out decades ago, with enormous effort and dedication. It is almost embarrassing to discover how quickly and cheaply they can be reproduced today! The change is due not only to the enormous speed and memory of modern computing machinery, but to major advances in the design of effective algorithms.

We start, in Section 10.2, with a consideration of boundary-value problems for ordinary differential equations, working up from linear, second-order equations of the type for which we have found closed-form solutions, to higher-order systems of nonlinear equations, such as are encountered in the famous self-similar solutions for flow near an infinite spinning disc. Special attention is paid to effective ways to deal with boundary conditions that are specified *at infinity*.

In Section 10.3, you will see how the techniques of Section 10.2 can be applied repeatedly, to solve linear partial differential equations of the *parabolic* type, and we indicate some of the fundamental difficulties that may be encountered with nonlinear parabolic equations. An introductory discussion of the *proper posing* of initial- and boundary-value problems is given. The accuracy of the method is illustrated by numerical solution of a problem, that of impulsively started flow outside a spinning cylinder, for which an analytical solution is given in Chapter 9. Two fairly ambitious problems, involving impulsively-started, three-dimensional stagnation-point flows, and spinning-disk flow, are then solved. Emphasis is placed on an appropriate magnification of the thin region occupied by vorticity at very early times. Extensive application of these techniques is postponed to Chapter 12, which deals with boundary-layer theory.

Section 10.4 starts with some discussion of the mathematical character of the full Navier–Stokes equations, and with the special problems that appear when one attempts to solve them numerically in a domain that contains only part of a flowfield. The issue of computational cost and computer capacity becomes serious; and the compromise between cost and accuracy is sketched. From that discussion emerges the realization that flows become increasingly difficult to compute as the value of the Reynolds number increases. The results of a few recently published calculations are shown, to give an impression of the state of the art.

10.2 FINITE-DIFFERENCE TECHNIQUES FOR ORDINARY DIFFERENTIAL EQUATIONS, WITH TWO-POINT BOUNDARY CONDITIONS

Any discussion of numerical methods per se in a text of this sort is bound to be seriously incomplete. Many full-length books have been devoted entirely to such discussions,[1] and the journal literature adds important new information each month. All that is intended here is an introduction to one method that works well, is easy to learn, and is so frequently used that the name of its principal algorithm has become a household word among students of fluid mechanics and heat transfer. The method represents derivatives by finite difference quotients, and hence is called a *finite-difference method*. Alternative ways of representing derivatives are mentioned in Chapter 14.

---

[1] See, for example, R. D. Richtmyer and K. W. Morton, *Difference Methods for Initial-Value Problems*. Interscience, New York, 1967.

Linear, Second-Order Equation

Consider the linear differential equation for a function $u(y)$,

$$u'' + r(y)u' + s(y)u = t(y) \tag{10.1}$$

to be solved in the interval $a \leq y \leq b$, subject to the linear boundary conditions

$$\alpha_1 u' + \beta_1 u = \gamma_1 \qquad \text{at} \qquad y = a \tag{10.2}$$

and $\qquad \alpha_N u' + \beta_N u = \gamma_N \qquad \text{at} \qquad y = b \tag{10.3}$

For the present, suppose that a and b are both finite numbers; procedures for boundary conditions at infinity will be explained later.

Subdivision of the interval, and approximation of derivatives. Subdivide the interval into $N - 1$ equal steps, each of length $h = (b - a)/(N - 1)$. introducing *nodal points* at the locations

$$y_n = (n - 1)h + a \tag{10.4}$$

Thus, $y = a$ corresponds to $n = 1$, while $y = b$ corresponds to $n = N$.

At the nth nodal point, approximate the derivatives by *three-point, centered, finite-difference quotients*:

$$u'_n = \frac{u_{n+1} - u_{n-1}}{2h} \tag{10.5}$$

and $\qquad u''_n = \dfrac{u_{n+1} - 2u_n + u_{n-1}}{h^2} \tag{10.6}$

If the function $u(y)$ may be represented by a Taylor series in the vicinity of each nodal point, these approximations may be derived. The largest terms omitted from the series expansion are of order h^2, compared to those retained.

The differential equation is now approximated, at the nth nodal point, by an algebraic equation, which can be written in the form

$$A_n u_{n-1} + B_n u_n + C_n u_{n+1} = D_n \tag{10.7}$$

which

$$A_n = \frac{1}{h^2} - \frac{r_n}{2h} \qquad C_n = \frac{1}{h^2} + \frac{r_n}{2h}$$

$$B_n = \frac{-2}{h^2} + s_n \qquad \text{and} \qquad D_n = t_n \tag{10.8}$$

There are N such equations, involving $N + 2$ unknown values of u_n. The r_n, s_n, and t_n are values of the known coefficient functions at the nth nodal point. Notice that two of the unknown values of u_n pertain to *image nodes*, for which $n = 0$ or $n = N + 1$. These extra unknowns can be eliminated by use of the boundary conditions, written with the same finite-difference approximations, as

$$\alpha_1 \frac{u_2 - u_0}{2h} + \beta_1 u_1 = \gamma_1 \tag{10.9}$$

and

$$\alpha_N \frac{u_{N+1} - u_{N-1}}{2h} + \beta_N u_N = \gamma_N \tag{10.10}$$

When these are used to eliminate u_0 and u_{N+1}, the first and last of the equations of form (10.7) may be replaced by the equations

$$B_1^* u_1 + C_1^* u_2 = D_1^* \tag{10.11}$$

and

$$A_N^* u_{N-1} + B_N^* u_N = D_N^* \tag{10.12}$$

in which

$$B_1^* = \beta_1 + \alpha_1 \frac{B_1}{2hA_1}$$

$$C_1^* = \alpha_1 \frac{A_1 + C_1}{2hA_1} \tag{10.13}$$

$$D_1^* = \gamma_1 + \alpha_1 \frac{D_1}{2hA_1}$$

and

$$A_N^* = -\alpha_N \frac{A_N + C_N}{2hC_N}$$

$$B_N^* = \beta_N - \alpha_N \frac{B_N}{2hC_N} \tag{10.14}$$

$$D_N^* = \gamma_N - \alpha_N \frac{D_N}{2hC_N}$$

The boundary-value problem is now approximated by a determinate set of linear algebraic equations, which can be readily solved by the *Thomas algorithm,* a very efficiently organized application of Gauss elimination. This algorithm has two elements:

1. Formulas for *triangularizing* the matrix of the coefficients:

$$E_1 = -\frac{C_1^*}{B_1^*} \qquad F_1 = \frac{D_1^*}{B_1^*}$$

and, for $n \geq 2$,

$$E_n = -\frac{C_n}{\text{DENO}} \quad \text{and} \quad F_n = \frac{D_n - A_n F_{n-1}}{\text{DENO}} \tag{10.15}$$

in which

$$\text{DENO} = A_n E_{n-1} + B_n$$

These can be evaluated sequentially, in a single FORTRAN DO-loop.

2. *Back-substitution* formulas, which give the u_n sequentially, starting with u_N. These are

$$u_N = \frac{D_N^* - A_N^* F_{N-1}}{B_N^* + A_N^* E_{N-1}}$$

and $\qquad\qquad u_n = E_n u_{n+1} + F_n$ $\qquad\qquad$ (10.16)

One of the advantages of the algorithm is that each value of a known coefficient, such as A_n, is needed only once, during the triangularization loop, and hence the coefficients do not require dimensioned storage arrays.

A test calculation, for flow of immiscible fluids down an inclined plane. As a test, we recompute one of the simplest solutions presented in Chapter 9, that for two superposed layers of immiscible liquid, sliding down an inclined plane. This example is not quite trivial, because the solution has discontinuities in slope and curvature at the liquid–liquid interface. It will interesting to see how well our numerical implementation of boundary conditions reproduces these discontinuities.

Let us choose the following values for the dimensionless parameters of the problem: $\beta = 1/2$, $R = 0.8$, $M = 100$. (See Section 9.3 and Fig. 9.10 for a reminder of definitions. The total depth of the layer is here called H, because h is used for the step length.) We solve for the dimensionless velocity, $\phi = v_1 u /(gH^2 \sin \alpha)$. With $\eta = y/H$, and with primes to denote differentiation with respect to η, the differential equation becomes

$$\phi'' = -1 \quad \text{in} \quad 0 \le \eta < \tfrac{1}{2} \qquad \phi'' = -\frac{R}{M} \quad \text{in} \quad \tfrac{1}{2} < \eta \le 1 \qquad (10.17)$$

The corresponding values of the coefficients are, except at the interface

$$A_n = C_n = 1 \qquad B_n = -2$$

$$D_n = -h^2 \quad \text{if} \quad 0 \le \eta \le \tfrac{1}{2} \quad \text{and} \quad D_n = -h^2 \frac{R}{M} \quad \text{if} \quad \tfrac{1}{2} < \eta \le 1$$

Let us take $N = 5$, with $n = 1$ corresponding to the solid plane, $n = 3$ to the liquid–liquid interface, and $n = N = 5$ to the upper surface. The no-slip boundary condition at the plane has the form given by Eq. (10.2) if $\alpha_1 = \gamma_1 = 0$; and the condition of vanishing tangential stress at the top surface has the form given by Eq. (10.3) if $\beta_N = \gamma_N = 0$.

From Eq. (10.13) we find $B_1^* = \beta_1$ and $C_1^* = D_1^* = 0$. It follows from Eq. (10.15) that

$$E_1 = F_1 = 0$$

From Eq. (10.14) we find $A_N^* = -\alpha_N / h$, $B_N^* = \alpha_N / h$, and $D_N^* = \alpha_N (R/M) h/2$. It follows from Eq. (10.16) that

$$\phi_N = \frac{F_{N-1} + 0.5 h^2 \dfrac{R}{M}}{1 - E_{N-1}}$$

Notice that both β_1 and α_N, whose values are physically meaningless in this problem, so long as they are finite and nonzero, have canceled out of the analysis.

To deal with the liquid–liquid interface, we must combine the conditions of continuity of velocity and stress with the differential equation, in the two forms assumed by the latter as the interface is approached from below or above. This will allow us to derive unambiguous values for the coefficients A_3, B_3, C_3, and D_3, which apply to a point where the two fluids meet. To do this, we invent image values ϕ_2^*, which belongs to the upper liquid, and ϕ_4^*, which belongs to the lower. The meaning of this belonging is that we write $\tau_{yx} = \mu_1(u_4^* - u_2)/(2h)$ on the lower side of the interface, and $\tau_{yx} = \mu_2(u_4 - u_2^*)/(2h)$ on the upper side.

Equating the two expressions for τ_{yx}, we get $\phi_4^* - \phi_2 = M(\phi_4 - \phi_2^*)$. Similarly, we represent the differential equation by $\phi_2 - 2\phi_3 + \phi_4^* = -h^2$ on the lower side, and by $\phi_2^* - 2\phi_3 + \phi_4 = -(R/M)h^2$ on the upper side. Combining these equations to eliminate ϕ_2^* and ϕ_4^*, we can derive the interface equation

$$\phi_2 - (1 + M)\phi_3 + M\phi_4 = -0.5(1 + R)h^2$$

from which we deduce the values $A_3 = 1$, $B_3 = -(1 + M)$, $C_3 = M$, and

$$D_3 = -0.5(1 + R)h^2$$

Note that these reduce to the expected expressions for a homogeneous layer of fluid when $R = 1$ and $M = 1$.

We are now ready to execute the Thomas algorithm. The triangulariza-tion routine generates the values tabulated below. We are using the step size $h = 0.25$.

n	1	2	3	4
E	0	0.5	0.995024876	0.995049505
F	0	0.03125	0.000870647	0.001363861

The free surface speed can now be calculated, giving $\phi_N = \phi_5 = 0.326000003$. The exact answer is $(4R + 2R/M + 2)h^2 = 0.326$. The other values of velocity are then calculated from the general form of Eq. (10.16), and are

n	4	3	2	1
ϕ	0.3257500003	0.325000003	0.193750002	0.0
exact ϕ	0.32575	0.325	0.19375	0.0

You should not be misled by the fact that such an accurate answer is found with so few nodal points; this will not usually happen, but does here because the exact answer in each liquid is a parabola, for which the error terms in our centered-difference formulas vanish. What the success of this example does illustrate is the appropriateness of the use of image values to approximate first derivatives at the edge of a computational domain, these values being eliminated by use of the difference form of the differential equation.

Nonlinear Second-Order Equation With a Boundary Condition at Infinity

Consider, as a slightly more complicated example, the differential equation

$$u'' + 2(1 - u^2) = 0 \tag{10.18}$$

with boundary conditions

$$u(0) = 0 \qquad u(\infty) = 1$$

The exact solution is

$$u(\eta) = 3 \tanh^2 (\eta + \beta) - 2 \tag{10.19}$$

in which $\qquad \beta = \text{arctanh} \, [\sqrt{(2/3)}] = 1.146215835$

The problem appears as the inner limit of a self-similar representation of plane flow in a wedge, along the apex of which there is a uniform line sink.

Two new features are encountered here: nonlinearity of the differential equation in the term u^2, and a boundary condition to be applied at infinity. The nonlinearity will be dealt with by an *iteration* scheme. We write

$$u = u^* + (u - u^*) \qquad \text{and hence} \qquad u^2 = u^{*2} + 2u^*(u - u^*) + (u - u^*)^2$$

from which we drop the last term, understanding that u^* represents a good estimate of the value of u, which will become more accurate as the iteration cycle is repeated. As this happens, the quadratic term $(u - u^*)^2$ will vanish more rapidly than will the linear term that precedes it. To start the cycle, we insert a convenient function that meets the boundary conditions, such as $u^* = 1 - \exp(-\eta)$. For the second iteration, u^* is given the value of u that came out of the first iteration, and so on. Thus, the equation that is approximated by finite differences is

$$u'' - 4u^*u = -2(1 + u^{*2}) \tag{10.20}$$

which is linear, because at any stage of the iteration u^* is a given quantity.

The boundary condition at infinity can be accommodated in a rough-and-ready way, by simply asserting that $u = 1$ at some suitably large, but finite, value of η, which we call η^*. In this case, $\eta^* = 6$ would be adequate. If the chosen value is too large, some calculations will be wasted, but no other harm done. If it is too small, there will be a loss of accuracy, and the symptoms of this loss become familiar to the experienced worker.

To proceed more systematically, with a better guarantee of accuracy, we can make an *asymptotic analysis* of the given problem, as $\eta \to \infty$. For this, define $\varepsilon = 1 - u$, and rewrite Eq. (10.18) as

$$\varepsilon'' - 4\varepsilon + 2\varepsilon^2 = 0$$

from which we drop the last term, because we are interested in the region where ε is very small. The solution of the truncated equation is

$$\varepsilon = A \exp(2\eta) + B \exp(-2\eta) \tag{10.21}$$

in which we must set $A = 0$, to meet the boundary condition at infinity. Now we have an expression which will, with the correct value of B, give high accuracy at any nominally large value of η. We use it to derive the equation $e' + 2\eta e = 0$, and rewrite this as

$$u' + 2\eta^* u = 2\eta^* \qquad (10.22)$$

which is in the form specified for the upper boundary condition by Eq. (10.3). We now have only to subdivide the interval $[0, \eta^*]$, approximate the second derivative in Eq. (10.20) by Eq. (10.6), and apply the Thomas algorithm to the set of linear algebraic equations of form (10.7), in which we have

$$B_1^* = \beta_1 \qquad C_1^* = 0 \qquad \text{and} \qquad D_1^* = 0$$

$$A_n = \frac{1}{h^2} \qquad C_n = \frac{1}{h^2}$$

$$B_n = -\frac{2}{h^2} - 4u_n^* \qquad \text{and} \qquad D_n = -2(1 + u_n^{*2}) \quad \text{for} \quad 2 \le n \le N - 1$$

$$A_N^* = -\frac{1}{h} \qquad B_N^* = 2(\eta^* + 1) + 4u_N^* h^2 \qquad \text{and} \qquad D_N^* = 2\eta^* + (1 + u_N^{*2})h$$

A FORTRAN program, HAMEL, to implement the Thomas algorithm for this problem is shown in Appendix B. The boundary condition was applied at $\eta = 4$, and four iterations were employed. Three more cycles of iteration produced no change in the first five digits after the decimal. Some results are tabulated below, for $N = 21$ and 41.

η	0	0.4	1.2	4.8
$u(21)$	0.0	0.50151	0.89126	0.99957
$u(41)$	0.0	0.50157	0.89183	0.99959
rich u	0.0	0.50159	0.89202	0.99959
exact u	0.0	0.501578	0.892016	0.999593

The value designated as "rich u" is calculated from $u(21)$ and $u(41)$ by a process called *Richardson extrapolation*, which increases the order of the error to h^4 if a numerical scheme is strictly second-order accurate. Calculated results for a discrete increment h are combined with those for $h/2$, the formula being rich $u = [4u(h/2) - u(h)]/3$.

The value of u' at $\eta = 0$ is often of interest. We call it SF, because it is often proportional to the shear stress on the wall, called *skin friction*. We eliminate the image value u_0 between the equations

$$2h\text{SF} = u_2 - u_0 \qquad \text{and} \qquad u_2 - 2u_1 + u_0 = -2h^2(1 - u_1^2)$$

recall that $u_1 = 0$, and find that $\text{SF} = u_2/h + h$. The computed, Richardson-extrapolated, and exact values of SF are compared below.

$$\text{SF}(21) = 1.63725, \qquad \text{SF}(41) = 1.63402,$$

$$\text{rich SF} = 1.63295, \qquad \text{exact SF} = 1.632993.$$

Coupled Equations

In Chapter 9, we saw some solutions, such as described the Ekman layer or the jet produced by a point force, which involved two coupled second-order equations. Systems of higher-order commonly appear in treatment of three-dimensional flows or flows with heat transfer. To isolate the complication introduced by the coupling of equations, we first study a numerical solution of the problem of the Ekman layer. We then present two new examples, which represent famous similarity solutions for viscous flow.

The steady Ekman layer. The problem of the Ekman layer can be formulated directly in the rotating frame of reference in which the flow is steady. To remove the need for specific values of v and ω in the numerical work, we use an independent variable $\zeta = (\omega/2v)^{1/2}z$. The dependent velocity components, divided by the specified surface velocity, are called u and v. The result is a pair of linear, coupled, second-order differential equations,

$$u'' = -2v \quad \text{and} \quad v'' = 2u \tag{10.23}$$

Replacing the derivatives with three-point centered-difference approximations, we obtain algebraic equations that may be presented in the form

$$A_{11}u_{n-1} + B_{11}u_n + C_{11}u_{n+1} + A_{12}v_{n-1} + B_{12}v_n + C_{12}v_{n+1} = D_1$$
and $\tag{10.24}$
$$A_{21}u_{n-1} + B_{21}u_n + C_{21}u_{n+1} + A_{22}v_{n-1} + B_{22}v_n + C_{22}v_{n+1} = D_2$$

The values of the coefficients are

$$A_{11} = \frac{1}{h^2} \quad B_{11} = -\frac{2}{h^2} \quad C_{11} = \frac{1}{h^2}$$

$$A_{12} = 0 \quad B_{12} = -2 \quad C_{12} = 0 \quad D_1 = 0$$

$$A_{21} = 0 \quad B_{21} = 2 \quad C_{21} = 0$$

$$A_{22} = \frac{1}{h^2} \quad B_{22} = -\frac{2}{h^2} \quad C_{22} = \frac{1}{h^2} \quad D_2 = 0$$

It may seem silly to write out an elaborate expression such as (10.24), in which so many of the coefficients turn out to be zero. The object is to introduce the *matrix notation*, with which the set of equations (10.24) is represented as

$$A\mathbf{y}_{n-1} + B\mathbf{y}_n + C\mathbf{y}_{n+1} = \mathbf{D} \tag{10.25}$$

In this, A, B, and C are 2×2 coefficient matrices, and ys and **D** are two-entry column vectors. For example:

$$A = \begin{bmatrix} A_{11} & A_{12} \\ A_{21} & A_{22} \end{bmatrix} \quad \mathbf{y}_n = \begin{bmatrix} u_n \\ v_n \end{bmatrix}$$

The Thomas algorithm will solve this matrix equation just as well as a single equation, with one essential complication. When we introduce the *back-substitution formula*

$$\mathbf{y}_n = E_n \mathbf{y}_{n+1} + \mathbf{F}_n \tag{10.26}$$

actually written for $\mathbf{y}_{n-1}$, to eliminate $\mathbf{y}_{n-1}$ from Eq. (10.25), we get

$$(A_n E_{n-1} + B_n)\mathbf{y}_n = -C_n \mathbf{y}_{n+1} + (\mathbf{D}_n - A_n \mathbf{F}_{n-1}) \tag{10.27}$$

Multiply Eq. (10.26) through by $(A_n E_{n-1} + B_n)$, to get

$$(A_n E_{n-1} + B_n)\mathbf{y}_n = (A_n E_{n-1} + B_n)E_n \mathbf{y}_{n+1} + (A_n E_{n-1} + B_n)\mathbf{F}_n \tag{10.28}$$

Compare Eqs. (10.27) and (10.28), and see that the E_n and the $\mathbf{F}_n$ are to be found by solving the equations

$$(A_n E_{n-1} + B_n)E_n = -C_n$$

and

$$(A_n E_{n-1} + B_n)\mathbf{F}_n = \mathbf{D}_n - A_n \mathbf{F}_{n-1} \tag{10.29}$$

This looks formidable, but is very easy when the order of the coefficient matrices is only two or three. Then Cramer's Rule may be used. For more complex problems, it may be more efficient to use a Gauss–Jordan *matrix inversion* routine. Let $G = A_n E_{n-1} + B_n$, and assign the *inverse* of G the symbol GI. Then the solutions of Eqs. (10.29) are

$$E_n = -GI\, C_n \quad \text{and} \quad \mathbf{F}_n = GI\,(\mathbf{D}_n - A_n \mathbf{F}_{n-1}) \tag{10.30}$$

The complication mentioned above is just that simple division must be replaced by this construction of an inverse matrix. For the simple case in hand, G has only four components, and the components of the inverse are

$$GI_{11} = \frac{G_{22}}{D} \quad GI_{12} = -\frac{G_{12}}{D} \quad GI_{21} = -\frac{G_{21}}{D} \quad GI_{22} = \frac{G_{11}}{D}$$

with

$$D = G_{11}G_{22} - G_{12}G_{21}$$

The boundary conditions for Eqs. (10.23) are

$$\text{at } z = 0: \quad u = 1, v = 0 \quad \text{while} \quad \text{as } z \to \infty: \quad u \to 0, v \to 0$$

This is not a case for which an asymptotic analysis helps with the boundary condition at infinity; we shall simply set u and v equal to zero at $\zeta = 12$, knowing that the envelope for the oscillations of u and v narrows in proportion to $\exp(-\zeta)$.

To evaluate $u'(0)$ and $v'(0)$, one must again account for the coupling of the equations. To evaluate the formula $u'(0) = (u_2 - u_0)/2h$, we use the difference approximation to the first Eq. (10.23), and the boundary condition $v_1 = 0$, to show that $u_0 = 2u_1 - u_2$. Then we set $u_1 = 1$ and find that $u'(0) = (u_2 - 1)/h$. A similar analysis shows that $v_0 = 2h^2 - v_2$, and hence that $v'(0) = v_2/h - h$.

A FORTRAN program, EKMAN, that embodies this analysis is given in Appendix B. Some results are tabulated in Table 10.1 for $N = 61$ and $N = 121$.

TABLE 10.1
Convergence of numerical results for the Ekman layer

ζ	0	1	2	4	8
$u(61)$	1.0	0.19915	−0.05513	−0.01200	−0.00004
$u(121)$	1.0	0.19886	−0.05602	−0.01198	−0.00005
rich u	1.0	0.19876	−0.05632	−0.01197	−0.00005
exact u	1.0	0.19877	−0.05632	−0.01197	−0.00005
$v(61)$	0.0	−0.30787	−0.12262	0.01352	−0.00032
$v(121)$	0.0	−0.30914	−0.12295	0.01378	−0.00033
rich v	0.0	−0.30956	−0.12306	0.01387	−0.00033
exact v	0.0	−0.30956	−0.12306	0.01386	−0.00033

$u'(0) = -0.99005 \quad v'(0) = -1.01005, \quad$ for $N = 61$;
$u'(0) = -0.99750 \quad v'(0) = -1.00250, \quad$ for $N = 121$;
$u'(0) = -0.99998 \quad v'(0) = -0.99998, \quad$ by Richardson extrapolation;
$u'(0) = -1.0 \qquad\quad v'(0) = -1.0, \qquad\;$ from exact solution.

Steady Three-Dimensional Stagnation-Point Flow

Suppose the potential flow $\mathbf{u}_p = Ax\mathbf{e}_x + By\mathbf{e}_y - (A + B)z\mathbf{e}_z$ appears in the half-space $z \geq 0$ at $t = 0$, and is maintained constant thereafter. If $A + B > 0$, as we shall assume, this flow will approach the plane $z = 0$, which is a solid, stationary wall. The no-slip condition will create a vortex sheet, which will diffuse out into the fluid but be converted back toward the wall. As time progresses, fresh vorticity will be created at the wall, as required to keep the no-slip condition in force. We expect a steady state to appear eventually.

In Sample Calculation D of Chapter 8, we have seen that a vortex sheet and its image, sufficient to cancel the slip velocity of this flow, induces no tangential velocity except in the region between sheet and image if it remains parallel to the wall. Since neither the velocity of diffusion away from the wall, nor that of convection back toward the wall, depends on x or y, it seems reasonable to seek a solution of the Navier–Stokes equations in which the vorticity is eventually confined to a layer of uniform thickness. This thickness can only be proportional to $(v/A)^{1/2}$, because this is the only quantity with dimension of length that can be formed from the data that specify the flow.[2] Thus, we define a dimensionless normal distance variable

$$\zeta = \frac{z}{\delta} \quad \text{with} \quad \delta = \left(\frac{v}{A}\right)^{1/2} \tag{10.31}$$

Let us now denote the physical velocity components by symbols with asterisks,

---

[2] This choice is a little arbitrary, in that one could just as logically use $A + B$ in the scaling of w. However, the present choice is more effective in the limit as $A + B \to 0$.

u^*, v^*, and w^*, and introduce *dimensionless, scaled* velocity components, u, v, and w. We set

$$u^*(x, y, z) = Axu(\zeta) \quad \text{and} \quad v^*(x, y, z) = Byv(\zeta) \qquad (10.32)$$

Then, to be consistent with the continuity equation, we set

$$w^*(x, y, z) = -A\delta w(\zeta) \qquad (10.33)$$

When we insert these into the momentum equation, we find from the z component that

$$\left(\frac{w^{*2}}{2} + \Pi - vw^*_{,z}\right)_{,z} = 0 \qquad (10.34)$$

From this, it is easy to show that $\Pi_{,x}$ and $\Pi_{,y}$ are independent of z, and can be given the values determined by the external potential flow. These are

$$\Pi_{,x} = -A^2 x \quad \text{and} \quad \Pi_{,y} = -B^2 y \qquad (10.35)$$

When Eqs. (10.24), (10.25), and (10.27) are inserted into the x- and y-momentum equations and the continuity equation, and nonzero common factors are canceled, we find the following coupled ordinary differential equations:

$$u'' + wu' + 1 - u^2 = 0$$
$$v'' + wv' + C(1 - v^2) = 0 \qquad (10.36)$$

and

$$-w' + u + Cv = 0$$

in which

$$C = \frac{B}{A}$$

The quantities u' and v' are of considerable physical interest, being proportional to components of vorticity, and they must be updated during the process of iteration; so it is convenient to assign them symbols. Those used here in the sample programs are $u' = R$, and $v' = S$. The working form of Eqs. (10.36) is then

$$u'' + wR + 1 - u^2 = 0$$
$$v'' + wS + C(1 - v^2) = 0,$$
$$-w' + u + Cv = 0 \qquad (10.37)$$
$$R = u' \qquad S = v'$$

The *nonlinearity* of these equations is treated in the usual way; for example, the term wR is approximated by the expression $wR^* + w^*R - w^*R^*$. This means that initial guesses are needed for u, v, w, R, and S. After the equations are linearized, the following difference approximations are introduced:

$$R_n = u'_n = \frac{u_{n+1} - u_{n-1}}{2h} \quad \text{and} \quad u''_n = \frac{u_{n+1} - 2u_n + u_{n-1}}{h^2}$$

with corresponding expressions for v' and v''. The continuity equation is differenced slightly differently, around the midpoint of the interval between ζ_{n-1} and ζ_n, as follows:

$$-\frac{w_n - w_{n-1}}{h} + \frac{u_n + u_{n-1}}{2} + C\frac{v_n + v_{n-1}}{2} = 0$$

This formula has the same order of accuracy as the three-point formulas, but is selected to reflect the fact that no condition on w is given at the upper boundary.

The resulting linear algebraic equations can be arranged in the matrix form of Eq. (10.25), with the matrices and vectors that follow. We see from this display that the matrix for **C** is very sparse. The fact that only C_{11} and C_{22} are nonzero implies, from the first Eq. (10.30), that $E_{ij} = 0$ whenever $j = 3$. Thus the expanded version of Eq. (10.26) reduces to

$$\begin{aligned} u_n &= E_{11n}u_{n+1} + E_{12n}v_{n+1} + F_{1n} \\ v_n &= E_{21n}u_{n+1} + E_{22n}v_{n+1} + F_{2n} \\ w_n &= E_{31n}u_{n+1} + E_{32n}v_{n+1} + F_{3n} \end{aligned} \tag{10.38}$$

The lack of reference to w_{n+1} confirms that no outer boundary value for w is needed.

The coefficient matrices are displayed below.

$$A = \begin{bmatrix} \dfrac{1}{h^2} - \dfrac{w_n^*}{2h} & 0 & 0 \\[2ex] 0 & \dfrac{1}{h^2} - \dfrac{w_n^*}{2h} & 0 \\[2ex] \dfrac{1}{2} & \dfrac{C}{2} & \dfrac{1}{h} \end{bmatrix}$$

$$B = \begin{bmatrix} -2\left(\dfrac{1}{h^2} + u_n^*\right) & 0 & R_n^* \\[2ex] 0 & -2\left(\dfrac{1}{h^2} + Cy_n^*\right) & S_n^* \\[2ex] \dfrac{1}{2} & \dfrac{C}{2} & -\dfrac{1}{h} \end{bmatrix}$$

$$C = \begin{bmatrix} \dfrac{1}{h^2} + \dfrac{w_n^*}{2h} & 0 & 0 \\[2ex] 0 & \dfrac{1}{h^2} + \dfrac{w_n^*}{2h} & 0 \\[2ex] 0 & 0 & 0 \end{bmatrix}$$

$$\mathbf{D} = \begin{bmatrix} -(1 + u_n^{*2}) + w_n^* R_n^* \\[1ex] -C(1 + y_n^{*2}) + w_n^* S_n^* \\[1ex] 0 \end{bmatrix}$$

The boundary conditions are:

$$w(0) = 0 \qquad \text{(impermeable wall)}$$
$$u(0) = v(0) = 0 \qquad \text{(no slip at wall)}$$
$$u(\infty) = v(\infty) = 1 \qquad \text{(vanishing vorticity far from the wall)}$$

The final approach to the limit can be analyzed as follows. Call $\varepsilon = 1 - u$, and $\phi = 1 - v$. If ε and ϕ vanish rapidly enough as $\zeta \to \infty$ so that their integrals from zero to infinity are finite, the continuity equation shows that $w \approx (1 + C)\zeta - \delta^*$ for very large values of ζ. The constant, δ^*, is called the *displacement integral*, and has the value

$$\delta^* = \int_0^\infty (\varepsilon + C\phi) \, d\zeta$$

Because w is, for large values of ζ, just a linear function of ζ, it can conveniently be used as the independent variable in the linearized equations that govern very small values of ε and ϕ. In the next few equations, a prime denotes differentiation with respect to w. Remember that $dw = (1 + C) \, d\zeta$.

The linearized equations are

$$\varepsilon'' + (1 + C)^{-1} w \varepsilon' - 2(1 + C)^{-2} \varepsilon = 0$$

and

$$\phi'' + (1 + C)^{-1} w \phi' - 2C(1 + C)^{-2} \phi = 0$$

Their most rapidly decaying asymptotic solutions for large values of w are[3]

$$\varepsilon \approx X \exp\left(\frac{-w^2}{2(1 + C)}\right) w^{-(3+C)/(1+C)}$$

and

$$\phi \approx Y \exp\left(\frac{-w^2}{2(1 + C)}\right) w^{-(3C+1)/(1+C)}$$

The parameter whose value determines the qualitative character of the solutions is the ratio $C = B/A$, which can vary throughout the range $-1 \le C \le 1$. We note that the algebraic factor in ϕ represents growth, rather than decay, when $C < -1/3$. Nevertheless, these formula predict that ϕ will decay to satisfactorily small values at reasonable values of ζ, for all values of $C > -1$.

When $C = -1$, $w \to w_\infty$, a constant, as $\zeta \to \infty$. Then the asymptotic expressions for ε and ϕ are

$$\varepsilon \approx X \exp\left(-\frac{\zeta}{2}[w_\infty + (w_\infty^2 + 8)^{1/2}]\right)$$

and

$$\phi \approx Y \exp\left(-\frac{\zeta}{2}[w_\infty + (w_\infty^2 - 8)^{1/2}]\right)$$

---

[3] The solutions are related to the *parabolic cylinder functions*. See A. Davey, "Boundary-layer flow at a saddle point," *J. Fluid Mech.* **10**: 503–610 (1961).

The asymptotic expressions can be used to derive outer boundary conditions of the form (10.3) for both u and v, with

$$\alpha_{uN} = \alpha_{vN} = 1$$

$$\beta_{uN} = \gamma_{uN} = w_N^* + \frac{3 + C}{w_N^*} \quad \text{and} \quad \beta_{vN} = \gamma_{vN} = w_N^* + \frac{3C + 1}{w_N^*}$$

For $C = -1$, we substitute the formulas

$$\beta_{uN} = \gamma_{uN} = \tfrac{1}{2}[w_\infty + (w_\infty^2 + 8)^{1/2}]$$

and

$$\beta_{vN} = \gamma_{vN} = \tfrac{1}{2}[w_\infty + (w_\infty^2 - 8)^{1/2}]$$

A straightforward generalization of the algebraic procedure used to get the first Eq. (10.16) can be carried out, allowing calculation of u_N and v_N when values of w_N^*, E_{N-1}, and $\mathbf{F}_{N-1}$ are known. The requisite formulas are given in the FORTRAN program "STAGPT" in Appendix B. Two more details require mention before the results of calculation are presented:

1. The calculation of R and S at $\eta = 0$ requires the introduction and elimination of image values u_0 and v_0. The resulting formulas are

$$R_1 = \frac{u_2}{h} + \frac{h}{2} \quad \text{and} \quad S_1 = \frac{v_2}{h} + \frac{Ch}{2} \tag{10.39}$$

2. The *displacement integral*, δ^*, is easily calculated from the formula

$$\delta^* = \lim_{\zeta \to \infty} [(1 + C)\zeta - w] \tag{10.40}$$

The variation of R_1, S_1, and δ^* with C is shown in Fig. 10.1. The results shown here agree closely with those of Davey.

A few typical velocity profiles are shown in Fig. 10.2. When $C > 0$, $v(z)$ exhibits a curvature of one sign only. An inflection point appears at

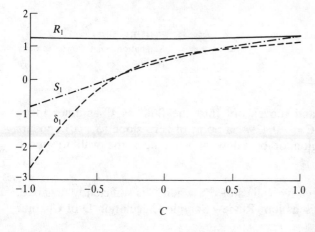

FIGURE 10.1
Three-dimensional stagnation-point flows. Wall shear and displacement integral.

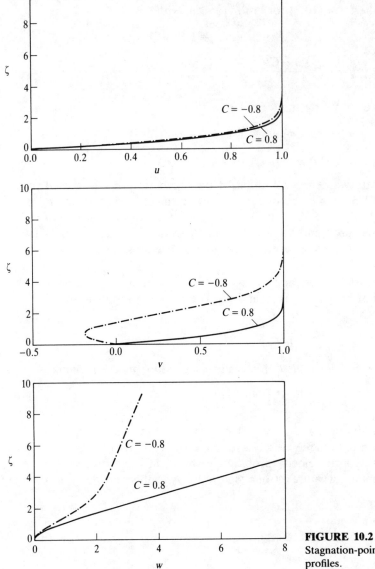

FIGURE 10.2
Stagnation-point flows: velocity profiles.

the wall when $C = 0$, and moves out into the fluid as C assumes more negative values. When $C = -0.4294$, a point of zero slope $(S_1 = 0)$ appears at the wall, and a region of backflow appears near the wall for more negative values of C.

The appearance of negative values of δ^*, when C is sufficiently negative, merits special notice and discussion. Review Sample Calculation D of Chapter

8. In it, we analyzed the velocity field due to a plane vortex sheet of strength $d\Gamma/dx^* = -Ax$, when it lies parallel to the wall $z = 0$, at a distance $z = h$. There was an image sheet at $z = -h$, and the two sheets together induced a velocity away from the wall that equalled Ah if $y > h$. This would be the effect of the y component of vorticity in the present problem if all that vorticity were concentrated at its average distance from the wall. When $B < 0$ in the present problem, we can use the same analysis to see that the x component of vorticity, which is proportional to our v', will have just the opposite effect; i.e., it will induce a velocity Bh toward the wall at points $z > h$. If the value of h were the same for the two vorticity components, the displacement effect of Ω_y would win out, because $A + B > 0$, and the combined effect would be a displacement flow away from the wall. This corresponds to our δ^* being positive. However, the two values of h are not equal! When $C < 0$, the vorticity of the v-profile lies, on the average, much farther from the wall than does that of the u-profile. The discrepancy increases as C becomes more negative, until the net displacement effect vanishes at $C \approx -0.38$, just before the first appearance of backflow in the v-profile. We note that it must be this negative displacement effect, induced by the vorticity itself, that allows the possibility of a solution when $C = -1$.

This solution has intrinsic interest as a model of flow near stagnation points, where flow from upstream divides to pass around a finite body. In that situation, the value of C depends largely upon the local curvatures of the body. Consider the sketch of a foot-long hotdog, shown in Fig. 10.3. At points A and A', we expect that $C > 0$, because the potential flow along the wall moves away from the stagnation point in all directions. At point B, the surface flow approaches the stagnation point along the local y axis, but departs along the x axis, corresponding to a negative value of C. We shall return to consideration of flow around the hotdog in Chapter 12.

Steady Flow Induced by a Spinning Disk

Another problem of three-dimensional flow governed by ordinary differential equations arises when an infinite solid plane spins steadily around an axis, say the z axis, that is normal to the plane. If this motion starts at time zero,

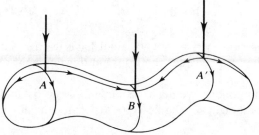

FIGURE 10.3
Hotdog in a breeze, illustrating different types of stagnation point.

adjacent to an initially stationary fluid that fills the entire half-space, $z \geq 0$, a vortex sheet is formed, with initially radial vortex lines. As this vorticity diffuses out into the fluid, the fluid moves circumferentially and radially outward. The radial motion, due to the absence of a radial pressure gradient sufficient to keep fluid particles on circular paths, induces an axial motion toward the plane. There is again a competition between diffusion and convection, leading us to expect that the vorticity will eventually be confined to a layer of finite thickness near the solid plane.

The only parameters to characterize the problem are ω and v, the angular velocity of the plane and the kinematic viscosity. From these we can form a characteristic length, $\delta = (v/\omega)^{1/2}$, and a characteristic velocity, $\omega\delta$. Let us again denote physical velocity components with an asterisk, saving the corresponding plain letters for dimensionless, scaled, quantities. If we apply dimensional analysis to constrain the possible relationships between the circumferential velocity component v^* and the independent variables and parameters, we find that

$$v^*(\omega\delta)^{-1} = F\left(\frac{r}{\delta}, \frac{z}{\delta}\right) \tag{10.41}$$

However, the no-slip condition requires that $v^* = \omega r$ where $z = 0$. This suggests that the dependence of the function F on the variable r/d is one of linear proportionality. Trying this, we find

$$v^*(\omega\delta)^{-1} = \frac{r}{\delta} v\left(\frac{z}{\delta}\right)$$

or simply
$$v^* = \omega r v(\zeta) \tag{10.42}$$

in which
$$\zeta = \frac{z}{\delta} = z\left(\frac{\omega}{v}\right)^{1/2} \tag{10.43}$$

To establish an appropriate scaling for the radial velocity component, we can make a crude force balance in the radial direction, noting that the centrifugal force, which acts on the layer of thickness δ in proportion to $\rho\delta v^{*2}/r$, must be comparable to the restraining stress exerted by the disk. If u^* is the peak value of radial velocity, this stress should be proportional to $\mu u^*/\delta$. Using the estimate, $v^* = \omega r$, and comparing these two forces, we find that u^* should be comparable to $\omega^2 r \delta^2/v$. Thus we write

$$u^* = \omega r\left(\frac{\omega\delta^2}{v}\right)u(\zeta) \tag{10.44}$$

Of course, in the present problem $\omega\delta^2/v = 1$, but we shall have use for the more general expression later, when we consider the transient development of this flow.

To scale the axial velocity component, we make a simple flowrate balance on a control volume of radius r and height δ. The flowrate that enters axially is

$\pi r^2 w^*$; that which leaves radially is $2\pi r\delta u^*$, where w^* and u^* are average values of the velocity components. This shows that w^* is typically proportional to $(\delta/r)u^*$, so that an appropriate scaling of w^* is given by the formula

$$w^* = -\frac{\omega^2\delta^3}{\nu}\,w(\zeta). \tag{10.45}$$

The corresponding form for the vorticity is

$$\Omega = -v^*_{,z}\mathbf{e}_r + u^*_{,z}\mathbf{e}_\theta + r^{-1}(rv^*)_{,r}\mathbf{e}_z$$

$$= \omega\left(\frac{r}{\delta}\right)\left(-v'\mathbf{e}_r + \frac{\omega\delta^2}{\nu}u'\mathbf{e}_\theta\right) + 2\omega v\mathbf{e}_z \tag{10.46}$$

To scale Π, we consult the axial momentum equation, noting that $\Pi_{,z}$ must be comparable to $\nu w^*_{,zz}$, and that, like w^*, it is independent of r. An appropriate scaling is thus

$$\Pi = (\omega\delta)^2\,P(\zeta) \tag{10.47}$$

What we have just executed is a procedure sometimes called *scaling*. It employs rough estimates, based on the equations of motion and the boundary conditions, to predict what we call the *order of magnitude* of the dependent variables, before any serious calculations are attempted. Specifically, it aims to express each dependent variable as a product of a factor that has the dimensions of the variable, and which is related in a specific way to the physical parameters, in this case w and u, the values of which specify the problem. It may, and often does, make use of some derived parameter, such as the δ of this case, which is easy to visualize and easy to relate, at least approximately, to the basic parameters. In the latter relationship, such as our estimate $\delta^2 = \nu/\omega$ for this steady flow, and the estimate used later for unsteady flow, constant numerical factors are irrelevant. It is only important that the dependence on the basic parameters be captured correctly: for example, $\delta^2 = 2\nu/\omega$ is just as good as our estimate, but $\delta = \nu/r\omega$ would not work at all. These ideas cannot be assimilated in one gulp, so we shall refer back to them as we present more examples. In the meanwhile, you will see how nicely the procedure works out mathematically and computationally.

Inserting the assumed forms for u^*, v^*, w^*, and Π into the continuity and momentum equations, we find that r, ω, and ν appear to the same power in every term of each equation, and can be canceled. The dimensionless, scaled variables u, v, and w are then related by the following equations:

From r-momentum (A54) $\quad u'' + wu' - u^2 + v^2 = 0$ $\hfill (10.48)$

From θ-momentum (A55) $\quad v'' + wv' - 2uv = 0$ $\hfill (10.49)$

From continuity (A53) $\quad\quad w' - 2u = 0$ $\hfill (10.50)$

From z-momentum (A56) $\quad P' + ww' + w'' = 0$ $\hfill (10.51)$

The equations are presented in this particular order, to look as much as possible like the equations for the stagnation-point flows. The last equation serves to determine P, after the first three have been solved for u, v, and w. It can be integrated by sight, into

$$P = P_0 - \frac{w^2}{2} - 2u \qquad (10.52)$$

in which w' has been replaced by $2u$, by virtue of the continuity equation. There is no specification of the pressure itself at any point, so we can only discuss pressure differences. For this purpose, we can set $P_0 = 0$.[4]

The boundary conditions translate into the following constraints on u, v, and w:

$$u(0) = 0 \qquad v(0) = 1 \qquad \text{(no-slip)}$$
$$w(0) = 0 \qquad \text{(impervious wall)}$$
$$u'(\infty) = 0 \quad \text{and} \quad v(\infty) = 0 \qquad \text{(vanishing vorticity as } \zeta \to \infty)$$

Note that there is no specified constraint on w as $\zeta \to \infty$, because the intensity of the induced axial inflow cannot be known in advance.

From Eq. (10.48), we see that it is consistent for u to vanish as $\zeta \to \infty$, since v vanishes. The only driving cause of u, the scaled radial velocity, is the circumferential motion. We drop both the quadratic terms from eqs. (10.48) and (10.50) and note, from Eq. (10.49) and the boundary condition on u', that w must approach a constant, say Δ, as $\zeta \to \infty$. This reveals the asymptotic forms

$$u'' + \Delta u' = 0 \quad \text{and} \quad v'' + \Delta v' = 0$$

with the decaying solutions $u = A \exp(-\Delta \zeta)$ and $v = B \exp(-\Delta \zeta)$. From these we can easily deduce upper boundary conditions of the type (10.3), with $\alpha_N = 1$, $\beta_N = \Delta^*$, and $\gamma_N = 0$ for both u and v. We use the symbol Δ^*, because the value of Δ is not known in advance and requires iterative refinement as successively better estimates of $w(\infty)$ are calculated. The formula to be used for this purpose is

$$\Delta^*(\text{new}) = w_{N-1} + 2 \frac{u_{N-1}}{\Delta^*(\text{old})} \qquad (10.53)$$

as can easily be shown for an exercise. It is also easy to show that the formulas for $u'(0)$ and $v'(0)$ are, for this problem,

$$u'(0) = \frac{u_2}{h} + \frac{h}{2} \quad \text{and} \quad v'(0) = \frac{v_2}{h} \qquad (10.54)$$

---

[4] This solution is quite famous, the reduction to ordinary differential equations having been discovered by T. von Kármán, and the first approximate solutions worked out by W. G. Cochran in 1934. The accuracy of the numerical work has been refined by several authors; the values in Table 3.6 of *Viscous Fluid Flow* by F. M. White are consistently accurate to the number of digits presented. Unfortunately, that and several other books give positive values for P, whereas the correct values are obviously negative.

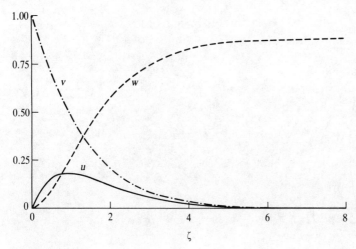

FIGURE 10.4
Velocity profiles for spinning disk.

The algebraic details of this calculation are much the same as for the stagnation-point flows. Computed profiles of u, v, and w are shown in Fig. 10.4. Profiles for Π and θ are shown in Fig. 10.5; θ is defined by the equation $\tan \theta = v/u$.

Finally, a plan view of the velocity vector, for comparison with an Ekman spiral, is shown in Fig. 10.6. A tangent to this curve at any point is normal to the horizontal component of vorticity at the corresponding value of z. The values of the limiting slopes at the ends of the curve are found to be:

$$\frac{S}{R} = -1.2071 \quad \text{at} \quad \zeta = 0 \qquad \frac{S}{R} \to 1.320 \quad \text{as} \quad \zeta \to \infty$$

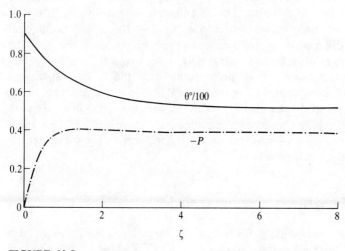

FIGURE 10.5
P and θ for the spinning disk.

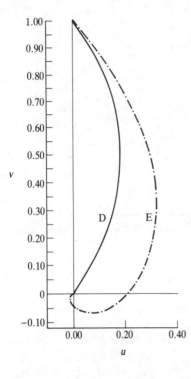

FIGURE 10.6

Plan view of velocity profiles: D = spinning disk, E = Ekman layer.

10.3 FLOWS GOVERNED BY PARABOLIC PARTIAL DIFFERENTIAL EQUATIONS

The next degree of computational complexity is illustrated in its simplest form by an example we studied analytically in Chapter 9, the flow outside an impulsively spun-up circular cylinder. This is an interesting example, in that it proves to be computationally easier to describe the flow by a direct numerical attack, using finite-difference approximations, than to evaluate the analytic solution, which involves an infinite integral with a very complicated integrand involving Bessel Functions. We shall present this example to demonstrate a successful scheme of computation, and then use the scheme on two more complex problems, to study the transient developments of the flow above a spinning disk when the plane acquires a constant angular velocity at time zero, and to study the three-dimensional stagnation-point flows, after the potential flow has been turned on at time zero.

Flow Outside an Impulsively Spun-up Cylinder

Two questions arise immediately when we contemplate a solution of this problem by finite-difference approximations: (1) What choice of dependent variable leads to the easiest formulation of the numerical scheme? (2) How can we cover, with a constant number of computational nodes, a temporally expanding region of motion?

It is clear that vorticity is a poor candidate for dependent variable, because its value is initially infinite at the cylinder surface, and is unknown thereafter. There is a boundary condition on $\Omega_{,r}$ at the cylinder surface, but it only specifies that $\Omega_{,r}$ vanishes there, and makes no reference to the angular speed of the cylinder. Velocity is a somewhat better candidate, but the product rv satisfies a slightly simpler differential equation and has a simpler asymptotic behavior as $r \to \infty$, so we shall choose it. Specifically, we shall use the dimensionless variable,

$$g = \frac{rv}{\omega a^2} \tag{10.55}$$

where ω and a are, as in Chapter 9, the angular velocity and radius of the cylinder.

The spreading domain, within which g decreases from 1 to 0, can be covered with a single mesh if the independent variables are combined properly. We know that the vorticity is confined, at very early times, to an annulus of width $r - a \approx (6vt)^{1/2}$; and that rv vanishes in proportion to $\exp(-r^2/4vt)$ at very late times and large values of r. These estimates are shown in Fig. 9.6. It follows that an independent variable

$$\eta = (r - a)(vt)^{-1/2} \tag{10.56}$$

will be appropriate, in the sense that the vorticity will be spread smoothly over the interval $0 \le \eta \le 8$, and the outer boundary condition, $g = 0$, can be applied at $\eta \approx 8$, for all times.

We could use, for the second independent variable, the dimensionless quantity vt/a^2, but the appearance of the square root of this quantity in the definition of η suggests that

$$T = \left(\frac{vt}{a^2}\right)^{1/2} \tag{10.57}$$

would be a slightly better choice. With this, we have $\eta = (r - a)/(aT)$.

The partial differential equation is the θ-momentum equation

$$g_{,t} = v(g_{,rr} - r^{-1}g_{,r})$$

with boundary conditions

$$g(a, t) = 1 \qquad t > 0 \qquad \text{(no-slip)}$$

and

$$g(\infty, t) = 0 \qquad t > 0 \qquad \text{(decay of vorticity)}$$

and the initial condition

$$g(0, r) = 0 \qquad r \ge a \qquad \text{(initially stationary fluid)}$$

Because g is a function of $\eta(r, t)$ and $T(t)$, the chain rule is needed for evaluation of the necessary partial derivatives. It is important to understand in the following analysis that the symbol $g_{,T}$ represents the partial derivative of g with respect to T while η is held constant, whereas $g_{,t}$ represents the partial derivative of g with respect to t, while r is held constant. Thus, we find

$$\eta_{,t} = -\frac{\eta}{2t} \qquad \eta_{,r} = \frac{\eta}{aT} \qquad \text{and} \qquad \frac{dT}{dt} = \frac{T}{2t}$$

When these are inserted into the chain-rule expressions

$$g_{,t} = g_{,T}\frac{dT}{dt} + g_{,\eta}\eta_{,t} \qquad g_{,r} = g_{,\eta}\eta_{,r} \qquad \text{and} \qquad g_{,rr} = g_{,\eta}(\eta_{,r})^2$$

the differential equation takes the final form, with primes to symbolize partial differentiation with respect to η,

$$g'' + \left[\frac{\eta}{2} - \frac{T}{1 + \eta T}\right]g' = \frac{T}{2}g_{,T} \qquad (10.58)$$

The boundary conditions are

$$g(0, T) = 1 \qquad \text{and} \qquad g(\infty, T) = 0$$

By our definition of η, which amounts to an appropriate *scaling of an independent variable*, we obtain an infinitely magnified view of the structure of the initial vortex sheet. This is revealed by setting $T = 0$ in Eq. (10.58), and leads us back, not surprisingly, to Stokes' First Problem and its solution:

$$g(\eta, 0) = \text{erfc}\left(\frac{\eta}{2}\right) \qquad (10.59)$$

It turns out that $g_{,T}$ vanishes so rapidly as $T \to \infty$ that the right-hand side of Eq. (10.58) vanishes with it, even though the factor $T/2$ is going to infinity. The resulting ordinary differential equation has the solution, $g = \exp(-\eta^2/4)$, which is not imposed as a constraint on the transient solution, but should automatically appear as we march the latter forward to very large values of T.

Marching Forward in Time. Computational Instability

Equation (10.58) is classified as a *parabolic* partial differential equation. Although a second derivative with respect to η appears, there is only a first derivative with respect to T. The main consequence, with respect to computation, is that information about the solution propagates forward in time, so that the boundary conditions imposed in the future have no effect on the solution in the present. On the other hand, at any given value of T, the value of g at one value of η affects the values of g at all other values of η. This effect is small when the two points are far apart in η, but it is there in principle.

It might seem easy to integrate Eq. (10.58) forward in time. We have an initial profile, and could simply approximate $g(\eta, T + dT)$ by the *forward difference* formula $g(\eta, T + dT) = g(\eta, T) + g_{,T}(\eta, T)\, dT$, using Eq. (10.58) to approximate $g_{,T}$ at time t. The η-derivatives of g, at time T, would be approximated by centered three-point difference formulas, as in our numeric work with ordinary differential equations. Unfortunately, this *explicit* procedure runs into a notorious difficulty of finite-difference schemes, called *computational instability*. The algebraic operations that produce $g(\eta, T + dT)$ from $g(\eta, T)$ amplify tiny differences between two almost identical initial profiles, and soon the computed solutions lose all their intended connection to real problem. The instability is not unconditional in the present case; it can be avoided by restricting dT to be less than $(d\eta)^2/2$. This restriction is unacceptable, however, because there is a computationally much less expensive approach.

The successful approach is called *implicit,* and involves the use of a *backward* difference formula

$$g(\eta, T + dT) = g(\eta, T) + g_{,T}(\eta, T + dT)\, dT$$

To implement an implicit scheme, one has to solve, at every value of T, the ordinary differential equation

$$g'' + \psi(\eta, t)g' - T\frac{g}{2dT} = -T\frac{g(T - dT)}{2dT} \tag{10.60}$$

in which g is the function of η at time T, and only $g(T - dT)$ is known from the earlier time. The coefficient of g',

$$\psi(\eta, T) = \frac{\eta}{2} - \frac{T}{1 + \eta T} \tag{10.61}$$

is a known function of the independent variables.

This scheme is unconditionally stable; the size of dT is constrained only by considerations of accuracy. As presented here, the implicit scheme is only accurate to the first order in dT; it is possible to gain accuracy by surrendering some margin of stability, but this scheme gives ample accuracy for present purposes, because the independent variables were designed so that changes of g with T are relatively small, and nearly linear for small ranges of T. In Chapter 12, a slightly more elaborate differencing scheme, giving second-order accuracy in dT, is introduced.

A FORTRAN program, "SPNCYL", to carry out the marching, is given in Appendix B. It is very little more complicated than the program needed to solve the ordinary differential equation by itself. As a demonstration of accuracy, we tabulate the vorticity at the surface of the cylinder as a function of time, comparing it with the result of the analytical solution.

To obtain Ω/ω from the discrete values of g, we approximate the equation

$$\Omega = r^{-1}(rv)_{,r}$$

by the centered-difference formula

$$\Omega_n = r_n^{-1}\omega a^2 \frac{g_{n+1} - g_{n-1}}{r_{n+1} - r_{n-1}}$$

and then use

$$\frac{r}{a} = 1 + T\eta$$

to get the final form

$$\frac{\Omega_n}{\omega} = \frac{g_{n+1} - g_{n-1}}{2hT(1 + T\eta_n)}$$

in which h is the finite increment of η. To evaluate Ω_1/ω, we employ the imaginary value g_0, and eliminate it with the centered-difference approximation to Eq. (10.58). This yields the formula

$$\frac{\Omega_1}{\omega} = \frac{g_2 - g_1}{Th(1 + Th/2)}$$

It is important, before claiming that such a calculation gives a high-precision solution to the Navier–Stokes equation, to refine the discretization parameters, h and dT, and to experiment with the value of η at which the

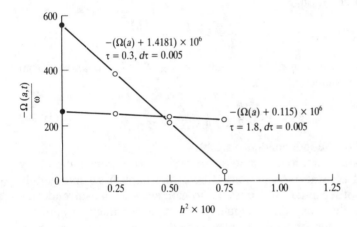

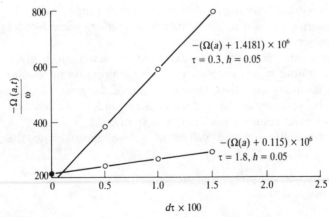

FIGURE 10.7
Truncation effects in finite-difference solution of the spinning cylinder equations.

outer boundary condition is applied. We must make sure that the reported results are, to the number of decimals presented, independent of the values chosen for these numerical parameters. In this process, it is valuable to verify that convergence is occurring at the expected rate, as is illustrated in Fig. 10.7.

If the *truncation errors* associated with finite values of h and dT are simply additive, so that

$$g(\text{computed}) = g(\text{true}) + Ah^2 + B\,dT$$

if h and dT are sufficiently small, one can perform a two-dimensional Richardson extrapolation to estimate $g(\text{true})$ from three values of $g(\text{computed})$. Specifically, let $g_1 = g(h, dT)$, $g_2 = g(2h, dT)$, and $g_3 = g(h, 2dT)$, for fixed values of η and T. Then the extrapolation formula is

$$g(\text{true}) = \tfrac{1}{3}(7g_1 - g_2 - 3g_3)$$

Using this formula, with $h = 0.05$ and $dT = 0.003125$, we find the values listed in Table 10.2 below as $-\Omega(a, t)/\omega$ (finite-difference). The outer boundary condition was applied at $\eta = 8$, after verifying that, for $h = 0.05$ and $dT = 0.003125$, the values of $-\Omega(a, t)/\omega$ were, to the number of decimals presented in the table, unaffected when the point of application of the outer boundary condition was moved out to $\eta = 10$.

The analytic solution was evaluated by setting $\xi = 1$ in Eq. (9.38); fixing the value of vt/a^2, which equals the τ of Eq. (9.38); employing computer library routines to evaluate the necessary Bessel functions; and executing the integral with Simpson's Rule. At $p = 0$, we use known approximations for the Bessel functions of very small argument, thus finding that the integrand equals 0.

TABLE 10.2
Accuracy tests for the finite-difference description of the flow around an impulsively spun-up cylinder

vt/a^2	$-\Omega(a, t)/\omega$	
	Finite-difference	Analytic
0.01000	−5.18073	−5.18079
0.04000	−2.39300	−2.39299
0.09000	−1.48137	−1.48135
0.16000	−1.03624	−1.03625
0.25000	−0.77626	−0.77626
1.00000	−0.29264	−0.29263
4.00000	−0.09651	−0.09651
9.00000	−0.04751	−0.04751

Transient Development of the Spinning-Disk Flow

This problem involves nonlinearity and coupling of equations, but is otherwise not more difficult than the one just presented. When the disk first starts to spin, the vorticity-containing layer grows thicker by diffusion alone, hence $\delta^2 \approx vt$. As $t \to \infty$, the thickness of the layer approaches a constant value, $\delta^2 \approx v/\omega$. We know these things from the simpler solutions that have been displayed in preceding sections. A function which goes smoothly from one of these limits to the other is

$$\delta^2(t) = \frac{v}{\omega}[1 - \exp(-\omega t)] \qquad (10.62)$$

As was mentioned in our discussion of the steady-flow limit of this problem, there is nothing sacred about this formula. The alternative formula $\delta^2(t) = (vt)/(1 + \omega t)$ would, in principle, work just as well. One formula may lead to slightly less-expensive computations for a given level of accuracy, but the results obtained by the use of either formula will be the same when expressed in terms of the original physical variables, providing that all calculations are done with sufficient control of truncation errors. We are dealing here with design choices, rather than mathematical theory.

Continuing with the choice (10.62), and using dimensionless independent variables

$$\tau = \frac{\omega t}{1 + \omega t} \qquad \text{and} \qquad \zeta = \frac{z}{\delta(t)} \qquad (10.63)$$

we transform the computational domain for the entire problem into a finite rectangle, $0 \le \tau \le 1$, and $0 \le \zeta \le 6$, the suggested upper limit on ζ being subject to confirmation by test computations, as explained in Section 10.2. The necessary partial derivatives are

$$\tau_{,z} = 0, \qquad \tau_{,t} = \frac{\omega}{(1 + \omega t)^2} = \omega(1 - \tau)^2$$

$$\zeta_{,z} = \left(\frac{\omega}{v}\right)^{1/2}[1 - \exp(-\omega t)]^{-1/2}$$

$$\zeta_{,t} = \frac{v\zeta}{2}\exp(-\omega t)[1 - \exp(-\omega t)]^{-1}$$

We go back now to Eqs. (10.43) to (10.47), treating δ as a function of t when we substitute those expressions into the momentum and continuity equations for unsteady flow. Again using asterisks to denote the physical values of the velocity components, we calculate, for example, the local time-derivative of u^*. Our scaling procedure gave

$$u^* = \omega r\left(\frac{\omega \delta^2(t)}{v}\right)u(\zeta(z, t), t(t))$$

From this we calculate

$$u^*_{,t} = \omega r\left(u\frac{d(\omega\delta^2/\nu)}{dt} + \frac{\omega\delta^2}{\nu}(u'\zeta_{,t} + u_{,\tau}\tau_{,t})\right)$$

$$= \omega^2 r\left[(1-T)\left(u - \frac{\zeta u'}{2}\right) + T(1-\tau)^2 u_{,\tau}\right]$$

in which we have introduced the symbol $T \equiv 1 - \exp(-\omega t)$. The equations to be integrated numerically are then:

From r-momentum (A54)

$$u'' + (1-T)\left(\frac{\zeta u'}{2} - u\right) + T^2 wu' - T^2 u^2 + v^2 = T(1-\tau)^2 u_{,\tau} \quad (10.64)$$

From θ-momentum (A55)

$$v'' + (1-T)\frac{\zeta v'}{2} + T^2 wv' - 2Tuv = T(1-\tau)^2 v_{,\tau} \quad (10.65)$$

From continuity (A53)

$$w' - 2u = 0 \quad (10.66)$$

From z-momentum (A56)

$$w'' + (1-T)\left(\frac{\zeta w'}{2} - \frac{3w}{2}\right) + T^2 ww' + P' = T(1-\tau)^2 w_{,\tau} \quad (10.67)$$

When $t \to \infty$, $\tau \to 1$ and $T \to 1$; we recover Eqs. (10.48) to (10.51). When $\tau = 0$, $T = 0$; we again encounter Stokes' First Problem for v, with solution

$$v(\zeta, 0) = \text{erfc}\left(\frac{\zeta}{2}\right) \quad (10.68)$$

For u, we find $u'' + \zeta u'/2 - u + v^2 = 0$, with boundary conditions $u(0) = u(\infty) = 0$.

This can easily be solved numerically, and the solution can be used numerically to integrate the equation

$$w(z, 0) = 2\int_0^\zeta u(\xi, 0)\, d\xi \quad (10.69)$$

Finally, the initial profile of P is computed by a similar quadrature:

$$P(\zeta, 0) = P(0, 0) - 2u(\zeta, 0) - \frac{\zeta w(\zeta, 0)}{2} + 2\int_0^\zeta w(\xi, 0)\, d\xi \quad (10.70)$$

In this last equation, we have used Eq. (10.62) to replace w' by $2u$. These initial profiles are shown in Fig. 10.8. Notice that they are qualitatively very similar to the final profiles shown in Fig. 10.4, and that there is only a modest change in the magnitudes of these scaled variables as τ increases from 0 to 1.

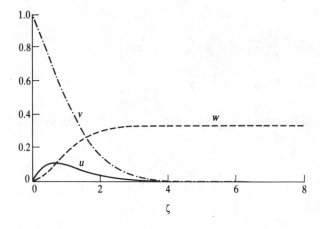

FIGURE 10.8
Scaled initial velocity profiles for the transient spinning-disk flow.

The solution is finally computed for increasing values of τ, by the marching technique shown in the previous example. The FORTRAN program SPNDSK is shown in Appendix B. Curves of $u'(0, t)$, $v'(0, t)$, and $w(\infty, t)$ are shown in Fig. 10.9. The smooth and modest changes of these quantities are evidence of the success of the scaling procedures. However, one cannot formulate a trustworthy physical interpretation of the solution until the scaling formulas have been inverted, to yield values of the physical variables. Consider, for example, the computed pressure distribution. The scaled pressure profile, shown in Fig. 10.10, is the one profile that changes qualitatively with increasing time. At early times, it features the pressure gradient which accelerates the external potential flow toward the disk. Of course, the value of P' that appears in this figure has to be multiplied by $\omega^2 \delta(t)$ to get the physical pressure gradient $\Pi_{,z}$. When this is done, we find that $\Pi_{,z}(\infty, t)$ starts at zero, increases at first in proportion to $t^{1/2}$ as the rate of fluid being centrifugally pumped increases, reaches a maximum value at about

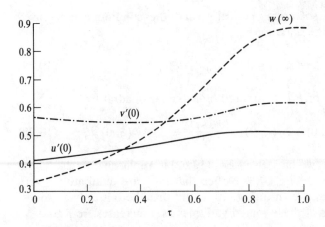

FIGURE 10.9
Temporal evolution of $u'(0)$, $v'(0)$, and $w(\infty)$ in the spinning-disk flow.

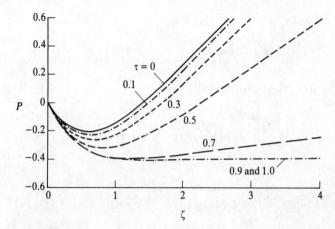

FIGURE 10.10
Temporal evolution of pressure profiles in the transient spinning-disk flow.

$\tau = 0.5$, and then decays to zero as the flow becomes steady. It is also interesting to calculate the physical inflow velocity w^* and the torque required to turn a finite portion of the disk, as functions of time. These calculations make good exercises, and will be left as such.

Transient three-dimensional stagnation-point flows

With some minor modifications of the analysis and computer program used for the transient flow over the impulsively spun-up disk, we can compute the transient establishment of the three-dimensional stagnation-point flows, after the potential flow is impulsively started. The velocity components are scaled as follows:

$$u^* = Ax\, u(\zeta, \tau) \qquad v^* = By\, v(\zeta, \tau) \qquad \text{and} \qquad w^* = -A\, \delta(t) w(\zeta, \tau) \quad (10.71)$$

$\Pi_{,x}$ and $\Pi_{,y}$ are the same as in the steady flow. A useful general expression for Π is

$$\Pi = -\tfrac{1}{2}[(Ax)^2 + (By)^2] + (A + B)\nu P(\zeta, \tau) \tag{10.72}$$

Note that this scaling factor for Π is independent of time. The thickness of the layer of rotational flow will again start to grow in proportion to $(\nu t)^{1/2}$, and will attain a steady value proportional to $\{\nu/A\}^{1/2}$. To keep all the vorticity within a more or less constant range of ζ, we try the scaling formula

$$\delta^2 = \frac{\nu}{A}T \qquad \text{with} \qquad T = 1 - \exp{(-At)}$$

The time-variable is taken to be

$$\tau = \frac{At}{1 + At}$$

The equations to be integrated numerically are:
From x-momentum (A48)

$$u'' + \left[Tw + (1 - T)\frac{\zeta}{2} \right] u' + T(1 - u^2) = T(1 - \tau)^2 u_{,\tau} \qquad (10.73)$$

From y-momentum (A49)

$$v'' + \left[Tw + (1 - T)\frac{\zeta}{2} \right] v' + CT(1 - v^2) = T(1 - \tau)^2 v_{,\tau} \qquad (10.74)$$

From continuity (A47)

$$w' - u - Cv = 0 \qquad (10.75)$$

From z-momentum (A50)

$$P' = -w'' + \tfrac{1}{2}(1 - T)(w - \zeta w') - T[ww' - (1 - \tau)^2 w_{,\tau}] \qquad (10.76)$$

This last equation can be integrated, again using the notation $\chi = \int_0^\zeta w \, d\zeta$, to give

$$P(z, t) = P(0, t) - u - Cv + (1 - T)\chi - \tfrac{1}{2}(1 - T)\zeta w - T\left[\frac{w^2}{2} - (1 - \tau)^2 \chi_{,\tau} \right] \qquad (10.77)$$

Again, the continuity equation has been used to eliminate w'. No condition is given on the pressure itself, so we can set $P(0, \tau) = 0$. The other boundary conditions are

$$u(0, \tau) = v(0, \tau) = w(0, \tau) = 0 \qquad u(\infty, \tau) = v(\infty, \tau) = 1 \qquad (10.78)$$

When $\tau = 0$, we once again encounter Stokes' First Problem, this time for both u and v. Thus,

$$u(\zeta, 0) = v(\zeta, 0) = \text{erf}\left(\frac{\zeta}{2}\right)$$

The integral for w has been evaluated in a different context in Chapter 9, and yields

$$w(\zeta, 0) = \zeta \, \text{erf}\left(\frac{\zeta}{2}\right) - \frac{2}{\sqrt{\pi}}\left[1 - \exp\left(-\frac{\zeta^2}{4}\right) \right]$$

Finally, the integral for χ can also be evaluated in closed form, with the final result

$$P(\zeta, 0) = -\frac{\zeta}{\sqrt{\pi}}$$

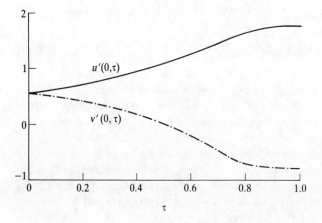

FIGURE 10.11
Impulsively-started flow at a saddle stagnation point. Normalized wall vorticities as functions of time; $C = -0.8$.

This pressure distribution arises because the sudden application of the no-slip condition and the resulting rapid growth of the boundary layer calls for a sudden reduction in the speed with which the fluid approaches the wall. As time goes by, this effect diminishes in importance.

The cases of greatest interest are those for which backflow near the wall eventually develops in the v-profile. The value $C = -0.8$ specifies a good example. The resulting histories of $u'(0, \tau)$, and $v'(0, \tau)$ are shown in Fig. 10.11.

We see that backflow first appears when $\tau = 0.270$. Figure 10.12 shows the profiles of R and S at $\tau = 0$, $\tau = 0.27$, and $\tau = 1$. As in the analysis of

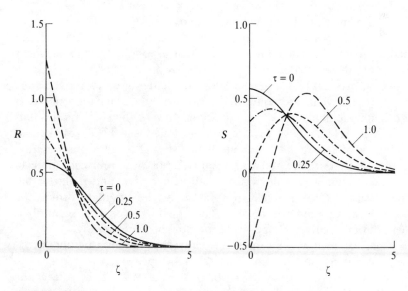

FIGURE 10.12
Evolution of vorticity profiles at a saddle stagnation point; $C = -0.8$.

steady flow, $R = u'$, $S = v'$, and the vorticity of this flow is represented by

$$\mathbf{\Omega} = -\frac{By}{\delta} S \mathbf{e}_x + \frac{Ax}{\delta} R \mathbf{e}_y \qquad (10.79)$$

To account for the different temporal evolutions of R and S, we can use the concepts of the Random-Vortex Method, which was introduced in Section 8.13. Specifically, we can imagine that the value of R at a given values of z and t is determined by the number of linear vorticity elements, parallel to the y-axis and having various signs and strengths, that appear in a small neighborhood of the specified value of z at time t. Each element is generated at the wall as required to annul a given slip velocity, and thereby acquires its plus or minus sign. We assign it an initial strength proportional to the slip velocity that was annulled at the time and place of its creation, and note that this strength is the same everywhere along the length of the element. There is a corresponding set of elements, parallel to the x axis, to account for the value of S. The effect of one set on a member of the other is only to modify the rate of stretching or of foreshortening caused by the potential flow. The elements responsible for the distribution of R are convected by u^* and w^*, and stretched by $v^*_{,y}$; corresponding statements can be made about the elements responsible for the distribution of S. Since $\nabla^2 \mathbf{\Omega}$ reduces to $\mathbf{\Omega}_{,zz}$, we can simulate diffusion by random displacements in the z direction only.

With the initiation of the potential flow, $\mathbf{u}^* = Ax\mathbf{e}_x + By\mathbf{e}_y - (A + B)\mathbf{e}_z$, vortex elements are created at the wall to cancel the slip-velocity components, Ax and By. We shall call these the *original* vortex elements.

An original element of $\mathbf{\Omega}_y$, which is created along the line $x = x_0$, $z = 0$, is assigned a strength proportional to x_0, so that the stronger elements are created farther from the stagnation point. Correspondingly, the stronger elements of $\mathbf{\Omega}_x$ are those created at larger values of y_0.

Since we always take $A > 0$, the original elements of $\mathbf{\Omega}_y$, once they diffuse away from the wall, are convected away from the stagnation point. If $B < 0$, the original elements of $\mathbf{\Omega}_x$ are convected toward the stagnation point. Simultaneously, the elements of $\mathbf{\Omega}_y$ are foreshortened and hence weakened, while the elements of $\mathbf{\Omega}_x$ are stretched and hence strengthened. Consequently, the elements of $\mathbf{\Omega}_y$ that are created at a given value of x are replaced, after an instant, by weaker elements. This occurs at all values of x, and requires that more vorticity with the same sense of rotation as the original vorticity be created at the wall, to keep the no-slip condition in force. This scenario is repeated, without qualitative change, until the flow becomes steady.

For this particular flow, we can readily show that the diffusive flux of $\mathbf{\Omega}_y$ through the wall is constant for $t > 0$. The relationship $-\nu\mathbf{\Omega}_{y,z} = A^2x$ at $z = 0$ can be transformed into the equation, $R'(0, t) = -\delta^2 A/\nu = -T/\alpha$, so that the slope of the R-profile at $z = 0$ becomes only gradually evident, as shown in Fig. 10.12.

The evolution of the S-profile is similarly explained, with the following important difference. If $B < 0$, the elements of Ω_x that are created at a given value of y are replaced by stronger elements. To keep the no-slip condition in force, new vorticity with the opposite sense of rotation must be created at the wall. The new elements interdiffuse with the original ones, gradually reducing the value of S at the wall.

The flux of new vorticity through the wall can be shown to imply the equation $S'(0, t) = -C(T/\alpha)$, a positive value since $C < 0$. Eventually, if C is sufficiently negative, the new vorticity overwhelms the original vorticity at the wall, and backflow is observed in the v-profile.

Much of this scenario, in particular the creation of vorticity of one sign that is rearranged by diffusion and convection so as to require the creation of vorticity of the opposite sign, eventually leading to backflow, appears when a viscous flow *separates* from a solid wall. A separating flow is, however, essentially more complicated than the present flow, in that the onset of backflow occurs along a particular locus on the wall, and is accompanied by a more or less severe local eruption of fluid away from the wall. The fact that backflow in the present case actually enhances flow toward the wall marks these stagnation-point flows as unusual.

The evolution of the scaled pressure profile is shown in Fig. 10.13.

An interesting observation about this last example is that the vorticity field starts out with one self-similar evolution, dominated by diffusion and the boundary conditions, and ends up in a second self-similar configuration, characterized by a steady balance between diffusion and convection. When the ordinary differential equations governing the steady flow were first analyzed, by Davey, it was discovered that the solution was not unique. A second solution was found, which exhibited more profound backflow and a velocity

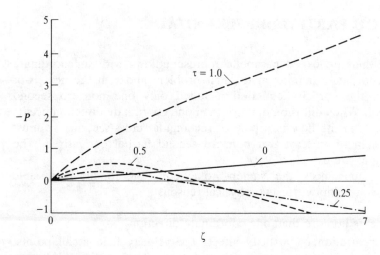

FIGURE 10.13
Evolution of pressure profile at a saddle stagnation point; $C = -0.8$.

overshoot. Davey argued that the solution without overshooting velocity was more likely to be physically relevant, and the present calculations confirm that Davey's solutions are the ones approached if the flow starts impulsively from rest.

You will learn much more about parabolic partial differential equations in Chapter 12, since they are the type of equation that arises naturally in boundary-layer theory. In particular, you will encounter examples in which a downstream-distance variable plays the role of the time variable in the present examples. Then, instead of marching forward in time, solving ordinary differential equations at each time step to account for unidirectional diffusion, the computation must march downstream, following the fluid. It is not always a straightforward matter to follow the fluid computationally, since the computation may reach a streamwise station at which local backflow exists, so that another step in the marching direction is locally a step backwards in time. Since diffusion has the effect of smoothing away the details of an initial distribution of diffusing substance, the initial distribution cannot be deduced by observing the distribution to which it leads. If we attempt to run either of the last two computational schemes backwards in time, the results will quickly blow up.

A final comment is in order about the computational costs of the calculations that have been described and presented in this section. These costs are extremely low, partly because the algorithms are very efficient and the preliminary scaling analysis has minimized the amount of work left to the computer, partly because of the ever-growing power, and the falling cost, of computers. By contrast, a computational study of a complex flow, in which the location of regions of concentrated vorticity cannot be known in advance, may still be extremely expensive.

10.4 ELLIPTIC PARTIAL DIFFERENTIAL EQUATIONS

The equations that are called parabolic represent flows with unicoordinate diffusion, the coordinate being r in the first problem and z in the next two. Mathematically, the vorticity equation included only one nonzero second partial derivative. When diffusion in more than one coordinate direction occurs simultaneously, as in the flow in a pipe of rectangular cross-section, we must deal with the sum of at least two nonzero second partial derivatives. The resulting equations are called *elliptic*.

In most circumstances, the Navier–Stokes equations for incompressible or anelastic flow are elliptic, for two physical reasons:

1. Diffusion occurs in more than one coordinate direction.
2. A local concentration of vorticity affects the velocity field at all points, whatever their direction away from the location of the vorticity. Alternatively, a local adjustment of pressure, needed to enforce the continuity equation, is felt instantaneously, everywhere in the flow.

The second kind of ellipticity is exhibited mathematically in the form of the equations that must be solved to compute velocity from vorticity, or pressure from velocity. The case of two-dimensional flow makes the point most simply; then the velocity components may be found from a streamfunction, Ψ, by the equations

$$u = \Psi,_y \quad \text{and} \quad v = -\Psi,_x$$

Since the single component of vorticity is given by $\Omega = v,_x - u,_y$, the distribution of Ψ is governed by the *Poisson equation*,

$$\nabla^2\Psi = \Psi,_{xx} + \Psi,_{yy} = -\Omega \tag{10.80}$$

Alternatively, we can take the divergence of the momentum equation and apply div $\mathbf{u} = 0$ in the resulting equation. What remains is a Poisson equation for the pressure,

$$\nabla^2\Pi = \Pi,_{xx} + \Pi,_{yy} = -\text{div}\,(\mathbf{u}\,\text{grad}\,\mathbf{u}) = -2(u,_x^2 + u,_y v,_x) \tag{10.81}$$

In special circumstances, one of the second derivatives in the expanded form of ∇^2 vanishes identically; this has been true in all of our examples so far. In more general circumstances, to be considered extensively in Chapters 11 and 12, one of the second derivatives in Eq. (10.80) is so much smaller than the other that it can be neglected without much loss of accuracy. In many cases, however, both second derivatives must be retained. When this is so, the solution at any point of the x–y plane affects the solution at every other point, and boundary conditions must be specified at every point of a curve that encloses the domain of interest. Values of Ψ itself, or of the rate of change of Ψ with distance normal to the boundary, must be given. A linear combination of these two quantities may be prescribed, but independent values of the two quantities must not be given at the same point.

Flows with Multidimensional Diffusion

Section 9.3 presents an analysis of a simple Couette flow between parallel plane walls, $y = 0$ and $y = H$, modified by the presence of stationary sidewalls, $z = \pm W$. The analysis came down to the solution of the Laplace equation $\nabla^2 u = 0$, with $u = 0$ on all bounding planes except $y = h$, where $u = U$. The solution was constructed by separation of variables and Fourier synthesis. Let us now consider a finite-difference approach to the same problem.

There are many effective ways to do this; we shall consider only two of the simplest. With this brief introduction, you should be able to read descriptions of more efficient methods, should you be interested. Both of the present methods start by covering the rectangular cross section of the channel by a uniform rectangular mesh, as shown in Fig. 10.14. The vertical lines of the mesh are given by $z = -W/2 + (i-1)h_z$, where $h_z = W/(M-1)$, and i runs from 1 to M. The horizontal lines are given by $y = (j-1)h_y$, where $h_y = H/(N-1)$, and j runs from 1 to N. The dimensionless velocity at the intersection of the ith vertical line and the jth horizontal line is called $u(i, j)$.

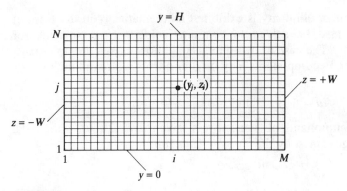

FIGURE 10.14
Simple mesh for finite-difference analysis.

With this terminology, the boundary conditions are written as

$$u(1, j) = u(M, j) = 0 \quad \text{for} \quad 2 \leq j \leq N - 1$$

$$u(i, 1) = 0 \quad \text{and} \quad u(i, N) = 1 \quad \text{for} \quad 2 \leq i \leq M - 1$$

The finite-difference approximation to the differential equation is derived by reference to the stencil of mesh points, shown in Fig. 10.15. The velocity at the center point, 0, is taken to be $u(i, j)$. At that point, $u_{,zz}$ is approximated by $[u(i + 1, j) - 2u(i, j) + u(i - 1, j)]/h_z^2$, while $u_{,yy}$ is approximated by $[u(i, j + 1) - 2u(i, j) + u(i, j - 1)]/h_y^2$. The approximation to the Laplace equation is then

$$\frac{u(i + 1, j) - 2u(i, j) + u(i - 1, j)}{h_z^2} + \frac{u(i, j) + 1) - 2u(i, j) + u(i, j - 1)}{h_y^2} = 0$$

$$(10.82)$$

There is a linear algebraic equation of this form for each interior mesh point, the number of such equations being $(M - 2)(N - 2)$. The number of unknowns

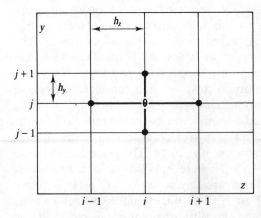

FIGURE 10.15
Finite-difference stencil for a Poisson-solver.

appearing in those equations equals $(M-2)(N-2) + 2[(M-2) + (N-2)]$, the first term accounting for unknowns at the interior points while the second accounts for unknowns on the boundary. Note that the corner points, where the velocity may not be uniquely defined, do not enter the calculation. For the present simple problem, the necessary boundary values are given in advance, so a determinate system of linear algebraic equations has been set up. Unfortunately, the coupling between equations in this set is more thorough than in the equations for one-dimensional diffusion, and more effort is required to get a solution.

A direct method of solution, using fast Fourier transforms. One popular method of solution of the linear algebra problem accomplishes the task with a single pass of calculations, and no trial-and-error iterations. Thus, it is called a *direct method*.

Discrete Fourier transform in z. In the present numerical approximation, the variation of u with z, at a fixed value of y (or j), is represented by a succession of discrete values of u, at equally-spaced values of z. By a straightforward calculation, a weighted linear sum of sine waves can be formed, so that the sum equals a specified value of u at each mesh point. The formula for this sum is written as

$$u(i,j) = \sum_{s=1}^{M} u_s(j) \sin\left[\pi(i-1)s/M\right), \qquad 2 \le i \le M \qquad (10.83)$$

while the formula for the coefficients in the sum is

$$u_s(j) = \frac{2}{M} \sum_{i=2}^{M} u(i,j) \sin\left[\pi s(i-1)/M\right] \qquad 1 \le s \le M-1 \qquad (10.84)$$

When (10.83) is used to represent each of the values of u that appear in (10.82), we get the equation

$$\sum_{s=1}^{M} \frac{u_s(j)}{h_z^2} \left\{ \sin\left[\pi i s/M\right] - 2\sin\left[\pi(i-1)s/M\right] + \sin\left[\pi(i-2)s/M\right\} \right.$$
$$+ \left(\frac{\sin\left[\pi(i-1)s/M\right]}{h_y^2}\right)[u_s(j+1) - 2u_s(j) + u_s(j-1)] = 0 \quad (10.85)$$

Using the trigonometric identity

$$\sin\left[\pi i s/M\right] + \sin\left[\pi(i-2)s/M\right] = 2\sin\left[\pi(i-1)s/M\right]\cos\left(s\pi/M\right)$$

we can deduce that (10.85) will be satisfied, provided that

$$u_s(j-1) + B(s)u_s(j) + u_s(j+1) = 0 \qquad (10.86)$$

In this,
$$B(s) = 2\left\{\left(\frac{h_y}{h_z}\right)^2 [\cos\left(s\pi/M\right) - 1] - 1\right\}$$

TABLE 10.3
Finite-difference results for Couette Flow in a rectangular
channel of aspect ratio 5

		Flow rate		Drag (bottom)		Drag (top)		Drag (sides)	
IM	JM	DIR	ADI	DIR	ADI	DIR	ADI	DIR	ADI
8	8	2.0793	2.0793	3.9548	3.9548	5.1266	5.1265	1.1717	1.1717
16	16	2.1803	2.1803	4.0799	4.0799	6.0467	6.0468	1.9668	1.9667
32	32	2.2139	2.2139	4.1084	4.1084	6.9382	6.9382	2.8298	2.8298
64	64	2.2243	2.2244	4.1152	4.1157	7.8230	7.8226	3.7078	3.7079

IM = number of equal increments in z, JM = number of equal increments in y.
Solution by separation of variables gives FLOW = 2.2285, DRAGB = 4.1175.

This manipulation has broken the algebraic coupling in the z-direction, and has led to a problem, represented by (10.86), that can be solved by the Thomas algorithm. After that has been done for each value of s, so that all the $u_s(j)$ are known, the unknown values of $u(i, j)$ can be found from (10.83). The boundary conditions at $i = 1$ and $i = M$ are automatically satisfied by the choice of sine functions in (10.83). The boundary values at $j = 1$ and $j = N$ are used to generate the $u_s(1)$ and $u_s(N)$, by application of (10.84). For the problem in hand, this leads to $u_s(1) = 0$ for all s, and to

$$u_s(N) = \frac{2}{M} \sum_{i=2}^{M} \sin\left[\pi s(i-1)/M\right] \qquad 1 \leq s \leq M \qquad (10.87)$$

At first sight, this seems to be a terribly expensive method, requiring the computation of an enormous number of sine functions. However, if $M = 2^m$, where m is an integer, the number of necessary calculations is greatly reduced by virtue of useful symmetries. Programs to do these *fast Fourier transforms* (FFT) are contained in most libraries of standard computational routines.[5] The result is that this is often the method of choice when the shape of the domain and the nature of the boundary conditions are sufficiently simple. A FORTRAN program that executes this algorithm literally, without calls to library FFT routines, is shown in Appendix B, under the name "DIRECOU". Numerical results are shown in Table 10.3.

An iterative method: Alternating-Direction-Implicit method (ADI). The second method is an attractive alternative when the time-dependent field of a

---

[5] For information on how to adapt this general technique to other types of boundary conditions, etc., see R. C. LeBail, "Use of fast Fourier Transforms for solving partial differential equations in physics," *J. Comput. Phys.* **9**: 440–465 (1972).

diffusing quantity is to be computed. Then we always know a good first approximation to the present field, given first by the initial conditions, and later by the field at the latest previous time level.

The method belongs to a general class of *time-splitting* methods, which employ the fact that it is permissible to compute simultaneous effects as though they occurred sequentially, providing that the time step is small enough. The Random-Vortex Method is another example.

The ADI method is perhaps easiest to understand as a method for solving the transient equation, $u_{,t} = v(u_{,yy} + u_{,zz})$. It prescribes the following algorithm with which to advance the solution from an initial field $u(y, z, t)$ to the new field $u(y, z, t + dt)$ that will exist after a small time increment. A spatial mesh is introduced, as in the direct method, and the initial field is represented by the discrete values of $u^n(i, j)$. The corresponding values at time $t + dt$ are symbolized by $u^{n+1}(i, j)$. The ADI method also employs imaginary intermediate values, $u^{n+1/2}(i, j)$, which have no direct physical meaning.

The intermediate values are first calculated, along the ith mesh column, using the tri-diagonal set of algebraic equations

$$u^{n+1/2}(i, j) = u^n(i, j) + \frac{dt}{2h_y^2}[u^{n+1/2}(i, j + 1) - 2u^{n+1/2}(i, j) + u^{n+1/2}(i, j - 1)]$$

$$+ \frac{dt}{2h_z^2}[u^n(i + 1, j) - 2u^n(i, j) + u^n(i - 1, j)] \qquad (10.88)$$

This half-step uses an implicit formulation of diffusion in the y direction, and an explicit formulation of diffusion in the z direction. The unknown quantities appear only on the left-hand side and in the middle group of terms on the right-hand side, so the set of such equations for each value of i can be solved by the Thomas algorithm. This solution naturally brings in the known boundary values for $j = 1$ and $j = N$. Equation (10.88) is solved for $2 \le i \le M - 1$, with the given boundary data for $i = 1$ and $i = M$ being used for the solutions at $i = 2$ and $i = M - 1$.

The solution is advanced through the second half-step with the analogous formula

$$u^{n+1}(i, j) = u^{n+1/2}(i, j) + \frac{dt}{2h_y^2}[u^{n+1/2}(i, j + 1) - 2u^{n+1/2}(i, j) + u^{n+1/2}(i, j - 1)]$$

$$+ \frac{dt}{2h_z^2}[u^{n+1}(i + 1, j) - 2u^{n+1}(i, j) + u^{n+1}(i - 1, j)] \qquad (10.89)$$

This treats diffusion in the z direction implicitly, and diffusion in the y direction explicitly.

It can be shown that this algorithm has just enough of an implicit character to make it unconditionally stable.[6] To use it for a steady-flow

---

[6] This explanation of the ADI method has been kindly shown to me by my colleague, Professor Philip Marcus.

problem, one needs to guess an initial field, hopefully not too different from the desired final field, and then to apply (10.88) and (10.89) successively, over and over, until the computed final field becomes substantially independent of further iterations. For this purpose, it is often effective to use a rather large value of *dt,* since the fields computed before convergence is obtained are of no interest. To simulate a physically unsteady flow with real initial conditions, one must use small enough values of *dt* to resolve the transient behavior.

A FORTRAN program, "ADICOU", to apply this method to the plane Couette flow with sidewalls, is shown in Appendix B. Results are compared with those of the direct method in Table 10.3. This is an interesting problem for numerical analysis, because only certain quantities, such as the flow rate and the drag of the stationary bottom wall, can be accurately computed. Other quantities, such as the drag of the moving top wall or that of the sidewalls, are intrinsically ill-defined, because the mathematical problem presupposes a physically impossible condition where the moving top wall and the stationary sidewalls meet. There the fluid is subjected to an infinite rate of deformation, and the local skin friction becomes so large that an infinite drag results. Thus, as the finite-difference mesh is refined, the values for flow rate and bottom-wall drag converge to the limits that can be computed from the analysis of Chapter 9. During the same process the top-wall drag diverges, again as predicted by the solution found in Chapter 9.

The ADI method requires an increasing number of iterations as the mesh refinement is increased. A slight tendency for oscillation was cured by use of an underrelaxation factor of 0.85. For a tolerance of 1 in the fifth decimal place of FLOW and DRAGB, the first three cases required 10, 11, and 24 iterations. The last case was not quite converged after 60 iterations.

It must be reemphasized at this point that this little discussion of numerical methods for elliptic equations has exposed only the tip of the iceberg. Many scientific software libraries contain a variety of *elliptic solvers,* which may be used to advantage once you understand how they work, and how to test them for accuracy in your particular application.

10.5   COMPUTATION OF ENCLOSED VISCOUS FLOWS

Numerical methods have been used extensively in recent years to explore flows much more complicated than the examples described in this chapter. In this section and the next, two rather impressive examples are exhibited, just to give an indication of the possibilities. Flows that are entirely enclosed by solid boundaries, and that are driven by motion or heating of the boundaries, are in principle the most cleanly defined, and will be exemplified by two-dimensional flow in a rectangular cavity, driven by a steady, or impulsively started, sliding motion of one wall.

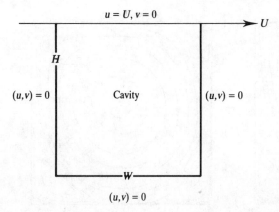

FIGURE 10.16
Defining sketch for driven-cavity flow.

Wall-Driven Cavity Flow

Imagine a two-dimensional flow in a rectangular cavity of width W and depth H, sketched in Fig. 10.16. Three walls are stationary, while the fourth slides with a constant speed U. There are no driving thermal effects, and ρ and μ are assumed to be constant. Interest in this problem centers on the possibility of a steady solution, and on the evolution of the streamline pattern as the Reynolds number, $\text{Re} = UW/\nu$, increases. The effect of the aspect ratio W/H is also interesting, but many studies are restricted to the case $W = H$.

This problem has served as a testing ground, almost a jousting ground, for competing numerical methods, the object of the competition being to prove that one's favorite method reveals, accurately, more realistic details of the flow at high values of Re than does any other method of comparable cost.[7]

Figure 10.17[8] shows some of the trends of two-dimensional steady flow that are now believed to be real, in the sense that they are revealed by many, quite different, computation schemes, and seem to be independent of further refinement of computational parameters such as mesh size. The striking feature is the appearance of an increasing number of eddies surrounded by closed streamlines, as Re increases. It will be shown in Chapter 11 that some of these may be expected even at very low values of Re.

---

[7] For a recent and critical introduction to the literature, and an indication of the present state of the art, see K. Gustafson and K. Halasi, "Cavity flow dynamics at higher Reynolds numbers and higher aspect ratios," *J. Comput. Phys.* **70**: 271–283 (1979). For an important warning about potential pitfalls, see R. Schreiber and H. B. Keller, "Spurious solutions in driven cavity calculations," *J. Comput. Phys.* **49**: 165–172 (1983).

[8] These results are from the very careful work of R. Schreiber and H. B. Keller, "Driven cavity flows by efficient numerical techniques," *J. Comput. Phys.* **49**: 310–333 (1983).

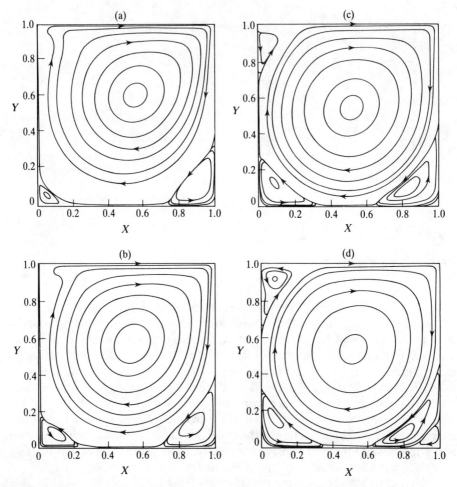

FIGURE 10.17
Steady flow in a driven cavity. Streamline patterns. (a) Re = 400; (b) Re = 1000; (c) Re = 4000; (d) Re = 10,000. (Reproduced, by permission, from R. Schreiber and H. B. Keller, 1983.[8])

10.6 COMPUTATION OF EXTERNAL VISCOUS FLOWS

Fundamental and interesting theoretical difficulties appear when the domain of interest is only part of a larger flow field, and one attempts to avoid computations in the external domain that is not of interest. Figure 10.18 illustrates some typical situations.

Computational schemes that cover the domain of interest with a computational grid or mesh ordinarily require specified values of unknown quantities and/or their gradients, all around the outermost set of mesh points. It is extremely difficult to specify these values in a way guaranteed to provide an accurate simulation of a particular experimental situation. One result of this

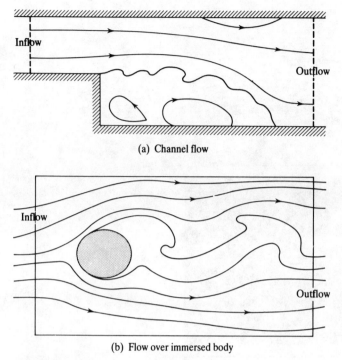

(a) Channel flow

(b) Flow over immersed body

FIGURE 10.18
Situation requiring inflow and outflow boundary conditions.

is that seemingly simple-posed tasks, such as "Calculate the streamline field for two-dimensional flow past a circular cylinder, given the fact that the stream in which the cylinder is immersed would be perfectly uniform if the cylinder were not there," never seem to be done to the complete satisfaction of all critics. Roughly speaking, the difficulty is that one cannot simply draw a large circle around the cylinder, to include the region on which one is willing to expend most of the computational resources, and say that the flow on that circle is not disturbed by the cylinder. This is particularly true where the wake of the cylinder cuts through the imaginary boundary cycle. Conditions in this region have come to be called *outflow boundary conditions,* and the best ways to handle these conditions are still a subject of active research.

Flow Past an Impulsively Accelerated Cylinder

In spite of the difficulties cited above, some very interesting and impressive numerical studies of external viscous flows have been made. The one used here for illustration shows the flow evolution shortly after a circular cylinder has been impulsively brought to a constant forward velocity in an infinite fluid that was initially at rest. Although the flow is highly unsteady and the value of $Re = Ua/v$ is fairly high, the computation is somewhat facilitated by the fact

FIGURE 10.19
Flow behind an impulsively started cylinder, Re = 3000. (Top) Computed by a discrete-vortex method. (Middle) Experimental flow visualization. (Bottom) Computed by finite differences. (Reproduced, by permission from P. A. Smith and P. K. Stansby, "Impulsively-started flow around a circular cylinder by the vortex method," *J. Fluid Mech.* **194**: 45–77 (1988).)

that the vorticity shed from the cylinder has not moved too far away, and so the computational domain can reasonably enclose it. Another reason for selection of this example is the existence of laboratory observations, a finite-difference calculation, and a random-vortex calculation, all showing very nearly the same intricate patterns of eddies at very nearly identical dimensionless times. These results are shown in Fig. 10.19.

10.7 SUMMARY

Numerical analysis has contributed greatly to our appreciation of the implications of the Navier–Stokes equations. These equations have a few, famous, self-similar, solutions that can be found by numerical integration of nonlinear ordinary differential equations. Typically, the solutions must satisfy boundary conditions at both ends of the interval of interest. One successful technique of solution, which is explored extensively in this chapter and used frequently in the remaining chapters, is built around the Thomas algorithm.

Successful computational analysis is often initiated by a clever scaling of independent and dependent variables, which reduce the range of variation of the numbers that need to be computed. This eases the computational effort required to obtain results of a given accuracy.

Parabolic partial differential equations, which appear in the next simplest class of flow analyses, can often be solved by a simple elaboration of the algorithms used to find the self-similar solutions. Cases for which this is not so simple are analyzed towards the end of Chapter 12.

In many important cases, the Navier–Stokes equations are fully elliptic, because diffusion proceeds in more than one direction, and because pressure changes in one part of the flow are felt instantaneously in all other parts. Algorithms for the numerical solution of the Poisson equation are thus important components of most schemes for numerical solution of the Navier–Stokes equations. There are two general classes of such algorithms, direct and iterative.

The formulation of boundary conditions for a numerical analysis of a portion of a larger flowfield is problematical, and still a subject of active research.

In spite of all difficulties, numerical simulations of viscous flows can reveal an astonishing richness of detail, often exceeding that which can be obtained by laboratory measurements. It seems inevitable that they will play an increasingly important role as witnesses of real flows, and that the student of such flows will have to become skilled and critical interrogators of such witnesses.

EXERCISES

10.1. Derive Eqs. (10.5) and (10.6) by the use of Taylor series expansions. Identify the first term omitted from each approximation, and verify that it is of order h^2 compared to the terms retained. Is this still true if the three points for the difference formulas are not equally spaced?

10.2. Fill in the algebra leading from Eq. (10.7) to (10.13) and (10.14).

10.3. Derive the formula for u_N given in Eq. (10.16).

10.4. An asymptotic analysis, such as that leading to Eq. (10.21), allows one to find an accurate solution of a differential equation in the finite domain, $0 \le \eta \le \eta^*$, although the outer boundary condition should in principle be applied at $\eta = \infty$.

Suppose such a numerical solution has been found. How do you then describe, quantitatively, the solution in the range $\eta > \eta^*$? Derive, for example, a formula for the error made when the quantity

$$\delta^* = \int_0^\infty (1 - u)\, d\eta \quad \text{is approximated} \quad \delta^* = \int_0^{\eta^*} (1 - u)\, d\eta$$

10.5. Evaluate the slope ϕ' of the velocity profile on each side of the interface between the two layers of immiscible fluids flowing down an inclined plane. (See Fig. 9.10.) Do this with the relevant finite-difference formulas and the numerical data given on page 182, and compare your answers with the analytic results of Chapter 9.

10.6. Verify that Eq. (10.19) is the desired solution of (10.18).

10.7. Prove that Eqs. (10.35) follow logically from (10.32) and (10.33).

10.8. Derive Eq. (10.53).

10.9 Calculate $w^*(\infty)$ and the torque required to turn the steadily spinning disk, using the dimensionless, scaled, values $w(\infty)$, $v'(0)$.

10.10. Verify that Eqs. (10.83) and (10.84) are mutually consistent by direct calculation for the case $M = 4$.

CHAPTER
11

CREEPING FLOWS

11.1 INTRODUCTION

The sample viscous flows presented in Chapters 9 and 10 could be comprehensively analyzed because their highly idealized initial and boundary conditions could be met by relatively simple solutions of the Navier–Stokes equations. In this chapter and the next, you will encounter important branches of viscous-flow theory, in which the possibility of a fruitful analysis depends on the value of the Reynolds number. This chapter deals with flows characterized by very small values of Re, so that the inertia term, $\rho\, D\mathbf{u}/Dt$, plays a relatively unimportant role in the momentum equation. Such flows are often called *creeping flows*. The distribution of vorticity in such flows is established entirely by diffusion, the role of convection being of minor importance and often negligible.

Engineering students are often told that almost all practically important flows are characterized by high values of Re: a fair statement if you are interested in airplanes, ships, rivers, or rocket motors. However, many practical devices, such as fluid-lubricated bearings, and many modern industrial processes, such as the manufacture of color film or magnetic recording tape, involve creeping flows. If your curiosity extends to subjects as diverse as the formation of rain drops, the flow of glaciers, or the feeding methods of the tiny organism that stand close to the base of the food chain, you will again need to understand creeping flows.[1]

Regardless of one's practical interests, creeping flows are intrinsically fascinating, exhibiting many phenomena that are strikingly different from those seen in flows with significant inertia. They also exhibit some phenomena, such as flow separation, that are frequently associated with inertia or convection, showing the need for care in one's attempts to identify the cause of a particular effect.

The mathematical theory of creeping flow can often be represented with linear differential equations and boundary conditions; therefore much of the theory is quite old and elaborate. It extends far beyond the scope appropriate for this text; this chapter is intended only to convey the main theoretical ideas and a few results that are of special conceptual or practical importance. References to more advanced treatments will be given with each section.

Section 11.2 presents some general experimental and theoretical results that highlight the unique character of this kind of flow. The theory of hydrodynamic lubrication, and of other creeping flows that occur in slender films, including films with free surfaces, is given in Sections 11.3 to 11.5. Even in its mathematically most idealized form, this theory suggests many practically important results. Few of the complications that arise in real applications are discussed, although the phenomena of cavitation and ventilation in journal bearings are mentioned as an indication of the care with which a student must learn to examine the implications of an idealized theory.

Section 11.6 presents a local analysis of flows near sharp corners. The boundary conditions for these idealized problems introduce no scale of length, and dimensional analysis dictates the form of the solution for most of them. The most interesting of the corner-flow examples introduces the concept of an eigensolution, and shows the practical utility of some classic mathematical formulas. It also introduces a very surprising flow phenomenon.

The second general topic is that of *external* creeping flows, typified by the flow over a sphere that is far from any other solid body. This example is treated fairly completely in Section 11.7. Some results for spheroids and for a spherical cap, are analyzed in Section 11.8. Singular solutions, and their superposition to approximate flows over bodies of complicated shape, are briefly analyzed in Section 11.9.

Section 11.10 treats small effects of inertia, first for bounded flows such as lubrication films, for which a regular perturbation series can be constructed, and then for external flows such as that over a sphere, for which a singular perturbation analysis can be made by the method of matched asymptotic expansions.

---

[1] This point is brilliantly made, and many of the phenomena of creeping flow are vividly illustrated in the educational film, "Low Reynolds-Number Flow", by Sir Geoffrey Taylor. Sponsored by NCFMF, distributed by Encyclopedia Britannica Films.

Finally, the creeping flow through porous media is briefly treated in Section 11.11.

11.2 GENERAL FEATURES OF CREEPING FLOW

Negligible inertia. In creeping flow, the inertia of a moving fluid is dynamically insignificant; whatever the acceleration, **a**, of each fluid particle, the quantity ρ**a** is negligibly small compared to the viscous force per unit volume, $\nabla \cdot \tau$. In such a flow, the viscous force is usually balanced by a combination of pressure and gravity forces, so that the momentum equation reduces to

$$\nabla \cdot \tau = \rho \nabla \Pi \tag{11.1}$$

where $\Pi = p/\rho + gh$, as in earlier chapters.

Suppose the fluid motion is caused by a moving body, with characteristic speed U, characteristic acceleration a, and characteristic length L. Then $\rho \mathbf{a} \ll \nabla \cdot \tau$ translates into constraints on the values of two Reynolds numbers:

$$\frac{UL}{\nu} \ll 1 \quad \text{and} \quad \frac{aL^2}{U\nu} \ll 1 \tag{11.2}$$

When these constraints are met, some astonishing things happen. The velocity of the fluid becomes precisely proportional to the velocity of the body. When the body starts or stops, the fluid does the same. If the body, having moved from A to B along a particular path, subsequently reverses itself and returns to A along the same path, every fluid particle returns to its initial position! Of course, Eqs. (11.2) must be continuously satisfied throughout the entire process.

Streamline patterns in steady flow become independent of the direction of flow along the streamlines, so that one cannot tell upstream or downstream just by looking at the streamlines. Figure 11.1 shows striking photographic evidence of this. The photos were obtained in steady flow, at Re ≈ 0.02, but the streamline pattern is independent of the value of Re, so long as that value is small enough.

When inertial forces and variations of density and viscosity can be neglected, the Navier–Stokes equations can be reduced to

$$\nabla \Pi = -\nu \nabla \times \Omega \tag{11.3}$$

and

$$\nabla \cdot \mathbf{u} = 0$$

We can take the divergence of (11.3) to get

$$\nabla^2 \Pi = 0 \tag{11.4}$$

and we can take the curl of it to get

$$\nabla \times (\nabla \times \Omega) = 0 \tag{11.5}$$

It is easy to show, especially for planar flows, that streamline patterns are independent of the value of Re, providing that (11.5) is justified. In such flows,

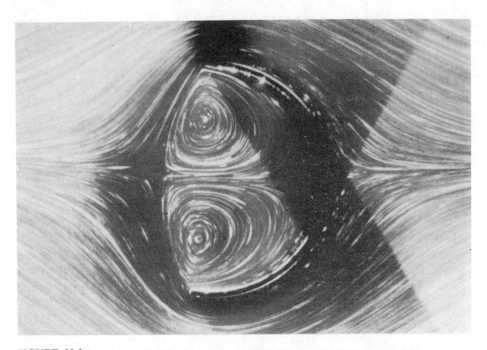

FIGURE 11.1
Creeping flow, from left to right. Flow from right to left looks the same when $\text{Re} \ll 1$. (Reproduced, by permission, from Taneda, 1979.[2])

Ω has a single scalar component that can be derived from a scalar streamfunction, ψ. Specifically,

$$u = \frac{\partial \psi}{\partial y} \qquad v = -\frac{\partial \psi}{\partial x}$$

and

$$\Omega = \Omega \mathbf{e}_z \qquad \text{where} \qquad \Omega = \frac{\partial u}{\partial y} - \frac{\partial v}{\partial x} = -\nabla^2 \psi$$

Then Eq. (11.5) implies that

$$\psi_{,xxxx} + 2\psi_{,xxyy} + \psi_{,yyyy} = \nabla^4 \psi = 0 \tag{11.6}$$

This is called the *biharmonic equation*.

Suppose the streamline pattern is observed from a frame of reference attached to the body, which is moving along the x axis. Then the boundary conditions on ψ are $\psi = Uy$ very far from the body, and $\partial \psi / \partial x = \partial \psi / \partial y = 0$ on the body. Neither the differential equation nor the boundary conditions, which govern the spatial distribution of the streamfunction, contain Re as a parameter, so the streamline pattern is independent of the value of Re.

---

[2] For the original publication of this and many other fascinating pictures of creeping flow, see S. Taneda, "Visualization of Separating Stokes Flows," *J. Physical Soc. Japan* **46**: 1935–1942 (1979).

It is also easy to show, either from Eq. (11.3) or from a simple dimensional analysis, that the net force acting on a finite body of length L, creeping along at speed U, through a fluid of viscosity μ, will be proportional to $\mu U L$, whatever the shape of the body. This result is true, however, only if no other physical parameters, such as surface tension, enter the problem through the boundary conditions.

Negligible convection. It is also true of creeping flows that the distribution of vorticity is determined only by diffusion: convection plays a negligible role. Another way of saying this is that vorticity can move through the fluid faster than the creeping body. Thus, vorticity is as likely to be found upstream of a body as in its wake; indeed, there is no such thing as a wake in creeping flow.

The notion of a pure creeping flow is somewhat like that of a pure incompressible flow; it contains some paradoxes and requires some careful thinking. It is, for example, not possible to think of the transient development of a creeping flow that starts from rest, because the transient effects are manifestations of inertia, which we have agreed to forget!

11.3 CREEPING FLOW IN SLENDER LAYERS

In Chapter 9, it was shown that special boundary conditions sometimes completely prevent the development of convective acceleration, even if the value of Re is large. If a steady flow is subject to such conditions, there will be no inertial forces, although there is plenty of inertia. The result may be very simple flow, such as Couette or Poiseuille flow.

In Section 9.4, the steady flow between plane porous walls was analyzed. A cross-flow velocity component, V, causes a convective redistribution of vorticity, but this effect is negligible if $\text{Re} = VH/\nu \ll 1$, where H is the distance between the walls. Then a theory in which V is completely neglected gives good approximations for the velocity component parallel to the walls, for the skin friction, and for the pressure field. Specifically, exact results differ from those given by the zero-V theory by amounts linearly proportional to Re, when $\text{Re} \ll 1$.

There are many other situations in which the boundary conditions on flow in a slender layer allow only a relatively small velocity component normal to the layer, so that a Reynolds number based on a typical value of this velocity component and the width of the layer has a small value. This happens very commonly in fluid-lubricated bearings and related devices, so that the theory of such flows is commonly called lubrication theory.

In Chapter 12, you will encounter a fairly systematic development of approximate equations for flow in slender layers. In this section, the basic ideas will be developed by means of a typical example.

Order-of-Magnitude Analysis: the Squeezed Film

Consider the flow produced when fluid is squeezed out of a narrow gap between two parallel walls that approach each other with constant relative speed V. Let the lower wall occupy the strip $-L/2 \le x \le L/2$ in the stationary plane $z = 0$. Let the upper wall lie directly above the lower wall, in the plane $z = H(t)$. The gap is slender, i.e. $H \ll L$.

To postpone the necessity of thinking about the pressure, we shall start with the vorticity equation and the equations that relate the vorticity and velocity components to the streamfunction. If the velocity is simply $\mathbf{u} = u(x, y, t)\mathbf{e}_x + (x, y, t)\mathbf{e}_y$, and the density and viscosity are constants, these equations are

$$\frac{\partial \Omega^*}{\partial t^*} + u^* \frac{\partial \Omega^*}{\partial x^*} + w^* \frac{\partial \Omega^*}{\partial z^*} = v\left(\frac{\partial^2 \Omega^*}{\partial x^{*2}} + \frac{\partial^2 \Omega^*}{\partial z^{*2}}\right) \tag{11.7}$$

$$\Omega^* = -\left(\frac{\partial^2 \psi^*}{\partial x^{*2}} + \frac{\partial^2 \psi^*}{\partial z^{*2}}\right) \tag{11.8}$$

$$u^* = \frac{\partial \psi^*}{\partial z^*} \quad \text{and} \quad w^* = -\frac{\partial \psi^*}{\partial x^*} \tag{11.9}$$

Here, as in Chapter 10, we use starred symbols to represent actual dimensional quantities, and now introduce scaled, dimensionless, equivalents to facilitate further analysis. The natural scale for x^* is the strip width, L; that for z^* is the initial gap height H_0; that for t^* is the time to close the gap, H_0/V. The natural scale for ψ^* is the rate, VL, at which fluid is displaced by the moving strip, per unit length of the strips. Thus, we define

$$\psi^*(x^*, z^*, t^*) = VL\psi(x, z, t),$$

in which

$$x = \frac{x^*}{L} \qquad z = \frac{z^*}{H_0} \qquad t = \frac{Vt^*}{H_0}$$

The corresponding representation of the vorticity is

$$\Omega^* = -\frac{VL}{H_0^2}\left[\left(\frac{H_0}{L}\right)^2 \frac{\partial^2 \psi}{\partial x^2} + \frac{\partial^2 \psi}{\partial z^2}\right] \tag{11.10}$$

Whereas scaling was employed in Chapter 10 just to facilitate computation, it will be used here to organize a systematic scheme of approximation. Thus, because $H_0/L \ll 1$, we neglect, for a first approximation, the first term on the right-hand side of Eq. (11.10). For the same reason, we neglect the first term on the right-hand side of Eq. (11.7). These two approximations profoundly simplify the mathematical character of the resulting theory. The first one significantly restricts the variety of kinematic phenomena that can be represented; the second simply asserts that streamwise diffusion of vorticity is negligible compared to cross-stream diffusion.

We use these approximations, and return to the subscript notation for partial derivatives to rewrite Eq. (11.7) in terms of the scaled variables, as

$$\frac{VH_0}{v}(\psi_{,zzt} + \psi_{,z}\psi_{,zzx} - \psi_{,x}\psi_{,zzz}) = \psi_{,zzzz} \tag{11.11}$$

Finally, we restrict attention to cases in which the cross-stream Reynolds number, VH_0/ν, is so small that one gets a good approximation by dropping the terms that are multiplied by this parameter.

The general solution for ψ, in the limit $VH_0/\nu \rightarrow 0$, is thus

$$\psi = A + Bz + Cz^2 + Dz^3$$

On the bottom plate, $z = 0$, we can set $\psi = 0$; and the no-slip condition requires that $\psi_z = 0$. Thus, $A = 0$, and $B = 0$.

On the top plate, the streamfunction can be found from the integral

$$\psi^*(x^*, H^*, t^*) = -\int_0^{x^*} v^* \, dx^* = Vx^*$$

if we specify that the plane $x^* = 0$ is another part of the stream surface $\psi^* = 0$. This means that $\psi(x, H/H_0, t) = x$. For future convenience, we introduce the notation

$$\xi = \frac{H}{H_0}$$

Thus,

$$C\xi^2 + D\xi^3 = x$$

When this is combined with the no-slip condition on the upper plate, which gives

$$2C + 3D\xi = 0$$

we find that $C = 3x\xi^3$, and $D = -2x\xi^2$. Finally, we get

$$\psi = x(3z^2\xi^{-2} - 2z^3\xi^{-3}), \tag{11.12}$$

or, in terms of the original physical variables,

$$\psi^* = Vx^*\left[3\left(\frac{z^*}{H}\right)^2 - 2\left(\frac{z^*}{H}\right)^3\right] \tag{11.13}$$

$$u^* = 6V\left(\frac{x^*}{H}\right)\left[\left(\frac{z^*}{H}\right) - \left(\frac{z^*}{H}\right)^2\right] \tag{11.14}$$

$$w^* = -V\left[3\left(\frac{z^*}{H}\right)^2 - 2\left(\frac{z^*}{H}\right)^3\right] \tag{11.15}$$

In the approximation we have been using to find the velocity field, the momentum equation reduces to

$$\frac{\partial \Pi^*}{\partial z^*} = \nu\frac{\partial^2 w^*}{\partial z^{*2}} \quad \text{and} \quad \frac{\partial \Pi^*}{\partial x^*} = \nu\frac{\partial^2 u^*}{\partial z^{*2}}$$

from which we find the pressure field

$$\Pi^*(x^*, z^*, t^*) = \Pi^*(0, 0, t^*) - 6\nu\left(\frac{V}{H}\right)\left[\left(\frac{z^*}{H}\right) - \left(\frac{z^*}{H}\right)^2 + \left(\frac{x^*}{H}\right)^2\right] \tag{11.16}$$

Velocity and Stress Fields for Creeping Flow in Slender Films

In the example just analyzed, the description given by the lowest approximation exhibits the following features, which are typical of many lubricating films and slender viscous films that appear in other applications.

1. The viscous stress tensor has only two dynamically significant components, τ_{zx} and τ_{zy}. These can be combined, for later convenience, into a two-component vector

$$\boldsymbol{\tau} = \tau_{zx}\mathbf{e}_x + \tau_{zy}\mathbf{e}_y \qquad (11.17)$$

2. The components of $\boldsymbol{\tau}$ are related to the velocity field in a particularly simple way:

$$\tau_{zx} = \mu u_{,z} \quad \text{and} \quad \tau_{zy} = \mu v_{,z} \qquad (11.18)$$

3. The viscous force per unit volume is due only to the variation of $\boldsymbol{\tau}$ with z.

$$\operatorname{div} \boldsymbol{\tau} = \boldsymbol{\tau}_{,z} \qquad (11.19)$$

Hence, it acts nearly parallel to the surfaces which form the top and bottom of the film.

4. The local viscous force per unit volume is independent of distance normal to the film. That is,

$$(\operatorname{div} \boldsymbol{\tau})_{,z} = \boldsymbol{\tau}_{,zz} = 0 \qquad (11.20)$$

5. There exists a local balance between viscous forces, pressure forces, and body forces. Thus, the resultant of pressure and body forces is also independent of z and very nearly parallel to the film-bounding surfaces.

The example from which these characteristics are deduced does not involve a film like that between closely-fitting circular cylinders, which is bounded by one or more curved surfaces. However, it is easy to show that such a curved film behaves as though it were flat, provided that its thickness is very small compared to the minimum radius of curvature of the actual bounding surface.

For further analysis, we shall thus assume that the bottom of the film coincides with the plane $z = 0$, while the top is found at $z = H(x, y, t)$. In typical bearing problems, the function H will be given; in film-application problems, it will be an unknown.

From Eq. (11.20), we see that $\boldsymbol{\tau}$ will be a linear function of z, namely

$$\boldsymbol{\tau} = \mathbf{A} + \mathbf{B}z \qquad (11.21)$$

in which $\mathbf{A}$ and $\mathbf{B}$ are independent of z, but may depend on x, y, and t. $\mathbf{B}$ is, of course, the viscous force per unit volume. In some applications, the boundary conditions will determine that $\mathbf{B}$ must vanish.

It is convenient to have a symbol for the velocity component tangent to the film, so we call

$$\mathbf{q} = u\mathbf{e}_x + v\mathbf{e}_y \tag{11.22}$$

From Eq. (11.21), and the assumption that μ is independent of z, it follows that u and v are quadratic functions of z, so that

$$\mathbf{q} = \mathbf{C} + \mathbf{D}z + \mathbf{E}z^2 \tag{11.23}$$

Another useful quantity is the integral of $\mathbf{q}$ over the local thickness of the film,

$$\mathbf{Q} = \int_0^H \mathbf{q} \, dz \tag{11.24}$$

Clearly, $$\mathbf{Q} = \mathbf{C}H + \mathbf{D}\frac{H^2}{2} + \mathbf{E}\frac{H^3}{3} \tag{11.25}$$

In specific applications, values of $\mathbf{C}$, $\mathbf{D}$, and $\mathbf{E}$ are determined by boundary conditions, and by considerations soon to be explained. For example, if the boundary $z = 0$ is a solid wall, to which is attached the frame of reference for the measurement of velocity, and at which a no-slip condition is enforced, $\mathbf{C}$ will be zero.

The principle of mass conservation provides the final, crucial connection between the pressure and velocity fields. Consider a control cylinder that extends across the film from $z = 0$ to $z = H$. It is shown in Fig. 11.2. Its lateral surface and its bottom are stationary, but its top is free to move up or down at the rate $\partial H/\partial t$.

Just as we have ignored variations of viscosity in the z direction, because the temperature is expected to vary only slightly with z, we shall now ignore variations of density in the z direction, because both temperature and pressure are expected to vary only slightly with z.

The rate at which mass enters the cylinder through its lateral surface can be represented by a line integral around the curve that outlines the bottom of the cylinder.

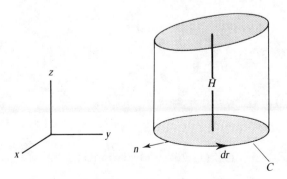

FIGURE 11.2
Control volume for derivation of the Reynolds equation.

$$\text{Rate of mass flow in} = -\int_C \rho\, \mathbf{Q} \cdot \mathbf{n}\, dr$$

The rate of change of mass stored within the cylinder is given by the equation

$$\frac{dm}{dt} = A\,\frac{\partial(\overline{\rho H})}{\partial t},$$

in which A is the base area of the cylinder, and the overbar signifies an average over A.

Applying the principle of conservation of mass, letting $A \to 0$, and recognizing the definition of the divergence of plane vector field, we obtain the equation

$$\frac{\partial(\rho H)}{\partial t} = -\mathbf{\nabla} \cdot (\rho \mathbf{Q}) \tag{11.26}$$

Reynolds' Equation for the Pressure.
Lubrication Theory

As a first application of this arsenal of equations, we show how the pressure distribution in a film-lubricated bearing is determined.

In the simplest bearing problems, one is given $H(x, y, t)$ and $\mathbf{U}(x, y, t)$, and is required to find $p(x, y, t)$. The body force is usually negligible, and will be excluded from this analysis.

Both surfaces are assumed to be rigid solids at which conditions of zero penetration and zero slip apply. The frame of reference for velocity is attached to the wall $z = 0$. Thus,

$$\mathbf{C} = 0 \quad \text{and} \quad \mathbf{D}H + \mathbf{E}H^2 = \mathbf{U}$$

Using the latter equation to eliminate $\mathbf{D}$ from Eq. (11.25), we find

$$\mathbf{Q} = \mathbf{U}\frac{H}{2} - \mathbf{E}\frac{H^3}{6}$$

The balance between viscous and pressure forces gives

$$\frac{\partial \mathbf{\tau}}{\partial z} = \mathbf{\nabla}p \tag{11.27}$$

from which we see that $\mathbf{E} = (2\mu)^{-1} \mathbf{\nabla}p$, and hence, that

$$\mathbf{Q} = \mathbf{U}H/2 - (12\mu)^{-1} \mathbf{\nabla}p\, H^3.$$

Placing this expression for $\mathbf{Q}$ into Eq. (11.26), and rearranging the result slightly, we get the *Reynolds equations for film lubrication*:

$$\nabla \cdot \left[\left(\frac{\rho H^3}{\mu} \right) \nabla p \right] = 12 \frac{\partial(\rho H)}{\partial t} + 6 \nabla \cdot (\rho H \mathbf{U}) \qquad (11.28)$$

Only for gas bearings is it important to account for variations of density. For other applications, ρ may simply be dropped from the equation.

11.4 SAMPLE LUBRICATION PROBLEMS

(a) The Squeezed Film

Let us start by confirming that the Reynolds equation gives the same result for pressure, in the initial illustrative example, as was obtained by direct calculation. We have $\partial H / \partial t = -V$, $\mathbf{U} = 0$, and $\nabla H = 0$. The density and viscosity are taken constant. Thus, Eq. (11.22) reduces to

$$-12\mu V = H^3 \nabla^2 p = H^3 \frac{\partial^2 p}{\partial x^2}$$

When this is integrated over x, with the symmetry condition that $\partial p / \partial x = 0$ at $x = 0$, and the boundary condition $p = 0$ at $x = L$, the result is

$$p(x, t) = (6\mu V / H^3)(L^2 - x^2)$$

Comparing this with Eq. (11.10), and allowing for the use of asterisks to denote physical quantities in the earlier equation, but not here, we notice a small difference. The earlier development predicts a dependence of pressure on z; the present development ignores that effect, anticipating that at most points in the slender film, $x/H \gg 1$, whereas z/H cannot exceed 1.

(b) A Slider Bearing

Consider the bearing sketched in Fig. 11.3. The upper boundary of the film is a flat, finite, slightly inclined plate. On this *slider*, $0 \le x \le L$ and $-W/2 \le y \le W/2$.

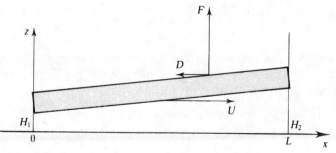

FIGURE 11.3
Slider bearing.

The trailing edge is at $z = H_1$, the leading edge at $z = H_2$. For shorthand, let $\alpha = (H_2 - H_1)/L$. The slider moves to the right, at constant speed U, so that H_1, H_2, and α are all independent of time. Clearly

$$\frac{\partial H}{\partial t} = -U \frac{\partial H}{\partial x} = -U\alpha$$

Because V, ∇U, and $\partial H/\partial y$ are all zero, and ρ and μ are assumed to be constants, Eq. (11.28) reduces to

$$\frac{\partial(H^3 \, \partial p/\partial x)}{\partial x} + H^3 \frac{\partial^2 p}{\partial y^2} = -6\mu U\alpha \qquad (11.29)$$

As it stands, this equation allows for the fact that lubricant may escape from the film between the slider and the stationary wall at the sides where $y = \pm W/2$, as well as from the leading and trailing edges. If the slider is very wide, $W/L \gg 1$, or if side guards prevent the sideways escape of lubricant, the velocity component v, and the corresponding dependence of p on y, can be neglected. We now explore this simplified case.

Because the frame of reference is attached to the lower, stationary, wall, both H and p depend on t as well as on x. However, the variation of p with t only enters, in a parametric way, through the known dependence of H on t. At a fixed instant, we have only to deal with an ordinary differential equation to find p as a function of x.[3]

The integration is facilitated by the use of H, rather than x, as the independent variable. At a fixed instant, we can set $dH = \alpha \, dx$, and write Eq. (11.23), without its middle term, as

$$\frac{\partial(H^3 \, \partial p/\partial H)}{\partial H} = -\frac{6\mu U}{\alpha} \qquad (11.30)$$

A first integration gives

$$\frac{\partial p}{\partial H} = \frac{\mu U}{\alpha}(-6H^{-2} + AH^{-3})$$

A second gives

$$p = \frac{\mu U}{\alpha}\left[6H^{-1} - \frac{A}{2}H^{-2}\right] + B$$

It is now assumed that the pressure at the leading and trailing edges of the slider is the same. We call this constant of integration p_A. Thus, the boundary conditions on p are taken to be

$$p(H_1) = p_A \quad \text{and} \quad p(H_2) = p_A$$

---

[3] If you have difficulty with this, try attaching the frame of reference to the slider. Then the flow appears steady.

The constants of integration have the values

$$A = 12H_1H_2(H_1 + H_2)^{-1} \quad \text{and} \quad B = p_A - \frac{6\mu U}{\alpha}(H_1 + H_2)^{-1}$$

The final form of the pressure distribution is then

$$p - p_A = \frac{6\mu U}{\alpha(H_1 + H_2)}\left[\frac{(H_2 - H)(H - H_1)}{H^2}\right] \tag{11.31}$$

It is convenient to have dimensionless variables for a presentation of these results, so we let

$$\frac{H}{H_1} = \phi \qquad \frac{H_2}{H_1} = R \quad \text{and} \quad x/L = \xi$$

These are related by the equation $\phi = 1 + (R - 1)\xi$. Equation (11.31) can now be written as

$$p - p_A = \frac{6\mu U}{\alpha H_1} F(R, \xi)$$

with

$$F(R, \xi) = \frac{(R - 1)^2}{R + 1} \cdot \frac{\xi(1 - \xi)}{[1 + (R - 1)\xi]^2}$$

The maximum value of $F(R, \xi)$ equals $(R - 1)^2/4R(R + 1)$, where $\xi = (R + 1)^{-1}$. $F(R, \xi)$ is plotted, for $R = 3/2$ and for $R = 4$, in Fig. 11.4(a).

Other features of the solution are worth examining. One is the distribution of skin friction along the slider surface. We have

$$\tau_{zx}(x, H) = \frac{\mu U}{H} + \frac{H}{2}\frac{\partial p}{\partial x} = \frac{\mu U}{H} + \frac{\alpha H}{2}\frac{\partial p}{\partial H}$$

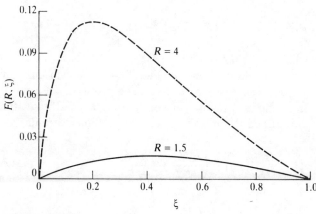

FIGURE 11.4(a)
Pressure on slider.

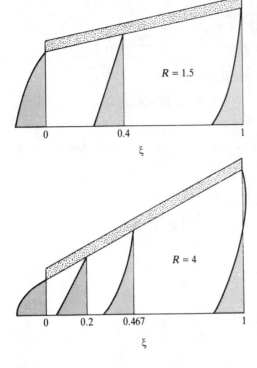

FIGURE 11.4(b)
Velocity profiles under slider.

and

$$\frac{\partial p}{\partial H} = \frac{6\mu U}{\alpha}[2H_1 H_2(H_1 + H_2)^{-1}H^{-3} - H^{-2}]$$

Thus

$$\tau_{zx}(x, H) = 2\mu U[3H_1 H_2(H_1 + H_2)^{-1}H^{-2} - H^{-1}]$$

or

$$\tau_{zx}(x, H) = \frac{2\mu U}{H_1}\{3R(R + 1)^{-1}[1 + (R - 1)\xi]^{-2} - [1 + (R - 1)\xi]^{-1}\}$$

If $R > 2$, τ_{zx} can vanish, where $\xi = (2R - 1)/(R^2 - 1)$.

The streamwise velocity profile is shown in Fig. 11.4(b), at various key values of ξ, and for $R = 3/2$ and for $R = 4$. Note that each profile is the sum of a shear-driven straight line and a pressure-driven parabola, and that the parabola bows out in the direction of falling pressure. The formula for $u(x, z, t)$ is

$$\frac{u}{U} = \left(4 - \frac{H^*}{H}\right)\left(\frac{z}{H}\right) + \left(\frac{H^*}{H} - 3\right)\left(\frac{z}{H}\right)^2 \tag{11.32}$$

in which

$$H^* = \frac{A}{2} = \frac{6H_1 H_2}{H_1 + H_2}$$

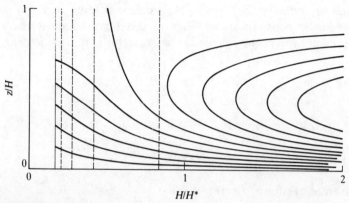

FIGURE 11.4(c)
Generalized streamlines of flow under slider, in frame of reference moving with slider. Area between dotted lines corresponds to $R = 4$. Area between dashed lines corresponds to $R = 1.5$.

In a frame of reference which moves with the slider, the flow is steady and exhibits an interesting streamline pattern. The formula for the streamfunction is

$$\frac{\psi}{UH^*} = \frac{1}{2}\left(4\frac{H}{H^*} - 1\right)\left(\frac{z}{H}\right)^2 + \frac{1}{3}\left(1 - 3\frac{H}{H^*}\right)\left(\frac{z}{H}\right)^3 - \left(\frac{H}{H^*}\right)\left(\frac{z}{H}\right) \quad (11.33)$$

The dependence of the flow on the design parameters R and α appears only when one converts from the variables z/H and H/H^* to more usual variables x/L and z/L. Figure 11.4(c) shows the generic flowfield defined by Eq. (11.33). The portion of this field which extends from $H/H^* = (1 + R)/6R$ to $H/H^* = (1 + R)/6$ would lie under the slider with a specified value of R. If $R > 2$, a dividing streamline appears, separating the flow that will pass under the trailing edge of the slider from the flow that turns back to pass around the leading edge. On the dividing streamline, $\psi = -UH^*/6$ and

$$z/H = \frac{1}{(6H/H^* - 2)} \quad (11.34)$$

Forces on the slider

It is usually assumed that the pressure p_A acts on the top surface of the slider. If this is true, the slider can support a load, in the z direction,

$$\text{Load} = (1 + \alpha^2)^{-1/2}\int_0^L (p - p_A)\, dx = (1 + \alpha^2)^{-1/2}\alpha^{-1}\int_{H_1}^{H_2} (p - p_A)\, dH$$

which gives

$$\text{Load} = (1 + \alpha^2)^{-1/2}\left(\frac{6\mu U}{\alpha^2}\right)\left(\ln R - 2\frac{R - 1}{R + 1}\right) \quad (11.35)$$

Typically, α is so small that the factor $(1 + \alpha^2)^{-1/2}$ can be omitted.

It is interesting to compare the load to the tangential force required to keep the slider moving at constant speed. Part of this force balances the x component of the pressure force; the remainder balances the skin friction drag. The first part is

$$D_p = (1 + \alpha^2)^{-1} \frac{6\mu U}{\alpha} \left[\ln R - 2 \frac{R-1}{R+1} \right]$$

The second part is given by the integral

$$D_f = \int_0^L \tau_{zx}(x, H) \, dx$$

Using the result given above for $\tau_{zx}(x, H)$, we find

$$D_f = \frac{2\mu U}{\alpha} \left(3 \frac{R-1}{R+1} - \ln R \right)$$

Finally, we add D_p and D_f to get the total drag

$$\text{Drag} = D_p + D_f = \frac{2\mu U}{\alpha} \left(2 \ln R - 3 \frac{R-1}{R+1} \right) \tag{11.36}$$

Remember that the skin friction vanishes at a point on the slider if $R > 2$. For larger values of R, the skin friction on the forward part of the slider acts in the direction of slider motion, and if $R > 13.14$, the net skin friction acts to abet, rather than to oppose, this motion. Of course, the total drag opposes the motion for all values of R; otherwise, we should have a perpetual motion machine!

A complete engineering analysis of the slider bearing would extend well beyond the scope of this text, to consider quantities such as the moment of the pressure around the trailing edge, and phenomena such as the response of the bearing to small perturbations in position, orientation, speed or load.[4] This important work would, however, add little to your understanding of the underlying fluid mechanics.

(c) A Journal Bearing

One of the most familiar applications of lubrication theory provides an idealized description of the operation of a *journal bearing,* in which a round shaft, the journal, spins inside a cylindrical sleeve, or bearing, of slightly larger radius. The gap between shaft and sleeve is partially or completely filled with a viscous lubricant, usually a liquid.

---

[4] A relatively short recent survey of lubrication mechanics is given by E. A. Saibel and N. A. Macken *Ann. Rev. Fluid Mech.,* **5**: 185–212 (1973). A more comprehensive text is D. D. Fuller, *Theory and Practice of Lubrication for Engineers,* 2d ed., Wiley-Interscience, 1984. Finally, there is a *Journal of Lubrication Technology.*

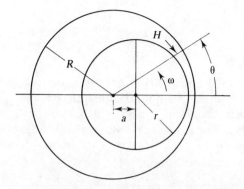

FIGURE 11.5(a)
Journal bearing.

The analysis starts with the assumption that the shaft, when spinning with a constant angular velocity ω and subject to a constant transverse load L, will find an equilibrium position relative to the sleeve, and that the resulting flow will be steady.

The separation of the parallel axes of shaft and sleeve in the equilibrium configuration is a, the journal radius is r, and the sleeve radius is R. These quantities are shown in Fig. 11.5a, along with an angular coordinate, θ, which is zero where the clearance between shaft and sleeve is smallest.

The radial clearance, $C = R - r$, is typically so small compared to R that the curvature of the streamlines has no significant dynamical effect. Thus, the analysis proceeds as though the gap were rolled out flat, and the local clearance is well approximated by the formula

$$H(\theta) = C - a \cos \theta \qquad (11.37)$$

One can represent the Reynolds equation in cartesian coordinates, with $z = 0$ on the stationary sleeve, and $z = H$ on the spinning shaft. The tangential speed of the shaft is then well approximated by $U = \omega R$. Throughout the analysis, no distinction is made between the values of R and r, except where their difference, C, appears.

If there is no axial flow or pressure gradient, and if the viscosity is constant, Eq. (11.28) reduces to

$$\frac{d(H^3 \, dp^*/d\theta)}{d\theta} = 6\mu\omega R^2 \frac{dH}{d\theta}$$

The first integration gives

$$\frac{dp^*}{d\theta} = 6\mu\omega R^2(H^{-2} - AH^{-3})$$

Let us consider only the simplest case, in which the gap between shaft and sleeve is entirely filled with the lubricating fluid. Then the value of the constant of integration, A, is determined by the requirement that the pressure be single-valued. Thus the integral of $dp^*/d\theta$ from $\theta = -\pi$ to $\theta = +\pi$ must vanish. This happens when

$$A = \frac{2C(1-\eta^2)}{2+\eta^2}$$

where $\eta = a/C$ is called the *eccentricity* of the shaft. It is easy to show that $A = 2Q/\omega R$, where Q is the volumetric flowrate per unit axial length, passing between shaft and sleeve at any angular location. A also equals the local clearance between shaft and sleeve at the angular locations where the velocity profile is a straight line.

A second integration gives the pressure distribution:

$$p(\theta) - p(0) = -6\mu\omega\left(\frac{R}{C}\right)^2 \frac{\eta \sin \theta(2 - \eta \cos \theta)}{(2 + \eta^2)(1 - \eta \cos \theta)^2} \tag{11.38}$$

This is shown, for three different values of the eccentricity, in Fig. 11.5b. In the limit as $\eta \to 1$, the pressure extrema approach the location of minimum H and become very large. The pressure gradient is largest where H is smallest, and acts to help the fluid move through the constriction in the direction in which it is dragged by the shaft motion.

Assuming that Eq. (11.38) holds for all values of θ, we can calculate the net pressure force on the shaft. Clearly, this acts in a direction normal to the line of centers of shaft and sleeve, specifically along the direction $\theta = \pi/2$ when $\omega > 0$. This force may be called the *lift L* of the bearing. A final integration gives

$$L = 2\int_0^P (p(0) - p(\theta))R \sin \theta \, d\theta$$

$$= 12\pi\mu\omega\left(\frac{R}{C}\right)^2 \eta(1 - \eta^2)^{-1/2}(2 + \eta^2)^{-1} \tag{11.39}$$

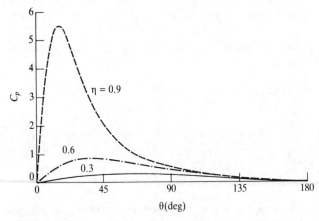

FIGURE 11.5(b)
Pressure in journal bearing. $C_p \equiv [p(\theta) - p(0)/6\mu\omega(R/C)^2$.

The torque, T, required to turn the shaft at constant ω, is also of interest. We use the approximation

$$T = R^2 \int_0^{2\pi} \tau_{xz}(\theta, H) \, d\theta$$

with
$$\tau_{xz}(\theta, H) = \left(\frac{H}{2R}\right) \frac{dp}{d\theta} + \mu\omega \frac{R}{H}$$

The result is

$$T = 2\pi\mu\omega\left(\frac{R^3}{C}\right)[3\eta^2 + (2 + \eta^2)](1 - \eta^2)^{-1/2}(2 + \eta^2)^{-1}$$

where the $3\eta^2$ inside [ ] comes from the pressure-driven flow, while the $(2 + \eta^2)$ comes from the shear-driven flow. Note that these terms are comparable when η is nearly equal to 1. When they added, we find the working formula

$$T = 4\pi\mu\omega\left(\frac{R^3}{C}\right)(1 + 2\eta^2)(1 - \eta^2)^{-1/2}(2 + \eta^2)^{-1} \qquad (11.40)$$

A figure of merit for the journal bearing is the dimensionless ratio LR/T, which has the value

$$\frac{LR}{T} = 3\frac{R}{C}\left(\frac{\eta}{1 + \eta^2}\right) \qquad (11.41)$$

This is analogous to the lift/drag ratio of the slider bearing. For a given value of R/C, the bearing is most efficient in the limit $\eta \to 1$. In this limit, however, the two solid surfaces touch; this is the very condition the bearing is designed to avoid! Another way to use (11.41) is to determine the value of η that will result from given values of T and L.

Equation (11.38) contains a warning of a possible limitation on the validity of this analysis. It shows the possibility of very low pressures, which could lead to the release of dissolved gases or even local boiling of a liquid lubricant, or to the displacement of liquid in the bearing by air or other gases which leak in from the ends of the bearing. We shall use the word *cavitation* for the local release of a gas phase, and the word *ventilation* for invasion of gas from the outside. Because the liquid oil is very nearly incompressible, a gas phase can only appear if the remaining liquid has somewhere to go. Thus, the result to be expected in practice depends on conditions imposed at the ends of the bearing, on the presence of pressure taps, etc. Some bearings are deliberately operated with the low-pressure side open to atmospheric pressure, in which case ventilation often consists of a finger-like intrusion of the meniscus, below the level it would occupy if the shaft were not rotating. A photograph of the phenomenon is shown in Fig. 11.6. The complexities of a real situation can be quite baffling;[5] an idealized one is shown in Sir Geoffrey Taylor's film.

There are many other fascinating practical problems associated with journal bearings, such as the possibility that the shaft may not settle down into a stable orientation, that are beyond the proper scope of this text.[6]

(d) A Simple Example of Forced Lubrication

In many film-lubricated bearings there is a velocity component that carries away fluid that has been heated by viscous dissipation, replacing it with fluid that has passed through an external cooler. Because of the linearity of the governing equations, this component of flow can be analyzed separately, and then superposed on the components that are responsible for the pressure distribution in the bearing. When lubricant is pumped into a bearing with the specific purpose of developing a load-supporting pressure field, we speak of *forced lubrication*.[7] A simple example is presented by the floating hockey pucks that are used to approximate frictionless motion and elastic collisions in introductory physics laboratories.

The device is a circular disk with a central hole through which a compressed gas passes from a bottle mounted on the disk into the gap between the bottom of the disk and a smooth horizontal table. The situation is sketched in Fig. 11.6, where the diameter of the feed hole and the clearance between disk and table are greatly exaggerated, relative to the radius of the disk. In typical real cases, $b \ll a$ and $H \ll a$.

We assume that disk and table remain parallel, even when the disk is sliding over the table. Then, if the fluid motion can be described by lubrication theory, the sliding motion does not affect the pressure field between disk and table. To calculate the pressure, one can use a frame of reference attached to the disk. Equation (11.22) reduces to

$$\frac{d(r\,dp/dr)}{dr} = 0$$

with solution $dp/dr = A/r$, and $p = A \ln r + B$. This solution will not be accurate where r is very close to zero, because the velocity field under the feed hole does not conform to the assumptions of a slender-film theory. For the moment, we ignore this difficulty, and evaluate A and B in terms of the pressure at the outer edge of the disk and the flowrate of lubricant through

---

[5] For a fascinating but sober introduction to these complexities, read G. I. Taylor, "Cavitation of a Viscous Fluid in Narrow Passages," *Jour. Fluid. Mech.* **16**: 595–619 (1963).

[6] See, for example, Fuller, *op. cit.*, Chapter 8.

[7] Fuller, *op. cit.*, Ch. 3, calls this *hydrostatic lubrication*, and gives many fascinating examples. He also describes the use of an electric analog, to optimize the placement of lubricant-feed holes.

FIGURE 11.6
Cavitation in a journal bearing. Shaft motion bottom to top. (Reproduced, by permission, from D. Dowson and C. M. Taylor, *Ann. Rev. Fluid Mech.* **11**, 35–66, 1979).

the feed hole. Thus $p(r) - p(a) = A \ln (r/a)$. The value of A is found by noting that at the edge of the disk $dp/dr = A/a$, so that the radial velocity component there is given by $u(a, z) = (A/2\mu a)z(z - H)$. The flowrate, Q, is given by

$$Q = 2\pi a \int_0^H u(a, z)\, dz = -\frac{\pi H^3 A}{6\mu}$$

and the pressure is related to the flowrate by

$$p(r) - p(a) = \frac{6\mu Q}{\pi H^3} \ln \left(\frac{a}{r}\right) \tag{11.42}$$

A final integration may be used to find the load, i.e., the weight of disk and gas bottle, that can be supported at a given clearance by a given flowrate. If the pressure acting on top of the disk and bottle is equal to $p(a)$, the result is

$$L = 2\pi \int_0^a r[p(r) - p(a)]\, dr = \frac{3\mu Q a^2}{H^3} \tag{11.43}$$

Since the purpose of the disk is to demonstrate low-friction motion, it is interesting to evaluate the resistance to sliding motion, and to form a corresponding friction coefficient. If the speed of sliding is V, the drag will be accurately approximated by $D = \pi a^2 \mu V / H$. The friction coefficient is then

$$f \equiv \frac{D}{L} = \frac{\pi}{3} \frac{VH^2}{Q} = \pi \left(\frac{\mu^2 a^4}{3L^2 Q} \right)^{1/3} V.$$

The second expression is written in case L and Q, rather than H and Q, are given.

(e) The Hele Shaw Cell

A final application of the Reynolds Equation has nothing to do with lubrication; instead, it provides a laboratory demonstration of streamline patterns to be expected in inviscid potential flow!

The cell is usually rectangular in planform, and is partially bounded by two accurately plane, parallel, and stationary plates, one of which is transparent. Two parallel edges are sealed; fluid enters at a uniform rate along one of the remaining edges and exits along the other. A wafer-like model that spans the gap may be placed in the cell, and extra inlets or outlets, which simulate the sources and sinks of potential-flow theory, may be installed in the large, nontransparent wall. When dye filaments are admitted with the entering flow, they trace out the patterns of streamlines in two-dimensional potential flow around the model or the sources and sinks. An example is shown in Fig. 11.7.[8]

That the apparatus works as claimed is easily seen from Eq. (11.28), which reduces to $\nabla^2 p = 0$, and by the corresponding equation for the velocity,

$$\mathbf{q} = (2\mu)^{-1} z(z - H) \nabla p \tag{11.44}$$

This shows that in plan view the streamlines are all parallel to the pressure gradient, while the pressure itself satisfies the Laplace equation. Thus, the pressure in a Hele-Shaw cell plays a role like that of the velocity potential in two-dimensional, incompressible, potential flow.

The boundary conditions on the model in the cell prevent flow through the model, but also require zero slip at the model surface. This can produce a local deviation from the state of stress assumed to exist in a slender-film flow, but if the model diameter is much greater than H, the effect of this on the streamline pattern is not noticeable.

---

[8] An effective demonstration of a Hele-Shaw cell in action may be seen in the short film *Hele Shaw Analog of Potential Flows*, Nos. FM-80 & FM-81 in the NCFMF series of films, distributed by Encyclopedia Britannica Educational Films.

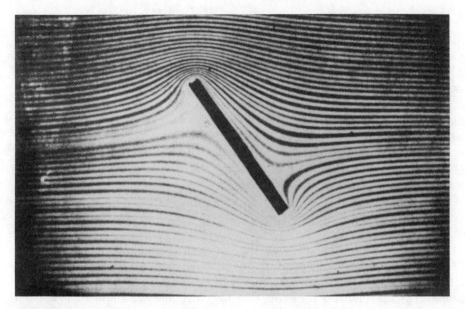

FIGURE 11.7
Hele–Shaw flow past an inclined plate. (Photograph by D. H. Peregrine.)

11.5 SLENDER VISCOUS FILMS WITH FREE SURFACES

A technically important class of slender viscous films involve free surfaces. The film thickness, H, is to be predicted or controlled. Motion within the film may be driven by the motion of a boundary, by an imposed stress at the free surface, or by a body force such as gravity. Surface tension may play a crucial role. By analyzing such films, one may gain some theoretical understanding of such diverse phenomena as the spreading of molten lava over flat ground, the merging of two tiny bubbles into a single one, or the withdrawal of a layer of viscous fluid on a solid surface which rises out of a pool of the fluid.[9]

The theory of viscous films with a free surface is somewhat analogous to that of water waves. In both cases, the flow is governed by a linear partial differential equation, but the boundary conditions at the free surface make the problem of prediction of $H(x, y, t)$ mathematically nonlinear. To see how this comes about for viscous films, consider the following very simple example.

---

[9] An industrially important subset, involved in the making of magnetic recording tape and color film, are called coating flows. A recent review of their theory is given by K. J. Ruschak, "Coating Flows," *Ann. Rev. Fluid Mech.* **17**: 65–90 (1985).

A Windblown Oil Film

A film of oil, of thickness $H = H(x, t)$, is dragged along a stationary plane, $z = 0$, by the action of a specified tangential stress $\tau(x, H, t)$. We may imagine that this stress is transmitted to the oil at the free surface by a co-flowing wind, and that the object of the study is to see whether the stress can be experimentally evaluated by measuring $H(x, t)$. We neglect effects of gravity, surface tension, and horizontal pressure gradients, and use the slender-layer creeping-flow approximations for the flow in the film.

We may start with Eq. (11.25) and (11.26). The first of these reduces to $Q = (\tau/2\mu)H^2$, and the second becomes $\partial H/\partial t = -\partial Q/\partial x$. These can be combined into a single equation for $h(x, t)$,

$$\frac{\partial H}{\partial t} + \frac{\tau H}{2\mu} \frac{\partial H}{\partial x} = -\frac{H^2}{2\mu} \frac{\partial \tau}{\partial x} \tag{11.45}$$

This equation shows that it would be necessary to measure H, $\partial H/\partial t$ and $\partial H/\partial x$ to pin down a value of τ, and that all one would get would be a relationship between τ and $\partial \tau/\partial x$. Setting aside this motivation, let us suppose, as a further simplification, that τ is independent of both x and t. Then the right-hand side of Eq. (11.45) vanishes, leaving

$$\frac{\partial H}{\partial t} + \frac{\tau H}{2\mu} \frac{\partial H}{\partial x} = 0 \tag{11.46}$$

This equation shows that H is constant along *characteristic trajectories* in an x–t plane, the inverse slope of these trajectories being $dx/dt = \tau H/2\mu$. To see this, we use the chain rule for differentiable functions,

$$dH = (\partial H/\partial t)\, dt + (\partial H/\partial x)\, dx,$$

and ask what value of dx/dt will make $dH = 0$. We see the answer in Eq. (11.46). The nature of this result is best seen graphically, as in Fig. 11.8. Because H is constant along each trajectory, the trajectories are straight lines.[10]

Consider the temporal evolution of the initial thickness distribution $H(x, 0) = F(x)$. Where $dF/dx > 0$, the film profile is destined to become flatter. Where $dF/dx > 0$, it is destined to become steeper, reaching an infinite slope at a time $t^* = -(2\mu/\tau)/(dF/dx)$. To see this, draw two neighboring characteristic trajectories that leave the x axis at points x_1 and x_2, in an interval where $dF/dx < 0$. Note that they will eventually intersect, where

$$x_1 + \frac{\tau H_1}{2\mu} t = x_2 + \frac{\tau H_2}{2\mu} t \quad \text{or when} \quad \frac{\tau}{2\mu} t = \frac{x_2 - x_1}{H_2 - H_1}$$

---

[10] If you have already studied inviscid compressible flow, you will recognize this analysis and its relation to the study of simple waves of compression or expansion in gases.

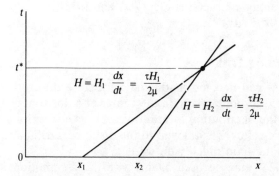

FIGURE 11.8
Characteristic lines for a wind-blown oil film.

The result for t^* appears in the limit when $x_2 - x_1 \to 0$. For any time just later than t^*, two different characteristics reach a given point, indicating that two different values of H exist at the same point and instant. The necessary inference is that $\partial H / \partial x$ has become infinite at time t^*, beyond which our simple theory does not make sense. Physically, any initial waviness in the film surface will develop breakers, like a surf, as shown in Fig. 11.9. Blowing over a pool of paint would not be a good way of spreading it into a film of uniform thickness!

Draining Down an Inclined Plane

In this very similar example, we suppose that x increases down the fall line of a stationary plane, inclined at an angle α to the horizon. We specify an initial distribution of film thickness, $H(x, 0) = F(x)$, and assume that H at all times depends only on x and t. We also assume that the only stress acting on the free surface is a uniform atmospheric pressure. Then Eq. (11.25) gives $Q = (g \sin \alpha / 3v)H^3$, and the differential equation for $H(x, t)$ is

$$\frac{\partial H}{\partial t} + \frac{g \sin \alpha H^2}{v} \frac{\partial H}{\partial x} = 0 \tag{11.47}$$

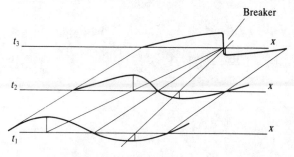

FIGURE 11.9
Evolution of a ripple in a wind-blown oil film.

Again, H will be constant along characteristic curves, and breakers will form in a time $t^* = -(v/g \sin \alpha)/(dF^2/dx)$.

Axisymmetric Pool on a Spinning Disk

To see how easily these problems can become mathematically more difficult, suppose a viscous liquid is supplied through a hole in the center of a horizontal spinning disk. We wish to know the distribution of film thickness as a function of radius r and time. In a frame of reference attached to the disk, a centrifugal force, $\rho\omega^2 r$, will appear to drive the fluid radially outward.

Equations (11.25) and (11.26) may be used again, with the assumption that no tangential stress, and only a uniform pressure, acts on the free surface of the liquid. In cylindrical polar coordinates, we have a radial velocity component $u = (\omega^2 r/2v)z(2H - z)$, so the radial flow rate is $Q = (\omega^2 r/3v)H^3$. For radial flow, $\nabla \cdot \mathbf{Q} = r^{-1} \partial(rQ)/\partial r$, so Eq. (11.26) then gives

$$\frac{\partial H}{\partial t} + \frac{\omega^2 r H^2}{v} \partial H/\partial r = -\frac{2\omega^2 H^3}{3v} \tag{11.48}$$

This equation again specifies the rate of change of H along a characteristic trajectory, but this rate depends strongly on H, and the characteristic trajectories are no longer straight.

Nevertheless, Eq. (11.48) is easily solved. from the analysis of the first example, it should be clear that (11.48) is equivalent to two coupled ordinary differential equations, to be integrated along each characteristic trajectory. These are

$$\frac{dr}{dt} = \frac{\omega^2 r H^2}{v} \tag{11.49}$$

and

$$\frac{dH}{dt} = -\frac{2\omega^2 H^3}{3v} \tag{11.50}$$

If we divide (11.49) by (11.50), we find $d(\ln r) = -(3/2)d(\ln H)$, so

$$\frac{r}{r_0} = \left(\frac{H}{F}\right)^{-3/2} \tag{11.51}$$

Here, F and r_0 are the values of H and r at the starting point of that particular characteristic.

Since (11.50) does not involve r, it can be integrated immediately, to get

$$\frac{H}{F} = \left[1 + \left(\frac{4\omega^2 F^2}{3v}\right)t\right]^{-1/2} \tag{11.52}$$

Combining this with (11.51), we find

$$\frac{r}{r_0} = \left[1 + \left(\frac{4\omega^2 F^2}{3v}\right)t\right]^{3/4} \tag{11.53}$$

With some analysis, which is left as an exercise, it can be shown again that breakers may form, providing that $d(\ln F)d(\ln r) < -2/3$, and that the time of formation is then

$$t^* = -\frac{v}{2\omega^2 F^2}[\tfrac{2}{3} + d(\ln F)d(\ln {}_r)]^{-1} \tag{11.54}$$

Gravity-Driven Spreading of a Circular Pool

Suppose there is initially an axisymmetric pool of viscous liquid on a stationary horizontal plane. The initial distribution of H is $H(r, 0) = F(r)$. We want to know how the pool spreads out and thins[11] under the action of gravity. We assume a uniform atmospheric pressure is the only stress at the free surface.

The driving horizontal force is now a z-independent pressure gradient, associated with the slope of the free surface, $\partial p/\partial r = \rho g \, \partial H/\partial r$. This drives a radial outflow $Q = -(gH^3/3v) \, \partial H/\partial r$. Equation (11.26) then becomes

$$\frac{\partial H}{\partial t} - \frac{gH^3}{3vr}\frac{\partial H}{\partial r} - \frac{gH^2}{v}\left(\frac{\partial H}{\partial r}\right)^2 - \frac{gH^3}{3v}\frac{\partial^2 H}{\partial r^2} = 0 \tag{11.55}$$

This is a much more complicated equation, highly nonlinear and second-order in r. The worst complication resides in the fact that one of the boundary conditions, that $H = 0$ at the outer rim of the pool, has to be applied at an unknown value of r.

There is a possibility for a *similarity solution*, which can be investigated by the method used to analyze the first example in Section 9.4. We assume that the pool has a constant volume, V, and a profile that can always be derived from a single curve, by multiplying H and r by suitable factors that depend only on time. Calling the radius of the pool $R(t)$, we set $H(r, t) = VR(t)^{-2}f(\xi)$, where $\xi = r/R$. Thus, we calculate

$$\frac{\partial H}{\partial t} = V\left[-2R^{-3}R'f + R^{-2}f'\left(-\frac{\xi R'}{R}\right)\right]$$

$$= -VR^{-3}R'(2f + \xi f')$$

$$\frac{\partial H}{\partial r} = VR^{-3}f' \quad \text{and} \quad \frac{\partial^2 H}{\partial r^2} = VR^{-4}f''$$

In these expressions, f' denotes $df/d\xi$; R' denotes dR/dt. Substitution of these expressions into Eq. (11.55) yields

$$R'(2f + \xi f') + \frac{gV^3}{3vR^7}[\xi^{-1}f^3f' + 3f^2(f')^2 + f^3f''] = 0$$

---

[11] Note that this not quite the problem of the spreading of oil on water, in which the motion of the water is coupled to that of the oil.

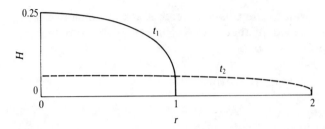

FIGURE 11.10
Circular pool of viscous liquid spreading on a stationary horizontal surface.

We rewrite this in the form

$$\frac{3\nu}{gV^3} R^7 R' = -\frac{\xi^{-1}f^3f' + 3f^2(f')^2 + f^3f''}{2f + \xi f'} = \lambda$$

The first term is independent of r, but is equal to the second term, which is independent of t. Thus both terms must equal the same constant, which we call λ. The self-similar profile must satisfy the ordinary differential equation

$$\xi f^3 f'' + 3\xi f^2(f')^2 + f^3 f' + \lambda(\xi^2 f' + 2\xi f) = 0 \qquad (11.56)$$

The boundary conditions for this equation are now imposed at known points, and are

$$f'(0) = 0 \qquad \text{and} \qquad f(1) = 0$$

By grouping terms and recognizing exact differentials, one discovers that the desired solution has the simple form $f(\xi) = [(3\lambda/2)(1 - \xi^2)]^{1/3}$. The value of λ is determined by the equation

$$V = 2\pi \int_0^R Hr \, dr = 2\pi V \left(\frac{3\lambda}{2}\right)^{1/3} \int_0^1 (1 - \xi^2)^{1/3} \xi \, d\xi = \frac{3\pi}{4} V \left(\frac{3\lambda}{2}\right)^{1/3}$$

which gives $\lambda = 128/(81\pi^3)$. The final answers are

$$f(\xi) = \frac{4}{3\pi}(1 - \xi^2)^{1/3} \qquad \text{and} \qquad (R/R_0)^8 = 1 + \frac{16}{9}\frac{H_0^3}{R_0^2}\frac{gt}{\nu} \qquad (11.57)$$

H_0 is the initial maximum pool depth. The profile is shown in Fig. 11.10.

We see that $\partial H/\partial r = -\infty$ at $r = R$, so the theory is not self-consistent there. In real cases, surface tension becomes an important factor near the outer rim. One must remember, too, that this solution is a very special one, and we have not shown that pools with an initial profile different from this one have any tendency to approach this profile as time goes on.

Effects of Surface Tension

Surface tension can affect the motion in a thin viscous film, in two ways.

1. It can impose a tangential pressure gradient as a result of spatial variations of the curvature of the interface. Imagine a simple case in which the bottom of the film is the plane $z = 0$, and the film thickness depends only on x and t. Suppose also that a uniform atmospheric pressure, p_a, acts on the top of the interface. The situation is drawn in Fig. 11.11. In a slender film, the slope of the interface is small, and the curvature is well approximated by $\partial^2 H / \partial x^2$. The pressure in the film just under the interface will equal $p_a - \Sigma \, \partial^2 H / \partial x^2$, where Σ is the surface tension. If Σ is independent of x, and if we neglect gravitationally induced pressure gradients, the pressure-driven velocity will be $u = (\Sigma / 2\mu) \, \partial^3 H / \partial x^3 z(2H - z)$. Equation (11.25) will yield $Q = (\Sigma / 3\mu) H^3 \, \partial^3 H / \partial x^3$, and Eq. (11.26) will yield

$$\frac{\partial H}{\partial t} = -\frac{\Sigma}{3\mu} \frac{\partial (H^3 \, \partial^3 H / \partial x^3)}{\partial x}$$

This equation is of fourth order in x and, in general, is highly nonlinear. To get some idea of its implications, suppose that the initial configuration of the interface exhibits a very small-amplitude sinusoidal variation in x, say $H_0 = H_m + A \sin kx$, where the wave amplitude A and wave number k are restricted so that $A \ll H_m$ and $kA \ll 1$. Then (11.58) may be approximated by the linear equation

$$\frac{\partial H}{\partial t} = -\frac{\Sigma H_m^3}{3\mu} \frac{\partial^4 H}{\partial x^4}$$

This equation has standing-wave solutions of the form

$$H(x, t) = H_m + A \exp(-\sigma t) \sin kx$$

When this is substituted into (11.58), the damping coefficient is found to be

$$\sigma = \Sigma \frac{H_m^3 k^4}{3\mu}$$

This is an interesting result, indiating that a high value of the viscosity leads to a slower damping of the waves. A little investigation shows that this is because the rate of deformation, and hence the rate of viscous dissipation

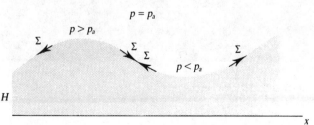

$p = p_a$

$p > p_a$

$p < p_a$

H

x

FIGURE 11.11
Pressure gradient induced by interfacial curvature and tension.

of energy, is inversely proportional to μ in a pressure-driven viscous film flow.

2. Surface tension can vary along the interface, owing to gradients of temperature or of surfactant concentration, and can thereby apply a tangential stress to the liquid below. To account for this in addition to the effects of variable curvature, we must reformulate the pressure gradient, into $\partial p/\partial x = -\partial(\Sigma\,\partial^2 H/\partial x^2)/\partial x$, and we must change the interfacial boundary condition on u from $\partial u/\partial y = 0$ to $\mu\,\partial u/\partial y = \partial\Sigma/\partial x$.[12] Thus, the velocity along the film becomes

$$u = -\frac{1}{2\mu}\frac{\partial(\Sigma\,\partial^2 H/\partial x^2)}{\partial x}z(2H-z) + \frac{1}{\mu}\frac{\partial\Sigma}{\partial x}z$$

the flowrate is

$$Q = -\frac{1}{3\mu}H^3\frac{\partial(\Sigma\,\partial^2 H/\partial x^2)}{\partial x} + \frac{1}{2\mu}\frac{\partial\Sigma}{\partial x}H^2$$

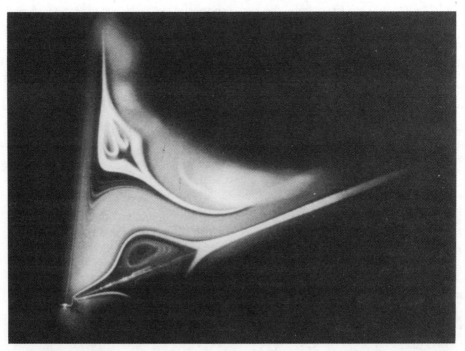

FIGURE 11.12
Flow in a thin film, with important effects of surface tension. (Reproduced, by permission, from P. M. Schweizer (1988)[13].)

---

[12] If you wonder about this, review the discussion leading up to Eq. (4.27).

and Eq. (11.26) becomes

$$\frac{\partial H}{\partial t} = \frac{1}{3\mu} \frac{\partial[H^3 \, \partial(\Sigma \, \partial^2 H/\partial x^2)/\partial x]}{\partial x} - \frac{1}{2\mu} \frac{\partial(H^2 \, \partial\Sigma/\partial x)}{\partial x}$$

To use this equation, one has to have a prescription of $\Sigma(x)$, or equations that will allow its prediction. This gets beyond the intended scope of this text, but the resulting flows are both industrially important and, as shown in Fig. 11.12, occasionally spectacular.[13]

11.6 TWO-DIMENSIONAL CREEPING FLOWS IN CORNERS

Many solutions of Eq. (11.6), the biharmonic equation, are known. To investigate creeping flows in corners, we examine solutions obtained by separation of variables in polar coordinates.[14] These have the form

$$\psi(r, \theta) = r^n \{ A \cos n\theta + B \sin n\theta + C \cos[(n - 2)\theta] + D \sin[(n - 2)\theta] \quad (11.58)$$

for almost any value of n. The exceptional cases are:

For $n = 0$ $\qquad\qquad \psi(r, \theta) = A + B\theta + C\theta^2 + D\theta^3$ $\qquad\qquad\qquad$ (11.59)

For $n = 1$ $\quad \psi(r, \theta) = r(A \cos \theta + B \sin \theta + C\theta \cos \theta + D\theta \sin \theta)$ $\qquad$ (11.60)

For $n = 2$ $\qquad \psi(r, \theta) = r^2(A \cos 2\theta + B \sin 2\theta + C\theta + D)$ $\qquad\qquad$ (11.61)

The velocity is $\mathbf{u} = u(r, \theta)\mathbf{e}_r + v(r, \theta)\mathbf{e}_\theta$, in which

$$u = r^{-1}\frac{\partial \psi}{\partial \theta} \quad \text{and} \quad v = -\frac{\partial \psi}{\partial r} \quad (11.62)$$

These solutions can represent an amazing variety of creeping flows that are bounded by intersecting planes. The planes may be solid, with specified radial or angular motion, or they may be idealizations of free surfaces.

Usually, we do not expect the creeping-flow approximations to be valid for all values of the radial distance r, because the relative magnitudes of viscous and inertial forces may depend on r. From eq. (11.62), it follows that u and v will be proportional to r^{n-1}, the viscous force per unit volume will be

---

[13] For interesting examples, see T. O. Oolman and H. E. Blanch, "Bubble Coalescence in Stagnant Liquids," *Chem. Eng. Commun.* **43**: 237–261 (1986); W. S. Overdiep, "The Levelling of Paints," in *Physicochemical Hydrodynamics*, Vol. 2, ed. B. Spalding, Advance Publications Limited, 1977, pp. 683–697; and P. M. Schweizer, "Visualization of coating flows," *Jour. Fluid Mech.* **193**: 285–302 (1988).

[14] The development of this section owes much to the original work and lucid account given by H. K. Moffatt, "Viscous and resistive eddies near a sharp corner," *Jour. Fluid Mech.* **18**: 1–18 (1964).

proportional to r^{n-3}, and the inertial force per unit volume will be proportional to r^{2n-3}. Hence the ratio of inertial to viscous force will be proportional to r^n. In some very interesting cases, n may be a complex number; then the real part of n is used in this discussion. In any event, when $n > 0$, the creeping-flow approximations will be valid for small values of r; when $n < 0$, they will be valid for large values of r. When $n = 0$, the approximations will be either good or bad for all values of r, depending on the value of $\mathrm{Re} = \psi_{max}/\nu$.

Flow in a Wedge with Line Source or Sink at the Apex

This very simple example makes a good transition between the study of lubricating films, and the analysis of the more complex flows that are the principal subject of this section. It is the limit, for $\mathrm{Re} \ll 1$, of an exact solution of the full Navier–Stokes equations. The limiting solution for $\mathrm{Re} \gg 1$ will be seen in Chapter 12.

Let the flow be bounded by stationary solid planes at $\theta = 0$ and $\theta = \alpha$. The source or sink of fluid, with volumetric flowrate per unit length equal to Q, lies at $r = 0$. For dimensional homogeneity, ψ must be proportional to Q, and since Q is independent of r, we deal with the case $n = 0$. The boundary conditions are:

$$\psi(0) = 0 \qquad \psi(\alpha) = Q \qquad u(r, 0) = 0 \qquad u(r, \alpha) = 0$$

Applying these conditions to the solution given by Eq. (11.39), we find

$$A = 0 \qquad B = 0 \qquad C = 3Q\alpha^{-2} \qquad D = -2Q^{-3}$$

so

$$\psi(\theta) = Q\left[3\left(\frac{\theta}{\alpha}\right)^2 - 2\left(\frac{\theta}{\alpha}\right)^3\right] \tag{11.63}$$

and

$$u(r, \theta) = 6\left(\frac{Q}{r\alpha}\right)\left[\left(\frac{\theta}{\alpha}\right) - \left(\frac{\theta}{\alpha}\right)^2\right] \tag{11.64}$$

This is the same parabolic velocity profile we found for the pressure-driven flow in a lubricating film. The viscous force per unit volume is $-12\mu(Q/r^3)\alpha^{-2}$, independent of θ. The maximum inertial force at a given radius occurs at $\theta = \alpha/2$, and equals $-(9/4)\rho(Q^2/r^3)$. Thus the ratio that must be small in order for the creeping-flow approximation to be accurate is $\mathrm{Re} = (3/16)(Q/\nu)\alpha^2$.

Effect of a Squeegee or a Submerging Belt

Suppose that one of two intersecting planes is stationary, or free of viscous stress, while the second plane slides with constant speed U toward or away from the line of intersection. The two situations of interest are shown in Fig. 11.13. Dimensional homogeneity now requires that ψ be proportional to Ur, so that the relevant exponent is $n = 1$.

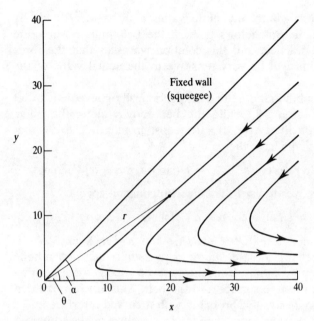

FIGURE 11.13
Streamlines under a squeezee.

For the window-wiping case, the boundary conditions are

$$\psi(r, 0) = 0 \qquad \psi(r, \alpha) = 0 \qquad u(r, 0) = U \qquad u(r, \alpha) = 0$$

When these are applied to the solution displayed in Eq. (11.60), we find

$$A = 0 \qquad \text{and} \qquad \begin{bmatrix} B \\ C \\ D \end{bmatrix} = \frac{U}{\alpha^2 - \sin^2 \alpha} \begin{bmatrix} \alpha^2 \\ -\sin^2 \alpha \\ \sin \alpha \cos \alpha - \alpha \end{bmatrix}$$

A few streamlines are sketched in Fig. 11.13. They look exactly the same, with just a reversal of flow direction, if the direction of motion of the sliding plane is reversed.[15]

It is interesting to calculate the viscous stresses that are exerted on the two walls. First we note that the fluid velocity is independent of r, so that the viscous stresses will be proportional to $1/r$. For example,

$$\tau_{\theta r}(r, 0) = \mu \left[r \frac{\partial(v/r)}{\partial r} + r^{-1} \frac{\partial u}{\partial \theta} \right]_{\theta=0} = 2\mu \frac{D}{r} \tag{11.65}$$

---

[15] Note that this is not thermodynamic reversibiity. Creeping flows are thermodynamically as irreversible as can be imagined. All of the work done to drive a creeping flow is dissipated by viscosity, rather than going into kinetic or gravitational potential energy.

The integral of this, from $r = 0$ to any finite value of r, is logarithmically infinite. As is pointed out in Batchelor's book,[16] the two walls never quite come together at $r = 0$ in practice, and this solution warns us that the force required to drive the motion will be very sensitive to the actual value of the clearance between them.

For the case of the submerging belt, it is algebraically convenient to let the stress-free surface be at $\theta = 0$, while the belt moves along the plane $\theta = -\alpha$. Then the condition that $\tau_{\theta r}(r, 0) = 0$ means, from Eq. (11.65), that $D = 0$. The nonzero coefficients have the values

$$B = -U(\sin \alpha \cos \alpha - \alpha)^{-1}\alpha \cos \alpha \qquad C = U(\sin \alpha \cos \alpha - \alpha)^{-1} \sin \alpha.$$

Fluid on the free surface moves toward the belt with constant speed

$$u(r, 0) = B + C = U(\sin \alpha \cos \alpha - \alpha)^{-1}(\sin \alpha - \alpha \cos \alpha) \qquad (11.66)$$

This varies from $-U/2$ when $\alpha \to 0$, to $-2U/\pi$ when $\alpha = \pi/2$, and to $-U$ when $\alpha = \pi$. In all but the last case, there is a requirement of infinite acceleration, and an even more infinite concentration of viscous force per unit volume, at the junction of the free surface and the belt. This again warns that the real situation must deviate locally, probably with a curved interface and a significant effect of surface tension, from the idealized picture analyzed here.

Flow in a Wedge with Swinging Walls

Imagine two plane walls, hinged at their line of intersection, and swinging so that the angle between them increases at a rate $d\alpha/dt = \omega$. Dimensional homogeneity now requires that ψ is proportional to ωr^2, so the relevant value of n is 2, and we work with Eq. (11.61). Let $\theta = 0$ denote the stationary bisecting plane, as shown in Fig. 11.14. The boundary conditions are $u(r, \pm\alpha/2) = 0$, $v(r, \pm\alpha/2) = \pm r\omega/2$. The resulting values of the coefficients are:

$$A = D = 0$$

$$B = -\frac{\omega}{4}(\sin \alpha - \alpha \cos \alpha)^{-1} \qquad C = \frac{\omega}{2}\cos \alpha(\sin \alpha - \alpha \cos \alpha)^{-1}$$

Streamlines are shown in Fig. 11.14.

Flow in a Wedge Driven by Circumferential Motion at Large r

The last, and most fascinating, of these examples involves flow in a wedge, driven by the motion of, say, a rotating cylinder, as shown in Fig. 11.15. The

---

[16] G. K. Batchelor (1967) *op. cit.*, p. 226.

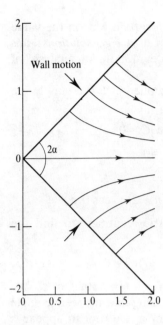

FIGURE 11.14
Creeping flow between swinging walls.

analysis does not directly account for the presence of the driving cylinder, but simply inquires as to the possible existence of solutions of the general form (11.38), under the homogeneous boundary conditions $u(r, -a) = 0$, $v(r, -\alpha) = 0$, $u(r, \alpha) = 0$, $v(r, \alpha) = 0$.

Moffatt was particularly interested in solutions in which ψ is an even function of θ, so that the radial inflow or outflow would be antisymmetric about the bisector. We get this from Eq. (11.59) by setting B and D equal to zero. The boundary conditions then need be imposed only at one wall, say $\theta = \alpha$. The result is a pair of homogeneous linear algebraic equations for A and C. Nonzero solutions exist only when the determinant of this set of equations vanishes, so that the equations are no longer linearly independent.

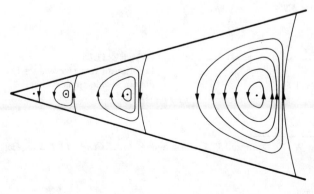

FIGURE 11.15
Corner eddies in creeping flow.

This happens only when the exponent n depends in a certain way on the value of the wedge angle α. The resulting solutions are called *eigensolutions*; this mathematical concept will be met again in Chapter 13, in an analysis of hydrodynamic stability.

In the present case, the boundary conditions applied at $\theta = \alpha$ supply the equations:

$$n \sin(n\alpha) A + (n-2) \sin[(n-2)\alpha]C = 0$$

and

$$\cos(n\alpha) A + \cos[(n-2)\alpha]C = 0$$

The determinant of the coefficients is then

$$\mathbf{D} = n \sin(n\alpha) \cos[(n-2)\alpha] - (n-2) \sin[(n-2)\alpha] \cos(n\alpha)$$

We equate this to zero, and use some trigonometric identities, to find the *eigenvalue equation*:

$$F(n, \alpha) = (n-1) \sin(2\alpha) + \sin[2(n-1)\alpha] = 0 \qquad (11.67)$$

Figure 11.16 shows $F(n, \alpha)$ as a function of $n - 1$, for a few values of α. We see that F will vanish for some real values of n, providing that the wedge angle 2α is sufficiently large. The minimum value of 2α for a real root to appear is 146.3°, corresponding to which $n = 1.5294$. As the value of α increases, the number of roots also increases.[17] If $2\alpha < 146.3°$, n must be a complex number, with fascinating consequences for the qualitative nature of the solution.

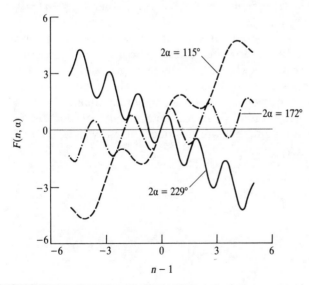

FIGURE 11.16
Location of real eigenvalues for flow in a wide-angle corner.

---

[17] The original analysis is due to W. R. Dean and P. E. Montagnon, *Proc. Camb. Phil. Soc.* **45**: 389–394 (1949).

To find the complex roots, we write $n - 1 = p + iq$, and use the identity

$$\sin[(p + iq)\alpha] = \sin(p\alpha)\cosh(q\alpha) + i\cos(p\alpha)\sinh(q\alpha)$$

Then Eq. (11.67) has real and imaginary parts

$$p\sin\alpha = \pm\sin(p\alpha)\cosh(q\alpha) \qquad q\sin\alpha = \pm\cos(p\alpha)\sinh(q\alpha)$$

Following Moffatt, we let $2p\alpha = \xi$, $2q\alpha = \eta$, and $(2\alpha)^{-1}\sin 2\alpha = k$, so that the working equations are

$$k\xi + \sin\xi\cosh\eta = 0 \qquad k\eta + \cos\xi\sinh\eta = 0 \tag{A}$$

A simple strategy for solving these is the following. Put them in the form

$$\left(\frac{k\xi}{\sin\xi}\right)^2 = \cosh^2\eta \qquad \left(\frac{k\eta}{\cos\xi}\right)^2 = \sinh^2\eta \tag{B}$$

Subtract the second from the first and rearrange, to get

$$\eta^2 = \cos^2\xi\left[\left(\frac{\xi}{\sin\xi}\right)^2 - k^{-2}\right] \tag{C}$$

Equation (C) can be used to eliminate η in Eq. (B), which can then be written as

$$F(\xi, k) = \frac{k\xi}{\sin\xi} + \cosh\left\{\cos\xi\left[\left(\frac{\xi}{\sin\xi}\right)^2 - k^{-2}\right]^{1/2}\right\} = 0$$

The function $F(\xi, k)$ is plotted versus ξ for $k = 2/\pi$ ($\alpha = 45°$) in Fig. 11.17. The first three roots that satisfy both of Eqs. (A) are indicated by the dots. The smallest root lies at $\xi = 4.303343$, $\eta = 1.757760$; the next at $\xi = 10.752313$, $\eta = 2.641506$, etc. The other roots are artifacts of the squaring performed in Eq. (B); the corresponding eigenfunctions do not satisfy the no-slip condition.

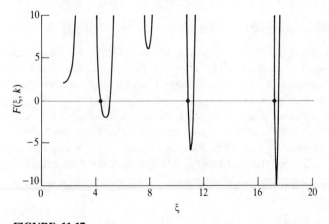

FIGURE 11.17
Location of real part of complex eigenvalues for flow in a 90° corner, $F(\xi, k)$ vs. ξ.

The fact that n can have complex values implies that the variation of the streamfunction with r can oscillate, rather than exhibiting a monotonic growth or decay. To see this, we write $r^n = r^{1+p+iq}$, and use the identity $r = \exp(\ln r)$. Then

$$r^n = \exp[(1+p+iq)\ln r] = \exp[(1+p)\ln r][\cos(q\ln r) + i\sin(q\ln r)]$$

For the angular variation of ψ, we must use expressions such as

$$\cos n\theta = \cos[(1+p+iq)\theta]$$
$$= \cos[(1+p)\theta]\cosh(q\theta) - i\sin[(1+p)\theta]\sinh(q\theta)$$

As in our previous use of complex numbers, we assume that the real part of quantities such as $r^n \cos n\theta$ is the physically relevant part. Thus, the final formula for the streamfunction can be written as

$$\psi(r, \theta) = A'\left(\frac{r}{r_0}\right)^{p+1}\left\{\cos\left[q\ln\left(\frac{r}{r_0}\right)\right]\mathscr{R}\,g(\theta) - \sin\left[q\ln\left(\frac{r}{r_0}\right)\right]\mathscr{I}\,g(\theta)\right\} \quad (11.68)$$

where r_0 is a reference value of r, and A' is a real-valued amplitude factor. For a general (real or complex) value of the original exponent n we can write

$$g(\theta) = \cos[(n-2)\alpha]\cos(n\theta) - \cos(n\alpha)\cos[(n-2)\theta] \quad (11.69)$$

In the present case, with $n = 1 + p + iq$, a decomposition into real and imaginary parts gives

$$\mathscr{R}\,g(\theta) = A(\alpha)W(\theta) - B(\alpha)X(\theta) - C(\alpha)Y(\theta) + D(\alpha)Z(\theta)$$
$$\mathscr{I}\,g(\theta) = -A(\alpha)X(\theta) - B(\alpha)W(\theta) + C(\alpha)Z(\theta) + D(\alpha)Y(\theta)$$

where $A = \cos[(p-1)\alpha]\cosh(q\alpha)$ $B = \sin[(p-1)\alpha]\sinh(q\alpha)$

 $C = \cos[(p+1)\alpha]\cosh(q\alpha)$ $D = \sin[(p+1)\alpha]\sinh(q\alpha)$

and $W = \cos[(p+1)\theta]\cosh(q\theta)$ $X = \sin[(p+1)\theta]\sinh(q\theta)$

 $Y = \cos[(p+1)\theta]\cosh(q\theta)$ $Z = \sin[(p-1)\theta]\sinh(q\theta)$

The real and imaginary parts of $g(\theta)$ are shown in Fig. 11.18(a), for $\alpha = 45°$, and for the lowest eigenvalue. The corresponding eigenfunctions for the second even mode are shown in Fig. 11.18(b). They describe a less likely flow with three eddies in each radial cell.

Two important features of these solutions should still be noticed. The first is the ratio of successive values of r, for which the oscillatory factor in r^n goes from $+1$ to -1 or vice versa. This has the value $r_{j+1}/r_j = \exp(p/q)$, showing that the radial extent of the eddies grows in geometrical proportion as r increases. Meanwhile, the exponential factor in r^{n-1}, to which the velocity is proportional, increases according to the formula

$$\frac{\exp(p\ln r_{j+1})}{\exp(p\ln r_j)} = \exp\left(\pi\frac{p}{q}\right)$$

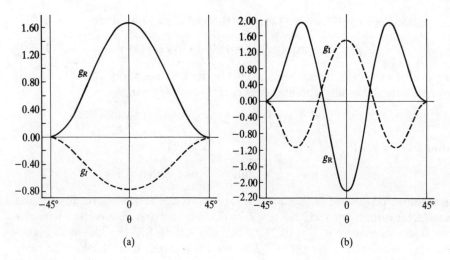

FIGURE 11.18
Even eigenfunctions for a Moffat eddy in a 90° corner: (a) lowest order; (b) second order.

For the lowest eigenvalues found for $\alpha = 90°$ ($p = 2.739594$, $q = 1.119026$), this factor equals 2,189; so the velocities in eddies described by that eigenfunction would die off drastically from the outer edge of an eddy to the inner edge. For the next even-function eigensolution, the factor is 30,291; for still higher-order solutions, the decay rates are even greater.

A special case of great interest is presented when $\alpha \rightarrow 0$. Imagine that this happens while two points on opposite walls, a distance $2H$ apart, are held fixed, the apex of the angle moving off to $x = -\infty$. The cause of the fluid motion may be supposed to be a moving belt at $x = 0$, as sketched in Fig. 11.19.

The streamfunction can be calculated from the limit of Eq. (11.68), or by a fresh analysis of the problem, using cartesian coordinates. For the latter

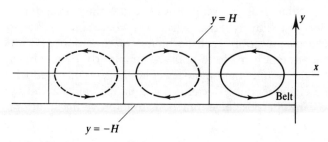

FIGURE 11.19
Moffatt eddies in a rectangle.

purpose, we take $y = 0$ halfway between the walls, and note that

$$\psi(x, y) = \exp(kx)[A \cos(ky) + B \sin(ky)] \qquad (11.70)$$

is an even function of y, satisfying the biharmonic equation $\nabla^4 \psi = 0$. It will also satisfy the boundary conditions $u(x, H) = v(x, H) = 0$, if

$$\sin(2kH) + 2kH = 0$$

Letting $2kH = \xi + i\eta$, we find

$$\sin \xi \cosh \eta = -\xi \quad \text{and} \quad \cos \xi \sinh \eta = -\eta$$

The solution procedure is as shown above, with the following results for the lowest eigenvalue: $\xi = 4.21239$, $\eta = 2.25073$. The second eigenvalue, with the same flow symmetry, is $\xi = 10.7125374$, $\eta = 3.1031487$. For the first eigenvalue, the decay rate $\exp(\pi\xi/\eta) = 358$; for the second, it is $51,290$. The eddy length, L, defined as the interval in ξ between successive appearances of $\psi = 0$, is given by $L/2H = \pi/\eta$. This has the value 1.396 for the first eigenvalue; 1.012 for the second.

This case has been investigated numerically and experimentally by Pan and Acrivos,[18] who solved the creeping-flow equations for flow in a rectangular cavity with a stationary end at $x = -10H$, and a moving end, with $v = U$, at $x = 0$. Their solution exhibited four large eddies, the middle two of which are quite well described by Eq. (11.70), and tiny corner eddies in the 90° corners of the closed end. They made a detailed numerical study of the corner eddies in a square cavity, and found extraordinary agreement with the properties of Moffatt's lowest eigenfunction for that case, too. For example, their computed values of the attenuation factor were 2110 and 2170, against the analytical value 2189. Presumably, their flowfields would closely resemble the first, window-wiper, example of this series very near the corners where the moving wall passes by a stationary one.

Summary

Solutions of the type exemplified by (a) to (d) above are not complete solutions to any realistic problem of creeping flow, because they prescribe a certain behavior of the flow at relatively large values of r, where some different behavior may be imposed by physically realistic boundary conditions. If, for example, an experimental attempt is made to realize example (d), by moving a wall across the opening of the wedge, the boundary conditions on that wall will not be met by any single eigenfunction. Because the equations and boundary conditions are linear, we can superpose eigenfunctions to improve the solution,

---

[18] F. Pan and A. Acrivos, "Steady Flows in Rectangular Cavities," *Jour. Fluid Mech.* **28**: 643–655 (1967).

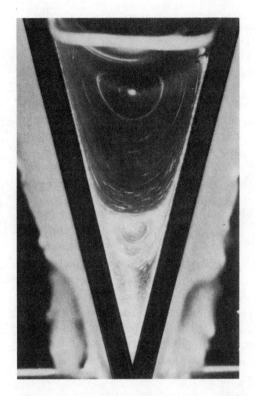

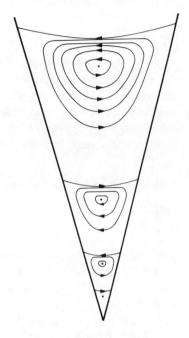

FIGURE 11.20(a)
Moffat's corner eddies: experimental (reproduced, by permission, from Taneda, 1979)[2].

FIGURE 11.20(b)
Moffat's corner eddies: computed
from Eq. (11.68).

but this would not be an easy task. On the other hand, the analysis shows that
variations in the boundary conditions at, say, $r = R$, will have little effect on
the flow where $r \ll R$. This happens because the higher-order eigenfunctions
decay much more rapidly than the lowest order one as r decreases. In fact,
experimental and computational demonstrations of the existence of the stacked
eddies of this example have become very common,[19] and it must be concluded
that they are an almost universal feature of laminar flow in corners. A sample
experimental demonstration is shown in Fig. 11.20.

11.7 EXTERNAL CREEPING FLOWS

Let us now consider the motion set up in an infinite body of fluid, when a finite
body moves steadily through that fluid. It is assumed that Re $\ll 1$, where Re is

---

[19] For an excellent, up-to-date review, see H. Hasimoto and O. Sano, "Stokeslets and Eddies in
Creeping Flow," *Ann. Rev. Fluid Mech.*, **12**: 335–363 (1980).

a Reynolds number based on the speed and characteristic length of the body. The body might be a tiny droplet of rain or fuel, a bit of dust or chaff, or a microscopic plant or animal.

In many applications, more than a single moving body is involved, or the fluid region is somewhere bounded by a wall. Since, as will be shown, the velocity perturbation created by a body dies off with only the first inverse power of distance when Re ≪ 1, problems involving the effect of one body on the flow around a second body are very important. It is also sometimes important to understand when and how the past motion of the body, if unsteady, affects the present flowfield. Only a very limited treatment of these complications is appropriate for this text; you should be mostly concerned to develop a thorough understanding of the simplest cases.

Stokes Flow Over a Sphere

For this classic and most useful special case, we assume that the sphere is not spinning, realizing that if it is, the flow due to spin can be calculated separately, and superposed on the flow due to translation. Spherical polar coordinates are used, in a frame of reference attached to the sphere. As in the study of axisymmetric flow produced by a point force, which was presented in great detail in Chapter 9, almost all steps of analysis will be shown. This will serve as further illustration of techniques you should master, and of results with which you should become familiar.

For axisymmetric, steady, non-spinning flow, the velocity is

$$\mathbf{u} = u(R, \theta)\mathbf{e}_R + v(R, \theta)\mathbf{e}_\theta \tag{11.71}$$

The vorticity has a single, azimuthal, component

$$\mathbf{\Omega} = \Omega(R, \theta)\mathbf{e}_\phi \quad \text{where} \quad \Omega = R^{-1}\left(\frac{\partial(vR)}{\partial R} - \frac{\partial u}{\partial \theta}\right) \tag{11.72}$$

The viscous force per unit volume is given by

$$\text{div } \tau = -\mu\nabla \times \mathbf{\Omega} = \frac{\mu}{R \sin \theta}\left(\frac{\partial(\Omega \sin \theta)}{\partial \theta}\right)\mathbf{e}_R + \frac{\mu}{R}\left(\frac{\partial(\Omega R)}{\partial R}\right)\mathbf{e}_\theta \tag{11.73}$$

Again, we introduce the shorthand $\Pi = p/\rho + gh$, where h is elevation. One can find $\Pi(R, \theta)$ by integrating the R-component of the momentum equation from a point at infinity, where Π is presumably constant and known, inward toward any finite value of R. The form of that equation is

$$\frac{\partial\Pi}{\partial R} = -\frac{v}{R \sin \theta}\frac{\partial(\Omega \sin \theta)}{\partial \theta} \tag{11.74}$$

The vorticity equation for creeping flow, Eq. (11.5), becomes

$$\frac{\partial^2(\Omega R \sin \theta)}{\partial R^2} + \frac{\sin \theta}{R^2}\frac{\partial}{\partial \theta}\left[\frac{1}{\sin \theta}\frac{\partial}{\partial \theta}(\Omega R \sin \theta)\right] = 0 \tag{11.75}$$

The velocity components can be derived from a streamfunction, $\psi(R, \theta)$, by the formulas

$$u = (R^2 \sin \theta)^{-1} \frac{\partial \psi}{\partial \theta} \quad \text{and} \quad v = -(R \sin \theta)^{-1} \frac{\partial \psi}{\partial R} \qquad (11.76)$$

The corresponding equation for the vorticity is

$$\Omega R \sin \theta = -\left\{ \frac{\partial^2 \psi}{\partial R^2} + \frac{\sin \theta}{R^2} \frac{\partial}{\partial \theta} \left[\frac{1}{\sin \theta} \frac{\partial \psi}{\partial \theta} \right] \right\}. \qquad (11.77)$$

Note that the same differential operator appears in (11.75) and (11.77).

Two boundary conditions on ψ can be stated immediately. The flow past the sphere should become a uniform stream, with velocity $\mathbf{u} = U\mathbf{e}_z$, as $R \rightarrow \infty$. The corresponding behavior of the streamfunction is

$$\psi \rightarrow \frac{UR^2}{2} \sin^2 \theta \quad \text{as} \quad R \rightarrow \infty \qquad (11.78)$$

The polar axis and the surface of the sphere can be assigned the value $\psi = 0$, i.e.,

$$\psi = 0 \quad \text{if} \quad \theta = 0 \text{ or } \pi \quad \text{for any value of } R$$

$$\psi = 0 \quad \text{if} \quad R = a \quad 0 \le \theta \le \pi$$

These conditions suggest the possibility of a simple solution in the form

$$\psi = f(R) \sin^2 \theta \qquad (11.79)$$

with $f(R)$ to be determined by the vorticity equation and the remaining boundary conditions. When (11.79) is substituted into (11.77), and the result is substituted into (11.75), one obtains the ordinary differential equation

$$\left(\frac{d^2}{dR^2} - \frac{2}{R^2} \right)^2 f(R) = 0 \qquad (11.80)$$

This obviously admits solutions of the form $f = R^n$, because

$$\left(\frac{d^2}{dR^2} - \frac{2}{R^2} \right)^2 R^n = [(n-2)(n-3) - 2][n(n-1) - 2]R^{n-4}$$

Hence (11.80) is satisfied for any of the exponents $n = -1, 1, 2,$ and 4, and one may write

$$f(R) = AR^{-1} + BR + CR^2 + DR^4 \qquad (11.81)$$

Equations (11.79) and (11.81) were derived to describe the flow outside the sphere, but they also work for $R < a$, providing that the spherical body is composed of a second Newtonian fluid. Only the values of A, B, C, and D will distinguish the solutions for the inner and outer flows. Before the boundary conditions are applied, let us gather some results that will be useful in further analysis. The velocity components are

$$u = 2 \cos \theta \, \{AR^{-3} + BR^{-1} + C + DR^2\} \tag{11.82}$$

and
$$v = \sin \theta \, \{AR^{-3} - BR^{-1} - 2C - 4DR^2\} \tag{11.83}$$

From Eq. (11.72), the vorticity is

$$\Omega = 2 \sin \theta \, (BR^{-2} - 5DR) \tag{11.84}$$

Integration of (11.74) gives

$$\Pi(R, \theta) = \Pi(R_0, \theta) + 2v \cos \theta \, [B(R^{-2} - R_0^{-2}) + 10D(R - R_0)] \tag{11.85}$$

The components of viscous stress on a spherical surface, $R = $ constant, are

$$\tau_{RR} = 2\mu \frac{\partial u}{\partial R} = -4\mu \cos \theta \, (3AR^{-4} + BR^{-2} - 2DR) \tag{11.86}$$

$$\tau_{R\theta} = \mu \left[R \frac{\partial}{\partial R} \left(\frac{v}{R} \right) + \frac{1}{R} \frac{\partial u}{\partial \theta} \right] = -6\mu \sin \theta \, (AR^{-4} + DR) \tag{11.87}$$

We now fix the values of the coefficients, for flow of one fluid, over a sphere of a second fluid. Results for a solid sphere may then be obtained by setting the viscosity of the inner fluid equal to infinity. To get the desired flow far from the sphere, one must set $C_0 = U/2$, and $D_0 = 0$. Note that this guarantees that Π approaches a constant value as $R \to \infty$. We call this constant Π_∞.

Since the velocity must be finite at the center of the inner sphere, we require that $A_i = B_i = 0$.

Since the surface of the sphere is a streamsurface, with $\psi(a, \theta) = 0$, we find that

$$A_0 + B_0 a^2 + \frac{U}{2} a^3 = 0 \quad \text{and} \quad C_i + D_i a^2 = 0$$

Of course, one gets the same results by setting $u = 0$ at $R = a$.

Continuity of tangential velocity at the interface between fluids requires that

$$v_0(a, \theta) = v_i(a, \theta) \quad \text{so that} \quad A_0 - B_0 a^2 - Ua^3 = -2C_i a^3 - 4D_i a^5$$

Surface tension is the agency that maintains the spherical form of the interface. If it is constant, there must be continuity of tangential stress at the interface. From

$$\tau_{R\theta o}(a, \theta) = \tau_{R\theta i}(a, \theta)$$

we find that

$$mA_o = D_i a^5 \quad \text{where} \quad m = \frac{\mu_o}{\mu_i}$$

There are now enough equations to determine values of all the coefficients. The nonzero values are:

$$A_o = \frac{Ua^3}{4} \frac{1}{1+m}, \qquad B_o = -\frac{Ua}{4} \frac{3+2m}{1+m},$$

$$C_i = -\frac{U}{4} \frac{m}{1+m}, \qquad D_i = \frac{U}{4a^2} \frac{m}{1+m},$$

The local stress that the outer fluid exerts on the inner fluid is

$$\sigma(a, \theta) = [-p_o(a, \theta) + \tau_{RRo}(a, \theta)]\mathbf{e}_R + \tau_{R\theta}(a, \theta)\mathbf{e}_\theta \qquad (11.88)$$

in which

$$p_0(a, \theta) = \rho_o \Pi_\infty - \rho_o g h(a, \theta) - \frac{\mu_o U}{2a} \frac{3+2m}{1+m} \cos \theta \qquad (11.89)$$

$$\tau_{RRo}(a, \theta) = \frac{2\mu_o U}{a} \frac{m}{1+m} \cos \theta \qquad (11.90)$$

$$\tau_{R\theta o}(a, \theta) = -\frac{3\mu_o U}{2a} \frac{1}{1+m} \sin \theta \qquad (11.91)$$

Substituting (11.89) to (11.91) into (11.88), and recognizing that $\cos \theta \, \mathbf{e}_R - \sin \theta \, \mathbf{e}_\theta = \mathbf{e}_z$, we can rewrite (11.88) in an interesting form. This is

$$\sigma(a, \theta) = \left\{ -\rho_o \Pi_\infty + \rho_o g h(a, \theta) + \frac{3\mu_o U}{a} \frac{m}{1+m} \cos \theta \right\} \mathbf{e}_R$$

$$+ \frac{3\mu_o U}{2a} \frac{1}{1+m} \mathbf{e}_z \qquad (11.92)$$

When this is multiplied by $dA = R^2 \sin \theta \, d\theta \, d\phi$, and integrated from $\phi = 0$ to 2π, and from $\theta = 0$ to π, we find that the net force exerted on the sphere by the outer fluid is

$$\mathbf{F} = \mathbf{B} + \pi \mu_o U a (6 + 4m)(1 + m)^{-1} \mathbf{e}_z \qquad (11.93)$$

In this result, **B** is the buoyancy, i.e., minus the weight of the outer fluid displaced by the sphere. The rest is the drag force acting on the sphere. To get

Stokes' famous result for the solid sphere,[20] we simply set $m = 0$. Then

$$\text{Drag}_{(m=0)} = 6\pi\mu_o Ua \qquad (11.94)$$

This would also be quite a good approximation for water droplets in air, for which the value of m is about $1/70$, making the constant 5.97π instead of 6π. For air bubbles in water, on the other hand, $m \approx 70$ and the constant drops to 4.03π. Experimental confirmation of (11.94) is excellent, provided that $Ua/v \leq 0.1$. On the other hand, experimental data for the drag of bubbles in liquids often gives values higher that one would deduce from (11.93). It is suspected that surfactant molecules diffuse to the interface, and are concentrated near the rear stagnation point by convection along the interface.[21] There they reduce the surface tension, as discussed in Section 6.7, and the resulting gradient of surface tension acts to immobilize, the interface partially or completely. The result is to make the bubble drag approach the value for a solid sphere. Because it is very difficult to control these surfactant effects, drag data for bubbles often exhibit a surprising amount of scatter.[22]

OBSERVATIONS AND CRITICISMS OF THE CREEPING-FLOW SOLUTION.
A number of features of the sphere-flow solution are of intrinsic interest, or help one to frame a logical criticism of the assumptions.

(a) Normal viscous stress at the interface.[23] Equation (11.90) shows that there is a nonvanishing value of $\tau_{RR_o}(a, \theta)$, provided that the interface is free to move, but that this component of viscous stress vanishes when $m = 0$. This result can be elucidated by combining the equation $\tau_{RR} = 2\mu \, \partial u/\partial R$ with the continuity equation in the form

$$\frac{\partial u}{\partial R} = -R^{-1}\left(2u + \csc\theta \frac{\partial(v \sin\theta)}{\partial\theta}\right)$$

This shows that at the interface, where $u = 0$, $\tau_{RR} = -2\mu(R \sin\theta)^{-1} \partial(v \sin\theta)/\partial\theta$. For a solid sphere, v would be zero at the interface, and so would τ_{RR}.

---

[20] The formula was originally derived by G. G. Stokes in 1851; the result for a fluid inner sphere was found independently in 1911 by Hadamard and Rybczynski. Equation (11.94), with a correction for rarefied-gas effects, was central to Robert Millikan's oil-drop experiment for measurement of the charge of an electron.

[21] Although convection of vorticity is negligible near a sphere in creeping flow, when compared with diffusion of the same quantity, the same result does not hold true for convection and diffusion of surfactant molecules, which tend to be large and to have diffusion coefficients much smaller than kinematic viscosity of the liquid through which they diffuse.

[22] An extensive discussion of this, with many references to the research literature, is given by R. Clift, J. R. Grace, and M. E. Weber, in *Bubbles, Drops, and Particles*, Academic Press, New York, 1978.

[23] This discussion is closely related to Exercise 5.4.

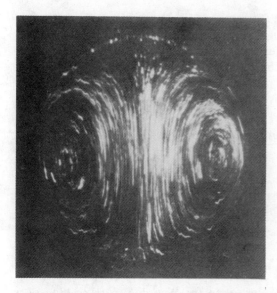

FIGURE 11.21(a)
Recirculating flow inside a droplet: clean surface. (Reproduced, by permission, from P. Savic, NRC (Canada) Rep. No. MT-22, 1953.)

(b) Maximum speed of the circulating flow inside a bubble. The maximum speed of the interface is easily shown to equal $(U/2)m(1+m)^{-1}$, which is nearly half the translational speed of an air bubble in water. Thus, if noncondensable light-scattering particles are in the air, it is possible to see a relatively vigorous internal circulation. Sample photographs are shown in Fig. 11.21.

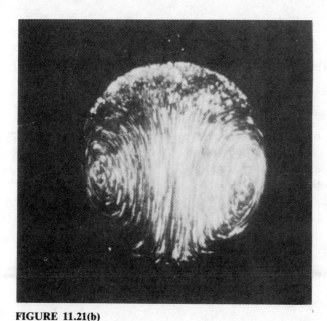

FIGURE 11.21(b)
Recirculating flow inside a droplet: contaminated surface. (Reproduced, by permission, from P. Savic, NRC (Canada) Rep. No. MT-22, 1953.)

(c) Decay and diffusion of vorticity as $R \to \infty$. Convection of vorticity is neglected in an analysis of creeping flow, but it is easily seen that convection is, in fact, more important than diffusion at sufficiently large distances from the sphere. Consider, for example, the rate of streamwise diffusion of vorticity. This is $\mathbf{e}_z \cdot (-\nu \, \nabla \Omega)$, which equals $3(\nu_0 \Omega / R) \cos \theta$. The rate of streamwise convection far from the sphere is simply $U\Omega$, so the ratio of convection to diffusion is

$$\frac{\text{Convection}}{\text{Diffusion}} = \frac{UR}{3\nu} \sec \theta \tag{11.95}$$

For any fixed value of the Reynolds number, Ua/ν, this ratio is large provided that $R \gg a$. This internal inconsistency of creeping-flow theory was noted in 1910 by C. W. Oseen, who proposed a remedy that will appear, in a modern disguise, in Section 11.8.

Flow Induced Outside a Spinning Sphere

This problem is even easier to analyze than the flow due to translation of a sphere. We suppose that a solid sphere is spinning steadily around the polar axis, with angular speed ω. One can readily imagine that the flow may have only an azimuthal component of velocity, i.e.,

$$\mathbf{u}(R, \theta) = w(R, \theta)\mathbf{e}_\phi \tag{11.96}$$

The corresponding vorticity is

$$\mathbf{\Omega}(R, \theta) = \frac{1}{R \sin \theta} \left(\frac{\partial(w \sin \theta)}{\partial \theta} \mathbf{e}_R - \frac{\partial(wR \sin \theta)}{\partial R} \mathbf{e}_\theta \right) \tag{11.97}$$

The viscous force per unit volume is

$$\text{div } \boldsymbol{\tau} = -\frac{\mu}{R^2 \sin \theta} \left\{ \frac{\partial^2(wR)}{\partial R^2} + \frac{\partial}{\partial \theta} \left[\frac{1}{R \sin \theta} \frac{\partial}{\partial \theta} (w \sin \theta) \right] \right\} \mathbf{e}_\phi$$

From the balance between viscous and pressure forces, we see that $\partial \Pi / \partial R = 0$, and hence that $\Pi = \Pi_\infty$ everywhere. In short, there is no pressure force in this flow, and the differential equation for w is simply obtained by setting $\text{div } \boldsymbol{\tau} = 0$. This gives

$$\frac{\partial^2(wR)}{\partial R^2} + \frac{\partial}{\partial \theta} \left[\frac{1}{R \sin \theta} \frac{\partial}{\partial \theta} (w \sin \theta) \right] \right\} = 0 \tag{11.98}$$

The boundary conditions on w are

$$w(a, \theta) = \omega a \sin \theta \qquad w \to 0 \quad \text{as} \quad R \to \infty$$

These suggest a trial solution in the form

$$w(R, \theta) = \omega a \sin \theta \, F(R) \tag{11.99}$$

Now we try $F(R) = R^n$, and easily find that $n = -2$ or 1. The latter value is excluded by the boundary condition at $R = \infty$, so that the final solution is

$$w(R, \theta) = \omega a^3 R^{-2} \sin \theta \tag{11.100}$$

The viscous stress on the surface of the sphere is

$$\tau_{R\phi}(a, \theta) = \mu \left(R \frac{\partial (w/R)}{\partial R} \right)_{R=a} = -3\mu\omega \sin \theta$$

and the torque required to keep the sphere spinning is

$$T = 6\pi\mu\omega a^3 \int_0^\pi \sin^3 \theta \, d\theta = 8\pi\mu\omega a^3 \tag{11.101}$$

11.8 EFFECTS OF BODY SHAPE ON EXTERNAL CREEPING FLOWS

The construction of creeping-flow solutions that meet the no-slip and no-penetration boundary conditions on bodies of more complex shape is usually very difficult. Nevertheless, some amazing closed-form solutions have been found by clever use of special coordinate systems. Fortunately, the linearity of the equations allows superposition of solutions, so that flow over a body like an ellipsoid, which possesses three axes of symmetry, need only be analyzed for cases of motion parallel to each of these axes. Then the forces and moments can easily be calculated for an arbitrary orientation.

Prolate Spheroids

Axisymmetric creeping flow over solid spheroids, which are bodies obtained by rotating an ellipse around its major or minor axis, has been analyzed by a number of authors. The solutions presented here employ elliptical, (ξ, η), coordinates. They are related to cylindrical polar coordinates by the equations

$$z = L \cosh \xi \cos \eta \qquad r = L \sinh \xi \sin \eta$$

where L is a reference length. The surface of the body is given by $\xi = \xi_0$. As shown in Fig. 11.22, the fluid fills the region $\xi > \xi_0$. The velocity component in the direction of increasing ξ is called u, that in the direction of increasing η is called v. A streamfunction is introduced, from which u and v can be derived. The formulas are

$$u(\xi, \eta) = -\frac{r}{h} \frac{\partial \psi}{\partial \eta} \qquad \text{and} \qquad v(\xi, \eta) = -\frac{r}{h} \frac{\partial \psi}{\partial \xi}$$

where the metric coefficient is given by $h = L(\cosh^2 \xi - \cos^2 \eta)^{1/2}$.

We use the shorthand notations $s = \cosh \xi$, and $c = \cos \eta$. The shape parameter, $s_0 = \cosh \xi_0$, is related to the ratio $E =$ body length/body diameter

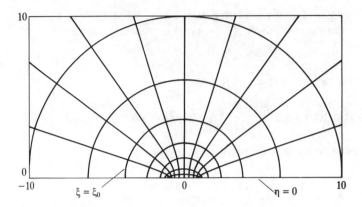

FIGURE 11.22
Prolate ellipsoid and elliptical coordinates.

by the equation

$$s_0 = \frac{E}{(E^2 - 1)^{1/2}} \tag{11.102}$$

The streamfunction is given by the formula[24]

$$\psi = \frac{UL^2}{2}(1 - c^2)\left[(s^2 - 1) - A(s^2 - 1)\ln\left(\frac{s+1}{s-1}\right) + Bs\right] \tag{11.103}$$

It must satisfy the boundary conditions $\psi = 0$ and $\partial\psi/\partial s = 0$, at $s = s_0$. Thus,

$$A = \frac{1}{2}\frac{s_0^2 + 1}{\text{Deno}} \quad \text{and} \quad B = \frac{s_0^2 - 1}{\text{Deno}}$$

where

$$\text{Deno} = \tfrac{1}{2}(s_0^2 + 1)\ln\left(\frac{s_0 + 1}{s_0 - 1}\right) - s_0$$

The vorticity is purely azimuthal, and has the magnitude

$$\Omega = -\frac{1}{h^2 r}\left((s^2 - 1)\frac{\partial^2\psi}{\partial s^2} + (1 - c^2)\frac{\partial^2\psi}{\partial c^2}\right)$$

$$= -\frac{2U}{L\,\text{Deno}}\frac{\text{ctnh }\xi \sin\eta}{\cosh^2\xi - \cos^2\eta} \tag{11.104}$$

The drag is

$$\text{Drag} = 8\pi\mu\frac{UL}{\text{Deno}} \tag{11.105}$$

---

[24] See L. E. Payne and W. H. Pell, "Stokes Flow for Axial Symmetric Bodies," *Jour. Fluid. Mech.*, **7**: 529–549 (1960) for the derivation of this solution.

The vorticity can again be set proportional to the drag, with the result

$$\Omega = \frac{\text{Drag}}{4\pi\mu L^2} \frac{\text{ctnh } \xi \sin \eta}{\cosh^2 \xi - \cos^2 \eta}$$

When $\xi \to \infty$, ctnh $\xi \to 1$, and $L^2(\cosh^2 \xi - \cos^2 \eta) \to R^2$, where R equals distance from the center of the spheroid. This shows that the far-field vorticity bears the same relationship to drag for a spheroid, as it does for a sphere.

Noting that tanh ξ_0 equals the (diameter/length) ratio of the spheroid, we check to see that the results for a sphere are recovered as $\xi_0 \to \infty$. We use the power series approximation for $s_0 \gg 1$,

$$\ln \left(\frac{s_0 + 1}{s_0 - 1} \right) \approx 2(s_0^{-1} + \tfrac{1}{3}s_0^{-3} + \tfrac{1}{5}s_0^{-5} + \cdots)$$

to find that

$$\text{Deno} \approx \tfrac{4}{3}s_0^{-1} + \tfrac{8}{15}s_0^{-3} + \cdots$$

so that

$$\text{Drag} \approx 6\pi\mu U L s_0 (1 - \tfrac{2}{5}s_0^{-2} + \cdots)$$

Since $L s_0$ equals the half-length of the spheroid, we see that the drag is just less than that of a sphere which contains the spheroid. On the other hand, since tanh $\xi_0 \approx (1 - \tfrac{1}{2}s_0^{-2} + \cdots)$ when $\xi_0 \gg 1$, we see that the drag is more than that of the largest sphere entirely contained inside the spheroid. It can be proved that this is a quite general result for drag in creeping flow.

In the opposite limit, when $E \gg 1$, we can use an approximation to Eq. (11.102), $s_0 \approx 1 + [1/(2E^2)]$, to find the corresponding approximations

$$\text{Deno} \approx 2 \ln (2E) - 1$$

and

$$\text{Drag} \approx \frac{2\pi\mu U(2L)}{\ln (2E) - \tfrac{1}{2}} \tag{11.106}$$

where now $2L$ is approximately the length of the body.

Oblate Spheroids

To obtain an *oblate* spheroid, one rotates the ellipse around its *minor* axis. Then

$$z = L \sinh \xi \cos \eta \qquad r = L \cosh \xi \sin \eta \qquad h = L(\sinh^2 \xi + \cos^2 \eta)^{1/2}$$

We use the shorthand $t = \sinh \xi$. The streamfunction is

$$\psi = \frac{Ur^2}{2} \left(1 - \frac{At}{1 + t^2} - B \text{ ctn}^{-1} t \right) \tag{11.107}$$

with

$$A = \frac{1 + t_0^2}{\text{Deno}} \qquad \text{and} \qquad B = \frac{1 - t_0^2}{\text{Deno}}$$

where

$$\text{Deno} = t_0 + (1 - t_0^2) \text{ ctn}^{-1} t_0$$

The magnitude of the vorticity is

$$\Omega = -\frac{1}{h^2 r}\left((1 + t^2)\frac{\partial^2 \psi}{\partial t^2} + (1 - c^2)\frac{\partial^2 \psi}{\partial c^2}\right)$$

$$= -\frac{2U}{L\,\mathrm{Deno}}\frac{\tanh \xi \sin \eta}{(\sinh^2 \xi + \cos^2 \eta)} \qquad (11.108)$$

The drag is again

$$\mathrm{Drag} = \frac{8\pi\mu UL}{\mathrm{Deno}} \qquad (11.109)$$

When $\xi_0 \gg 1$, $\mathrm{Deno} \approx \frac{4}{3}t_0^{-1} + \frac{8}{15}t_0^{-3} + \cdots$, so we again recover the result for a solid sphere.

When $\xi_0 = 0$, $\mathrm{Deno} = \pi/2$, and we get the result for a *circular disk*:

$$\psi = \frac{Ur^2}{2}\left[1 - \frac{\pi}{2}\left(\frac{t}{1 + t^2} - \mathrm{ctn}^{-1}t\right)\right] \qquad (11.110)$$

$$\Omega = -\frac{4U}{\pi L}\frac{\tanh^2 \xi \sin \eta}{\sinh^2 \xi + \cos^2 \eta} \qquad (11.111)$$

and
$$\mathrm{Drag} = 16\mu UL \qquad (11.112)$$

In this limit, L is the radius of the disk.

Spheroid of Minimum Drag for a Fixed Volume

A quantity important to the survival of microorganisms is the drag to be overcome while propelling a body of a certain volume V; thus it makes sense to define a shape-dependent drag coefficient for creeping flow by the formula

$$C = \frac{\mathrm{Drag}}{\mu UV^{1/3}} \qquad (11.113)$$

For a solid sphere, $C = (162\pi^2)^{1/3} = 11.69333$.

The volume of a prolate spheroid equals $(4\pi L^3/3)s_0(s_0^2 - 1)$, so that the formula for C is

$$C = (384\pi^2)^{1/3}[s_0(s_0^2 - 1)]^{-1/3}\left[\tfrac{1}{2}(s_0^2 + 1)\ln\left(\frac{s_0 + 1}{s_0 - 1}\right) - s_0\right]^{-1}$$

As a spheroid becomes increasingly prolate, the value of C decreases slightly, reaching a minimum of 11.17310 when $E = 1.953$. For still more slender bodies, C increases again.

There is good theoretical reason to believe that this spheroid is very nearly the body for which C has the smallest possible value.[25] Its resemblance

---

[25] See J.-M. Bourot, "On the Numerical Computation of the Optimum Profile in Stokes Flow," *Jour. Fluid Mech.*, **65**: 513–515 (1974).

to the shape of some tiny organisms, such as mammalian sperm cells, which propel themselves by waving an attached flagellum, is really striking.

The Spherical Cap

All of the bodies considered so far possess fore-and-aft symmetry, which is shared by the streamline and vorticity patterns. A very different, but geometrically simple, axisymmetric body is the spherical cap, shown in Fig. 11.23. It is characterized by a radius a and the angle α subtended by the circular arc.

A closed-form solution for the streamfunction has been found, but the formula requires ten lines of type![26] The drag is given by the simple formula

$$\text{Drag} = \mu U a (6\alpha + 8 \sin \alpha + \sin 2\alpha) \qquad (11.114)$$

When $\alpha = \pi$, we have a complete solid sphere, with the familiar drag formula.

The striking qualitative feature of this flow is a separation streamsurface, which joins the solid cap at its rim. The angle between the separation surface and the extended surface of the cap, called λ and shown in Figure 11.23, is

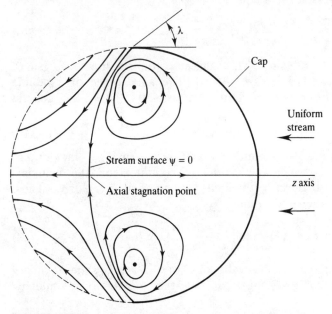

FIGURE 11.23
Flow around a spherical cap. (Reproduced, by permission, from J. M. Dorrepaal, M. E. O'Neill, and K. B. Ranger, 1976.[26])

---

[26] See J. M. Dorrepaal, M. E. O'Neill, and K. B. Ranger, "Axisymmetric Stokes Flow Past a Spherical Cap," *Jour. Fluid Mech.*, **75**: 273–286 (1976).

given by the formula

$$3 \tan\left(\frac{\lambda}{2}\right) = \mathrm{ctn}\left(\frac{\alpha}{2}\right) \qquad (11.115)$$

Thus, when $\alpha = 90°$, $\lambda = 36.87°$. As $\alpha \to 0$, $\lambda \to 180°$. In this limit, the cap approaches a plane circular disk, and the eddy between the cap and the separation surface vanishes.

The solution shows that there is always a single eddy between the cap and the separation surface. It also illustrates the striking proposition, well confirmed by experiment, that the eddy always stands on the concave side of the cap, whether that side faces downstream or upstream!

The Circular Cylinder Broadside to the Flow.
The Stokes Paradox

To investigate the creeping flow normal to a long slender cylinder, it is natural to start with the idealization of an infinitely long circular cylinder. We use cylindrical polar coordinates, with velocity $\mathbf{u} = u(r, \theta)\mathbf{e}_r + v(r, \theta)\mathbf{e}_\theta$. A streamfunction ψ is introduced, such that $ru = \partial\psi/\partial\theta$ and $v = -\partial\psi/\partial r$. The vorticity equation, expressed as an equation for ψ, is

$$\left[r^{-1}\frac{\partial(r\,\partial/\partial r)}{\partial r} + r^{-2}\frac{\partial^2}{\partial\theta^2} \right]^2 \psi = 0 \qquad (11.116)$$

When $r \to \infty$, we want $\psi \to Ur \sin\theta$. This suggests a trial solution $\psi = r^n \sin\theta$. This works if $(n+1)(n-1)(n-1)(n-3) = 0$. The appearance of a double root means that $\psi = r \ln r \sin\theta$ is another possibility. Thus, the general solution is

$$\psi = (Ar^{-1} + Br + Cr\ln r + Dr^3)\sin\theta \qquad (11.117)$$

The boundary condition at $r = \infty$ requires that $D = 0$. Leaving C indeterminate for the moment, consider the boundary condition on the cylinder: $u(a, \theta) = v(a, \theta) = 0$. Use these to eliminate A and B in favor of C. Thus, $A = (a^2/2)C$, and $B = -(\ln a + \frac{1}{2})C$, and

$$\psi = C\left(\frac{a}{2}\right)\left\{ 2\left(\frac{r}{a}\right)\ln\left(\frac{r}{a}\right) - \frac{r}{a} - \frac{a}{r} \right\}\sin\theta \qquad (11.118)$$

Here is Stokes' Paradox. There is no value of C that will satisfy the boundary condition at the free stream. We must conclude that our theory is somehow fundamentally flawed.

Remembering the seemingly harmless inconsistency in the theory of creeping flow around a sphere, that convection is not really negligible compared to diffusion in the very far field, we compare these two agencies in the present solution.

The vorticity that corresponds to the solution (11.1.19) is

$$\Omega = -2Cr^{-1}\sin\theta \qquad (11.119)$$

We compare convection to streamwise diffusion in the far field, just as in the study of flow around the sphere, and find a similar result; the ratio is proportional to Ur/v. For the cylinder, the failure to account somehow for a bit of convection makes it impossible to complete the analysis. We shall return to Stokes' Paradox in Section 11.9, but first consider the possibility of circumventing the difficulty by considering creeping flow across a very slender body of finite length.

The Prolate Spheroid Broadside to the Flow

In 1876, Oberbeck analyzed creeping flow past a general ellipsoid.[27] His analysis is too complicated to duplicate here, but it leads to fairly simple formulas for the drag of a prolate spheroid, moving normal to its axis of rotation.[28]

$$\text{Drag} = \frac{32\pi\mu UL}{(3 - s_0^2) \ln \left(\dfrac{s_0 + 1}{s_0 - 1}\right) + 2s_0} \tag{11.120}$$

The approximation for a very slender spheroid, with $E \gg 1$, is

$$\text{Drag} \approx \frac{4\pi\mu U(2L)}{\ln (2E) + \frac{1}{2}} \tag{11.121}$$

When this is compared with Eq. (11.106), we find that, for very slender prolate spheroids

$$\frac{\text{Drag (normal)}}{\text{Drag (axial)}} \approx \frac{2[2 \ln (2E) - 1]}{2 \ln (2E) + 1}$$

a value which slowly approaches 2 as $E \to \infty$.

11.9 SINGULAR SOLUTIONS FOR THE CREEPING-FLOW EQUATIONS

In the theory of inviscid, incompressible, irrotational flow, which is also governed by linear differential equations, much good use is made of *singular* or *fundamental* solutions, such as sources, doublets, and vortices. In that theory there is a velocity potential, governed by the Laplace equation, $\nabla^2\Phi = 0$. In creeping flows the pressure exess, Π, is governed by the same equation; hence various singular solutions for the pressure, with their associated velocity fields, are useful for the analysis of geometrically complex creeping flows. This

---

[27] A. Oberbeck, "Uber stationäre Flüssigkeitsbewegungen mit Berücksichtigung der inneren Reibung," *J. reine angew. Math.* **81**: 62–80 (1876).

[28] For a derivation, see Articles 335, 336 of H. Lamb, *Hydrodynamics*, 6th ed., Dover Press, 1932.

technique has been particularly fruitful in the study of the swimming and feeding motions of microorganisms.[29]

The Stokeslet

If the results for drag of the sphere and for the pressure in the outer flow are compared, an interesting result emerges.

$$\Pi - \Pi_\infty = -\frac{\text{Drag}}{4\pi\rho R^2}\cos\theta \tag{11.122}$$

Except for the minus sign, this is identical to Eq. (9.89), which describes the pressure field generated by the action of a point force, $F\mathbf{e}_z$ on a stationary fluid when the Reynolds number $F/\rho v^2 \ll 1$.

A meridional cross section of surfaces of constant Π and a view of curves tangent to $\nabla\Pi$ are shown in Fig. 11.24. The picture might just as well represent the equipotential surfaces and streamlines for an axially-oriented mass-flow *doublet* in incompressible potential flow.

For applications in which a flow field may be constructed by a superposition of singular solutions, a coordinate-free representation of (11.122) is useful. It is

$$\Pi - \Pi_\infty = -\nabla \cdot \frac{\mathbf{F}}{4\pi\rho r} \tag{11.123}$$

in which r is distance from the point of application of the force $\mathbf{F}$. This particular creeping-flow solution is called a *Stokeslet,* in honor of Sir G. G. Stokes, who did so much of the early study of viscous flow.

The velocity field of a Stokeslet is readily derived from the streamfunction given in Eq. (9.88). In spherical polar coordinates, with polar axis pointing in the direction of $\mathbf{F}$, the velocity components are

$$\mathbf{u} = \frac{F}{8\pi\mu R}(2\cos\theta\,\mathbf{e}_R - \sin\theta\,\mathbf{e}_\theta) \tag{11.124}$$

A coordinate-free representation is

$$\mathbf{u} = (6\pi\mu r)^{-1}\mathbf{F} - \frac{r^2}{6v}\nabla\Pi \tag{11.125}$$

in which Π is represented by Eq. (11.123).

---

[29] See M. J. Lighthill, "Flagellar Hydrodynamics," *SIAM Review* **18** (No. 2) (April 1976), for a fascinating presentation of both the biological and mathematical aspects of this study. The discussion in the text is strongly influenced by Lighthill's paper.

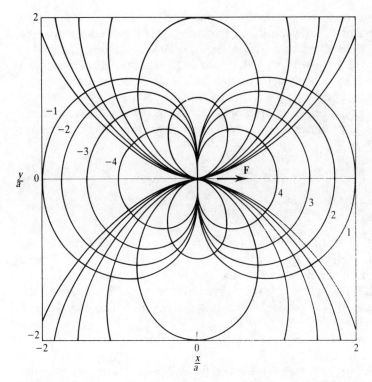

FIGURE 11.24
Isobaric surfaces for a Stokeslet. Isobars are marked with values of the quantity $P = (16\pi a^2/F) \times (\Pi - \Pi_\infty)$.

Unlike the doublet of potential flow, the Stokeslet has a vorticity field

$$\mathbf{\Omega} = -\frac{\text{Drag}}{4\pi\mu R^2} \sin\theta \, \mathbf{e}_\phi \tag{11.126}$$

Again, this is identical, except for sign, to the result given in Eq. (9.90). The vortex lines encircle the axis of the Stokeslet, the direction of the vorticity being given by a right-hand rule. Let your right thumb point in the direction of the force, **F** or −**Drag**, that engenders the Stokeslet; your curled fingers will then point in the direction of $\mathbf{\Omega}$. A coordinate-free representation of the vorticity is

$$\mathbf{\Omega} = (6\pi\mu r^2)^{-1}(\mathbf{F} - 2r^3 \rho\nabla\Pi) \times \mathbf{e}_r \tag{11.127}$$

Other Singular Solutions

One could construct a variety of other singular solutions for Π by partial differentiation of the Stokeslet solution. One particular example is of special interest to the theory of self-propulsion of microorganisms, because in that process the propulsive force generated by one part of the organism at any given time is just equal and opposite to the drag. This happens because the organism, like the fluid it displaces, has negligible inertia, and hence negligible capacity to store kinetic energy. To the fluid at a distance, the kinematical effect of the organism is equivalent to that of a pair of equal-but-opposite, slightly displaced Stokeslets. If we suppose that the two Stokeslets share a common axis, the polar axis, and are applied a distance L apart, the velocity and vorticity fields at distances $R \gg L$ can be represented in polar coordinates by

$$\mathbf{u} = \frac{FL}{8\pi\mu R^2}(1 - 3\cos^2\theta)\mathbf{e}_R \tag{11.128}$$

$$\mathbf{\Omega} = -\frac{3FL}{4\pi\mu R^3}\sin^2\theta\,\mathbf{e}_\phi \tag{11.129}$$

The most interesting feature of these results is the accelerated decay of the velocity with distance from the organism. Only a finite amount of kinetic energy is stored in all the fluid outside the sphere $R = R_0$, where R_0 is large enough so that (11.128) gives a good approximation to $\mathbf{u}$. In contrast, the corresponding quantity for a single Stokeslet is infinite. The interference between the swimming motions of neighbouring microorganisms is also much smaller than one would deduce by looking only at the formulas for a single Stokeslet.

For actual computations of creeping flows, another kind of singular solution is particularly useful. This is the mass-flow doublet of potential-flow theory. In the creeping-flow regime, this kind of doublet makes no contribution to the pressure field, because it has no vorticity, and hence no viscous force per unit volume to balance a pressure gradient. It does, however, help one to construct solutions that satisfy the boundary conditions of zero penetration and zero slip. Like the Stokeslet pair, it has an axis, and must be characterized by a vector strength, which we may call $\mathbf{G}$. Its velocity field is

$$\mathbf{u} = -\mathbf{G} \cdot \nabla\left(\frac{1}{4\pi}r\right) \tag{11.130}$$

To see, by example, how it can be useful, we can construct the flowfield around the sphere, by setting $\mathbf{G} = -(a^2/6\mu)\mathbf{F}$ and adding the velocity field of this doublet to that of the Stokeslet (11.125). The second term in the velocity field of the Stokeslet is precisely canceled, at $r = a$, by the field of the doublet. The remaining velocity on that surface is a constant $\mathbf{u} = \mathbf{F}/(6\pi\mu a)$, which we recognize as the velocity of the solid sphere.

Techniques of Superposition

There are many ways to construct approximate solutions to boundary-value problems by superposition of Stokeslets, doublets, and other singular solutions. Typically, a finite number of singularities are disposed at predetermined

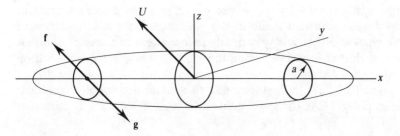

FIGURE 11.25
Approximation of the effect of a translating slender spheroid by distributed Stokeslets, **f**, and doublets, **g**.

points outside the fluid-filled domain, and the boundary conditions are satisfied at a correspondingly finite number of points on the boundary. This leads to a set, usually quite large, of linear algebraic equations, to be solved for the vector strengths of the singularities. Sometimes a careful preliminary analysis indicates particular combinations of singularities that will give good accuracy for a given computational effort. For an example, we consider the treatment of slender bodies, developed for the study of microorganisms that swim by waving flagella,[30] but we apply it to the simpler problem of estimating the drag of a very slender prolate spheroid in crossflow, for which we already have an exact answer.

The situation to be imagined is shown in Fig. 11.25. Distance along the axis of the spheroid is denoted by x; the local radius of the spheroid is called $a(x)$. A continuous distribution of Stokeslets, with strength $\mathbf{f}(x)$ per unit length, and an accompanying distribution of mass-flow doublets, with strength $\mathbf{g}(x)$ per unit length, is laid down along the axis. The directions of $\mathbf{f}$ and $\mathbf{g}$ are taken to be parallel or antiparallel to the direction of motion of the spheroid; this is a major simplifying feature of our example that is not carried over to the study of swimming.

We use a theorem presented by Lighthill that says for our problem that if $\mathbf{g}(x) = -(a^2/4\pi)\mathbf{f}(x)$, the combination of the Stokeslets and doublets will produce an approximately uniform velocity over the entire circle of radius a, which girdles the spheroid at a specified value of x. This velocity has the value

$$\mathbf{u}(x) = (8\pi\mu)^{-1}\left\{2\mathbf{f}(x) + \int [(x - \xi)^2 + a^2]^{-1/2}\mathbf{f}(\xi)\,d\xi\right\} \qquad (11.131)$$

where the integration is carried out over the entire length of the spheroid, excluding a local interval in which $|x - \xi| < \delta$, where $\delta = (a/2)\sqrt{e} = 0.82a$.

To see that this formula works for a simple but fairly representative case,

---

[30] M. J. Lighthill (1976), *op. cit.*, pp. 190–196.

we can evaluate **u** at a point $(0, y, z)$, where $y^2 + z^2 = a^2$, by direct integration, supposing that **f** has a constant value in the interval $-L \leq x \leq L$, and is zero elsewhere. We shall find that **u** is very nearly independent of y and z, and that its value agrees with that found by taking the limit $E \to \infty$ in the exact solution for the spheroid.

At the point $(0, y, z)$, the velocity due to a Stokeslet-doublet combination located at point $(\xi, 0, 0)$ and oriented so that $\mathbf{f} = F\mathbf{e}_z$, has cartesian components

$$(u, v, w) = \frac{F}{8\pi\mu} \left[\frac{-\xi z,\, yz,\, r^2 + z^2}{r^3} + \frac{a^2}{2} \left(\frac{3\xi z,\, -3yz,\, r^2 - 3z^2}{r^5} \right) \right]$$

Here $r^2 = \xi^2 + y^2 + z^2$. The contribution of all such singularities is then obtained by integration from $\xi = -L$ to $\xi = L$. The x component of the integrand is an odd function of ξ, so there will be no resultant x component of the velocity. What remains on the boundary, where $y^2 + z^2 = a^2$, and $r^2 = a^2 + \xi^2$, is

$$(v, w) = \frac{F}{8\pi\mu} \int_{-L}^{L} \left[\frac{yz,\, a^2 + z^2 + \xi^2}{(a^2 + \xi^2)^{3/2}} + \frac{a^2}{2} \left(\frac{-3yz,\, a^2 - 3z^2 + \xi^2}{(a^2 + \xi^2)^{5/2}} \right) \right] d\xi$$

With the aid of integral tables, this can be reduced to

$$v = -\frac{F}{4\pi\mu} \frac{yzL}{(a^2 + L^2)^{3/2}}$$

$$w = \frac{F}{8\pi\mu} \left[-\frac{z^2 L}{(a^2 + L^2)^{3/2}} + 2\sinh^{-1}\left(\frac{L}{a}\right) - \frac{L}{(a^2 + L^2)^{1/2}} \right]$$

We now notice that all of v, and the first term in w, i.e., all the terms that represent variation of velocity over the bounding circle, are smaller than the last term in w, by a factor proportional to $(a/L)^2$. Since this factor is assumed to be very small, we are left with a uniform velocity, which we now call **U**. Inverting our calculation to solve for **f**, the presumed force exerted on the fluid per unit length of the slender body, we find

$$f = \frac{8\pi\mu\mathbf{U}}{\text{Deno}} \tag{11.132}$$

where

$$\text{Deno} = 2\sinh^{-1}\left(\frac{L}{a}\right) + \frac{L}{(a^2 + L^2)^{1/2}}$$

Finally, we return to the notation of the discussion of spheroid drag, $E \equiv L/a$, and make a further approximation of Deno, dropping only terms of order E^{-2}, to get

$$\text{Deno} \approx 2\ln(2E) + 1$$

Noting that the total force equals $2Lf$, we find that the result of this analysis is identical, to order E^{-2}, to the result given in Eq. (11.121)!

As a last move, we return to Eq. (11.131) and integrate it for constant **f**, now taking note of the excluded interval. The result is

$$\mathbf{U} = \frac{\mathbf{F}}{4\pi\mu} \left[1 - \sinh^{-1}\left(\frac{\delta}{a}\right) + \sinh^{-1}\left(\frac{L}{a}\right) \right]$$

$$\approx \frac{\mathbf{F}}{4\pi\mu} \left[1 - \ln\left(\frac{2\delta}{a}\right) + \ln\left(\frac{2L}{a}\right) \right]$$

With the value specified for δ, $2\delta/a = e^{1/2}$; so $\ln(2\delta/a) = 1/2$, and the value of w from the more involved integration is recovered.

11.10 SMALL EFFECTS OF INERTIA

As was mentioned in the last section, equations that omit any reference to inertia or convection sometimes have no solution that meets all the required boundary conditions, no matter how small the value of Re. Even when they do have solutions, these may be insufficiently accurate for the successful prediction of the final destination of a small body that moves through the fluid for a long time, or for an adequate description of the mutual interactions between neighboring particles, or between a particle and a wall. There are many practical concerns that demand an accounting for small effects of inertia. This section will address only three topics in this large field:[31] (1) the systematic improvement of lubrication theory; (2) Oseen's resolution of Whitehead's paradox, by an improved description of flow far from a creeping sphere; (3) the development of lift on a sphere creeping through a shear flow.

Small Inertial Effects in Slender, Bounded Flows

The solutions found in Sections 11.3 to 11.6 to describe creeping flow in slender, bounded regions, involved approximations that could be shown to be reasonable everywhere in the flowfield. Specifically, if the terms dropped from the full governing equations are evaluated by appropriate manipulations of the approximate solution, they are found to be smaller than the terms retained, everywhere, provided only that the crossflow Reynolds number has a sufficiently small value. Such approximate solutions are said to be *uniformly valid*, and there is a straightforward, if sometimes tedious, way of improving them to allow for somewhat larger values of the Reynolds number. To see how this works, consider the introductory example of the two-dimensional squeezed film, analyzed in Section 11.3.

---

[31] These and other limitations of the simplest creeping-flow theory are described and analyzed in L. G. Leal, "Particle Motions in a Viscous Fluid," *Ann. Rev. Fluid Mech.* **12**: 435–476 (1980).

For convenience, the definitions of nondimensional variables are repeated here.

$$x = \frac{x^*}{L} \qquad z = \frac{z^*}{H_0} \qquad t = \frac{t^*V}{L} \qquad \psi = \frac{\psi^*}{VL} \qquad \Omega = \Omega^* \frac{H_0^2}{VL} \qquad \Pi = \Pi^* \frac{H_0}{\nu V}$$

The flow is characterized by two small parameters, $\mathrm{Re} = VH_0/\nu$, and $\alpha = (H_0/L)^2$. The full equations for the vorticity and the streamfunction are

$$\mathrm{Re}\{\Omega_{,t} + \psi_{,z}\Omega_{,x} - \psi_{,x}\Omega_{,z}\} - \alpha\Omega_{,xx} + \Omega_{,zz} = 0 \qquad (11.133)$$

and

$$\Omega - \alpha\psi_{,xx} + \psi_{,zz} = 0 \qquad (11.134)$$

Just as in the treatment of small variations of viscosity, given in Sample Calculation A of Chapter 7, one may now postulate regular perturbation expansions for ψ and Ω, allowing for the fact that there are two small parameters in the present problem. Thus, we write[32]

$$\psi(x, z, t; \mathrm{Re}, \alpha) = {}^{0,0}\psi(x, z, t) + \mathrm{Re}\,{}^{1,0}\psi(x, z, t) + \alpha^{0,1}\psi(x, z, t)$$
$$+ \mathrm{Re}\,\alpha^{1,1}\psi(x, z, t) + \cdots$$

and

$$\Omega(x, z, t; \mathrm{Re}, \alpha) = {}^{0,0}\Omega(x, z, t) + \mathrm{Re}\,{}^{1,0}\Omega(x, z, t) + \alpha^{0,1}\Omega(x, z, t)$$
$$+ \mathrm{Re}\,\alpha^{1,1}\Omega(x, z, t) + \cdots$$

When the series for ψ is substituted into (11.134), and the result is compared with the series for Ω, we find that

$$ {}^{0,0}\Omega = -{}^{0,0}\psi_{,zz} \qquad {}^{1,0}\Omega = -{}^{1,0}\psi_{,zz} $$

$$ {}^{0,1}\Omega = -({}^{0,0}\psi_{,xx} + {}^{0,1}\psi_{,zz}) \qquad \text{and so on} $$

When both series are substituted into (11.133), and the cofactor of each power of Re and of a is equated to zero, we find

$$ {}^{0,0}\Omega_{,zz} = 0 $$

$$ {}^{1,0}\Omega_{,zz} = {}^{0,0}\Omega_{,t} + {}^{0,0}\psi_{,z}\,{}^{0,0}\Omega_{,x} - {}^{0,0}\psi_{,x}\,{}^{0,0}\Omega_{,z} $$

$$ {}^{0,1}\Omega_{,zz} = {}^{0,0}\Omega_{,xx} $$

and so on. From Section 11.3, we draw the result

$$ {}^{0,0}\psi(x, z, t) = x(3z^2\xi^{-2} - 2z^3\xi^{-3}) $$

in which the quantity $\xi = H(t)/H_0$ carries the dependence on time. We easily calculate that

$$ {}^{0,0}\Omega(x, z, t) = -6x(\xi^{-2} - 2z\xi^{-3}) $$

---

[32] The preceding-superscript notation, as in ${}^{i,j}\psi(x, z, t)$, identifies the coefficient of $\mathrm{Re}^i\alpha^j$ in the double power series expansion.

It follows that $\quad {}^{1,0}\Omega_{,zz} = -12x(\xi^{-3} - 6z^2\xi^{-5} + 4z^3\xi^{-6})$

and that $\quad {}^{0,1}\Omega_{,zz} = 0$

We can integrate the last two of these twice over z, to get general solutions for ${}^{1,0}\Omega$ and ${}^{0,1}\Omega$. Those are put into the appropriate equations for ${}^{1,0}\psi_{,zz}$ and ${}^{0,1}\psi_{,zz}$, and integrated twice again. There will be four constants of integration in each final expression. These are fixed by the boundary conditions ${}^{1,0}\psi = 0$ at $z = 0$ and 1, and ${}^{1,0}\psi_{,z} = 0$ at $z = 0$ and 1, and corresponding conditions on ${}^{0,1}\psi$. The results are

$$
\begin{aligned}
{}^{1,0}\psi = 2xz^2(1-z)^2 [\tfrac{1}{4}\xi^{-3} &- \tfrac{1}{10}(z^2 + 2z + 3)\xi^{-5} \\
&+ \tfrac{1}{35}(z^3 + 2z^2 + 3z + 4)\xi^{-6}]
\end{aligned}
$$

and $\quad {}^{0,1}\psi = 0.$

A direct way to see the practical implications of the higher-order perturbations is to calculate the modification of the force required to keep the upper plate advancing at speed V. Applying the perturbation procedure to the x-momentum equation, we get

$$
{}^{1,0}\Pi_{,x} = -{}^{1,0}\Omega_{,y} - ({}^{0,0}\psi_{,tz} + {}^{0,0}\psi_{,z}\,{}^{0,0}\psi_{,xz} - {}^{0,0}\psi_{,x}\,{}^{0,0}\psi_{,zz}
$$

Using the known expressions for ${}^{0,0}\psi$ and ${}^{1,0}\psi$, one can reduce this to

$$
{}^{1,0}\Pi_{,x} = -6x(\xi^{-3} - \tfrac{4}{5}\xi^{-5} + \tfrac{2}{7}\xi^{-6})
$$

Integrating over x, with the boundary condition $\Pi = 0$ at $x = 1$, one finds that

$$
{}^{1,0}\Pi = 3(1 - x^2)(\xi^{-3} - \tfrac{4}{5}\xi^{-5} + \tfrac{2}{7}\xi^{-6})
$$

This compares with

$$
{}^{0,0}\Pi = 6(1 - x^2)\xi^{-3}
$$

If we are content to leave the analysis at this order, and hence there is no need to calculate more values of ${}^{i,j}\psi_{,t}$ and ${}^{i,j}\Omega_{,t}$, we can now set $\xi = 1$ in final presentations of the results. This gives

$$
{}^{0,0}\psi(x, z, 0) = x(3z^2 - 2z^3)
$$

$$
{}^{0,0}\Omega(x, z, 0) = -6x(1 - 2z)
$$

$$
{}^{0,0}\Pi = 6(1 - x^2)
$$

$$
{}^{1,0}\psi = \tfrac{1}{70}xz^2(1 - z)^2(9 - 16z - 2z^2 + 4z^3)
$$

$$
{}^{0,1}\psi = 0
$$

$$
{}^{1,0}\Pi = \tfrac{51}{35}(1 - x^2)
$$

$$
{}^{0,1}\Pi = 0
$$

The dimensionless force is just twice the integral of $P\,dx$ from $x = 0$ to $x = 1$. This gives the result

$$
F = 8 + \tfrac{108}{35}\,\mathrm{Re} \tag{11.135}
$$

which shows that the error in F, due to complete neglect of inertia, is less than 1% if $\mathrm{Re} < 0.026$.

This procedure can be continued, particularly if one has access to a symbol-manipulating computer system, and in some cases series of this sort have been carried out to a very large number of terms, so that the results can be used at moderately large values of the nominally small parameters.

Improved Description of Flow Far From a Body in Creeping Flow

A century ago, Whitehead[33] attempted to improve on Stokes solution for creeping flow over a solid sphere, following a procedure much like the one we have just demonstrated for the squeezed film. He used Stokes' solution to evaluate the inertia terms that were dropped from the first approximation, and was able to find a correction to Stokes' streamfunction that was proportional to the Reynolds number. Unfortunately, there was no way to make this solution meet the boundary conditions both on the sphere and at infinity. This became known as *Whitehead's paradox*. The difficulty arose because the creeping-flow approximations are not uniformly valid when bodies move in an unbounded fluid. Sufficiently far from the body, the neglected inertia terms are not small compared to the viscous terms.

This paradox was resolved in 1910 by Oseen;[34] the treatment that appears here is a modern one that closely follows the beautiful presentation by Van Dyke.[35] Just as we used two kinds of series expansion to analyze, in Chapter 9, the solution for flow due to a point force when $F/\rho v^2 \gg 1$, we analyze the flow over a sphere when $\mathrm{Re} \ll 1$ by use of an inner expansion and an outer expansion. In all that follows, the symbols R and ψ will denote dimensionless quantities, $R = R^*/a^*$, and $\psi = \psi^*/U^*a^{*2}$. The Reynolds number will be $\mathrm{Re} = U^*a^*/v^*$.

The inner (Stokes) expansion. This is built on the concept of a *Stokes limit* for each dependent quantity, such as the streamfunction. With spherical polar coordinates, we symbolize this concept as follows:

$$\mathrm{Lim}_S \{\psi(R, \theta; \mathrm{Re})\} = \lim_{\mathrm{Re} \to 0} [\psi(R, \theta; \mathrm{Re})] \quad \text{with } R \text{ and } \theta \text{ fixed.}$$

Another concept is that of the *first Stokes deviation*, which is, for the

[33] A. N. Whitehead, "Second Approximations to Viscous Fluid Motion," *Quart. Jour. Math.*, **23**: 143–152 (1889).

[34] C. W. Oseen, "Über die Stokes'che Formel, und über eine verwandte Aufgabe in der Hydrodynamik," *Ark. Math. Astronom. Fys.*, **6** (No. 29) (1910).

[35] M. Van Dyke, *Perturbation Methods in Fluid Mechanics* (Annotated edition), The Parabolic Press, Stanford, Calif. (1975).

streamfunction,

$$\psi_{S1}(R, \theta) = \text{Lim}_S\, (\text{Re}^{-1}\, \{\psi(R, \theta; \text{Re}) - \text{Lim}_S\, [\psi(R, \theta; \text{Re})]\}).$$

The Stokes limit is what we calculated in Section 11.6; the first Stokes deviation is what Whitehead tried, but failed, to calculate.

When Whitehead adjusted constants in his expression for the first Stokes deviation, to satisfy the boundary conditions at the surface of the sphere, the resulting expression could not be made to satisfy the condition that $\psi \to \frac{1}{2}R^2 \sin^2 \theta$ as $R \to \infty$. The way out of this bind is to construct a special description of the flow far from the sphere, in the form of the Oseen expansion.

The outer (Oseen) expansion. This is built on the concept of an *Oseen limit* for each dependent quantity. This concept, for the streamfunction, is symbolized as follows:

$$\text{Lim}_O\, \{\psi(R, \theta; \text{Re})\} = \lim_{\text{Re} \to 0}\, [\psi(\rho, \theta; \text{Re})] \quad \text{with } \rho \text{ and } \theta \text{ fixed.}$$

Here $\rho = \text{Re}\, R$, so that in the Oseen limit the point of observation moves infinitely many radii away from the sphere. Thus, $\text{Lim}_O\, \{\psi(R, \theta; \text{Re})\} = \frac{1}{2}\,\text{Re}^{-2}\,\rho^2 \sin^2 \theta$, the streamfunction of the undisturbed free stream.

The *first Oseen deviation* is defined,[36] for the streamfunction, by

$$\psi_{O1}(\rho, \theta) = \text{Lim}_O\, (\text{Re}^{-1}\, \{\psi(R, \theta; \text{Re}) - \text{Lim}_O\, [\psi(R, \theta; \text{Re})]\}).$$

To calculate it, we need first to derive the governing differential equation. For this purpose, we insert the two-term Oseen expansion,

$$\psi(R, \theta; \text{Re}) = \frac{1}{2}\,\text{Re}^{-2}\,\rho^2 \sin^2 \theta + \text{Re}^{-1}\,\psi_{O1}(\rho, \theta) + \cdots$$

into the full equation for the streamfunction,

$$D^4\psi = \frac{\text{Re}}{\rho^2 \sin \theta}\left(\psi_{,\theta}\,\frac{\partial}{\partial R} - \psi_{,R}\,\frac{\partial}{\partial \theta} + 2\,\text{ctn}\,\theta\,\psi_{,R} - \frac{2}{R}\,\psi_{,\theta}\right)D^2\psi$$

to find that the terms that remain in the Oseen limit are

$$D^4\psi_{O1} = \left(\cos \theta\,\frac{\partial}{\partial \rho} - \frac{\sin \theta}{\rho}\,\frac{\partial}{\partial \theta}\right)D^2\psi_{O1} \tag{11.136}$$

In these last two equations, the operator D^2 is defined by

$$D^2 = \frac{\partial^2}{\partial R^2} + \frac{\sin \theta}{R^2}\,\frac{\partial}{\partial \theta}\left[\frac{1}{\sin \theta}\,\frac{\partial}{\partial \theta}\right]$$

---

[36] The higher-order Stokes and Oseen deviations contain more complicated functions of Re in the place of Re^{-1}. See Van Dyke (1975), *op. cit.*, for the explanation of this.

if R appears elsewhere in the equation. If ρ appears elsewhere, as in Eq. (11.136), then it replaces R in the definition of D^2. In either case, D^4 symbolizes two successive operations by D^2.

The solution of Eq. (11.136) involves some clever moves;[37] the end result can be checked by substitution. For the vorticity, we get

$$\Omega_{O1} = -\text{Re} \, (\rho \sin \theta)^{-1} D^2 \psi_{O1}$$

$$= -A \left(\frac{\text{Re}}{\rho}\right)\left(1 + \frac{2}{\rho}\right) \exp\left(-\frac{\rho}{2}(1 - \cos \theta)\right) \sin \theta \qquad (11.137)$$

and for the streamfunction

$$\psi_{O1} = -2A(1 + \cos \theta)\left[1 - \exp\left(-\frac{\rho}{2}(1 - \cos \theta)\right)\right]. \qquad (11.138)$$

These are not the only possible solutions of (11.136), but all others are flawed in some essential way; they misbehave either as $\rho \to \infty$, or as $\rho \to 0$.

To evaluate the constant A, we require that, if possible, the two-term Oseen expansion be made to give the same result, where $\rho \to 0$, as we find from the one-term Stokes expansion, where $R \to \infty$. As a preliminary step, we expand the right-hand side of (11.138) for small values of ρ, getting

$$\psi_{O1} \approx -A\rho \sin^2 \theta + \cdots$$

We now replace ρ by $R \, \text{Re}$ in the two-term Oseen expansion, using this approximation to ψ_{O1}. The result is

$$\psi_O \approx (\tfrac{1}{2}R^2 - AR) \sin^2 \theta + \cdots$$

The dominant terms in the Stokes solution, for large values of R, are

$$\psi_S \approx (\tfrac{1}{2}R^2 - \tfrac{3}{4}R) \sin^2 \theta + \cdots$$

The two representations agree if $A = \tfrac{3}{4}$.

The second term in the Stokes expansion. Whitehead's paradox resolved. If we now insert a two-term Stokes expansion,

$$\psi_S(R, \, \theta; \text{Re}) = \tfrac{1}{4}(2R^2 - 3R + R^{-1}) \sin^2 \theta + \text{Re} \, \psi_{S1}(R, \, \theta)$$

into the full equation for the streamfunction, we find that

$$D^4 \psi_{S1} = -\tfrac{9}{4}(2R^{-2} - 3R^{-3} + R^{-5}) \sin^2 \theta \cos \theta \qquad (11.139)$$

Whitehead found the particular integral,

$$-\tfrac{3}{32}(2R^2 - 3R + 1 - R^{-1} + R^{-2}) \sin^2 \theta \cos \theta$$

---

[37] See Van Dyke (1975), *op. cit.*, pp. 156–158.

to which we can add the complementary function,

$$B(2R^2 - 3R + R^{-1}) \sin^2 \theta$$

Both parts satisfy the boundary conditions on the sphere, and are dominated by the same power of R as $R \to \infty$. Thus, we propose the two-term Stokes expansion

$$\psi_S(R, \theta; \text{Re}) = (\tfrac{1}{4} + B \text{ Re})(2R^2 - 3R + R^{-1}) \sin^2 \theta$$

$$-\frac{3 \text{ Re}}{32}(2R^2 - 3R + 1 - R^{-1} + R^{-2}) \sin^2 \theta \cos \theta$$

To evaluate B, we replace R by ρ/Re in this, and re-expand the result for $\text{Re} \to 0$. The first two terms of this expansion are

$$\psi_S(\rho, \theta; \text{Re}) \approx \text{Re}^{-2}(\tfrac{1}{2}\rho^2 \sin^2 \theta) + \text{Re}^{-1}\left[\tfrac{3}{16}\rho^2 \sin^2 \theta\left(\frac{32B}{3} - \cos \theta\right) - \tfrac{3}{4}\rho \sin^2 \theta\right]$$

This is to be compared with the first two terms of the Oseen expansion, in the limit as $\rho \to 0$. These two terms are

$$\psi_O(\rho, \theta; \text{Re}) \approx \text{Re}^{-2} (\tfrac{1}{2}\rho^2 \sin^2 \theta) + \text{Re}^{-1} [\tfrac{3}{16}\rho^2 \sin^2 \theta(1 - \cos \theta) - \tfrac{3}{4}\rho \sin^2 \theta)]$$

The two representations agree if $B = \tfrac{3}{32}$. With this value of B, the two-term Stokes expansion can be regrouped into the form

$$\psi_S(R, \theta; \text{Re}) = \tfrac{1}{4}(R - 1)^2 \sin^2 \theta\left[\left(1 + \frac{3 \text{ Re}}{8}\right)(2 + R^{-1})\right.$$

$$\left. -\frac{3 \text{ Re}}{8}(2 + R^{-1} + R^{-2}) \cos \theta\right] \qquad (11.140)$$

The two-term Stokes expansion for the vorticity can be calculate⁻ from Eq. (11-140), with the result

$$\Omega_S(R, \theta; \text{Re})$$

$$= -(3 \sin \theta/2R^2)\{(1 + 3 \text{ Re}/8) + (\text{Re } R/8) \cos \theta(4 - 9R^{-1} + 3R^{-2} - 2R^{-3})\}. \qquad (11.141)$$

For comparison, the two-term Oseen expansion for vorticity may be rewritten from Eq. (11-137), with $A = 3/4$ and the substitution $\rho = \text{Re } R$. The result is

$$\Omega_O(R, \theta; \text{Re}) = -(3 \sin \theta/2R^2)(1 + \text{Re } R/2) \exp \{-(\text{Re } R/2)(1 - \cos \theta)\} \qquad (11.142)$$

The composite expansion, giving a uniformly valid approximation. We have constructed two approximate pictures of the flow, one designed to be most accurate near the sphere, the other most accurate far away. What is needed is a single picture, with good accuracy everywhere. Van Dyke gives some recipes for the construction of such a composite picture; the simplest involves adding the two representations at each point, and then substracting what they have in common. Let us do that for the vorticity field.

To find what $\Omega_S(R, \theta; \text{Re})$ and $\Omega_O(R, \theta; \text{Re})$ have in common, we take either the Stokes limit of Ω_O, or the Oseen limit of Ω_S, retaining the first two terms in ascending powers of Re. The result in either case is

$$\Omega_{SO}(R, \theta; \text{Re}) = \Omega_{OS}(R, \theta; \text{Re}) = -(3 \sin \theta/2R^2)(1 + \text{Re } R \cos \theta/2) \quad (11\text{-}143)$$

The two-term composite expansion is then

$$\Omega(R, \theta; \text{Re}) \approx -(3 \sin \theta/2R^2)\{3 \text{ Re}/8 - (\text{Re}/8) \cos \theta(9 + 3R^{-1} - 2R^{-2})$$
$$+ (1 + \text{Re } R/2) \exp \{-(\text{Re } R/2)(1 - \cos \theta)\} \quad (11\text{-}144)$$

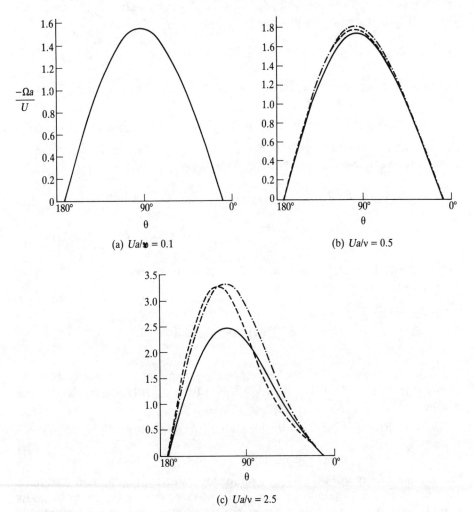

(a) $Ua/\nu = 0.1$

(b) $Ua/\nu = 0.5$

(c) $Ua/\nu = 2.5$

FIGURE 11.26
Surface vorticity when —— full Navier–Stokes equations (Dennis and Walker, 1971[38]), · · · · two-term Stokes expansion (Eq. 11.141), – – – two-term composite expansion (Eq. 11.144).

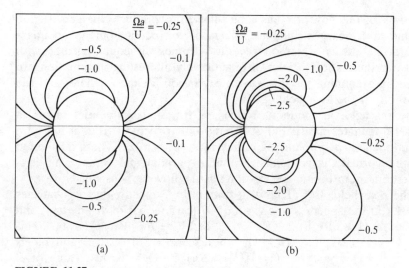

FIGURE 11.27

Comparison of full Navier–Stokes solution to two-term Stokes expansion. Vorticity field. Flow left to right. Top: Dennis and Walker, 1971.[38] Bottom: two-term Stokes expansion. (a) $va/v = 0.5$; (b) $va/v = 5.0$.

To see how the accuracy of this approximation depends on the value of Re, we can make various comparisons with a very highly regarded numerical solution of the full Navier–Stokes equations.[38] Figure 11.26 shows the vorticity at the surface of the sphere, $R = 1$. When Re ≤ 0.5, the values obtained from the composite expansion and from the two-term Stokes expansion are virtually indistinguishable and agree well with the results from the full Navier–Stokes equations. However, this agreement rapidly worsens as Re increases much beyond that value. When Re $= 5$, the curve obtained from the two-term Stokes expansion looks qualitatively correct, but that from the composite expansion has gone wild. Figure 11.27 shows contours of constant Ω, using the two-term Stokes expansion for the lower half of the figure, and the full Navier–Stokes results for the upper half. Again, the agreement is fine when Re $= 0.5$, and there is a qualitative resemblance when Re $= 5$. The corresponding map for the composite expansion bears no resemblance to either top or bottom of Fig. 11.27. This probably happens because the Oseen expansion satisfies neither the no-penetration condition nor the no-slip condition, and gives an exaggerated picture of the effect of convection near the sphere that is not adequately corrected by the procedure used to form the composite expansion.

The quantitative results obtained by matching asymptotic inner and outer

[38] S. C. R. Dennis and J. D. A. Walker, "Calculation of the Steady Flow Past a Sphere at Low and Moderate Reynolds Numbers," *J. Fluid Mech.* **48**: 771–790 (1971). (Note that their Reynolds number is $2Ua/u$.)

representations of this flow are thus somewhat disappointing. One gets a good picure of the first, nearly negligible, effects of convection, but no adequate picture of these effects when convection becomes an equal partner with diffusion. Nevertheless, quite a lot of qualitatively interesting information can be obtained by examining just the Oseen expansion for the far field, and the Stokes solution for the near field.

In the far field, at moderate values of Re, the exponential factor in Eq. (11-143) dominates the spatial distribution of vorticity. It is constant on a downstream-opening paraboloid of radius r, $z + z_0 = r^2/z_0$. Thus, it is often said that the vorticity is contained within a paraboloidal wake. This is a valuable description of the far field as long as the flow remains steady.

For the near field, Eq. (11-140) factors nicely, to show that ψ_S vanishes where $R = 1$ (on the sphere), where $\sin \theta = 0$ (on the axis of symmetry), and finally where $(1 + 3\,\mathrm{Re}/8)(2 + R^{-1}) = (3\,\mathrm{Re}/8)(2 + R^{-1} + R^{-2}) \cos \theta$. This streamline leaves the sphere $(R = 1)$, where $\cos \theta = 3/4 + 2/\mathrm{Re}$, and meets the axis $(\cos \theta = 1)$ where $R = (1/4)\{(1 + 3\,\mathrm{Re})^{1/2} - 1\}$. This last value of R must equal at least 1 for the eddy to exist; this happens first when $\mathrm{Re} = 8$.

The calculations of Dennis and Walker show that an eddy qualitatively very similar to that described by two-term Stokes expansion appears first when $\mathrm{Re} \approx 10.1$. This value of Re is far out of the range for which the procedures of asymptotic analysis can be justified.

Higher-order approximations. Hoping to extend the range of Re for which the matched expansions give accurate estimates, various workers have continued the expansions to higher orders. However, the results are quite disappointing. Each expansion apparently constitutes an asymptotic series, rather than a convergent one, so that the accuracy of representation at a given value of Re is not much improved by adding more terms to each series. This is demonstrated by predictions of the drag. The two-term approximation that has been presented here gives the estimate

$$\text{Drag} = 6\pi\mu U a \left(1 + \frac{3\,\mathrm{Re}}{8}\right)$$

The highest-order analytical estimate of this type gives the formula

$$\text{Drag} = 6\pi\mu U a \left[1 + \frac{3\,\mathrm{Re}}{8} + \frac{9\,\mathrm{Re}^2}{40} \left[\ln\left(\frac{\mathrm{Re}}{2}\right) + 0.8302\right] + \frac{27}{80}\,\mathrm{Re}^3 \ln(\mathrm{Re})\right]$$

A careful comparison with experimental and numerical data[39] indicates that this is better than the first term alone when $\mathrm{Re} < 0.3$, but only marginally better when $\mathrm{Re} = 1$. Thus, even for quite low values of Re, we have to turn to numerical solution of the full Navier–Stokes equations to get an accurate

---

[39] See Dennis and Walker (1971) *op. cit.*, Fig. 1.

description. Even this is easier said than done, as is shown by the lack of precise agreement between the results produced by different numerical algorithms.

The Circular Cylinder; Stokes' Paradox Resolved

A similar analysis can be made for the circular cylinder in crossflow, but the practical reward is even smaller, because the expansion parameter turns out to be $\varepsilon = [\ln (3.703/\text{Re})]^{-1}$, which approaches zero much more slowly than Re itself.[40] The most extended analysis of this type gives the following prediction for the drag per unit length:

$$\frac{\text{Drag}}{\text{Length}} = 4\pi\mu U[\varepsilon - 0.87\varepsilon^3 + o(\varepsilon^4)] \tag{11.142}$$

This fits experimental data rather well if $\text{Re} < 0.4$, but misses by almost a factor of 2 when $\text{Re} = 1$.

An interesting question may be posed at this point. Suppose you wish to predict the drag per unit length at the midsection of a finite cylinder with the following specifications: $\text{Re} = Ua/v = 0.1$, $E = \text{length/diameter} = 500$. Which formula should you use: (11.121) for a finite, but very long, cylinder, or (11.142) for an infinite cylinder? Note that the two formulas make very different predictions, the ratio of which is

$$\frac{\text{Drag (finite)}}{\text{Drag (infinite)}} \approx \frac{\ln (3.703/\text{Re})}{\{\ln (2E) + 0.5\}}$$

For this example, the ratio equals 0.301, so the choice is not just a matter of taste! We are used to thinking that a very long cylinder can be conveniently idealized as a piece of an infinite cylinder, but this is clearly not so in creeping flow, in which the freedom of the fluid to escape around the ends of a finite cylinder has an essential effect. You would have to think carefully about other constraints on the flow. For example, is the cylinder of interest stretched between two large, parallel, walls, or is it falling freely through a much larger region of fluid?

Lift on a Sphere in Shear Flow

It can be shown theoretically that a solid sphere, creeping along a streamline of a steady shear flow such as laminar, fully-developed pipe flow, may experience a side force or lift as a consequence of small inertial effects. If its density is

---

[40] See Van Dyke (1975), *op. cit.*, pp. 161–165, for the analysis and criticism.

nearly the same as that of the fluid, the sphere, like a swimming microorganism, will move so that the net force on it will be very nearly zero. This means that it will drift sideways, crossing the streamlines of the carrier flow. This effect was first deduced from experiments,[41] which demonstrated a tendency for small, neutrally buoyant spheres in a Poiseuille flow to migrate towards a preferred radial position $r \approx 0.6 r_p$ in the pipe.

Many theoretical analyses have been constructed, each focusing on some particular set of circumstances that offered some simplification of an impossibly complicated general problem, and each trying to isolate and quantify the effect that is dominant in those circumstances. It has been shown, for example, that if there is a substantial differential speed between the sphere and the carrier fluid, the sphere will tend to move sideways, so as to increase the absolute magnitude of that differential speed. This situation might be encountered when the carrier flow is vertical and the sphere is positively or negatively buoyant. The relevant analysis[42] involves construction of matched inner and outer expansions, and is too complicated to be reproduced here, even though it takes no explicit notice of the presence of the tube or channel walls.

Briefly, the motion of the sphere was characterized by the relative speed, V, and by an angular velocity, ω, parallel to the vorticity of the carrier flow. The carrier flow was characterized by a constant rate of shear, $k = dU/dy$. There are three relevant Reynolds numbers, all based on the sphere radius a. They are $\mathrm{Re}_V = Va/\nu$, $\mathrm{Re}_\omega = \omega a^2/\nu$, and $\mathrm{Re}_k = ka^2/\nu$. All were required to be $\ll 1$, and there was a subsidiary requirement that $\mathrm{Re}_V \ll \mathrm{Re}_k^{1/2}$. The theoretical estimate of the side force is then

$$L = 6.46 \mu V a^2 \left(\frac{k}{\nu}\right)^{1/2} \tag{11.143}$$

The direction of the force is shown in Fig. 11.28. To this level of approximation, there is no effect of the sphere's rate of rotation.

Another important analysis,[43] of the motion of neutrally buoyant spheres, exploited the assumption that the Reynolds number of the carrier flow itself is $\ll 1$. This seems very restrictive, but it avoids the necessity for constructing matched asymptotic expansions, because even the walls of the pipe or channel lie within the domain of validity of the inner, Stokes, expansion for the flow perturbation caused by the sphere. The reward was quite spectacular: a clearcut prediction that these spheres should migrate in a

---

[41] G. Segré and A. Silberberg, "Behavior of Macroscopic Rigid Spheres in Poiseuille Flow. Part 2. Experimental Results and Interpretation," *J. Fluid Mech.* **14**: 136–157 (1962).

[42] P. G. Saffman, "The Lift on a Small Sphere in a Slow Shear Flow," *J. Fluid Mech.* **22**: 385–400 (1965).

[43] B. P. Ho and L. G. Leal, "Inertial Migration of Rigid Spheres in Two-dimensional Unidirectional Flows," *J. Fluid Mech.* **65**: 365–400 (1974).

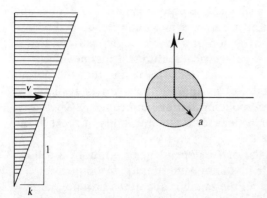

FIGURE 11.28
Side force on a sphere creeping through a
shear flow.

Poiseuille flow in just the way they are observed to! Under these conditions, the side force deduced by Saffman is relatively unimportant, because the only cause of a relative speed V is a rather subtle increase of sphere drag, due to the presence of the walls.

11.11 CREEPING FLOW THROUGH A POROUS MEDIUM

When a fluid seeps through a porous rock formation, or through a rigidly packed granular solid, individual flow passages are often tiny, the velocity of flow in them is small, and inertia forces are negligible. To study such a flow in the large, so as to predict, for example, the rate of seepage under a dam, one quickly abandons any hope of predicting the details of flow in any given interstice, and attempts to construct a simple statistical picture. One imagines a plane slicing through the porous material and oriented by a unit normal vector **n**. A *superficial velocity* **u** is defined by saying that $\mathbf{n} \cdot \mathbf{u}$ is the volumetric flow rate across unit area of that plane. As in the construction of a continuum picture of a fluid, one hopes to find that the flow rate per unit area comes to a definite limit as the test area is reduced, before statistical fluctuations due to the finite size of individual flow passages appear.

In the most idealized porous media, like a bed of tightly-packed, single-sized, spheres, the viscous drags of all the solid particles in a given volume add up to an average force per unit volume that is proportional to $-\mu\mathbf{u}$, where μ is the fluid viscosity. Since there is no significant inertia, this force is balanced against gravity and an average pressure force per unit volume. We symbolize the latter combination, as before, by $-\rho\boldsymbol{\nabla}\Pi$. The result of this model is an equation called Darcy's law:

$$\mathbf{u} = -\frac{k}{\nu}\boldsymbol{\nabla}\Pi \tag{11.144}$$

The new factor of proportionality in this, k, is called the *permeability* of the medium. It is measured by placing a plug of the porous material in a tube, and measuring the pressure drop required to drive a measured flowrate. Presumably, the value of k depends on the fraction of a macroscopic test area that is open to flow, and on other averaged geometrical features of the medium.

By letting the permeability be a scalar, one is assuming that the porous medium is isotropic. If a good model of anisotropic behavior is available, it may be possible to construct a corresponding tensor permeability. Obviously, it may be very difficult to know good values for the permeability of rocks found in the field.

If the factor (k/v) is both scalar and independent of position, it follows that the field of **u** is irrotational. If in addition the fluid is incompressible, Π will obey the Laplace equation and all the familiar apparatus of potential-flow theory will be available.

Actually, flow through porous media is often a great deal more complicated than this introduction may suggest. A typical practical problem, secondary oil recovery by steam injection, involves two-phase flow, large temperature gradients, exquisitely complicated effects of surface tension, and important chemical reactions between the solid and the fluid. Roughly speaking, the practical problem is that the steam may not effectively drive the oil ahead of it, but may find a path of least resistance that bypasses the oil.[44] To understand some of these effects, idealized problems of creeping flow in individual interstices are studied.

Another important deviation from idealized flow occurs near impermeable boundaries of a porous medium, particularly if there is an important transfer of heat or mass from these boundaries to the fluid creeping through the porous medium. There are two readily identified deviations from ideal behavior near the impervious wall.

1. A no-slip condition will be imposed naturally there, but there is no room for such a condition in the theory of irrotational flow.
2. The permeability of the medium, especially if it is granular or fibrous, may change appreciably very near the wall, which constrains how closely the solid elements can be packed.

11.12 SUMMARY

Creeping flows exhibit negligible, or almost negligible, effects of inertia and/or convection. Thus, each fluid particle moves under a nearly exact balance of surface and body forces, and the vorticity field is established almost exclusively

---

[44] A good starting point for further study is J. O. Richardson, *Flow Through Porous Media*, Section 16 of *Handbook of Fluid Dynamics*, Editor-in-Chief V. L. Streeter, McGraw-Hill, New York, 1961.

by diffusion. This state of motion appears when an appropriate characteristic Reynolds number has a very small value.

Flows without inertia exhibit effects, such as fore-and aft symmetry and instantaneous response to changes in the motion of solid boundaries, that are startling when first seen, because we are so used to watching flows with lots of inertia.

When the viscous fluid is constrained, by the presence of nearly parallel walls, or by the act. In of a body force, to move in a slender layer, the appropriate Reynolds number is based on the thickness of the layer, and upon a typical value of the velocity component normal to the layer. The typical distribution of the velocity component along the layer is the sum of a shear-driven straight-line profile, and a pressure-driven parabolic profile. If, as in lubrication applications, the amplitude of the shear-driven component is specified, the continuity equation then determines the amplitude of the pressure-driven component. This permits calculation of the unknown pressure field. Sometimes the viscous film has a free surface, the position and shape of which determines the pressure field. Then the continuity equation determines the evolution of the film thickness.

Except in problems involving a free surface of unknown location, the equations of creeping flow are linear, and useful solutions can be found by separation of variables or superposition of fundamental singular solutions. The first technique is used to find several interesting solutions for creeping flow in a region partially bounded by intersecting planes, and to find the flow around isolated spheres and spheroids. Two memorable conclusions about the drag of isolated rigid bodies are that (1) the drag is always larger than that of any sphere that can be fitted inside the body, and less than that of any sphere that will contain the body; (2) the drag of a long, slender body of revolution moving normal to its axis is about twice as large as that of the same body moving parallel to its axis. The technique of superposition of singular solutions is especially useful for analysis of the motion around long, slender bodies, bodies near other bodies, and bodies near walls.

Whereas the smallness of an appropriate Reynolds number guarantees that inertia and/or convection can be safely ignored everywhere in a bounded region, such as the fluid-filled region of a bearing; it does not give the same guarantee for the unbounded region outside an isolated, translating body. For the first class of problems, small effects of inertia may be predicted by a *regular* perturbation analysis, in which dependent variables are ordinarily described by convergent series in increasing powers of the Reynolds number. The second class of problems must be dealt with by a *singular* perturbation analysis, in which different asymptotic series may be used to represent the dependent variables in regions close to the body or far away. By an appropriate matching procedure, the far-field boundary values that cannot be directly accommodated by the inner, or Stokes, expansions are made accessible by means of an outer, Oseen, expansion.

Many technically important creeping flows involve a multitude of closely neighboring, moving bodies, or flow through a geometrically complex matrix of stationary, possibly interconnected, bodies. The former type of flow, often called the flow of suspensions, is of great interest to chemical engineers, but its theory is too complicated for this text.[45] The latter, usually called flow through porous media, occasionally submits to a very simple macroscopic theory, embodied in Darcy's law. In this theory, the quantity $\Pi = p/\rho + gh$ acts as a potential for the superficial velocity, and obeys the Laplace equation.

11.13 SAMPLE CALCULATIONS

A. The cone-plate viscosimeter. The axis of a rotating cone is normal to the plane of the stationary plate; the apex of the cone very nearly touches the plate. The semi-apex angle of the cone is $\beta = \pi/2 - \alpha$, where $\alpha \ll 1$. The angular velocity of the cone is ω. Thus, in spherical polar coordinates, with polar axis coincident with the cone axis, the boundary conditions on the velocity of the fluid that fills the space between plate and cone are well approximated by

$$u(R, \pi/2) = v(R, \pi/2) = w(R, \pi/2) = 0;$$

$$u(R, \beta) = v(R, \beta) = 0; \quad w(R, \beta) = \omega R \sin \beta \approx \omega R.$$

This is rather like the example of a pool of viscous liquid on a spinning disk, in which the radial flow is driven by the centrifugal pressure gradient, $\partial \Pi / \partial r \approx \rho \omega^2 R$. This results in a typical value, U, of the radial velocity component, which is proportional to $(H^2/\mu) \, \partial \Pi / \partial r$. Noting that $H = \alpha R$, we write this for future use as $U = \omega^2 \alpha^2 R^3 / \nu$. The continuity equation then determines that a typical value, V, of the latitudinal velocity component, on which we wish to base the Reynolds number, will be proportional to αU. Defining $\text{Re} = VH/\nu$, we draw all this together and see that the appropriate Reynolds number is $\text{Re} = (\omega/\nu)^2(\alpha R)^4$. We note that this gets rapidly larger as R increases, so that creeping flow will exist only near the axis. Nevertheless, we shall proceed as in the examples of the corner flows, to see whether a useful solution for small values of R can be obtained without explicit consideration of conditions far from the axis.

Let us now define dimensioness variables $u = u^*/U$, $v = v^*/V$, $w = w^*/(\omega R)$, $\Pi = \Pi^*/(\omega R)^2$. For convenience, we introduce a magnified angular variable, $\xi = (\pi/2 - \theta)/\alpha$. For dimensional homogeneity, u, v, w, and Π will depend on ξ, and on some dimensionless form of R, such as Re. For small

[45] An important advanced treatise that deals extensively with these problems, and with many others that are beyond the scope of this text, is J. Happell and H. Brenner, *Low Reynolds Number Hydrodynamics*, 2d ed., Nordhoff Int. Publ., Leyden, 1973.

values of Re, we postulate the regular perturbation expansions

$$u(\xi, \text{Re}) = u_1(\xi) + \text{Re} \, u_2(\xi) + \cdots$$
$$v(\xi, \text{Re}) = v_1(\xi) + \text{Re} \, v_2(\xi) + \cdots$$
$$w(\xi, \text{Re}) = w_0(\xi) + \text{Re} \, w_1(\xi) + \cdots$$
$$\Pi(\xi, \text{Re}) = \Pi_0(\xi) + \text{Re} \, \Pi_1(\xi) + \cdots$$

and substitute them into the full Navier–Stokes equations in spherical polar coordinates. In this substitution, we shall consistently neglect quantities of order α^2. For example, the exact continuity equation gives $5u + v' + \alpha v \, \text{ctn} \, \theta = 0$. Because $\alpha \ll 1$, $\text{ctn} \, \theta \approx \alpha \xi$. Then the last term in the continuity equation is of order α^2 compared to the first, and is dropped. This leaves

$$5u_1 + v_1' = 0 \quad \text{and} \quad 9u_2 + v_2' = 0$$

where a prime denotes $d/d\xi$.

After some algebra, which is left as an exercise, the ϕ-momentum equation yields

$$w_0'' = 0 \quad \text{and} \quad w_1'' = 2u_1 w_0 + v_1 w_0'.$$

The θ-momentum equation yields

$$\Pi_0' = 0 \quad \text{and} \quad \Pi_1' = 0$$

The R-momentum equation yields

$$u_1'' = 2\Pi_0 - w_0^2 \quad \text{and} \quad u_2'' = 6\Pi_1 - 2w_0 w_1 + 3u_1^2 - v_1 u_1'$$

In terms of the new variables, the boundary conditions become

$$u_1(0) = v_1(0) = w_0(0) = 0; \quad u_1(1) = v_1(1) = 0, \quad w_0(1) = 1$$
$$u_2(0) = v_2(0) = w_1(0) = 0; \quad u_2(1) = v_2(1) = w_1(1) = 0.$$

Finally, we need a condition to determine the numerical values of Π_0 and Π_1. These come from the fact that there is no net volumetric flux across any surface of constant R, so that

$$\int_0^1 u_1 \, d\xi = \int_0^1 u_2 \, d\xi = 0$$

These equations can now be solved sequentially, to determine, in the following order, that $w_0 = \xi$; $\Pi_0 = \text{constant}$; and $u_1 = \Pi_0(\xi^2 - \xi) + \frac{1}{12}(\xi - \xi^4)$. The integral constraint now gives $\Pi_0 = \frac{3}{20}$, so that $u_1 = -\frac{1}{60}\xi(5\xi^3 - 9\xi + 4)$, and $v_1 = \frac{1}{12}\xi^2(\xi^3 - 3\xi + 2)$. This completes one cycle of calculation. The second cycle starts out with $w_1 = -\frac{1}{2520}\xi(5\xi^6 - 63\xi^4 + 7\xi^3 + 51)$.

From these results, one can compute the torque T required to turn the cone steadily. If the radius of the cone is a (see Fig. 11.29(a)), we find

$$T = 2\pi \frac{\mu\omega}{\alpha} \int_0^a \left[w_0'(1)R^2 + \left(\frac{\omega}{\nu}\right)^2 \alpha^4 R^6 w'(1) \right] dR$$

which gives
$$T = \frac{2\pi}{3} \frac{\mu \omega a^3}{\alpha} \left[1 + \left(\frac{201}{2520} \right) \text{Re}\,(a) + \cdots \right]$$

Another measurable quantity is the angle, γ, between a circle of constant R, and the skin friction vector on the plate. (See Figure 11.29(b).) This is given by

$$\tan \gamma = \frac{\tau_{\theta R}}{\tau_{\theta \phi}} = \left(\frac{u^*_{,\theta}}{w^*_{,\theta}} \right)_{\theta = \pi/2} = -\tfrac{1}{15} \text{Re}^{1/2} + \cdots$$

One special feature of this analysis merits a comment. The result shows that flow moves radially inward near the plate, and returns outward near the cone. However, the solution was obtained without any specification of boundary conditions at the edge of the cone. Presumably, this flow is somewhat like the flows in corners, which were analyzed in Section 11.6, in that actual boundary conditions at a specified value of R cause changes that die out fairly rapidly as R decreases, so that the present solution gives a good approximation at points not too close to the edge of the cone.

B. Trajectory of a small droplet, in a nonuniform flow. A small water droplet, of radius a, is entrained in an airflow that passes around a circular cylinder. Assume that any relative motion between the droplet and the air is resisted by

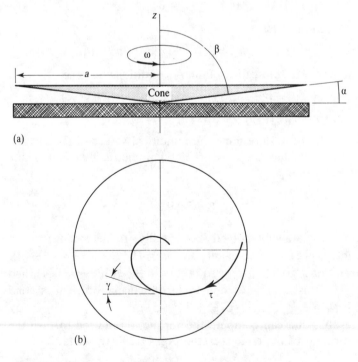

(a)

(b)

FIGURE 11.29
Spinning-cone viscosimeter: (a) definition of geometry; (b) orientation of skin friction on plate.

a Stokes drag, $\mathbf{F} = 6\pi\mu a(\mathbf{u} - d\mathbf{r}/dt)$, where $\mathbf{u}$ is the local air velocity and $\mathbf{r}$ is the instantaneous position vector of the droplet, relative to the center of the cylinder. Assume also that F is the only force acting on the droplet, and derive differential equations for $r(t)$ and $\theta(t)$, where r and θ define the vector $\mathbf{r}(t)$. Let $\mathbf{u} = u(r, \theta)\mathbf{e}_r + v(r, \theta)\mathbf{e}_\theta$.

We apply Newton's second law to the droplet, getting $m\, d^2\mathbf{r}/dt^2 = \mathbf{F}$. Since

$$\mathbf{r}(t) = r(t)\mathbf{e}_r(\theta(t))$$

$$\frac{d\mathbf{r}}{dt} = \left(\frac{dr}{dt}\right)\mathbf{e}_r + r\left(\frac{d\mathbf{e}_r}{d\theta}\right)\left(\frac{d\theta}{dt}\right) = \left(\frac{dr}{dt}\right)\mathbf{e}_r + r\left(\frac{d\theta}{dt}\right)\mathbf{e}_\theta$$

and

$$\frac{d^2\mathbf{r}}{dt^2} = \left(\frac{d^2 r}{dt^2}\right)\mathbf{e}_r + \left(\frac{dr}{dt}\right)\left(\frac{d\mathbf{e}_r}{d\theta}\right) + \left(\frac{dr}{dt}\right)\left(\frac{d\theta}{dt}\right)\mathbf{e}_\theta$$

$$+ \left(\frac{d^2\theta}{dt^2}\right)\mathbf{e}_\theta + r\left(\frac{d\theta}{dt}\right)\left(\frac{d\mathbf{e}_\theta}{d\theta}\right)\left(\frac{d\theta}{dt}\right)$$

$$= \left[\frac{d^2 r}{dt^2} - r\left(\frac{d\theta}{dt}\right)^2\right]\mathbf{e}_r + \left\{r\left(\frac{d^2\theta}{dt^2}\right) + 2\left(\frac{dr}{dt}\right)\left(\frac{d\theta}{dt}\right)\right\}\mathbf{e}_\theta$$

Inserting these expressions for the velocity and acceleration of the droplet, we get the coupled equations

$$m\left[\frac{d^2 r}{dt^2} - r\left(\frac{d\theta}{dt}\right)^2\right] + 6\pi\mu a\left(\frac{dr}{dt}\right) = 6\pi\mu a u(r, \theta)$$

$$m\left[r\left(\frac{d^2\theta}{dt^2}\right) + 2\left(\frac{dr}{dt}\right)\left(\frac{d\theta}{dt}\right)\right] + 6\pi\mu a r\left(\frac{d\theta}{dt}\right) = 6\pi\mu a v(r, \theta)$$

Before integration, it is convenient to introduce dimensionless variables, $\xi = r/A$, $\tau = Ut/A$, $f(\xi, \theta) = u/U$, and $g(\xi, \theta) = v/U$; and a dimensionless parameter $K = 6\pi\mu aA/(mU)$. Then the trajectory equations become

$$\frac{d^2\xi}{d\tau^2} - \xi\left(\frac{d\theta}{d\tau}\right)^2 + K\left(\frac{d\xi}{d\tau}\right) = Kf(\xi, \theta) \tag{11.145}$$

$$\xi\left(\frac{d^2\theta}{d\tau^2}\right) + 2\left(\frac{d\xi}{d\tau}\right)\left(\frac{d\theta}{d\tau}\right) + K\xi\left(\frac{d\theta}{d\tau}\right) = Kg(\xi, \theta) \tag{11.146}$$

The value of the parameter K determines how closely the droplet follows the streamlines of the air flow. When it is large, the droplet follows the flow very closely; this is the condition needed for laser Doppler anemometry. When it is small, droplets starting in front of the cylinder run right into it; this is the beginning of the problem of aircraft icing.

C. A free-falling prolate spheroid. A prolate spheroid, with $E = 50$, is falling at its terminal velocity, without rotation. The angle between the major axis and the vertical is θ. At what angle to the vertical, β, will its center of gravity

FIGURE 11.30
Falling prolate spheroid.

move? What is the maximum value of $\beta(\theta)$, and how rapidly will the spheroid lose altitude in this orientation, compared to its rate of fall when $\beta = 0$? Refer to Fig. 11.30.

From Eq. (11.121), we calculate that $K = D_n U_t / D_t U_n = 1.609$. Since the normal and tangential components of hydrodynamic force must add vectorially to cancel the weight of the body, we get $\tan \theta = D_n / D_t$. From the figure, we see that $\tan (\theta - \beta) = U_n / U_t$. Thus $\tan \theta = K \tan (\theta - \beta)$, which is easily solved to get

$$\tan \beta = \frac{(K - 1) \tan \theta}{K + \tan^2 \theta}$$

By differentiation, we find that $\tan \beta$ is maximum when $\tan^2 \theta = K$, and that

$$\beta_{max} = \arctan \left(\frac{K - 1}{2\sqrt{K}} \right)$$

For the given spheroid, $\beta_{max} = 13.499°$, when $\theta = 51.75°$.

The speed of the inclined spheroid may be called U_1, that of the vertically aligned one U_2. Then the ratio of rates of fall is $R = U_1 \cos \beta / U_2$. The weight of the spheroid is the same in either case, so $(D_n^2 + D_t^2)_1 = D_{t2}^2$.

From this, together with $\tan \theta = (D_n / D_t)_1$, we get $D_{t1} / D_{t2} = U_{t1} / U_2 = \cos \theta$. But

$$\frac{U_{t1}}{U_1} = \cos (\theta - \beta)$$

so

$$R = \frac{\cos \theta \cos \beta}{\cos (\theta - \beta)} = (1 + \tan \theta \tan \beta)^{-1} = \cos^2 \theta (1 + K^{-1} \tan \theta)$$

For the value of θ that maximizes β, this gives $R = 2/(1 + K)$. For our spheroid this value equals 0.767.

EXERCISES

11.1. Prove that in a plane creeping flow, curves of constant Π (isobars) are orthogonal to curves of constant Ω (isovorticity contours.) Is this also true of axisymmetric flows in which $\Omega = \Omega \mathbf{e}_\phi$?

11.2. Use lubrication theory to estimate the pressure variation along a tube, the radius of which is specified by $a(z) = a_0(1 + (\alpha z)^2)^{1/2}$, where z is distance along the axis of the tube. Express your answer in the form $\Pi(Z) - \Pi(0) = Pf(\xi)$, where $P = 4\mu Q/(\pi \alpha a_0^4)$ and $\xi = \alpha z$. Calculate the pressure drop between $z = -L/2$ and $z = +L/2$, and find the radius, a^*, of a straight-walled tube which would cause the same pressure drop. Express this answer as $a^*/a_0 = g(\alpha L)$. Compare these values of a^* with a simple arithmetic average of $a(L)$ and a_0.

11.3. To judge the applicability of lubrication theory to flow in a tube of variable radius, $a(z)$, we can evaluate the crossflow Reynolds number, $\text{Re} = (Q/\pi va) \, da/dz$. If the longitudinal section of a tube is a hyperbola, so that $a(z) = a_0(1 + (\alpha z)^2)^{1/2}$, the maximum value of Re, as a function of z, is given by $\text{Re}_{\max} = (Q\alpha/2\pi v)$. Prove this, and find the value of z at which the maximum occurs. Show that $\alpha^2 = (a_0 R_0)^{-1}$, where R_0 is the radius of curvature, at $z = 0$, of the longitudinal section.

11.4. Calculate the pressure distribution, lift, and drag for a stepped slider bearing, for which H is constant over a length L, and then drops abruptly to zero at the trailing edge. Admitting that there must be some leakage of lubricant under the trailing edge, suppose that this edge has length L^*, and clearance H^*. Assume that ambient pressure exists at the rear end of this trailing passage and estimate the fractional reduction of lift as a function of H^*/H and L^*/L.

11.5. Calculate from Eq. (11.31) the moment about the trailing edge, $x = 0$, produced by the distributed pressure on the inclined slider. Locate the center of pressure.
Answer:

$$M = \frac{3\mu U H_1}{\alpha^3} \frac{(5 + R)(R - 1) - (2 + 4R) \ln R}{R + 1}$$

11.6. Repeat the analysis of lubrication theory for a squeezed film when the moving plate is a circular disc of radius a. Suppose that the axis of symmetry is vertical; that the force pushing the disk toward the stationary plane is just the weight, W, of the disk; and that changes in inertia of the disk are negligible. Show that the Reynolds number based on the speed of the disk and the thickness of the film can be calculated from $\text{Re} = (2\pi/3)(W/\rho v^2)(H/a)^4$.

11.7. Using Eq. (11.34), show that the slope of the dividing streamline in the flowfield under a slider bearing for which $R > 2$ is given by $dz/dx = -2\alpha$ at the point where the streamline meets the slider.

11.8. A slider bearing loses lift when fluid is free to escape past the side edges of the slider. Applying the technique of separation of variables (used to analyze the effect of sidewalls on a Couette flow, Section 9.3), estimate this loss of lift for a slider of length L and width W. You may assume that $W \gg L$ and that the pressure equals p_a all around the edge of the slider.

11.9. In his movie, Sir Geoffrey Taylor demonstrates a *teetotum*, made of three cardboard blades which are attached to a central shaft, rather like a three-bladed propeller. The blades rest on a smooth table, and are inclined to it, so that they act as the sliders of a slider bearing when a torque is applied to the shaft. Then the teetotum rises slightly off the table, and spins for a long time, until the drag of the air film beneath it finally brings it to rest.

Make a theoretical model of the performance of the teetotum, assuming that its weight is exactly balanced against the lift of the three blades, aiming to predict its angular speed as a function of time. Assume that the only torque acting on it after it is launched is due to the drag of the blades, and that this and the lift of the blades can be approximated by Eqs. (11.36) and (11.35). These equations give force per unit width of the blade; for a first approximation you may simply multiply them by the width of the blade and assume that the forces are applied at the geometrical center of the blade. To be specific, you may assume that each blade surface is a square, $L \times L$, and that its center is at a distance $2L$ from the axis.

11.10. Suppose the angular velocity and the load on a journal bearing are held constant, while the oil warms up as a result of the work done on it. How does the resulting decrease of viscosity affect the rate of work input? Don't forget that the eccentricity of the shaft position will change as a result of the heating.

11.11. From Eq. (11.47), show that DH/Dt, the rate of normal displacement of a fluid particle on the interface, equals $\frac{1}{2}\,\partial H/\partial t$ in a viscous film draining down an inclined plane.

11.12. In the text, the gravity-driven flow down an inclined plane was treated approximately, in that the additional pressure-driven flow due to $\partial H/\partial x$ was neglected. Derive the partial differential equation that governs $H(x, t)$ when this effect is added to that of the slope of the plane.

11.13. Derive a formula for the Reynolds number of a pool of viscous liquid that is thinning because of centrifugal forces. Use the definition $\mathrm{Re} = -(H/v)DH/Dt$, where H is evaluated at the center of the pool.

11.14. Derive Eq. (11.54) for the time of first formation of breakers in a film thinned by centrifugal force.

11.15. For the circular pool spreading under the effect of gravity, define a Reynolds number, $\mathrm{Re} = -H(\partial H/\partial t)/v$. Show that the maximum value of this appears at $t = 0$ and $r = 0$, and that the maximum value is given by the formula $\mathrm{Re} = (4/9)(g/v^2)(H_0^5/R_0^2)$. Evaluate Re for the pool of glycerol described in Exercise 11.16. Would a creeping-flow analysis be appropriate for a pool of water of the same initial dimensions?

11.16. A pool of glycerol, with initial radius 5 cm and initial maximum depth 1 cm, is spilled on a flat table. If the analysis culminating in Eq. (11.57) is applicable, at what time will the pool radius be 10 cm?

11.17. Verify Eq. (11.55).

11.18. Check, by differentiation and substitution, that Eq. (11.57) is a solution of Eq. (11.56).

11.19. Calculate the vorticity and pressure fields associated with the streamfunction given in Eq. (11.63). Verify that when $\alpha \ll 1$ the streamwise component of $\nabla \Pi$ is independent of θ, and larger, by a factor α^{-1}, than the cross-stream component.

11.20. Calculate the vorticity and pressure fields set up by the submerging belt (Section 11.6(b)). Can the proposed solution apply to a situation in which the pressure is uniform in the region above the interface?

11.21. Show that the pressure is independent of θ in the flow between swinging walls. (Section 11.6(c)).

11.22. Verify that the streamfunction given in Eq. (11.70) satisfies the biharmonic equation. Show that the vorticity is given by $\Omega = 2Bk \exp{(kx)} \cos{(ky)}$. Remembering that k is a complex number, evaluate the real part of this expression for Ω, letting B be a real number. Calculate and plot $\Omega(x, 0)$ over a range of x for which it is continuously positive. At the value of x, say x^*, for which the curve reached its maximum, calculate and plot $\Omega(x^*, y)$ for $-H \leq y \leq H$. From these two plots, sketch a map of isovorticity lines.

11.23. Show that the Reynolds number of a rigid sphere of radius a and net weight W, falling at its terminal velocity U, tends to the limit $\mathrm{Re} \equiv Ua/v = (1/6\pi)(W/\rho v^2)$ as $a \to 0$. (*Net weight* is weight of sphere minus weight of fluid displaced.)

11.24. Find the terminal velocity and radius of a water droplet falling through STP air with a Reynolds number $Ua/v = 0.1$.

11.25. Assume that $D = 6\pi\mu Ua$ for a solid sphere in creeping flow, even when U is changing with time. Find $U(t)$ for a sphere of density ρ_i, which is released at time $t = 0$ to fall under gravity through a fluid of density ρ. Calculate L/a, where L is the distance the sphere falls before attaining 99% of its terminal speed, as a function of the terminal Reynolds number and the density ratio.

11.26. Show that a fluid sphere of radius a, creeping with speed U through an immiscible fluid, produces the same velocity field in the surrounding fluid as would a solid sphere of radius b, moving at speed V, where

$$b = a\left(1 + \frac{2m}{3}\right)^{-1/2} \quad \text{and} \quad V = U\left(1 + \frac{2m}{3}\right)^{3/2}(1 + m)^{-1}$$

11.27. In the analysis of creeping flow over a fluid sphere, assume a uniform surface tension, Σ, at the interface, and calculate the value of $\Pi - \Pi_\infty$ at the center of the sphere.

11.28. Calculate the flow speed at the center of the inner sphere, in the problem of Stokes flow over a fluid sphere. Show that it equals the maximum speed of a particle on the interface.

11.29. Show that the velocity field of a Stokeslet, whose force is directed along the positive z axis, is

$$(u, v, w) = \frac{F}{8\pi\mu R^3}(zx, zy, z^2 + R^2)$$

where

$$R^2 = x^2 + y^2 + z^2$$

11.30. Show that the velocity field of a double-Stokeslet (two Stokeslets, nose-to-nose) is purely radial, and find the angular variation of the speed. The solution can be obtained by differentiation of the velocity field of a single Stokeslet, shown in Exercise 11.29, with respect to z.

11.31. Verify Eq. (11.125).

11.32. Verify the statements made in the paragraph immediately following Eq. (11.130).

11.33. Complete the calculation set up in Sample Calculation A, to determine $u_1(\xi)$.

11.34. Given the dimensionless airflow velocity components $f(\xi, \theta) = (1 - \xi^{-2}) \cos \theta$, $g(\xi, \theta) = (1 + \xi^{-2}) \sin \theta$, carry out numerical integrations of the differential equations derived in Sample Calculation B, (11.145) and (11.146) to find the value of the parameter K for which a droplet started at $\xi = 10$, $\xi \sin \theta = 0.5$, with initial velocity $d\xi/d\tau = f$, $\xi \, d\theta/d\tau = g$, just hits the cylinder. What is the value of θ where it hits? Assume that a hit is scored when $\xi = 1$.

11.35. Noting that the creeping-flow solution for flow outside a spinning, but nontranslating, sphere is uniformly valid, postulate a regular perturbation expansion for small values of the Reynolds number, $\omega a^2/\nu$. Note that even though there is flow in the ϕ direction, $\partial/\partial\phi = 0$. Thus the continuity equation will be the same as though there were no spinning, and u and v can be derived from a streamfunction, just as in the analysis of flow past a nonspinning sphere.

Define nondimensional variables

$$R = R^*/a, \quad \mathbf{u} = \mathbf{u}^*/\omega a, \quad \Pi = (\Pi^* - \Pi_\infty^*)/\rho\omega^2 a^2, \quad \psi = \psi^*/\omega a^3,$$

and let the expansions take the form $u(R, \theta; \mathrm{Re}) = \mathrm{Re}\, u_1(R, \theta) + \cdots$ and $v(R, \theta; \mathrm{Re}) = \mathrm{Re}\, v_1(R, \theta) + \cdots$, so that $\psi(R, \theta; \mathrm{Re}) = \mathrm{Re}\, \psi_1(R, \theta) + \cdots$. Also use $\Pi(R, \theta; \mathrm{Re}) = \mathrm{Re}\, \Pi_1(R, \theta) + \cdots$, and finally

$$w(R, \theta; \mathrm{Re}) = w_0(R, \theta) + \mathrm{Re}\, u_1(R, \theta) + \cdots.$$

Show that the ϕ component of the vorticity equation gives, in the limit as $\mathrm{Re} \to 0$, the equation

$$D^4\psi_1(R, \theta) = 2w_0\left[\cos\theta \frac{\partial w_0}{\partial R} - \left(\frac{\sin\theta}{R}\right)\frac{\partial w_0}{\partial\theta}\right]$$

Here D^4 is the operator defined just below Eq. (11.136).

Now use $w_0 = R^{-2}\sin\theta$ (Eq. (11.100) in dimensionless variables), and try a solution of the form $\psi_1(R, \theta) = R^n \sin\theta \sin 2\theta$. You should find that $n = -1$ gives the particular integral, while $n = 0$ and $n = -2$ identify the solutions of the homogeneous equation, $D^4\psi_1(R, \theta) = 0$, that behave realistically as $R \to \infty$. Form a linear sum of these solutions and apply the no-slip and no-penetration boundary conditions. The result should be

$$\psi_1(R, \theta) = -\tfrac{1}{32}(1 - R^{-1})^2 \sin\theta \sin 2\theta$$

Finally, show that the vorticity associated with this secondary flow is given by

$$\Omega_1(R, \theta) = \tfrac{3}{16}R^{-2}(1 - R^{-1})^2 \sin\theta \sin 2\theta$$

11.36. Analyze the effects of slight inertia in the axisymmetric squeezed-film problem, when the disk executes the harmonic displacement, $z = z_0[1 + 0.2 \cos(\omega t)]$. Specifically, assume that the Reynolds number, $\text{Re} = \omega z_0^2/\nu$, is very small, and calculate the time-dependent pressure force on the disk, correct to order Re. In particular, obtain a result for the phase shift between the displacement of the disk and the resulting pressure force.

CHAPTER 12

LAMINAR
BOUNDARY
LAYERS

12.1 INTRODUCTION

In Chapter 10, we examined some exact solutions of the Navier–Stokes equations in which vorticity was confined, usually by the effects of convection, to a layer of uniform thickness. The components of vorticity tangent to the layer were proportional simply to the rate of change of an appropriate tangential velocity component with distance normal to the layer. For example, if x and y are the tangential coordinates and z the normal coordinate, the full equations $\Omega_y = u_{,z} - w_{,x}$ and $\Omega_x = w_{,y} - v_{,z}$ were reduced exactly to $\Omega_y = u_{,z}$ and $\Omega_x = -v_{,z}$. This allowed us to calculate u and v from the formulas

$$u(x, y, z) = u(x, y, 0) + \int_0^z \Omega_y \, dz \quad \text{and} \quad v(x, y, z) = v(x, y, 0) - \int_0^z \Omega_x \, dz$$

which are much simpler than the general Biot–Savart law. Also, in these exact solutions, the effects of viscosity are the same as though vorticity had diffused only in the direction normal to the layer.

Whenever vorticity is confined to a *slender* region, either by the effects of convection or because the entire flow moves through a slender passage, these useful simplifications may be recovered as *good approximations*. For the approximations to be good everywhere in the slender region, and at all times, the boundary values imposed on the flow must vary only gradually with distance in the tangent directions. Also, the flow must not spontaneously generate steep tangential variations that are not required by the boundary conditions, i.e., it must remain *laminar*. The corresponding approximate theory is ordinarily called *laminar boundary-layer theory*. Its original publication, by L. Prandtl in 1904, was one of the most important events in the history of our understanding of fluid mechanics, because it offered the first rational explanation of the profound differences between real flows in fluids with very little viscosity, and hypothetical flows in fluids with no viscosity at all.

For most applications of boundary-layer theory, we must know in advance the general location of the region of rotational flow. Specifically, we usually know that the region is wrapped closely around a *reference surface*, such as that of a ball, or a nozzle; or that it surrounds a *reference line*, such as the axis of the jet analyzed in Chapter 9. The reference surface is usually just called the *wall*. The usual object of boundary-layer theory is to predict the thickness of the region as a function of tangential position and/or time, the variation of tangential velocity components in the normal direction, and the tangential stress exerted by the flow on the reference surface. If variable temperature or chemical composition is involved, we wish to predict its profile and the transfer of heat and mass to the wall.

In this chapter, we start by considering, in Section 12.2, a coordinate system that fits a slender region of known location. The boundary-layer equations are first derived in this coordinate system for the case in which the region is wrapped closely around a reference surface whose principal radii of curvature are finite. Then there is a well-defined normal direction at every point of the reference surface, and the local thickness of the layer, δ, is measured along that direction. The boundary-layer equations are derived by requiring that δ be vanishingly small compared to both radii of curvature and to a characteristic tangential length, L. The length L may be recognized in different ways, but is always independent of viscosity.

The modified equations appropriate for a region, like a round jet, which is wrapped around a reference line, rather than a reference surface, are also derived.

Section 12.3 presents the *momentum-integral* equations of the boundary layer, which are obtained by integrating the momentum and continuity equations over the normal coordinate. The unknowns of the momentum-integral theory are δ and a few parameters that characterize, in an integral sense, the shapes of the profiles of the tangential velocity components. In the simplest case, there are two of these *shape factors*. The theory is indeterminate at this stage, there being three unknowns and only one equation.

Numerical integration of the differential boundary-layer equations is now easy for many cases, but there were many years during which elaborate and ingenious schemes for *closing* the momentum-integral theory were invented. We present only one of these, *Thwaites' Method,* which is so easy to use that it has many very useful closed-form solutions.[1] We use the basic equation of Thwaites' method to make some important observations about the response of boundary layers to typical sorts of initial and boundary conditions. We also present a collection of solutions for interesting and varied boundary conditions, and subsequently find them to be quite accurate by comparison with accurate numerical integrations of the partial differential equations.

Section 12.4 deals systematically with the *scaling* of the independent and dependent variables of laminar boundary-layer theory, the ultimate objective being to prepare the equations for efficient numerical solution, as we did for the transient development of the spinning-disk flow and the stagnation-point flows in Section 10.3. An immediate byproduct of this analysis is a carefully framed discussion of *self-similar* solutions of the boundary-layer equations. Several of the most famous and useful self-similar solutions are presented.

Section 12.5 completes the development of an efficient algorithm for numerical analysis of the plane or axisymmetric, steady, boundary layer on a reference surface of finite curvature. Sample calculations[2] illustrate the design of successful transformation formulas, which render the subsequent calculations relatively easy, accurate, and extremely inexpensive. A variety of interesting solutions are presented and some rules of thumb are devised to summarize some of the results.

Section 12.6 shows how distributed suction through a porous wall can modify boundary-layer development, and presents a simple modification of the basic finite-difference scheme, which allows stepwise determination of the distribution of suction velocity that is necessary to control some specified aspect of the boundary-layer development.

Section 12.7 deals with the development of boundary layers within boundary layers, as exemplified by the effects of suddenly-applied suction, and the early development of a wake.

Section 12.8 deals with the complications of boundary-layer behavior that are due to a third velocity component or to unsteadiness. This requires, for the first time, a serious consideration of how information is passed through a boundary layer by convection. The concepts of domain of dependence and of region of influence are introduced and applied.

Section 12.9 presents some examples of two-dimensional, unsteady boundary layers, and Section 12.10 does the same for three-dimensional boundary layers. These examples introduce numerical methods that can be used for a wide class of problems.

---

[1] This term, *closing*, is used here and in Chapter 14 to describe the invention of equations that must be added to equations already in hand to arrive at a mathematically determinate system. Thus, for example, the Newtonian-fluid relation between stress and rate of deformation was invented to help close an indeterminate system of conservation equations.

[2] Many of the examples in Section 12.5 were first worked out, as homework exercises, by graduate students at the University of California, Berkeley. All numerical results shown here were freshly computed for this text.

The behavior of a laminar boundary layer that separates from the wall is discussed in Section 12.11. The effect of separation on the validity of the boundary-layer approximations is discussed for steady, two-dimensional flow. The concept of separation in unsteady or three-dimensional flows is also discussed and exemplified.

Finally, Section 12.12 hints at some of the wealth of analysis aimed at determining the mathematical status of boundary-layer theory as an approximation to the theory of the full Navier–Stokes equations. In particular, it suggests how one may develop a higher-order boundary-layer theory that would be capable of determining the magnitude of errors made when boundary-layer theory is applied in slightly inappropriate circumstances.

12.2 BOUNDARY-LAYER COORDINATES AND EQUATIONS

Coordinate Systems

Consider a plane or axisymmetric body or flow channel, such as the examples sketched in Fig. 12.1. Call the orthogonal coordinates α, β, and γ, with the coordinate surface $\gamma = 0$ coinciding with the wall. Let β increase in the downstream direction, along the wall, and let α increase in the remaining tangential direction, transverse to the flow. We shall initially treat only either bodies that are cylinders, such as the wing, in which case the surfaces of

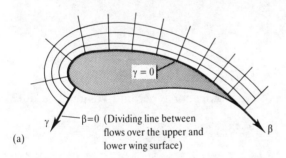

(a)

γ — $\beta = 0$ (Dividing line between flows over the upper and lower wing surface) β

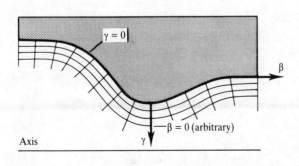

(b)

Axis γ

$\beta = 0$ (arbitrary)

FIGURE 12.1
Boundary-layer coordinate meshes: (a) plane flow over a wing; (b) axisymmetric flow through a nozzle.

constant α are parallel planes, or bodies of revolution, such as the nozzle, for which case the α-surfaces are planes that intersect in the axis of symmetry. For the present, assume that the velocity has the representation

$$\mathbf{u} = u(\beta, \gamma, t)\mathbf{e}_\beta + v(\beta, \gamma, t)\mathbf{e}_\gamma \tag{12.1}$$

A third velocity component, and variation with α, will be restored in Section 12.8.

Kinematics. Even in curvilinear coordinates, the basic expressions for div $\mathbf{u}$, def $\mathbf{u}$, and $\mathbf{u}$ grad $\mathbf{u}$ are much simplified by the restrictions inherent in Eq. (12.1), and by the companion assumptions that no scalar quantity depends on α. Making the corresponding simplifications in Eqs. (A.35), (A.36) and (A.39), we find that

$$\operatorname{div} \mathbf{u} = \tfrac{1}{2}\{d_{\alpha\alpha} + d_{\beta\beta} + d_{\gamma\gamma}\}$$

and

$$\operatorname{def} \mathbf{u} = \begin{bmatrix} d_{\alpha\alpha} & 0 & 0 \\ 0 & d_{\beta\beta} & d_{\gamma\beta} \\ 0 & d_{\beta\gamma} & d_{\gamma\gamma} \end{bmatrix}$$

in which

$$d_{\alpha\alpha} = 2\left[\left(\frac{u}{h_\alpha h_\beta}\right)h_{\alpha,\beta} + \left(\frac{v}{h_\gamma h_\alpha}\right)h_{\alpha,\gamma}\right]$$

$$d_{\beta\beta} = 2\left[\left(\frac{1}{h_\beta}\right)u_{,\beta} + \left(\frac{v}{h_\gamma h_\beta}\right)h_{\beta,\gamma}\right]$$

$$d_{\gamma\gamma} = 2\left[\left(\frac{1}{h_\gamma}\right)v_{,\gamma} + \left(\frac{u}{h_\gamma h_\beta}\right)h_{\gamma,\beta}\right]$$

and

$$d_{\beta\gamma} = d_{\gamma\beta} = \left(\frac{h_\gamma}{h_\beta}\right)\left(\frac{v}{h_\gamma}\right)_{,\beta} + \left(\frac{h_\beta}{h_\gamma}\right)\left(\frac{u}{h_\beta}\right)_{,\gamma}$$

To evaluate the vorticity, and the viscous force per unit volume, we note that

$$\mathbf{\Omega} = \Omega(\beta, \gamma)\mathbf{e}_\alpha$$

with

$$\Omega = \left(\frac{1}{h_\beta h_\gamma}\right)[(vh_\gamma)_{,\beta} - (uh_\beta)_{,\gamma}]$$

and that

$$\operatorname{curl} \mathbf{\Omega} = \left(\frac{1}{h_\alpha h_\gamma}\right)(\Omega h_\alpha)_{,\gamma}\mathbf{e}_\beta$$

$$- \left(\frac{1}{h_\alpha h_\beta}\right)(\Omega h_\alpha)_{,\beta}\mathbf{e}_\gamma$$

Finally,

$$\mathbf{u}\,\operatorname{grad}\mathbf{u} = \left\{\left(\frac{u}{h_\beta}\right)\left[u_{,\beta} + \left(\frac{v}{h_\gamma}\right)h_{\beta,\gamma}\right] + \left(\frac{v}{h_\gamma}\right)\left[u_{,\gamma} - \left(\frac{v}{h_\beta}\right)h_{\gamma,\beta}\right]\right\}\mathbf{e}_\beta$$

$$+ \left\{\left(\frac{u}{h_\beta}\right)\left[v_{,\beta} - \left(\frac{u}{h_\gamma}\right)h_{\beta,\gamma}\right] + \left(\frac{v}{h_\gamma}\right)\left[v_{,\gamma} + \left(\frac{u}{h_\beta}\right)h_{\gamma,\beta}\right]\right\}\mathbf{e}_\gamma$$

Metric coefficients for a boundary layer. It is possible to generate a strictly orthogonal coordinate mesh to fit a wall of very general shape; to do so, one has to relate the metric coefficients[3] of the mesh precisely to the local values of the coordinates, taking into account the shape of the body. The shape is characterized locally by the values of the two principal radii of curvature of the wall: the *longitudinal* radius of curvature, R_L, appears in the intersection of the wall with a plane of constant α; the *transverse* radius of curvature, R_T, is infinite in the case of plane flow, and equal to the local radius of the body of revolution in the case of axisymmetric flow.

Because only a very thin region, proportionally much thinner than is suggested by Fig. 12.1, must be covered, a slightly nonorthogonal coordinate system will serve perfectly well.

It also helps to clarify the physical meaning of the equations if we set $h_\beta = 1$ at the wall, i.e., where $\gamma = 0$. When the wall has *longitudinal curvature*, h_β will depend on the value of γ. For small values of γ, an adequate approximation is $h_\beta = 1 + \gamma/R_L$. If the flow along the body surface starts at some obvious point, such as the forward stagnation point on a wing, the β-surface that contains this point is the origin for β. Otherwise, as in the nozzle, the origin for β is arbitrary, and must play no role in the results.

The third coordinate, α, is distance in the spanwise direction for plane flow, or azimuthal angle for axisymmetric flow. Correspondingly, h_α equals 1 or r, where r is distance to the axis of symmetry. In the latter case, h_α will depend on γ. An adequate approximation for thin layers will be $h_\alpha = R_T + \gamma \cos \phi$ where ϕ is the angle between a normal to the axis, and a normal to the wall, as shown in Fig. 12.2.

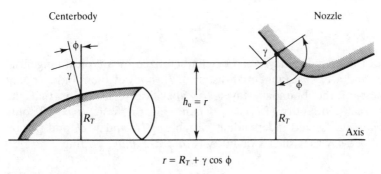

$$r = R_T + \gamma \cos \phi$$

FIGURE 12.2
h_α for axisymmetric flows.

---

[3] A metric coefficient, say h_γ, is defined by the equation $ds_\gamma = h_\gamma \, d_\gamma$, where ds_γ is the local distance between coordinate surfaces $\gamma = \gamma_0$ and $\gamma = \gamma_0 + d\gamma$. See Appendix A for more details. Thus we see $h_\gamma = 1$, so that γ itself is distance from the wall.

We can analyze plane and axisymmetric flows simultaneously, by use of an index j, which equals 0 for plane flow and equals 1 for axisymmetric flow. This involves writing

$$h_\alpha = r^j \approx R_T^j \left[1 + j\left(\frac{\gamma}{R_T}\right) \cos \phi \right]$$

Only the first two terms of a Taylor series are retained, because we shall require that $(\gamma/R_T) \ll 1$ for all points in the boundary layer.

Using these choices to evaluate the metric coefficients and their derivatives, setting $\gamma = 0$ in the resulting formulas, and then inserting the results into the general expressions for plane or axisymmetric flow, we find

$$d_{\alpha\alpha} = \frac{2j}{R_T} \{ uR_{T,\beta} + v \cos \phi \} \qquad d_{\beta\beta} = 2\left(u_{,\beta} + \frac{v}{R_L} \right)$$

$$d_{\gamma\gamma} = 2v_{,\gamma} \qquad d_{\beta\gamma} = d_{\gamma\beta} = u_{,\gamma} + v_{,\beta} - u/R_L$$

$$\Omega = v_{,\beta} - u_{,\gamma} + u/R_L$$

$$-\text{curl } \Omega = \left[\Omega_{,\gamma} + \left(\frac{j\Omega}{R_T}\right) \cos \phi \right] \mathbf{e}_\beta - \left[\Omega_{,\beta} + \left(\frac{j\Omega}{R_T}\right) R_{T,\beta} \right] \mathbf{e}_\gamma$$

$$\mathbf{u} \text{ grad } \mathbf{u} = \left[u\left(u_{,\beta} + \frac{v}{R_L} \right) + vu_{,\gamma} \right] \mathbf{e}_\beta + \left[u\left(v_{,\beta} - \frac{u}{R_L} \right) + vv_{,\gamma} \right] \mathbf{e}_\gamma$$

Scaling. To scale the velocity components and their derivatives in this coordinate system, we start with the statement that $u_{,\beta}$ is of *order of magnitude* U/L in the boundary layer. This is written as

$$u_{,\beta} = O\left(\frac{U}{L}\right)$$

and means that the boundary-layer can be characterized by a typical tangential flow speed, U, and a typical tangential distance, L, both of them independent of the thickness of the boundary layer, δ. Specifically, it means that the dimensionless, scaled quantity $(L/U)u_{,\beta}$ depends, in the limit as $\delta \to 0$, only on the dimensionless, scaled independent variables β/L and γ/δ, and not on the value of δ. We say that such a quantity is *of order unity*, writing

$$\left(\frac{L}{U}\right) u_{,\beta} = O(1)$$

Typical choices of U, L, and δ are:

U = flight speed of a vehicle, or flow speed at the throat of a nozzle

L = body length, or diameter of a nozzle throat

δ = boundary-layer thickness at a forward stagnation point, or at a nozzle throat

Since the shape of the body itself is independent of δ, it follows that R_L/L, R_T/L, and $\cos \phi$ are all of order unity.

We look now at the continuity equation, restricting our interest to incompressible flow, so that div $\mathbf{u} = 0$. Suppose for the moment that the wall is flat, so that the terms involving R_L and R_T do not appear. Then, if $\mu_{,\beta} = O(U/L)$, it follows that $v_{,\gamma} = O(U/L)$. Writing

$$v(\beta, \gamma) = v(\beta, 0) - \int_0^\gamma u_{,\beta} \, d\gamma$$

and noting that, in the boundary layer, $\gamma = O(\delta)$, we see that v is the sum of two terms: the first either equal to zero, for an impermeable wall, or of indeterminate order, if there is suction or blowing through the wall.[4] The second term is $O(U\delta/L)$. With the warning that we must reconsider our conclusions if there is suction or blowing, we forget the first term for the present, and proceed with

$$v = O\left(\frac{U\delta}{L}\right)$$

We now use these ideas to continue reducing the complexity of the kinematic description of the boundary layer. In the following equations, the order of magnitude of each term appears below the term.

$$d_{\alpha\alpha} = \frac{2j}{R_T} \{uR_{T,\beta} + v \cos \phi\} \qquad d_{\beta\beta} = 2\left(u_{,\beta} + \frac{v}{R_L}\right)$$
$$(U/L) \quad (U\delta/L^2) (U/L) \quad (U\delta/L^2)$$

$$d_{\gamma\gamma} = 2v_{,\gamma} \qquad d_{\beta\gamma} = d_{\gamma\beta} = u_{,\gamma} + v_{,\beta} - u/R_L$$
$$(U/L) (U/\delta) \quad (U\delta/L^2) \quad (U/L)$$

$$\Omega = v_{,\beta} - u_{,\gamma} + u/R_L$$
$$(U\delta/L^2) \quad (U/\delta) \quad (U/L)$$

$$-\text{curl } \Omega = \left[\Omega_{,\gamma} + \left(\frac{j\Omega}{R_T}\right) \cos \phi \right] \mathbf{e}_\beta - \left[\Omega_{,\beta} + \left(\frac{j\Omega}{R_T}\right) R_{T,\beta} \right] \mathbf{e}_\gamma$$
$$(U/\delta^2) (U/\delta L) (U/\delta L) (U/\delta L)$$

$$\mathbf{u} \text{ grad } \mathbf{u} = \left[u\left(u_{,\beta} + \frac{v}{R_L} \right) + v u_{,\gamma} \right] \mathbf{e}_\beta$$
$$(U^2/L) \quad (U^2\delta/L^2) (U^2/L)$$

$$+ \left[u\left(v_{,\beta} - \frac{u}{R_L} \right) + vv_{,\gamma} \right] \mathbf{e}_\gamma$$
$$(U^2\delta/L^2) (U^2/L) (U^2\delta/L^2)$$

---

[4] Suction or blowing of even minute flows through a nominally solid wall can have profound effects on boundary-layer behavior, as we shall see later. Suction only improves the accuracy of thin-layer approximations. Blowing may ruin them.

Most of these equations contain terms of two different orders of magnitude, usually differing by a factor δ/L. However, in the expressions for $d_{\beta\gamma}$ and Ω the second term is smaller by a factor $(\delta/L)^2$; thus, as $\delta/L \to 0$, these terms are the first to become negligible.

The Boundary-Layer Equations

When these expressions for $\mathbf{u}\,\mathrm{grad}\,\mathbf{u}$ and $-\mathrm{curl}\,\Omega$ are introduced into the momentum equation for constant-property flow, Eq. (7.18), and all but the largest term in each group of terms are dropped, we find

$$uu,_\beta + vu,_\gamma + \Pi,_\beta = -v\Omega,_\gamma = vu,_{\gamma\gamma}$$
$$(U^2/L) \quad (U^2/L) \qquad\quad (vU/\delta^2) \tag{12.2}$$

$$-\frac{u^2}{R_L} + \Pi,_\gamma = v\left[\Omega,_\beta + \left(\frac{j\Omega}{R_T}\right)R_T,_\beta\right] \tag{12.3}$$
$$(U^2/L) \qquad\qquad (vU/\delta L)$$

We need to conclude from Eq. (12.3) that it is permissible to treat $\Pi,_\beta$ as a function only of β in Eq. (12.2). Using a Taylor series for small values of γ, we can set $\Pi,_\beta(\beta, \gamma) \approx \Pi,_\beta(\beta, 0) + \Pi,_{\beta\gamma}(\beta, 0)\gamma$. We can differentiate Eq. (12.3) with respect to β, and note from the result that $\Pi,_{\beta\gamma}$ is at most of order $(U/L)^2$. Then, since γ is of order δ, the part of $\Pi,_\beta$ that varies with δ is at most of order $\delta(U/L)^2$, which is negligible by comparison with the acceleration terms already in Eq. (12.2). Thus, it is permissible to replace (12.3) with the simple equation

$$\Pi,_\gamma = 0 \tag{12.4}$$

At this level of approximation, the effects of wall curvature have dropped entirely out of the momentum equation, but they remain in the continuity equation, which has become

$$u,_\beta + \left(\frac{ju}{R_T}\right)R_T,_\beta + v,_\gamma = 0$$

A more convenient form of this is

$$(uR_T^j),_\beta + (vR_T^j),_\gamma = 0 \tag{12.5}$$

The boundary-layer approximation to the vorticity equation will often be useful, and is easily obtained by differentiating Eq. (12.2) with respect to γ. Because of Eq. (12.4), this eliminates Π. When the continuity equation is used in its first form to simplify the result, we find that

$$u\left[\quad\Omega,_\beta \quad -\left(\frac{j\Omega}{R_T}\right)R_T,_\beta\right] + \quad v\Omega,_\gamma \quad = \quad v\Omega,_{\gamma\gamma} \tag{12.6}$$
$$(U^2/L\delta) \quad (U^2/L\delta) \qquad (U^2/L\delta) \quad (vU/L\delta^3)$$
$$(\text{------------- convection -----------}) \quad (\text{diffusion})$$

which is easily transformed into

$$\left(\frac{\Omega}{R_T^i}\right)_{,\beta} + v\left(\frac{\Omega}{R_T^i}\right)_{,\gamma} = v\left(\frac{\Omega}{R_T^i}\right)_{,\gamma\gamma} \tag{12.7}$$

The dependence of boundary-layer thickness on viscosity. A boundary layer is, by its very essence, a place where convection and diffusion of vorticity compete on equal terms. Thus, the orders of magnitude indicated for the two types of terms in Eq. (12.6) must be equal. This enforces a relationship between δ, v, U, and L, which is

$$\delta^2 = v\frac{L}{U} \tag{12.8}$$

This is the same relationship found in the exact solutions for the steady spinning-disk flow and the three-dimensional stagnation point flows, in each of which there was neither an L nor a U, but just a characteristic frequency, ω or A.

Flow Equations for a Slender Axisymmetric Jet

When an axisymmetric flowfield includes the axis itself, terms of order δ/R_T cannot be neglected. It is best to go back to the beginning, and introduce cylindrical polar coordinates, with $\beta = z$, $\gamma = r$, and $\alpha = -\theta$. Then $h_\alpha = r$, while $h_\beta = h_\gamma = 1$. It is still true that $v/u = O(\delta/L)$, and that $\Omega = -u_{,\gamma}$. However, the dominant term in curl Ω must be retained in the form $(1/r)(r\Omega)_{,\gamma}$. The β-momentum equation is then

$$uu_{,\beta} + vu_{,\gamma} + \Pi_{,\beta} = \frac{v}{r}(ru_{,\gamma})_{,\gamma} \tag{12.9}$$

while the continuity equation is

$$(ru)_{,\beta} + (rv)_{,\gamma} = 0 \tag{12.10}$$

Boundary conditions. The values of both velocity components are ordinarily prescribed at the wall. Since the coordinate system for the boundary-layer description is attached to the wall, this leads to the conditions

$$u(\beta, 0) = 0 \qquad v(\beta, 0) = v_w(\beta) \tag{12.11}$$

As was mentioned before, controlled suction or blowing through a slightly porous wall may be applied, to manipulate the boundary layer. In such cases, it is assumed that $v_w(\beta)$ is a given function.

In the usual applications of boundary-layer theory, the tangential pressure gradient acting in the boundary layer is also a given quantity, related to the tangential speed of an *external potential flow* by the Euler equation

$$\Pi_{,\beta} = -(U_{,t} + UU_{,\beta}) \tag{12.12}$$

From our study of the velocity induced by a vortex sheet next to a wall, including the effect of the image sheet behind the wall, it is clear that the effect of the boundary-layer vorticity on the potential flow vanishes as $\delta \to 0$, providing that the boundary layer remains attached to the wall. The derivation of the boundary-layer equations shows that they are intended for use in this limit, so U denotes the tangential velocity which the potential flow would reach, at the wall itself.

The *outer boundary condition* on the boundary-layer equations is a statement that

$$u(\beta, \gamma) \to U(\beta) \qquad \text{as} \qquad \left(\frac{\gamma}{\delta}\right) \to \infty \tag{12.13}$$

To understand the meaning of this, review the discussion in Section 9.4 of the outer and inner limits to the exact solution for the axisymmetric jet produced by a point force. Consider also the series of sketches shown in Fig. 12.3. They show profiles of circumferential velocity in the steady flow past a circular cylinder, for which the potential, outer, flow profile is, in cylindrical polar coordinates,

$$U(r, \theta) = \frac{U(a, \theta)}{2}\left[1 + \left(\frac{a}{r}\right)^2\right]$$

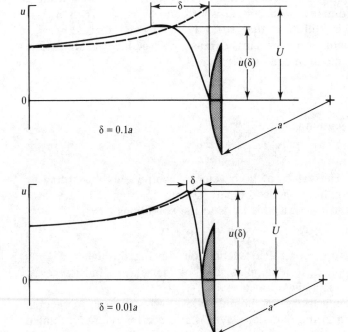

FIGURE 12.3
The approach of u to U as $\gamma/\delta \to \infty$.

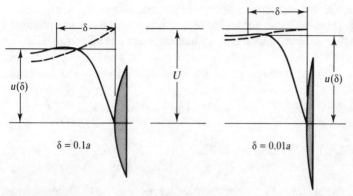

FIGURE 12.4
Magnified view of the boundary layer.

This profile is shown as a dotted line, while the profile for a given value of δ/a is shown by a solid line. The thickness of the boundary layer is much exaggerated in the sketches, to make it easier to see how, as $(\delta/a) \to 0$, the velocity at the outer limit of the boundary layer approaches $U(a, \theta)$.

In Fig. 12.3, the radius of the cylinder is held fixed as $(\delta/a) \to 0$. Figure 12.4 shows a close-up of the boundary-layer region, with δ held fixed as $(\delta/a) \to 0$. Here we notice how the curvature of the wall and the slope of the potential-flow profile become less visible as $(\delta/a) \to 0$. Note that, in the limit as $(\delta/a) \to 0$, we can expect that $u_{,\gamma} \to 0$ as $(\gamma/\delta) \to \infty$.

Notice that no condition is given on the normal velocity component, v, at the outer edge of the boundary layer. If $\Pi_{,\beta}$ is a given quantity, the situation is precisely what was encountered in the study of the stagnation-point flows; another condition on v would only overdetermine the solution

Initial conditions and/or upstream conditions. Conditions in a boundary layer depend not only on conditions imposed at the edges of the layer, but on the initial field of velocity, and on the properties convected into the layer at its upstream end. Boundary-layer theory leads to relatively simple calculations only when there is no ambiguity about the direction to be named upstream, i.e., only when the sign of the tangential velocity component, u, is the same everywhere in the layer. Then the upstream condition consists of a prescription of the distribution of the incoming tangential velocity profile,

$$u(\beta_0, \gamma, t) = u_0(\gamma, t) \tag{12.14}$$

NEW SYMBOLS FOR THE INDEPENDENT VARIABLES. Up to this point, symbols β and γ have been used to denote distance along and across the boundary layer, to emphasize the concept of a curvilinear, body-fitted, coordinate system. We have concluded, however, that the shape of the body is only represented in the final boundary-layer equations by the factor R_T^i, which appears in the continuity equation and vorticity equation. A much better accord with other texts will be achieved if we now switch nomenclatures, subsequently replacing β by x, and γ by y. Thus, at least until Section 12.8,

> x denotes distance along the wall, in the direction taken by the potential flow, and
>
> y denotes distance normal to the wall, with $y = 0$ at the wall.

12.3 THE MOMENTUM-INTEGRAL EQUATION

To save rewriting in Section 12.8, the possibility of unsteady flow is here included in the derivation of the momentum-integral equation. For this, one needs only to add the term $u_{,t}$ to the representation of tangential acceleration in Eq. (12.2). We also use (12.11) to replace $\Pi_{,x}$, and thus start with the equation

$$u_{,t} + uu_{,x} + vu_{,y} - (U_{,t} + UU_{,x}) = vu_{,yy} \qquad (12.15)$$

Multiply the continuity equation (12.5) by $(u - U)$; multiply (12.15) by R_T^j; and add the resulting equations. The sum can be rearranged into the equation

$$[R_T^j(u - U)]_{,t} + [R_T^j u(u - U)]_{,x} + R_T^j(u - U)U_{,x} + [R_T^j v(u - U)]_{,y} = vR_T^j u_{,yy}$$

Integrate this from $y = 0$ to $y = \infty$, assuming that $u - U$ and $u_{,yy}$ approach zero fast enough so that all the integrals have finite values. This assumption will be supported by subsequent analysis and calculations. Name the following integrals:

$$\int_0^\infty \left(1 - \frac{u}{U}\right) dy = \delta^* \qquad \text{the } \textit{displacement thickness} \qquad (12.16)$$

$$\int_0^\infty \frac{u}{U}\left(1 - \frac{u}{U}\right) dy = \theta \qquad \text{the } \textit{momentum thickness} \qquad (12.17)$$

and assign a symbol to the tangential stress exerted by the fluid on the wall:

$$\mu u_{,y}(x, 0) = \tau_w \qquad \text{the } \textit{skin friction} \qquad (12.18)$$

After the integration, apply the boundary conditions (12.11) and the condition that $u_{,y} \to 0$ as $(y/\delta) \to \infty$. What remains will be

$$(U\delta^*)_{,t} + (U^2\theta)_{,x} + jU^2(\ln R_T)_{,x}\theta + UU_{,x}\delta^* = Uv_w + \frac{\tau_w}{\rho} \qquad (12.19)$$

This single equation relates the three unknowns, δ^*, θ, and τ_w. Everything else in the equation is given data that specifies the problem at hand.

Equation (12.17) can be rewritten as an equation for θ, in which the dimensionless *shape factors,*

$$H = \frac{\delta^*}{\theta} \quad \text{and} \quad T = \frac{\tau_w \theta}{\mu U} \tag{12.20}$$

appear. The values of these depend only on the shape of the velocity profile, and not on the thickness of the boundary layer. For example, suppose the dimensionless, scaled boundary-layer profile is $u/U = 2y/\delta - (y/\delta)^2$. The integration for δ^* and θ is easily done, and we find $\delta^* = \delta/3$, $\theta = 2\delta/15$, and $\tau_w = 2mU/\delta$. Thus $H = 5/2$, $T = 4/15$.

In general, H and T will be functions of x and t, and the equation for $\theta(x, t)$ will be

$$\left(\frac{H}{U}\frac{\partial}{\partial t} + \frac{\partial}{\partial x}\right)\left(\frac{\theta^2}{\nu}\right) = -2[(2 + H)(\ln U)_{,x} + j(\ln R_T)_{,x} + U^{-2}(UH)_{,t}]\left(\frac{\theta^2}{\nu}\right)$$

$$+ 2\left(\frac{v_w}{U}\right)\frac{\theta}{u} + \frac{2T}{U} \tag{12.21}$$

Steady, Plane Flow over a Solid Wall. Thwaites' Method

When $\partial/\partial t = 0$, $j = 0$, and $v_w = 0$, Eq. (12.19) reduces to

$$\left(\frac{\theta^2}{\nu}\right)_{,x} + \frac{2}{U}\left[(2 + H)U_{,x}\left(\frac{\theta^2}{\nu}\right) - T\right] = 0 \tag{12.22}$$

This is easily integrated if H and T are constants, but the resulting restriction to velocity profiles of constant shape robs the result of much potential interest. However, several people noticed, after many special velocity profiles had been accurately calculated, that values of H and T depend almost entirely upon values of a dimensionless quantity λ, which is defined by

$$\lambda = U_{,x}\frac{\theta^2}{\nu} \tag{12.23}$$

The significance of this parameter is seen geometrically in a suitably scaled plot of the vorticity profile. When $-\Omega\theta/U$ is plotted against y/θ, as in Fig. 12.5, the value of T appears as the height of the profile at $y = 0$; the slope of the profile at that point has the value $-\lambda$; and the value of the abscissa at the centroid of the area under the profile equals H. The shape of the curve is subject to three physically imposed constraints, namely:

1. The ordinate must go to zero as the abscissa goes to infinity.
2. It and the axes must enclose unit area.
3. It must have zero curvature at $y = 0$.
4. In the absence of unusual upstream perturbations, the curve never crosses the axis $\Omega = 0$.

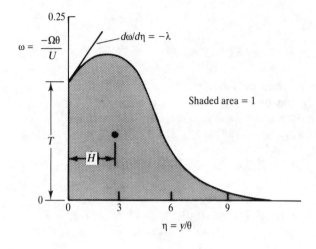

FIGURE 12.5
Nondimensional vorticity profile showing the shape factors λ, T, and H.

The proof of these statements makes an excellent series of exercises, which appear at the end of the chapter. The implication of the constraints is that the shape of the profile is nearly determined once the value of λ is given. Positive values of λ lead to relatively high values of T and correspondingly low values of H. As the value of λ decreases, the value of T also decreases, while that of H increases.

We now concentrate on the work of B. Thwaites,[5] whose correlation of data known in 1949 admits quite a bit of scatter but is neverless very effective. It is shown in Fig. 12.6. Most importantly, Thwaites notices that, to a good

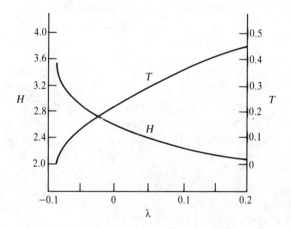

FIGURE 12.6
Shape factors for Thwaites' method.

[5] B. Thwaites, "Approximate Calculation of the Laminar Boundary Layer. *Aeronautical Quarterly,* **1**:245–280 (1949).

approximation,

$$2T - 2(H + 2)\lambda = 0.45 - 6.0\lambda \tag{12.24}$$

When this is inserted into Eq. (12.20), that equation becomes as easy to integrate as though H and T were constant. The result is

$$\frac{\theta^2(x)}{\nu} = C\left(\frac{U(x_0)}{U(x)}\right)^6 + \frac{0.45}{U(x)}\int_{x_0}^{x}\left(\frac{U(x')}{U(x)}\right)^5 dx' \tag{12.25}$$

in which x_0 is the upstream, starting, value of x, and x' is a dummy variable of integration. If the value of θ is specified at $x = x_0$, so that $\theta(x_0) = \theta_0$, the value of the constant of integration is $C = \theta_0^2/\nu$.

Equation (12.25) can be used as it stands, to show semiquantitatively how the local value of θ depends on the upstream history of the boundary layer. Because of the large exponent on the velocity ratio $[U(x_0)/U(x)]$, the contribution of the first term on the right of (12.23) grows rapidly with increasing x if $U_{,x} < 0$; i.e., if the potential flow is decelerating. Simultaneously, the contribution of the integral is built up most strongly at the smallest values of x. We say that

A boundary layer developing under a decelerating potential flow is strongly affected by its upstream history.

This fact guides the strategy that must be followed in design of flow systems if separation of the boundary layer is to be delayed.

On the other hand, when the potential flow is accelerating, the value of θ_0 is multiplied by a factor that rapidly diminuishes as x increases, and the value of the integral is dominated by contributions from the region immediately upstream, where $x' \approx x$. We say that

A boundary layer under an accelerating potential flow rapidly forgets its early history

The fact makes accurate predictions of θ possible, in some situations where values of x_0 and θ_0 are not even known. This is the case, for example, with the boundary layer in a strongly convergent flow-metering nozzle.

Sample Applications of Thwaites' Method

For many simple but varied potential flows, Eq. (12.25) leads to results in closed form. We present a representative sample here, and leave others as exercises. For final calculations of skin friction and displacement thickness, the following fitting formula, provided by White,[6] is handy:

$$T \approx (\lambda + 0.09)^{0.62} \tag{12.26}$$

---

[6] F. M. White, *Fluid Mechanics*, 2nd ed., McGraw-Hill, New York, 1986, p. 408.

Flat plate with constant-speed potential flow. A flat plate set parallel to a uniform stream would not, except for viscosity, disturb the stream. Thus $U(x) = U_0$, a constant. the boundlary layer starts to grow, with $\theta_0 = 0$, at the leading edge of the plate, denoted by $x = 0$. Equation (12.25) then gives

$$\frac{\theta^2}{v} = \frac{0.45x}{U_0} \tag{12.27}$$

From Eq. (12.26), with $\lambda = 0$, we calculate $T = 0.225$. The corresponding value of H can be found from Eq. (12.24), by application of L'Hospital's rule, and is $H = 2.55$. The exact values for this boundary layer are $T = 0.220$ and $H = 2.59$, values that were known to Thwaites and were used as part of the data for his correlation. The exact result for momentum thickness is $\theta^2/v = 0.441x/U_0$.

Plane stagnation-point flow. Suppose $U(x) = Ax$, so that $x = 0$ identifies a forward stagnation point. For this case, $U_0 = 0$, so the first term in (12.25) drops out, whatever the value of θ_0. Actually, the value of θ_0 is now *predicted*, since the integral contribution gives

$$\frac{\theta^2}{v} = \frac{0.45}{6A} = \frac{0.075}{A} \tag{12.28}$$

This result corresponds to a value $\lambda = 0.075$, for which the correlation gives $H = 2.35$, $T = 0.327$. These values, taken from the case $C = 0$ of a stagnation-point flow, computed as an exact solution in Section 10.2, were used to anchor Thwaite's correlation, so we learn nothing quantitatively new here.

Flat plate with a line sink along the trailing edge. Suppose that a flat plate lies along the x axis, from $x = 0$ to $x = a$, whre there is a line sink. If the sink is the only cause of the potential flow, we have

$$U(x) = \frac{m}{2\pi r} \tag{12.29}$$

in which m is the strength of the sink, and $r = a - x$. We then find

$$\frac{\theta^2}{v} = 0.1125\left(\frac{2\pi}{m}\right)r^2\left[1 - \left(\frac{r}{a}\right)^4\right] \tag{12.30}$$

This solution is shown graphically in Fig. 12.7.

This solution shows one important general result of laminar boundary-layer theory, and one special result, important to an interesting class of problems. The general result is that every measure of the thickness of the boundary layer is proportional to $v^{1/2}$, while the streamwise location of any characteristic feature of the thickness distribution, for example, the maximum of θ, is independent of v. The specific result is that the parameter that determines the values of shape factors has a very nearly constant value over

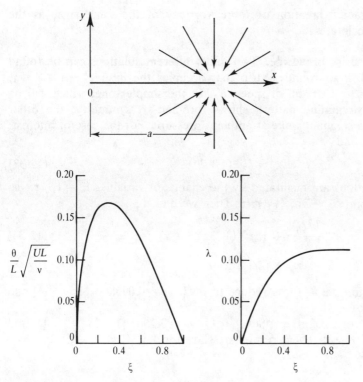

FIGURE 12.7
Momentum thickness and acceleration parameter. Flat plate with sink at trailing edge.

the last quarter of the plate length. This is shown by the result

$$\lambda = 0.1125\left[1 - \left(\frac{r}{a}\right)^4\right] \qquad (12.31)$$

Correspondingly, within the range $0 \le r \le a/4$, the value of λ is very nearly independent of the value of a. The acceleration of the potential flow between the leading edge of the plate and the $\frac{3}{4}$-chord point has been sufficient to wipe out memory of the fact that there was a leading edge.

A simple power-series treatment of (12.30) for $x \ll a$, with the substitution $U_0 a = m/2\pi$, shows that near the leading edge of the plate

$$\frac{\theta^2}{v} = \frac{0.45x}{U_0}\left(1 - \frac{7}{2}\frac{x}{a} + \cdots\right) \qquad (12.32)$$

This starts out just as though there were no acceleration of the potential flow. When the potential flow of speed U_0 first encounteres the plate, the no-slip condition requires the immediate generation of a vortex sheet of strength U_0. The initial growth of the boundary layer reflects the diffusion of that sheet into a stream of speed U_0; the next correction term in (12.32) starts to account for the necessary gradual creation of more vorticity of the same sign, as the potential flow accelerates.

Flow over a flat strip, broadside. Two quite different solutions can be found for the potential flow around a strip that occupies the interval $-a \le Y \le a$, $X = 0$, broadside to a stream of speed U_0. In the simpler one, which will be used here, the streamline pattern shows fore-and-aft symmetry; the other pattern features a stagnant wake at ambient pressure. For the case of interest, with $\xi = x/a$, we find

$$U(x) = U_0\xi(1 + \xi^2)^{-1/2} \tag{12.33}$$

The calculations are facilitated by the change of variables $\xi^2 = 1 - \tau^2$, so that $U^5 \, dx = -U_0^5 a(\tau^{-4} - 2\tau^{-2} + 1) \, d\tau$. Then we find

$$\frac{\theta^2 U_0}{va} = 0.15\tau^3(1 + 3\tau)(1 + \tau)^{-3} \tag{12.34}$$

and

$$\lambda = 0.15(1 + 3\tau)(1 + \tau)^{-3} \tag{12.35}$$

Near the stagnation point, these reduce to $\theta^2 U_0/va \approx 0.075[1 - \frac{9}{8}\xi^2 + \cdots]$ and $\lambda \approx 0.075[1 + \frac{3}{8}\xi^2 + \cdots]$.

Near the edge of the plate, $\theta^2 U_0/va \approx 0.150\tau^3[1 - 3\tau^2 + \cdots]$, and $\lambda \approx 0.150[1 - 3\tau^2 + \cdots]$. These results are shown in Fig. 12.8.

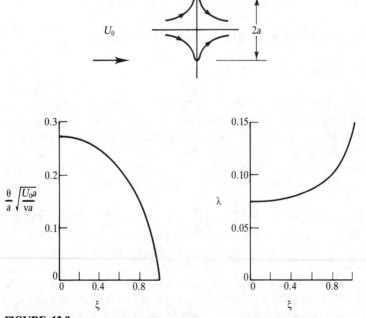

FIGURE 12.8
Momentum thickness and acceleration parameter. Flat plate broadside to a uniform stream.

Flow over a circular cylinder. Suppose that flow from a uniform stream of speed U_0 passes around a circular cylinder of radius a. The surface velocity of the potential flow is

$$U(x) = 2U_0 \sin \phi \qquad (12.36)$$

if $\phi = 0$ identifies the forward stagnation point. Since $U(0) = 0$, only the integral term of (12.25) need be evaluated. The integration is facilitated by the change of variables $t = \cos \phi$, so that $U^5 \, dx = -32U_0^5 a(1 - t^2)^2 \, dt$. We then find

$$\frac{\theta^2 U_0}{\nu a} = 0.03 (\sin \phi)^{-6} (8 - 15t + 10t^3 - 3t^5)$$

The polynomial can be factored, and trigonometric identities used, to put this in the more transparent form

$$\frac{\theta^2 U_0}{\nu a} = 0.0375 \left(1 - \frac{3\xi}{2} + \frac{3\xi^2}{5} \right) (1 - \xi)^{-3} \qquad (12.37)$$

in which $\xi = \sin^2(\phi/2)$. The corresponding expression for λ is

$$\lambda = 0.075 \left(1 - \frac{7\xi}{2} + \frac{18\xi^2}{5} - \frac{6\xi^3}{5} \right) (1 - \xi)^{-3} \qquad (12.38)$$

Equations (12.35) and (12.36) are represented graphically in Fig. 12.9.

One interesting observation is that neither θ or λ change much from their inital values until $\xi = 0.03$, about 20° away from the stagnation point. This suggests that the stagnation-point solution, derived as an idealized flow against an infinite plane wall, describes the boundary layer on a considerable portion of the upstream face of a circular cylinder quite accurately. It is also interesting to compare the boundary layer on the cylinder with that on the front side of the flat strip.

The other interesting feature of the distribution (12.38) is the value of ξ at which $\lambda = -0.09$, indicating separation of the boundary layer, according to the data represented in Fig. 12.6. This value is $\xi = 0.613$, corresponding to an angle of 103° from the stagnation point. We note that separation is predicted to occur where $U(x)$ has decreased to 0.974 times its maximum value, which was reached at an angle of 90°. Thus, this boundary layer has very little tolerance for deceleration of the potential flow, separating very soon after the flow starts to slow down.

Boundary layer in a plane nozzle. Consider the potential flow through a gap between two orthogonal walls, one the plane $y = 0$, $-\infty \le x \le \infty$; the other the half-plane $x = 0$, $a \le y \le \infty$. It has hyperbolic streamlines and elliptic equipotential lines. The speed along the infinite wall is

$$\frac{U(x)}{U_0} = (1 + \xi^2)^{-1/2} \qquad \text{in which} \qquad \xi = \frac{x}{a} \qquad (12.39)$$

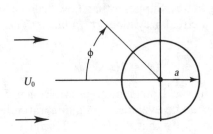

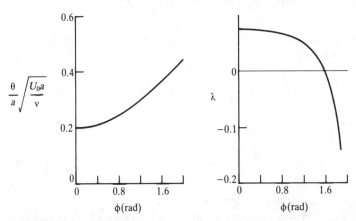

FIGURE 12.9
Momentum thickness and acceleration parameter. Circular cylinder in uniform stream.

To evaluate the integral in Eq. (12.23), we set $\xi^2 = \eta^2(1 - \eta^2)^{-1}$. Then $U^5\, dx = U_0^5 a(1 - \eta^2)\, d\eta$. We also note that the boundary layer starts growing at $x = -\infty$, where $U = 0$, so that only the integral contributes to the local value of θ. The result is

$$\frac{\theta^2 U_0}{va} = 0.15(2 - \eta)(1 - \eta)^{-2}(1 - \eta^2)^{-1} \tag{12.40}$$

To clarify the behavior of this far upstream, where $\eta \to -1$, we rewrite (12.38) in terms of ξ, and find that

$$\frac{\theta^2 U_0}{va} \approx 0.1125\xi^2 \tag{12.41}$$

Note that this is essentially the same behavior we found for the plate in the potential flow due to a line sink at the trailing edge when the leading edge of the plate is far upstream of the point of observation. Far upstream of its throat, i.e., where $(-\xi) \gg 1$, the nozzle looks just like a wedge with a sink at the apex. The complete expression for λ can be reduced to

$$\lambda = -0.15\eta(2 - \eta)(1 - \eta)^{-2} \tag{12.42}$$

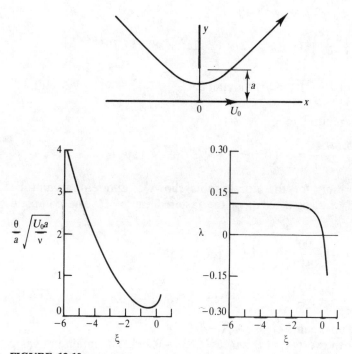

FIGURE 12.10
Momentum thickness and acceleration parameter. Nozzle boundary layer.

Far upstream, this is well approximated by $\lambda \approx 0.1125[1 - \frac{8}{3}\xi^2]$. At the throat of the nozzle, we find $\theta^2 U_0/va = 0.30$ and $\lambda = 0$. At $\eta = 0.2095$, which corresponds to $\xi = 0.211$ and to $U/U_0 = 0.978$, we predict separation. Results are shown in Fig. 12.10.

It turns out to be equally easy to calculate the boundary layer development on the other wall, or to insert a wall along the streamline that is asymptotic to the ray $y = x$. See the Exercises at the end of the chapter.

Flow along a wall under a stationary line vortex. Suppose a line vortex of circulation Γ stands at the point $(0, a)$ over an infinite plane wall, $y = 0$. The tangential velocity at the wall, induced by the vortex and its image at $(0, -a)$, is

$$U(x) = \frac{\Gamma}{\pi a}(1 + \xi^2)^{-1} \tag{12.43}$$

in which $\xi = x/a$. The boundary layer starts again at $x = -\infty$, where $U(x) = 0$, so that only the integral part of Eq. (12.25) is needed.

At first glance, this problem seems almost identical to the nozzle problem, but in the present problem $U(x)$ decreases much faster as $|x|$ increases. The integration is tedious, and the result is rather unwieldy. Using the reference speed $U_0 = \Gamma/\pi a$, we find

$$\frac{\theta^2 U_0}{va} = \left(\frac{0.45}{384}\right)\xi(1 + \xi^2)^2\left[279 + 511\xi^2 + 385\xi^4 + 105\xi^6\right.$$

$$\left. + \left(\frac{105}{\xi}\right)(1 + \xi^2)^4\left(\arctan \xi + \frac{\pi}{2}\right)\right] \tag{12.44}$$

Where $\xi \ll -1$, this reduces to

$$\frac{\theta^2 U_0}{va} \approx -0.050\xi^3[1 + \tfrac{21}{11}\xi^{-2} + \cdots]$$

Compared to the result for the nozzle, this shows a different asymptotic exponent, 3 instead of 2, as a result of the faster decay of U. The complete formula for λ is

$$\lambda = -\left(\frac{0.90}{384}\right)\xi^2\left[279 + 511\xi^2 + 385\xi^4 + 105\xi^6\right.$$

$$\left. + \left(\frac{105}{\xi}\right)(1 + \xi^2)^4\left(\arctan \xi + \frac{\pi}{2}\right)\right] \tag{12.45}$$

which becomes, far upstream, $\lambda = 0.100[1 - \tfrac{1}{11}\xi^{-2} + \cdots]$.

Directly under the vortex, at $\xi = 0$, $\theta^2 U_0/va = 0.194$. The minimum value of θ is reached at $\xi = -0.429$, where $U(x) = 0.845U(0)$. Separation is predicted for $\xi = 0.157$, where $U(x) = 0.976U(0)$. Results are plotted in Fig. 12.11.

Some Analyses, Assuming Constant Shape Factors

Thwaites' method allows us to study boundary layers in which variation of the shape factors is essential, but was not designed to permit study of axisymmetric or unsteady flows, or flows with suction through the wall. If separation of the boundary layer is not to be expected, a simple analysis which uses constant shape factors can be used fruitfully for a wide variety of situations. Here are three examples.

Flate plate parallel to uniform flow, with uniform suction. Suppose $U(x) = U_0$ and $v_w(x) = v_0$, where U_0 and v_0 are constants. Let us assume that constant profile shape $u/U_0 = 1 - \exp(-y/\delta^*)$, which is actually the correct profile far downstream. The corresponding values of the shape factors are $H = 2$, $T = 1/2$. We choose to recast Eq. (12.21) as an equation for δ^*, which is

$$\left(\frac{\delta^*}{v}\right)\delta^*_{,x} = \frac{2}{U_0}\left(1 + \frac{v_0\delta^*}{v}\right) \tag{12.46}$$

Before proceeding, we notice that the line $\delta^* = -v/v_0$ divides the $\delta^* - x$

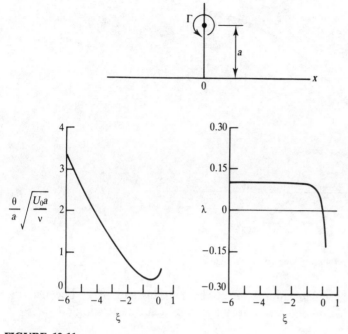

FIGURE 12.11
Momentum thickness and acceleration parameter. Plane wall under stationary vortex.

plane into two regions, in one of which $\delta^*_{,x} \geq 0$, while in the other $\delta^*_{,x} \leq 0$. Noting that $v_0 < 0$ if there is suction, and that $\delta^* = 0$ at $x = 0$ we expect our solution to approach the dividing line asymptotically, from below. We define dimensionless variables

$$\phi = -\frac{v_0 \delta^*}{\nu} \quad \text{and} \quad \xi = -v_0 \left(\frac{x}{2\nu U_0}\right)^{1/2}$$

Then Eq. (12.44) becomes $\phi \, d\phi/(1 - \phi) = 4\xi \, d\xi$, which is nonlinear but *separable*. It is easily solved for ξ as a function of ϕ; the answer that satisfies the initial condition is

$$\xi^2 = -\tfrac{1}{2}[\phi + \ln(1 - \phi)] \tag{12.47}$$

This is plotted in Fig. 12.12, along with a curve from an accurate integration of the full *slender-layer* equations. The latter term is used here, in preference to the term boundary-layer equations, because the typical scaling of a laminar boundary layer, which dictates that $\delta^* = \nu^{1/2}$ while the value of x for which $\phi = 1/2$ would be independent of ν, is not obeyed by this flow.

Very close to the leading edge, $\phi \ll 1$. We use the corresponding series expansion for the logarithm to find $\delta^* \approx 2(\nu x/U_0)^{1/2}$. This recovers the standard boundary-layer scaling, because diffusion effectively outraces the convection due to suction when the layer is very thin. We miss the correct

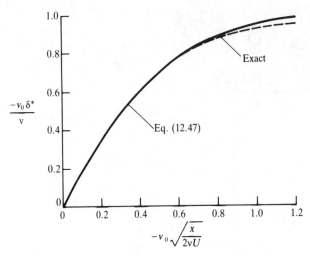

FIGURE 12.12
Displacement thickness on a plate with uniform suction.

value of the factor, which should be about 1.72 instead of 2, because the assumed values of H and T are not very accurate near the leading edge.

Axisymmetric flow on a circular disk, with a point sink at the center. If r denotes distance from the sink and m is the strength of the sink, the velocity along the wall will be

$$U(x) = \frac{m}{4\pi r^2}$$

Let a denote the value of r at which the boundary layer starts to grow, and let $U_0 = U(a)$. Then use $m = 4\pi a^2 U_0$ and $\xi = x/a = 1 - r/a$, to write

$$U(x) = U_0(1 - \xi)^{-2} \tag{12.48}$$

If $v_w = 0$, and H and T are assigned constant values, Eq. (12.21) has the solution

$$\frac{\theta^2}{v} = C\left(\frac{U(x_0)}{U(x)}\right)^{2H+4}\left(\frac{R_T(x_0)}{R_T(x)}\right)^{2j} + \frac{2T}{U(x)}\int_{x_0}^{x}\left(\frac{U(x')}{U(x)}\right)^{2H+3}\left(\frac{R_T(x')}{R_T(x)}\right)^{2j} dx' \tag{12.49}$$

For the case in hand, $R_T = r$. We can set $C = 0$, because $\theta = 0$ at $x = x_0 = 0$. Setting $j = 1$, and introducing $\eta = r/a$, we can integrate and write the result as

$$\frac{\theta^2 U_0}{va} = \frac{2T}{\beta}\eta^3(1 - \eta^\beta) \tag{12.50}$$

in which $$\beta = 4H + 3$$

We can pick values for H and T, so as to get the best fit with an exact

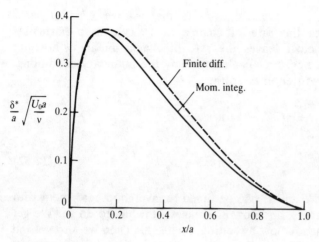

FIGURE 12.13
Displacement thickness on a disk with a point sink at the center.

calculation, but there is little point in this. To get the general behavior, we can use typical values for boundary layers under accelerating potential flows, say $H = 2.25$ and $T = 0.3$, to get

$$\frac{\theta^2 U_0}{va} = 0.05\eta^3(1 - \eta^{12})$$

This is shown in Fig. 12.13, along with an accurate result from the partial differential equations.

Boundary layer on a semi-infinite plate with impulsively started potential flow. When a uniform potential flow of speed U starts at time zero, over a parallel plate that extends from $x = 0$ to $x = \infty$, the convection of vorticity has no net effect at positions $x > Ut$. Where $x < Ut$, fluid in the outer portion of the boundary layer has passed part of the time since the beginning of flow at x positions where there is no plate, and hence no initial charge of vorticity. Thus, there will be a qualitative difference between the velocity profiles in these two regions. Steady-flow conditions will appear first very near the leading edge; farther downstream, a gradual transition from the velocity profile of Stokes' first problem to the profile for steady flow will begin at time x/U. The momentum-integral theory, dealing only with y-averaged properties of the boundary layer, replaces this gradual transition with a sudden jump. To see this, we start with Eq. (12.21), setting $j = 0$, $v_w = 0$, $U_{,t} = 0$, and $U_{,x} = 0$. What remains is

$$\left(\frac{H}{U}\frac{\partial}{\partial t} + \frac{\partial}{\partial x}\right)\frac{\theta^2}{v} = 2\frac{T}{U} \tag{12.51}$$

This equation is of *hyperbolic* type. We need a solution for the first quadrant of the x–t plane that vanishes along the positive x and t axes. Equation (12.49) specifies the rate of change of (θ^2/v), in a particular direction in the x–t plane. It leaves the rate of change in an orthogonal direction unspecified, and hence available to satisfy the boundary conditions. The solution can thus be written in the form

$$\frac{\theta^2}{v} = 2T\frac{x}{U} \qquad \text{if } x \le \frac{U}{H}t$$

and
$$\frac{\theta^2}{v} = 2T\frac{t}{H} \qquad \text{if } x \ge \frac{U}{H}t \tag{12.52}$$

Again, the values of T and H can be tuned to give good agreement with exact values for limiting cases, and now it seems worthwhile to do so. We get the exact value of θ for steady flow by setting $T = 0.220$. Then we also get the exact value of θ for Stokes' first problem if we choose $H = 2.01$. This makes the frontier between purely transient flow and steady flow move at about half the speed of the potential flow. Figure 12.14 summarizes the key features of this analysis.

This analysis, and a much deeper discussion of the problem itself, was first given by K. Stewartson.[7]

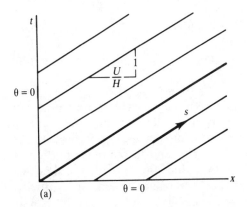

(a)

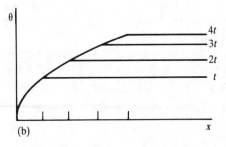

(b)

FIGURE 12.14
Momentum-integral analysis of impulsively-started Blasius boundary $d(\theta^2/v)/ds = 2T/U$. (a) Domain of the solution. (b) Evolution of the solution.

[7] K. Stewartson, "On the Impulsive Motion of a Flat Plate in a Viscous Fluid," *Quart. Jour. Mech.* **4**:182–198 (1951).

12.4 TRANSFORMATION OF THE BOUNDARY-LAYER EQUATIONS. SIMILAR SOLUTIONS

We now return to the differential form of the boundary-layer equations, and begin to prepare them for efficient solution by finite-difference methods. In the process, we shall discover various circumstances that allow *self-similar* solutions. The basic ideas and procedures have already been presented in Section 10.3.

Initially considering only steady, plane or axisymmetric flow, we start with the equations

$$uu,_x + vu,_y - UU,_x = vu,_{yy} \tag{12.53}$$

and

$$(uR_T^j),_x + (vR_T^j),_y = 0 \tag{12.54}$$

It is convenient to introduce a streamfunction $\Psi(x, y)$, to satisfy the continuity equation. Thus we let

$$uR_T^j = \Psi,_y \quad \text{and} \quad vR_T^j = -\Psi,_x \tag{12.55}$$

Scaling again. Our first exercise in scaling the variables that describe flow in a slender layer was intended to identify terms in the full Navier–Stokes equations that would be negligible in a thin boundary layer. The limiting process for that exercise was one in which $(\delta/L) \to 0$, δ and L being values characteristic of the flow as a whole. We now take the next step, intended to design variables that retain unit order of magnitude as x varies, by factoring out of the streamfunction those given or independently calculated quantities that determine its typical magnitude, and by stretching y appropriately. Specifically, we introduce dimensionless, scaled independent variables

$$\xi = \xi(x) \quad \text{and} \quad \eta = \eta(x, y) = \frac{y}{g(x)} \tag{12.56}$$

in which $g(x)$ is some *local* measure of the boundary-layer thickness. Simultaneously, we introduce the dimensionless, scaled streamfunction, $f(\xi, \eta)$, by the equation

$$\Psi(x, y) = Q(x)f(\xi, \eta) \tag{12.57}$$

in which

$$Q(x) = R_T^j(x)U(x)g(x)$$

This is written with the expectation that the factor $Q(x) = R_T^j(x)U(x)g(x)$ will be approximately equal to the flowrate passing through the boundary layer, at a station identified by the value of x.

Calculations of the type carried out in Section 10.3 are now made. We use f' as shorthand for $f_{,\eta}$, and also use a prime to denote the x-derivative of any quantity, such as ξ or g, that depends only upon x. Finally, physical velocity components are again identified with an asterisk. The results are:

$$u^*(x, y) = U(x)f' \tag{12.58}$$

$$v^*(x, y) = -Ug[(\ln Q)'f + \xi'f_{,x} + \eta_{,x}f'] \tag{12.59}$$

$$u^*_{,x} = U'f' + Ug[\xi'f'_{,x} + \eta_{,x}f'']$$

$$u^*_{,y} = \frac{Uf''}{g} \quad \text{and} \quad u^*_{,yy} = \frac{Uf'''}{g^2}$$

The x-momentum equation is transformed into

$$u'' + \gamma(\xi)fu' + \delta(\xi)(1 - u^2) = \psi(\xi)[uu_{,\xi} - u'f_{,\xi}] \tag{12.60}$$

in which we have again reintroduced the symbol u, to represent the scaled tangential velocity component. Thus

$$f' = u \tag{12.61}$$

The ξ-dependent coefficients are calculated from the formulas

$$\gamma = \frac{g^2 U}{\nu}(\ln Q)' \qquad \delta = \frac{g^2 U}{\nu}(\ln U)' \qquad \psi = \frac{g^2 U}{\nu}\xi' \tag{12.62}$$

It is helpful to notice that the coefficient γ is proportional to the percentage rate of increase of the nominal flowrate in the boundary layer, while δ is similarly proportional to the percentage rate of increase of U, both taken as ξ increases. It may be surprising to discover that boundary layers sometimes disgorge fluid, so that negative values of γ are possible.

The *boundary conditions* are

$$u(\xi, 0) = 0 \quad \text{(no slip)}$$

and
$$u(\xi, \infty) = 1 \quad \text{(matching with the potential flow)}$$

If the wall is impermeable, we simply set $f(\xi, 0) = 0$. If there is suction or blowing, we can calculate

$$\Psi(x, 0) = -\int_{x_0}^{x} R_T^i v_w(x)\, dx \tag{12.63}$$

and then specify the function

$$f(\xi, 0) = f_0(\xi) = \frac{\Psi(x, 0)}{Q(x)} \tag{12.64}$$

We note again, however, that this boundary value will only be independent of Reynolds number if a characteristic value of v_w, say v_{w0}, is related to a characteristic value of U, say U_0, as follows:

$$\frac{v_{w0}}{U_0} = A\left(\frac{U_0 L}{v}\right)^{-1/2}$$

When this is satisfied, one calculation solves the laminar boundary-layer problem for all values of the Reynolds number for which the flow remains laminar and vorticity remains confined to an adequately slender layer. If it is not satisfied, the calculation has to be done anew for each value of the Reynolds number.

The initial, or upstream, condition specifies an initial profile for u,

$$u(0, \eta) = F(\eta) \qquad \text{for} \qquad 0 \le \eta \le \infty$$

A Transformation Based on the Momentum-Integral Equation

We say that the boundary layer equations have been *transformed* by these changes of variables. The invention of an effective transformation formula is an exercise in design, in that matters of choice and purpose are involved. The element of choice enters now, as we come to decide how best to pick g and ξ, as functions of x.

Before finite-difference solutions of the full boundary-layer equations became relatively easy and inexpensive, the art of designing transformations was focused on the possibility of showing that two seemingly different boundary-layer problems can be transformed into identical mathematical form. This was particularly important for boundary layers exhibiting variable density and viscosity. Now the natural focus is on transformations that accurately capture the natural growth of the boundary layer, leaving the computer to work out relatively minor changes in profile shape.

A design choice that seems to combine effectiveness with ease of programming sets g proportional to the momentum thickness, and ξ simply proportional to x. That is, we set

$$g(x) = A\theta(x) \qquad \text{and} \qquad \xi(x) = \frac{x}{a} \tag{12.65}$$

For plane flow without suction, we then use Thwaites' approximation to the momentum-integral equation to generate a differential equation for g^2/v,

$$\left(\frac{g^2}{v}\right)' = \frac{0.45A^2 - 6(g^2/v)U'}{U} \tag{12.66}$$

This equation can be used to eliminate g' from the definition of γ, giving the simple result

$$\gamma = 0.255A^2 - 2\delta \tag{12.67}$$

Since the definition of δ is equivalent to

$$\delta = \frac{g^2}{\nu} U' \qquad (12.68)$$

we recognize that $\delta = A^2 \lambda$, where λ is the shape-determining parameter of Thwaites' method.

The value of the constant A determines the range of η, which must be accommodated by the computer program. Since the initial profile, $F(\eta)$, often corresponds to a familiar self-similar solution, it is convenient to use the value of A to give the coefficients γ and δ initial values that will make the initial profile numerically identical to the corresponding, previously published, profile. Examples will be given in Section 12.5.

Self-Similar Boundary Layers

The mathematical equivalent of the notion of a shape-preserving boundary-layer profile is the statement that $u(\xi, \eta)$ reduces to a function of η alone. You can now see, systematically, what is required for this to happen. Although the arguments are mathematical, the requirements for similarity are common-sense physical restrictions, quite independent of mathematical transformations.

1. The most obvious requirement is that the coefficients γ and δ must be independent of x. If they are not, u will certainly depend on ξ as well as η, even if ψ is zero.
2. If there is suction or blowing, $v_w(x)$ must be constrained so that f_0 is independent of ξ.
3. The initial profile, $F(\eta)$, must be identical to the solution of the ordinary differential equation that appears when the right-hand member of Eq. (12.60) is removed. For an accurate approximation to similarity, a misfit initial profile may be tolerated, if followed by a long interval of accelerating potential flow, which causes initial conditions to be forgotten.

To understand the implications of requirements (1) and (2), we use Eqs. (12.62) and (12.64). Let us first treat the usual case, in which $v_w = 0$. Noting that the shape factors H and T are independent of x when the boundary-layer profile is shape-preserving, we see that the key requirement is that δ, or λ, be independent of ξ.

In our sample problems for Thwaites' Method, we found many cases in which λ was exactly constant, or in which it was almost constant for a considerable range of x. In all of these cases, $U(x)$ varies as a constant power of x, when the origin of x is suitably located. If $j = 1$, so that R_T is involved, it is also required that $R_T(x)$ vary as a constant power of x. Examples were found in which self-similar behavior arises naturally in the evolution of a boundary layer, either as a feature of the birth of the boundary layer, or as an asymptotic condition approached after initial conditions are forgotten.

If we explore the consequences of assuming that $U(x) = Ax^m$, we find that, for δ to be constant, $g^2/v = Bx^{(1-m)}$. Further, if there is suction or blowing, γ will be constant only if $v_w = Cx^{(m-1)/2}$. If we then require that $f(\xi, 0)$ be independent of ξ on an axisymmetric body, we discover the final restriction, that $R_T = Dx^n$. There is no necessary connection between the values of m and n, but the suction or blowing must have been applied consistently since the birth of the boundary layer.

None of the cases in which λ was found to be independent of x involved decelerating potential flows, and it is indeed hard to imagine a physically realistic scenario in which self-similar development could be established in a laminar boundary layer under such a potential flow. The difficulty is that the required power laws for U and g are not compatible with realistic initial conditions, but initial conditions are crucially important to the development of the boundary layer when $U' \leq 0$.

Nevertheless, it is possible to investigate the solutions of the ordinary differential equations that are obtained when δ is assigned a negative value, and a great deal of attention has been given to these solutions in the literature. We review some of this material now.

To organize the presentation, we recognize two particularly important classes of self-similar boundary layers, which we shall call the *Falkner–Skan flows* and the *sink-driven flows*.

Falkner–Skan Flows

Named for two English investigators, these are boundary layers that start to grow at $x = 0$, and which grow continuously thereafter, so that the quantity Q increases with increasing x, and the value of γ is positive. We recall that an arbitrary constant factor can be included in the values of γ and δ, to adjust the range of η occupied by the boundary layer. We shall employ this factor to make $\gamma = 1$ for all cases. To find the ratio of γ to δ, which determines the shape of the velocity profile, we return to the general definitions and calculate

$$\frac{\gamma}{\delta} = \frac{(\ln Q)'}{(\ln U)'} = 1 + \frac{jUR'_T}{R_T U'} + \frac{Ug'}{gU'}$$

For this family of flows,

$$U(x) = Ax^m \qquad g = Bx^{(1-m)/2} \qquad \text{and} \qquad R_T = Dx^n$$

so that

$$\delta = \frac{2m}{1 + m + 2jn} \qquad \text{if} \qquad \gamma = 1$$

Classic examples are:

1. The *Blasius* boundary layer on a flat plate, with $m = 0$; so that $\gamma = 1$, $\delta = 0$.
2. The *plane stagnation-point flow*, with $m = 1$, $j = 0$; so that $\gamma = 1$, $\delta = 1$.

3. The *axisymmetric stagnation-point flow*, with $m = 1$, $j = 1$, $n = 1$; so that $\gamma = 1$, $\delta = \frac{1}{2}$.

4. The boundary layer on an infinite wedge of semi-apex angle ϕ, pointing symmetrically into a uniform stream, with $m = \phi/(\pi - \phi)$, $j = 0$; so that $\gamma = 1$, $\delta = 2\phi/\pi$. An example of two physically different boundary layers being described by the same mathematical function appears when $\phi = \pi/4$. then $\gamma = 1$ and $\delta = \frac{1}{2}$, just as for the axisymmetric stagnation-point flow.

5. The boundary layer that starts to grow at a corner, where a wall suddenly bends away from a uniform stream, with $m = -\phi/(\pi + \phi)$, $j = 0$; so that $\gamma = 1$, $\delta = -2\phi/\pi$. Here ϕ is the angle by which the downstream wall is deflected.

Numerical solutions of the boundary-value problem

$$u'' + fu' + \delta(1 - u^2) = 0$$

$$f' = u \tag{12.69}$$

$$u(0) = 0 \qquad u(\infty) = 1 \qquad f(0) = f_0$$

are easily generated by the techniques explained in Section 10.2, but a special feature arises when δ is negative. It is revealed by an analysis of the way in which solutions of the Falkner–Skan equations approach the specified outer boundary condition. We have seen a special case of this analysis in Chapter 10, in connection with the stagnation-point boundary layers.[8] Applied to the Falkner–Skan equations, the analysis shows that

$$1 - u \approx A\zeta^{-(2\delta+1)} \exp(-0.5\zeta^2) + B\zeta^{2\delta} \tag{12.70}$$

in which $\zeta = \eta - \delta_1 - f(0)$, δ_1 being the dimensionless displacement integral. We see that the boundary condition, $u \to 1$ as $\eta \to \infty$, can be satisfied only if $B = 0$, when $\delta \geq 0$. The permissible solution then approaches its asymptote very rapidly, once ζ is sufficiently large. However, when $\delta < 0$, both terms in (12.68) die away eventually, although the one containing B may do so very slowly. There is then a possibility that two solutions of the system (12.67) can be found numerically, although the numerical methods introduced in Chapter 10 will always product the solution with $B = 0$, if the first-guess profile is at all close to that solution.

Sink-Driven Boundary Layers

These are boundary layers that are born at $x = -\infty$, and that just hold their own, or actually disgorge fluid as they move downstream, because $Q' \leq 0$. The

---

[8] A good review of this somewhat advanced topic is given on pages 247 and 248, and the Appendix at the end of Chapter 5 L. Rosenhead (ed.), *Laminar Boundary Layers*, Oxford University Press, 1963.

corresponding values of γ are either zero or negative. Some examples are:

1. Boundary layer driven by a line sink in a flat wall. $U = A(-x)^{-1}$, $g = -Bx$, $j = 0$; hence $\gamma = 0$. We can use the disposable factor in g to make $\delta = 2$. The resulting equation was used as an example in Section 10.2, and has an exact, closed-form solution, given by Eq. (10.19).
2. Axisymmetric boundary layer driven by a line sink normal to a flat wall. Again $U = A(-x)^{-1}$ and $g = -Bx$, but $j = 1$ and $R_T = -x$. Since $(\ln Q)' = 1/x$, which is negative, γ will be negative. Meanwhile, $(\ln U)' = -1/x$, so $\delta = -\gamma$. We can use the disposable scale factor to make $\gamma = -1$, $\delta = 1$.
3. Axisymmetric boundary layer driven by a point sink in a flat wall. Now $U = A(-x)^{-2}$, $g = B(-x)^{3/2}$, $j = 1$ and $R_T = -x$. Since $(\ln Q)' = 1/2x$, while $(\ln (U))' = -2/x$, we find $\delta = -4\gamma$, and set $\gamma = -1$, $\delta = 4$.

When $\gamma = -1$, as in the two axisymmetric sink-flow boundary layers, the decay of $1 - u$ as $\eta \to \infty$ is comparatively slow, being governed by the formula

$$1 - u \approx A\zeta^{(2\delta-1)} \exp(0.5\zeta^2) + B\zeta^{-2\delta} \tag{12.71}$$

We see that δ must be positive, and that A must equal zero, in order for the boundary condition to be met. For the two sink-driven boundary layers, $1 - u$ dies away only in proportion to ζ^{-2} (in Case 2), or to ζ^{-8} (in Case 3). To calculate such solutions accurately, without wasting a large number of mesh points in the slowly-decaying tail of the profile, we can derive an outer boundary condition of the form (10.3). Unfortunately, it is not clear how to do this when the shape of the boundary-layer profile is not self-preserving, as in some of the examples which will soon be analyzed, and a less elegant approach must then be taken.

The Vortex-Driven Boundary layer

This relatively unfamiliar case is a sort of hybrid of the previous two; the boundary layer is born at $x = -\infty$, but it ingests fluid as x increases, so γ is positive. It involves the plane boundary layer on a flat wall, driven by a stationary line vortex above the wall. This layer, like that on the nozzle wall, is only asymptotically self-similar, where $(-x/a) \gg 1$. There we find $U = A(-x)^{-2}$, $g = B(-x)^{3/2}$, $j = 0$. Now $(\ln Q)' = -1/2_x$, while $(\ln U)' = -2/x$, so $\delta = 4\gamma$, and we can set $\gamma = 1$, $\delta = 4$. This makes it mathematically a member of the Falkner–Skan family, although its physical origins are completely different.

The Effects of Pressure Gradient and Suction on Profile Shapes

Figure 12.15 shows normalized distributions of vorticity for some of the Falkner–Skan family of profiles, including the vortex-driven boundary layer. We see that favorable pressure gradient, $\delta \geq 0$, and suction, $f(0) \geq 0$, have

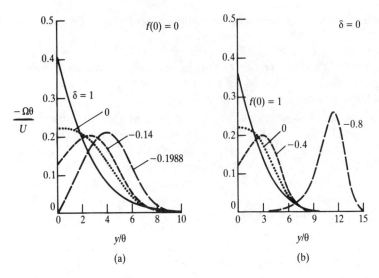

FIGURE 12.15
Effect of pressure gradient and flow through the wall on Falkner–Skan boundary layers.

qualitatively very similar effects on the shape of the profile. The former requires a continuous source of vorticity, of the same sign as that already in the boundary layer, to keep the no-slip condition in force while U increases. The latter requires the same kind of source, to replace vorticity which is being continuously withdrawn through the wall.

There is also some resemblance between the effects of adverse pressure gradient, $\delta \leq 0$, and of blowing, $f(0) \leq 0$, but there are fundamental differences, too. Deceleration of the potential flow calls forth a wall source of opposite-signed vorticity, which dilutes the preexisting vorticity, and can eventually reduce the value of Ω to zero at the wall. For the solid wall, this happens to the Falkner–Skan flow when $\delta = -0.19884$. On the other hand, blowing through the wall adds no new vorticity of either sign, but displaces the vorticity so that the maximum value occurs out in the fluid.

Both suction and blowing may be used to achieve design objectives.[9] Blowing is an effective means of reducing heat transfer to the wall, and is used to cool turbine blades and to protect high speed aircraft such as the space shuttle. Suction is usually employed to offset the effect of adverse pressure gradient, to delay *separation* and *instability* of laminar boundary layers. Both of these phenomena are intimately associated with the displacement, away

---

[9] These possibilities are shown dramatically in the film *Boundary-Layer Control*, narrated by D. Hazen, in the NCFMF series. Film 21614 available from Britannica Films and Video, 425 No. Michigan Ave, Chicago, IL, 60611.

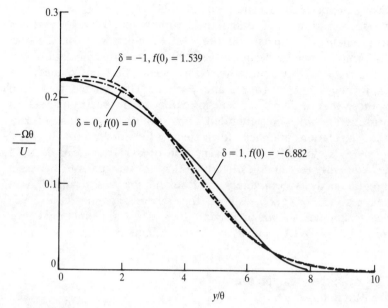

FIGURE 12.16
Result of matching δ and $f(0)$ to maintain a constant value of the shape factor T.

from the wall, of the peak of the vorticity profile. We shall show, in a later section, a computed example of this use of suction. For the present, this suggestion is strengthened by the appearance, in Fig. 12.16, of three normalized vorticity profiles of the Falkner–Skan family that all share the value $T = 0.220$, which characterizes the Blasius boundary layer. One is the Blasius ($\delta = 0$, $f(0) = 0$) layer; one balances adverse pressure gradient against suction; and the third balances favorable pressure gradient against blowing. The profile shapes are not quite identical, and we shall see in Chapter 13 that the one with a slight maximum of vorticity away from the wall may be a little less likely to remain laminar if slightly disturbed.

The calculations represented by Fig. 12.16 involve the use of a simple outer iteration loop, which determines the correct value of $f(0)$ for a given value of δ, by the Newton–Raphson method. The calculations are started with $\delta = 0$, and then proceed gradually through a range of increasing or decreasing values of δ. Since the result from each calculation can be used as the first-guess profile after the next increment of δ, a small number of iterations, usually about five, suffices to give five-decimal convergence.

12.5 NUMERICAL PREDICTION OF THE EVOLUTION OF NONSIMILAR BOUNDARY LAYERS

The similarity solutions provide valuable insights, but actual predictions for cases that do not admit of self-similar solutions must rely on a numerical scheme for solving the partial differential equation (12.60). In this section, you will learn one successful way of doing this, with a number of variations appropriate to problems with special features. In particular, you will see how transformation formulas may be designed to facilitate accurate calculation. We begin with a description of a general-purpose scheme that implements the transformation formulas $g(x) = A\theta(x)$, and $\xi = x/a$. The scheme is a little different from those that ordinarily appear in textbooks and survey papers, in that the transformation may be generated numerically, by solution of Eq. (12.66), as the calculation marches downstream. Of course, if Thwaites' method easily yields a closed-form representation of $\theta(x)$, that can be used directly. Later, you will see examples in which it is advantageous to bypass a momentum-integral analysis completely, and to employ a shrewdly-chosen closed-form expression for $g(x)$. All of the schemes employ the basic finite-difference techniques shown in Section 10.3. They are embodied in a FORTRAN program, NEWBL, which appears in Appendix B.

Example: The Boundary Layer on a Wall Under a Line Vortex

To demonstrate Program NEWBL, we work out in detail the case of the boundary layer on a flat wall, under a stationary line vortex. The potential flow first accelerates, and then decelerates, so prominent effects of both favorable and unfavorable pressure gradient are expected. By allowing the wall to be a finite flat plate halfway between two equally strong, counterrotating vortices, we can study the effect of upstream history on the flow in both kinds of pressure gradient. This is done by moving the leading edge of the plate closer to the vortices.

The work to be done, with the help of NEWBL, can be broken down into four main tasks:

1. Establishment, at some starting value of x, of appropriate initial values of the quantities that depend only on x. These are g itself, and the coefficients γ, δ, and ψ. This task is usually done as part of a preliminary analysis.
2. Calculation of starting profiles of $u(\xi, \eta)$ and (ξ, η). This is usually accomplished in NEWBL, where appropriate ordinary differential equations are solved.
3. Application of Eqs. (12.66, 67, 68, and 62), to determine values of g, γ, δ, and ψ at the next, nearby, downstream station.
4. Solution, at the new value of x, of the ordinary differential equations for u and f, which result when the derivatives $u_{,\xi}$ and $f_{,\xi}$ are approximated by trailing differences, as in the examples of Section 10.3. The necessary calculations for this step and step (2) are usually assigned to a subroutine, called THOMAS.

Let us now examine these tasks one by one, for the flow under the line vortex.

Starting conditions. Two cases will be analyzed: one in which the wall is imagined to extend to infinity, and one in which it has a leading edge at a finite value of x. In both cases, we use closed-form solutions from Thwaites' method: Eq. (12.44) for the infinite wall, and a simple modification of it for the wall with a leading edge.

(a) CASE WITH AN INFINITE WALL. THE BLASIUS SERIES. Having noticed, in the momentum-integral analysis, that the shape-determining parameter, λ, approaches a constant value when x is large and negative, we suspect that the boundary layer at some large but finite distance upstream of the vortex can be closely approximated by the Falkner–Skan type profile, with $\gamma = 1$ and $\delta = 4$. To confirm this idea, and to get a guaranteed level of accuracy at, say, $\xi = -10$, we employ a power-series expansion, which we may call a *Blasius series*. It was originally invented as a technique with which to extend solutions of the boundary-layer equations away from the stagnation region of a body such as a circular cylinder, while using numerical methods to solve only ordinary differential equations.[10]

The Blasius series. We wish to prove that it is appropriate to start the calculation with a similarity solution, for the example of the vortex-driven boundary layer. Recall, from Section 12.4, that $\theta^2 U'/v \approx 0.100[1 - \frac{1}{11}\xi^{-2} + \cdots]$ where $\xi = x/a$ is large and negative. Picking $A^2 = 40$, we obtain

$$\delta = 4 - \tfrac{4}{11}\xi^{-2} + \cdots \qquad \gamma = 1 + \tfrac{8}{11}\xi^{-2} + \cdots$$

and

$$\psi = -2\xi - \tfrac{20}{11}\xi^{-1} + \cdots$$

Far upstream, γ and δ approach constant values, as is necessary for a self-similar solution, but the fact that $\psi \to \infty$ in this same limit is worrisome. It must be shown that the right-hand side of Eq. (12.60) vanishes as $\xi \to -\infty$, even though it contains ψ as a factor. Only if that is true will it be permissible to calculate starting profiles of u and f by solving ordinary differential equations. To make the required demonstration, we attempt a solution to the partial differential equation in the form of the series

$$f(\xi, \eta) = f_0(\eta) + f_1(\eta)\xi^{-2} + \cdots \qquad (12.72)$$

which implies $\qquad u(\xi, \eta) = u_0(\eta) + u_1(\eta)\xi^{-2} + \cdots$

$$f_{,\xi} = -2f_1(\eta)\xi^{-3} + \cdots \qquad \text{and} \qquad u_{,\xi} = -2u_1(\eta)\xi^{-3} + \cdots$$

---

[10] The series method is named for H. Blasius, and is extensively described in Chapter 9 of H. Schlichting, *Boundary-Layer Theory*, 6th Edn, McGraw-Hill, New York, 1968.

all valid for $\xi \ll -1$. When these, along with the corresponding expansions for γ, δ, and ψ, are introduced into (12.60), (12.61), and the boundary conditions, it proves possible to collect terms according to their power of ξ^{-2} and to examine first the equations that arise in the limit $\xi \to -\infty$. These turn out to be exactly the equations that define the self-similar solution. For easy reference, we repeat these:

$$u_0'' + f_0 u_0' + 4(1 - u_0^2) = 0 \qquad f_0' = u_0$$
$$f_0(0) = 0 \qquad u_0(0) = 0 \qquad u_0(\infty) = 1 \tag{12.73}$$

The next group of terms, all containing the factor ξ^{-2}, define the boundary-value problem

$$u_1'' + f_0 u_1' - 12 u_0 u_1 + 5 u_0' f_1 = -\tfrac{8}{11} f_0 u_0' + \tfrac{4}{11}(1 - u_0^2)$$
$$f_1' = u_1 \tag{12.74}$$
$$f_1(0) = 0 \qquad u_1(0) = 0 \qquad u_1(\infty) = 0$$

The appearance of a well-defined sequence of boundary-value problems, associated with Eqs. (12.73) and (12.74), suggests that the guessed form of the series (12.72) is successful, and provides a straightforward way to generate the upstream data needed for the integration of (12.60). We can now pick a finite starting value of ξ, say $\xi = -10$, where the first two terms of the series are expected to represent the complete series to the desired level of accuracy. Numerical experimentation, to see whether this expectation is correct, will be a part of the subsequent program of computation.

Second-order streamwise differencing. In Chapter 10, none of the examples of a time-dependent solution presented much challenge to the accuracy of the finite-difference representation of the time-derivative. Now, in preparing a program suitable for boundary layers that develop under decelerating potential flows, and which are consequently very sensitive to their upstream history, it is good to use second-order difference formulas to represent $u_{,\xi}$ and $f_{,\xi}$. One can choose from a variety of second-order formulas; a choice that is easily combined with the linearization procedures shown here is the three-point trailing-difference formula,

$$u_{,\xi}(\xi, \eta) = au(\xi, \eta) + bu(\xi - d\xi_1, \eta) + cu(\xi - d\xi_1 - d\xi_2, \eta)$$

For routine calculations, $d\xi$ can be taken constant; then $(a, b, c) = (3, -4, 1)/(2\,d\xi)$. As the boundary layer approaches separation, it is occasionally advisable to cut $d\xi$ in half. Then, with $d\xi_1 = \tfrac{1}{2} d\xi_2$, we find that $(a, b, c) = (8, -9, 1)/(6\,d\xi_1)$.

In program NEWBL, storage arrays for $u(\xi, \eta)$, $u(\xi - d\xi_1, \eta)$, and $u(\xi - d\xi_1 - d\xi_2, \eta)$ are named UM(N), UMN1(N), and UMN2(N). A corresponding nomenclature is used for $f(\xi, \eta)$, etc. Whenever dx is halved, an integer flag is set, to warn subroutine THOMAS to use the appropriate values of a, b, and c.

Linearization of the right-hand side of Eq. (12.60). In Chapter 10, the right-hand side of the transformed partial differential equations was a remnant of the operator $\partial/\partial t$, and required no linearization. For Eq. (12.60) we proceed in the following way.

$$uu_{,\xi} \approx u(\xi, \eta)[au(\xi, \eta) + bu(\xi - d\xi_1, \eta) + cu(\xi - d\xi_1 - d\xi_2, \eta)]$$
$$\approx u(\xi, \eta)[2au^*(\xi, \eta) + bu(\xi - d\xi_1, \eta) + cu(\xi - d\xi_1 - d\xi_2, \eta)]$$
$$- au^{*2}(\xi, \eta)$$
$$Sf_{,\xi} \approx S(\xi, \eta)[af(\xi, \eta) + bf(\xi - d\xi_1, \eta) + cf(\xi - d\xi_1 - d\xi_2, \eta)]$$
$$\approx S(\xi, \eta)[af^*(\xi, \eta) + bf(\xi - d\xi_1, \eta) + cf(\xi - d\xi_1 - d\xi_2, \eta)]$$
$$+ aS^*(\xi, \eta)f(\xi, \eta) - aS^*(\xi, \eta)f^*(\xi, \eta)$$

Here, as in Chapter 10, we are using the symbol S as shorthand for u'.

The algorithm embodied in program NEWBL has the following steps, some special to this problem but most common to a wide range of problems;

1. The self-similar profiles for $\xi = -\infty$, $f_0(\eta)$, $u_0(\eta)$, and $S_0(\eta)$, are computed iteratively, by repeated calls to subroutine THOMAS. This requires an input of the values $\gamma = 1$, $\delta = 4$, and $\psi = 0$, and of first-guess profiles $u^*(\eta)$, $f^*(\eta)$ and $S^*(\eta)$. The value assigned to ψ causes THOMAS to solve the set of Eqs. (12.73). The programming logic assures that the output from each execution of THOMAS provides the necessary input of u^*, f^*, and S^* for the next execution. Seven iterations is usually more than enough to give convergence to five decimals. When the iteration is complete, dimensionless integrals proportional to δ^* and θ are computed, as are the values of the shape factors H and T. The values of f, u and S are saved, under the names F0, U0, and S0.

2. The coefficients in Eq. (12.74) are then defined, and THOMAS is used to calculate f_1, u_1, and S_1. No iteration is required, because (12.74) is linear.

3. A starting position, $\xi = \xi_0$, is chosen and the two-term series approximation, (12.72), is used to set up starting profiles of f, u and S at ξ_0 and $\xi_0 + d\xi$. These profiles are stored as UMN2(N) and UMN1(N) and used for evaluations of skin friction, displacement and momentum thicknesses, and shape factors.

4. A numerical value is selected for the streamwise step, $d\xi$, and the numerical marching procedure is started at the position $\xi_0 + 2\,d\xi$. The necessary values of g^2, δ, γ, and ψ are calculated, using Eqs. (12.44), (12.68), (12.67), and (12.64). At very large negative values of ξ, the values of g^2 must be given by the series expansion of the exact result, (12.44), because roundoff error in the computer will prevent accurate evaluation of the closed-form expression!

5. These new values are submitted to THOMAS, which provides the first calculated estimates of f, u, and S at $\xi_0 + 2\,d\xi$. Two cycles of iteration are usually ample to deal with the nonlinearity of the equations to be solved,

because the design of the transformation formulas assures that the first-guess values prepared in step (4) will be very close to the converged values, if reasonably small values are chosen for $d\xi$. For the first cycle of iteration, f, u, and S are approximated by their values at $\xi_0 + d\xi$.

6. The shape factors, H and T, and the locally-scaled displacement thickness, momentum thickness, and skin friction are calculated from the converged values of u and f. The locally-scaled values are then converted back to globally-scaled values such as $(U_0/va)^{1/2}\delta^*(x/a)$, which should be demonstrably independent of the particular transformation formulas used to ease the calculations. In the program listing, this step is called *inverting the transformation*.

7. In preparation for the next step downstream, the values stored as FMN1 are moved to FMN2, and so on, and the most recent values are stored as FMN1, etc.

8. Another increment is added to ξ, and steps (4) through (7) are repeated, until a target value of ξ is reached, or until separation of the boundary layer is indicated by negative skin friction.

The calculation of the boundary layer on the wall under a vortex illustrates certain fairly typical results, which are best appreciated by running program NEWBL with various choices of $d\xi$, $d\eta$, and the value of η at which u is set equal to 1. These are mentioned now, not only to impress you with the necessity for a disciplined approach to the use of numerical methods, but because the calculation strongly confirms a conclusion we had drawn from simple momentum-integral formulas, which certainly has significant implications for the control of boundary-layer behavior.

This conclusion is that a boundary layer fairly quickly forgets its upstream history, as long as it develops in a favorable pressure gradient, i.e., under an accelerating potential flow. Thus, it is quite possible, for example, to make an error in the calculation of the two-term Blasius series, perhaps a simple sign error in the combination of u_0 and u_1, at $\xi = -10$, and to discover that all calculated quantities at $\xi = -5$ are unchanged to five decimals when the error is corrected. One implication of this is that fairly large values of $d\xi$ may be used, without noticeable effect on accuracy, in regions of favorable pressure gradient.

The companion conclusion, that a boundary layer in an adverse pressure gradient is very sensitive to its past history, is even more strongly emphasized by the calculations. The experiments you can perform with program NEWBL will quickly convince you that much care is required to maintain accuracy of the computation in such regions. In particular, smaller and smaller values of $d\xi$ are required as the point of separation is approached. Even errors accumulated in computing the history of the boundary layer under an accelerating flow, in this case the history for $\xi < 0$, may be remembered and amplified when the pressure gradient turns adverse.

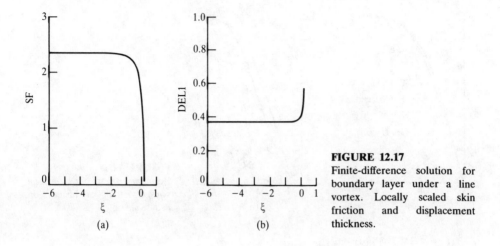

FIGURE 12.17
Finite-difference solution for boundary layer under a line vortex. Locally scaled skin friction and displacement thickness.

Evidence of the success of the transformation based on Thwaites' estimate of $\theta(x)$ appears in several forms. Figure 12.17 shows the variation of the locally-scaled skin friction and displacement thickness. The fact that only small changes are observed means that the computed profiles change only slightly; this makes life easy for the computer.

Because it is not obvious how to impose a softened outer boundary condition in a calculation which involves ξ-derivatives, the condition $u(\xi, \eta_{max}) = 1$ is imposed directly. Thus, it is important to examine vorticity profiles for representative values of ξ, to make sure that the value of η_{max} is everywhere sufficient, so that the vorticity at the outer boundary is negligible. The standard imposed in the sample calculation is that $S(\xi, \eta_{max}) \leq 0.00001$.

To check whether the two-term Blasius series gives an accurate start to the computation at $\xi = -10$, we can compare the results of that series to the results of finite-difference calculation, at points closer to the vortex. The Blasius series gives the following predictions:

$$SF = 2.34746 - 0.0738\xi^{-2} \qquad DEL1 = 0.36905 + 0.00226\xi^{-2}$$

Even where $x = -4$, these formulas fit the values from the step-by-step integration with an error that is only 0.00004 in SF, and 0.00000 in DEL1.

Figure 12.18 shows the effect of this particular flow history on the shape of the velocity profile at two locations where a comparison can be made with self-similar profiles. One is at $\xi = 0$, where there is zero pressure gradient, and the other is at the point of zero skin friction.

(b) THE EFFECT OF A LEADING EDGE AT FINITE ξ. Suppose now that the boundary layer is born at some finite negative value of ξ, where the wall begins. Our first step is to redo the momentum-integral analysis, requiring that

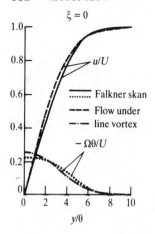

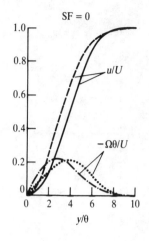

FIGURE 12.18
Boundary layer under a line vortex. Effect of history on profile shape.

$\theta = 0$ at $\xi = \xi_0$. The result is

$$\frac{\theta^2 U_0}{va} = 0.45(1 + \xi^2)^6[F(\xi) - F(\xi_0)]$$

with

$$F(\xi) = \tfrac{1}{384}[48\xi(1 + \xi^2)^{-4} + 56\xi(1 + \xi^2)^{-3} + 70\xi(1 + \xi^2)^{-2} + 105\xi(1 + \xi^2)^{-1} + 105 \arctan (\xi)]$$

Where $(\xi - \xi_0) \ll 1$, the leading approximation is

$$\frac{\theta^2 U_0}{va} = 0.45(1 + \xi_0^2)(\xi - \xi_0)$$

With the choice $A^2 = \frac{40}{9}$, this leads to

$$\delta = -4\xi_0(1 + \xi_0^2)^{-1}(\xi - \xi_0) \qquad \gamma = 1 + 8\xi_0(1 + \xi_0^2)^{-1}(\xi - \xi_0)$$

and

$$\psi = 2(\xi - \xi_0)$$

To be completely consistent with the procedure used when the leading edge is at upstream infinity, we should generate a new two-term Blasius series. It turns out that a successful form is

$$f(\xi, \eta) = f_0(\eta) + f_1(\eta)(\xi - \xi_0) + \cdots \tag{12.75}$$

The leading term will now be the Blasius function. The second term is easily found, and the marching calculation can be carried forward just as before.

We are most interested to see how the separation point moves as the leading edge is brought closer to the vortex. According to Thwaites' correlation, a thinner boundary layer can support a larger deceleration of the flow before it separates; however, the thickness of the boundary layer at $\xi = 0$ is not much influenced by the position of the leading edge, if $\xi_0 \ll -1$. Table 12.1 shows $\theta(0)$, ξ_{sep}, and the value of Thawites' λ at separation, as functions of ξ_0.

TABLE 12.1
Effect of leading-edge location on the boundary layer under a line vortex

ξ_0	$\theta(0)$	ξ_{sep}	λ_{sep}
−10.0	0.47001	0.173	−0.122
−4.0	0.46999	0.174	−0.122
−2.0	0.46947	0.174	−0.122
−1.0	0.45931	0.176	−0.119
−0.5	0.40800	0.182	−0.111
−0.2	0.28902	0.223	−0.099
−0.0	0.00000	0.288	−0.091

No appreciable effect of the value of ξ_0 is seen until $\xi_0 > -2$, indicating that the strong favorable pressure gradients upstream of the vortex cause almost total loss of memory of the birth of the boundary layer when the leading edge is farther upstream. The point of separation moves from $\xi = 0.173$ for distant leading edges, to $\xi = 0.288$ when $\xi_0 = 0$. The corresponding values of U at the point of separation are $0.970U_0$, and $0.923U_0$, indicating that not much reduction of the potential-flow speed occurs before separation, even when the boundary layer has a completely fresh start at the beginning of the interval of deceleration.

Remember that the value $\lambda = -0.090$ corresponds to separation in Thwaites' method. This value is very close to what we find here when $\xi_0 = 0$. The other values are a little higher, but not so much as to reduce the value of the method for quick predictions.

The variation of skin friction, τ_w, near the separation point is shown in Fig. 12.19. Note that τ_w^2 is plotted versus x, because a local analysis by S. Goldstein[11] predicts that τ_w should vanish in proportion to $(\xi - \xi_{sep})^{1/2}$.

Conditions at, and After Separation

If you run Program NEWBL for problems like these, you may find that the calculation seems to step right over the separation point, and to continue for a few, rather wobbly, steps before actually indicting a negative skin friction. It turns out that the calculation scheme breaks down as soon as it has to deal with negative values of u. This may seem surprising, since a nearly identical method had no trouble describing the development of backflow in the transient analysis of the three-dimensional stagnation-point flows. There is, however, a crucial

---

[11] S. Goldstein, "Concerning Some Solutions of the Boundary-Layer Equations in Hydrodynamics," *Proc. Camb. Phil. Soc.* **26**:1–30 (1930).

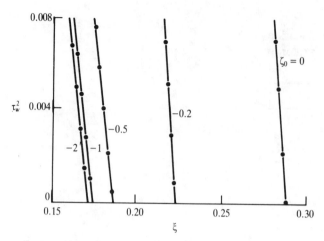

FIGURE 12.19
Boundary layer under a line vortex. Wall shear near separation point.

logical difference between the two types of problem. For the stagnation-point flows, the calculation marches unambiguously forward in time, never letting the prediction be influenced by something that has not yet happened. In the present problem, however, the calculation marches in the direction of increasing x, and when it encounters reverse flow we are attempting to predict the state of fluid particles whose past state is unknown, because they arrive from a region which we have not yet examined. This will be discussed in greater depth in Section 12.11.

Other Examples

Having described one example in some detail, we now show more briefly how some others are analyzed.

Boundary layer on a circular cylinder, and on a sphere. To show how to start a calculation at a stagnation point, we now work out two classic and closely related examples, those of the boundary-layer development on a circular cylinder and on a sphere.

For the cylinder, we have already done the momentum-integral analysis, finding Eq. (12.37). We wish to choose the constant in $g = A\theta$ so that the corresponding values of γ and δ match those for the plane stagnation-point flow. At the stagnation point, Eq. (12.38) gives the value $\lambda(0) = 0.075$, and we know that $\delta = A^2\lambda$. To make $\delta(0) = 1$, we set $A^2 = \frac{40}{3}$. According to Eq. (12.67), this value of A makes $\gamma(x) = 3 - 2\,\delta(x)$, which yields the value $\gamma(0) = 1$, as desired. The starting value of ψ is zero.

The corresponding general expression for $g(x)$ is then

$$\frac{g^2 U_0}{va} = \frac{1}{2}\left(1 - \frac{3z}{2} + \frac{3z^2}{5}\right)(1-z)^{-3}$$

with $z = \sin^2(\xi/2)$ and $\xi = x/a$.

For the *sphere*, the potential flow along the surface has speed $U(\xi) = (3U_0/2)\sin\xi$, where again $\xi = x/a$, and U_0 is the speed of the uniform flow in which the sphere is immersed. Thwaites' method was not designed for axisymmetric flow, but we can use the momentum-integral equation with constant shape factors, to provide an estimate of $\theta(\xi)$. Furthermore, we shall pick a value of H, namely $H = 2$, that makes the integration easy. Noting that $R_T = a\sin\xi$, integrating, and using the identity $1 - \cos\xi = 2\sin^2(\xi/2)$, we find

$$\frac{\theta^2 U_0}{va} = \frac{2T}{15}\frac{(1 - (10/3)z + (30/7)z^2 - (5/2)z^3 + (5/9)z^4)}{(1-z)^5}$$

where again $z = \sin^2(\xi/2)$.

From the definition of δ and the equation for $U(\xi)$, we find that $\delta(0) = 2A^2T/15$. This can be made to agree with the value $\delta = \frac{1}{2}$ of the axisymmetric stagnation-point solution, by seting $A^2T = \frac{5}{2}$. Note that it is unnecessary to guess a value of T, if θ is to be used only for scaling purposes.

Combining the definition of γ with the momentum-integral equation for axisymmeric flow, and using the assumed value $H = 2$, we can derive the relation $\gamma = A^2T - 3\delta$. With $A^2T = \frac{5}{2}$, this gives $\gamma = \frac{5}{2} - 3\delta$, which yields $\gamma = 1$ at the stagnation point, as desired. Again $\psi = 0$ at the stagnation point, and the calculation proceeds just as for the cylinder.

Calculations were carried out with $d\xi = \pi/800$ up to $\xi = \pi/2$. Half this value of $d\xi$, then a quarter, then an eighth, were used as the separation point was approached. The values of η^* and h were varied to make sure that final answers are independent of the exact choices made. Separation was found at nearly the same angle for both bodies: $104.39°$ for the cylinder, $104.82°$ for the sphere. The angle for maximum skin friction was $57.81°$ for the cylinder, $57.83°$ for the sphere; the maximum ordinate value in Fig. 12.20 being 2.2573 for the cylinder, and 1.5595 for the sphere. The displacement and momentum thicknesses are almost equal on the sphere and the cylinder.

There is good reason to believe that these calculations are accurate, but they turn out, for reasons with which we shall deal later, to be unrealistic. Briefly, they do not include an adequate description of the vorticity distribution downstream of the point of separation. The distribution found in experiments is unsteady, unless certain artificial precautions are taken. Even if it were steady, the vorticity in the separated flow, being far from the wall and thus not accompanied by nearby image vorticity, would induce large changes in the velocity field upstream of the computed separation point. This effect causes the boundary layer to develop under a potential flow which decelerates even before it reaches the widest cross section of the body. If the boundary-layer

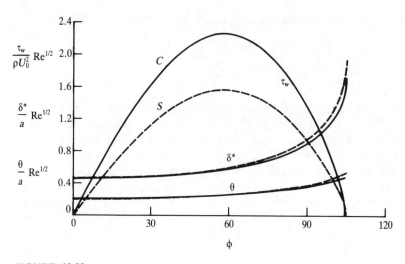

FIGURE 12.20
Skin friction, displacement thickness, and momentum thickness on a sphere (dotted curves) and a circular cylinder (solid curves).

calculation is repeated, with the function $U(x)$ obtained from the experimental time-averaged pressure distribution, the computed separation point moves forward to about $\phi = 81°$, in good agreement with time-averaged measurements. The tentative conclusion is that the phenomenon of separation does not, except very locally, adversely affect the accuracy of the boundary-layer equations; it just complicates the decision as to what boundary-layer problem to solve.

Boundary layer on a Rankine half-body. Görtler's transformation. When the velocity field of a uniform steady stream is added to that of the flow from a stationary line source, oriented normal to the stream, the combined velocity potential is

$$\Phi = U_0\left(a - R\cos\phi - a\ln\frac{a}{R}\right)$$

The flow from the source is separated from that of the stream by the streamline $R = a\phi \csc \phi$, which is sketched in Fig. 12.21. A solid body that would just fit inside that streamline is called a *Rankine half-body*.

The speed of potential flow over such a body, given by the equation

$$\frac{U}{U_0} = (1 - \phi^{-1}\sin 2\phi + \phi^{-2}\sin^2\phi)^{1/2}$$

starts at zero at a forward stagnation point, reaches a maximum of about $1.26U_0$ at an angle $\phi \approx 117°$, and asymptotically returns to U_0, the speed of the uniform stream.

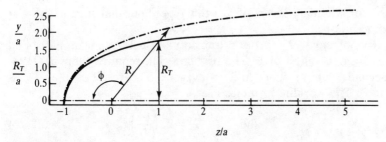

FIGURE 12.21
Rankine half-bodies: —— plane; ····· axisymmetric.

Separation of the boundary layer is threatened, but never quite occurs, and the velocity profile asymptotically approaches that for a flat-plate boundary layer under an undisturbed potential flow. It is interesting to see how far from the leading edge one must look, to find a good approximation to the classic flat-plate flow. Figure 12.22 shows the potential-flow speed, $U(\phi)$, both for this body and for the axisymmetric body that fits the dividing streamline when we have a point source instead of a line source.

For these potential flows, it is difficult to get a closed-form evaluation of the integral in Thwaites' method. We can, however, easily evaluate it numerically, and have done so for the case of plane flow. The result shows that the shape-determining parameter reaches a minimum value, $\lambda = -0.060$ at $\phi = 155°$, thus predicting that the boundary layer will not separate. To get a more definitive prediction, we take the opportunity to introduce a differently designed transformation of the partial differential equations. It was invented

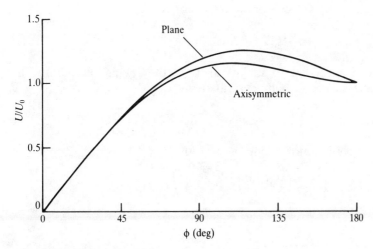

FIGURE 12.22
Speed of potential flow along Rankine half-bodies.

by H. Görtler[12] for boundary layers of just this type: ones that start to grow at a well-defined origin of x.

The design objective of Görtler's transformation is to make $\gamma = 1$ and $\psi = 2\xi$, for all positive values of ξ. The first criterion requires that $(\ln Q)' = v/g^2U$; the second requires that $(\ln \xi)' = 2v/g^2U$. The result is that Q^2 is proportional to ξ. The working formulas are

$$\frac{d\xi}{dx} = \frac{U(x)}{U_0 a} \quad \text{and} \quad \frac{g^2 U_0}{va} = 2\xi \left(\frac{U_0}{U}\right)^2 \tag{12.76}$$

These formulas look a little ominous but it is rather easy, for the Rankine body, to evaluate ξ, which is just a dimensionless velocity potential. The necessary formula is

$$\xi = \ln \left(\frac{\phi}{\sin \phi}\right) - \phi \operatorname{ctn} \phi + 1$$

To calculate δ we combine its definition with Eq. (12.76) to get

$$\delta = 2\xi \frac{d(\ln U)}{d\xi} = 2\xi \frac{[d(\ln U)/d\phi]}{d\xi/d\phi}$$

To evaluate this, we can differentiate the expressions for $x(\phi)$ and $U(\phi)$. Small-angle approximations can be used for points close to the stagnation point, to show that $\xi \approx \frac{1}{2}\phi^2 + \cdots$, while $U/U_0 \approx \phi + \cdots$, so that the starting value of δ equals 1. Thus, this calculation, like that for the circular cylinder, starts out with the classic profiles for stagnation-point flow.

Axisymmetric source-shaped body. Mangler's transformation. Before showing the results for a plane source-shaped body, let us analyze the corresponding axisymmetric flow. If a point source, rather than a line source, is superposed on a uniform stream, the velocity potential becomes

$$\Phi = U_0 \left(2a - R \cos \phi - \frac{a^2}{R}\right)$$

The constant term is added to make $\Phi = 0$ at the stagnation point. The dividing streamline outlines a body described by

$$R = a \sec \left(\frac{\phi}{2}\right) \quad \text{so that} \quad R_T = 2a \sin \left(\frac{\phi}{2}\right)$$

The speed of the potential flow along the wall is

$$\frac{U}{U_0} = \left[4 \sin^2 \left(\frac{\phi}{2}\right) - 3 \sin^4 \left(\frac{\phi}{2}\right)\right]^{1/2}$$

---

[12] H. Görtler, "A New Series for the Calculation of Steady Laminar Boundary Layer Flows," *Jour. Math. Mech.* **6**:1–66 (1957).

These results are shown in Figs. 12.21 and 12.22.

If we again design a transformation to make $\gamma = 1$ and $\psi = 2\xi$, we can follow the analysis of the plane-flow case, only remembering that Q now contains the factor R_T. The result is a combination of Görtler's transformation and a transformation due to W. Mangler.[13]

$$\frac{d\xi}{dx} = \frac{U(x)R_T^2}{U_0 a^3} \quad \text{and} \quad \frac{g^2 U_0}{va} = 2\xi\left(\frac{U_0 a}{UR_T}\right)^2 \tag{12.77}$$

The integration to evaluate ξ is easy, if we note that

$$U\frac{dx}{U_0 a} = \frac{d\Phi}{U_0 a} = -(3 + p^{-2})\,dp \quad \text{if} \quad p = \cos\left(\frac{\phi}{2}\right)$$

The result is

$$\xi = 4\sin^4\left(\frac{\phi}{2}\right)\sec\left(\frac{\phi}{2}\right)$$

The corresponding expression for δ is

$$\delta = \tfrac{1}{2}(1 - \sin^2\phi)(1 - \tfrac{3}{2}\sin^2\phi)(1 - \tfrac{3}{4}\sin^2\phi)^{-2}$$

As expected, at the stagnation point of an axisymmetric body, $\delta(0) = \tfrac{1}{2}$.

Program NEWBL can now be used, just as for the cylinder and the sphere. Figure 12.23 shows the evolution of the shape factors, H and T, versus

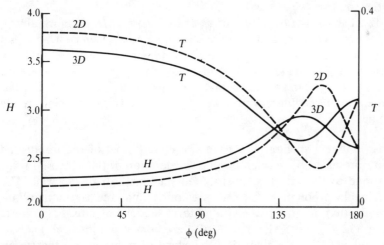

FIGURE 12.23
Shape factors for the boundary layers on Rankine half-bodies.

[13] Mangler's original paper is in German. See H. Schlichting, *Boundary Layer Theory*, 6th ed. pp. 235–237.

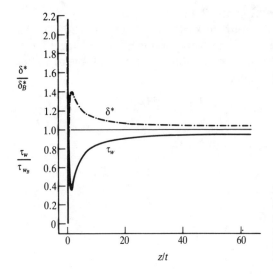

FIGURE 12.24
Effect of a rounded nose on skin friction and displacement thickness far downstream on a flat plate: z = distance from leading edge; δ_B^* and τ_{w_B} are values for a Blasius boundary layer.

ϕ. This presentation emphasizes the development of the profile near the nose of the body, and compresses the region far downstream. A more detailed examination of the shape factors far from the nose shows that T, the slower to approach its asymptote, attains 99% of its final value when z/t exceeds about 80 on the plane body, or about 10 on the axisymmetric body. In these figures, t represents the asymptotic thickness of the plate ($t = 2\pi a$), or the asymptotic diameter of the rod ($t = 4a$).

Figure 12.24 shows the evolution of τ_w and δ^* versus z/t, the normalizing factors being the values of τ_w and δ^* in a Blasius boundary layer, at an equal distance from the leading edge of an infinitely thin plate. We see that the adverse pressure gradient makes the boundary layer thicker than the reference Blasius layer. The effect can be approximated by displacing the origin of z, but the displacement for τ_w is different from that for δ^*, and is therefore not really worth reporting.

Flat plate with uniform suction. We have considered most of the interesting effects that are produced by acceleration or deceleration of the potential flow; let us now attend to the effects of distributed suction through the wall. The simplest imaginable problem is the one previously analyzed by momentum-integral methods: that of a flat plate with uniform suction velocity, $v_w(x) = v_0$.

Equation (12.47) gave the result of the momentum-integral analysis inversely, i.e., as $x(\delta^*)$ rather than $\delta^*(x)$. Instead of trying to use that result explicitly, we shall base our transformation on a simple interpolation formula, somewhat analogous to the ones used for the transient flows of Section 10.3. We adopt the following design criteria.

1. $\gamma(x)$ should be such that $\gamma(0) = 1$, so the initial profile will be the classic Blasius profile. This will happen if $g^2 \to 2vx/U_0$ as $x \to 0$.

2. $g(x)$ should approach a constant value as $x \to \infty$, such that $\eta_{max} = 6$ will guarantee that u differs from 1 by less than one part in 10^5. We can use this criterion because we expect the solution to approach that for an infinite flat plate with uniform suction, which is $u/U_0 = 1 - \exp{(v_0 y/v)}$. Setting $y = 6g$, and $u/U_0 = 1 - 10^{-5}$, we find that the desired asymptotic value for g is $g = \frac{5}{6}(\ln 10)(v/-v_0) = 1.92(v/-v_0)$. Remember that $v_0 < 0$ for the case of suction. Let us round off the 1.92 to 2.

3. $\xi(x)$ should be proportional to $x^{1/2}$, at least for small values of x. The reason for this appears when we see how $f(\xi, 0)$ varies with ξ where $\xi \ll 1$. From the definition of $f(\xi, \eta)$ and the fact that v_0 is independent of x, we calculate that $f(\xi, 0) \equiv \sigma = -v_0 x/gU_0$. Since we intend that g should start out in proportion to $x^{1/2}$, it is clear that s will start out the same way. This boundary condition will force $f(\xi, \eta) - f(0, \eta)$ to be proportional to $x^{1/2}$ for small values of x. Such a variation can be most accurately captured by a finite-difference scheme if we design the function $\xi(x)$ so that $f(\xi, \eta)$ starts out as a linear function of ξ.

A simple formula that meets all these criteria is

$$g = \frac{-2v}{v_0}[1 - \exp{(-\xi)}] \qquad \text{with} \qquad \xi = -v_0\left(\frac{x}{2vU_0}\right)^{1/2}$$

The working formulas are then

$$\gamma = \exp{(-\xi)}\frac{1 - \exp{(-\xi)}}{\xi}, \qquad \delta = 0$$

$$\psi = \frac{[1 - \exp{(-\xi)}]^2}{\xi} \qquad \text{and} \qquad \sigma = \frac{\xi^2}{1 - \exp{(-\xi)}}$$

Most problems involving suction require the specification of a dimensionless suction strength, but this one is special, in that the effect of that parameter can be absorbed into the relationship between ξ and x.

The calculation with NEWBL starts out with the Blasius profile, because both ψ and σ vanish at the leading edge. It is interesting to analyze what happens as $\xi \to \infty$. In that limit, γ vanishes, $\psi \to 1/\xi$, and $\sigma \to \xi^2$. We expect that $u(\xi, \eta)$ should become independent of ξ, which implies that $f(\xi, \eta)$ will have the asymptotic form

$$f(\xi, \eta) \approx \sigma(\xi) + F(\eta) \qquad \text{where} \qquad u = F'(\eta)$$

Thus, Eq. (12.60) will approach the form $u'' = -\sigma' u'/\xi = -2u'$, with solution $u = 1 - \exp{(-2\eta)}$.

Care must be taken with the calculation of both SF and DEL1, because the value of σ influences both quantities. To see the effect on SF, it is easiest to return to the unscaled x-momentum equation. At $y = 0$, it reduces to $v_0 u^*_{,y} = vu^*_{,yy}$, the scaled version of which is $u''(\xi, 0) = g(v_0/u)u'(\xi, 0) = -2[1 - \exp{(-\xi)}]u'(\xi, 0)$.

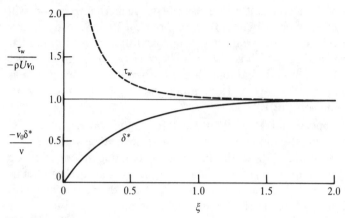

FIGURE 12.25
Flat plate with uniform suction. Skin friction and displacement thickness.

Using the second-order centered difference formulas to approximate both $u''(\xi, 0)$ and $u'(\xi, 0)$, we find $SF \equiv u'(\xi, 0) = (u_2/h)\{1 - h[1 - \exp(-\xi)]\}^{-1}$. The effect on DEL1 is simply additive:

$$\text{DEL1} \equiv \int_0^\infty (1 - u) \, d\eta = \eta_{\max} - f(\xi, \eta_{\max}) + \sigma$$

Figure 12.25 shows the distribution of skin friction and displacement thickness versus ξ, while Fig. 12.26 shows the evolution of the velocity and vorticity profiles.

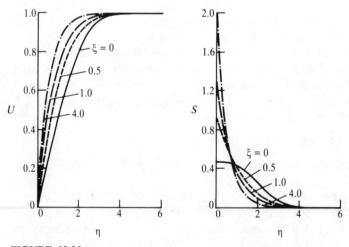

FIGURE 12.26
Flat plate with uniform suction. Velocity and vorticity profiles.

This problem was originally solved, in the days before fast computers and efficient algorithms, by R. Iglisch.[14] Considering the difficulties faced by Iglisch, the accuracy of his results is remarkable.

The Boundary-Layer Equations with Pressure Eliminated

There are flows, such s in slender channels, in which all the approximations of boundary-layer theory are expected to be appropriate, but in which the pressure is not known beforehand. To deal with such cases, one can use the boundary-layer approximations on the vorticity equation. If a streamfunction is used to satisfy the continuity equation, we have, for steady flow,

$$\psi_{,y}\Omega_{,x} - \psi_{,x}\Omega_{,y} = \nu\Omega_{,yy} \quad \text{and} \quad \Omega = -\Psi_{,yy} \quad (12.78)$$

The boundary conditions will be $\Psi(x, 0) = F(x)$ and $\Psi_{,y}(x, 0) = 0$ at the wall, and $\Omega = 0$ at the outer edge of a boundary layer, or at the midplane of a symmetric channel flow. Because the system of equations is now of fourth order in y, an additional condition is needed, outside the boundary layer. In the cases presented heretofore, that condition was simply $\psi_{,y} = U(x)$.

An Idealized Entry-Length Problem

Consider the following example, which will clarify the principle. Figure 12.27 shows the inlet to a channel between plane parallel walls, at $y = 0$ and $y = 2H$. We suppose, without asking how it could be true, that flow enters the channel at $x = 0$ with a uniform velocity, $\mathbf{u} = (Q/H)\mathbf{e}_x$. The volumetric flowrate per unit span of channel is $2Q$, and this quantity will be independent of x. In this problem, the second upper boundary condition specifies the value of a linear combination of Ψ and $\Psi_{,y}$, for reasons that will presently be shown.

Two slightly different transformations of variables will be used: one for Region 1, in which relatively thin boundary layers grow towards each other,

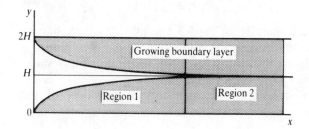

FIGURE 12.27
Inlet to channel between plane, parallel walls.

---

[14] R. Iglisch, "Exact Calculation of the Laminar Boundary Layer on a Flat Plate in Slow Flow, with Homogeneous Suction." Translated from the German in NACA TM1205, 1949.

the other for the Region 2, in which vorticity is found from wall to wall and the flow is acquiring a fully-developed character. In both regions, the streamwise distance variable is defined by

$$\xi = \left(\frac{2vx}{QH}\right)^{1/2} \tag{12.79}$$

a choice suggested by the simple notion that the entry length will be proportional to the distance over which a Blasius boundary layer grows to thickness H. That distance is approximately QH/v; the square root is introduced because it improves the accuracy of the finite-difference approximations of the operator $\partial/\partial\xi$.

For the region $0 \leq \xi \leq \xi_0$, the dimensionless streamfunction is introduced by the transformation

$$\Psi(x, y) = Q\xi f(\xi, \eta) \quad \text{in which} \quad \eta = \xi^{-1}\left(\frac{y}{H}\right) \tag{12.80}$$

The dimensionless vorticity is introduced by $\Omega(x, y) = (Q/H^2)\xi^{-1}S(\xi, \eta)$. For $\xi \geq \xi_0$, these are simply changed to

$$\Psi(x, y) = Q\xi_0 f(\xi, \eta) \quad \text{in which} \quad \eta = \xi_0^{-1}\left(\frac{y}{H}\right) \tag{12.81}$$

and the vorticity is $\Omega(x, y) = (Q/H^2)\xi_0^{-1}S(\xi, \eta)$.

Procedure for the upstream region. The transformed differential equations for $0 \leq \xi \leq \xi_0$ are

$$S'' + fS' + Sf' = \xi(f'S_{,\xi} - S'f_{,\xi}) \tag{12.82}$$

and $$f'' + S = 0$$

The boundary conditions are

$$f(\xi, 0) = 0 \quad f'(\xi, 0) = 0 \quad S(\xi, \eta_{max}) = 0$$

and $$\xi f(\xi, \eta_{max}) + (1 - \xi\eta_{max})f'(\xi, \eta_{max}) = 1 \tag{12.83}$$

As in previous numerical analyses, η_{max} is a constant, large enough so that its exact value makes no difference to the computed results, representing a point outside the boundary layer. The third boundary condition simply states that there is no vorticity there.

The fourth boundary condition is the transformed equivalent of the integral statement

$$Q = \int_0^H u(x, y)\, dy$$

in which the range of integration has been broken into two intervals, $0 \leq y \leq H\xi\eta_{max}$, and $H\xi\eta_{max} \leq y \leq H$. The first interval, corresponding to the boundary layer, contributes the first term of Eq. (12.83). In the second interval, the velocity is assumed to be independent of y, hence equal to $(Q/H)f'(\xi, \eta_{max})$. This assumption is consistent with *first-order boundary-layer theory*, which is the theory discussed in all but the last section of this chapter.

The calculation is started at $\xi = 0$, by solving the ordinary differential equations to which Eqs. (12.82) reduce there; the result is just the Blasius profile. It is stopped where $\xi_0 = 1/\eta_{max}$, which identifies the section at which $y_{max} = H\xi\eta_{max} = H$. The downstream marching procedure employs three-point trailing differences with constant $d\xi$, preceded by a predictor step, in which the first guesses for f and S at a new value of ξ are given by

$$f^*(\xi, \eta) = 2f(\xi - d\xi, \eta) - f(\xi - 2\,d\xi, \eta)$$

and $\qquad S^*(\xi, \eta) = 2S(\xi - d\xi, \eta) - S(\xi - 2\,d\xi, \eta)$

A couple of iterations at each new value of ξ give convergence to about five decimals. The necessary linear algebra is the same as in program EKMAN, which dealt with two coupled second-order equations. Details may be seen in Appendix B, in the first main section of program ENTRY.

Procedure for the downstream region. The transformed differential equations for $\xi_0 \le \xi$ are

$$S'' = \xi_0^2 \xi^{-1}(f's_{,\xi} - S'f_{,\xi}) \tag{12.84}$$

and $\qquad\qquad\qquad f'' + S = 0$

The boundary conditions are

$$f(\xi, 0) = 0 \qquad f'(\xi, 0) \qquad S(\xi, \eta_{max}) = 0$$

and $\qquad\qquad\qquad \xi_0 f(\xi, \eta_{max}) = 1$

Now $\eta = \eta_{max}$ represents a point halfway between the walls. The vorticity is zero there because of symmetry. The last condition simply states that $\Psi = Q$ on the midplane.

The coefficients in the two sets of differential equations and boundary conditions are continuous at $\xi = \xi_0$, and so are the solutions. The second main section of program ENTRY continues the downstream marching until a test of fully-developed flow is met.

Once the velocity field is known, the pressure field can be found from the x-momentum equation, by the quadrature

$$\Pi(x) - \Pi(0) = \int_0^x \{vu_{,yy} - (uu_{,x} + vu_{,y})\}\, dx.$$

Because Π is independent of y, the quadrature may be evaluated at any convenient value of y. To display the midplane velocity and viscous force per unit volume as interesting flow features, one can evaluate the quadrature at

$y = H$. The transformed equation is then

$$CP(\xi) = 2 \int_0^\xi S'(\xi, \eta_{max})\left(\frac{1}{\xi}\right) d\xi - [f'(\xi, \eta_{max})]^2 \qquad \text{for } 0 \le \xi \le \xi_0$$

and

$$CP(\xi) = 2 \int_0^\xi S'(\xi, \eta_{max})\left(\frac{\xi}{\xi_0^2}\right) d\xi - [f'(\xi, \eta_{max})]^2 \qquad \text{for } \xi_0 \le \xi$$

In these, $CP = 2[\Pi(x) - \Pi(0)]/(U^2)$, where $U = Q/H$ is the inlet speed.

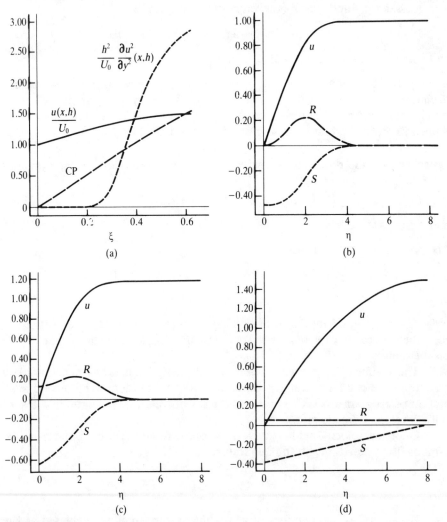

FIGURE 12.28
Entry flow between parallel walls: (a) midplane velocity, viscous force, and pressure; (b) inlet (Blasius) profiles of velocity (u), vorticity (S), and viscous force (R); (c) profiles where $\eta = 8$ means $y = h$; (d) profiles where $\xi = 0.6$.

Some results are shown in Fig. 12.28. If the criterion for fully-developed flow is that $f'(\xi, \eta_{\max}) = 0.99$ times its asymptotic value for $\xi = \infty$, it is met at $\xi = 0.306$. the alternative criterion, that $S'(\xi, \eta_{\max}) = 0.99$ times its asymptotic value, is met at $\xi = 0.395$.

12.6 BOUNDARY-LAYER CONTROL BY DISTRIBUTED SUCTION

It is quite easy to add a control loop to program NEWBL, with which to calculate the distribution of suction velocity that would modify boundary layer behavior in some desired way. To illustrate this possibility, and to exhibit some of the features of boundary layers that have been controlled in this manner, we calculate the distribution of suction velocity, required to prevent the growth of the displacement thickness on a circular cylinder. It turns out that this control objective also prevents separation of the boundary layer, and would make the layer likely to remain laminar and steady at quite large values of the Reynolds number.

Since this control strategy, if successful, keeps at least one measure of boundary-layer thickness constant, we use the very simple transformation formulas

$$g^2 = \frac{va}{2U_0} \qquad \text{and} \qquad \xi = \frac{x}{a}$$

The potential-flow velocity is, as before, $U = 2U_0 \sin \xi$, so the coefficients in the transformed momentum equation are $\gamma = \delta = \cos \xi$, and $\psi = \sin \xi$.

The boundary value, $f(\xi, 0) = \sigma(\xi)$, will be determined, step-by-step. To be sure that the calculation gets off to a proper start, and to obtain a guide to the expected behavior of σ for small values of ξ, we construct a Blasius series. Nothing that, for $\xi \ll 1$,

$$\gamma \approx 1 - \tfrac{1}{2}\xi^2 \qquad \delta \approx 1 - \tfrac{1}{2}\xi^2 \qquad \text{and} \qquad \psi \approx \xi$$

we seek a solution in the form

$$f(\xi, \eta) = f_0(\eta) + \xi^2 f_1(\eta) + \xi^4 f_2(\eta) + \cdots$$

with the corresponding expectation that $\sigma(\xi) = \xi^2 \sigma_1 + \xi^4 \sigma_2 + \cdots$. The fact that we let σ be zero at $\xi = 0$, means that δ^* will be kept constant, at the value it has at the stagnation point on a *solid* cylinder.

The problem to be solved for f_0 is the familar one for a plane, stagnation-point flow. The problem for f_1 is prescribed by the equations

$$u_1'' + f_0 u_1' - 4u_0 u_1 + 3u_0' f_1 = \tfrac{1}{2}(f_0 u_0' + 1 - u_0^2)$$

$$f_1' = u_1 \tag{12.85}$$

$$f_1(0) = \sigma_1 \qquad u_1(0) = 0 \qquad u_1(\infty) = 0$$

The required value of σ_1 is that which makes

$$\int_0^\infty u_1 \, d\eta = f_1(\infty) - \sigma_1 = 0 \tag{12.86}$$

Since the problem for f_1 is linear, we have only to solve it for any two values of σ_1, determine the corresponding values of $f_1(\infty) - \sigma_1$, and then determine the value of σ_1 that satisfies Eq. (12.86), by linear interpolation or extrapolation. The value turns out to be $\sigma_1 = 0.22054$.

The problem for f_2 is prescribed by the equations

$$u_2'' + f_0 u_2' - 6u_0 u_2 + 5u_0' f_2 = -\tfrac{1}{24}(f_0 u_0' + 1 - u_0^2) - \tfrac{4}{3}u_0 u_1$$

$$+ \tfrac{5}{6}u_0' f_1 + \tfrac{1}{2}u_1' f_0 - 3(f_1 u_1' - u_1^2) \qquad (12.87)$$

$$f_2' = u_2$$

$$f_2(0) = \sigma_2 \qquad u_2(0) = 0 \qquad u_2(\infty) = 0$$

The value of σ_2 is found in the same way as that of σ_1, and equals 0.02917.

The Blasius series suggests that the finite-difference approximation of streamwise rates of change will be most accurate if ξ^2, rather than ξ, is used as a streamwise variable. Thus we define $\zeta = \xi^2$, and make the corresponding adjustment of the formula for ψ, which becomes

$$\chi = 2\xi \sin \xi = 2\zeta^{1/2} \sin (\zeta^{1/2})$$

After a value is selected for the finite increment $d\zeta$, the Blasius series is used to establish the profiles of u and f at $\zeta = 0$, and at $\zeta = d\zeta$. The finite-difference approximation of ζ-derivatives is used to calculate the profiles at $\zeta = 2\,d\zeta$, $3\,d\zeta$, and so on.

Before leaving the Blasius series, we use it to evaluate the local sensitivity of the value of DEL1 to changes in the value of σ. A sufficient approximation is given by the formula $\partial\delta^*(\zeta, \sigma)/\partial\sigma \approx \delta^*(\sigma_1 = 1) - \delta^*(\sigma_1 = 0)$. In the listing of program BLCTRL in Appendix B, this quantity is named DERSIG.

The control strategy is implemented by an outer iteration loop, which uses the Newton–Raphson method of successive approximations. The procedure includes the following steps.

1. The streamwise variable is incremented to the value, ζ, at which u and f are to be determined. The corresponding local values of γ, δ, and ψ are computed from the formulas cited above. A first guess at the local value of σ is made by linear extrapolation, using the formula $\sigma(\zeta) = 2\sigma(\zeta - d\zeta) - \sigma(\zeta - 2\,d\zeta)$. A corresponding extrapolation is made, for the values of $f^*(\zeta, \eta)$ which appear in the inner iteration loop, which is used to deal with the nonlinearity of the finite-difference equations. It is not necessary to extrapolate the values for $u^*(\zeta, \eta)$, because the boundary conditions force u^* to change slowly with ζ. These extrapolations are not an essential feature of the algorithm, but they greatly reduce the required number of iterations. With all this input, the new profiles of u and f are computed, and a value of DEL1 is established.

2. Ordinarily, this value of DEL1 will not be what is wanted; we compare it to the target value and define an error, DEL1–TARG. We call this ERROLD, and call the corresponding value of σ SIGOLD. An improved approximation to the value of σ can now be calculated from the formula SIG = SIGOLD – ERROLD/DERSIG. For the sensitivity, DERSIG, we use a value carried over from calculations at $\zeta - d\zeta$. For the first finite-difference step, this value was prepared from the Blasius series.

3. Using the new value of σ, we recalculate DEL1, and find a new value of the error, calling it ERR. An improved value of the sensitivity can now be formed, from the equation DERSIG = (ERR – ERROLD)/(SIG – SIGOLD). After this is done, SIG and ERR are renamed, SIGOLD and ERROLD, and a further-improved value of σ is calculated, again using the formula SIG = SIGOLD – ERROLD/DERSIG.

4. Step (3) is repeated until the absolute value of ERR is satisfactorily small, or until a maximum permissible number of iterations is reached. The necessary number of repetitions is coupled to the size of the step, $d\zeta$, so that reductions in step size do not automatically incur proportional increases of computing effort.

Calculations were made with the program listed in Appendix B, with $d\zeta = \pi^2/400$. The outer iterations were usually converged to an error tolerance of 1 in the fifth decimal, before step (4) needed to be applied. As the value of ξ exceeded about 170°, the number of necessary applications of step (4) began to increase rapidly, reaching 6 or 7 in the last few steps before a spuriously negative value of skin friction was computed, terminating the program.

Perhaps the most interesting feature of the results is the evolution of the vorticity distribution, particularly in the region where the potential flow is decelerating. Figure 12.29 shows that the vorticity remains everywhere positive, with the largest value at the wall. The deceleration of the potential flow, which would call for the generation of negative vorticity to maintain the no-slip condition at an impervious wall, is accommodated in this case by the removal of surplus positive vorticity, by convection through the porous wall. For a large initial range of ξ, this occurs without any dramatic effect on the shape of the vorticity profile. Eventually, however, a striking two-layer structure appears, with most of the vorticity plastered right up against the wall, and a slowly decaying tail of the distribution extending out to very large values of η. This causes accuracy problems for the calculation method, which is not well suited either to resolve the structure of the very thin layer, or to encompass the broad layer. However, we recognize that very little vorticity is located in this region, where the potential-flow velocity is very small, and suspect that the local loss of accuracy has little effect on integral quantities, such as the drag of the cylinder, or the total amount of fluid withdrawn into the cylinder.

Figure 12.30 shows the necessary distribution of suction velocity, which

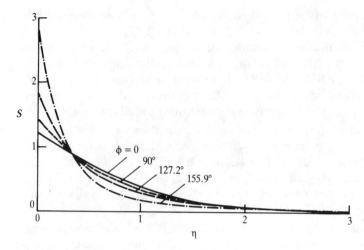

FIGURE 12.29
Boundary-layer control by suction. Vorticity profiles on circular cylinder.

may not be very accurate for $\xi \geq 170°$. The total suction flow rate for the upper half cylinder is found to be

$$Q \approx 10.0 U_0 a \left(\frac{\nu}{U_0 a} \right)^{1/2}$$

while the integrated skin-friction drag for the upper half-cylinder is

$$D \approx 9.91 \rho U_0^2 a \left(\frac{\nu}{U_0 a} \right)^{1/2}$$

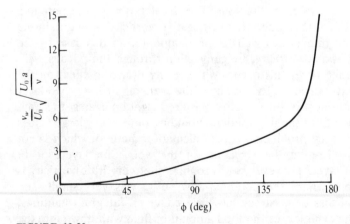

FIGURE 12.30
Boundary-layer control by suction. Suction velocity required to maintain constant displacement thickness on a circular cylinder.

These results are conveniently represented by the nondimensional flow coefficient

$$C_Q \equiv \frac{Q}{U_0 a} \approx 14.1 \, \text{Re}^{-1/2}$$

and the drag coefficient

$$C_D \equiv \frac{2D}{\rho U_0^2 a} \approx 28.0 \, \text{Re}^{-1/2}$$

In both of these, the Reynolds number is based on the diameter, so $\text{Re} = 2U_0 a / \nu$.

Considerations that are presented in Chapter 13 suggest that this flow may remain steady and laminar at Reynolds numbers as large as a million, at which value $C_Q = 0.007$, and $C_D = 0.028$. When δ^* is constant, the vorticity has very little effect on the potential flow; in particular, it should not materially disrupt the fore-and-aft symmetry of the pressure distribution, so there should be little or no pressure drag. In assessing the practical benefits of this sort of boundary-layer control, one should make allowance for the cost of operation of the suction system, and one should be aware of the difficulty of fabricating a cylinder through which the specified flow rate can be withdrawn in the required manner.

12.7 BOUNDARY LAYERS WITHIN BOUNDARY LAYERS

Sudden changes in one or more of the boundary conditions that apply at the wall may produce an effect most easily described as a new boundary layer growing inside an old one. For example, an impervious wall may suddenly give way to a porous one, through which fluid is withdrawn, or the boundary layer may simply arrive at the end of the wall, as at the trailing edge of a flat plate.

Accurate solutions of the boundary-layer equations can be produced for each of these cases, but the associated analysis is extremely complex and not really necessary to a description and understanding of the events. For the description, we can again rely on program NEWBL; the understanding draws on the concepts of vorticity transport developed in Chapter 8. We start with the example of a boundary layer that first develops along a solid wall, then encounters a section of wall through which fluid is withdrawn with a uniform value of v_w, and finally returns to flow along a solid wall. For simplicity, the potential flow outside the boundary layer is taken to be uniform. The problem was first analyzed by W. Rheinboldt,[15] following up ideas that had been

---

[15] W. Rheinboldt, "Zur Berechnung stationärer Grenzschichten bei kontinuerlicher Absaugung mit unstetig veränderlicher Absaugegeschwindigkeit," *J. Rational Mech. Anal.* **5**:539–604 (1956). For a translation into English, see NASA Translation TT-F-29 (1961).

systematically developed by S. Goldstein.[16] What we use from these analyses is a warning that conditions change very rapidly after the change in boundary conditions, in a small region close to the wall and just downstream of the point at which the boundary conditions change. Thus, the computation employs very small steps in ξ and η.

The Flat Plate With Discontinuous Suction

We employ a simple transformation that ignores the effects of suction,

$$g^2 = \frac{2vx}{U_0} \quad \text{and} \quad \xi = \frac{x}{a}$$

The first section of solid plate extends from $x = 0$ to $x = a$; the second solid section begins at $x = 1.5a$. In the interval $1 < \xi \le 1.5$, the suction velocity has the value

$$v_w = -U_0 \left(\frac{v}{U_0 a} \right)^{1/2}$$

The coefficients in the transformed momentum equation are

$$\gamma = 1 \qquad \delta = 0 \qquad \text{and} \qquad \psi = 2\xi$$

The transformed streamfunction assumes the following values at the wall:

$$f(\xi, 0) \equiv \sigma(\xi) = \begin{matrix} 0 & 0 \le \xi \le 1 \\ (\xi - 1)(2\xi)^{-1/2} & \text{if} \quad 1 < \xi \le 1.5 \\ 0.5(2\xi)^{-1/2} & 1.5 < \xi \end{matrix}$$

The profile at $\xi = 1$ is the classic Blasius profile. The calculation marches downstream with tiny steps, $d\xi = 0.001$ until $\xi = 1.1$; then the step is increased to $d\xi = 0.01$ until $\xi = 1.5$. The step size is returned to $d\xi = 0.001$ until $\xi = 1.6$; and finally is increased again to $d\xi = 0.01$ until $\xi = 2.0$, where the calculation stops.

The most striking feature of the result appears in the vorticity profiles, shown in Fig. 12.31. We see that the profile for $\xi = 1.001$ is just like that for $\xi = 1$, displaced slightly towards the wall, *except very close to the wall*, where it rises suddenly. As ξ is increased, this pattern is repeated, with a larger shift of the outer profile toward the wall, and a higher and broader rise at the wall.

The explanation of these curves is simple. As the vorticity of the Blasius profile is carried with the fluid, to pass over the porous section of the wall, the suddenly encountered suction velocity convects some of the vorticity through the wall. This leaves the boundary layer without enough vorticity to satisfy the

[16] S. Goldstein, "On the Two-Dimensional Steady Flow of a Viscous Fluid Behind a Solid Body, Part 1," *Proc. Roy. Soc.* **A142**: 545–562 (1933).

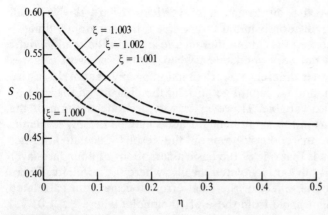

FIGURE 12.31
Vorticity generated at the wall to replace vorticity convected through a suddenly porous wall. Close-up view of the region very near the wall.

no-slip condition, so new vorticity must be generated. This can only occur at the wall itself, with the fresh vorticity subsequently diffusing back out into the fluid. Rheinboldt predicts that the peak value of this newly-generated vorticity will grow in proportion to $(\xi - 1)^{1/3}$, and that the thickness of the region into which it has diffused is also proportional to $(\xi - 1)^{1/3}$. In Fig. 12.32, we see how the skin friction and displacement thickness respond to the sudden application of suction. Recalling that the value of the displacement thickness locates the center-of-vorticity of the boundary-layer profile, i.e., that

$$U\delta^* = -\int_0^\infty y\Omega \, dy$$

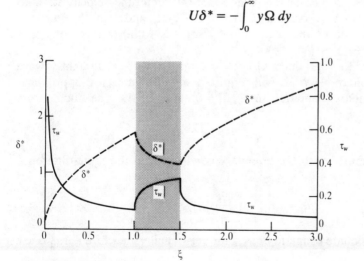

FIGURE 12.32
Effect of uniform suction (in the shaded strip) on skin friction and displacement thickness. $(v_w/U_0)(v/U_0 a)^{-1/2} = 1.0$.

we deduce that a convective displacement of vorticity toward the wall will reduce δ^*. The rate of reduction is initially very steep, but apparently finite.

The agreement between the finite-difference results and the semi-analytic results of Rheinboldt is not very good, even when we take steps as small as $d\xi = 0.0001$. The first-order difference approximation simply cannot do justice to the $\frac{1}{3}$-power variation, and we cannot get around the difficulty in this case by redefining the streamwise variable. However, a very accurate solution of the boundary-layer equations in this small neighborhood may be of only academic interest if errors made there do not poison the results calculated farther downstream. The reason is that one of the basic assumptions of boundary-layer theory, that $v_{,x} \ll u_{,y}$, is obviously violated locally, where $v_{,x}$ is infinite. There seems to be no serious after-effect of the local errors, because our calculated results become undistinguishable from those of Rheinboldt where $\xi > 1.01$.

Just downstream of the end of the porous section, another boundary layer within a boundary layer appears, as a result of the sudden removal of the source of vorticity at the wall.

The Wake Behind a Finite Flat Plate

At the trailing edge of a finite flat plate, the vorticity from the boundary layer beneath the plate is suddenly freed to diffuse into the fluid above the plate, and vice versa. The solution will exhibit a discontinuity in the first derivative, $u_{,y}$, at the trailing edge itself. This contrasts with the discontinuity in the second derivative, $u_{,yy}$, which appears at the sudden beginning or end of suction. The solution of the boundary-layer equations for this case was first given by S. Goldstein,[17] who demonstrated that the centerline velocity starts to recover according to a formula $u(x, 0) = A(\xi - 1)^{1/3}$, and that the vorticity from the opposite side of the plate initially spreads within a layer of depth proportional to $(\xi - 1)^{1/3}$.

Any attempt to reproduce Goldstein's results with a straightforward application of program NEWBL will require a suitable numerical approximation to the boundary condition $u'(\xi, 0) = 0$, for $\xi > 1$. The momentum equation gives

$$u''(\xi, 0) = \psi u(\xi, 0) u(\xi, 0)_{,\xi}$$

which can be combined with the boundary condition into the finite-difference statement

$$u_2 = u_1 \left[1 + \left(\frac{\psi h^2}{2} \right) u_{1,\xi} \right]$$

To get an adequate estimate of $u_{1,\xi}$ for very small values of $\xi - 1$, we use

---

[17] S. Goldstein (1930), *op. cit.*

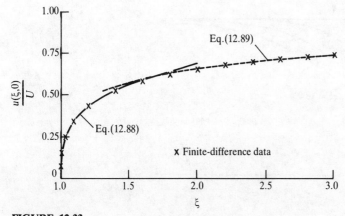

FIGURE 12.33
Recovery of symmetry-plane velocity in the wake of a flat plate.

the approximation $u_{1,\xi} = u_1/[3(\xi - 1)]$, which embodies the $\frac{1}{3}$-power variation disclosed by Goldstein's analysis. When ξ exceeds 1.004, we switch to the usual finite-difference representation of this derivative. With either representation of the derivative, the boundary conditons are converted into the following statements:

$$E_{11}(1) = \left[1 + \left(\frac{\psi h^2}{2}\right)u_{1,\xi}^*\right]^{-1} \qquad F_1(1) = 0$$

This procedure produced oscillating iterative estimates during the first few steps; convergence is much accelerated by a slight degree of *under-relaxation*. This was achieved by adding 0.9 times the latest estimate of $E_{11}(1)$ to 0.1 times the next-to-latest estimate, before recomputing the velocity profile.

Figure 12.33 shows the computed distribution of $u(x, 0)$, compared to the near-wake formula of Goldstein, which is

$$u(\xi, 0) = 0.77246(\xi - 1)^{1/3} - 0.18600(\xi - 1)^{4/3} + 0.10650(\xi - 1)^{7/3} + \cdots$$

$$(12.88)$$

and to the far-wake formula of Stewartson,[18]

$$u(x, 0) = 1 - 0.37469(\xi - 0.42)^{-1/2} - 0.07020(\xi - 0.42)^{-1} + \cdots \quad (12.89)$$

Figure 12.34 shows how the vorticity profile very close to the trailing edge adjusts to the sudden change in boundary conditions.

---

[18] K. Stewartson, "On Asymptotic Expansions in the Theory of Boundary Layers." *Jour. Math. Phys.* **36**:173–191 (1957).

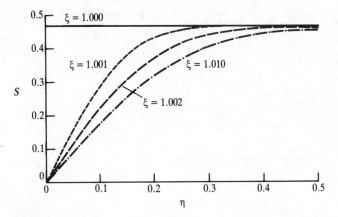

FIGURE 12.34
Vorticity profiles close behind a flat plate. Close-up view of the region near the symmetry plane.

Wakes in an Accelerating or Decelerating Stream

The procedure outlined above can readily be modified, to allow for acceleration or deceleration of the flow outside the wake. The results shown in Fig. 12.35 are for potential flows produced by a line sink placed one platelength downstream of the trailing edge, and by a line source placed 20 platelengths upstream of the leading edge. The results are of practical interest, in showing how quickly the effect of the plate is forgotten when the external flow accelerates, and how the effect of the plate is magnified when the external flow decelerates. Thus, flow-straightening vanes can be located almost with

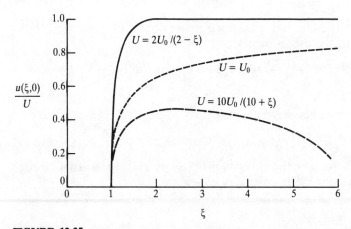

FIGURE 12.35
Recovery of symmetry-plane velocity in the wake of a flat plate, with accelerating, uniform, and decelerating potential flow.

impunity upstream of a rapid area contraction in a duct or wind tunnel, whereas the vanes in a vaned diffuser may improve pressure recovery, but will guarantee very nonuniform flow.

Before leaving this topic, you should be warned that the evolution of wake profiles, and of their companions, jet profiles, is decidedly more difficult to compute with guaranteed accuracy, than is the evolution of a boundary layer. The reason is that the symmetry condition at $\eta = 0$ does not pin down the *range* of the velocity profile, which may then drift slowly away from the physically correct value through an accumulation of truncation error. There is, however, a test for accuracy, determined by the momentum-integral equation when τ_w is set equal to zero. This is

$$(U^2\theta)_{,x} + \left(\frac{U^2}{2}\right)_{,x} \delta^* = 0 \qquad (12.90)$$

When the potential flow is unaccelerated, this simply requires that θ remain constant. For the calculations represented by Figs. 12.31 and 12.32, θ increased from an initial value of 0.66414 to a maximum of 0.66482 at $\xi = 1.004$, and subsequently fell to 0.6639 at $\xi = 3.0$.

The calculations of wakes in accelerating and decelerating potential flows satisfied Eq. (12.90) with a tolerance of about 2%. Greater accuracy was not sought, because such laminar flows are very unstable, as we shall see in Chapter 13. Thus, a very accurate solution seems of only academic interest.

12.8 UNSTEADY AND/OR THREE-DIMENSIONAL BOUNDARY LAYERS

Many of the boundary layers with which one must deal in practice are both unsteady and three-dimensional. The boundary layers on helicopter rotors or turbocompressor blades provide important examples. The concept of a boundary layer is little changed by the fact that the flow within it may be unsteady or three-dimensional. The layer is still supposed to be slender, and to hug the body like a glove until it separates.

Wherever there is a well-defined direction normal to the surface of the body, i.e., away from corners and edges, that is the dominant direction for diffusion. The slender-layer approximations allow us to neglect diffusion in any direction tangent to the layer; the pressure still varies hydrostatically across the layer. The process that is fundamentally more complex than in the examples we have studied is convection. Kinematic phenomena such as separation and the displacement of the potential flow become much more complex, and require careful description. Computational schemes that involve a fixed Eulerian grid require careful formulation if they are to retain the modest requirements for computer speed and storage, characteristic of the simple algorithms we have been using.

Propagation of Information in a Boundary Layer

Serious and successful computational analysis of unsteady and three-dimensional boundary layers was greatly facilitated by the work of G. S. Raetz,[19] whose ideas form the base for much of the following discussion.

Examining solutions of the simple heat equation, $u_{,t} = vu_{,yy}$, one learns that a disturbance that is introduced at any value of y is, in principle, felt instantaneously at any other point, whatever its value of y. As a practical matter, the disturbance diffuses away from its point of origin at a finite rate, but we shall ignore this for the moment and assert that:

1. At a given instant, the values of vorticity and temperature at each point on a given line normal to the boundary layer affect, by diffusion, the corresponding values at every other point which lies on that same line, inside the boundary layer.

   Region of influence. It follows from assertion (1) that

2. The solution of the boundary-layer equations along a specified normal line, at a specified instant, i.e., at a given point P in the space of the *time-like variables*,[20] influences the solution in a specific region of that space, which lies downstream, or in the future, with respect to the point P.

This *region of influence* of P cannot be predicted in advance, except in a small neighborhood of P, and then only after the distribution of tangential velocity components and local accelerations is prescribed along the normal line through P. This local description is, however, sufficient for the design of computational procedures. Examples involving only two time-like variables will illustrate this concept.

Three-dimensional steady flow. In typical three-dimensional boundary-layer flows, the tangential component of **u** changes direction as a function of distance along the normal line. This is illustrated in Fig. 12.36, for which the line of sight is along the normal, by the fan of arrows radiating downstream from P. The figure differs from similar views of the velocity profile of the Ekman layer, or of the spinning-disk flow, in that only the direction, and not the magnitude, of **u** is shown. The ray PS might represent the direction of the external potential flow; and PW might represent the limiting direction as the wall is approached. These directions need not be at the edges of the fan, but they often are.

---

[19] G. S. Raetz, Northrop Aircraft, Inc., Report No. NAI-58-73 (BLC-114), 1957.

[20] These variables include time itself, and by analogy, distance downstream.

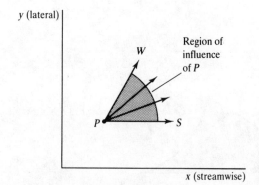

FIGURE 12.36
Convective spread of influence in a steady, three-dimensional boundary layer. The slope of the arrows is $dy/dx = v/u$.

We interpret Fig. 12.36 by saying that the solution to the boundary-layer equations along the normal through point P influences the solution at all neighboring points within the fan. The influence is spread first by nearly-tangential convection to some point in the fan, and then by purely-normal diffusion to other points along the normal through the point reached by convection.

Unsteady, two-dimensional, flow. Figure 12.37 shows the analogous sketch for unsteady, two-dimensional flow. Forward flow is depicted at some levels in the boundary layer, the maximum forward speed being represented by the arrow PF. Backward flow is depicted at other levels, the maximum backward speed being represented by PB. If the wall is stationary, the ray that represents a particle on the wall runs straight up.

For three-dimensional unsteady flow, P lies in the space of x, y, and t, and its region of influence fills out a sort of cone in that space.

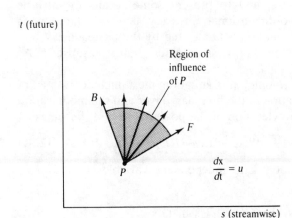

FIGURE 12.37
Convective spread of influence in unsteady, two-dimensional flow.

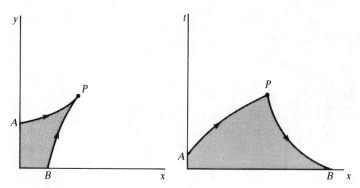

FIGURE 12.38
Domain of dependence (shaded area) and associated initial-data arc (AB).

Domain of dependence

3. Just as the solution at P influences the solution in a certain downstream-future region, so must the solution at P depend on the solution everywhere in some upstream-past domain, throughout which information is transmitted by convection and diffusion to reach P. Figure 12.38 illustrates this concept.

The term, *domain of dependence* of P, is used to identify either the entire shaded region in the relevant sketch or the *initial-data arc*, along which profiles of vorticity and temperature must be specified if they are to be computed at P. The precise initial-data arc that corresponds to a given point P is not ordinarily known in advance, but we do not often require this information.

12.9 UNSTEADY, TWO-DIMENSIONAL, BOUNDARY LAYERS

To illustrate the foregoing ideas, and to present some results of intrinsic interest, we first consider two-dimensional unsteady flow. If the flow is incompressible, the continuity equation is unaffected by the unsteadiness, and we can again use a streamfunction $\Psi(x, y, t)$, to assure that it is satisfied. As before, $u = \Psi,_y$ and $v = -\Psi,_x$.

The x-component of the momentum equation must include two extra terms, to describe local acceleration in the boundary layer and the modification of the pressure force by local acceleration of the potential flow. It becomes

$$\Psi,_{yt} + \Psi,_y \Psi,_{yx} - \Psi,_x \Psi,_{yy} - U,_t - UU,_x = \nu\Psi,_{yyy} \qquad (12.91)$$

We transform the dependent and independent variables in the same general way as before, setting

$$\Psi(x, y, t) = Q(x, t)f(\xi, \eta, \tau)$$

with
$$Q(x, t) = U(x, t)g(x, t)$$

$$\xi = \xi(x) \qquad \tau = \tau(x, t) \qquad \text{and} \qquad \eta = \frac{y}{g(x, t)}$$

The design of a specific transformation amounts to specifying the three functions $\xi(x)$, $\tau(x, t)$, and $g(x, t)$ so as to achieve some desired effect. The function $g(x, t)$ should embody a reasonable a priori estimate of the local, instantaneous, boundary thickness. The transformed equation is

$$u'' + (\gamma f + \varepsilon \eta)u' + \delta(1 - u^2) + \kappa(1 - u)$$
$$= \beta u_{,\tau} + \psi(uu_{,\xi} - u'f_{,\xi}) + \theta(uu_{,\tau} - u'f_{,\tau}) \quad (12.92)$$

The coefficients have the following definitions:

$$\gamma = \frac{g^2 U}{\nu}(\ln Q)_{,x} \qquad \varepsilon = \frac{1}{2\nu}g^2_{,t}$$

$$\delta = \frac{g^2 U}{\nu}(\ln U)_{,x} \qquad \kappa = \frac{g^2}{\nu}(\ln U)_{,t}$$

$$\beta = \frac{g^2}{\nu}\tau_{,t} \qquad \psi = \frac{g^2 U}{\nu}\xi_{,x} \qquad \theta = \frac{g^2 U}{\nu}\tau_{,x}$$

Impulsively-Started Two-Dimensional Boundary Layer On a Wedge

For a first example, consider the development of the boundary layer on a stationary infinite wedge, after the potential flow $U(x, t) = Ax^m$ is impulsively started at time $t = 0$. The transient development of the boundary layer should lead to the steady, self-similar solution of the Falkner–Skan family that is appropriate for the given wedge angle. In fact, we have already computed, in Section 10.3, the solution for $m = 1$.

A special feature of this example is that the initial and boundary conditions introduce no natural scales of time or distance. Thus x and t do not play entirely independent roles, and the degree of progress through the transient development of the flow is determined by the value of a combination of these variables, namely the quantity $\chi = Ut/x$.

As a consequence, it should be possible to construct a function $G(\chi)$, such that the expression $g^2 = \nu t G(\chi)$ exhibits the limiting behavior $g^2 \approx \nu t$ when $\chi \to 0$, and $g^2 \approx \nu x/U$ when $\chi \to \infty$. We can invent many functions G that produce this effect, and will present calculations done with the choice $G = [2/(m + 1)]\chi^{-1}\,\text{erf}\,(\alpha\chi)$, which makes

$$g^2 = \nu t \left(\frac{2}{m + 1}\right)\chi^{-1}\,\text{erf}\,(\alpha\chi) \qquad (12.93)$$

in which
$$\alpha = \frac{(m + 1)\sqrt{\pi}}{4}$$

The choice of constants was made to bring forth familiar solutions of the boundary-layer equations at the extreme limits of χ. To see whether this has a chance to work, we try the transformation

$$\psi(x, y, t) = Q(x, t) f(\chi, \eta)$$

Sure enough, the resulting coefficients in the transformed momentum equation are functions only of χ. Specifically, the equation becomes

$$u'' + (\gamma f + \varepsilon \eta) u' + \delta(1 - u^2) = \psi(u u_{,x} - u' f_{,x}) + \theta u_{,x} \qquad (12.94)$$

with coefficients

$$\gamma = \mathrm{erf}\,(\alpha \chi) - \left(\frac{m-1}{2}\right) \chi \exp\,(-\alpha^2 \chi^2) \qquad \varepsilon = \tfrac{1}{2} \exp\,(-\alpha^2 \chi^2)$$

$$\delta = \frac{2m}{m+1} \, \mathrm{erf}\,(\alpha \chi)$$

$$\psi = \frac{2m}{m+1} \chi \, \mathrm{erf}\,(\alpha \chi) \qquad \theta = \frac{2}{m+1} \, \mathrm{erf}\,(\alpha \chi)$$

The boundary conditions are

$$f(\chi, 0) = 0 \quad \text{(impermeable wall)} \qquad u(\chi, 0) = 0 \quad \text{(no slip)}$$

$$u(\chi, \infty) = 1 \quad \text{(matching to potential flow)}$$

Since the transformed differential equations and boundary conditions contain only two independent variables, χ and η, the computational problem seems little different from that for steady flow. There is, however, a pitfall, which we discover by examining the initial conditions and the region of influence of those conditions.

We see that $\chi = 0$ when $t = 0$, for all $x > 0$. Putting $\chi = 0$ into the expressions for the coefficient functions, we find that the differential equation reduces to

$$u'' + \tfrac{1}{2} \eta u' = 0$$

The solution that satisfies the boundary conditions is our old friend,

$$u(0, \eta) = \mathrm{erf}\left(\frac{\eta}{2}\right)$$

At the leading edge, $x = 0$. When $t > 0$, the corresponding value of χ is either 0, if $m > 1$, or ∞, if $m < 1$. In the former case, a straightforward computation that marches forward from $\chi = 0$ towards $\chi = \infty$ will follow the natural convective flow of information, downstream and into the future. If $\chi \gg 1$, the coefficient functions approach the limits

$$\gamma = 1 \qquad \delta = \frac{2m}{m+1} \qquad \alpha = \frac{2}{m+1} \qquad \varepsilon = 0 \qquad \psi = \frac{2(m-1)}{m+1} \chi$$

If the derivatives $u_{,\chi}$ and $f_{,\chi}$ become sufficiently small in that limit, and our calculations show that they do,[21] the terms on the right-hand side of Eq. (12.94) become negligible and the solutions approach the familiar Falkner–Skan solutions for steady boundary layers on wedges.

Trouble appears when we consider cases for which $m < 1$. Its source can best be appreciated by inspection of Figs. 12.39 and 12.40.

Figure 12.39(a) shows the domain of dependence of two points in the x–t plane for the case $m < 1$. For $P2$, this domain intersects only the x axis, on which $\chi = 0$, so that a solution can be found at $P2$ by marching forward from $\chi = 0$. For $P1$, however, the domain of dependence includes part of the t axis, on which $\chi = \infty$. Correspondingly, the solution at $P1$ is influenced by data that have no role in the marching calculation. Figure 12.39(b) shows that there are no points like $P1$ if $m \geq 1$.

Figure 12.40 shows the curves of constant χ, which correspond to the (a) and (b) portions of Fig. 12.39. The curve labeled χ_c in Fig. 12.40(a) coincides with the locus along which the first signals arrive from the leading edge. The curves are numbered in order of increasing values of χ. The considerations of domain of dependence which are illuminated by these figures suggest that the boundary-layer development can be described by a computation that marches forward in χ; for all values of χ if $m \geq 1$, and for $\chi \leq \chi_c$ if $m < 1$. The techniques available for the calculation include a Blasius series,[22] which is of

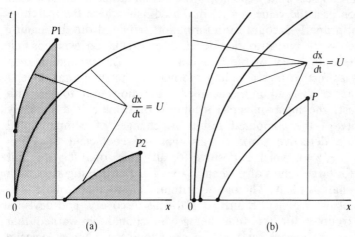

FIGURE 12.39
Domains of dependence for $U = Ax^m$: (a) $m < 1$; (b) $m \geq 1$.

---

[21] For the case $m = 0$, the analysis of Stewartson (1951) *op. cit.* suggests that these derivatives become smaller than any inverse power of χ.

[22] The series technique is well documented in Schlichting's *Boundary-layer Theory*, Chapter 15, and in Rosenhead's *Laminar Boundary Layers*, Chapter 7.

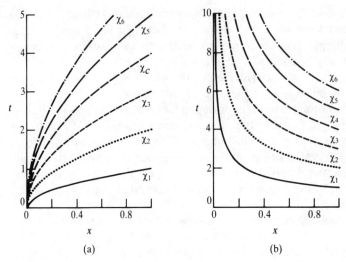

FIGURE 12.40
Loci of constant $c = Ut/x$ when $U = Ax^m$: (a) $m < 1$, $\chi_c = 1/(1 - m)$; (b) $m > 1$.

special interest because some of the resulting ordinary differential equations can be solved analytically, and a finite-difference technique such as we have been using in the preceding section.

It is interesting to see what happens to the computation when $m \leq 1$, if we try to carry it on past the value $\chi = \chi_c$, into a domain where the marching procedure does not properly account for information arriving from the leading edge. The computed vorticity distribution develops an oscillation, which is first noticeable near the outer edge of the boundary layer. The oscillation then grows very rapidly as further increases in χ are made. The symptoms depend somewhat on the size of $d\chi$, and on the number of iterations used to deal with nonlinearity. In fact, the most universal symptom of trouble is a suddenly increased sensitivity of the computed results to changes of computational parameters. A little detective work reveals that the coefficient of $u_{,\chi}$ is proportional to $1 - (\chi/\chi_c)u$, which is positive for all values of η if $\chi \leq \chi_c$ but becomes negative for large values of η when $\chi > \chi_c$. Further investigation, with the aid of a basic analysis of the Thomas algorithm,[23] shows that trouble is to be expected as a result. We shall return soon to this discovery, and devise a way around the trouble; for the time being you should be warned that computed results do not always send such a clear message when a logically erroneous procedure is attempted. You must be well-informed, and attentive to even the faintest warning of suspicious results, if you hope to do reliable computations.

---

[23] For the algorithm in its simplest form, the requirement is that $A_n + B_n + C_n \leq 0$ for all values of n. See Richtmyer and Morton, *op. cit.* p. 200.

Impulsively-Started Flows Within the Region of Influence of a Leading Edge

To avoid the difficulties encountered above when $m < 1$, we need to treat x and t as truly independent variables while we are solving the boundary-layer equations, although the answer we obtain may subsequently be represented as a function of the single combined variable, Ut/x. The procedures described here apply equally well to cases in which x and t cannot be combined into a single variable, even for representation.

We start with Eq. (12.80). To compute a solution for $0 \le x \le L$, and $0 \le t \le \infty$, we employ the simple transformation formulas

$$\xi = \frac{x}{L} \quad \text{and} \quad \tau = \frac{Ut}{L}$$

We use the same function for $g(x, t)$, given by Eq. (12.93), and we continue our examination of the family of flows for which $U(x, t) = Ax^m$ for $t > 0$. The resulting equations for the nonzero coefficients in Eq. (12.92) are

$$\gamma = \text{erf}(\alpha\chi) - \frac{(m-1)}{2}\chi \exp(-\alpha^2\chi^2) \quad \varepsilon = \tfrac{1}{2}\exp(-\alpha^2\chi^2)$$

$$\delta = \frac{2m}{m+1}\text{erf}(\alpha\chi) \quad \kappa = 0 \quad \beta = \frac{2}{m+1}\xi\,\text{erf}(\alpha\chi)$$

$$\psi = \frac{2}{m+1}\xi\,\text{erf}(\alpha\chi) \quad \theta = \frac{2m}{m+1}\tau\,\text{erf}(\alpha\chi).$$

Since the potential flows past wedges impose no adverse pressure gradients on the boundary layer if $m \ge 0$, we know that $u(\xi, \tau, \eta)$ will be nonnegative for all values of the independent variables. Thus the trailing finite-difference approximations

$$u_{,\xi}(\xi, \tau, \eta) \approx \frac{u(\xi, \tau, \eta) - u(\xi - d\xi, \tau, \eta)}{d\xi}$$

and

$$u_{,\tau}(\xi, \tau, \eta) \approx \frac{u(\xi, \tau, \eta) - u(\xi, \tau - d\tau, \eta)}{d\tau}$$

will bring information from the upstream-past to the point (ξ, τ) at which we seek to compute u. For the approximation of spatial derivatives, this procedure is called *upwind differencing*. We shall see a more interesting example of it later, in a calculation of bi-directional flow.

Figure 12.41 illustrates the computational domain, and the flow of information, for the program, TRANBL, that implements this analysis. This program, which appears in Appendix B, has two sections. The first carries out the marching calculation, from $\chi = 0$ to $\chi = \chi_c$, setting up initial data on the diagonal edge of the rectangularly gridded region in the figure. The resulting profiles of u and f are stored as UI and FI.

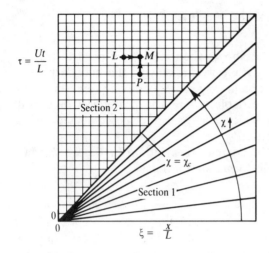

FIGURE 12.41
Flow of information in program TRANBL.

The second section of TRANBL starts with the calculation of the appropriate Falkner–Skan profiles, which are subsequently used as boundary data along the axis $\xi = 0$. These profiles are named UO and FO.

The calculation for $\chi > \chi_c$ starts in the lower left corner of the rectangularly gridded region, sweeping across each line of constant τ from left to right. When the new profiles at a typical interior point P have been found, they are stored under two auxiliary names: UL and FL, which are needed for the next increment of ξ at the current value of τ; and UP and FP, which will be needed for the new sweep at the next value of τ. Only UP and FP require storage arrays for all discrete values of ξ, and that for only one value of τ, so the storage requirements of the calculation are very modest.

When the calculation has been carried out to the desired final value of t, the results can be examined to see whether u and f really depend only on χ and η, rather than on ξ, τ, and η. Results from points near the lower left corner, where only a very few grid points represent the entire range from $\chi = \chi_c$ to $\chi = \infty$, show a lot of scatter, but the rest confirm that ξ and τ are only important in the combination that equals χ.

Figure 12.42 shows how the shape factor, H, varies with χ, for four values of the exponent m. For $m = 1.5$ or 1.0, the straightforward marching calculation covers the whole range of χ; for $m = 0.5$, the transient evolution is almost complete when $\chi = \chi_c = 2$. The most dramatic case is that for $m = 0$, in which no change of profile shape occurs until $\chi = 1$, when the first signal arrives from the leading edge. Without that signal, the velocity profile would be an error function forever.

Figure 12.43 shows the variation of displacement thickness along the flat plate in the case $m = 0$, when the first signal from the leading edge has just reached the point $x = L$. We see that steady conditions are very nearly established from $\xi = 0$ to $\xi = 0.25$, i.e., over a distance equal to $\frac{1}{4}$ the distance moved by a particle in the potential flow. The two straight lines, which are

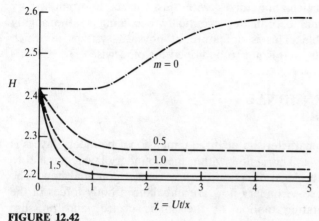

FIGURE 12.42
Evolution of the shape factor H for impulsively started Falkner–Skan boundary layers.

asymptotes to the finite-difference solution, would represent the two pieces of the solution, (12.52), given by the momentum-integral method, if that solution were converted to a prediction of δ^*, and if the shape factors were tuned to the values $H = 2.324$, $T = 0.274$.

Although this numerical treatment of impulsively-started flow over a flat plate with a leading edge seems quite successful, an element of doubt remains, because of the poor accuracy of finite-difference aproximations in the lower left corner of the computational domain. The values of u and f in this corner influence the values everywhere in the rectangularly gridded region; if they are wrong, the entire calculation may be poisoned! The hope that this has not happened is based on the fact that correct information is constantly being fed into the calculation from the boundaries, and that erroneous information

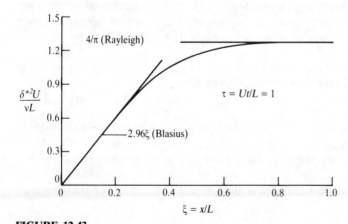

FIGURE 12.43
Finite-difference analysis of impulsively started Blasius boundary layer. Spatial variation of displacement thickness.

generated early in the calculation mixes with this correct information, by diffusion along the normal coordinate, eventually becoming so thoroughly diluted as to be negligible. This is a feature of physical systems in which diffusion has strong effects, and is not to be hoped for otherwise.

12.10 THREE-DIMENSIONAL BOUNDARY LAYERS

The most distinguishing characteristic of three-dimensional boundary layers is often called *secondary flow*. This is flow within the boundary layer, parallel to the wall and normal to the local direction of the external potential flow. The principal factors that cause secondary flow are unbalanced body forces in the secondary direction, rotating motion of the solid surface, and pressure gradients associated with *geodesic curvature*[24] of the surface streamlines of the potential flow. These causes need only have acted somewhere upstream, in the domain of dependence of the station of interest; secondary flows die away quite slowly after the causes have been removed.

A full exploration of three-dimensional boundary layers would carry us far beyond the scope of this text, so we must be content with just a few key ideas and results. We shall start with the boundary-layer equations, written in an (α, β, γ) orthogonal coordinate system. The surface $\gamma = 0$ coincides with the wall on which the boundary layer grows, and the coordinate γ is taken to be distance along a local normal to the wall; hence the metric coefficient $h_\gamma = 1$. The design of the coordinate mesh that covers the wall will be discussed later.

If the boundary-layer thickness is very small, compared to the minimum radius of curvature of the wall, the actual shape of the wall influences the boundary-layer flow only insofar as it determines the potential flow outside the boundary layer. We have already seen that we may encounter an essentially two-dimensional boundary layer on a curved body like a sphere, and a three-dimensional boundary layer on a flat body like a spinning disk.

Design of a Coordinate System to Cover the Wall

There are two popular ways to cover the wall with an orthogonal coordinate mesh. If the body surface is *developable,* so that it can be rolled out flat without any wrinkles, we may choose a simple $x-y$ rectangular mesh, with x being distance normal to the generators of the surface, and y being distance along the generators. Alternatively, if the surface streamlines of the potential

---

[24] Geodesic curvature is defined in Fig. 12.44. It is a property of a curve on a surface, seen when one looks down on the surface, along the local normal to the surface.

flow can readily be traced, we may prefer to use the streamlines as one set of coordinate curves, and the equipotentials as the other. The appearance of the boundary-layer equations depends somewhat on the choice of coordinates, so we start by exhibiting the equations for both systems, taking special care to define symbols clearly.

Boundary-layer equations with rectangular surface coordinates. We let u^* be the dimensional velocity component in the direction of increasing x, inside the boundary layer; the corresponding component of the potential flow is called U. Similarly, v^* and V are associated with y, the other tangential coordinate; w^* is associated with the normal coordinate, z. The x and y components of the momentum equation are then

$$u_{,t}^* + u^* u_{,x}^* + v^* u_{,y}^* + w^* u_{,z}^* - (U_{,t} + UU_{,x} + VU_{,y}) = vu_{,yy}^*$$

and $\quad v_{,t}^* + u^* v_{,x}^* + v^* v_{,y}^* + w^* v_{,z}^* - (V_{,t} + UV_{,x} + VV_{,y}) = vv_{,yy}^*$

while the continuity equation is

$$u_{,x}^* + v_{,y}^* + w_{,z}^* = 0$$

The wall- and external-flow boundary conditions are typically

$$u^*(x, y, 0) = v^*(x, y, 0) = 0 \quad \text{(no-slip)} \qquad w^*(x, y, 0) = 0 \quad \text{(impervious wall)},$$

$$u^*(x, y, z) \to U(x, y) \qquad \text{and} \qquad v^*(x, y, z) \to V(x, y) \qquad \text{as } y/\delta \to \infty$$

(matching the potential flow)

Because the continuity equation now has three terms, there is not much to be gained by the use of streamfunctions. The alternative is to find a reasonable way to transform w^* directly. Following the alternative, we introduce dimensionless variables u, v, w; ξ, η, ζ and τ, as follows.

$$u^*(x, y, z, t) = U(x, y, t) u(\xi, \eta, \zeta, \tau)$$

$$v^*(x, y, z, t) = V(x, y, t) v(\xi, \eta, \zeta, \tau),$$

$$w^*(x, y, z, t) = -\frac{v}{g} w(\xi, \eta, \zeta, \tau)$$

$$\xi = \xi(x) \qquad \eta = \eta(y) \qquad \tau = t(t) \qquad \zeta = \frac{z}{g(x, y, t)}$$

This way of defining a dimensionless normal component of velocity seems to contrast strongly with the way adopted for our analysis of the impulsively-started three-dimensional stagnation-point flows in Section 10.3. Notice that the boundary conditions do not prescribe a natural scale for w^*, and that w^* is never differentiated with respect to x, y, or t. The present choice is motivated by tidiness; the corresponding coefficient of w in the transformed equations is just 1. The transformed equations are

$$u'' + (\gamma_1 \zeta u + \gamma_2 \zeta v + w + \varepsilon \zeta)u' + \delta_{x1}(1 - u^2) + \delta_{x2}(1 - uv) + \omega_x(1 - u)$$
$$= \psi_1 uu_{,\xi} + \psi_2 vu_{,\eta} + \theta u_{,\tau} \quad (12.95)$$

$$v'' + (\gamma_1 \zeta u + \gamma_2 \zeta v + w + \varepsilon \zeta)v' + \delta_{y1}(1 - uv) + \delta_{y2}(1 - v^2) + \omega_y(1 - v)$$
$$= \psi_1 uv_{,\xi} + \psi_2 vv_{,\eta} + \theta v_{,\tau} \quad (12.96)$$

and
$$\gamma_1 \zeta u' + \gamma_2 \zeta v' + w' - \delta_{x1}u - \delta_{y2}v = \psi_1 u_{,\xi} + \psi_2 v_{,\eta} \quad (12.97)$$

The coefficients have the following definitions:

$$\gamma_1 = \frac{U}{2v} g_{,x}^2 \qquad \gamma_2 = \frac{V}{2v} g_{,y}^2 \qquad \varepsilon = \frac{1}{2v} g_{,t}^2$$

$$\delta_{x1} = \frac{g^2 U}{v}(\ln U)_{,x} \qquad \delta_{y1} = \frac{g^2 V}{v}(\ln U)_{,y}$$

$$\delta_{x2} = \frac{g^2 U}{v}(\ln V)_{,x} \qquad \delta_{y2} = \frac{g^2 V}{v}(\ln V)_{,y}$$

$$\omega_x = \frac{g^2}{v}(\ln U)_{,t} \qquad \omega_y = \frac{g^2}{v}(\ln V)_{,t}$$

$$\psi_1 = \frac{g^2 U}{v}\xi_{,x} \qquad \psi_2 = \frac{g^2 V}{v}\eta_{,y} \qquad \theta = \frac{g^2}{v}\tau_{,t}$$

The number of terms in these equations makes them seem formidable, but leads to no novel algebraic difficulties. The complexity is partly due to the fact that allowance has been made for unsteadiness, as well as three-dimensionality, of the flow. This has been done to prepare the way for an interesting sample calculation, which will follow the presentation of equations for the other popular coordinate system.

Coordinates based on streamlines and equipotentials of the potential flow. Let us now use x to denote distance along the surface streamlines of the potential flow, and y to denote distance along the corresponding equipotential curves. Distance along the local normal to the wall is again called z. This usage is conventional, but we must not forget that the basic coordinate curves are now not curves of constant x and y, because the distance between a given pair of streamlines or equipotentials may vary with position along the curves.

The velocity components along the directions of increasing x, y, and z are again called u^*, v^*, and w^* inside the boundary layer. The potential-flow velocity component corresponding to u^* is the surface speed of the potential flow, U; by definition of the coordinate system, the value of V is zero.

The geodesic curvature of the coordinate curves will now appear in our representations of $\mathbf{u} \, \text{grad} \, \mathbf{u}$ and $\text{div} \, \mathbf{u}$. Figure 12.44 shows how these curvatures are defined, in a view looking directly down the local normal to the wall. The x

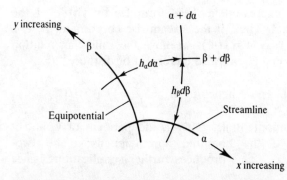

FIGURE 12.44
Definition of geodesic curvatures: $\kappa_x = (\ln h_\alpha),_y$ is the curvature of the streamline; $\kappa_y = (\ln h_\beta),_x$ is the curvature of the equipotential.

and y components of the momentum equation now appear as

$$u^*,_t + u^*u^*,_x + v^*u^*,_y + w^*u^*,_z + \kappa_x u^*v^* - \kappa_y v^{*2} - (U,_t + UU,_x) = vu^*,_{yy}$$

and $\quad v^*,_t + u^*v^*,_x + v^*v^*,_y + w^*v^*,_z + \kappa_y u^*v^* - \kappa_x u^{*2} + \kappa_x U^2 = vv^*,_{yy}$

while the continuity equation is

$$u^*,_x + v^*,_y + w^*,_z + \kappa_y u^* + \kappa_x v^* = 0$$

To transform these, we use the formulas

$$u^*(x, y, z, t) = U(x, y, t)u(\xi, \eta, \zeta, \tau) \qquad v^*(x, y, z, t) = U(x, y, t)v(\xi, \eta, \zeta, \tau)$$

$$w^*(x, y, z, t) = -\left(\frac{v}{g}\right)w(\xi, \eta, \zeta, \tau)$$

$$\xi = \xi(x) \qquad \eta = \eta(y) \qquad \tau = \tau(t) \qquad \zeta = \frac{z}{g(x, y, t)}$$

The transformed equations are then

$$u'' + (\gamma_1 \zeta u + \gamma_2 \zeta v + w + \varepsilon\zeta)u' + \delta_{x1}(1 - u^2) + \delta_{x2}v^2 + \omega(1 - u)$$
$$= \psi_1 uu,_\xi + \psi_2 vu,_\eta + \theta u,_\tau \quad (12.98)$$

$$v'' + \{\gamma_1 \zeta u + \gamma_2 \zeta v + w + \varepsilon\zeta\}v' - (\delta_{x1} + \delta_{x2})uv + \delta_{y1}(1 - u^2 - v^2) - \omega v$$
$$= \psi_1 uv,_\xi + \psi_2 vv,_\eta + \theta v,_\tau \quad (12.99)$$

and $\qquad \gamma_1 \zeta u' + \gamma_2 \zeta v' + w' - (\delta_{x1} + \delta_{x2})u = \psi_1 u,_\xi + \psi_2 v,_\eta \quad (12.100)$

The coefficients have the following definitions:

$$\gamma_1 = \frac{U}{2v}g^2,_x \qquad \gamma_2 = \frac{U}{2v}g^2,_y \qquad \varepsilon = \frac{1}{2v}g^2,_t$$

$$\delta_{x1} = \frac{g^2 U}{v}(\ln U),_x \qquad \delta_{x2} = \frac{g^2 U}{v}\kappa_y \qquad \delta_{y1} = \frac{g^2 U}{v}(\ln U),_y$$

$$\omega = \frac{g^2}{v}(\ln U),_t \qquad \theta = \frac{g^2}{v}\tau,_t$$

$$\psi_1 = \frac{g^2 U}{v}\xi,_x \qquad \psi_2 = \frac{g^2 U}{v}\eta,_y$$

Some simplifications are expressed in these formulas by virtue of the properties of a velocity potential. Thus, if the streamwise coordinate, α, is a function only of the potential, say $\alpha = f(\Phi)$, we have the following relation between $d\alpha$ and the corresponding streamwise increment of distance, dx.

$$d\alpha = f'(\Phi)\, d\Phi = f'(\Phi)\, U\, dx \qquad \text{hence} \qquad h_\alpha = \frac{dx}{d\alpha} = [f'(\Phi)U]^{-1}$$

Since y increases along the equipotentials, $f'(\Phi)$ is independent of y, and it follows that $\kappa_x = -(\ln U)_{,y}$. If the potential flow happens to be two-dimensional, then $\kappa_y = -(\ln U)_{,x}$. This produces further simplifications, since then $\delta_{x2} = -\delta_{x1}$.

The boundary conditions at the wall are ordinarily taken to be

$$u(\xi, \eta, 0, \tau) = 0 \qquad v(\xi, \eta, 0, \tau) = 0 \qquad \text{and} \qquad w(\xi, \eta, 0, \tau) = 0$$

while the matching conditions with the potential flow are

$$u(\xi, \eta, \zeta, \tau) \to 1 \qquad \text{and} \qquad v(\xi, \eta, \zeta, \tau) \to 0 \qquad \text{as} \qquad \zeta \to \infty$$

We notice that there is only one term in Eq. (12.99) that does not vanish if v is everywhere zero. This term, $\delta_{y1}(1 - u^2)$, is proportional to the geodesic curvature of the streamlines of the potential flow. Thus, if this curvature is everywhere zero, and if no secondary flow is demanded by local boundary conditions or inherited from the past or upstream conditions, there will never be any secondary flow. These equations, evidently, do not account for possible generation of secondary flow by local action of transverse body forces, but they could be easily modified to do so.

Special equations applicable in a plane of symmetry. Many of the simpler three-dimensional boundary-layer flows involve a plane of symmetry, in which the boundary-layer equations assume important special forms. This can happen when the boundary layer develops on a geometrically symmetric body, such as an ellipsoid, that is oriented with a plane of symmetry parallel to an oncoming uniform stream. This configuration is suggested by the lefthand sketch in Fig. 12.45; a more elaborate example is sketched in Fig. 10.3. Another symmetric configuration, for an internal flow, appears in the righthand sketch of Fig. 12.45.

The symmetry-plane solutions for three-dimensional boundary layers play a role somewhat like that of the stagnation-point solutions for two-dimensional boundary layers. Whereas the latter are governed by ordinary differential equations if the flow is steady, the former are governed by equations that involve only one streamwise variable, distance along the intersection of the symmetry plane and the wall.

Symmetry-plane equations in streamline-equipotential coordinates. In this system of coordinates, U is an even function of y, and v^* is an odd function; thus, the normalized velocity, v, will be an odd function. In order to retain the

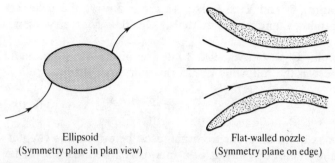

Ellipsoid
(Symmetry plane in plan view)

Flat-walled nozzle
(Symmetry plane on edge)

FIGURE 12.45
Three-dimensional flows with symmetry planes. The symmetric ellipsoid is at an angle of attack. The nozzle layer of interest is on the flat sidewalls.

information inherent in Eq. (12.99), we must first differentiate that equation with respect to η, and then strike out those quantities that are zero on the symmetry plane. To start, we note the following properties of the coefficients:

Even functions of y; γ_1, ε, δ_{x1}, δ_{x2}, ω, θ, ψ_1, ψ_2
Odd functions of y: γ_2, δ_{y1}

With the notation $\phi = v_{,\eta}$, the resulting symmetry-plane equations become

$$u'' + (\gamma_1 \zeta u + w + \varepsilon \zeta)u' + \delta_{x1}(1 - u^2) + \omega(1 - u) = \psi_1 u u_{,\xi} + \theta u_{,\tau} \quad (12.101)$$

$$\phi'' + (\gamma_1 \zeta u + w + \varepsilon \zeta)\phi' - (\delta_{x1} + \delta_{x2})u\phi + (\delta_{y1})_{,\eta}(1 - u^2) - \omega\phi - \psi_2\phi^2$$
$$= \psi_1 u\phi_{,\xi} + \theta\phi_{,\tau} \quad (12.102)$$

and
$$\gamma_1 \zeta u' + w' - (\delta_{x1} + \delta_{x2})u - \psi_2\phi = \psi_1 u_{,\xi} \quad (12.103)$$

The boundary conditions at the wall ordinarily require that $u(\xi, \eta, 0) = 0$, $\phi(\xi, \eta, 0) = 0$, and $w(\xi, \eta, 0) = 0$. The conditions of matching with the potential flow are $u(\xi, \eta, \zeta) \to 1$ and $\phi(\xi, \eta, \zeta) \to 0$ as $\zeta \to \infty$. The last condition is satisfied because v itself approaches zero when $\zeta \to \infty$, by definition, for all values of y.

Nozzle Sidewall Boundary Layer

For a numerical example, consider the boundary layer on the flat sidewall of a nozzle, as sketched in Fig. 12.45. If the contoured walls of the nozzle are hyperbolas, the streamlines of the potential flow will be other hyperbolas of the same family. The speed of the potential flow is related to the velocity potential Φ and streamfunction Ψ by the formula

$$\frac{U}{U_0} = \left[\cosh^2\left(\frac{\Phi}{U_0 a}\right) - \sin^2\left(\frac{\Psi}{U_0 a}\right)\right]^{-1/2} \quad (12.104)$$

in which U_0 is the speed where the symmetry plane passes though the throat of the nozzle. At this point, Φ and Ψ are zero, as are x and y, the distances measured along streamlines and equipotentials. On the symmetry plane, $x/a = \sinh(\Phi/U_0 a)$.

The coefficients in Eqs. (12.98, 99, 100) are given by simple and well-behaved expressions if the following transformation functions are used:

$$\xi = \frac{x}{a} \qquad \eta = \frac{y}{a} \qquad g^2 = va\frac{U_0}{U^2} \qquad (12.105)$$

This formula for g looks quite different from any used before in this chapter, but it agrees with the asymptotic behavior of a sink-driven boundary layer far upstream, and shows a smooth decrease of g to a finite minimum value at the throat of the nozzle, which is what we have found in studies of a two-dimensional boundary layer under the same distribution of U. The resulting expressions for the coefficients are

$$\gamma_1 = \tanh\left(\frac{\Phi}{U_0 a}\right)$$

$$\delta_{x1} = -\delta_{x2} = -\tanh\left(\frac{\Phi}{U_0 a}\right) \qquad (\delta_{y1})_{,\eta} = \operatorname{sech}^3\left(\frac{\Phi}{U_0 a}\right)$$

$$\psi_1 = \psi_2 = \cosh\left(\frac{\Phi}{U_0 a}\right)$$

$$\varepsilon = 0 \qquad \omega = 0 \qquad \theta = 0$$

These can readily be expressed in terms of ξ. When this is done, we see that a Blasius series of the form

$$u(\xi, \zeta) = u_0(\zeta) + \xi^{-2}u_1(\zeta) + \cdots \qquad w(\xi, \zeta) = w_0(\zeta) + \xi^{-2}w_1(\zeta) + \cdots$$

and

$$\phi(\xi, \zeta) = \xi^{-3}\phi_1(\zeta) + \cdots$$

can be used to get the calculation started at a suitably large negative value of ξ. Then a straightforward downstream-marching calculation can be carried out, up to the point at which $u'(\xi, 0)$ becomes negative. The FORTRAN program SYMMBL is listed in Appendix B.

The principal feature of interest in this calculation is the effect of lateral divergence of the secondary flow, on the streamwise variation of skin friction and of the normal velocity component. To highlight this, SYMMBL was run twice: once with the real values of $(\delta_{y1})_{,\eta}$, and once with $(\delta_{y1})_{,\eta}$ artificially set equal to zero. The strategy of the second run removes the term that drives the secondary flow, so that the program generates the result $\phi(\xi, \zeta) = 0$. Figure 12.46 compares the results of the two runs, and shows that the secondary flow significantly reduces the normal component of velocity, particularly in the region where the favorable streamwise pressure gradient is weakening. The skin friction is enhanced in the same region, and the point where it vanishes is moved downstream, from about $\xi = 0.2$, where $U = 0.98U_0$, to $\xi = 0.375$, where $U = 0.936U_0$.

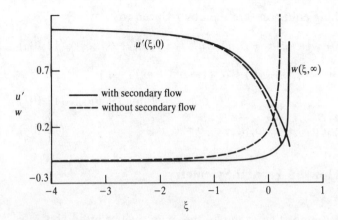

FIGURE 12.46
Effect of secondary flow on skin friction and normal velocity, in symmetry plane of nozzle sidewall.

Figure 12.47 shows the profiles of primary and secondary velocity components at the point where $u'(\xi, 0) = 0$, in the real case. The dimensional quantities are $u^* = Uu$, and $v^*_{,y} = U\phi/a$.

Rectangular coordinates may be especially convenient if the symmetry line is straight and the inviscid velocity components are simple functions of x and y. The symmetry plane may then be either $x = 0$ or $y = 0$. For the example to be presented, a helpful reminiscence of problems solved in Chapter 10 arises if we let the symmetry plane be $x = 0$. Then V is an even function of x, and U is odd. By their definitions, *both* u and v are even. The scale of thickness, g, is also even. From this, it follows that γ_1, δ_{x2}, ω_x, ψ_1, $u_{,\xi}$ and $v_{,\xi}$ all vanish

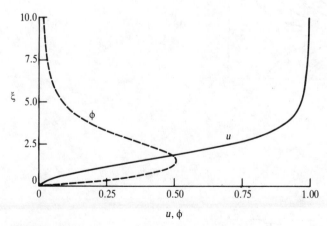

FIGURE 12.47
Streamwise velocity and lateral divergence of secondary velocity on symmetry plane of nozzle sidewall, at point where skin friction vanishes.

when $\xi = 0$. The governing equations are thereby reduced to

$$u'' + (\gamma_2 \xi v + w + \varepsilon \zeta)u' + \delta_{x1}(1 - u^2) = \psi_2 vu_{,\eta} + \theta u_{,\tau} \qquad (12.106)$$

$$v'' + (\gamma_2 \zeta v + w + \varepsilon \zeta)v' + \delta_{y1}(1 - uv) + \delta_{y2}(1 - v^2) + \omega_y(1 - v)$$
$$= \psi_2 vv_{,\eta} + \theta v_{,\tau} \qquad (12.107)$$

and
$$\gamma_2 \zeta v' + w' - \delta_{x1}u - \delta_{y2}v = \psi_2 v_{,\eta} \qquad (12.108)$$

Impulsively Started Flow on Frontal Symmetry Plane of a Hot Dog

Let us solve these for an interesting example, the impulsively-started flow in a symmetry plane that contains two stagnation points, as shown in Fig. 12.48. We specify the potential flow

$$\mathbf{U} = \left[Ax\mathbf{e}_x + By\left(1 - \frac{y}{L}\right)\mathbf{e}_y - \left(A + B - \frac{2By}{L}\right)z\mathbf{e}_z \right] \qquad (12.109)$$

which is turned on at $t = 0$. Let the dimensionless wall-mesh coordinates be $\xi = x/L$ and $\eta = y/L$, and let the dimensionless time be $\tau = At/(1 + At)$. The constant $C = B/A$ is the parameter used to classify stagnation-point solutions in Chapter 10.

For this flow, ω_y and δ_{y1} vanish. To complete the transformation, we again use the formula employed before in the analysis of impulsively started

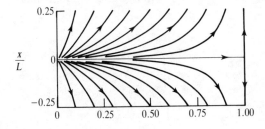

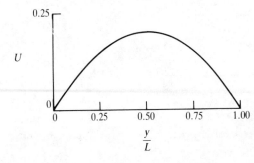

FIGURE 12.48
Symmetry-plane flow with adjacent stagnation points. $U(y) = 0.8U_0(y/L)(1 - y/L)$.

stagnation-point flows:

$$g^2 = \frac{vT}{A} \tag{12.110}$$

in which
$$T = 1 - \exp{(-At)}$$

The corresponding expressions for the coefficients are

$$\gamma_2 = 0 \qquad \delta_{y1} = 0$$

$$\delta_{x1} = T \qquad \delta_{y2} = CT(1 - 2\eta)$$

$$\varepsilon = \tfrac{1}{2}(1 - T) \qquad \psi_2 = C\eta(1 - \eta)T \qquad \theta = T(1 - \tau)^2$$

To gain a more complete resemblance to the equations solved in Section 10.5, we now rescale the normal velocity component, setting

$$w^*(x, y, z, t) = -Agw(\xi, \eta, \zeta, \tau)$$

The only effect of this is to change the coefficient of w in the transformed equations, from unity to T. The final equations to be integrated are then

$$u'' + \left(Tw + (1 - T)\frac{\zeta}{2}\right)u' + T(1 - u^2) = CT\eta(1 - \eta)vu_{,\eta} + T(1 - \tau)^2 u_{,\tau}$$

$$v'' + \left(Tw + (1 - T)\frac{\zeta}{2}\right)v' + CT(1 - 2\eta)(1 - v^2)$$

$$= CT\eta(1 - \eta)vv_{,\eta} + T(1 - \tau)^2 v_{,\tau}$$

$$w' - u - C(1 - 2\eta)v = C\eta(1 - \eta)v_{,\eta}$$

At each of the two stagnation points, where $\eta = 0$ or $\eta = 1$, these equations become identical to Eqs. (10.73) to (10.75). For a positive value of C, the equations at $\eta = 0$ are those for the stagnation point flow with that value of C; those at $\eta = 1$ correspond to the flow with coefficient $-C$. At the initial time, when $T = 0$, the momentum equations reduce to

$$u'' + \left(\frac{\zeta}{2}\right)u' = 0 \qquad v'' + \left(\frac{\zeta}{2}\right)v' = 0$$

The solutions that meet the boundary conditions are $u = v = \mathrm{erf}\,(\zeta/2)$, so that $v_{,\eta} = 0$ and the continuity equation reduces to

$$w' = [1 + C(1 - 2\eta)]\,\mathrm{erf}\left(\frac{\zeta}{2}\right)$$

and the initial distribution of w is

$$w = [1 + C(1 - 2\eta)]\left\{\zeta\,\mathrm{erf}\left(\frac{\zeta}{2}\right) - \frac{2}{\sqrt{\pi}}\left[1 - \exp\left(-\frac{\zeta^2}{4}\right)\right]\right\}$$

In the opposite limit, when $At \to \infty$ so that $T \to 1$ and $\tau \to 1$, the equations reduce to

$$u'' + wu' + 1 - u^2 = C\eta(1-\eta)vu,_\eta$$

$$v'' + wv' + C(1-2\eta)(1-v^2) = C\eta(1-\eta)vv,_\eta$$

$$-w' + u + C(1-2\eta)v = -C\eta(1-\eta)v,_\eta$$

These equations were studied intensively by Cooke and Robins,[24a] who easily obtained solutions that smoothly connected the two stagnation-point solutions, for C and $-C$, providing that v remained everywhere positive. As expected, this is the case if $C \le 0.4294$. For larger values of C, the y component of skinfriction vanishes at some point between the stagnation points, in a rather weakly singular way. Computations done with an adaptation of program SYMMBL confirm their results. For example, if $C = 0.8$, $v'(0, \eta, 0)$ vanishes and the displacement thickness

$$\delta^* = \lim_{\zeta \to \infty} [(1 + C(1-2\eta))\zeta - w]$$

seems to approach infinity, as $\eta \to 0.7594$. The simple numerical marching procedure that follows the flow from $\eta = 0$ to this point cannot, however, be continued to reach the second stagnation point, because it is unstable in the region where $v < 0$.

For this problem, a suitably designed algorithm that mimics the natural convection of information in the region of bidirectional flow avoids this instability and permits a better look at the singularity.[25] The key to the algorithm is an upwind-differencing scheme portrayed in Fig. 12.49. In Fig. 12.49, **M** marks the station at which ordinary differential equations will be solved to determine the profiles of u, v, and w. At that station, partial derivatives with respect to τ and η are first approximated by the formulas

$$u,_\tau = \frac{3u(\mathbf{M}) - 4u(\mathbf{P}) + u(\mathbf{PP})}{2\,d\tau}$$

$$u,_\eta = \frac{3u(\mathbf{P}) - 4u(\mathbf{L^*}) + u(\mathbf{LL^*})}{2\,d\eta} \quad \text{if} \quad u(\mathbf{P}) \ge 0 \qquad (12.111)$$

and $\quad u,_\eta = -\dfrac{3u(\mathbf{P}) - 4u(\mathbf{R^*}) + u(\mathbf{RR^*})}{2\,d\eta} \quad \text{if} \quad u(\mathbf{P}) \le 0$

[24a] J. C. Cooke and A. J. Robins, "Boundary-Layer Flow Between Nodal and Saddle Points of Attachment," *J. Fluid Mech.* **41**:823–836 (1970).

[25] The possibility of doing this was first demonstrated in a calculation that was somewhat flawed by a poorly designed transformation formula. See H. A. Dwyer and F. S. Sherman, "Some Characteristics of Unsteady Two- and Three-Dimensional Reversed Boundary-Layer Flows," in *Numerical and Physical Aspects of Aerodynamic Flows*, edited by T. Cebeci, Springer Verlag, New York, (1982), Chap. 18.

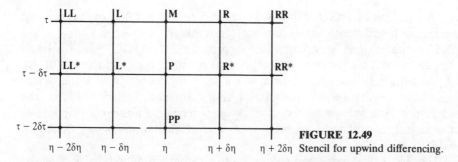

FIGURE 12.49
Stencil for upwind differencing.

The errors in these approximations are proportional to $(d\tau)^2$ and $(d\eta)^2$, but the formulas cannot be used for time $\tau = d\tau$, or for the positions $\eta = d\eta$ or $\eta = 1 - d\eta$, where they refer to undefined quantities. At these points we can use the second-order, two-point formulas

$$u_{,\eta} = \frac{2}{d\eta}[u(\mathbf{M}) - u(\mathbf{L})] - u_{,\eta}(\mathbf{L}) \qquad \text{at } \eta = d\eta$$

and

$$u_{,\eta} = \frac{2}{d\eta}[u(\mathbf{R}) - u(\mathbf{M})] - u_{,\eta}(\mathbf{R}) \qquad \text{at } \eta = 1 - d\eta$$

At first glance, these do not appear useful, because they refer to end-point values of $u_{,\eta}$. However, for this problem these values are easily obtained. We have only to differentiate the equations governing u, v, and w with respect to η, and to set $\eta = 0$ or $\eta = 1$ in the resulting equations. This gives equations for $u_{,\eta}$, $v_{,\eta}$, and $w_{,\eta}$, which contain u, v, and w as coefficients. In each time step of a numerical marching procedure, the values of u, v, and w at the stagnation points are first determined, as was done in Chapter 10; these values are then used to advance the values of $u_{,\eta}$, $v_{,\eta}$ and $w_{,\eta}$ through that same time step.

Notice, however, that the approximations of $u_{,\eta}$ *at internal points* are additionally inaccurate, in that they involve quantities at time $\tau - d\tau$ instead of τ. Remembering that some iteration is required to account for the nonlinearity of the finite-difference equations, we design the iteration to improve the accuracy of $u_{,\eta}$ at the same time, by the following procedure.

1. In the first sweep of calculations at time τ, $u(\mathbf{M})$, $v(\mathbf{M})$ and $w(\mathbf{M})$ are approximated by $u(\mathbf{P})$, $v(\mathbf{P})$, and $w(\mathbf{P})$, wherever first-guess values are needed for the evaluation of coefficients. Equations (12.111) are used to approximate $u_{,\eta}$.

2. In a second sweep at the same time level, the results of the first sweep are used in the evaluation of coefficients, and to provide estimates of $u(\mathbf{L})$ and $u(\mathbf{R})$ for the formulas

$$u_{,\eta} = \frac{3u(\mathbf{M}) - 4u(\mathbf{L}) + u(\mathbf{LL})}{2\,d\eta} \qquad \text{if} \qquad u(\mathbf{P}) \geq 0 \qquad (12.112)$$

but

$$u_{,\eta} = -\frac{3u(\mathbf{M}) - 4u(\mathbf{R}) + u(\mathbf{RR})}{2\,d\eta} \qquad \text{if} \qquad u(\mathbf{P}) \leq 0$$

Step (2) can be recycled, but a single execution of it will provide excellent convergence if sufficiently small time steps are taken.

This analysis has been embodied in program HOTDOG, which is listed in Appendix B. This is the most complicated of the programs included in this book, and the results obtained from it will now be described in some detail. Some of the results are of immediate physical interest; others reinforce the message that one cannot be too careful when trying to answer a physical or mathematical question by numerical means. Calculations were done for $C = 0.8$.

The first calculations were done with a rather crude subdivision of the interval between the stagnation points, $\delta\eta = 0.1$, and made the program look very good. In particular, the flow seemed to become nicely steady at all points after the time $\tau \approx 0.9$, and in the steady flow the η component of skin friction varied smoothly, with no sign of singular behavior, from the positive value given by the stagnation-point solution for $C = 0.8$ at $\eta = 0$, to the value for $C = -0.8$ at $\eta = 1$. This was, however, a bit suspicious, because it did not confirm the finding, by Cooke and Robins and by our own direct, steady-flow calculation, of a weak singularity at the point of first flow reversal. Other trial calculations confirmed that $\zeta_{max} = 7$, and $\delta\zeta = 0.10$ would be adequate for about 4-decimal accuracy. The calculation was then repeated, with $\delta\eta = 0.05$, 0.025, 0.0125, and 0.00625.

Constraints on the ratio of δ_τ to δ_η. A second type of computational instability. In all other illustrative calculations done for this book, this procedure of refining the discrete computational steps leads uneventfully to a converged solution, as accurate as may be required. The finite-difference procedures are stable, because of the implicit formulation. However, these calculations, with the exception of those done with program TRANBL, involve only a single timelike independent variable: either time itself, or distance in an unambiguously downstream direction. This time, there are two timelike variables, η and τ, and a second type of computational instability appears where there is backflow, if the ratio of $\delta\eta$ to $\delta\tau$ becomes too small. For example, when $\tau = 0.8$, by which time a substantial region of bidirectional flow had appeared, the successive halvings of $\delta\eta$ were first carried out with $\delta\tau = 0.01$. All went well as $\delta\eta$ was reduced to 0.0125, but suspicious wiggles appeared in the results for $\delta\eta = 0.00625$. These wiggles disappeared when $\delta\tau$ was reduced to 0.005 after $\tau = 0.7$. The wiggles are confined to the region of bidirectional flow, and are most conspicuous in plots of the displacement thickness, δ^*. Further trial computations showed that the wiggles can be avoided, and properly converged solutions can be obtained, for times as late as $\tau = 0.9$, providing that the finite increment of dimensionless time, $\delta(At)$, does not exceed about $20\delta\eta$. The results of converged calculations are shown in Fig. 12.50(a) for the scaled η component of skin friction, and Fig. 12.50(b) for the displacement thickness.

As can be seen in Fig. 12.50(b), the computed flow first becomes steady

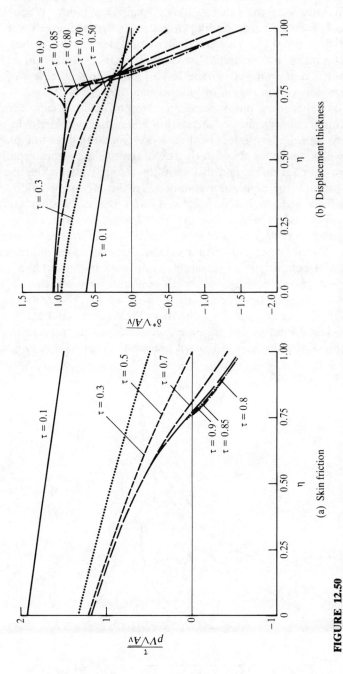

(a) Skin friction

(b) Displacement thickness

FIGURE 12.50

Impulsively started flow in the symmetry plane between nodal and saddle stagnation points: $\tau = \mu v_{,z}(0, y, 0)$; $A\delta^* = \lim_{z \to \infty}(w - zW_{,z})$.

at $\eta = 0$, where the nodal stagnation point is. By the time $\tau = 0.8$ ($At = 4$) the domain of steady flow has extended out to about $\eta = 0.6$; and the flow has become nearly steady at the other, saddle, stagnation point. When $\tau = 0.9$ ($At = 9$) the solution has become independent of τ everywhere except in the small range $0.75 < \eta < 0.85$. The steady-state profiles of the velocity component that becomes bidirectional are shown in Fig. 12.51 for the two stagnation points, a station just to the left of the first flow reversal, and a station in the middle of the region of bidirectional flow.

As $\tau \to 1$ it becomes more and more expensive to obtain converged results in the immediate vicinity of the station where the η-component of skin friction changes sign. Even the rule of thumb $\delta(At) < 20\delta\eta$ does not guarantee convergence when τ exceeds about 0.91. The derivative $v_{,\eta}$ varies rapidly with η near the flow-reversal point, and this variation strongly affects the computed values of w, through the continuity equation. In the region of unidirectional flow, these effects are carried downstream, and the calculation converges nicely for all values of τ. Where the flow is bidirectional, a sort of feedback loop exists, causing a change in w at one value of η to affect, at the next time step, the flow at both larger and smaller values of η. This does not necessarily provoke a violent computational instability, nor does it prevent the solution from converging and becoming steady throughout most of the region of bidirectional flow. It does, however, cause the solution in a rather narrow band of η to drift aimlessly after a certain value of τ. This makes it difficult to decide with real confidence whether the displacement thickness at the station of first flow reversal remains finite as $\tau \to 1$. As a practical matter, one may not much

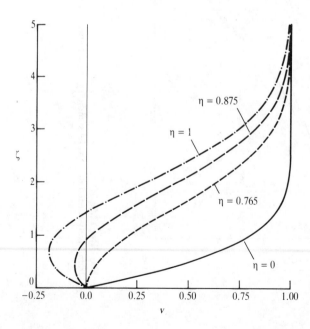

FIGURE 12.51
Steady-state distributions of v.

care about such fine points, but one is bound to be impressed by the care with which this seemingly straightforward calculation must be done in order to obtain results that do any justice at all to the boundary-layer equations.

12.11 BOUNDARY-LAYER SEPARATION

Boundary-layer separation is still imperfectly understood; it is even difficult to know how to describe it when the flow is unsteady or three-dimensional. This section will only hint at the difficulties, by a brief discussion and a few examples.

Steady, Two-Dimensional Flow

The universally accepted picture of separation in steady, two-dimensional flow is shown in Fig. 12.52. Two streamlines approach the separation point, S, along the wall. At S they merge into a single streamline that divides the flow field, at least locally, into a part that comes from upstream, and a part that comes from downstream. In some circumstances, as in the concave bend of a wind tunnel contraction, the flow which separates at S reattaches at point R. In such cases the separation streamline, SR, divides the main, outside, flow from the recirculating flow inside the *separation bubble*. As the separation stream-line approaches the reattachment point, it divides into the two streamlines that lead upstream and downstream along the wall.

It will ease further discussions to have a formal nomenclature for the splitting and merging of streamlines and streamsurfaces. Where one streamline or streamsurface, followed in the direction of flow along it, splits into two, as at R, there is a *positive bifurcation*.[26] Where two streamlines or streamsurfaces

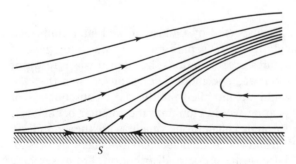

FIGURE 12.52
Steady, two-dimensional, separating flow.

---

[26] This nomenclature is taken from H. Hornung and A. E. Perry, "Some Aspects of Three-Dimensional Separation, Part I: Streamsurface Bifurcations. *Zeit. f. Flugwissenschaften u. Weltraumforschung* **8**:77–87 (1984). I agree with these authors that it is important to develop a precise vocabulary when attempting to describe complex things.

merge to form a single streamline or streamsurface, as at *S,* there is a *negative bifurcation.* Bifurcations of streamsurfaces can occur out in the fluid, as well as at the wall. In plane flow, what we draw as a streamline is just a streamsurface seen in edge view.

In plane flow, the skin friction is zero at a point of separation or attachment. That is what has been used to identify the separation point in the examples shown previously in this chapter.

It was mentioned in Section 12.5 that our marching calculations of steady two-dimensional boundary layers could not be carried on successfully past the point of separation. This is not an anomaly associated with the particular algorithm used, but a common affliction of all marching schemes for the full boundary-layer equations. There appear to be two reasons for this, both of which can be circumvented without a need to abandon any of the approximations of boundary-layer theory.[27]

1. In calculations for which the pressure distribution is obtained from potential-flow theory or from a smooth fit to experimental data, all the computed quantities change very rapidly as the separation point is approached. The skin friction, for example, seems to vanish in proportion to $(x - x_S)^{1/2}$. These changes are not accurately described by the usual sort of finite-difference approximations, and so accuracy is lost. This is often called the problem of *Goldstein's singularity.*

2. If local accuracy is not desired, the calculation can be adjusted to step over the Goldstein singularilty, to presumably safe territory on the other side. There, however, there is bidirectional flow; and in the backflow region the marching scheme is trying to project information in a direction opposite to the way it is being carried by the fluid. This makes no sense, and most schemes for solving the finite-difference equations promptly become unstable and blow up.

The Goldstein singularity. It was suspected by Goldstein, and later confirmed by many workers, that the singularity at the separation point was not a general feature of solutions to the boundary-layer equations, but that it was a result of the way the pressure distribution was specified. The suspicion was that the actual pressure distribution would be modified subtly, as a consequence of separation, and that this modification would eliminate the singular behavior.

The confirmation came through calculations, made with the boundary-layer equations, but with a distribution of displacement thickness or of skin friction being specified in the place of the pressure distribution. Let us see how that might be done, when displacement thickness is specified.

---

[27] For an excellent discussion, with references to the research literature, see J. C. Williams, III, "Incompressible Boundary-Layer Separation," *Ann. Rev. Fluid Mech.* **9**:113–144 (1977).

The starting point is the vorticity-streamfunction formulation of the boundary-layer equations, used above to study entry flow in a channel.

$$\psi_{,y}\Omega_{,x} - \psi_{,}\,\Omega_{,y} = \nu\Omega_{,yy} \quad \text{with} \quad \Omega = -\psi_{,yy}$$

To introduce the displacement thickness into the equations, convert these to equations for the perturbation streamfunction $\psi' = \psi - (y - \delta^*)u^*$, where $u^* = \psi_{,y}$ is the streamwise velocity component. This is constructed so that ψ' will vanish outside the boundary layer.

We introduce the following scaling, in which U and a are constant reference values for velocity and streamwise distance.

$$x = \xi a \quad y = \eta\delta^* \quad u^* = Uu(\xi, \eta) \quad \psi' = U\delta^* f(\xi, \eta)$$

$$\Omega = \left(\frac{U}{\delta^*}\right)S(\xi, \eta) \quad \delta^* = \left(\frac{\nu a}{U}\right)^{1/2} D(\xi)$$

The vorticity equation is transformed into

$$S'' + \gamma(\eta - 1)uS' + \gamma uS = \chi\left\{u\frac{\partial S}{\partial \xi} - (\eta - 1)S'\frac{\partial u}{\partial \xi}\right\} \tag{12.113}$$

in which the coefficients

$$\gamma(\xi) = \frac{U}{2\nu}\frac{d\delta^{*2}}{dx} \quad \text{and} \quad \chi(\xi) = \frac{U}{a\nu}\delta^{*2} \tag{12.114}$$

bring in the information about the assumed variation of displacement thickness.

From the relation between velocity and vorticity in a boundary layer, $\Omega = -\partial u^*/\partial y$, we get another transformed equation,

$$u' + S = 0 \tag{12.115}$$

Finally, differentiating the definition of ψ' with respect to y, and remembering that $u^* = \partial\psi/\partial y$, we get $\partial\psi/\partial y = u^* = \partial\psi'/\partial y + u^* - \delta^* \partial u/\partial y = \partial\psi'/\partial y + u^* - \delta^*\Omega$. The u^*s cancel, and the remaining terms transform into

$$f' - (\eta - 1)S = 0 \tag{12.116}$$

Equations (12.113, 114, 115, and 116) require four boundary conditions in η, which are

$$u(\xi, 0) = 0 \quad \text{(no slip)} \quad f(\xi, 0) = 0 \quad \text{(impervious wall)}$$

$$S(\xi, \infty) = 0 \quad \text{(no vorticity outside the boundary layer)}$$

and $\quad f(\xi, \infty) = 0 \quad$ (by the definition of displacement thickness)

We take $\xi = 0$ to denote a station somewhat upstream of the point where separation would be predicted by the usual boundary computations, and use those computations to provide initial profiles of u, S, and F. They also provide initial values for the coefficients γ and χ.

Marching in a region of bidirectional flow. Suppose for the moment that this reformulation of the calculation will remove the Goldstein singularity, and that one can implement it with a wise guess of the function $D(\xi)$. What is to be done about the instability of the forward-marching procedure in a region with backflow?

The simplest answer is inspired by the fact that the backflow velocity in many small separation bubbles is a very small fraction, often only a few percent, of the forward speed outside the boundary layer. This suggests that the important path by which information reaches a point in the backflow region follows the fluid downstream in the region of forward flow, and then cuts across the streamlines by diffusion, in toward the wall. The weak convective transport by the horizontal component of the backflow is, by comparison, unimportant.[28]

To test this idea, one has only to insert a flag in the computer program that will substitute a very small positive number for u whenever the computed value is negative, in the first term on the right-hand side of Eq. (12.113). This has been done by Carter and Wornum, using distributions of displacement thickness for which accurate numerical solution of both the full Navier–Stokes equations and the boundary-layer equations had previously been computed. The more accurate treatment of the boundary-layer equations employed an iterative calculation of the recirculating region, with a procedure that looks rather like a time-marching scheme with upwind differencing.

An alternative approach, in which a regular distribution of skin friction along the wall is prescribed in the place of the displacement thickness, works just as well. By these means, it has been shown that the solutions of the boundary-layer equations match well with those of the full Navier–Stokes equations, when the separation bubble is really slender.[29] The boundary-layer calculations, even when iteration is employed, are computationally much less demanding than the Navier–Stokes calculations.

Conclusion. These results allow the conclusion that the boundary-layer equations are not invalidated by the phenomenon of separation. On the other hand, it is still clear that much more than a simple boundary-layer theory is required to predict and describe separation in circumstances where separation grossly perturbs the external inviscid flow. We shall also see, in Chapter 13, that the whole topic of laminar, steady, flow separation is a bit academic, because the separated flow is notoriously unstable in a physical sense, and gives way to unsteady or turbulent flow at quite small values of Reynolds number.

---

[28] These ideas were once really offensive to me, but the evidence of their success is irresistible. See J. E. Carter and S. F. Wornum (1975) "Forward-Marching Procedure for Separated Boundary-Layer Flows." *AIAA Journal* **13**, No. 8, 1101–1103.

[29] See Williams (1977) op. cit. for more discussion, and details of the results.

Unsteady Two-Dimensional Flow

The examination of simple examples, such as the oscillating shear flow of Stokes' Second Problem, shows that vanishing skin friction is an insufficient criterion for separation of an unsteady boundary layer. Many efforts have been made to find a suitable criterion, but it is not easy to do so, because of the temporal variation of the streamline pattern. This section presents a single example, which illustrates both a reversal of skin friction without separation and a second reversal that is accompanied by separation. Separation in this example seems to conform to the concept worked out by Moore, Rott, and Sears.[30]

LAMINAR FLOW AROUND THE EDGE OF AN IMPULSIVELY ACCELERATED PLATE. When a flat plate is suddenly moved forward, in a direction normal to its plane, the vorticity generated on the windward side, and to some extent on the leeward side, is swept up into a concentrated and growing vortex which moves away from the sharp edge of the plate. You can see something like this when you suddenly move a spoon near the surface of a cup of coffee; more well-defined experiments are depicted in Van Dyke's *Album of Fluid Motion*,[31] and in Fig. 12.53.

Observations like these led Prandtl[32] to draw attention to the possibility of an inviscid, self-similar flow pattern in which a growing vortex sheet, coiled into a spiral of infinitely many turns, develops at the edge. He deduced that distance from the edge to the center of the spiral is proportional to $t^{2/3}$, and that the total circulation in the spiral grows in proportion to $t^{1/3}$. For the purposes of this section, the spiral is represented by a single line vortex that moves downstream on a line normal to the plate. The imaginary situation at a given time is shown in Fig. 12.54.

The boundary layer on the back, or leeward, side of the plate is of special interest. It develops under a transient potential flow, composed partly of the flow that would exist in the absence of the cast-off vortex, and partly of the flow induced by that vortex. The resulting velocity on the plate is given by the formula

$$\frac{U(x, t)}{U_0} = \left(\frac{2}{\xi}\right)^{1/2} \frac{1 - \chi}{1 + (1 - \chi)^2} \tag{12.117}$$

in which

$$\xi = \frac{x}{H} \quad \text{and} \quad \chi = \tau^{1/3}\left(\frac{2}{\xi}\right)^{1/2} \tag{12.118}$$

---

[30] Williams (1977) op. cit. is especially good on this topic. Another good reference is D. P. Telionis, *Unsteady Viscous Flows*. Springer series in computational physics, Springer-Verlag, New York, 1981

[31] See photos 80–83 of the *Album*.

[32] L. Prandtl, "Über die Entstehung von Wirbeln in der idealen Flüssigkeit," in *Vorträge aus den Gebiet der Hydro- und Aerodynamik,* edited by Th. v. Kármán and T. Levi-Civita, pp. 19–34 (1922).

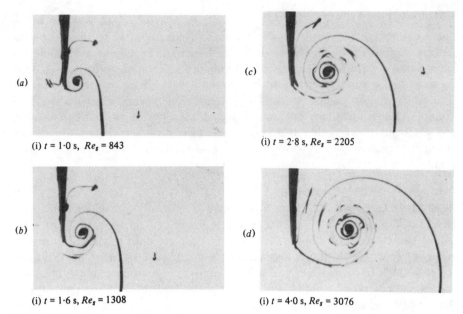

FIGURE 12.53

Suddenly started flow past a sharp edge. Flow approaches a 5° wedge from the left, with a speed proportional to $t^{0.45}$. The smoothly curved dye streak is a time line that initially extended straight down from the bottom of the wedge. The ripoled streakline that comes off the bottom of the wedge is fed by dye washed off the solid surface. This is a self-similar flow, but not quite the one analyzed below. (Reproduced, by permission, from D. I. Pullin and A. E. Perry, *J. Fluid Mech.* **97**: 239–255, 1980).

U_0 and H are reference values of velocity and length, which would be proportional to the forward speed and width of a finite plate. For present purposes, there is no need to be more precise about them. Note that the potential flow exhibits a moving stagnation point, which moves away from the edge with a speed proportional to the speed of the vortex. The stagnation

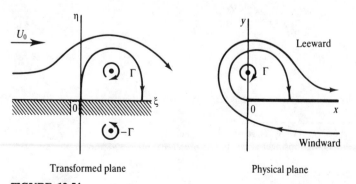

Transformed plane   Physical plane

FIGURE 12.54

Impulsively-started flow around the edge of a plate. Theoretical model.

point corresponds to $\chi = 1$, the edge to $\chi = \infty$, and the point where $x = \infty$ to $\chi = 0$. It can be shown, from (12.117), that $\partial U/\partial x > 0$ in the range $2 - \sqrt{2} < \chi < 2 + \sqrt{2}$, so that the potential flow will, at least here, encourage the boundary layer to stay close to the wall.

The lack of a physically defined reference length, which led Prandtl to the idea that the vortex spiral should grow in a self-similar way, may be used again for the boundary-layer analysis. As will be shown soon, a successful transformation for the streamfunction in the boundary layer is

$$\Psi^* = U_0\left(\frac{vtH}{x}\right)^{1/2} f(\chi, \eta) \tag{12.119}$$

in which

$$\eta = \frac{y}{(vt)^{1/2}} \tag{12.120}$$

Thus, we expect that this problem will be computationally no more complex than one of steady, two-dimensional flow.

The velocity components derived from the streamfunction (12.119) are

$$u^* = \Psi^*_{,y} = U_0\xi^{-1/2}f'(\chi, \eta)$$

and

$$v^* = -\Psi^*_{,x} = U_0\left(\frac{v}{U_0H}\right)^{1/2} \tau^{1/2}\xi^{-3/2}[\chi f(\chi, \eta)]_{,x}$$

The equation for streamwise momentum in the boundary-layer,

$$u^*_{,t} + u^*u^*_{,x} + v^*u^*_{,y} - (U_{,t} + UU_{,x}) = vu^*_{,yy}$$

now appears as

$$u'' + \left(\frac{\eta}{2} - \gamma f\right)u' + \gamma u^2 + \delta = \psi(uu_{,\chi} - u'f_{,\chi}) + \phi u_{,\chi} \tag{12.121}$$

with

$$f' - u = 0$$

The primes denote partial differentiation with respect to η, and the coefficients are all functions only of χ, to wit:

$$\gamma = \frac{\sqrt{2}}{8}\chi^3 \qquad \psi = -\frac{\sqrt{2}}{8}\chi^4 \qquad \phi = \frac{\chi}{3}$$

and

$$\delta = -\frac{\sqrt{2}}{12}\left(\frac{\chi^2}{D^3}\right)(3\chi^4 - 11\chi^3 + 2\chi^2 + 18\chi - 16)$$

in which

$$D = 1 + (1 - \chi)^2$$

The boundary conditions are

$$u(\chi, 0) = 0 \quad \text{(no slip)} \qquad f(\chi, 0) = 0 \quad \text{(impervious wall)},$$

and

$$u(\chi, \infty) = \frac{\sqrt{2}(1 - \chi)}{D} \quad \text{(matching with potential flow)}$$

The initial condition is

$$u(0, \eta) = \frac{1}{\sqrt{2}} \operatorname{erf}\left(\frac{\eta}{2}\right) \tag{12.122}$$

which is simply the solution of (12.121) at $\chi = 0$. This corresponds to the fact that diffusion always dominates the initial development of an impulsively-started boundary layer.

Figure 12.55 shows an overall view of the x–t plane, as a guide to important features of the inviscid flow along the wall. The $\frac{3}{2}$-power parabolas issuing from the origin are loci of constant χ that bound regions in which the flow shows some special feature. The curve AB is the path followed by a particle of the potential flow that lies at $x = H$ when the flow is started. It starts moving away from the edge, but is soon overtaken by the effect of the vortex, which arrests it and moves it eventually back to the edge. The acceleration of the particle is toward the edge, until $\chi = 2.9634$, where the quantity δ, which is proportional to the pressure gradient, changes sign. Where $\chi < 2.9634$, the vorticity continuously added at the wall has the opposite sign from that in the boundary layer at $\zeta = 0^+$. We might expect this to lead eventually to flow reversal near the wall. The dotted region is that where the potential flow outside the boundary layer is approaching the wall.

Equation (12.121) can be solved numerically by marching forward in the variable χ, from the initial condition (12.122), providing that the coefficient of $\partial f'/\partial \chi$ remains everywhere positive. As was discussed in Sections 12.8 and 12.9, this requirement is the Courant–Friedrichs–Lewy (CFL) condition, that the domain of dependence of the finite-difference scheme should include the corresponding domain for the differential equation. If the coefficient becomes negative, the marching calculation becomes analogous to an effort to solve the heat equation with negative time increments, and blows up. In the present

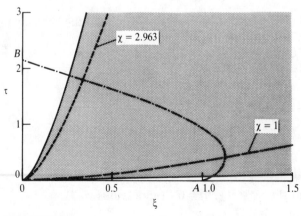

FIGURE 12.55
Inviscid flow domains.

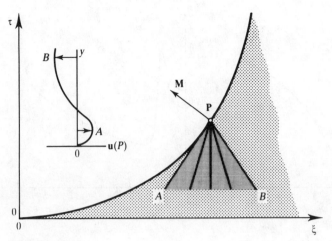

FIGURE 12.56
The Courant–Friedrichs–Levy condition.

case, the coefficient equals $\phi + \psi u$, which is positive if $\chi^3 u < \sqrt{(32/9)}$. The CFL requirement is shown graphically in Fig. 12.56.

For an (x, t) station P, the marching calculation projects information forward in the direction of increasing χ, along the vector **PM**. The domain of dependence of that calculation scheme is the lightly dotted region, in which $\chi < \chi_p$. Information is convected physically to the station P along rays such as BP, the slope of which is $1/u_B$, and AP, the slope of which $1/u_A$. The CFL requirement is that none of the convection vectors, such as **BP**, should have a negative dot product with the marching vector **PM**. If that happened, the calculation would be trying to project information into a region out of which information would really be coming. Remember this drawing, when you come to the description of the Moore–Rott–Sears (MRS) criterion for separation.

The necessary calculations can be done with a simple adaptation of program NEWBL. Since the effectiveness of the transformation of ψ is not known a priori, special attention must be paid to the choice of η^*, but $\eta^* = 8$ proves to be adequate throughout the entire range of χ. Because the matching condition at $\eta = \eta^*$ introduces a stronger dependence on the marching variable than is typical of most calculations described here, a second-order difference formula for χ-derivatives is almost a necessity.

Figure 12.57 shows computed distributions of the scaled skin friction, $u'(\chi, 0)$, and a scaled vorticity thickness, δ_Ω, as functions of χ. The vorticity thickness is defined by an unconventional formula,

$$\delta_\Omega = \frac{\int_0^\infty \eta \, |\Omega| \, d\eta}{\int_0^\infty |\Omega| \, d\eta} \tag{12.123}$$

which yields the conventional scaled displacement thickness if there is no bidirectional flow in the boundary layer.

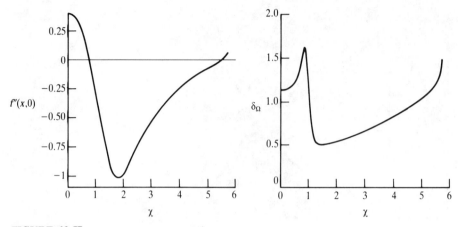

FIGURE 12.57
Scaled quantities versus χ. (a) Skin friction; (b) vorticity thickness.

A first reversal of skin friction occurs where $\chi = 0.76705$; a second where $\chi = 5.44732$. Unlike the situation in steady flow with prescribed $U(x)$, the skin friction at both of these points passes through zero with a finite slope, and there is no difficulty in carrying the marching solution past the point of reversal. The marching scheme blows up, however, at about $\chi = 5.706$, where the CFL requirement is violated. More about that later.

The scale factor by which δ_Ω is multiplied to get a physical thickness is simply $(\nu t)^{1/2}$, so the spatial variation of boundary-layer thickness at any given time is surprisingly small. The fairly sudden decrease between $\chi = 1$ and $\chi = 1.5$ is the result of cancellation of the original charge of vorticity by the outward diffusion of vorticity of the opposite sign. Presumably this decrease

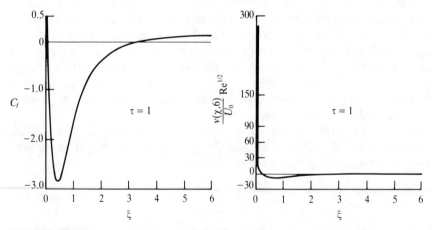

FIGURE 12.58
Spatial distributions at $\tau = 1$. (a) Skin friction; (b) normal velocity where $\eta = 6$.

would not be shown by a boundary-layer thickness based on a temperature profile, if the wall temperature were slightly increased at time zero, and thereafter held constant.

Figure 12.58 shows spatial distributions of skin friction and of the y component of velocity at $\eta = 6$, at nondimensional time $\tau = 1$. The shape of the latter distribution differs little from that of $-\partial U/\partial x$, except as $\chi \to 5.706$, the value at which the CFL criterion is violated and at which the calculation ends. It is difficult to draw definite conclusions about the possibility of a singularity in the distribution of velocity normal to the wall, but the present numeric data are quite well fitted by the formula

$$\mathrm{Re}^{1/2}\frac{v(\chi, 6)}{U_0} = 2.675(5.706 - \chi)^{-1.15} \tag{12.124}$$

which certainly implies a singularity. There seems to be no corresponding singularity for the skin friction.

The Moore–Rott–Sears separation criterion. The MRS criterion for unsteady separation evolved from experience with steady flows over moving walls. Streamwise stations had been found in these flows, where the velocity and the vorticity vanished simultaneously at a point out in the boundary layer, and where the velocity normal to the wall seemed singularly large. This was not a station at which the skin friction vanished, but it was one where flows from upstream and downstream appeared to meet and to erupt away from the wall. It seemed logical to say that separation occurred at such a station, and it seemed intuitively plausible that a similar situation might be seen in an unsteady boundary layer, if one were in the correct moving frame of reference. Just as a surfer who moves with just the right speed gets an impressive opportunity to observe the curl of a large breaking wave, so an experimentalist who could arrange to move with the moving separation point in an unsteady boundary layer would get a clear view of the phenomenon of separation. According to MRS, the streamline pattern that would be observed from this frame of reference would look, in our case, like the picture sketched in Fig. 12.59. Far from the wall, it looks just like the familiar sketch of steady separation, but close to the wall there is a distinctive sort of undertow.

Semisimilar unsteady boundary layers, such as the one being used here as an example, provide a unique opportunity to know just how fast one should move to lock on to the moving separation phenomenon. This is because the shape of the boundary-layer profile depends only on the value of χ, and we know precisely how fast to move to keep that value constant. In dimensionless terms, the necessary speed is $d\xi/d\tau = (\frac{4}{3})\tau^{-1/3}\chi^{-2}$. From this we can calculate, by setting $u/U_0 = \xi^{-1/2}f'(\chi, \eta) = d\xi/d\tau = (\frac{4}{3})\tau^{-1/3}\chi^{-2}$, that we should find a local maximum of f', given by $\chi^3 f'_{\max} = \sqrt{(\frac{32}{9})}$, at a station where the MRS criterion for separation is met. This is exactly the critical condition for violation of the CFL condition, just as it is for a simple downstream-marching calculation of a steady boundary layer that approaches separation.

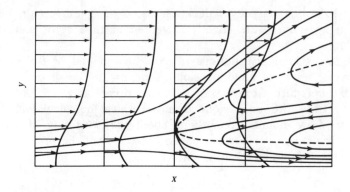

FIGURE 12.59
Streamline pattern for unsteady (MRS) separation.

This concurrence of tri-authored criteria makes perfect sense. The CFL rule says that the calculation cannot be successfully marched past the point where fluid appears, bearing information about conditions up ahead. The MRS picture shows how that foreign fluid arrives at the separation point, sidles up to the fluid that has been followed by the marching calculation, and moves off abruptly away from the wall. In a way, this concurrence of criteria is unfortunate, because it prevents the use of boundary-layer theory for the generation of a properly computed version of Fig. 12.59. This is not a case, like that of the impulsive sliding motion of the flat plate, for which a clever upwind-differencing scheme can circumvent the CFL constraints, because we have no way to deduce boundary conditions along the $x = 0$ axis. We simply do not know anything about the flow that sneaks into the boundary layer from the edge of the plate under a potential flow that is moving the other way.

Figure 12.60 gives a pictorial summary of the flow in this example, showing the principal flow features at dimensionless time $\tau = 1$.

The points to be emphasized before we leave this example are the following.

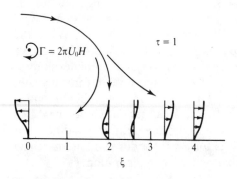

FIGURE 12.60
Pictorial sketch of boundary-layer behavior, $\tau = 1$.

1. It is perfectly possible that the skin friction may pass through zero at a point under an unsteady boundary layer, without any eruption of flow away from the wall. This happens in this flow, at $\chi = 0.747$. The situation then seems qualitatively more akin to Stokes' Second Problem than to steady separation.

2. When unsteady separation does occur, the point where flows from left and right collide and move away from the wall is displaced away from the wall, and does not lie directly over the point of vanishing skin friction.[33]

3. This example is special, in the sense that separation appears immediately at $\tau = 0^+$, right at the edge of the plate. Quite a different mode of unsteady separation may occur when, as in the case of an impulsively accelerated circular cylinder, the boundary layer develops for some time before separation appears anywhere. This other mode, analogous to the *open mode* of separation of steady, three-dimensional boundary layers, is not necessarily well described by the MRS criterion.[34]

Three-Dimensional Steady Flow Separation

When a body such as a prolate spheroid is placed at an angle of attack in a wind tunnel, with its surface lightly coated with oil; or if the same body is placed in a water tunnel, with its surface initially coated with a soluble dye, one sees a remarkably complex pattern of streaks in the oil film, and an even more complex evolution of the sheet of dyed water which was once the attached boundary layer.[35] The vorticity originally in the boundary layer is carried away from the body in sheets, which roll up into concentrated tubes and trail off downstream.

In many cases, the velocity fields induced by the vortex sheets and rolled-up vortex tubes have a prominent effect on the distribution of aerodynamic forces and heat transfer, on the performance of turbomachinery and fluid distribution systems, on erosion and sediment transport, and a host of other practically important phenomena. In some cases, the location of the places where vorticity leaves the surface of a body is determined by geometric

---

[33] Several fascinating calculations of streamline patterns in separating unsteady boundary layers have recently been made. See, for a fine example, T. L. Doligalski and J. D. A. Walker, "The Boundary layer Induced by a Convected Two-Dimensional Vortex," *J. Fluid Mech.* **139**:1–28 (1984).

[34] See K. C. Wang, "On the Current Controversy About Unsteady Separation," in *Numerical and Physical Aspects of Aerodynamic Flows,* edited by T. Cebeci, Springer-Verlag. New York, 1982, Chap. 16.

[35] Picture 74 of Van Dyke's *Album* hints at this. You may also want to see the film loops FM 54 and FM 66 in the NCFMF/EDC film series, distributed by Encyclopedia Britannica Educational Corporation.

features such as sharp edges, but in many others it is not. In any case, it is likely to be hard to predict, or even to describe, the spatial configuration the shed vorticity will assume. As a result, this section deals with a field of active research in which there is still a significant diversity of opinion as to the most promising lines of approach. If you become interested in it, you will certainly want to keep abreast of both theoretical and experimental research.

The part of this subject that fits most naturally into this chapter concerns the prospects for theoretical calculation of the development of three-dimensional boundary layers in the vicinity of separation. Because of the importance of the concept of domain of dependence, not just as a theoretical abstraction but as a practical constraint on what can be calculated by standard methods of boundary-layer theory, it is first necessary to gain, primarily from experiment, some idea of the nature of three-dimensional separation.

The most mature effort concerns the analysis and classification of oil-streak patterns, or of the pattern of the *skin friction lines* that they are thought to approximate. These are lines on the surface of the body, everywhere tangent to the skin friction vector τ. Like streamlines in the body of the fluid, they cannot cross each other, except at points, like stagnation points, where the length of τ vanishes and its direction is therefore indeterminate. The first key to a general understanding of the skin friction field is the classification of the various local fields in the vicinity of such singular points.[36] Figures 12.61 and 12.62 show examples of the sort commonly seen in oil-streak patterns. In Fig. 12.61(a), point N is called a *nodal point* of the skin friction field; the point S is a *saddle point*. Infinitely many skin friction lines pass through a nodal point with the same slope; one distinguished line passes through with a different slope. Only two skin friction lines pass through a saddle point; the rest swerve away from it.

Figure 12.61(b) shows another saddle point, and the other type of singular point, called a *focus* or a *vortical node,* through which infinitely many skin friction lines may spiral, or around which they may circle.

Figure 12.62 shows a pair of singular lines without definable ends, one seeming to swallow up skin friction lines, and the other seeming to give birth to them. Following the nomenclature of Hornung and Perry, we call lines like these *bifurcation lines*: *positive bifurcations* if skin friction lines run out of

---

[36] An accessible, and very influential, presentation of this point of view is given by M. J. Lighthill, in *Laminar Boundary Layers,* edited by L. Rosenhead, Oxford University Press, 1963, Chap. 11, Sections 26–29. You would be well rewarded by study of this entire chapter.

For a more recent view, see M. Tobak and D. J. Peake, "Three-Dimensional Separated Flows," *Ann. Rev. Fluid Mech.* **14**:61–85 (1982). These authors begin to direct attention to the flow field off the wall, above a singular point of the skin-friction field.

Finally, you should read Hornung and Perry (1984), op. cit., and their companion article "Some Aspects of Three-Dimensional Separation, Part II: Vortex Skeletons," *Zeitschrift für Flugwissenschaften und Weltraumforschung* **8**:155–160 (1984).

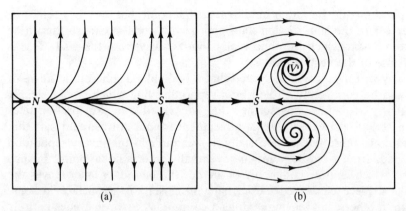

FIGURE 12.61
Three-dimensional steady boundary layers. Singular points of the skin-friction field.

them, *negative bifurcations* if skin friction lines seem to run into them. Many authors call the former *lines of attachment*, and the latter *lines of separation*.

Lighthill showed how the skin friction field could help one determine whether the streamlines just off the surface would stay nearly parallel to the wall, or would move sharply away from it. He assumed, as we have throughout this section, that the essential feature of separating flow is this pronounced motion away from the surface.

Consider a tiny streamtube, bounded by the wall, by two normal fences rising out of adjacent skin friction lines, and by a top surface located at a very small distance $z(s)$ from the wall. Let h be the local distance between the skin friction lines, and suppose that z is so small that the speed of flow parallel to the wall is approximately $(\tau/\mu)z$. Then the flowrate through the streamtube is approximately $(\tau/\mu)hz^2/2$. In steady flow this is independent of s, the distance along the streamtube. We see that the elevation of the top surface is proportional to $(\tau h)^{-1/2}$. Thus, the flow along the top moves away from the wall either because τ is approaching zero, as the skin friction lines approach a singular point, or because the skin friction lines are converging. Note that the two effects partly cancel where the skin friction lines approach a saddle point,

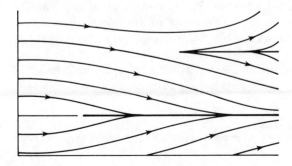

FIGURE 12.62
Open bifurcation lines of the skin-friction field (positive bifurcation above, negative below).

so that separation may be rather mild near such a point, but that they reinforce each other where the skin friction lines approach a nodal point, so that these are the points near which the fluid really squirts away from the wall. This is especially true of the vortical nodes.

In early writings, it was suggested that a body like a spheroid at angle of attack would be completely girdled by a separation line which would cleanly separate that part of the body surface that is covered with skin friction lines that originated at upstream points of attachment from a downstream part that is inaccessible to these lines. Specifically, it was thought that every separation line must issue from a saddle point and eventually enter a nodal point. In such a situation, which is quite commonly observed, the boundary layer leaves the surface like a piece of draped fabric, hanging closer to the surface in some places than in others, but leaving it along a continuous curve that goes entirely around the body, meeting itself on the other side. This is commonly called *closed separation,* with a *closed negative bifuraction line.*

Now it is known that this description does not fit all the experimental facts, but that *open bifurcations,* such as are shown in Fig. 12.62, are quite common. Then some skin friction lines from a particular upstream origin approach a negative bifurcation line from one side, while others, from the same origin, slip around the end of the separation line and approach it from the other side. The boundary layer then leaves the surface in a way analogous to the fold in a draped fabric, which often forms just below the shoulder of a garment. The excess fabric can be rolled up, so that the remaining fabric clings to the arm, well below the tip of the fold. The vortex sheet that springs from an open bifurcation line usually rolls up in much this same way, into a concentrated trailing vortex. The positive open bifurcation shown in Fig. 12.62 would then be an effect of the velocity field induced by that vortex.[37]

It would obviously be asking a lot of a marching solution of the boundary-layer equations, to reveal much of the detail shown in Fig. 12.61(b). Aside from the great difficulty of organizing the calculations so as to abide, if possible, by the constraint that the calculation should nowhere try to march against the local flow, there is likely to be a reenactment of the difficulties encountered in two-dimensional flow when the pressure field is prescribed in advance. It may be necessary to specify the problem, near separation, in some way analogous to the specification of a distribution of displacement thickness or skin friction in the two-dimensional case. This anticipates that key features of the observed skin friction fields may be the result of the velocity induced by the vorticity after it has left the wall. This case is made very convincingly in Part II of Hornung and Perry.

It may be that local solutions of the Navier–Stokes equations, designed

---

[37] If this seems hopelessly abstract or hard to visualize, consult Part II of Hornung and Perry (1984) op. cit.

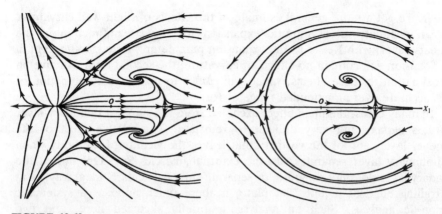

FIGURE 12.63
Skin-friction fields generated by series expansions. (Reproduced, by permission, from A. E. Perry and M. S. Chong, 1986.[38])

to display specific observed kinds of separation, may somehow be fitted, by adjustment of parameters, into an overall scheme of calculation that employs conventional boundary-layer theory where that is possible, and a theory of inviscid vortical flow, where that is appropriate. Partly motivated by this possibility, a very remarkable task of computer-aided analysis has been undertaken.[38] It aims to construct solutions of the Navier–Stokes equations that can reveal the complexity of flow near singular points by power-series expansions in the space coordinates. Such expansions, carried only to second-order, are the basic tools by which the singular points are classified; the aim now is to carry the expansions to sufficiently high order that the accurately-described domain encompasses more than one singular point. Figure 12.63 exhibits some results of this work.

12.12 HIGH-ORDER BOUNDARY-LAYER THEORY

What happens to a boundary layer when the Reynolds number is not very high, and the layer is not very slender? Some of the effects that are left out of the simplest boundary-layer theory are *local*, and would be relatively easy to reinstate. The curvature of the wall, which has been neglected, produces such an effect. In fact, if one intends to integrate the boundary-layer equations numerically, for a specific value of the Reynolds number, a few minor modifications of the program will reinstate the wall-curvature effects. However, each such calculation is useful for only one value of the Reynolds

---

[38] A. E. Perry and M. S. Chong "A Series-Expansion Study of the Navier–Stokes Equations with Applications to Three-Dimensional Separation Patterns," *J. Fluid Mech.* **173**:207–223 (1986).

number. To get a more general estimate of the effects of slight wall curvature, one can postulate an asymptotic expansion for the streamfunction in the boundary layer, with $\mathrm{Re}^{-1/2}$ as the expansion parameter.[39] The leading term in the expansion will contain no effects of curvature; the second term will give an approximation to these effects, which will suffice to make the combination of terms accurate over an extended range of Re.

In many cases, the main effect excluded from the simplest boundary-layer theory is *global*, and very difficult to reinstate. This is the effect of the boundary layer, and of the vorticity that leaves the vicinity of the body when the boundary layer separates, on the external inviscid flow. Only when the boundary layer does not separate, or separates in a particularly ideal way, as at the trailing edge of a finite flat plate, is there much hope of a systematic asymptotic analysis. Such an analysis is briefly sketched here, and two examples by Van Dyke are presented as illustrations.

As in Chapter 9, where the structure of flow due to a point force was reanalyzed for high values of the Reynolds number $F/\rho v^2$, we introduce inner (Prandtl) variables and an inner expansion to describe the region containing vorticity. The resulting equations lack boundary conditions at the outer edge of the boundary layer. These are supplied by matching the inner representation of the solution to an asymptotic representation of the outer flow, which is formulated as another expansion in terms of outer (Euler) variables.

The matching for the tangential component of velocity u is depicted in Fig. 12.64. At the lowest level of approximation, the level at which we have been working heretofore in this chapter, the inner representation of u approaches a constant value U_1, as the inner variable, $N = y/\delta(x)$, approaches

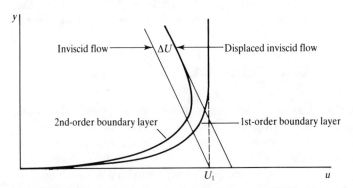

FIGURE 12.64
Matching of inner and outer velocity profiles.

[39] See M. Van Dyke, "Higher Approximations in Boundary-Layer Theory. Part I. General Analysis," *J. Fluid Mech.* **14**:161–177 (1962). In his book *Perturbation Methods*, referred to in Chapter 11, Van Dyke verifies this choice of expansion parameter.

infinity. The outer representation of u approaches U_1 as the outer variable, y/L, approaches zero.

At the second level of approximation, a two-term inner representation of u will approach a straight-line representation of the outer flow. This line has the slope of the lowest-order outer velocity profile, evaluated at $y/L = 0$. Its intercept at the wall differs from U_1 by an amount that is due to the displacement effect of the lowest-order boundary layer.

The slope of the second-order matching line is easy to calculate in cases where the lowest-order external flow is described in closed form. For example, consider flow over a parabolic cylinder. The complete lowest-order analysis of the boundary layer is worked out as a Sample Calculation at the end of this chapter. In parabolic coordinates (ξ, η) the streamfunction for the lowest-order inviscid flow is given by $\Psi = U_\infty a\xi(\eta - 1)$. Here U_∞ is the flow speed far upstream from the cylinder, and a is the radius of curvature of the nose. Curves of constant η are parabolas; $\eta = 1$ denotes the body surface. The distance between such parabolas is given by $d\eta = h(\xi, \eta)\, d\eta$, where $h = a(\xi^2 + \eta^2)^{1/2}$ is the metric factor for the coordinate system. The ξ-component of velocity is $U = h^{-1}\, \partial\Psi/\partial\eta = U_\infty\xi(\xi^2 + \eta^2)^{-1/2}$, and the derivative of this with respect to distance from the wall is $\partial U/\partial n = h^{-1}\, \partial U/\partial\eta = (U_\infty/a)\xi\eta(\xi^2 + \eta^2)^{-2}$. Thus, the quantities needed for the matching are

$$U_1(\xi) = U_\infty\xi(\xi^2 + 1)^{-1/2} \qquad \text{(the first-order intercept)}$$

and

$$S_1(\xi) = (U_\infty/a)\xi(\xi^2 + 1)^{-2} \qquad \text{(the slope of the second-order matching line)}$$

In a plane, potential flow, $S_1 = -\kappa U_1$, where κ is the longitudinal curvature of the wall.

The displacement of the intercept of the matching line, ΔU in Fig. 12.64, is relatively hard to evaluate. It is obtained by re-solving the equations for inviscid flow, with a wall boundary condition that specifies a normal velocity component there, equal to the normal velocity component at the outer edge of the first-order boundary layer. The result cannot ordinarily be obtained in closed form, but we shall present a numerical procedure that works for the parabolic cylinder.

Solutions of the first-order boundary-layer equations do not have the property that u approaches a straight line of nonzero slope as $y/\delta \to \infty$, so it is clear that it will be up to the second-order inner solution to accommodate this feature of the matching. In principle, one can imagine how this procedure might be carried on to third order, in which the matching curve for u would have a curvature determined by the first-order outer solution, a slope as determined above but modified by the second-order outer solution, and an intercept with an additional displacement associated with the additional v-velocity induced by the second-order inner solution. However, enough is enough: few people would care to try it, feeling that their time would be better invested in a brute-force numerical assault on the full Navier–Stokes equations.

To put this picture into equations, we write out the postulated inner and outer expansions.

Inner

$$\psi(x/L, y/\delta; \text{Re}) \approx \text{Re}^{-1/2}\,\psi_1(x/L, y/\delta) + \text{Re}^{-1}\,\psi_2(x/L, y/\delta) + \cdots$$
$$u(x/L, y/\delta; \text{Re}) \approx u_1(x/L, y/\delta) + \text{Re}^{-1/2}\,u_2(x/L, y/\delta) + \cdots$$
$$v(x/L, y/\delta; \text{Re}) \approx \text{Re}^{-1/2}\,v_1(x/L, y/\delta) + \text{Re}^{-1}\,v_2(x/L, y/\delta) + \cdots$$
$$\Omega(x/L, y/\delta; \text{Re}) \approx \text{Re}^{1/2}\,\Omega_1(x/L, y/\delta) + \Omega_2(x/L, Y/\delta) + \cdots$$

Outer

$$\Psi(x/L, y/L; \text{Re}) \approx \Psi_1(x/L, y/L) + \text{Re}^{-1/2}\,\Psi_2(x/L, y/L) + \cdots$$
$$U(x/L, y/L; \text{Re}) \approx U_1(x/L, y/L) + \text{Re}^{-1/2}\,U_2(x/L, y/L) + \cdots$$
$$V(x/L, y/L; \text{Re}) \approx V_1(x/L, y/L) + \text{Re}^{-1/2}\,V_2(x/L, y/L) + \cdots$$

We insert these into the full Navier–Stokes equations and make successive applications of the Prandtl and Euler limiting processes, as appropriate.[40] If the outer flow is irrotational far upstream, it turns out that both Ψ_1 and Ψ_2 satisfy the Laplace equation. The first-order inner equations are just the ordinary boundary-layer equations (12.2) and (12.5).

At the second-order approximation to the inner flow, streamwise diffusion is still negligible, but the equations connecting velocity and vorticity to streamfunction show the effects of wall curvature. Both transverse curvature of an axisymmetric body and longitudinal curvature κ are important, but we shall limit attention here to plane flow. That is enough to demonstrate the general idea; it turns out that effects of transverse curvature can be separately calculated and superposed, since the perturbation analysis produces linear equations for the second-order quantities.

We continue the analysis for plane flow, using N to denote the inner variable y/δ, S to denote x/L, and using $U_0 a$ to make ψ dimensionless, without changing the symbol for ψ. If there is no vorticity in the flow far upstream, the governing equation for ψ_2 is

$$\psi_{2,NNN} - [u_1\psi_{2,NS} + v_1\psi_{2,NN}] - [\psi_{2,N}u_{1,S} + \psi_{2,S}u_{1,N}]$$
$$= \frac{d(U_1(s,0)U_2(s,0))}{ds} + \kappa\{Nu_{1,NN} + u_{1,N} - u_1 v_1\}$$
$$+ \{\kappa[NU_1^2(s,0) + M(s,N)]\}_{,s} \tag{12.125}$$

where
$$M(s,N) = \int_N^\infty (U_1^2(s,0) - u_1^2)\, dN$$

---

[40] For the Prandtl limit, hold y/δ constant as $\text{Re} \to \infty$; for the Euler limit, hold y/L constant. When necessary, remember that δ/L is proportional to $\text{Re}^{-1/2}$.

Note that $U_2(s, 0)$, the velocity due to first-order displacement effects, acts as a forcing function in the differential equation for ψ_2, as well as appearing in a boundary condition. The boundary conditions on ψ_2 are

$$\psi_{2,N}(s, 0) = 0 \quad \text{(no slip)} \qquad \psi_2(s, 0) = 0 \quad \text{(impervious wall)}$$

$$\psi_{2,N}(s, \infty) = U_2(s, 0) + NS_1 = U_2(s, 0) - N\kappa U_1(s, 0) \quad \text{(matching to outer flow)}$$

Calculation of flow due to displacement. To discover the wall boundary condition for the second-order outer flow, we match representations of the normal velocity, taking two terms of the outer expansion and one of the inner. Anticipating that we need the behavior of V close to the wall, and that $V_1 = 0$ on the wall, we insert $V_1(s, y) \approx (y/L)V_{1,y}(s, 0) + \cdots$ into the first term of the outer expansion. To express the result in inner variables, we use $y = \text{Re}^{-1/2} N$, and find that the outer expansion begins with $\text{Re}^{-1/2}[NV_{1,y}(s, 0) + V_2(s, \text{Re}^{-1/2} N)] + \cdots$.

The first term in the inner expansion, expressed in terms of the outer variable, is just $\text{Re}^{-1/2} v_1(s, \text{Re}^{1/2} y) + \cdots$. Equating these two expressions in the limit $\text{Re} \to \infty$, we get $NV_{1,y}(s, 0) + V_2(s, 0) = v_1(s, \infty)$. Finally, we note that $V_1, y(s, 0)$ and $v_{1,N}(s, \infty)$ represent the same quantity, so we can write

$$V_2(s, 0) = \lim_{N \to \infty} [v_1(s, N) - Nv_{1,N}(s, N)] = \frac{d[U_1(s, 0)\delta_1^*(s)]}{ds} \quad (12.126)$$

Here δ_1^* is just the displacement thickness of the first-order boundary layer. If we recognize $V_2(s, 0)\, ds$ as an apparent increment of volumetric flow passing out of the wall, through a strip of length ds, we can think of the effect of the first-order boundary layer on the outer flow as being equivalent to a continuous distribution of sources along the wall, the source density per unit length being $2d[U_1(s, 0)\,\delta_1^*(s)]/ds$. The factor of 2 appears because half the flow from each source goes back into the parabola. This quantity can easily be added to the output of program NEWBL; we have only to deduce a way of using it to determine $U_2(s, 0)$.

Second-Order Theory for a Parabolic Cylinder

For a parabolic cylinder, a numerical procedure for the calculation of $U_2(s, 0)$, given input from program NEWBL, can easily be devised. We start by deducing the flow speed induced at one point on a parabolic cylinder, say at parabolic coordinate ξ, by a source of strength m at another point ξ'. This is easily done in two steps. First we find the speed induced at point $(\xi, 1)$ on the plane $\eta = 1$, by a source of strength m located at $(\xi', 1)$, on the same plane. The answer is simply $U(\xi, \xi') = (m/2\pi)(\xi - \xi')^{-1}$.

Next, we transform the plane $\eta = 1$ into the parabola $2X = Y^2$ with the conformal mapping $2(X + iY)/a = (\xi + i\eta)^2 + 1$. This kind of mapping carries streamlines into streamlines, and equipotential lines into equipotential lines. Specifically, if the mapping carries point P, where the velocity potential is Φ_p, into point P'; the potential at point P' will also equal Φ_p. We use this to deduce that speed on the parabola will equal that on the plane times the ratio of the distance between points at ξ and $\xi + d\xi$ on the plane and the distance between the corresponding points on the parabola. The former distance is simply $d\xi$, the latter is $h\,d\xi$, where h is the metric coefficient for the transformation, with value $h = a(\xi^2 + 1)^{1/2}$. Therefore, on the parabola, $U(\xi, \xi') = [m(\xi')/2\pi h(\xi)](\xi - \xi')^{-1}$. Since there are sources all over the surface of the parabola, we express the final desired result as the integral

$$U_2(s, 0) = \frac{1}{2\pi h(\xi)} \int_{-\infty}^{\infty} m(\xi')(\xi - \xi')^{-1} d\xi' \qquad (12.127)$$

in which

$$m(\xi') = \frac{d[U_1(\xi', 0)\delta_1^*(\xi')]}{d\xi'}$$

Since $m(\xi')$ is an even function of ξ', we can convert (12.127) to an integral from 0 to ∞, in the form $U_2(s, 0) = I(\xi)/(2\pi h(\xi))$, where

$$I(\xi) = \int_0^{\infty} m(\xi')[(\xi - \xi')^{-1} + (\xi + \xi')^{-1}] d\xi'$$

This integral is not easy to evaluate numerically as it stands, because of the singularity where $\xi = \xi'$. It is helpful to integrate once by parts. To avoid logarithms of a negative number, we break up the integral so that the resulting integrands are positive everywhere. This gives

$$I(\xi) = \int_0^{\xi} m(\xi')[(\xi - \xi')^{-1} + (\xi + \xi')^{-1}] d\xi'$$

$$- \int_{\xi}^{\infty} m(\xi')[(\xi' - \xi)^{-1} - (\xi + \xi')^{-1}] d\xi$$

$$= \int_0^{\xi} m(\xi') \, d \ln \frac{\xi + \xi'}{\xi - \xi'} + \int_{\xi}^{\infty} m(\xi') \, d \ln \frac{\xi' + \xi}{\xi' - \xi}$$

$$= -\int_0^{\xi} \ln \frac{\xi + \xi'}{\xi - \xi'} m'(\xi') \, d\xi' - \int_{\xi}^{\infty} \ln \frac{\xi' + \xi}{\xi' - \xi} m'(\xi') \, d\xi'$$

Here m' denotes $dm/d\xi'$. You can easily show that the integrated portions add to zero. We now give separate treatment to the small intervals $\xi - \varepsilon < \xi' < \xi$ and $\xi < \xi' < \xi + \varepsilon$, by expanding $m'(\xi')$ in a Taylor series in x, where $\xi' = \xi - x$ in the first integral, and $\xi' = \xi + x$ in the second. The final result is

$$I(\xi) = \int_0^{\xi - \varepsilon} \ln \frac{\xi - \xi'}{\xi + \xi'} m'(\xi') \, d\xi' + \int_{\xi + \varepsilon}^{\infty} \ln \frac{\xi' - \xi}{\xi' + \xi} m'(\xi') \, d\xi'$$

$$+ 2\varepsilon m'(\xi)\left[\ln\left(\frac{\varepsilon}{2\xi}\right) - 1 \right] + o(\varepsilon^3 \ln \varepsilon)$$

Now the trapezoidal rule can be applied to the remaining integrals. This quadrature must be truncated at some finite value of ξ, but it was easy to show, using the gradually-expanding steps described in the sample calculation,

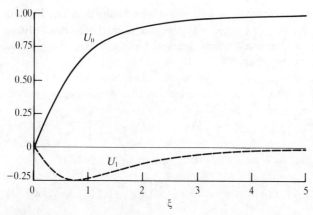

FIGURE 12.65
Inviscid surface speed on a parabolic cylinder; $U = U_0 + \mathrm{Re}^{-1/2}U_1$.

that no change appeared in the first five decimals if the quadrature was carried out to $\xi' = 980$, instead of being stopped at $\xi' = 200$.

The results of a calculation embodying these formulas are shown in Fig. 12.65. Near the nose of the parabola, the numerical results are well fitted by the formula

$$U_2(\xi, 0) \approx -\xi(0.6034 - 0.818\chi^2 + \cdots) \qquad \text{where} \qquad \chi^2 = \xi^2(1 + \xi^2)^{-1}$$

In his study of this problem, Van Dyke worked with extended Blasius series, and found an estimate of 0.61 for the first coefficient in this expression.

SOME PROPERTIES OF THE SECOND-ORDER BOUNDARY LAYER To complete the analysis of the second-order boundary layer, it is interesting to follow Van Dyke and to separate effects due to wall curvature from those due to modifications of the external flow. This is possible because of the linearity of the equations and boundary conditions giverning ψ_2. We substitute

$$\psi_2 = \kappa\psi_{2\kappa} + U_2(s, 0)\psi_{2d} \tag{12.128}$$

into Eq. (12.125) and the boundary conditions, and then let all the terms containing κ or $d\kappa/ds$ define one boundary-value problem, while all those containing $U_2(s, 0)$ or $dU_2(s, 0)/ds$ define another. It would be very tedious to carry this out for the complete range of s; Van Dyke has calculated the leading term in a solution of these problems by a Blasius-series approximation, and we simply cite his results for the second-order skin friction near the nose of the parabola:

$$\tfrac{1}{2} \mathrm{Re}^{1/2} c_f \approx (1.232588 - 3.03\,\mathrm{Re}^{-1/2})\frac{s}{a}$$

Of the second-order coefficient, curvature contributes -1.91, displacement -1.12.

This prediction has been compared with numerical solutions of the full Navier–Stokes equations, by R. T. Davis.[41] The result is fairly disappointing, in that the second-order correction proves accurate only when $\text{Re} > 4000$, where the correction itself is less than 5%.

Davis also presents results for the pressure drag, and we can easily compute the effect of boundary-layer displacement on that quantity. We have

$$D_p = 2 \int_0^\infty p \, dY = 2 \int_0^\infty \left[p_0 - \frac{\rho}{2} (U_1^2 + 2 \, \text{Re}^{-1/2} \, U_1 U_2 + \cdots) \right] dY \quad (12.129)$$

Using $U_1 = U_\infty \chi$, $Y = a\xi$, and $p_0 = p_\infty + (\rho/2)U_\infty^2$, one easily finds that

$$D_p = 2 \int_0^\infty p_\infty \, dY + \rho U_\infty^2 a \left[\frac{\pi}{2} - 2 \, \text{Re}^{-1/2} \int_0^\infty \chi U_2 \, d\xi \right] \quad (12.130)$$

To get a finite result, we can imagine that the parabolic cylinder ends at some very large value of x, and that the free stream pressure, p_∞, acts over the base of the cylinder. This removes the first integral. The second integral in (12.130) can easily be evaluated numerically, since we already have $U_2(\xi)$. This gives the result

$$C_{Dp} = \frac{D_p}{\rho U_\infty^2 a} = \frac{\pi}{2} + 0.8128 \, \text{Re}^{-1/2} + \cdots \quad (12.131)$$

Davis gives a corresponding result[42]

$$C_{Dp} = 1.57 + 2.6 \, \text{Re}^{-1/2} + \cdots$$

The discrepancy is partly due to the fact that our estimate does not yet include the effect of curvature on the wall pressure. The pressure that we have computed is that which would exist at the wall if the outer inviscid flow persisted all the way to the wall. It thus includes the effect of a centrifugal pressure gradient that is somewhat lessened by the reduction of tangential velocity in the boundary layer. Thus the pressure at the wall is actually higher than we have estimated, by an amount

$$\Delta p = \rho \kappa \int_0^\infty [U_1^2(s, 0) - u_1^2(s, N)] \, dN = \rho \kappa U_1^2(s, 0)(\delta_1^* + \theta_1)$$

where δ_1^* and θ_1 are the displacement and momentum thicknesses of the first-order boundary layer. For the parabola, $\kappa a = (1 + \xi^2)^{-3/2}$, and the increment of drag due to curvature is

$$\Delta C_{Dp} = 2 \, \text{Re}^{-1/2} \int_0^\infty \xi^2 (1 + \xi^2)^{-5/2} (\text{DISPL} + \text{MOM}) \, d\xi$$

---

[41] R. T. Davis,. "Numerical Solution of the Navier–Stokes Equations for Symmetric Laminar Incompressible Flow Past a Parabola," *J. Fluid Mech.* **51**:417–433. (1972).

[42] R. T. Davis (1972) op. cit. his equation (6.4), expressed with our nomenclature.

Again, the integral is easy to evaluate from the data already calculated, and we get

$$\Delta C_{Dp} = 2.1271 \, Re^{-1/2}$$

This makes our final estimate of the present drag

$$C_{Dp} = \frac{\pi}{2} + 2.929 \, Re^{-1/2} + \cdots \qquad (12.132)$$

which is in fair agreement with the result of Davis. Actually, it is hard to see how a 10% discrepancy in the coefficients of $Re^{-1/2}$ could be explained. In any event, it is interesting to note that the effect of curvature on the pressure drag is about 2.5 times that of displacement, and of the same sign. Thus, second-order effects appear to reduce the skin-friction drag, and definitely increase the pressure drag, for a parabolic cylinder.

Second-Order Analysis of Entry Flow into a Channel

At the end of Section 12.5, a first-order analysis of boundary-layer development near the inlet of a channel between plane, parallel walls was given. The velocity at the inlet plane was assumed to be uniform, and the velocity of the inviscid flow between the encroaching boundary layers was assumed to depend only on distance from the inlet plane. In reality, the potential flow due to displacement by the boundary layers cannot be this simple. As part of a larger study,[43] Van Dyke determined the second-order inviscid flow for an infinite *cascade* of identical channels. For this arrangement, sketched in Fig. 12.66(a),

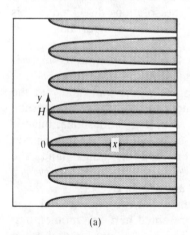

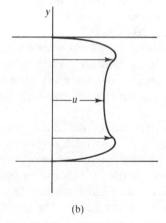

(a)   (b)

FIGURE 12.66
Entry flow in an infinite cascade.

[43] M. Van Dyke, "Entry Flow in a Channel," *J. Fluid Mech.* **44**:813–823 (1970).

the flow is a periodic function of y, with $v = 0$ on the imaginary upstream extensions of walls of zero thickness. The problem for the second-order inviscid flow is that of flow through an infinite cascade of parabolic cylinders, that being the shape into which the walls are transformed by the first-order, Blasius, boundary layers.

A solution composed from the first-order boundary-layer and the sum of the first- and second-order inviscid flows exhibits an interesting overshoot of u at the edge of the boundary layer, sketched in Fig. 12.66(b). This feature has been noted experimentally, and appears in finite-difference solutions of the full Navier–Stokes equations. Details are given in Van Dyke's paper, and this appears to be a very successful application of higher-order boundary-layer theory.

12.13 SUMMARY

In this chapter, you have encountered the idea of a laminar boundary layer as a thin region adjacent to a wall or other interface at which the no-slip condition acts to introduce vorticity into the flow. Specifically, the boundary layer is the region within which this vorticity is confined. Under favorable circumstances, which are usually encountered on the upstream side of a body immersed in a uniform stream, or in the converging portion of a nozzle that conducts flow out of a large reservoir, the boundary layer remains attached to the wall, and the flow outside the boundary layer is a potential flow. A useful distinction between external potential flow and boundary layer appears only when a Reynolds number based on a typical streamwise dimension and potential-flow speed is large, but a laminar boundary layer will be found only if the Reynolds number is not too large. Most of the solutions discussed in this chapter are tolerably accurate when such a Reynolds number has values between about 10^3 and 10^6.

The distinguishing facts of boundary-layer flow that facilitate its analysis and description are that the vectors representing velocity, acceleration, and net viscous force per unit volume are all nearly tangent to the wall, and that the tangential component of the pressure force is independent of distance normal to the layer. These facts apply equally to the flow in lubrication films, but the resultant acceleration of the fluid, which is usually negligible in lubrication films, is crucially important in boundary layers. On the other hand, the pressure distribution, which is the principal unknown of lubrication theory, is a known quantity in the usual forms of boundary-layer theory.

Flow in the simplest boundary layers is steady and two-dimensional, and can quite easily be predicted. The main tools presented here are momentum-integral analysis and finite-difference analysis. The momentum-integral equation must be supplemented by some information about the shape of the tangential velocity profile, such as is provided by the correlations used in Thwaites' method. According to these correlations, the shape of the profile is fairly well determined by the value of a parameter $\lambda = (\theta^2/\nu)\, dU/dx$. The value of λ is in turn affected by both the upstream history of the boundary layer, which determines the value of θ, and by the local acceleration of the external potential flow.

From the momentum-integral method, one learns that the local state of a boundary layer that has developed under an accelerating potential flow is only weakly influenced by its upstream history, whereas this history is very important to the local state when the external flow is decelerating. One also discovers that a laminar boundary layer cannot long stay attached to a wall once the potential flow starts to decelerate. Separation usually occurs before U has decreased more than about 15% from its maximum value; often this decrease is only 5 or 10%.

Finite-difference methods can be used when accurate predictions of the shape of the velocity profile are needed, as they are for predictions of the stability of the laminar flow. For the range of examples shown in this chapter, the resulting computational task is, by modern standards, relatively trivial and certainly inexpensive. The key to this relative ease of computation seems to be the design of appropriate transformations of the independent variables, especially the design of the function $g(x, t)$ that is used to scale the distance normal to the wall. You should review each example, aiming to understand the reason for the design employed in that example.

The finite-difference method is fairly easily adapted, to allow study of strategies for controlling boundary-layer development by the suction of small quantities of fluid through a porous or slotted wall. The aim of this control is often to delay or prevent separation of the boundary layer or transition from laminar to turbulent flow, by direct removal of "counterrotating" vorticity.

When flow in a boundary layer is unsteady or three-dimensional, a correct computational description of convection requires some careful preliminary thought about regions of influence and domains of dependence. In either case, the concept of boundary-layer separation requires extensive rethinking.

Separation is the main phenomenon that limits the usefulness of boundary-layer theory. When vorticity moves away from the body, into a configuration that is no longer roughly known in advance of detailed calculation, it is not only harder to describe, but it affects the potential flow much more extensively.

In cases not involving separation, boundary-layer theory can be systematically improved by calculation of the next terms in asymptotic expansions for large values of the Reynolds number. The first correction terms, for the simple case of incompressible flow past a parabolic cylinder in a uniform stream, account for the longitudinal curvature of the wall and for the displacement effect of the first-order boundary layer. The second-order corrections amount to about 1% of the values predicted by first-order theory when, for the parabola, a Reynolds number based on nose radius equals about 50,000.

SAMPLE CALCULATION

Requirements. Calculate the development of a steady boundary layer on a parabolic cylinder, first by Thwaites' method, and then by application of Program NEWBL. Compare results with the results of an extended Blasius series, as given by Van Dyke, *J. Fluid Mech.* **19**: 145–159 (1964).

Given data. The profile of a parabolic cylinder is described parametrically in cartesian (X, Y) coordinates by $X = (a/2)\xi^2$, and $Y = a\xi$, where a denotes the radius of curvature of the nose, which is located at the origin. Arclength along its surface, x, may be found from the differential relation $dx = a(1 + \xi^2)^{1/2} d\xi$. The speed of potential flow along the surface is then $U(x) = U_0\xi(1 + \xi^2)^{-1/2}$, where U_0 is the speed of the approaching free stream.

Thwaites' method. We start with Eq. (12.23), noting that only the integral term is needed, because the boundary layer starts at a stagnation point. Substitution of the given formulas for U and dx gives

$$\frac{\theta^2 U_0}{\nu a} = 0.45\xi^{-6}(1 + \xi^2)^3 \int_0^\xi \zeta^5(1 + \zeta^2)^{-2} \, d\zeta$$

The integral is easily evaluated by aid of the substitution $1 + \zeta^2 = \phi$, with the result

$$\frac{\theta^2 U_0}{\nu a} = 0.225\xi^{-6}(1 + \xi^2)^3[1 + \xi^2 - 2\ln(1 + \xi^2) - (1 + \xi^2)^{-1}]$$

To evaluate Thwaites' shape-determining factor λ we need to calculate

$$\frac{dU}{dx} = \left(\frac{dU}{d\xi}\right)\left(\frac{d\xi}{dx}\right) = U_0(1 + \xi^2)^{-2}$$

Then $\quad \lambda = \dfrac{\theta^2}{\nu}\dfrac{dU}{dx} = 0.225\xi^{-6}(1 + \xi^2)[1 + \zeta^2 - 2\ln(1 + \xi^2) - (1 + \xi^2)^{-1}]$

We note that $\lambda \to 0$ as $\xi \to \infty$, so we expect the boundary-layer behavior to approach that of a boundary layer under a uniform potential flow (Blasius layer) far downstream.

Preparation for numerical integration. We could proceed with a transformation based on this momentum-integral analysis, but there is an attractive simple alternative. One might guess that the edge of the boundary layer will roughly coincide with a parabola that lies just outside the one that defines the solid wall. The distance from the wall to this parabola is approximately $(1 + \xi^2)^{1/2}(\eta - 1)$. This suggests the choice

$$\frac{g^2 U_0}{\nu a} = 1 + \xi^2$$

which leads to the very simple expressions

$$\gamma(\xi) = 1 \qquad \delta(\xi) = (1 + \xi^2)^{-1} \qquad \text{and} \qquad \psi(\xi) = \xi$$

The Blasius series. For $\xi \ll 1$, $\delta = 1 - \xi^2 + \cdots$. Thus, the coefficients needed for the Blasius series are

$$\gamma_0 = 1 \qquad \delta_0 = 1 \qquad \psi_0 = 0 \qquad \text{and} \qquad \gamma_1 = 0 \qquad \delta_1 = -1 \qquad \psi_1 = 1$$

The marching calculation with NEWBL. To complete the necessary input for NEWBL, we have only to enter the general formulas for G2, U, $\gamma(\xi)$, $\delta(\xi)$, and $\psi(\xi)$.

The only disadvantage in the use of ξ as a marching variable is that significant changes in the boundary-layer profile still occur at quite large values of ξ, whereas relatively small steps in ξ are required for an accurate description of the relatively rapid changes near the nose of the parabola. This can be managed by use of gradually-increasing values of $d\xi$. If, for example, one uses $d\xi_m = \alpha^{-1} d\xi_{m-1}$, wtih $\alpha = 0.95$, the first increment in ξ can be as small as 0.002, while it takes only 200 of the gradually increasing increments to reach $\xi = 978$. Of course, the three-point trailing-difference formulas for $\partial u/\partial \xi$ and $\partial f/\partial \xi$ must be appropriately modified, becoming

$$\frac{\partial u}{\partial \xi} \approx \frac{A u_m + B_{m-1} + C_{m-2}}{d\xi}$$

with

$$A = \frac{2 + \alpha}{1 + \alpha} \qquad B = -\frac{1 + \alpha}{\alpha} \qquad C = \frac{1}{\alpha(1 + \alpha)} \qquad \text{and} \qquad d\xi = \xi_m - \xi_{m-1}$$

We set $\eta^* = 6$ for the initial try, and find that to be adequately large. The interval $0 \le \eta \le 6$ is divided into 240 equal steps, with a double-check at 480. These data allow Richardson extrapolation, to obtain five-decimal accuracy.

Comparison with Van Dyke's series. Van Dyke presents series approximations for skin friction and displacement thickness in his equations (2.10) and (2.17). In terms of our symbols, his series becomes

$$c_f = \left(\frac{U_0 a}{\nu}\right)^{-1/2} 2(1 - \chi^2)^{1/2}(1.2325877\chi - 0.4938405\chi^3 - 0.106505\chi^5$$
$$- 0.047331\chi^7 - 0.02675\chi^9 - 0.0172\chi^{11} - \cdots)$$

When the transformations are unwound in NEWBL, a quantity called TAU is produced. It is proportional to c_f, the relationship being $c_f = (U_0 a/\nu)^{-1/2} 2$ TAU. Thus, we can convert Van Dyke's series to a prediction of our TAU, and add to NEWBL a statement defining that prediction. A corresponding series-prediction can be made of our DISPL. A comparison of values is shown in Table 12.2. Incidentally, this part of the sample calculation is intended to illustrate the necessity of a careful unwinding of transformation formulas when comparing one's results with those of others.

TABLE 12.2

	TAU		DISPL	
ξ	(NEWBL)	(Van Dyke)	(NEWBL)	(Van Dyke)
0.03992	0.04910	0.04910	0.64871	0.64871
0.20284	0.23631	0.23631	0.66867	0.66867
0.40771	0.40549	0.40549	0.73027	0.73027
0.60024	0.48243	0.48242	0.82177	0.82178
1.02798	0.47058	0.47074	1.11842	1.11739
2.03855	0.28265	0.28603	2.11773	2.07745
4.00716	0.13323	0.14098	4.45113	4.15617
7.07374	0.07057	0.07777	8.23710	7.41009
14.54495	0.03292	0.03735	17.43993	15.30292
33.09553	0.01426	0.01636	40.11970	34.86045
209.96133	0.00224	0.00258	255.44021	221.21875

The final value of DISPL equals 1.21661ξ; the corresponding value for a Blasius boundary layer is 1.21677ξ. For $\xi > 20$, the displacement thickness on the parabola is less than 1% different from that on a flat plate of length $X = (a/2)\xi^2$. The discrepancy in local skin friction is even smaller.

EXERCISES

12.1. Given that the local skin friction on a flat plate is $\tau_w = 0.332\rho U^2 (\nu/Ux)^{1/2}$, find the total drag of a plate of length L, wetted on both sides. U, ρ, and ν are all constants.

12.2. Given that the velocity profile is $u = U[1 - \exp(v_w y/\nu)]$ in a boundary layer with suction, calculate the local skin friction. Remember that v_w is negative when there is suction.

12.3. Given that the boundary-layer thickness on a sphere, at a station 45° up from the forward stagnation point, is $\delta = 3.8(\nu D/U)$, where D is the sphere diameter and U is the speed of the approaching stream. Calculate δ when $D = 10$ cm and $U = 1$ m/s, for water and for air. Assume $T = 20°C$, $p = 1$ atm.

12.4. The discharge coefficient of a metering nozzle is defined as the ratio of the actual flowrate through the nozzle at a specified pressure drop, to the ideal flowrate, which would correspond to uniform flow through the throat at the same pressure drop.

Suppose that the flow in the throat would be uniform, except for the presence of a boundary layer. The displacement thickness of the boundary layer equals 10% of the radius of the nozzle throat. What is the value of the discharge coefficient?

12.5. Prove that the displacement thickness of a two-dimensional, steady boundary layer equals the average distance from the wall to the vorticity, i.e., that

$$\delta^* \int_0^\delta \Omega \, dy = \int_0^\delta y\Omega \, dy$$

where $y = \delta$ marks the outer edge of the boundary layer, beyond which there is no more vorticity.

12.6. Show that the normal velocity at the outer edge of a two-dimensional boundary layer is given by the equation

$$v(x, \delta) = v_w(x) - \delta U_{,x} + (U\delta^*)_{,x}$$

and give a physical interpretation to each term in the equation.

12.7. Show that the displacement effect of a jet, i.e., the velocity induced in the region outside of the jet, by the distribution of vorticity inside the jet, must be equivalent to a distribution of sinks along the axis or centerplane of the jet. (Recall the analysis of the Landau–Squire jet in Chapter 9 to get started.)

12.8. Verify the steps of analysis that lead from Eqs. (12.5) and (12.14) to the Momentum-Integral Equation in the form (12.17).

12.9. Derive the Momentum-Integral Equation by analyzing the conservation of mass and momentum for a fixed, finite, control volume bounded by planes $x = x_0$, $x = x_0 + dx$, $y = 0$, and $y = h$. Let the plane $y = 0$ be a fixed wall, where $u = 0$ and $v = v_w$. The plane $y = h$ lies just outside the boundary layer, where $u = U$ and where the vorticity and viscous stress are negligible. Start with Eqs. (4.24) and (4.25), and show how the surface integrals are simplified by approximations valid in boundary layers. To keep the analysis simple, assume that $\mathbf{u} = u(x, y)\mathbf{e}_x + v(x, y)\mathbf{e}_y$.

12.10. (a) Evaluate the coefficients in the polynomial

$$u = a + b\eta + c\eta^2 + d\eta^3$$

so that $\quad u = 0 \quad$ and $\quad u'' = 0 \quad$ at $\quad \eta = 0 \quad$ (the wall)

and that $\quad u = 1 \quad$ and $\quad u' = 0 \quad$ at $\quad \eta = 1 \quad$ (the outer edge)

Explain why each condition is appropriate if the polynomial is to approximate the velocity profile in a steady boundary layer under a uniform potential flow.

(b) With your final form of the polynomial, evaluate the integrals

$$\delta^* = \int_0^1 (1 - u)\, d\eta \qquad \text{and} \qquad \theta = \int_0^1 u(1 - u)\, d\eta$$

and the shape factors

$$H = \frac{\delta^*}{\theta} \qquad \text{and} \qquad T = \theta u'(0)$$

12.11. Do Exercise 12.10(b), for the velocity profille $u = \text{erf}\,(\eta/2)$.
(*Hint*: Integration by parts is effective; for this you need
$$u' = (2/\sqrt{\pi}) \exp(-\eta^2/4).)$$

12.12. (a) Calculate the rate of dissipation of kinetic energy in the boundary layer on a flat plate of length L, using the approximate velocity profile

$$u(x, y) = \frac{U}{2}(3\eta - \eta^3) \quad \text{if } 0 \le \eta \le 1 \qquad u = U \quad \text{if } \eta \ge 1$$

Here $\eta = y/\delta$. The boundary-layer approximation for the rate of dissipation per unit volume is

$$\Phi = \mu(u_{,y})^2$$

(b) Calculate the rate at which the plate does work on the fluid, if the plate is moving through the fluid at speed U.

(c) Compare your answers to part (a) and part (b) and discuss the comparison.

12.13. Explain the four constraints on the shape of the normalized vorticity profile shown in Fig. 12.5. The constraints are listed in the paragraph following Eq. (12.21).

12.14. Prove that the sketch of the normalized vorticity profile shown in Fig. 12.5 is correctly dimensioned.

12.15. Verify that Eq. (12.47) is a solution to the Momentum Integral Equation (12.9) for steady flow *if* the shape factors H and T are constant.

12.16. Show that Eq. (12.47) can be generalized to include the effect of suction through the wall, provided that either

$$v_w = S\theta U_{,x} \qquad \text{or} \qquad v_w = \frac{Su}{\theta}$$

S being another constant. Find the form of the solution in each case.

12.17. A potential flow is established by a line source of strength m, located at $x = -a$. A boundary layer grows under this flow, on a flat plate that lies in the plane $y = 0$, $x \geq 0$. The speed of the potential flow, at points on the wall, is

$$U(x) = \frac{U_0}{\xi} \qquad \text{where} \qquad \xi = \frac{x}{a}$$

Noting that the boundary layer will start with zero thickness at $x = 0$, analyse its growth and separation, using Thwaites' method. Evaluate the ratio of U/U_0 at the separation point.

(*Answer:* $\xi = 1.158$ at separation.)

12.18. A potential flow is established by a line source of strength m, parallel to a plane wall, and a distance a from the wall. the speed of this flow, at points on the wall, is

$$U(x) = \frac{m}{\pi a} \frac{\xi}{1 + \xi^2} \qquad \text{where} \qquad \xi = \frac{x}{a}$$

and $x = 0$ lies directly under the source.

Noting that the boundary layer will start at a stagnation point at $x = 0$, analyse its growth and separation, using Thwaites' method. Evaluate the ratio of U at the separation point to the maximum value of U.

(*Answer:* $\xi = 1.351$ at separation.)

12.19. A potential flow is established by a line sink of strength m, parallel to a plane wall, and a distance a from the wall. The speed of this flow, at points on the wall, is

$$U(x) = -\frac{m}{\pi a} \frac{\xi}{1 + \xi^2} \qquad \text{where} \qquad \xi = \frac{x}{a}$$

and $x = 0$ lies directly under the sink.

Noting that the boundary layer will start at $x = \pm\infty$, analyse its growth and separation, using Thwaites' method. Show carefully why only the integral term in Eq. (12.23) is needed. Evaluate the ratio of U at the separation point to the maximum value of U.

(*Answer:* $\xi = 0.8572$ at separation.)

12.20. A potential flow is established by two parallel line sinks, each of strength m, located at $x = \pm a$, in a plane wall. The resulting flow speed along the wall is

$$U(x) = \frac{m}{\pi a} \frac{\xi}{1 - \xi^2} \qquad \text{where} \qquad \xi = \frac{x}{a}$$

Noting that the boundary layer found between the sinks will start at a stagnation point at $x = 0$, analyse its growth, using Thwaites' method. Noting that the boundary layer found at $x^2 > a^2$ will start at $x = \pm\infty$, analyse its growth using Thwaites' method. Show carefully why only the integral term in Eq. (12.23) is needed.

(*Answers*:

$$\frac{\theta^2 m}{\pi v a^2} = 0.3(1 - \xi^2)^2(4 - \xi^2) \qquad \text{if} \qquad x^2 < a^2$$

$$\frac{\theta^2 m}{\pi v a^2} = 0.01875\xi^{-6}(1 - \xi^2)^2(1 - 4\xi^2 + 6\xi^4) \qquad \text{if} \qquad x^2 > a^2)$$

12.21. A two-dimensional flow passage is bounded by the plane $Y = 0$ and the hyperbola $Y^2 = X^2 + a^2$. The steady, irrotational, flow through the passage has velocity potential Φ, and streamfunction Ψ, and is most easily described when these are used as independent variables. Thus the speed of flow is given by

$$\left(\frac{U}{U_0}\right)^2 = \left[\sinh^2\left(\frac{\Phi}{U_0 a}\right) + \cos^2\left(\frac{\Psi}{U_0 a}\right)\right]^{-1}$$

and points are located by

$$\frac{X}{a} = \sinh\left(\frac{\Phi}{U_0 a}\right)\cos\left(\frac{\Psi}{U_0 a}\right)$$

$$\frac{Y}{a} = \cosh\left(\frac{\Phi}{U_0 a}\right)\sin\left(\frac{\Psi}{U_0 a}\right)$$

The curved wall is then a streamline on which $\Psi/U_0 a = \alpha$, where α is the asymptotic angle of the hyperbola.

(*a*) Noting that $d\Phi = U\,dx$, where dx is an increment of distance along a streamline, convert Thwaites' quadrature expression for $\theta^2(x)/v$ into an expression in which the independent variable is the dimensionless velocity potential, $\xi = \Phi/U_0 a$, and the integration is carried over the range of ξ, rather than x. You should get

$$\frac{\theta^2 U_0}{ua} = 0.45(\sinh^2\xi + \cos^2\alpha)^3 \int_{-\infty}^{\xi} (\sinh^2\xi' + \cos^2\alpha)^{-2}\,d\xi'$$

(*b*) Carry out the integration for the three cases, $\alpha = 0°$, $45°$, and $90°$.

(*Answers*:

$$\alpha = 0° \qquad \frac{\theta^2 U_0}{va} = 0.15\cosh^6\xi\,(3\tanh\xi - \tanh^3\xi + 2)$$

$$\alpha = 45 \qquad \frac{\theta^2 U_0}{va} = 0.1125\cosh^3(2\xi)(\tanh(2\,\xi) + 1)$$

$$\alpha = 90° \qquad \frac{\theta^2 U_0}{va} = 0.15\sinh^6\xi\,(3\,\mathrm{ctnh}\,\xi - \mathrm{ctnh}^3\,\xi + 2))$$

(c) Derive the corresponding formulas for $\lambda(\xi)$.

(d) Verify that your results for $\alpha = 0°$ agree with Eqs. (12.38) and (12.40). (Note that the η of Eq. (12.38) equals the tanh ξ of this analysis.)

12.22. Suppose the speed of potential flow just outside the boundary layer can be represented by $U(x) = U_0\{1 - x/a\}^m$, in which m is a nonnegative constant. Apply Thwaites' method, with the initial condition $\theta(0) = 0$, to locate the separation point as a function of the value of the exponent m.

12.23. Using Eqs. (12.27, 28, 29) in Eqs. (12.60), derive formulas for the coefficients, $\gamma(\xi)$, $\delta(\xi)$, and $\psi(\xi)$, needed for integration of the full boundary-layer equations for the flow driven by a line sink at the trailing edge of a flat plate. Choose the value of A to make $\gamma(0) = 2$.

(*Answers*: $\gamma = 2(1 - \xi)^4$, $\delta = 1 - (1 - \xi)^4$, $\psi = (1 - \xi)[1 - (1 - \xi)^4]$.)

12.24. Using Eqs. (12.31, 32) in Eqs. (12.60), derive formulas for the coefficients, $\gamma(\xi)$, $\delta(\xi)$, and $\psi(\xi)$, needed for integration of the full boundary-layer equations for the flow around a flat plate, oriented normal to the oncoming stream. Choose the value of A to make $\delta(0) = 1$.

(*Answers*: $\delta = 2(1 + 3\tau)(1 + \tau)^{-3}$, $\gamma = 3 - 2\delta$, $\psi = 2\xi\tau^2(1 + 3\tau)(1 + \tau)^{-3}$, in which $\tau^2 = 1 - \xi^2$.)

12.25. Using Eqs. (12.35, 36) in Eq. (12.60), derive formulas for the coefficients, $\gamma(\xi)$, $\delta(\xi)$, and $\psi(\xi)$, need for integration of the full boundary-layer equations for the flow around a circular cylinder. Choose the value A to make $\delta(0) = 1$.

12.26. Using the formulas given in Problem 12.21, derive formulas for the coefficients, $\gamma(\xi)$, $\delta(\xi)$, and $\psi(\xi)$, needed for integration of the full boundary-layer equations for the flow through a two-dimensional passage with a hyperbolic wall. Do the case $\alpha = 0$, choosing the value of A to make $\delta \rightarrow 2$ as $\xi \rightarrow -\infty$.

12.27. Using the formulas given in Problem 12.23, derive the equations for the first two terms of a Blasius series, in increasing powers of ξ, for the flow near the leading edge of the plate with a line sink on its trailing edge. Adapt program NEWBL, and solve these equations. Calculate, with five-decimal precision, the values of the normalized skin frictions, $S_0(0)$ and $S_1(0)$.

(*Answers*: $S_0(0) = 0.66412$, $S_1(0) = 2.25584$.)

12.28. Using the formulas given in Problem 12.24, derive the equations for the first two terms of a Blasius series, in increasing powers of ξ^2, for the flow near the stagnation point of the flat strip normal to the flow. Adapt program NEWBL, and solve these equations. Calculate, with five-decimal precision, the values of the normalized skin frictions, $S_0(0)$ and $S_1(0)$.

(*Answers*: $S_0(0) = 1.23259$, $S_1(0) = 0.13927$.)

12.29. (a) Using the formulas given in Problem 12.21, derive formulas for the coefficients, $\gamma(\xi)$, $\delta(\xi)$, and $\psi(\xi)$, for the flow near upstream infinity on the curved hyperbolic wall of a plane-flow nozzle. Let the asymptotic angle of the nozzle be $\alpha = 45°$. Choose the value of A to make $\delta \rightarrow 2$ as $\xi \rightarrow -\infty$.

(*Answers*: $\delta = 2[1 - \exp(4\xi)]$, $\gamma = 4\exp(4\xi)$, $\psi = 2[1 + \exp(4\xi)]$.)

For large negative values of ξ, derive the equations for the first two terms of a Blasius series, in increasing powers of the small quantity $\exp(4\xi)$. Adapt program NEWBL, and solve these equations. Calculate, with five-decimal precision, the values of the normalized skin frictions, $S_0(0)$ and $S_1(0)$.

(*Answers*: $S_0(0) = 1.63299$, $S_1(0) = -0.56179$.)

12.30. In cases for which the integral in Thwaites' method must be evaluated numerically, it is often more accurate and/or convenient to work directly with the parent differential equation,

$$U\left(\frac{\theta^2}{v}\right)_{,x} = 0.45 - 6\left(\frac{\theta^2}{v}\right)U_{,x}$$

Transform this into an equation in which the dependent variable is $Y \equiv \theta^2 U_0/va$, while the independent variable is the dimensionless velocity potential, $\xi = \Phi/U_0 a$. Recall that in a potential flow, $d\Phi = U\,dx$.

12.31. The boundary-layer growth on a two-dimensional wing is to be calculated, but the only information available about $U(x)$ is a collection of wall pressure measurements, available at equal increments of arclength in the x direction. Devise a computer program that will

(a) Exploit Thwaites' method to calculate $\theta^2 U_0/va$ as a function of x. U_0 is the flight speed; a is the chord of the wing.

(b) Use the numerical data obtained in (a) to generate the coefficients $\gamma(\xi)$, $\delta(\xi)$, and $\psi(\xi)$, where $\xi = x/a$.

12.32. (*More a project, than an exercise.*) Select one of the boundary-layer problems presented in the text or in preceding exercises, or a problem of your own choice, and do a complete finite-difference analysis. Pay particular attention to a demonstration of numerical accuracy. Write a comprehensive report, to document your preliminary analysis, your computer program, your results and conclusions.

CHAPTER 13

INSTABILITY
OF VISCOUS
FLOWS

13.1 INTRODUCTION

The idealized flows we have analyzed in the last four chapters all represented the simplest possible response of the fluid to the specified motions of bounding surfaces, or to other causes of motion, such as initially unbalanced body forces. We have now to confront a fascinating fact that has great practical importance. Under certain circumstances, to be identified in this chapter, the simplest possible flow is *unstable*. If it is perturbed, ever so slightly, the perturbations grow spontaneously, and the simplest possible flow is replaced by a more complex one. A broad review of instability phenomena, and a revealing experimental study of a typical one, has been given in an educational film[1] that will greatly help you to anticipate and understand the arguments and conclusions to be presented here.

The practical import of instability can be either positive or negative. On the positive side, an unstable flow is one in which potentially large, desired, effects may be produced just by tickling the flow, with very little expenditure

---

[1] *Flow Instabilities*, with E. L. Mollo-Christensen. Produced by NCFMF/EDC. Film No. 2169, distributed by Encyclopedia Britannica Educational Corp. 27 minutes, color.

of energy. For example, under certain circumstances the separated flow in a wide-angle diffuser forms a jet that tends to cling to one wall of the diffuser but does not prefer one wall to the other. The action of a fairly weak side jet may cause it to switch from one wall to the other, an effect for which there may be many practical uses.

On the negative side, an unstable flow is unpredictable, teetering on the edge between conditions that may imply, for example, very different distributions of aerodynamic loading. Unless the simplest flow is very selective, in the sense that only very special, controllable, disturbances may trigger an instability, the practical impossibility of knowing what flow to expect may be a barrier to successful design. Of course, in some cases the simplest flow may be desired for some inherent property, such as low skin friction or heat transfer, and the loss of this property, because the desired flow is unstable, defeats the intent of a design.

For many engineers, the most important manifestation of flow instability is the transition from laminar to turbulent flow, and for many years almost all knowledge of transition was empirical. The conceptual relevance of stability theory for an understanding of transition was conceded, but there was an enormous gap between the most complicated scenarios that could be constructed theoretically and the simplest ones that were observed experimentally. Aeronautical engineers made rough predictions of the location of transition with formulas that were fathered by stability theory, mothered by necessity, and regarded without much respect by either parent.

The gap between theory and experiment has narrowed greatly in recent years, and modern survey discussions of transition make exciting, if still sobering, reading.[2] This excitement is, however, only available to readers who are familiar with some basic terminology, techniques of analysis, and elementary results. This chapter should provide you with most of these necessary fundamentals.

The subject of hydrodynamic stability, and the literature devoted to it, are enormous. Anyone who hopes to deal with it in a single chapter has to adopt a highly selective set of goals.[3] The main goal of this chapter, besides the pedagogical goal announced in the preceding paragraph, is to illustrate the concept of a favored mode of instability, or of a most dangerous disturbance, since an understanding of these things seems essential to any effort at deliberate manipulation of instability.

---

[2] See, for example, M. V. Morkovin, "Recent Insights into Instability and Transition to Turbulence in Open-Flow Systems," AIAA Paper-88-3675 (1988) and T. Herbert, "Secondary Instability of Boundary Layers," *Ann. Rev. Fluid Mech.* **20**:487–526 (1988).

[3] More extended treatments may be found in R. Betchov and W. Criminale, *Stability of Parallel Flows*, Academic Press, New York, 1967, and in P. G. Drazin and W. H. Reid, *Hydrodynamic Stability*, Cambridge University Press, 1981. There is also an excellent chapter on stability in C-S. Yih, *Fluid Mechanics*, West River Press, Ann Arbor, Mich. 1977.

Section 13.2 presents a preliminary discussion of the concept of instability, and of the roles played by vorticity and viscosity, in determining how a given flow will respond to an accidental disturbance. Section 13.3 attempts to identify, without formal mathematics, the typical characteristics of unstable distributions of velocity and vorticity. Section 13.4 presents the general outline of a mathematical inquiry into the stability of a steady flow that is subject to infinitesimal perturbations, and shows when and how attention can be focused on elementary disturbances called *normal modes*. It also presents a significant classification of normal-mode instabilities, introducing the concepts of *convective, absolute,* and *global instability*.

Section 13.5 introduces the normal-mode analysis of *parallel shear flows*, based on the *Rayleigh equation* and the *Orr–Sommerfeld equation*. It presents some classic integral theorems, based on the Rayleigh equation, which refine one's ability to identify basic flows that are likely to be unstable, and that limit the possible phase velocities and amplification rates of unstable normal modes. It ends with Squire's theorem, based on the Orr–Sommerfeld equation, which further limits the parameter space in which one must search to find the most dangerous disturbances.

Section 13.6 illustrates, with an approximate analysis due to Lord Rayleigh, how the wavelength of a normal-mode disturbance affects its amplification. It introduces the important concepts of a *cutoff wavelength* and a *most dangerous disturbance*.

Section 13.7 shows how the Rayleigh and Orr–Sommerfeld equations may be solved numerically with a simple adaptation of the program used to solve the laminar boundary-layer equations, provided that the velocity profile of the basic flow exhibits certain symmetries. The most dangerous disturbances of free shear layers, jets, and plane Poiseuille flow are analyzed with this computational tool, with special emphasis on understanding both the stabilizing and destabilizing effects of viscosity. References are given for other numerical techniques that are complicated but more powerful.

Section 13.8 describes corresponding theoretical results for a number of flows, especially boundary layers, and discusses the relevance of the theory to an understanding of experimental observations.

Section 13.9 briefly considers the instability of basic flows with curved streamlines. A different class of most dangerous disturbance, the *Taylor–Görtler vortex*, is described. The simplest example of Taylor vortex flow is analyzed in detail. A fairly extensive introduction to the modern literature on Görtler vortices is given. The section ends with a discussion of crossflow instability in three-dimensional boundary layers.

Section 13.10 briefly introduces a very different but important kind of instability, called *viscous fingering*, which appears when a less viscous fluid is used to drive a more viscous one out of a porous medium or Hele–Shaw apparatus.

In Section 13.11 a series of questions left unanswered by the simplest applications of linear theory is addressed, and the concepts of *subcritical* instability and *secondary instability* are introduced. The *Floquet analysis* of secondary instabilities in boundary layers is sketched.

Finally, Section 13.12 examines the relevance of stability theory for the prediction of transition. The concept of the *receptivity* of a convectively unstable flow to various kinds of incoming disturbances is presented. Typical measured characteristics of flow in transition to turbulence are described, as are numerical simulations based on the full Navier–Stokes equations. The chapter ends with an extremely brief introduction to some of the concepts and terminology of *chaos theory*,[4] and a guess as to the future role of that theory in analysis of the fate of unstable laminar flows.

13.2 VORTICITY AND INSTABILITY

One of the themes of this chapter is that instability of the simple flows that we have analyzed can be blamed on the presence of vorticity, and understood by studying the evolution of the vorticity field. The question of instability becomes this: Will the future evolution of a specified initial distribution of vorticity be disproportionately affected if that initial distribution is infinitesimally altered?

The potential usefulness of this approach is suggested by the Biot–Savart law,[5] which is repeated here for convenient reference:

$$\mathbf{u}_V(\mathbf{r}, t) = \frac{1}{4\pi} \iiint_V \xi^{-3} \mathbf{\Omega}(\mathbf{r}', t) \times \xi \, dV(\mathbf{r}')$$

in which $\xi = \mathbf{r} - \mathbf{r}'$. This law shows explicitly how the velocity induced on one particle of vorticity-bearing fluid depends on the precise location of other vorticity-bearing volume elements. In most of the sample flows presented in earlier chapters, vorticity has been quite widely distributed throughout space, and the simplicity of the resulting velocity field is made possible by a very precise cancellation of the contributions to $\mathbf{u}_V(\mathbf{r}, t)$ that come from symmetrically disposed volume elements $dV(\mathbf{r}')$. Thus, when the distribution of vorticity is just right, the contributions to the Biot–Savart integral that would tend to move a particle sideways in a fully-developed laminar pipe flow, although perhaps individually large, add up to zero. Thus, the particle is left to move straight down the pipe.

It is easy to imagine that the precise balance that allows a simple flow is

---

[4] See J. Glieck, *Chaos, The Making of a New Science,* (1987) for a well-written account intended for nonspecialized, but scientifically informed, readers. Just be warned that he totally disregards the connection between the new science and the older theory of instability. He also completely ignores the contribution of engineers to our knowledge of instability and turbulence. This may seen a little narrow-minded, but perhaps it is good for us to be kept humble.

[5] You may wish to review Sections 3.4 and 8.11, especially Fig. 8.3.

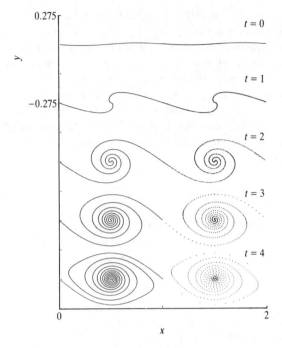

FIGURE 13.1
Rollup of an unstable vortex sheet. (Reproduced, by permission, from R. Krasny, (1986).[6])

precarious, and that any small disturbance of the vorticity field will lead to an aggravated convective rearrangement of the vorticity. It is not easy to imagine what happens next; whether there will be a tendency for the vorticity to return to its unperturbed configuration, or whether the initial perturbations will tend to grow. That is, briefly, what this chapter is all about.

Figures 13.1 and 13.2 exhibit computed examples of these ideas. The stability problem illustrated in Fig. 13.1 is an old classic; that illustrated in Fig. 13.2 will be unfamiliar to most readers, even quite experienced ones.

For the first example, the unperturbed distribution of vorticity is represented by a rectilinear array of equally spaced, equally strong vortex blobs. Each blob has a radially symmetric concentration of vorticity, which falls to zero outside some small radius δ. The array of blobs goes off to $x = \pm\infty$, so that in the unperturbed state the velocity induced on each blob by its neighbor on the right is exactly canceled by its neighbor on the left, and so on. Clearly, there is no tendency for any blob to move. To initialize the calculation, the blobs are rearranged slightly, to fall approximately on a sine curve of an arbitrarily chosen length.[6] For the case shown, this wavelength greatly exceeds the radius of a blob.

---

[6] The details are not essential to us yet, but they are crucially important to the success of the calculation. See R. Krasny, "Desingularization of Periodic Vortex Sheet Roll-Up," *J. Comp. Phys.* **65**:292–313 (1986).

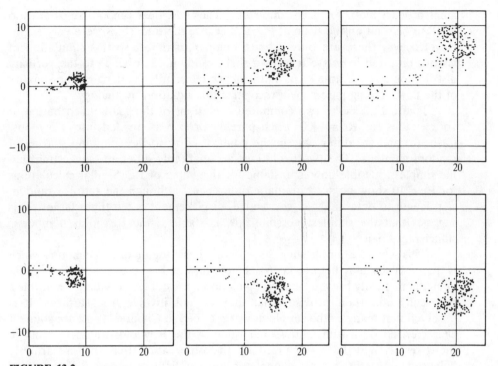

FIGURE 13.2
Wandering of an unstable vortex pair. (Reproduced, by permission, from M. J. Dooley, 1988.[8])

As time advances, each blob is displaced convectively by the velocity induced at its center by all the remaining blobs. No attempt is made to simulate diffusive effects associated with viscosity. As we see, the sine wave grows in amplitude and each cycle of the wave evolves into a jelly roll. We would say that the initial configuration is unstable to the specified initial disturbance, and you will learn here how to characterize, at least for small times, the growth of that disturbance.

The second example involves both convection and diffusion. Initially, the vorticity is concentrated into two equally strong, counterrotating line vortices. In the absence of diffusion, the two vortices would move steadily along parallel lines as along a train track, at a speed $U = \Gamma/4\pi H$, where Γ is the circulation of the positive vortex, and $2H$ is the initial distance between vortices. When diffusion is allowed, so that each vortex spreads out as it moves, the simplest possible flow is fairly complicated, and probably cannot be described in terms of simple functions. However, we can safely assume that it would show certain symmetries about a line halfway between the railroad tracks. The flow is characterized by the Reynolds number, $Re = \Gamma/\nu$, and its evolution and stability presumably depend on the value of Re.

For the computation, the strength of each line vortex is equipartitioned among a large number of vortex blobs, and viscous diffusion of vorticity is simulated by adding a small random displacement to the convective displacement of each blob, for each time step.[7] The individual random steps are too small to be seen on the scale of Fig. 2, because the time steps were very small. Nevertheless, there are only a finite number of blobs, so this simulation of viscous diffusion is *noisy*. It gives a slight random disturbance to the vorticity distribution at each time step, and is hence constantly challenging the stability of the flow. Young parents will recognize this situation by analogy!

Figure 13.2 shows two computed evolutions of the spatial distribution of vortex blobs for Re = 1000. Each parent vortex was divided into 125 blobs. Cases (a) and (b) differ only in the point of entry to the computer's string of random numbers. The two evolutions are markedly different, suggesting that the simplest possible flow is unstable for this value of Re. Similar calculations for Re = 10 show no such dramatic differences, although the ratio of random step length to deterministic, convective, displacement is ten times larger. This suggests that the simplest possible flow is stable if the Reynolds number is sufficiently small.[8]

This second example suggests a second basic theme of this chapter, which is that viscosity may help to keep a simple flow stable. It is observed, and can be mathematically proved, that creeping flows are stable, providing—and this a practically important proviso—that they do not involve free surfaces. It is observed that many of the simple flows described in Chapters 9–12 are stable if characterized by a sufficiently small value of the Reynolds number. Most of these results may be rationalized by the observation that the precariously balanced convective rearrangements of vorticity fields, the potential for which is seen in the Biot–Savart law, may be overwhelmed by diffusive rearrangements when the Reynolds number is small, especially if the spatial scale of the rearrangement is small. You will find this idea expressed in different ways as the chapter develops, and you will find exceptions, in boundary layers and channel flows, where the diffusive rearrangement of vorticity aids and abets the convective rearrangement, and hence is destabilizing.

13.3 EXAMPLES OF UNSTABLE ARRANGEMENTS OF VORTICITY

Disturbed Equilibrium of a Row of Vortices

In the first example shown above, the instability of the original arrangement of vortex blobs is easy to understand. As a preliminary step, consider the velocity field induced by a single array of line vortices, placed at $y = 0$ and at $x = \pm n\lambda$, where $n = 0, 1, 2, 3, \ldots$. Each vortex has the same circulation, Γ. The

---

[7] You may want to review Section 8.13, where the random-vortex simulation is described.

[8] Figures taken from thesis work by M. J. Dooley, Department of Mechanical Engineering, University of California, Berkeley, 1988.

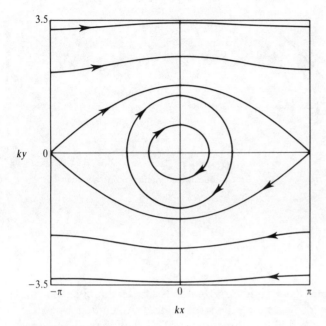

FIGURE 13.3
Streamlines associated with an infinite array of identical, equally spaced line vortices. Note the "cat's eye" pattern.

streamfunction of the flow induced at point $P(x, y)$ by the nth vortex is

$$\Psi_n(x, y) = -\frac{\Gamma}{4\pi} \ln \left[(x - n\lambda)^2 + y^2\right]$$

Superposing the contributions of all the vortices in the array, we get an infinite sum that can be represented by the closed-form expression

$$\Psi(x, y) = -\frac{\Gamma}{4\pi} \ln \left\{\cosh(ky) - \cos(kx)\right\}$$

where $k = 2\pi/\lambda$ is the *wave number* of the array. Figure 13.3 shows one wavelength of the corresponding steamline pattern.

Note that there is just one way in which a second array of vortices, just like the first, can be placed so as to remain stationary vis-à-vis the first array. Each vortex must be placed at one of the stagnation points midway between the original vortices. You can see from the streamline pattern that this equilibrium configuration is unstable; the slightest displacement will cause the two arrays to begin to shift their relative positions. This is shown in Fig. 13.4, a closeup view of the vicinity of a stagnation point, where the streamfunction is

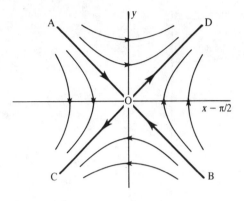

FIGURE 13.4
Closeup of stagnation point between corotating vortices.

given locally by

$$\Psi(x, y) \approx -\frac{\Gamma}{4\pi}[(ky)^2 - (kx')^2] \quad \text{where} \quad x' = x - n\frac{\lambda}{2}$$

The concept of a *most dangerous disturbance* is also exemplified in this picture. From the direction of flow along the streamlines, we see that a particle displaced precisely along the dividing streamline AOB would tend to come home, but that one displaced along COD would tend to move directly away from home. Thus, we call the latter displacement or disturbance more dangerous than the former.

It is quite easy to show that for small times,

$$R^{-1} dR/dt = -(k^2\Gamma/2\pi) \sin 2\theta,$$

where R and θ represent the particle position in polar coordinates centered at the stagnation point. For the most dangerous initial displacement, $d\theta/dt = 0$, so the displacement from equilibrium, R, grows exponentially with time, $R(t) = R(0) \exp(\sigma t)$. The *amplification rate*, $\sigma = k^2\Gamma/2\pi$, is larger, the stronger the vortices and the closer they are together.

This little analysis is, except for its simplicity, fairly typical of what you will encounter in this chapter. It identifies the possibility of instability, the most dangerous initial disturbance, and an initial exponential growth rate for that disturbance. It also exemplifies the common fact that growing departures from the equilibrium state soon depart from a simple exponential growth law.

A Sinusoidally Rippled Vortex Sheet

Consider now a vortex sheet, separating two initially uniform, counterflowing streams. When undisturbed, it occupies the plane $y = 0$, and each material particle of it is at rest. We impose on it an initial disturbance in which each material particle of the sheet is displaced from its initial position $(x_0, 0)$ to a new position $x = x_0 - A \sin kx$, $y = A \sin kx$. This displacement along 45° lines is suggested by the discovery we made above about the most dangerous initial

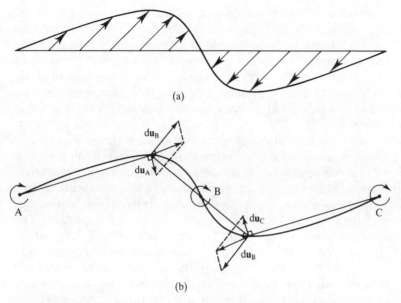

FIGURE 13.5
Perturbed vortex sheet: (a) most dangerous initial displacements; (b) kinematic demonstration of instability.

perturbation of a row of distinct vortices. The perturbed situation is sketched in Fig. 13.5(a).

Note that this disturbance not only corrugates the sheet, but that it changes the distribution of circulation per unit length along the x axis. In a mild sense, it causes the vorticity to bunch up around alternate nodal points of the sine wave, and to be depleted around the intervening nodal points. The effect of this disturbance on the subsequent motion of the sheet is illustrated in Fig. 13.5(b), for a point at the crest of the ripple. The vorticity at node A induces a downward velocity, with a small component to the right. That at node B induces an upward component, also with a small component to the right. Because the crest is slighly closer to the vorticity distributed along PA than to that distributed along PB, the fluid at P is induced to move to the right, and slightly upward. This argument can be extended to include, two-by-two, the contributions of all bits of circulation attached to material elements that were, before the disturbance, at equal distances left and right from the undisturbed position of the crest point, and the integrated result is the same. For very small times, each material particle continues to move along a 45° line, with a speed proportional to distance from its undisturbed position.

The dependence of the initial amplification rate on the parameters that describe the base flow and the disturbance can be obtained partially, for this very simple case, by dimensional analysis. The undisturbed base flow is characterized only by the velocity jump, U, across the vortex sheet. The initial disturbance is characterized by its wavelength, λ, and by the angle, θ, of the trajectory along which each particle is displaced. In a linear theory, the amplitude A of the initial disturbance has no effect on the value of the amplification factor σ, so it is left out of the dimensional analysis. The value of σ is then dependent on the values of U, λ, and θ. The dimensions of σ are T^{-1}, and the only reference time presented by the given parameters is λ/U. Then, we must find $\sigma = (\lambda/U)f(\theta)$. From a detailed analysis,[9] it turns out that $f(\theta)$ is largest when $\theta = 45°$.

From this simple exercise in picture-drawing, we have almost proved that a flat vortex sheet is unconditionally unstable, in that there is no range of values of U or λ in which the initial disturbance will not grow. Of course, infinitely thin vortex sheets are only an idealized abstraction; the vorticity in a real shear layer is spread out over a finite y-distance, say δ. Also, diffusion of vorticity, characterized by the kinematic viscosity ν, may be important. The result of dimensional analysis will then be $\sigma = (\lambda/U)f(\theta, \delta/\lambda, \text{Re})$, where $\text{Re} = U\delta/\nu$.[10] It turns out, as you will learn in subsequent sections, that there is then a *short-wave cutoff* and a *low-Re cutoff*, meaning that disturbances will not grow if λ/δ is too small, or if Re is too small. However, any flow in which the vorticity is concentrated in a manner that even vaguely resembles our vortex sheet is likely to be conspicuously unstable. This is why a designer worries about the stability of boundary layers in adverse pressure gradients; they exhibit a maximum of vorticity away from the wall, and do therefore vaguely resemble our sheet.

If you have absorbed the implications of this analysis of the infinitely thin sheet, you may well wonder how the computation that produced Fig. 13.1 was carried out. It shows the smooth evolution of an initial sinusoidal disturbance of the type specified above, with a finite value of the disturbance wavelength. Why is the sheet not wrinkled by the growth of disturbances of shorter wavelength, which have, according to dimensional analysis, a larger amplification rate?

The answer is partly inherent in the basic computational simulation; the elements of which the sheet is composed have a finite radius, so that the sheet itself has an effective finite thickness. This imposes a short-wave cutoff, but not one that would guarantee the sort of smooth behavior we see. The rest of the answer has to do with meticulous care to avoid the introduction of short-wave disturbances through truncation and roundoff errors inherent in the numerical determination of the trajectory of each sheet element. This is the sort of thing that can be done in a computational experiment, but is very hard to reproduce in a laboratory experiment.

---

[9] R. E. Conte, "Etude Analytique des Nappes Tourbillonnaires," Doctoral thesis, l'Université Pierre et Marie Curie, Paris, 1975. See pp. V.6–V.8.

[10] By convention, the Reynolds number is almost always based on a characteristic length of the base flow, rather than of the disturbance.

Vorticity Distributed Between Coaxial Rotating Cylinders

Only rarely is the vorticity of the basic flow so simply distributed that one can perform a back-of-the-envelope evaluation of implications of the Biot–Savart law. Often it is easier to work with concepts of force and momentum, as in the famous heuristic argument to be presented now. The question at hand is the stability of flow with continuously distributed vorticity, and with curved streamlines. Viscosity is again neglected, and the base flow is assumed steady, with velocity $\mathbf{u} = v(r)\mathbf{e}_\theta$ in cylindrical polar coordinates. This could represent flow between coaxial rotating cylinders, with the function $v(r)$ depending on the rates and directions of rotation of the inner and outer cylinders. The fluid particles are kept on circular streamlines by the radial pressure gradient

$$\frac{\partial p}{\partial r} = \rho\frac{v^2}{r} = \rho\frac{\Gamma^2}{r^3} \qquad \text{where} \qquad \Gamma \equiv rv$$

Imagine, skeptically, that the fluid initially occupying a small toroidal streamtube at radius r can move outward to fill a new torus at radius $r + dr$ without perturbing the pressure field or altering the circulation around the material loop that is the axis of the torus. The latter stipulation is suggested by Kelvin's theorem. See Fig. 13.6 for the mental picture of this process.

At the new position of the torus, the radial pressure gradient will be $\rho[\Gamma(r + dr)]^2/(r + dr)^3$, and the centrifugal force on unit volume of the torus will be $\rho[\Gamma(r)]^2/(r + dr)^3$. The inward pressure force will exceed the outward centrifugal force, and the torus will tend to go home, if

$$\frac{d(\Gamma^2)}{dr} > 0 \tag{13.1}$$

The distribution of Γ in the basic flow is

$$\Gamma = \frac{\Gamma_1(r_2^2 - r^2) - \Gamma_2(r^2 - r_1^2)}{r_2^2 - r_1^2}$$

where r_1 is the inner radius, and r_2 is the outer radius. The criterion for the

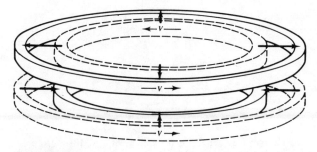

FIGURE 13.6
Radially-displaced rings of fluid, testing for centrifugal instability. Two rings are shown, as a reminder of the continuity equation.

existence of a restoring force is then

$$(\Gamma_2 - \Gamma_1)\{\Gamma_2(r^2 - r_1^2) + \Gamma_1(r_2^2 - r^2)\} \geq 0 \tag{13.2}$$

This will be satisfied, and the analysis suggests that the flow will be stable against this sort of disturbance, if the inner cylinder is stationary and the outer is rotating. The opposite case is predicted to be unstable, and there are interesting predictions for cases in which both cylinders are spinning. For example, if the cylinders are counterrotating, with $\Gamma_2 = -\Gamma_1$, the flow should be stable where r^2 is greater than $(r_2^2 + r_1^2)/2$, and unstable where r^2 is less than this value.

Considering the rough-and-ready nature of this argument, it is astonishingly successful. All the results it suggests are observed, although no account is taken of viscosity, and no precise specification of the nature of the disturbance is given. It turns out that the initally most dangerous disturbances develop into another steady and rather spectacular flow, which is shown in Fig. 13.7. The cellular motions that seem to be stacked like doughnuts are called *Taylor vortices,* in honor of Sir Geoffrey Taylor. Similar streamwise vortices are

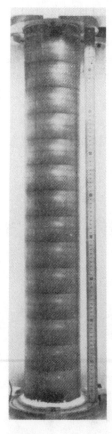

FIGURE 13.7
Taylor vortices shown by reflection of light from tiny disks. (Reproduced, by permission, from J. E. Burkhalter and E. L. Koschmieder, *J. Fluid Mech.* **58**: 547–560, 1973.)

observed in laminar boundary layers along concave walls. They are called *Görtler vortices*, and the phenomenon is often called *Taylor-Görtler instability*, or *centrifugal instability*, wherever it is observed. It will be discussed at somewhat greater length in Sections 13.8 and 13.9. In the meanwhile, we simply note that a slight longitudinal wall curvature, far too small to affect a noticeable change in the simplest possible flow along that wall, may very substantially affect the stability of that simple flow.

13.4 MATHEMATICAL ANALYSIS OF THE INSTABILITY OF STEADY, PLANE FLOWS

Many of the main ideas of stability theory appear in the simple context of a flow in which the velocity has only two scalar components, both before and after perturbation. The unperturbed flow is often called the *basic flow*; its velocity and vorticity are symbolized by $\mathbf{U}$ and $\boldsymbol{\Omega}$. If it is a plane, steady flow, it may be represented by $\mathbf{U} = U(x, y)\mathbf{e}_x + V(x, y)\mathbf{e}_y$, so that $\boldsymbol{\Omega} = \Omega(x, y)\mathbf{e}_z$. It is supposed that $\mathbf{U}$ and $\boldsymbol{\Omega}$ satisfy the equations of motion, specifically the vorticity equation

$$U\frac{\partial\Omega}{\partial x} + V\frac{\partial\Omega}{\partial y} = \nu\,\nabla^2\Omega \tag{13.3}$$

and the continuity equation

$$\frac{\partial U}{\partial x} + \frac{\partial V}{\partial y} = 0 \tag{13.4}$$

To investigate the stability of the basic flow, one asks what will happen to a second flow that has, at time zero, slightly different distributions of velocity $\mathbf{U}' = \mathbf{U} + \mathbf{u}(x, y, 0)$, and vorticity $\boldsymbol{\Omega}' = \boldsymbol{\Omega} + \zeta(x, y, 0)$. Specifically, one studies the evolution of the perturbations $\mathbf{u}$ and ζ, requiring that $\mathbf{U}'$ and $\boldsymbol{\Omega}'$ satisfy the equations of motion and that they are constrained by the same flow-forcing boundary conditions as are $\mathbf{U}$ and $\boldsymbol{\Omega}$. The precise meaning of these statements will be made clear as the analysis is developed. For now, you have only to imagine that the perturbations may propagate like a ripple on a pond, or that they may stay where they are initiated, and that in either case they may amplify or decay. In some situations, it may be quite tricky to identify the boundary between amplification and decay, but you can worry about that later.

The vorticity and continuity equations for the perturbed flow are

$$\frac{\partial\Omega'}{\partial t} + U'\frac{\partial\Omega'}{\partial x} + V'\frac{\partial\Omega'}{\partial y} = \nu\,\nabla^2\Omega'$$

and

$$\frac{\partial U'}{\partial x} + \frac{\partial V'}{\partial y} = 0$$

Subtracting Eqs. (13.3) and (13.4) from these, we get equations for the perturbations,

$$\frac{\partial \zeta}{\partial t} + U \frac{\partial \zeta}{\partial x} + V \frac{\partial \zeta}{\partial y} + u \frac{\partial \Omega}{\partial x} + v \frac{\partial \Omega}{\partial y} - v \nabla^2 \zeta = -\left(u \frac{\partial \zeta}{\partial x} + v \frac{\partial \zeta}{\partial y} \right) \qquad (13.5)$$

and

$$\frac{\partial u}{\partial x} + \frac{\partial v}{\partial y} = 0 \qquad (13.6)$$

Noting that the presumably small quantities u, v, and ζ appear only linearly on the left-hand side of (13.5), but that they appear in products on the right-hand side, we start the theoretical inquiry with linearized equations, replacing the right-hand side of (13.5) by zero. In the resulting *linear theory of stability*, much use is made of the notion that the functional form of an arbitrary initial distribution of z can be constructed by superposition of simpler, building-block functions, each having its own temporal evolution. If it can be shown that any one of the building-block, or elementary, perturbations amplifies spontaneously, one concludes that the basic flow is unstable. The term, *most dangerous disturbance*, can now be defined mathematically. It may signify either the first elementary disturbance that exhibits amplification as the basic flow is brought out of a stable regime by gradual increase of the Reynolds number, or the most rapidly growing elementary disturbance at some higher value of Re.

Normal-Mode Analysis

Suppose that the basic flow is steady. Then the coefficients, U, $\partial \Omega / \partial x$, etc, in the linearized version of Eq. (13.5) are independent of t, and solutions of the form $\zeta(x, y, t) = Z(x, y) \exp(-i\omega t)$, $\mathbf{u}(x, y, t) = \mathbf{Y}(x, y) \exp(-i\omega t)$ are possible. In these, ω is a constant, which may be real, imaginary, or complex.[11] These particular solutions, or special varieties of them, are called *normal-mode solutions*. Other solutions to an arbitrary initial-value problem can be found by use of Laplace transforms; the solutions so constructed usually evolve into, or are dominated by, normal modes after a relatively brief starting transient.

It is expeditious to introduce a disturbance streamfunction, $\psi(x, y, t)$ such that

$$\mathbf{u} = \mathbf{e}_z \times \nabla \psi \qquad \text{and} \qquad \zeta = -\nabla^2 \psi$$

We then set $\psi(x, y, t) = \phi(x, y) \exp(-i\omega t)$, whereupon the equation for the disturbance vorticity becomes

$$\left(-i\omega + U \frac{\partial}{\partial x} + V \frac{\partial}{\partial y} - v\nabla^2 \right) \nabla^2 \phi = \left(\frac{\partial \Omega}{\partial x} \right) \left(\frac{\partial \phi}{\partial y} \right) - \left(\frac{\partial \Omega}{\partial y} \right) \left(\frac{\partial \phi}{\partial x} \right) \qquad (13.7)$$

---

[11] The inclusion of $-i$ along with ω is a convention, having no immediate physical meaning.

The boundary conditions on ϕ are $\phi = 0$ and $\mathbf{n} \cdot \nabla \phi = 0$ on the complete boundary of the (x, y) domain of interest, if that boundary is a solid wall. This would focus the investigation on the question, "Can perturbations grow inside a bounded region, while vanishing on its boundary?" The answer would be yes, if we could find solutions for which the imaginary part of ω were positive.

Unfortunately, the solution of Eq. (13.7) for such a situation would be, in general, very difficult. It is a fourth-order partial differential equation, considerably more complicated than the equation for the Oseen approximation to creeping flow. To make things worse, the desired solutions are eigensolutions, which exist only for special values of parameters that characterize the building-block solution and the basic flow. The search for all possible eigenvalues can be very difficult. Consequently, the vast bulk of stability theory deals with basic flows that are imagined to extend from $x = -\infty$ to $x = +\infty$, and in which the velocity field is either rigorously or approximately independent of x. These basic flows are called *parallel flows*.

In a steady, rectilinear, parallel flow, we can align coordinate axes so that $V = 0$, and will then find that $\partial \Omega / \partial x = 0$. U and Ω will depend only upon y. Equation (13.7) reduces to

$$\left(-i\omega + U\frac{\partial}{\partial x} - v\nabla^2\right)\nabla^2\phi = -\left(\frac{d\Omega}{dy}\right)\left(\frac{\partial\phi}{\partial x}\right) \tag{13.8}$$

Since the coefficients of this equation do not depend on x, solutions that depend on x as $\phi(x, y) = f(y) \exp(ikx)$ can be found.[12] For them, $\nabla^2\phi = (-k^2 + D^2)f \exp(ikx)$, where the symbol D denotes the operation d/dy. The remaining unknown function, f, is governed by an ordinary differential equation

$$[-i\omega + ikU + v(k^2 - D^2)](-k^2 + D^2)f + ik\left(\frac{d\Omega}{dy}\right)f = 0 \tag{13.9}$$

Let U_0 and H be the characteristic velocity and length of the basic flow. For example, for flow between parallel walls, they might be the mean velocity and the distance between the walls. The base-flow Reynolds number will be $\mathrm{Re} = U_0 H / v$. We define dimensionless disturbance parameters $\alpha = kH$, $\sigma = \omega H / U_0$; dimensionless independent variable $\eta = y/H$; and dimensionless variable coefficients $U = U^*/U_0$, $U'' = -(H^2/U_0)(d\Omega/dy)$. In terms of these dimensionless quantities, and with primes to denote $d/d\eta$, we have[13]

$$(\alpha U - \sigma)(f'' - \alpha^2 f) - \alpha U''f + i\,\mathrm{Re}^{-1}[f^{\mathrm{iv}} - 2\alpha^2 f'' + \alpha^4 f] = 0 \tag{13.10}$$

This is called the Orr–Sommerfeld equation, in honor of two of the early investigators of stability theory. Without the viscous terms, i.e., in the limit

---

[12] Again, we include the factor i along with k, to be loyal to convention.

[13] Traditionally, these equations are divided through by α, because in much of the early theory, α was taken to be real. This is no longer the case, so we leave it in the numerator for now.

Re→∞, it is called the Rayleigh equation.[14] For subsequent analysis, it is convenient to employ a dimensionless vorticity perturbation, $g = \alpha^2 f - f''$, and to replace (13.10) by two, equivalent, coupled second-order equations.

$$(D^2 - \alpha^2)g - i\alpha \, \text{Re} \, \{(U - c)g + U''f\} = 0 \qquad (13.11)$$

and
$$(D^2 - \alpha^2)f + g = 0 \qquad (13.12)$$

in which $c = \sigma/\alpha$, and D denotes $d/d\eta$.

For a stability investigation, we seek solutions of (13.10) that vanish at the boundaries of the domain of interest. That is, we set $f(y_1) = f(y_2) = 0$; $f'(y_1) = f'(y_2) = 0$, where y_1 and y_2 may be finite or infinite.

In several interesting cases, y_1 and/or y_2 are infinite, but U is constant outside of a limited interval of y. Then there will be an asymptotic solution

$$f \approx A \exp(\lambda_1 y) + B \exp(\lambda_2 y) + C \exp(\lambda_3 y) + D \exp(\lambda_4 y) \quad (13.13)$$

when y is in a domain of constant U. The four lambdas are roots of the quartic equation

$$(\lambda^2 - \alpha^2)[(\lambda^2 - \alpha^2 - i\alpha \, \text{Re} \, (U_\infty - C)] = 0 \qquad (13.14)$$

Each root will ordinarily be a complex number, but the value of its real part will show whether the corresponding term in (13.13) vanishes with increasing distance away from the region where there is vorticity. Only terms that do so can be retained for the solution.

Eigensolutions. The differential equation and the boundary conditions imposed on f are all homogeneous, so that the solution for arbitrary combinations of the parameters α, c and Re is just $f = 0$. To avoid this, there must be a very special relationship between the parameters. If values are specified for two of the parameters, a nonzero solution for f will exist only if the third parameter has a special value or values. The special values are called *eigenvalues*; the corresponding solutions are called *eigenfunctions*. We shall be especially interested in how the eigenvalues of α and c depend on the value of Re, and on the shape of the velocity profile $U(y)$.

Definition and Classification of Instabilities

An elementary disturbance of the form

$$\psi(x, y, t) = f(y) \exp[i(kx - \omega t)] \qquad (3.15)$$

can be rewritten, for complex values of k and ω, in a form that exhibits both amplification and propagation. Using subscripts R and I to denote real and

---

[14] In honor of Lord Rayleigh, who proved important theorems about its solutions.

imaginary parts, we get

$$\psi(x, y, t) = \exp(\omega_I t - k_I x)[F(y) \cos(k_R x - \omega_R t - \theta(y)] \qquad (13.16)$$

Here, F and θ are the modulus and argument of the presumably complex quantity $f(y)$. If all of the disturbance parameters, ω_R, ω_I, k_R, and k_I, are nonzero, the disturbance is rather hard to visualize. At a fixed point, its dependence on time would be that of a growing or decaying oscillation. The same is true of the variation with x at fixed values of y and t. In laboratory work, one must contend with the fact that many disturbances of this type are likely to be superposed at any given time, or in any given region. To minimize the resulting confusion, special efforts are often made to reduce the accidental seeding of disturbances, while attention is focused on the evolution of artifically created disturbances of special types.

A special, idealized, disturbance that is very useful for conceptual discussions and for the classification of different kinds of instability is rather analogous to that produced by dropping a pebble into a pond. Think, if you will, of the point force that produced the self-similar jet motion of Chapter 9. Make that force infinitely strong, but let it act only for an infinitesimal instant of time. Apply it at a point in a boundary layer, or in some other flow whose stability is in question, and see what happens.

One thing that may happen is directly analogous to what you see when you toss a stone into a river. A set of waves radiates away from the splash point, and is washed downstream if the river is fast enough. After a short time, everything is back to normal. If you were curious, you could fly over the river in just the right way to keep your eye on some patch of the spreading, convecting wave pattern, to see what happened to it as it moved away from the origin of the disturbance. If the river current were uniform, the waves seen in your patch would gradually die away, as a result of viscous dissipation of the extra water motions associated with them. A similar thing might happen after the application of our point impulse to a laminar boundary layer. If so, we would call the layer stable, believing that such an impulse would excite every possible elementary disturbance of the form given by (13.13).

Convective Instability and Absolute Instability

A close approximation to this thought-experiment has been carried out for the unstable boundary layers in an otherwise very quiet environment.[15] The disturbance was caused by a sudden pressure pulse, applied through a tiny hole in the wall. The result was like the spreading ripples on the river, except that the ripples that moved downstream amplified as they went along. The other

---

[15] M. Gaster and I. Grant, "An Experimental Investigation of the Formation and Development of the Wave Packet in a Laminar Boundary Layer," *Proc. Roy. Soc. London* **A347**:253–269 (1975).

difference is that the dominant wavelength of the growing ripples in the boundary layer was determined by selective amplification and interference of component elementary disturbances, and had nothing to do with the size of the pressure hole. After a short time, the boundary layer over the hole recovered its orignal state, while the growing disturbance moved away downstream. In modern terminology, we say that the experiment demonstrated *convective instability*.[16]

Finally, if the disturbance were to be injected into the flow between coaxial rotating cylinders, it might either die away everywhere, indicating stability of the basic flow, or it might spread throughout the entire body of fluid, eventually changing the state of motion everywhere. In the latter case, we speak of *absolute instability*.

The distinction between these types of instability can be remembered with the aid of Fig. 13.8, which shows how a disturbance spreads in the $x-t$ plane. In either kind of instability, the growing response to a point impulse is initially confined to a wedge in this plane. If both edges of the wedge move downstream (lean to the right in the figure), the instability is convective, and the undisturbed state will eventually reappear at any value of x. If the two edges have opposite slopes, the instability is absolute; the disturbance will eventually appear at any value of x. It turns out that this classification scheme has important implications for the experimental study of instabilities; one has, for example, to be very cautious when introducing any sort of measuring probe into an absolutely unstable flow. On the other hand, the role of impressed disturbances, from noise, free-stream vorticity, wall roughness, etc., is much more clearly seen in convectively unstable flows.[17]

Global Instability

In certain circumstances, a convective instability is associated with a feedback mechanism, in which the amplifying wave packet interacts with an obstacle downstream, so that a transient signal is returned to the point of origin of the disturbance. This may be sufficient to trigger a new disturbance, which moves downstream, interacts with the obstacle, sends back another triggering signal, and so on. The result may be a very clean, self-sustaining oscillation that looks rather like a deliberately-stimulated convective instability, but which would continue after any deliberate stimulation was ended. This has come to be called a *global instability*. Again, the distinction is important to efforts to understand how, if at all, instabilities and transition may be controlled for some desired effect. The nature of the interaction, and of the process by which

---

[16] See Drazin and Reid (1981) op cit., p. 152, where the term is *convected* instability.

[17] Much of this discussion comes from P. Huerre and P. A. Monkewitz, "Absolute and Convective Instabilities in Free Shear Layers," *J. Fluid Mech.* **159**:151–168 (1985).

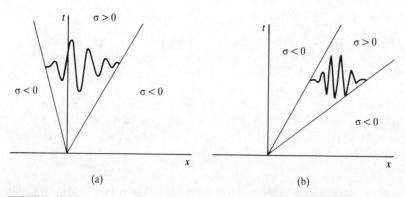

FIGURE 13.8
Definition of convective and absolute instability: (a) absolute; (b) convective. (Reproduced, by permission, from P. Huerre and P. Monkewitz 1985.[17])

the returning signal triggers a new instability, is an interesting subject of current research.[18]

Temporal Instability and Spatial Instability

There are five real parameters, characterizing the mean flow and the normal-mode disturbance. If the disturbance amplitude is not to be identically zero, these parameters must be connected by a *dispersion equation*, $F(\text{Re}, \alpha, \sigma) = 0$. Since this is a relationship involving complex quantities, it is actually two equations connecting the five real parameters. Thus, a search for eigenvalues always starts with an assignment of values to three of the parameters, and ends with the determination of those values of the other two parameters that cause the real and imaginary parts of the function F to vanish.

Much of the early work on stability theory starts with the choice $\alpha_I = 0$, so that the normal-mode disturbance is a simple sine wave in x, the amplitude of which may grow with time. The other parameters to which values are assigned are α_R and Re, and one seeks the eigenvalues of c_R and c_I. If the value of c_I proves to be positive, the flow is said to exhibit *temporal instability*. This is the obvious choice for the analysis of an absolutely unstable flow. This choice is also often made for simplicity, even when the flow is convectively unstable, but then the results of the analysis need to be translated before they can be compared to experiment. We shall return to this point in a moment.

In experimental studies of convectively unstable flows, such as boundary layers, controlled disturbances are often introduced along a line of constant x, by vibrating a tiny ribbon sinusoidally in time. To model this theoretically, it is proper to set $\sigma_I = 0$, and to assign values to σ_R and Re, seeking the eigenvalues

---

[18] See Morkovin (1988) op cit. for a fascinating discussion of this issue.

of α_R and α_I. If the value of α_I proves to be negative, the flow is said to exhibit *spatial instability*. You will also encounter these words as adjectives, when authors write of a *temporal-instability analysis* or a *spatial-instability analysis*.

As is hinted above, it is somewhat easier to analyze the Orr–Sommerfeld equation, or the Rayleigh equation, for temporal instability. You might guess this by noting that α appears more frequently in the equation than does σ; thus, it is relatively advantageous if α is purely real. When this is done for a convectively unstable flow, for which spatially growing modes are physically more relevant, the translation is made as follows.

1. We note that for temporal-instability modes, the disturbance described by (13.16) looks like a traveling wave with *phase speed* $c = \omega_R/k_R$. However, the solutions to the dispersion equation show that c depends on k_R. Waves for which this is true are called *dispersive*; water waves from a splash in a pond are a good example. The splash excites both short and long elementary wave components; the long ones move more quickly away from the place where they were excited. Waves with different wavelengths interfere with one another in such a way that what you see is a spreading circular band, with individual wave crests appearing at the inside, moving through the band, and disappearing at the outside. It is clear that the perturbation energy added by the splash moves outward more slowly than do the crests of the visible waves. The velocity with which the energy moves is called the *group velocity*, c_g. For a wave component of the form (13.15) the group velocity is given by the formula $c_g = \partial\omega_R/\partial k_R$.[19]

2. Consider what would be a corresponding case of spatial instability, with $k_I = 0$. We argue that the time t that appears in the exponential prefactor of (13.16) is the time during which the energy has had a chance to grow since initiation of the disturbance. Further, we argue that during that time, the location where we should look for that energy has moved downstream a distance $x = c_g t$ from the point where the disturbance is initiated. Thus, (13.16) is translated into

$$\psi(x, y, t) = \exp\left(\frac{\omega_I x}{c_g}\right)[F(y)\cos(k_R x - \omega_R t - \theta(y))] \qquad (13.17)$$

Thus, we have assumed that, corresponding to a temporally-growing mode with $\omega = \omega_R + i\omega_I$, $k = k_R$, there will be a spatially-growing mode with $\omega = \omega_R$, $k = k_R - i\omega_I/c_g$. This is admittedly only an approximation,[19a] but it

---

[19] For a much more systematic introduction to the concept of group velocity, see M. J. Lighthill, *Waves in Fluids*, pp. 237–245, Cambridge University Press, 1978.

[19a] M. Gaster, "A note on the relation between temporally-increasing and spatially-increasing disturbances in hydrodynamic stability," *J. Fluid Mech.* **14**:222–224 (1962).

served a very useful purpose in the days before trustworthy numerical methods for the analysis of spatial instability were available, and the notion that a calculation of temporal instability may provide a useful model of spatial instability is still frequently used.

13.5 ANALYSIS OF TEMPORAL INSTABILITY. THEOREMS THAT NARROW THE SEARCH FOR UNSTABLE BASIC FLOWS, AND FOR THE PARAMETERS OF DANGEROUS NORMAL MODES

The analysis of temporal instability is very highly developed; all that is intended in this section is the presentation of some famous theorems that sharpen one's intuition about the kinds of velocity profiles that are particularly unstable, and that narrow the domain in the space of the disturbance parameters within which one may expect to find unstable eigenvalues. These theorems take no account of viscosity, except in the implicit recognition that viscosity was probably responsible for the establishment of the basic vorticity distribution. As will be mentioned again in Chapter 14, viscosity often acts over a relatively long time to establish the distribution of vorticity in a basic flow but may have very little effect on the rapidly-developing events that follow a sudden perturbation of that flow.

Rayleigh's and Fjørtoft's Theorems Concerning Inflection Points in the Velocity Profile

In Eq. (13.10), set $\text{Re} = \infty$ and $\alpha = \alpha_R$. Divide through by α_R, and introduce the notation $c = \sigma/\alpha_R$, noting that c may have both a real and an imaginary part. The result is Rayleigh's equation:

$$(U - c)(f'' - \alpha^2 f) - U''f = 0 \tag{13.18}$$

Because this is only a second-order differential equation, its solutions are subjected only to the zero-penetration boundary conditions, $f(y_1) = 0$, $f(y_2) = 0$.

Recalling that f is a complex quantity, call its complex conjugate f^*. Multiply (13.18) through by $f^*/(U - c)$, and integrate the resulting equation from $y = y_1$ to $y = y_2$.

$$\int_{y_1}^{y_2} \left((f''f^* - \alpha^2 ff^*) - \frac{U''}{U - c} ff^* \right) dy = 0$$

Integrate the first term by parts, getting

$$[f'f^*]_{y_1}^{y_2} - \int_{y_1}^{y_2} (f'f^{*'} + \alpha^2 ff^*) \, dy = \int_{y_1}^{y_2} \frac{U''}{U - c} ff^* \, dy$$

The boundary condition $f(y_1) = 0$ implies that both the real and imaginary parts of f must vanish at the boundary; therefore $f^*(y_1) = 0$. The same is true at y_2, so the first term on the left-hand side vanishes. Now note that any complex quantity times its conjugate is a real, positive-definite, quantity. By assumption, α is real, so α^2 is also positive-definite. This means that

$$\int_{y_1}^{y_2} \frac{U''}{U - c} ff^* \, dy = -A^2 \qquad (13.19)$$

where A is real. Remembering that $c = c_R + ic_I$, we break (13.19) into its real and imaginary parts. The latter is

$$c_I \int_{y_1}^{y_2} \frac{U''}{(U - c_R)^2 + c_I^2} ff^* \, dy = 0 \qquad (13.20)$$

This leads to two possible conclusions, either that $c_I = 0$, in which case the basic flow is neutrally stable; or that the integral in (13.20) vanishes. In the latter case, we note that every factor in the integrand, except for U'', is everywhere positive; thus, the integral can only vanish if U'' is somewhere positive and elsewhere negative. That can only happen if there is at least one inflection point in the velocity profile.

This theorem, originally proved by Lord Rayleigh, establishes that the presence of an inflection point is *necessary* for instability. It is not, however, sufficient; there are stable profiles with inflection points. Another necessary condition was proved, in a rather similar way, by Fjørtoft. He showed that

$$\int_{y_1}^{y_2} \frac{U''(U - U_s)}{(U - c_R)^2 + c_I^2} ff^* \, dy \leq 0$$

This means that a profile with an inflection point where the velocity equals U_s must satisfy the condition $U''(U - U_s) \leq 0$ in some part of the range $y_1 < y < y_2$. Figure 13.9 illustrates this result for a profile with only one inflection point. Note that a boundary-layer profile in an adverse pressure gradient, a diffused free shear-layer profile (as at the edge of a jet), and the profile in a wake of a flat plate, all satisfy Fjørtoft's condition for instability.[20]

When you read about modern studies of the events leading up to transition from laminar to turbulent flow, you will often encounter comments about inflection-point instability, or simply *inflectional instability*. The authors are referring to the results just presented.

Howard's Semicircle Theorem

Before starting a search for unstable eigenvalues, it is helpful to have some idea where to look for them. The following theorem, originally proved by

---

[20] See Drazin and Reid (1981) op cit., Section 22, for more information on theorems of this sort.

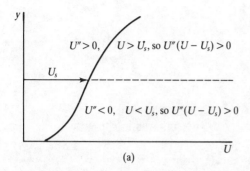

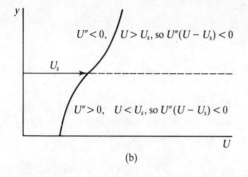

FIGURE 13.9
Fjørtoff's theorem: (a) stable flow; (b) possibly unstable flow.

L. N. Howard for stably stratified flows, is presented here for a fluid with a uniform density field. Like Rayleigh's and Fjørtoft's theorems, it take no account of viscosity. The object will be to prove the following inequality:

$$(U_m - c_R)^2 + c_I^2 \le \left(\frac{\Delta U}{2}\right)^2 \tag{13.21}$$

in which $\qquad U_m = \dfrac{U_{max} + U_{min}}{2} \qquad$ and $\qquad \Delta U = U_{max} - U_{min}$

The meaning of this is illustrated in Fig. 13.10; any possible complex eigenvalues must fall on or within a semicircle in the complex c plane. The diameter of the semicircle runs along the real axis from U_{min} to U_{max}. Thus the value of c_R must fall somewhere between the extremes of U, and the value of c_I cannot exceed $\Delta U/2$. All this is shown in Fig. 13.10; possible unstable eigenvalues must lie within the shaded region.

Proof of theorem. We adopt the following symbols: $W \equiv U - c$, $F \equiv f/W$, in terms of which the Rayleigh equation becomes

$$(W^2 F')' - \alpha^2 W^2 F = 0 \tag{13.22}$$

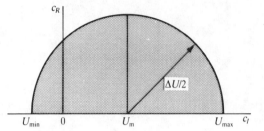

FIGURE 13.10
Howard's semicircle theorem.

Multiply (13.20) through by F^*, and integrate from $y = y_1$ to $y = y_2$:

$$\int_{y_1}^{y_2} \{F^*(W^2F')' - \alpha^2 W^2 FF^*\} \, dy = 0$$

Integrate the first term by parts, getting

$$[F^*W^2F']_{y_1}^{y_2} - \int_{y_1}^{y_2} W^2 F'F^{*\prime} \, dy - \alpha^2 \int_{y_1}^{y_2} W^2 FF^* \, dy = 0 \qquad (13.23)$$

Again, the first term on the left-hand side vanishes, because of the boundary conditions. Now introduce the shorthand notation $Q = F'F^{*\prime} - \alpha^2 FF^*$; this quantity is positive-definite. Then (13.23) is reduced to

$$\int_{y_1}^{y_2} W^2 Q \, dy = 0 \qquad (13.24)$$

Now take the real and imaginary parts of (13.24), noting that

$$W^2 = (U - c_R)^2 - c_I^2 - i2c_I(U - c_R)$$

This gives

$$\int_{y_1}^{y_2} [(U - c_R)^2 - c_I^2]Q \, dy = 0 \qquad (13.25)$$

and

$$c_I \int_{y_1}^{y_2} (U - c_R)Q \, dy = 0 \qquad (13.26)$$

From (13.26), we draw two alternative conclusions: *Either $c_I = 0$, or*

$$\int_{y_1}^{y_2} UQ \, dy = c_R \int_{y_1}^{y_2} Q \, dy \qquad (13.27)$$

We want to pursue the second conclusion, $c_i \neq 0$. Expanding

$$(U - c_R)^2 - c_I^2 = U^2 - 2c_R U + (c_R^2 - c_I^2)$$

we can put (13.24) into the form

$$\int_{y_1}^{y_2} U^2 Q \, dy = 2c_R \int_{y_1}^{y_2} UQ \, dy - (c_R^2 - c_I^2) \int_{y_1}^{y_2} Q \, dy$$

Using (13.27) to eliminate the middle integral, we get

$$\int_{y_1}^{y_2} U^2 Q \, dy = -(c_R^2 - c_I^2) \int_{y_1}^{y_2} Q \, dy \qquad (13.28)$$

Finally, we introduce the maximum and minimum value of U, by noting that the quantity $(U - U_{\min})(U - U_{\max})$ is intrinsically negative. Since Q is positive, we see that

$$\int_{y_1}^{y_2} (U - U_{\min})(U - U_{\max}) Q \, dy \le 0$$

We can now expand the product, and use (13.27) and (13.28) to eliminate the integrals of $U^2 Q$ and UQ in favor of the positive-definite integral of Q alone. The inequality shown just above then demands the inequality

$$(c_R^2 + c_I^2) + U_{\min} U_{\max} - c_R (U_{\max} - U_{\min}) \le 0$$

This is easily rearranged into the inequality that defines the semicircle.

Why a semicircle, instead of a whole circle? If we write out the real and imaginary parts of the Rayleigh equation, we get

$$(U - c_R)(f_R'' - \alpha^2 f_R) + c_I(f_I'' - \alpha^2 f_I) - U'' f_R = 0$$

and

$$(U - c_R)(f_I'' - \alpha^2 f_I) - c_I(f_R'' - \alpha^2 f_R) - U'' f_I = 0$$

We see that c and f can be replaced by c^* and f^*, without changing these equations. Nevertheless, it turns out that the solutions with negative values of c_I have no physical significance, because they represent a particular motion of a hypothetical inviscid fluid that is not the limit, as $\mathrm{Re} \to \infty$, of any motion of a real, viscous, fluid. Solutions of the Rayleigh equation for positive values of c_I can be shown to have this necessary property. Thus, the semicircle, rather than the circle.

Squire's Equivalence Theorem. The Development of Obliquely Propagating Normal Modes

Consider, for a moment, a slightly more general situation than we have been analyzing. Let the base flow be $\mathbf{U} = U(y)\mathbf{e}_x + W(y)\mathbf{e}_z$, so that $\mathbf{\Omega} = -U'(y)\mathbf{e}_z + W'(y)\mathbf{e}_x$. This could model a three-dimensional boundary layer, an Ekman layer, or the flow on a spinning disk. Correspondingly, let the disturbance velocity have a third scalar component, $w(x, y, z, t)$.

The linearized disturbance equations are now

$$\frac{\partial u}{\partial x} + \frac{\partial v}{\partial y} + \frac{\partial w}{\partial z} = 0 \tag{A}$$

$$\frac{\partial u}{\partial t} + U\frac{\partial u}{\partial x} + W\frac{\partial u}{\partial z} + vU' + \frac{1}{\rho}\frac{\partial p}{\partial x} = v\,\nabla^2 u \tag{B}$$

$$\frac{\partial v}{\partial t} + U\frac{\partial v}{\partial x} + W\frac{\partial v}{\partial z} + \frac{1}{\rho}\frac{\partial p}{\partial y} = v\,\nabla^2 v \tag{C}$$

$$\frac{\partial w}{\partial t} + U\frac{\partial w}{\partial x} + W\frac{\partial w}{\partial z} + vW' + \frac{1}{\rho}\frac{\partial p}{\partial z} = v\,\nabla^2 w \tag{D}$$

We aim to combine these into a single equation for v alone.

1. Add the x-derivative of equation (B) to the z-derivative of Eq. (D). Introduce the operator $D = \partial/\partial t + U\,\partial/\partial x + W\,\partial/\partial z - v\nabla^2$, noting that it commutes with $\partial/\partial x$ and $\partial/\partial z$. The result is

$$D\left(\frac{\partial u}{\partial x} + \frac{\partial w}{\partial z}\right) + U'\frac{\partial v}{\partial x} + W'\frac{\partial v}{\partial z} + (1/\rho)\left(\frac{\partial^2}{\partial x^2} + \frac{\partial^2}{\partial z^2}\right)p = 0 \tag{E}$$

2. Use (A) to eliminate u and w from (E). The resulting equation and Eq. (C) involve only v and p.
3. Take $\partial/\partial y$ of the result of step (2), remembering to differentiate U and W where they appear in the operator D.
4. Take $(\partial^2/\partial x^2 + \partial^2/\partial z^2)$ of Eq. (C).
5. Subtract the result of step (4) from that of step (5). The final result can be written as

$$\left(\frac{\partial}{\partial t} + U\frac{\partial}{\partial x} + W\frac{\partial}{\partial z}\right)\nabla^2 v - \left(U''\frac{\partial v}{\partial x} + W''\frac{\partial v}{\partial z}\right) = v\,\nabla^4 v$$

6. Let v be a normal mode, with $v = f(y)\exp[i(kx + mz - \omega t)]$, to obtain the ordinary differential equation

$$(\mathbf{k}\cdot\mathbf{U} - \omega)(D^2 - \mathbf{k}\cdot\mathbf{k})f - (\mathbf{k}\cdot\mathbf{U})'' + iv(D^2 - \mathbf{k}\cdot\mathbf{k})^2 f = 0 \tag{13.29}$$

This introduces the *wave-number vector*, $\mathbf{k} = k\mathbf{e}_x + m\mathbf{e}_z$. We see that $\mathbf{U}$ enters the problem only in the combination $\mathbf{k}\cdot\mathbf{U}$. Only the magnitude of k and the component of basic velocity that is normal to the disturbance wave fronts affects the eigenfunction f.

When $\mathbf{U}$ has only one scalar component, a normal mode for which $\mathbf{k}$ is not parallel to $\mathbf{U}$ is called *oblique*, or *skewed*. All such oblique modes are affected by basic velocity profiles with a common shape, but with a maximum speed that is smaller the more oblique the wave. Remember that the maximum

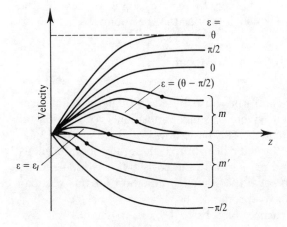

FIGURE 13.11
Projections of the velocity profile of a three-dimensional boundary layer. (Reproduced, by permission, from D. I. A. Poll, *J. Fluid Mech.* **150**: 329–356, 1985.)

basic-flow speed, together with the thickness of the basic-flow layer, provide a natural timescale for amplification or decay of the disturbance. If viscosity is unimportant, oblique disturbances will then develop more slowly than do disturbances that run directly downstream. If viscosity is important, the relevant Reynolds number for the oblique wave will be smaller than that for the direct one. This means that a search for the critical value of Re, below which all infinitesimal disturbances decay, does not have to consider oblique waves. These general considerations are often referred to as *Squire's Theorem.* The theorem itself asserts the precise equivalence of the eigenvalue problem for an oblique wave of a given wavelength, and that for a direct wave of the same length, in a slower basic flow.

It is good to know of Squire's theorem, but you must not let it lull you into forgetting all about oblique disturbances. There are at least two circumstances in which they appear to be very important. The first involves basic flows in which $W(y)$ is not zero, and presents a different profile than does $U(y)$. Then the profile of $\mathbf{k} \cdot \mathbf{U}$ depends on the obliqueness of the disturbance wave. This is especially significant in three-dimensional boundary layers, where there is always some range of the direction of $\mathbf{k}$ for which the effective basic-flow profile has an inflection point. This is illustrated in Fig. 13.11. In such cases, the most dangerous disturbance is often oblique, relative to the direction of flow outside the boundary layer. Three-dimensional boundary layers are expected, as a consequence, to be less stable than two-dimensional ones.

The second circumstance in which oblique waves seem very important need not concern you yet, as it has to do with the later, nonlinear, evolution of disturbances. It will be mentioned in Section 13.10.

13.6 EFFECTS OF WAVELENGTH ON STABILITY. SHORT-WAVE CUTOFFS

We return now to consider strictly two-dimensional disturbances of a unidirectional basic flow. It would be very helpful to know, before undertaking a search for unstable eigenvalues, whether there are some ranges of the nondimensional wave number α for which they cannot be found.

Layer of Uniform Vorticity Between Uniform Streams

Suppose that vorticity is initially uniformly distributed throughout a flat layer of thickness 2δ, which is then given a small sinusoidal corrugation of wavelength λ not vastly greater than δ. This basic flow is defined in Fig. 13.12. We have already seen how the vorticity distributed along any infinitesimally thin sheet of this layer would tend to aggravate the corrugation of that sheet.

Actually, it is quite easy to solve the Rayleigh equation for this vorticity distribution. We define the velocity profile to be $U = 1$ if $\eta > 1$, $U = -1$ if $\eta < -1$, $U = \eta$ if $\eta^2 < 1$. In each subinterval, $U'' = 0$, and the Rayleigh equation reduces to $f'' - \alpha^2 = 0$. A convenient form of the solution, constructed so that $f \to 0$ as $\eta \to \pm\infty$ for positive α, and so that f, and hence v, is continuous at $\eta = \pm 1$, is

$$f = A \cosh(\alpha\eta) + iB \sinh(\alpha\eta) \quad \text{if} \quad \eta^2 \leq 1$$

with $\qquad f = [A \cosh(\alpha) + iB \sinh(\alpha)] \exp[\alpha(1 - \eta)] \quad \text{if} \quad \eta > 1$

and $\qquad f = [A \cosh(\alpha) - iB \sinh(\alpha)] \exp[\alpha(\eta + 1)] \quad \text{if} \quad \eta < -1$

At $\eta = \pm 1$ the pressure must also be continuous. The linearized x-momentum equation shows that

$$\frac{\partial p}{\partial x} = -\rho\left(\frac{\partial u}{\partial t} + U\frac{\partial u}{\partial x} + v\frac{dU}{dy}\right)$$

If p is continuous at $\eta = \pm 1$, $\partial p/\partial x$ will also be continuous. For a normal-mode disturbance, the equivalent requirement is that $(U - c)f' - fU'$ be continuous.

Since the basic velocity profile is an odd function of η, there is no reason to expect that a normal-mode disturbance should propagate either to the right

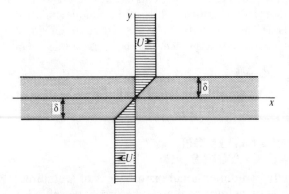

FIGURE 13.12
Rayleigh's uniform-vorticity layer.

or the left, there being no logical way to decide between the two directions. We thus assume that $c_R = 0$. Demanding that the pressure be continuous at $\eta = 1$, where $U = 1$, we get

$$\alpha(1 - ic_I)[A \sinh(\alpha) + iB \cosh(\alpha)] - (1)[A \cosh(\alpha) + iB \sinh(\alpha)]$$

$$= -\alpha(1 - ic_I)[A \cosh(\alpha) + iB \sinh(\alpha)] + 0$$

The real and imaginary parts of this provide two homogeneous linear algebraic equations for A and B. For a nonzero solution, the determinant of the coefficients must vanish, giving the final result

$$c_I^2 = -\left(\frac{1}{2\alpha}\right)^2 [1 - 2\alpha + \exp(-2\alpha)][1 - 2\alpha - \exp(-2\alpha)] \qquad (13.30)$$

For small values of α (long-wave disturbances), we use the Taylor series for $\exp(-2\alpha)$ to show that

$$c_I^2 = (1 - \tfrac{8}{3}\alpha + \tfrac{1}{3}\alpha^2 + \cdots)$$

For these long waves, c_I^2 is positive, and the disturbance is a growing standing wave. This situation persists until the middle factor on the right of (13.30) vanishes, at $\alpha \approx 0.639$. This is an example of a *short-wave cutoff*. The value $\alpha \approx 0.639$ corresponds to a disturbance wavelength equal to 4.91 times the total thickness of the vortex layer. The fastest-growing, and hence most dangerous, normal mode is the one for which αc_I is largest. It corresponds to $\alpha = 0.398$, $c_I = 0.505$, about half the maximum value allowed by the semicircle theorem. The variation of c_I versus α is shown in Fig. 13.13.

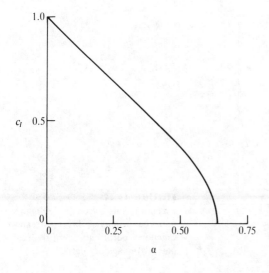

FIGURE 13.13
Shortwave cutoff of instability in Rayleigh's uniform-vorticity layer.

Calculation of the Streamfunction and of Streamlines

To get a bit more familiar with this kind of analysis, and with a fairly typical result, substitute the equation found for c_i into either of the equations relating B to A. The result is

$$B/A = \{[1 - 2\alpha + \exp(-2\alpha)]/[2\alpha - 1 + \exp(-2\alpha)]\}^{1/2}$$

As $\alpha \to 0$, $B/A \to \infty$ as $1/\alpha$, and the eigenfunction f becomes purely imaginary. As $\alpha \to 0.639$, $B/A \to 0$, and the eigenfunction f becomes purely real. When $\alpha = \frac{1}{2}$, corresponding very nearly to the most dangerous disturbance, $B/A = 1$. The real and imaginary parts of f are shown in Fig. 13.14 for this case.

For calculation of the streamlines of the perturbed flow, it is convenient to represent the complex function f by its *modulus* $F = (f_R^2 + f_I^2)^{1/2}$, and *argument* $\theta = \arctan(f_I/f_R)$. Then the perturbed streamfunction for a single normal mode can be represented by the equation

$$\psi(\xi, \eta, \tau) = \Psi(\eta) + \text{AMP}(\tau)F(\eta) \cos[\alpha\xi + \theta(\eta)] \qquad (13.31)$$

For the plotting of streamlines, this can be easily inverted, to evaluate ξ for given values of ψ, η, and τ, and for a specified basic-flow streamfunction $\Psi(\eta)$. The formula is

$$\xi = \alpha^{-1}\{\arccos[(\psi(\xi, \eta, \tau) - \Psi(\eta))/(\text{AMP}(\tau)F(\eta))] - \theta(\eta)\} \qquad (13.32)$$

In the case at hand, because f_R is an even function of η, while f_I is an odd function, each point (ξ, η) of a given streamline is matched to a second point $(2\pi/\alpha - \xi, -\eta)$ on the same streamline. These formulas have been used to prepare Fig. 13.15 for the case $\alpha = \frac{1}{2}$, with the value AMP = 0.1. The characteristic qualitative features of the streamline field are the regions of closed streamlines, called *cat's eyes*. They often appear when the perturbed

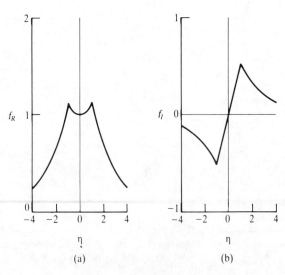

FIGURE 13.14
Eigenfunctions for Rayleigh's uniform-vorticity layer—streamfunction perturbation: (a) real part; (b) imaginary part.

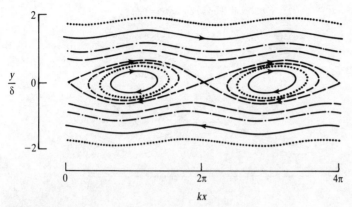

FIGURE 13.15
Streamlines of a perturbed uniform-vorticity layer, AMP = 0.1.

streamline field is seen, as in this case, from a frame of reference in which the real phase velocity of the normal-mode disturbance, c_R, is zero. The eyeballs are located where the perturbation has built up a local concentration of vorticity, the extreme case of total concentration being shown in Fig. 13.3.

Evolution of Material Surfaces

Because flow visualization is often used to study flow instabilities, it is very important to proceed carefully when analyzing the motion of material particles or surfaces in a disturbed flow. The procedure is straightforward, but the results are hard to anticipate intuitively, because small effects have a chance to accumulate over a finite period of time.

Taking a Lagrangian viewpoint, let the current position of a material particle be specified by dimensionless coordinates $\xi(\xi_0, \eta_0, \tau)$ and $\eta(\xi_0, \eta_0, \tau)$, where ξ_0 and η_0 specify its position at some initial time $\tau = 0$. The ordinary differential equations to be solved for ξ and η are

$$\frac{d\xi}{d\tau} = U(y) + \text{AMP}\,(\tau)[f_R'(\eta)\cos(\alpha\xi) - f_I'(\eta)\sin(\alpha\xi)] \qquad (13.33)$$

and

$$\frac{d\eta}{d\tau} = \alpha\text{AMP}\,(\tau)[f_R(\eta)\sin(\alpha\xi) + f_I(\eta)\cos(\alpha\xi)] \qquad (13.34)$$

where

$$\text{AMP}\,(\tau) = \text{AMP}\,(0)\exp(\alpha c_I\tau)$$

Figure 13.16 shows the computed evolution of two material surfaces, one initially at $y_0 = 0$, the other initially at $y_0 = \delta/2$. Two curves, distinguished by the time at which the material surface was marked, are shown for each initial location. The streamline pattern for the time at which the material surfaces are shown is that shown in Fig. 13.15. You see that the configuration of the surface at the time when AMP has a given value depends not only on that

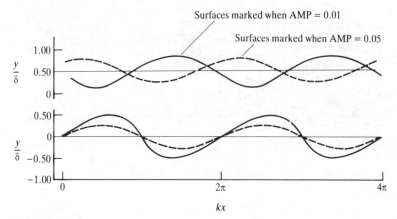

FIGURE 13.16
Distortion of material surfaces in a perturbed uniform-vorticity layer, AMP = 0.1.

value, but on the value AMP (0), corresponding to the time at which the surface was marked.

13.7 NUMERICAL SEARCHES FOR EIGENSOLUTIONS

Numerical analysis of the Orr–Sommerfeld and Rayleigh equations is quite tricky in many cases. There are two principal causes of difficulty.

1. One of the important objectives of the analysis is to define curves, in a Re–α plane, that define the boundary between stable and unstable behavior of the basic flow. Along these curves, c is purely real and lies between the extrema of $U(y)$. At the *critical layer* where $U - c$ vanishes, the Rayleigh equation often has a singularity. Even when this is somewhat smoothed out by viscosity, the eigenfunction varies steeply and numerical accuracy may suffer.

2. If the base flow extends to infinity in one or both directions, as does a boundary layer, jet or wake, in a region of uniform U, the general solution of the Orr–Sommerfeld equation is given by the sum of four linearly independent functions, as shown by Eq. (13.13). There is no difficulty in computing the evolution of a particular solution that grows in the direction of integration, so one can integrate easily from an initial point in one region of uniform basic flow, in towards the region where the basic vorticity is concentrated. When one attempts to continue the integration into a region where the desired solution dies away exponentially, the numerical integration scheme becomes unstable. Inevitable numerical errors, whether from truncation or round-off, initiate a rapid, often explosive, growth of one of the unwanted, exponentially amplifying solutions. The computed solution

fails to approach the required boundary value, no matter how accurate the approximation to the eigenvalue. In the case of the Orr–Sommerfeld equation, the dangerous unwanted solution has the exponential amplification rate $\mathcal{R}(\lambda_3)$ or $\mathcal{R}(\lambda_4)$, depending on the direction of integration. This becomes very large when the Reynolds number is very large, so the search for accurate eigenvalues of the Orr–Sommerfeld equation when $\alpha\,\mathrm{Re} \gg 1$ is particularly difficult.

The most popular numerical methods are shooting methods, executed with a Runge–Kutta algorithm. The parasitic growth of unwanted solutions is overcome by a technique called *orthonormalization*, or avoided entirely by converting the linear Orr–Sommerfeld equation into a nonlinear *Riccati* equation. The Riccati technique is very powerful and somewhat easier to program. Another very effective technique involves representation of the eigenfunction by a sum of *Chebyshev polynomials*. The differential equation is converted to a large set of linear algebraic equations that have a nontrivial solution only when the unknown normal-mode parameters are eigenvalues.

When the basic velocity profile is either an even or an odd function of y, eigensolutions of the Rayleigh and Orr–Sommerfeld equations are readily computed with slightly modified versions of the finite-difference method that we have used to solve the boundary-layer equations. In fact, our old friend the Thomas algorithm got its name from a noteworthy application to solution of the Orr–Sommerfeld equation.[21]

Stability of the Velocity Profile $U = \mathrm{erf}\,(\sqrt{\pi}\,\eta/2)$

A smooth velocity profile that has the same maximum vorticity and the same total shear as Rayleigh's layer of uniform vorticity, is described by $U = \mathrm{erf}(\sqrt{\pi}\,\eta/2)$ in $-\infty < \eta < \infty$. The characteristic thickness of the layer is taken to be $\delta = U/(dU/dy)_{\max}$. We shall first solve the Rayleigh equation, and then the Orr–Sommerfeld equation, using the Thomas algorithm. The first step is to establish that the desired eigenfunctions are even functions of η.

---

[21] Drazin and Reid (1981) op. cit., Article 30, give a very helpful review of alternative numerical methods. Betchov and Criminale (1967) op. cit. give a good explanation of the shooting method, and of the details of an iterative search for eigenvalues, particularly in Chapter 1, article 5, and in Appendix III. L. M. Mack gives a superb tutorial review in "Boundary-Layer Linear Stability Theory," in *Special Course on Stability and Transition of Laminar Flow*, AGARD Report No. 709, Chap. 3 (available from National Technical Information Service, Port Royal Rd., Springfield, VA 22161.) For the Riccati method, see A. Davey, "On the Numerical Solution of Difficult Eigenvalue Problems," *J. Comput. Phys.* **24**:331–338 (1977). For the use of Chebyshev polynomials, see S. A. Orszag, "Accurate solutions of the Orr–Sommerfeld Stability Equation," *J. Fluid Mech.* **50**:689–704 (1971). The simple method used here is inspired by the classic paper of L. H. Thomas, "The Stability of Plane Poiseuille Flow," *Phys. Rev.* (2)**91**:780–783 (1953).

To investigate whether the eigenfunction has such properties, one breaks the Rayleigh equation into real and imaginary parts. These are

$$f_R'' - \left(\alpha^2 + U''\frac{U - c_R}{D}\right)f_R + \frac{U''c_I}{D}f_I = 0 \tag{13.35}$$

and

$$f_I'' - \left(\alpha^2 + U''\frac{U - c_R}{D}\right)f_I - \frac{U''c_I}{D}f_R = 0 \tag{13.36}$$

in which $D = (U - c_R)^2 + C_I^2$.

When U, and hence U'', are odd functions of η, we might suspect the existence of unstable normal modes that do not travel, i.e., for which $c_R = 0$. We have seen them in the example of Section 13.6. For such modes, the coefficients of f_R in (13.35) and of f_I in (13.36) are even functions of η, while the coefficients of f_I in (13.35) and of f_R in (13.36) are odd functions of η. Thus, if we assume that f_R is an even function and f_I an odd function, every term in (13.36) will be even, and every term in (13.37) will be odd, confirming the assumption.

To find the eigenvalue and eigensolution for an unstable normal mode, say for $\alpha = \frac{1}{2}$, we subdivide the interval $-\eta^* < \eta < 0$ into equal subintervals for finite-difference approximation to f''. The value of η^* is selected so that U is essentially constant and U'' is negligible outside of the subdivided range.

For this example, $U'' = -(\pi/2)\eta \exp(-\pi\eta^2/4)$, and we take $\eta^* = 4$. The boundary conditions on f will be:

$$f = A \exp(\alpha\eta) \qquad \text{for} \qquad \eta \le -\eta^*$$

$$f_R'(0) = 0 \qquad \text{(from the evenness of } f_R)$$

$$f_I(0) = 0 \qquad \text{(from the oddness of } f_I)$$

A is a complex constant that plays no role in the solution procedure.

Using three-point centered differences, we convert the Rayleigh equation to the algebraic system

$$f_{n+1} - 2f_n + f_{n-1} + (d\eta)^2 G(\eta_n)f_n = 0,$$

in which

$$G = -\left(\frac{\alpha^2 + U''}{U - c}\right)$$

In our usual notation for the Thomas algorithm, we have

$$A_n f_{n-1} + B_n f_n + C_n f_{n+1} = D_n$$

with $\qquad A_n = C_n = 1 \qquad B_n = -2 + (d\eta)^2 G_n \qquad$ and $\qquad D_n = 0$

Applying the back-substitution formula, $f_n = E_n f_{n+1} + F_n$, in the region $\eta < -\eta^*$ where $f_n = A \exp(\alpha\eta_n)$ and $f_{n+1} = A \exp\{\alpha(\eta_n + d\eta)\}$, we find that it is satisfied by $F_n = 0$ and $E_n = \exp(-\alpha\, d\eta)$. These are used as the starting values for E and F, when $n = 1$.

From the recursion relation (10.15), $F_n = (D_n - A_n F_{n-1})/(B_n + A_n E_{n-1})$, and the knowledge that $D_n = 0$ and $F_1 = 0$, we see that $F_n = 0$ for all n. This means that the general back-substitution formula is reduced to

$$f_n = E_n f_{n+1} \qquad (13.37)$$

At $\eta = 0$, where U and U'' are zero, the boundary conditions are supplemented by the reduced form of Eq. (13.36):

$$f''_R - \alpha^2 f_R = 0 \qquad (13.38)$$

Suppose $n = m$ identifies the point $\eta = 0$, and $n = m - 1$ the point where $\eta = -d\eta$. The centred-difference approximation for f''_R can be simplified because $f_{R\,m+1} = f_{R\,m-1}$. Then (13.39) leads to

$$f_{R\,m-1} = [1 + 0.5(\alpha\,d\eta)^2]f_{R\,m}$$

When the back-substitution formula (13.37) is expanded into real and imaginary parts, we find

$$f_{R\,m-1} = E_{R\,m-1}f_{R\,m} - E_{I\,m-1}f_{I\,m}$$

However, we know that $f_{I\,m} = 0$, because f_I is an odd function of η. This leaves the equation

$$E_{R\,m-1} = 1 + 0.5(\alpha\,d\eta)^2$$

or $$2E_{R\,m-1} + B_{R\,m} = 0 \qquad (13.39)$$

which must be satisfied if the value of the parameter c is an eigenvalue.

To find an eigenvalue of c for a given value of α, we can proceed iteratively, as follows.

1. Make a first guess at c, using whatever information may be at hand. Store the value under the name C. For the present example, the value found analytically in the preceding section could be used. If eigenvalues are being computed for a range of values of α, the value of c found for a neighboring value of α makes a good first guess.
2. Starting with $E_1 = \exp(-\alpha\,d\eta)$, apply the recursion formula

$$E_n = -\frac{1}{B_n + E_{n-1}} \qquad (13.40)$$

until a value is found for $E_{R\,m-1}$. This will not, unless you are very lucky, be the required value. Form an *error quantity* ERR $= 2E_{R\,m-1} + B_{R\,m}$. Store its value, calling it ERROLD. Give the corresponding value of c the name COLD.
3. Make a slight change in the value of c, and repeat step (2), calling the resulting error ERR. The corresponding value of c is still called C.
4. Using the two values of c, and the two values of the error, make a linear extrapolation towards the value of c that should cause the error to vanish.

This done with the formulas

$$CORR = -ERR(C - COLD)/(ERR - ERROLD)$$

and
$$CNEW = C + GAM * CORR$$

Here GAM is called an *underrelaxion* factor. It helps to avoid oscillations in the successive estimates of c, but does not affect the final answer. An effective value is about 3/4.

5. Using the newly estimated value of c, repeat steps (2) and (4). Continue until the error becomes, and stays, satisfactorily small.

6. Assigning f an arbitrary real value at $\eta = 0$, say $f_m = 1 + 0i$, apply the back-substitution formula to calculate f_n for the remaining value of n. At this point the constant A could be evaluated, if it were of any interest. Values of f for $\eta > 0$ are obtained from the computed values by use of the even-and-oddness properties.

A FORTRAN program, EIGEN, to carry out this algorithm is shown in Appendix B. The computer handles the arithmetic of complex numbers, providing that storage space is provided for both the real and imaginary parts, and that the numbers that are complex are clearly identified as such.

Figure 13.17 shows c_i as a function of α, both from the Rayleigh equation and from the Orr–Sommerfeld equation. For the Rayleigh equation, the short-wave cutoff is found at $\alpha \approx 0.9149$. The eigenfunctions for stream function (f) and for vorticity (g) are shown in Figs. 13.18 and 13.19 for $\alpha = 0.50$, which gives a relatively high rate of amplification.

Stabilizing Effects of Viscosity

A simple extension of the work done on the Rayleigh equation produces an algorithm that works for the Orr–Sommerfeld equation.

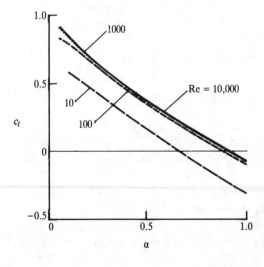

FIGURE 13.17
Eigenvalues for the shear layer $U = \text{erf}(\sqrt{\pi}y/2\delta)$. Temporal instability. Effect of wavenumber and Reynolds number.

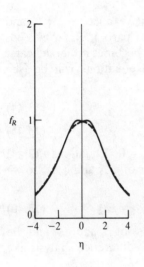

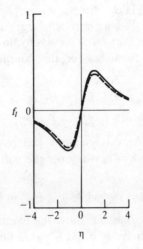

FIGURE 13.18
Eigenfunctions for the shear layer $U = \mathrm{erf}(\sqrt{\pi}y/2\delta)$. Streamfunction perturbation: —— $Re = \infty$; ---- $Re = 40$; $\alpha = 0.5$.

As a preliminary, we write Eq. (13.10) as a pair of coupled second-order equations

$$f'' - \alpha^2 f + g = 0 \tag{13.41}$$

and
$$g'' - \{\alpha^2 + i\,Re\,\alpha(U - c)\}g - i\,Re\,\alpha U''f = 0 \tag{13.42}$$

For the case of a free shear layer between infinite uniform streams, the boundary conditions are essentially the same as in the inviscid case:

$$\text{at} \quad \eta = -\eta^* \quad f = B\exp(\alpha\eta^*) \quad \text{and} \quad g = 0 \tag{13.43}$$

These result from inspection of the comparative values of the λs that satisfy Eq. (13.14). The first two, $\lambda_1 = \alpha$ and $\lambda_2 = -\alpha$, correspond to a

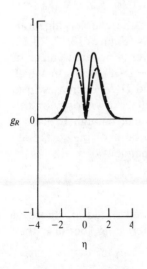

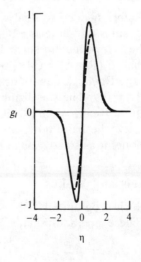

FIGURE 13.19
Eigenfunctions for the shear layer $U = \mathrm{erf}(\sqrt{\pi}y/2\delta)$. Vorticity perturbation: —— $Re = \infty$; ---- $Re = 40$; $\alpha = 0.5$.

disturbed potential flow. The others, $\lambda_3 = \alpha[1 + i(\mathrm{Re}/\alpha)(U_\infty - c)]^{1/2}$ and $\lambda_4 = -\alpha[1 + i(\mathrm{Re}/\alpha)(U_\infty - c)]^{1/2}$, have much larger real parts if Re/α exceeds about 10. If Re/α is not large, the search for eigenvalues becomes rather delicate.

The back-substitution formulas for the coupled equations will be [see (10.38)]

$$f_n = E_{11n}f_{n+1} + E_{12n}g_{n+1} \tag{13.44}$$

and

$$g_n = E_{21n}f_{n+1} + E_{22n}g_{n+1} \tag{13.45}$$

The calculation starts out at $\eta = -\eta^*$, where $g_n = 0$. Equations (13.44) and (13.45) must be true for $n = 1$, whatever the values of f_{n+1} and g_{n+1}, so we find that

$$E_{111} = \exp(-\alpha\,d\eta) \qquad \text{while} \qquad E_{121} = E_{211} = E_{221} = 0 \tag{13.46}$$

As in the inviscid case, one can show the evenness of f_R and g_R, and the oddness of f_I and g_I, as functions of η. At $\eta = 0$, a finite-difference analysis like the one performed for inviscid flow gives

$$2f_{R\,m-1} = B_{11\,R\,m}f_{R\,m} - B_{12\,R\,m}g_{R\,m}$$

and

$$2g_{R\,m-1} = B_{21\,R\,m}f_{R\,m} - B_{22\,R\,m}g_{R\,m}$$

The real parts of the back-substitution formulas reduce to

$$f_{R\,m-1} = E_{11\,R\,m-1}f_{R\,m} + E_{12\,R\,m-1}g_{R\,m}$$

and

$$g_{R\,m-1} = E_{21\,R\,m-1}f_{R\,m} + E_{22\,R\,m-1}g_{R\,m}$$

Elimination of $f_{R\,m-1}$ and $g_{R\,m-1}$ between these two pairs of formulas yields a two linear, homogeneous equations for $f_{R\,m}$ and $g_{R\,m}$. In order that nonzero solutions may exist, the determinant must vanish. Thus, the eigenvalue criterion is

$$(2E_{11\,R\,m-1} + B_{11\,R\,m})(2E_{22\,R\,m-1} + B_{22\,R\,m})$$
$$- (2E_{21\,R\,m-1} + B_{21\,R\,m})(2E_{12\,R\,m-1} + B_{12\,R\,m}) = 0 \tag{13.47}$$

Aside from the extra algebra needed to deal with the coupled algebraic equations, the rest of the program is the same as EIGEN. Figures 13.17, 13.18, and 13.19 show results for the error-function shear layer. For large values of Re, the calculations show that $C_i(\mathrm{Re}, \alpha) = C_i(\infty, \alpha) - D(\alpha)\,\mathrm{Re}^{-1}$, in which $C_i(\infty, \alpha)$ is the eigenvalue of the Rayleigh equation, and $D(\alpha)$ decreases from about 7.2 at $\alpha = 0.1$ to 2.9 at $\alpha = 0.9$. The stabilizing effect of viscosity on this kind of unbounded flow is quite weak; when Re is small enough to have a substantial effect on the stability, the basic flow would be evolving so rapidly that it makes little sense to model it as steady, or independent of x.

Other Symmetric Flows. Jets and Wakes

Other velocity profiles for which program EIGEN is well suited are those of symmetric jets and wakes. These are additionally interesting, because they

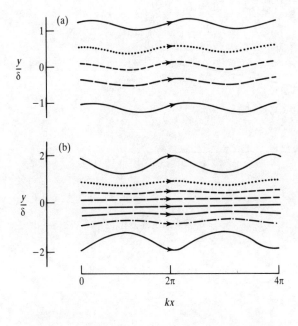

FIGURE 13.20
Disturbed streamlines of the Bickley jet: (a) sinuous mode, $\alpha = 1.0$; (b) varicose mode, $\alpha = 0.5$.

exhibit two dangerous modes of instability, one called *sinuous*, the other *varicose*. In a sinuous mode, f is an even function of y; in a varicose mode, it is an odd function. Figure 13.20 shows streamline patterns for the two kinds of mode, and explains their names.

The Bickley jet, $U = \text{sech}^2(\eta)$. Sinuous modes. Because this velocity profile is an even function of η, the eigenfunctions can have properties of evenness or oddness, even when $c_R \neq 0$. Both the real and imaginary parts of $\alpha^2 + U''/(U - c)$ will be even functions.

Program EIGEN can be used again, with the following small modification, for the sinuous modes. Combination of the Rayleigh equation and the requirement that f be an even function of η leads to the eigenvalue condition

$$2E_{m-1} + B_m = 0 \tag{13.48}$$

The derivation of this is left as an exercise. For the Orr–Sommerfeld equation, the function g shares the evenness or oddness of f, and Eq. (13.48) is replaced by the requirement $\|2\mathbf{E}_{m-1} + \mathbf{B}_m\| = 0$, i.e., that the determinant of the matrix should vanish. For a convergence criterion, the absolute values of these complex quantities must fall below, and stay below, a specified tolerance.

Values of c_I versus α for various values of Re are shown in Fig. 13.21. Eigenfunctions are shown for streamfunction and vorticity in Figs. 13.22 and 13.23.

Varicose modes for the Bickley jet. To calculate the varicose modes, for which f is an odd function of η, it is helpful to differentiate the Rayleigh equation

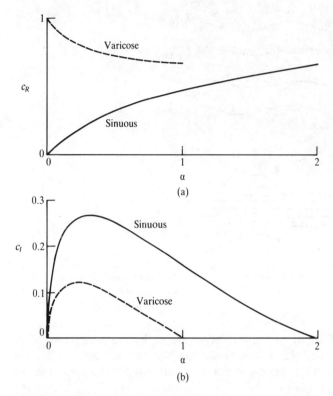

FIGURE 13.21
Eigenvalues for the Bickley jet, $U = \text{sech}^2(\eta)$: (a) c_R; (b) c_I.

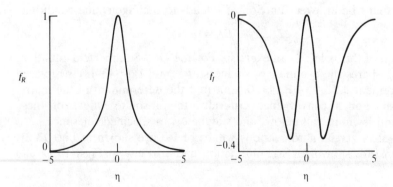

FIGURE 13.22
Sinuous-mode perturbations of the Bickley jet. Eigenfunctions for streamfunction. $\alpha = 1.0$.

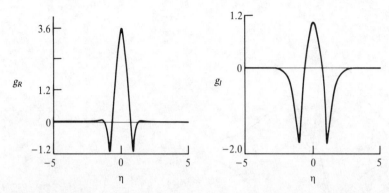

FIGURE 13.23
Sinuous-mode perturbations of the Bickey jet. Eigenfunctions for vorticity. $\alpha = 1.0$.

with respect to η, and then solve for f', which is an even function. If the Rayleigh equation is written as

$$f'' + G(\eta)f = 0$$

the new equation set to be solved is

$$u'' + G(\eta)u + G'(\eta)f = 0 \tag{13.49}$$

and $$f' - u = 0 \tag{13.50}$$

The differencing scheme is the same as used for the Falkner–Skan equations in Chapter 12. The matrix for **E** has only two nonzero members, the starting values of which are

$$E_{11\,1} = \exp\left(-\alpha\,d\eta\right) \qquad \text{and} \qquad E_{21\,1} = (1/\alpha)\exp\left(-\alpha\,d\eta\right)$$

The condition for an eigenvalue is $2E_{11\,m-1} + B_{11\,m} = 0$. Eigenvalues are shown in Fig. 13.21; eigenfunctions in Figs. 13.24 and 13.25.

Stability of Wall-Bounded Shear Flows

Suppose the shear flow is, like a boundary layer, bounded on one side by a solid wall. Its stability will be affected, even if viscosity is still disregarded and only a no-penetration boundary condition is enforced. If the no-slip condition is added, so that viscosity cannot be neglected, there may be further effects on the stability, some of them surprising. Let us first consider the effect of a zero-penetration condition.

Stabilizing effect of a no-penetration condition. In the absence of viscosity, the effect of a solid plane wall next to an unstable distribution of vorticity is equivalent to the presence of an image distribution of vorticity on the other side of the wall. The result can be qualitatively anticipated by comparing results for the sinuous and varicose modes of the Bickley jet. In fact, the

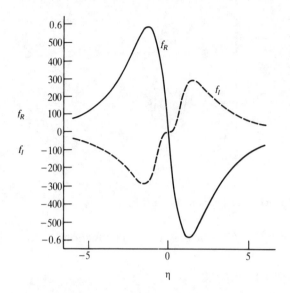

FIGURE 13.24
Varicose-mode perturbations of the Bickley jet. Eigenfunctions for streamfunction. $\alpha = 0.5$.

analysis for the varicose modes is indistinguishable from that of the instability of a *wall jet*, for which $U = \operatorname{sech}^2 \eta$ for $\eta > 0$, and $v = 0$ on $\eta = 0$. In Fig. 13.26 we see that the growth rate for a varicose disturbance of a given value of α is considerably smaller than that of a sinuous disturbance with the same value of α.

For a more general case the reduction of disturbance growth rates will depend on the value of the parameter kH, where H is the distance from the

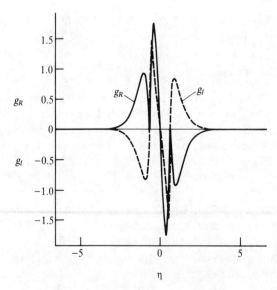

FIGURE 13.25
Varicose-mode perturbations of the Bickley jet. Eigenfunctions for vorticity. $\alpha = 0.5$.

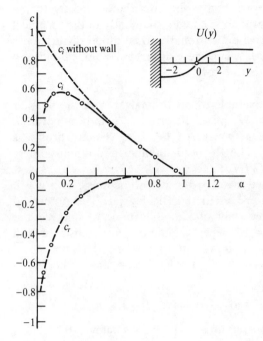

FIGURE 13.26
Effect of wall on inviscid instability. (Reproduced, by permission, from R. Betchov and W. Criminale, 1967.[3])

wall to the undisturbed position of the vortex sheet. Figure 13.26, from Betchov and Criminale, shows this effect.[22] Note that the effect of the wall is most stabilizing for long waves, whereas there is an intrinsic short-wave cutoff associated with the finite thickness of the layer. Hence the range of α for which there can be an unstable normal mode shrinks as kH decreases.

This effect of a neighboring wall becomes practically important in many applications of free shear layers, for reasons which are explained more fully in Chapter 14.

The Destabilizing Effect of Viscosity. TS Waves

When early investigators of stability theory contrasted the implications of Rayleigh's inflection-point theorem, which suggests that a laminar boundary layer developing under a favorable pressure gradient would be stable, with the observed fact that such boundary layers can become turbulent, they had to search for a way by which viscosity and/or the no-slip condition could destablize a flow. What they found to resolve the apparent paradox came to be

---

[22] R. Betchov and W. Criminale, op. cit., Fig. 5.6.

known as *TS waves*, in honor of four men[23] who contributed greatly to the understanding of this puzzle. The term appears very frequently in the technical literature, as a nickname for the two-dimensional normal-mode disturbances we have been considering, particularly when these disturbances appear close to a wall on which a no-slip condition is in force.

Plane Poiseuille flow. The easiest example for us to analyze with the tools at hand is that of plane Poiseuille flow. We place the origin of y midway between the flow-bounding walls, which lie at $y = \pm H$. The basic velocity profile is $U(y) = U_0(1 - \eta^2)$, where $\eta = y/H$. The Reynolds number and dimensionless wave number are defined by $\mathrm{Re} = U_0 H/\nu$, and $\alpha = kH$.

Since U is again an even function of η, the eigenfunctions f and g will both be either even or odd functions, as in the Bickley jet. The numerical algorithm used for the jet requires only one modification, to enforce the desired boundary conditions at $\eta = \pm 1$. These conditions are translated into initial values for the matrix $\mathbf{E}$, as follows. In the finite-difference representation, $n = 1$ now corresponds to $\eta = -1$, the lower wall. The back-substitution formula at the point is

$$f_1 = E_{11\,1}f_2 + E_{12\,1}g_2 \qquad \text{and} \qquad g_1 = E_{21\,1}f_2 + E_{22\,1}g_2$$

The zero-penetration condition requires that $f_1 = 0$, whatever the values of f_2 and g_2. Thus,

$$E_{11\,1} = E_{12\,1} = 0$$

The zero-slip condition, $f'(-1) = 0$, means that $f_2 = f_0$, where the imaginary value f_0 can be eliminated by use of the finite-difference representation of Eq. (13.41). This yields the equation $g_1 = (-2/B_{12\,1})f_2$, which must also be true, whatever the value of f_2. Thus,

$$E_{21\,1} = \frac{-2}{B_{12\,1}} \qquad \text{and} \qquad E_{22\,1} = 0 \qquad \text{(zero slip)}$$

Later, it will prove to be of interest to experiment with an artificial boundary condition that turns off the surface source of vorticity that is needed to keep the no-slip condition in effect. This condition is $g'(-1) = 0$. It requires that $g_2 = g_0$, where the imaginary value g_0 can be eliminated by use of the finite-difference representation of Eq. (13.42). This leads to $g_1 = (-2/B_{22\,1})g_2$, and hence to

$$E_{21\,1} = 0 \qquad \text{and} \qquad E_{22\,1} = \frac{-2}{B_{22\,1}} \qquad \text{(zero flux of vorticity)}$$

---

[23] W. Tollmien and H. Schlichting, who developed the successful theory, and G. B. Schubauer and H. K. Skramstad, who produced definitive experimental confirmation.

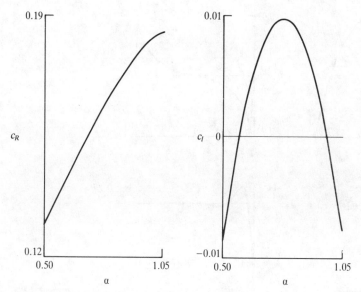

FIGURE 13.27
Eigenvalues for plane Poiseuille flow, Re = 40,000.

This is a very famous stability problem, and our search for eigenvalues would be a blind and stumbling one, except for the enormous amount of past analytical and numerical work, well reviewed by Drazin and Reid, which tells us approximately where to look. Because the basic flow is stable in the absence of viscosity, there are no easily found inviscid eigenfunctions to serve as a guide. It is known that the flow is stable against all disturbances of the form governed by the Orr–Sommerfeld equation if Re < 5772, and that the normal mode that is first amplified is a sinuous (even-function) mode with $\alpha = 1.021$.

Some results for the most dangerous sinuous modes at Re = 40,000 were computed with the algorithm described above, using 2000, 3000, and 4000 equal steps in the interval $-1 \le \eta \le 0$. These seem like large numbers, but the calculations converge in about 7 to 15 iterations, and are still very fast and cheap. Figure 13.27 shows the distributions of c_I and c_R versus α for this family of modes; the eigenfunctions for streamfunction and perturbation vorticity are shown for the most amplified mode ($\alpha = 0.775$) and for two neighboring damped modes ($\alpha = 0.5$ and 1.05) in Figs. 13.28 and 13.29.

The search for these eigenvalues is delicate and a bit of an adventure, because of the existence of many damped modes of various families, whose eigenvalues lie close to those of the family of interest.[24] It is quite possible to

---

[24] Figure 4.19 of Drazin and Reid shows a map of these neighboring eigenvalues, originally computed by L. Mack.

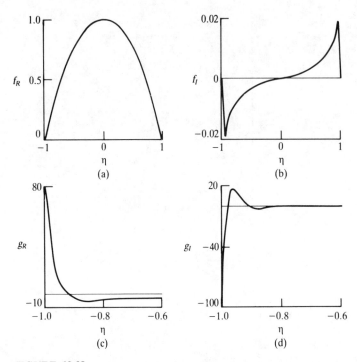

FIGURE 13.28
Eigenfunctions for the most rapidly growing disturbance of plane Poiseuille flow, $\alpha = 0.775$, $Re = 40,000$: (a) f_R; (b) f_I; (c) g_R; (d) g_I. Note expanded scale for η in (c) and (d).

find one of these other modes by accident, depending on the starting point of the search, and the value of the underrelaxation factor.

Notice that the instability of plane Poiseuille flow is very slight; the maximum amplification rates are very small by comparison to those for free shear layers and jets. It is not easy to give a persuasive verbal picture of cause and effect for this instability, and to illuminate the role of viscosity.

As an illustration of this difficulty, we present, and then criticize an expanded version of an argument originally given by Lighthill.[25] Suppose that the convection associated with an inviscid normal-mode disturbance produces a traveling-wave concentration of perturbation vorticity, particularly intense near the critical layer, where the wave travels with the speed of the basic flow. You might imagine a freight train, running along the critical layer. Cars $1, 3, 5, \ldots$ carry positive vorticity; cars $2, 4, 6, \ldots$ carry negative vorticity. As each car passes over a fixed point on the wall, its vorticity and the image of that vorticity behind the wall induce an illegal slip velocity on the wall. To

---

[25] M. J. Lighthill (1963) "Aerodynamical Background", op. cit., Art. 3.2.

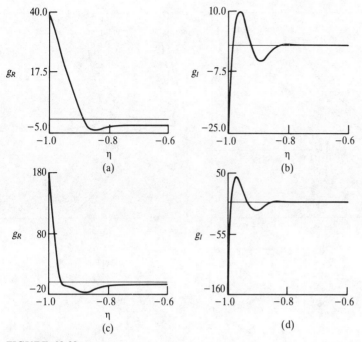

FIGURE 13.29

Vorticity eigenfunctions for neighboring damped disturbances of plane Poiseuille flow, $\alpha = 0.5$, Re = 40,000: (a) f_R; (b) f_I; (c) g_R; (d) g_I.

keep the no-slip condition in force, vorticity of the opposite sign is generated at the wall. This vorticity diffuses outward, toward the critical layer, and is simultaneously convected by the basic flow. Convection by the sheared basic flow may diminish the streamwise modulation of the outward-diffusing vorticity, but this modulation is still significant when the vorticity reaches the critical layer. At this point it will interfere, either constructively or destructively, with the convectively induced vorticity concentration which gave it birth. In the train analogy, positive vorticity induced by the passage of a car full of negative vorticity arrives at the critical layer just at the right time and place to board the following car, adding to its load of positive vorticity.

This sounds plausible, and suggested computation of the maps of perturbation vorticity shown in Fig. 13.30(a, b, c). These are drawn for the most strongly amplified wave, and for a longer (a) and a shorter (c) wave of the same family, each of which is fairly strongly damped. Each figure shows the contours of a ridge of positive perturbation vorticity, which adjoins a valley of negative vorticity on either side. A cross-section of the ridge-valley system at any value of η is a sine wave. The elevation of the ridge line drops precipitously from a maximum at the wall to a broad saddle at about the level of the critical layer, which is indicated by a horizontal line. Beyond the saddle, there is a low hill, and then the ridge continues to the midplane of the channel

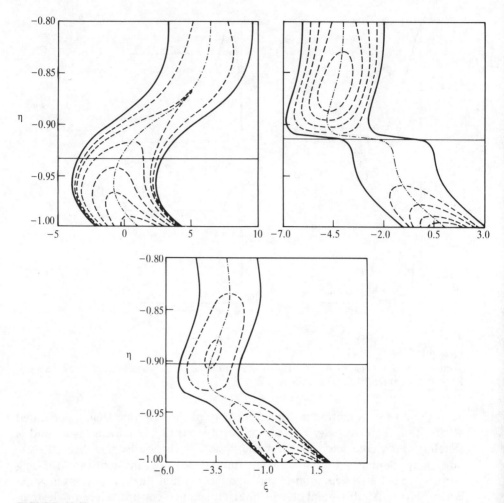

FIGURE 13.30
Contour maps for perturbation vorticity in plane Poiseuille flow, Re = 40,000: (a) $\alpha = 0.5$ (stable); (b) $\alpha = 0.775$ (unstable); (c) $\alpha = 1.05$ (stable).

at a nearly constant, low elevation. We try to understand why the perturbation shown in (b) is destined to grow, while those shown in (a) and (c) are destined to die away.

This has not been a very helpful exercise, because the real distributions of perturbation vorticity do not at all resemble the one imagined in the word picture. Where we imagined a great concentration of vorticity near the critical layer, we find only the saddle in the ridge. Even more disappointing, the contour maps for the amplifying wave and for the shorter decaying wave seem qualitatively very much alike. It is, however, easy to turn off the no-slip condition in the numerical solution, substituting the condition that there be no flux of perturbation vorticity through the wall. If the word picture has any merit at all, this should have a crucial effect. Sure enough, when this is done to the most dangerous disturbance ($\alpha = 0.775$, Re = 40,000), the amplification ceases and the disturbance decays.

What can you have learned from this inconclusive discourse? Take delight in a persuasive word picture, and try to construct them yourself because you may thereby enhance your understanding. However, retain some scepticism, and never cease trying to check the picture against the real thing.

13.8 COLLECTED RESULTS OF NORMAL-MODE THEORY FOR PARALLEL SHEAR FLOWS

Calculations of the sort described above have been made for very many parallel, or nearly parallel, basic flows. Because the instability often proves to be of the convective type, with disturbances that would wash away if not constantly replenished at the upstream boundary of the region under observation, the calculations have also been done with the assumption that the normal-mode frequency, ω, is purely real, and that the wave number, k, is complex.

In this section, a survey of some of the computed results is given, to identify the trends that one might expect to observe experimentally or to exploit in design. A brief description of the way one might attempt to check the theory experimentally is given, in the hope that this will help you to recognize the implications and limitations of the theory, especially when the basic flow is not strictly parallel. Finally, the success or failure of the theory, as a tool for the prediction of the observed breakdown of various laminar flows, will be briefly assessed.

Free shear layers have been found to be very unstable at any value of Re for which the basic flow is even approximately parallel. The shear layer that is often observed experimentally appears at the edge of a jet, and the disturbances propagate downstream at the average of the jet speed and the speed of the surrounding fluid. If this propagation speed is called U, and the shear is called ΔU, the disturbance amplitude grows by a factor $\exp(2\pi c_i/U)$ while the disturbance travels one wavelength. A typical value of c_i may be $0.3\,\Delta U$, and if the jet is discharging into a fluid at rest, $\Delta U \approx 2U$. For this case, the disturbance amplifies by a factor of about 43 in one wavelength! The consequence is that a snapshot of the disturbed flow is likely to show only two or three wiggles that look vaguely sinusoidal, and then an explosive growth into a disturbance that looks very little like a normal mode. Figure 13.31 shows an example.

Many calculations have been made, to see how the stability characteristics of a laminar boundary layer depend on the shape of the velocity profile.[26]

---

[26] For an impressive collection of results, with thoughtful interpretation, see H. J. Obremski, et al., *A Portfolio of Stability Characteristics of Incompressible Boundary Layers*, Agardograph 134, (1969). Available from NTIS, Springfield. Va. 22161.

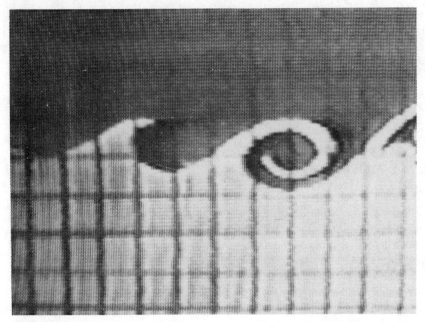

FIGURE 13.31
Rapid development of normal-mode disturbance of free shear layer. $U_{top} = 11.0$ cm/s; $U_{bof} = 2.88$ cm/s. Laser-induced fluorescence in water of lower layer. (Photo courtesy of D. I. Polinsky)

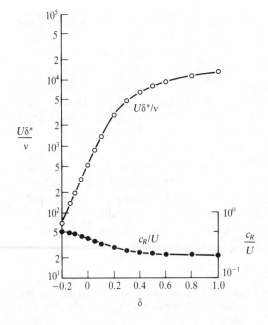

FIGURE 13.32
Effect of pressure gradient on the critical Reynolds number and speed of critical disturbances. Falkner–Skan boundary layers. (Adapted, by permission, from H. J. Obremski et al., 1969.[26])

Quantities such as the minimum critical Reynolds number, maximum frequency of unstable disturbances, and maximum spatial amplification rate can be tolerably well correlated with the value of the shape factor, $H = \delta^*/\theta$. Figures 13.31 and 13.32, based on the calculations of Wazzan, Okamura, and Smith, show these correlations for the Falkner–Skan family of similarity profiles. When data from nonsimilar profiles are included, a scatter band appears, but the general trends survive. High values of Re_c, low values of the high-frequency cutoff, and low values of maximum amplification rate, all denoting relatively stable behavior, correspond to low values of H. Remember that H increases as a result of adverse pressure gradient or other factors, such as blowing through the wall, which displace the point of maximum vorticity away from the wall.

The TS waves found in boundary layers are long waves, relative to the thickness of the boundary layer. The length of the first wave to become unstable, when $\mathrm{Re} = \mathrm{Re}_c$, equals about $10\delta^*$ for the separating Falkner–Skan profile, and about $37\delta^*$ for the stagnation-point boundary layer. The most highly amplified waves at higher values of Re are even longer.

TS waves move downstream with a phase speed that ranges from about 0.1 to 0.5 times the fluid speed at the outer edge of the boundary layer, depending on H and Re, The trends for the Falkner–Skan boundary layers are shown in Fig. 13.32.

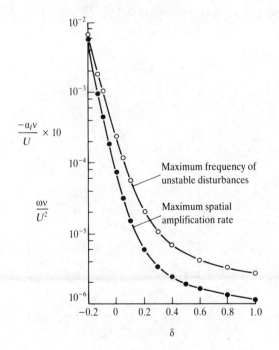

$\dfrac{-\alpha_i \nu}{U} \times 10$

$\dfrac{\omega \nu}{U^2}$

Maximum frequency of unstable disturbances

Maximum spatial amplification rate

δ

FIGURE 13.33
Effect of pressure gradient on the maximum spatial amplification rate and frequency of unstable disturbances. Falkner–Skan boundary layers.

Evolution of a TS Wave as it Moves Downstream

Figure 13.34 shows contours of equal spatial amplification rate for TS waves in a Blasius boundary layer, in a plane of wave frequency versus Reynolds number. We note, in passing, that the maximum spatial amplification rates occur when $Re = 2Re_c$.

Figure 13.34 is very helpful when one tries to calculate the evolution of a TS wave of a certain frequency as it propagates downstream. In many experiments, such waves are deliberately excited by use of a vibrating ribbon at some upstream location. To make this calculation, one must confront, for the first time, that fact that the boundary layer is not really a parallel flow, but that the characteristic length and velocity scales, δ^* and U_∞, change as the wave moves downstream. In the most literal application of the parallel-flow theory, one assumes that the spatial amplification rate, $-k_I$, changes slowly with x to keep up with the changes in the basic flow. Thus, the local amplitude of the disturbance is to be predicted by integrating the equation $(1/A) \, dA/dx = -k_I(x)$, to get

$$A(x) = A(0) \exp\left(-\int_0^x k_I(x) \, dx\right) \tag{13.51}$$

For an example of such a calculation, consider a wave of frequency $\omega = 10^{-4}U^2/\nu$, growing in a Blasius boundary layer. From Fig. 13.34 we see that the wave initially loses amplitude, until $U\delta^*/\nu$ attains a value of about 720. It then grows, with an increasing rate of amplification, until $U\delta^*/\nu \approx 930$,

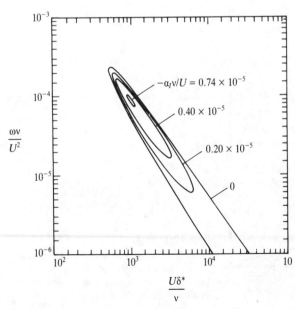

FIGURE 13.34
Spatial amplification map for Blasius boundary layer. (Adapted, by permission, from N. A. Jaffe, T. T. Okamura, and A. M. O. Smith, *AIAA Jour.* **8**: 301–308, 1970.)

and continues to grow, with decreasing amplification rate, until $U\delta^*/v \approx 1250$. Thereafter, it again decays. This all assumes, of course, that it never reaches an amplitude that invalidates the assumptions of a linearized theory.

It turns out that the streamwise length of the region of amplification corresponds to about 15 times the average wavelength of the wave, and that the integral in (13.51) amounts only to about 1.72. Thus, the amplitude of this wave has grown by about a factor of 5.6 while the wave passes through the entire region of amplification. This result is, however, a very strong function of wave frequency, as is shown in Fig. 13.35. For lower values of $\omega v/U^2$, the extent of the region of amplification corresponds to more wavelengths, and the total amplification between branches of the neutral stability curve can be much larger.[27] Thus, TS waves can easily attain amplitudes for which nonlinear interactions between waves become important. A brief introduction to the theory of such interactions will be presented in Section 13.12.

13.9 CENTRIFUGAL INSTABILITY. TAYLOR–GÖRTLER VORTICES

Before we review experimental studies of the instabilities of parallel shear flows, let us consider a second important class of instabilities, in which the curvature of the streamlines of the basic flow plays an essential part. The classic example is sketched in Fig. 13.36.

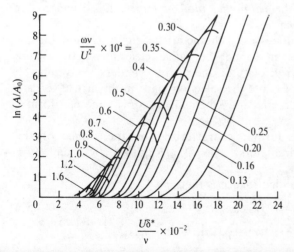

FIGURE 13.35
Total amplification of Tollmien–Schlichting waves of various frequencies. $A_0 =$ amplitude at branch I of neutral curve. Basic flow is Blasius boundary layer. (Reproduced by permission from L. M. Mack, 1984)[21]

---

[27] More examples of this kind of calculation are given by L. M. Mack (1984) op cit., Sec. 6.

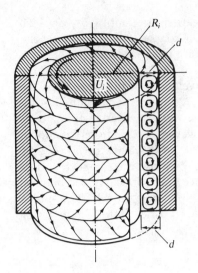

FIGURE 13.36
Taylor vortices. (Reproduced, by permission, from Schlichting, "Boundary-Layer Theory" McGraw-Hill, New York, 1968.)

The Instability of Circular Couette Flow. Taylor Vortices

Consider a basic steady flow with circular streamlines, so that the perturbed velocity field has the form $\mathbf{u} = u(r, \theta, z, t)\mathbf{e}_r + [V(r) + v(r, \theta, z, t)]\mathbf{e}_\theta + w(r, \theta, z, t)\mathbf{e}_z$. The corresponding pressure field is $p = P(r) + p'(r, \theta, z, t)$. Since the basic velocity and pressure profiles depend only on r, one can imagine a normal-mode perturbation that depends on t, z, and θ only through complex exponential factors. The dependence on θ must involve the factor $\exp(in\theta)$, where n is called the *azimuthal mode number*. Disturbances for which $n = 0$ are called *axisymmetric*.

A particular basic flow will possess a characteristic length, H, and angular velocity, Ω. Often, H is chosen to be the radial clearance between the cylinders, and Ω is the angular velocity of one of the cylinders. You must be careful, in using the literature, to note the selection made in the article you are reading. From these parameters and the kinematic viscosity, one can form a Reynolds number, $\text{Re} = \Omega H^2 / v$.

The basic flow will be further characterized by the values of the geometric and kinematic parameters, $\eta \equiv R_{\text{inner}}/R_{\text{outer}} = R_1/R_2$, and $\mu = \Omega_{\text{outer}}/\Omega_{\text{inner}} = \Omega_2/\Omega_1$.

The assumed form of the disturbance is then

$$\mathbf{u}(r, \theta, z, t) = \mathbf{u}^*(r) \exp[i(n\theta + kz - \omega t)] \tag{13.52}$$

$$p'(r, \theta, z, t) = p^*(r) \exp[i(n\theta + kz - \omega t)]$$

Disturbances that are independent of z, and in which $w = 0$, are called *two-dimensional* disturbances. They are cousins to the Tollmien–Schlichting waves, but not particularly dangerous.

It is helpful to introduce a dimensionless radial coordinate, $\xi = (r - r_m)/H$, where $r_m = (R_1 + R_2)/2$, and dimensionless disturbance parameters $\alpha = kH$, and $\sigma = \omega/\Omega$.

The disturbance equations. The general procedure used to derive the Orr–Sommerfeld equation can be applied to this problem, starting with the momentum and continuity equations in cylindrical polar coordinates. With velocity components measured in units of ΩH and pressure in units of $\rho(\Omega H)^2$, the first results are

r-momentum

$$[i(nF - \sigma) - \text{Re}^{-1} E]u^* - [2F - 2in\,\text{Re}^{-1}\,\xi^{-2}]v^* = \frac{-dp^*}{d\xi} \qquad (13.53)$$

q-momentum

$$[i(nF - \sigma) - \text{Re}^{-1} E]v^* + [G - 2in\,\text{Re}^{-1}\,\xi^{-2}]u^* = -np^* \qquad (13.54)$$

z-momentum

$$[i(nF - \sigma) - \text{Re}^{-1}(E + \xi^{-2})]w^* = -i\alpha p^* \qquad (13.55)$$

continuity

$$\frac{du^*}{d\xi} + \frac{u^*}{\xi} + \frac{inv^*}{\xi} + i\alpha w^* = 0 \qquad (13.56)$$

The dimensionless coefficient functions F and G represent the angular velocity $V/\Omega r$ and vorticity $\Omega^{-1}[dV/dr + V/r]$ of the basic flow. Recalling that $V = Ar + B/r$, we see that G is just the constant $2A/\Omega$, while $F = (A/\Omega)[1 - (B/A)r^{-2}]$. For future reference, we recall that

$$A = \Omega_1 \frac{\mu - \eta^2}{1 - \eta^2} \qquad \text{and} \qquad B = \Omega_1 R_1^2 \frac{1 - \mu}{1 - \eta^2}$$

The symbol E represents the differential operator

$$E = \frac{d^2}{d\xi^2} + \xi^{-1}\frac{d}{d\xi} - n^2 - \alpha^2$$

Equation (13.56) can be used to eliminate w^* from (13.55), and the resulting equation can be used to eliminate p^* from (13.54) and (13.53). This produces a pair of equations for u^* and v^*. They include $d^4u^*/d\xi^4$ and $d^2v^*/d\xi^2$ as the highest-order derivatives, and hence constitute a sixth-order system. All three disturbance velocity components vanish on both cylinders. Equivalently, u^*, v^* and $du^*/d\xi$ must vanish there. The problem is to find eigenvalues of σ that allow the boundary conditions to be satisfied for specified values of μ, η, Re, n and α. By now, efficient numerical procedures for solving the problem for rather general combinations of parameters are available.

Axisymmetric disturbances in a small gap, with only the inner cylinder rotating. With so many parameters and such complicated disturbance equations, it is hard to survey all the interesting possibilities. To convey the essence of the subject, we shall continue the analysis for only one case, in which $\eta \approx 1$ and $\Omega_2 = 0$. Taylor discovered that the first unstable disturbances to be encountered in this case, when the speed of the inner cylinder is gradually increased from zero, are axisymmetric and nonpropagating.[28] Although the basic velocity profile approaches that of plane Couette flow as $\eta \to 1$, the essential centrifugal instability does not go away, as long as η is not exactly equal to 1. The smallness of $1 - \eta$ does allow one to drop terms like u^*/ξ when they are compared to $du^*/d\xi$, and it allows replacement of the variable coefficient F by a constant. When only the inner cylinder is rotating, the natural scale for perturbations of the circumferential velocity component, v, is $\Omega_1 R_1$. On the other hand, a convenient scale for the radial and axial components, u and w, turns out to be $\Omega_1 (R_1 \delta)^{1/2}$, where $\delta = R_2 - R_1$. A natural definition of a Reynolds number, analogous to the one we have used for slender viscous layers in Chapter 11, is $\mathrm{Re} = \Omega_1 (R_1 \delta)^{1/2} \delta / \nu$. If we then define $u = u^{**} \Omega_1 (R_1 \delta)^{1/2}$, $v = v^{**} \Omega_1 R_1$, and $w = w^{**} \Omega_1 (R_1 \delta)^{1/2}$, and take $n = 0$, the disturbance equations become, after elimination of p' and w^{**},

$$(\mathrm{DLD} - \alpha^2) u^{**} = 2\alpha^2 \,\mathrm{Re}\, F v^{**} \tag{13.57}$$

and
$$L v^{**} = \mathrm{Re}\, G u^{**} \tag{13.58}$$

where D and L are the operators $D = d/d\xi$ and $L = D^2 - \alpha^2 - i\,\mathrm{Re}\,\sigma$. In the limit as $\eta \to 1$, $F \to 1$ and $G \to -1$.

A further simplification occurs when one searches only for neutral, nonpropagating normal modes, for which $\sigma = 0$. Then (13.57) and (13.58) can be combined into a single equation with constant coefficients:

$$[(D^2 - \alpha^2)^3 + 2(\alpha \,\mathrm{Re})^2] u^* = 0 \tag{13.59}$$

The quantity $2\,\mathrm{Re}^2 = 2\Omega_1^2 R_1 \delta^3 / \nu^2$ is often given the symbol Ta, and is called a *Taylor number.*

In principle, the solution to (13.59) is easily found in closed form, as

$$u^* = \sum_{i=1}^{6} A_i \exp(\lambda_i \xi)$$

---

[28] See G. I. Taylor, "Experiments with Rotating Fluids," *Proc. Camb. Phil. Soc.* **20**:326–9 (1921), and G. I. Taylor, "Stability of a Viscous Liquid Contained Between Two Rotating Cylinders," *Phil. Trans. Roy. Soc.* **A223**:289–343 (1923). Results for nonaxisymmetric disturbances were first obtained by R. C. DiPrima, "Stability of Nonrotationally Symmetric Disturbances in Viscous Flow between Rotating Cylinders," *Phys. Fluids* **4**:751–755 (1961).

provided that the six λ_i are solutions of the equation $(\lambda_i^2 - \alpha^2)^3 + \alpha^2 Ta = 0$. With $H = \delta$, the boundary conditions are applied at $\xi = \pm\frac{1}{2}$. They produce the six equations

$$0 = \sum_{i=1}^{6} A_i \exp\frac{\lambda_i}{2} \qquad 0 = \sum_{i=1}^{6} A_i \exp\frac{-\lambda_i}{2} \qquad \text{(from } u^* = 0)$$

$$0 = \sum_{i=1}^{6} A_i\lambda_i \exp\frac{\lambda_i}{2} \qquad 0 = \sum_{i=1}^{6} A_i\lambda_i \exp\frac{-\lambda_i}{2} \qquad \text{(from } Du^* = 0)$$

$$0 = \sum_{i=1}^{6} A_i(\lambda_i^2 - \alpha^2)^3 \exp\frac{\lambda_i}{2} \qquad 0 = \sum_{i=1}^{6} A_i(\lambda_i^2 - \alpha^2)^3 \exp\frac{-\lambda_i}{2} \qquad \text{(from } v^* = 0)$$

The eigenvalue relationship is then found by setting the determinant of this set of equations equal to zero. Obviously, this would be a rather monstrous algebraic task, and it is usual to seek numerical values in some other way, for example, by finite-difference analysis. Other important techniques are presented by Drazin and Reid, on pages 94–97 of their text.

For finite-difference work, it is convenient to convert (13.57) and (13.58) into three coupled second-order equations

$$(D^2 - \alpha^2)s^{**} = 2\alpha^2 \operatorname{Re} v^{**}$$

$$(D^2 - \alpha^2)v^{**} = -\operatorname{Re} u^{**}$$

$$(D^2 - \alpha^2)u^{**} = s^{**}$$

These apply to the search for the locus $T(\alpha)$ along which $\sigma = 0$.

To employ techniques with which we are familar, we replace the differential equations by three-point finite-difference equations, of the form $\mathbf{A}y_{n-1} + \mathbf{B}y_n + \mathbf{C}y_{n+1} = 0$, in which $\mathbf{A}$ and $\mathbf{C}$ are unit matrices and $\mathbf{y}$ is the solution vector with components $y_1 = s$, $y_2 = v$, and $y_3 = u$. The other coefficient matrix is

$$\mathbf{B} = -\begin{bmatrix} 2 + (\alpha\, d\xi)^2 & 2(\alpha\, d\xi)^2 \operatorname{Re} & 0 \\ 0 & 2 + (\alpha\, d\xi)^2 & -d\xi^2 \operatorname{Re} \\ d\xi^2 & 0 & 2 + (\alpha\, d\xi)^2 \end{bmatrix}$$

From the back-substitution formula, $\mathbf{y}_n = \mathbf{E}_n\mathbf{y}_{n+1}$, and the boundary conditions, one finds that $E_{31\,1} = 2/d\xi^2$, and that all other members of $\mathbf{E}_1$ are zero.

The eigensolution vector $\mathbf{y}$ will be either an even or an odd function of ξ. Taylor found that the even functions are more dangerous disturbances, and so we search for them. As in the case of the even nodes of disturbance of plane Poiseuille flow, the eigenvalue condition is

$$\|2\mathbf{E}_m + \mathbf{B}_{m-1}\| = 0 \tag{13.60}$$

where $n = m$ denotes the middle of the gap. The eigenvalues search is made by

TABLE 13.1

α	Ta	α	Ta	α	Ta
2.0	2176.9	3.0	1710.9	4.0	1878.9
2.2	1993.7	3.2	1709.1	4.2	1960.4
2.4	1868.1	3.4	1726.6	4.4	2057.0
2.6	1785.1	3.6	1761.5	4.6	2168.4
2.8	1734.8	3.8	1812.5	4.8	2295.8

The minimum Taylor numer is Ta = 1707.76 at $\alpha = 3.117$
This corresponds to Re = 29.221

the secant method, as in our other stability calculations. Program TAYLOR, shown in Appendix B, executes this algorithm. Some results that agree with those obtained by other accurate methods are shown in Table 13.1.

The eigenfunctions u^*, v^*, and w^* are shown in Fig. 13.37 for the critical case. The projection of the disturbance velocity on a plane of constant θ is shown in Figure 13.38. Note that the Taylor vortices come in matched, counter-rotating pairs. The fact that α very nearly equals π for the most dangerous disturbance implies that the disturbance motion forms a cellular vortex which is nearly square in cross section.

For a pair of cylinders with $R_1 = 10$ cm and $R_2 = 11$ cm, containing water with kinematic viscosity $v = 0.01$ cm^2/sec, the critical Taylor number is reached when $\Omega_1 \approx 0.00385$ rad/sec, or about 0.23 RPM. For obvious reasons, most experimental work is done with more viscous fluids.

It turns out that the Taylor-vortex flow becomes fairly vigorous, but retains its steadiness and general appearance if Re is gradually increased somewhat beyond the value at which the vortices first appear. Then, at a higher critical value of Re, which depends greatly on the value of η, they

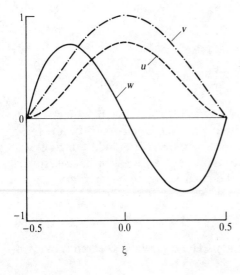

FIGURE 13.37
Eigenfunctions for the Taylor vortex in a narrow gap: Ta = 1707.76; $\alpha = 3.117$.

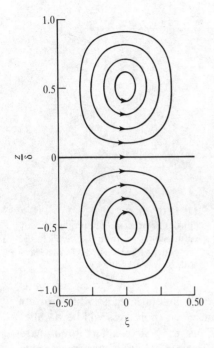

FIGURE 13.38
Streamlines of secondary flow for Taylor vortices in a narrow gap: Ta = 1707.76; $\alpha = 3.117$.

become unstable and give way in turn to another, somewhat more complicated, flow. We shall say more about these fascinating phenomena in a later section, but first direct attention to a related instability that appears in a somewhat greater variety of flows.

Instability of Boundary-Layer Flow Along a Concave Wall. Görtler Vortices

Another important example of centrifugal instability appears in a laminar boundary layer that flows along a wall with concave longitudinal curvature. Figure 13.39 illustrates the situation. Clearly the circulation around the local center of wall curvature, uR, decreases with distance from the center. According to Rayleigh's criterion, the flow is centrifugally unstable, and if viscosity does not stabilize it we may expect to see something akin to Taylor vortices growing in the boundary layer. Such vortices are indeed observed, but they are comparatively difficult to study quantitatively. This study was started by H. Görtler[29] in 1940, and continues to be challenging to this day.

---

[29] H. Görtler, "On the Three-Dimensional Instability of Laminar Boundary Layers on Concave Walls," 1940. (English translation in NACA Tech. Memo. 1375)

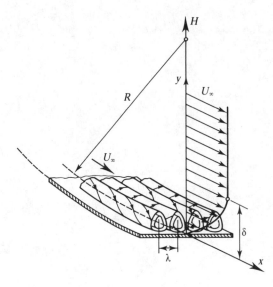

FIGURE 13.39
Görtler vortices. (Reproduced, by permission, from Schlichting, "Boundary-Layer Theory" McGraw-Hill, New York, 1968.)

Observations in two-dimensional boundary layers suggest that Görtler vortices, given a chance, will be steady. In plan view, their axes will be parallel to the flow outside the boundary layer. Thus they are called *streamwise vortices*. Once they can be seen, they tend to preserve whatever transverse size and spacing they acquired at birth.

The theoretical study of such vortices is quite subtle, and the literature is full of competing formulations of a linearized theory. The assumptions that the theories all embody include the following.

1. The curvature of the wall is gentle, in the sense that the radius of curvature, R, is very large compared to the streamwise extent, L, of the region in which the vortices develop.
2. The boundary layer is thin, in the sense that its thickness, δ, is very small compared to L. Specifically, δ/L is of order $(U_0L/v)^{-1/2}$, where U_0 is a typical speed of the flow outside the boundary layer.
3. The two small parameters, L/R and δ/L, are of comparable order of magnitude, so that their ratio, $L^2/\delta R$, is to be held constant while either parameter tends toward zero. This ratio, or its square root, is called the *Görtler number*.
4. The normal and transverse components of the velocity perturbation due to a Görtler vortex are systematically smaller, roughly by a factor δ/L, than the streamwise component.
5. The perturbed flow is steady, and varies sinusoidally in the transverse direction.
6. Streamwise diffusion of the vorticity of the perturbation is negligible, but transverse diffusion is important.

7. The gradient of the pressure perturbation has a negligible streamwise component. Note that assumptions (6) and (7) imply that the streamwise development of the vortex is governed by partial differential equations that can, if necessary, be solved by a downstream-marching numerical procedure. Since the variation with the transverse coordinate is given analytically, the equations become parabolic, and the task of integration is not essentially more complicated than that required to compute the evolution of the basic boundary-layer flow.

This view of the problem does not, however, do justice to the historical development of its analysis. That development is very closely tied to the ideas and procedures developed in the study of the Taylor vortices between rotating cylinders, and the study of TS waves in parallel or nearly parallel flows. All of these studies lead, either by natural symmetries or by grim necessity and good luck, to an eigenvalue problem for a set of ordinary differential equations. The diversity of linearized treatments of the Görtler instability arose from efforts to reduce the partial differential equations, on which all authors agreed, to a representative set of ordinary differential equations that captures the essence of the phenomenon and identifies a minimal set of important parameters. It has been difficult to achieve agreement on how to do this, because it turns out that the Görtler vortices grow rather slowly, in a direction in which the boundary layer itself is growing. A successful set of ordinary differential equations must embody, in its representation of the basic flow, just the right amount of information about the growth of the boundary layer.

For the purposes of this chapter, a demonstration of the numerical integration of the parabolic equations would represent overkill; we shall only exhibit the equations, and then discuss the results of one of the most recent reductions to a set of ordinary differential equations.

The perturbed velocity and pressure fields are taken to be

$$\frac{u}{U_\infty} = U(x, y) + u'(x, y, z)$$

$$\frac{v}{U_\infty} = \frac{v}{U_\infty \delta} [V(x, y) + v'(x, y, z)]$$

$$\frac{w}{U_\infty} = \frac{v}{U_\infty \delta} w'(x, y, z)$$

$$\frac{p}{\rho U_\infty^2} = P(x, y) + \left(\frac{v}{U_\infty \delta}\right)^2 p'(x, y, z)$$

Capital letters denote properties of the basic flow; x is arc length downstream along the wall, y is distance normal to the wall, z is transverse distance along the wall. Dimensionless coordinates are introduced by $\xi = x/L$, $\eta = y/\delta$,

$\zeta = z/\delta$. The reason for the scaling of z is that the vortices usually seen are fairly square in cross-section and lie mostly within the boundary layer. The Navier–Stokes equations are written in terms of these variables, and all terms multiplied by any of the small factors δ/L, δ/R, or powers thereof, including the factor $(\delta/L)^3 d(L/R)/dx$, are dropped when compared to unity. What remains are the familiar boundary-layer equations for the two-dimensional basic flow, and the following partial differential equations for the perturbations.

$$Uu'_{,\xi} + Vu'_{,\eta} + u'U_{,\xi} + v'U_{,\eta} = u'_{,\eta\eta} + u'_{,\zeta\zeta} \tag{13.61}$$

$$Uv'_{,\xi} + Vv'_{,\eta} + u'V_{,\xi} + v'V_{,\eta} - 2G^2\left(\frac{L}{R(\xi)}\right)Uu' = -p'_{,\eta} + v'_{,\eta\eta} + v'_{,\zeta\zeta} \tag{13.62}$$

$$Uw'_{,\xi} + Vw'_{,\eta} = -p'_{,\zeta} + w'_{,\eta\eta} + w'_{,\zeta\zeta} \tag{13.63}$$

and

$$u'_{,\xi} + v'_{,\eta} + w'_{,\zeta} = 0 \tag{13.64}$$

Since none of the coefficients in these linear equations depends on z, we search for solutions of the form

$$u'(x, y, z) = u^*(x, y)\cos\left(\frac{2\pi z}{\lambda}\right) \qquad v'(x, y, z) = v^*(x, y)\cos\left(\frac{2\pi z}{\lambda}\right)$$

$$w'(x, y, z) = w^*(x, y)\sin\left(\frac{2\pi z}{\lambda}\right) \qquad p'(x, y, z) = p^*(x, y)\cos\left(\frac{2\pi z}{\lambda}\right)$$

A useful dimensionless parameter, $\Lambda = (U_0\lambda/v)(\lambda/R_0)^{1/2}$, can be formed from the transverse wavelength. Here R_0 is a constant, average, value of R.

The equations that descend from (13.63) and (13.64) can be combined to eliminate w^*, yielding

$$p = -\left(\frac{\lambda}{2\pi}\right)^2 [U(u^*_{,\xi\xi} + v^*_{,\xi\eta}) + V(u^*_{,\xi\eta} + v^*_{,\eta\eta}) - (u^*_{,\xi\eta\eta} + v^*_{,\eta\eta\eta})] + (u^*_{,\xi} + v^*_{,\eta})$$

This would seem to introduce second derivatives with respect to ξ, but we can use the descendant of (13.61) to eliminate $u^*_{,\xi}$ in favor of quantities which contain no unknown ξ-derivatives. When the resulting expression for p^* is used in (13.62), we get an equation in which the highest-order streamwise derivatives of the unknowns are $u^*_{,\xi}$ and $v^*_{,\xi}$. This equation, together with (13.61), can be solved, with suitable care, by a downstream-marching numerical procedure. Initial profiles of u^* and v^* must be provided at the upstream end of the region of interest. There is no logically inevitable way to choose these initial profiles; one simply hopes that there is sufficient natural reshaping of the profiles as ξ increases, so that the details of the presumed upstream conditions have become unimportant at downstream locations where the disturbances attain an easily measurable amplitude. The ξ-dependence of basic-flow quantities can be routinely fed into the integration as it is marched downstream.[30]

The variety of published predictions of the growth of Görtler vortices arises from various efforts to reduce these partial differential equations to ordinary differential equations. The usual procedure is to ignore the ξ-variability of the coefficients, U, V, and their derivatives, and to represent the corresponding variability of u^* and v^* by the equations

$$u^*(\xi, \eta) = u^{**}(\eta) \exp (\sigma\xi) \quad \text{and} \quad v^*(\xi, \eta) = v^{**}(\eta) \exp (\sigma\xi)$$

The resulting ordinary differential equations, together with the boundary conditions that u^{**}, v^{**}, and $du^{**}/d\eta$ should vanish both at $\eta = 0$ and $\eta = \infty$, constitute an eigenvalue problem very much like the Taylor-vortex problem. Solution of the problem yields the amplification rate σ for each mode of motion that is possible for given values of the Görtler number G and transverse wavelength parameter Λ. Numerical results vary from author to author, even when there is no doubt of the accuracy of numerical work, for at least two reasons:

1. Retention or discarding of various terms in the ordinary differential equation. Thus, Görtler discarded all terms involving $U_{,\xi}$, V, or $V_{,\xi}$, treating the basic flow as a parallel flow except for the term that explicitly brings in the longitudinal curvature. His equations thus reduced to those of the Taylor problem, with a different function for $U(\eta)$. At the other extreme, Floryan and Saric[31] retain all the terms exhibited in (13.61) to (13.64)

2. Treatment of the variation of basic-flow streamline curvature, as a function of η. In an interesting review paper,[32] it is shown that this can have a strong effect on numerical results, thus warning that any set of results may apply only to a rather specifically-defined potential flow.

After the eigenvalue problem is solved, it is recognized that G will vary with ξ, and the amplitude growth of a given vortex mode is calculated from an integral formula similar to (13.51). This procedure, which works fairly well for TS waves, is based on the hope that a physically important amplification of an

---

[30] For such integrations, see P. Hall, "The Nonlinear Development of Görtler Vortices in Growing Boundary Layers," *J. Fluid Mech.* **193**:243–266 (1988). Hall has been the principal advocate of this approach to the Görtler-vortex problem.

[31] J. M. Floryan and W. S. Saric, "Stability of Görtler–Vortices in Boundary Layers," *AIAA Jour.* **20**:316–324 (1982).

[32] Th. Herbert, "On the Stability of the Boundary Layer Along a Concave Wall," *Arch. Mech. Stosowanej,* **28** (5–6):1039–1055 (1976).

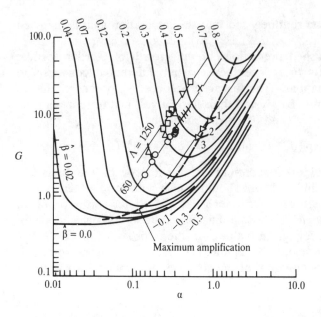

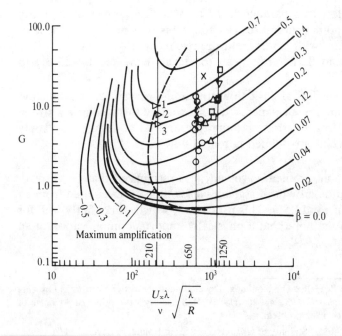

FIGURE 13.40
Amplification map for Görtler vortices. (Reproduced, by permission, from J. M. Floryan and W. S. Saric, 1982.[31])

unstable disturbance will occur over a short interval of ξ, throughout which the coefficient functions in equations such as (13.61) and (13.62) remain fairly constant.

Figure 13.40 shows contours of constant temporal amplification rate, on a plot of Görtler number versus the wavelength parameter Λ, from the paper of Floryan and Saric. It has two interesting features.

1. When G is only slightly above the predicted minimum value for any instability, growth rates are nearly equal for a wide range of values of Λ. This may help to explain the apparent lack of any strong wavelength selectivity in experimenal realizations of the instability. In experiments during which no attempt is made to control the wavelength artificially, the tiniest imperfections of the experimental apparatus seem to determine the spacing and location of the vortices.

2. Once the value of G exceeds the minimum critical value by about a factor of 2, the locus of maximum amplification rate stays fairly close to the line $\Lambda = 200$. In one experiment in which an effort was made, by the use of screens, to provide a broad spectrum of initial disturbances,[33] the resulting Görtler vortices did appear at about $\Lambda = 210$.

13.10 INSTABILITY OF THREE-DIMENSIONAL BOUNDARY LAYERS

As was hinted in Chapter 12, three-dimensional laminar boundary-layer flows, such as are found on swept wings and on spinning bodies, are likely to be particularly unstable. Recall, from our discussion of Squire's theorem, that the profile of $\mathbf{k} \cdot \mathbf{U}$ governs the development of the TS waves that have wave number $\mathbf{k}$. Because of the secondary flow, or crossflow, in three-dimensional boundary layers, there is always a range of directions of $\mathbf{k}$ for which the profile of $\mathbf{k} \cdot \mathbf{U}$ exhibits an inflection point, indicating the likelihood of a strong inviscid instability. The resulting phenomenon, often called *crossflow instability*, is well documented in Van Dyke's *Album*.[34]

The most frequently observed disturbance in this kind of flow appears to be an array of co-rotating stationary vortices, with axes nearly perpendicular to that $\mathbf{k}$ for which the profile of $\mathbf{k} \cdot \mathbf{U}$ has an inflection point and a zero-crossing point at the same distance from the wall. This rule of thumb allows a fairly successful prediction of the orientation of the vortices, but a full stability analysis is again mathematically fairly complicated.

---

[33] H. Bippes and H. Görtler, "Driedimensionale Störungen in der Grenzschicht an einer konkaven Wand," *Acta Mech.* **16**;251–267 (1972).

[34] M. D. Van Dyke, *An Album of Fluid Motion*, Parabolic Press, Stanford, CA, 1982. See Figs. 132–135.

Stability of Flow on a Spinning Disk

The classic example of crossflow instability causes the breakdown of steady laminar flow induced by a spinning circular disk.[35] The basic flow is described by that exact, self-similar, solution of the Navier–Stokes equations first found by von Kármán, and studied in Chapter 10. The disturbance equations, written in cylindrical polar coordinates, have coefficients that are independent of t and θ, so that each elementary disturbance contains the complex exponential factor $\exp[i(n\theta + \sigma t)]$. Thus, for example,

$$\mathbf{u}'(r, \theta, z, t) = \mathbf{u}^*(r, z) \exp[i(n\theta + \sigma t)]$$

Let us decide that the frame of reference in which $\mathbf{u}'$ is observed is stationary. To explore disturbances that appear stationary in a frame of reference that spins with the disk, at angular speed Ω, one can take the exponential factor to be $\cos[n(\theta - \Omega t)]$. Thus, n cycles of oscillation of the flow will be observed at a fixed point for each rotation of the disk.

The resulting disturbance equations will be partial differential equations in r and z. The analytical situation looks superficially similar to that encountered in the study of Görtler vortices, but differs in that radial diffusion and radial gradients of the perturbation pressure are important in the present problem, because the vortices are more nearly aligned with circles of constant r than with rays of constant θ. Thus, the partial differential equations cannot be solved, even for high values of the local Reynolds number, $\mathrm{Re} = r(\Omega/v)^{1/2}$, by a numerical procedure which marches outward in r. The experimental observations suggest, however, that the disturbances amplify by a factor of several thousand while the important r-dependent coefficients of the disturbance equations vary by only about 30%. Thus, there is a strong case for the further approximation

$$\mathbf{u}^*(r, z) = \mathbf{u}^{**}(z) \exp(i\alpha r)$$

where α is taken to be a complex constant. The real part of α, together with the value of the azimuthal mode number, n, determines the angle between the spiralling vortices and a circle of constant r. The imaginary part determines the rate of radial growth of disturbance amplitude. The eigenfunctions $\mathbf{u}^{**}$ and p^{**} are then found by solving ordinary differential equations that can be organized as a sixth-order system rather similar to that used to analyze the growth of Taylor or Görtler vortices.

The earliest analyses of this stability problem stripped terms away from the full linear disturbance equations until they were reduced to the Orr–Sommerfeld equation. The resulting theory predicted a minimum critical value

---

[35] N. Gregory, J. T. Stuart, and W. S. Walker, "On the Stability of Three-Dimensional Boundary Layers, with Application to the Flow due to a Rotating Disk," *Phil. Trans. Roy. Soc., London* **A248**: 155–199 (1955).

of the Reynolds number, $\text{Re} \approx 176$, which was somewhat lower than the observed value. The angle of the vortex spiral, which is observed to be about 14°, corresponded fairly well with the computed value for greatest radial amplification rate. Modern calculations,[36] based on a very slightly reduced version of the sixth-order differential equations, raise the predicted critical value to $\text{Re} = 285.36$, which seems quite well in accord with most experiments. At the critical point, the predicted values of disturbance parameters are $\alpha_R = 0.38482$ and $n = 22.14$, so that the angle between vortex axis and a circle of constant r is 11.40°. We shall call this angle ε.

The neutral curve is sketched in Fig. 13.41 for a wide range of Re, first as resultant wavenumber k versus Re; then as ε versus Re. Along the branch with the higher values of k, called the upper branch, the characteristics of the neutral disturbance tend, as Re increases, towards those predicted by applying Rayleigh's equation to the profile of $\mathbf{k} \cdot \mathbf{U}$. Along the lower branch, the value of ε approaches 39.64° as $\text{Re} \to \infty$. The axes of the vortices become normal to the direction of maximum basic-flow skin friction.

The radial amplification rate, α_I, has been computed[37] as a function of n and Re, with results that are shown in Fig. 13.42. The same computations yield values of α_R from which we can compute $\varepsilon = \arctan(n/r\,\alpha_R)$. Figure 13.43 shows the integrated amplification $A(\text{Re}, n)/A(250, n) = \exp(\int_{250}^{\text{Re}} \sigma\, d\text{Re})$, where $\sigma = -\alpha_I (v/\Omega)^{1/2}$, and the path of integration is one of constant n. We see that the values of n and ε at the peaks of the amplification curves increase as r, and hence Re, increases. It is hard to find visual evidence of this trend; pictures suggest rather that n remains fairly independent of r. This warns us that there is no reason to expect, in general, that visible disturbances of an unstable flow will have the appearance of any single eigenfunction, unless that eigenfunction has been artificially excited to a level at which it overwhelms all other eigenfunctions that may accidentally be present. This topic will be revisited in Section 13.12.

Flow on Other Spinning Bodies

There are many close cousins to the flow on a spinning disk, and fascinating studies have been made of their instabilities. Flow induced by spinning a cone in a fluid at rest, or with its axis aligned to the flow in a wind tunnel, has been studied intensively; so has the flow induced by spinning a sphere. All these flows exhibit regular spiral vortices, or sometimes annular vortices, as an initial feature of the transition to turbulent flow. The vortices on a disk, on relatively blunt cones, and on a sphere all seem to have the same sense of rotation. As

---

[36] M. R. Malik, "The Neutral Curve for Stationary Disturbances in Rotating-Disk Flow," *Jour. Fluid Mech.* **164**:275–287 (1986).

[37] L. M. Mack "The Wave Pattern Produced by Point Source on a Rotating Disk," AIAA Paper 85-0490. (1985).

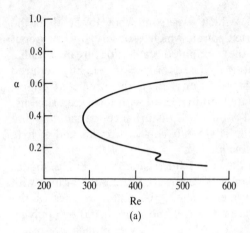

(a)

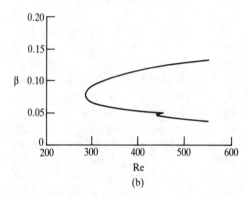

(b)

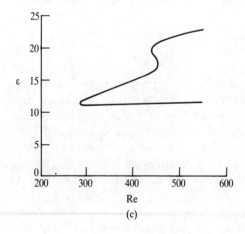

(c)

FIGURE 13.41
Neutral stability curve for a spinning disk. (Reproduced, by permission, from M. R. Malik, 1986.[36])

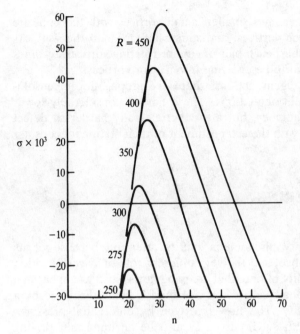

FIGURE 13.42
Radial amplification rate for cross-flow instability. (Reproduced, by permission, from L. M. Mack, 1985.[37])

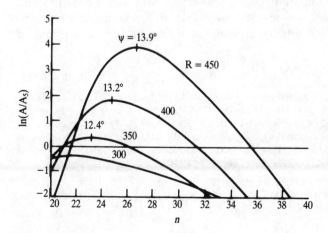

FIGURE 13.43
Integrated amplification as functions of azimuthal wave number and Reynolds number. (Reproduced by permission from L. M. Mack, 1985.[37])

the apex angle of the cone becomes smaller, small vortices with the opposite rotation appear between these vortices, and somewhat closer to the wall. As the cone angle decreases further, each pair of counterrotating vortices becomes more symmetric, as they seem to be evolving into Taylor vortices.[38]

Besides the stationary eigenfunctions, there are propagating eigensolutions of the crossflow instability equations. These have a considerably lower minimum critical Reynolds number, but are experimentally harder to detect and do not seem to interfere with the establishment of their stationary cousins.

13.11 VISCOUS FINGERING

One more class of instability phenomena merits attention because of its widespread practical importance, and the very different role of viscosity within it. This is basically an instability of bounded creeping flows, such as appear in a porous medium or a Hele–Shaw cell, when one fluid is being displaced by a second fluid of lower viscosity. The flow is driven by differential pressures imposed at the open boundaries. Gravity may also be an important driving force, and surface tension provides an important stabilizing effect if the two fluids are immiscible. This instability greatly affects the efficiency of secondary recovery of petroleum, when water is injected into the oil-bearing rocks or sands, in an attempt to expel the oil.

Figure 13.44 shows an example of the results of this instability in a Hele–Shaw cell. As you can imagine from the appearance of this picture, it is extremely difficult to obtain a comprehensive theoretical explanation, especially of the nonlinear developments of the flow. Serious study of the phenomena appears to have started in 1952, with the work of S. Hill, who was concerned with the displacement of a concentrated sugar solution by fresh water, and vice versa.[39] This study continues very actively at the present time, and many important questions are still unanswered.[40] All that is presented here are the simplest imaginable linear analyses of fingering instability, and an

---

[38] R. Kobayashi, Y. Kohama, and M. Kurosawa, "Boundary-Layer Transition on a Rotating Cone in Axial Flow," *J. Fluid Mech.* **127**:341–352 (1983). R. Kobayashi and H. Izumi, "Boundary-Layer Transition on a Rotating Cone in Still Fluid," *J. Fluid Mech.* **127**:353–364 (1983). Y. Kohama and R. Kobayashi, "Boundary-Layer Transition and the Behaviour of Spiral Vortices on Rotating Spheres," *J. Fluid Mech.* **137**:153–164 (1983). Y. Kohama, "Some Expectations on the Mechanism of Crossflow Instability," *Acta Mech.* **66**:21–38 (1987).

[39] S. Hill, "Channeling in packed columns," *Chem. Engr. Sci.* **1**:247–253 (1952).

[40] A fascinating overview is given in G. M. Homsy, "Viscous Fingering in Porous Media," *Ann. Rev. Fluid Mech.* **19**:271–311 (1987).

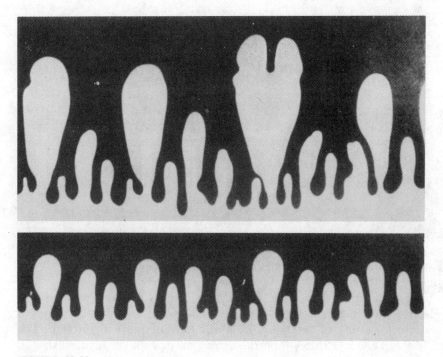

FIGURE 13.44
Example of viscous fingering. (Reproduced, by permission, from T. Maxworthy, *Phys. Fluids* **2B,** No. 9: Gallery of Fluid Motion, page 2637, 1985.)

invitation to you to keep an eye on this fascinating problem in the future. The linear analysis is originally due to Saffman and Taylor, and to Chouke, et al.[41]

Linear Analysis of the Stability of an Initially Plane Interface Between Immiscible Fluids

Let the basic state be one of rectilinear creeping flow of two incompressible fluids that make contact with one another across a plane interface normal to the x axis, see Fig. 13.45. In a porous medium, the contact is necessarily made throughout an interfacial zone of finite thickness, but we assume that any macroscopic ripple superposed on the interface will have a wavelength much greater than this thickness. Also, we recognize that the effects of surface tension in a porous medium may be extraordinarily complex, while in this

---

[41] P. Saffman and G. I. Taylor, "The Penetration of a Fluid into a Porous Medium or Hele–Shaw Cell Containing a More Viscous Liquid," *Proc. Roy. Soc.* **A245**:312–329 (1958). R. L. Chouke, P. von Meurs, and C. van der Poel, "The Instability of Slow, Immiscible Viscous Liquid–Liquid Displacements in Permeable Media," *Pet. Trans. AIME* **216**:188–194 (1959).

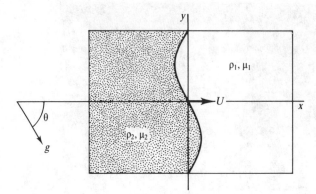

FIGURE 13.45
Definition sketch for viscous fingering.

analysis we conceive only of a surface tension associated with the macroscopic curvature of the disturbed interface. Both of these concerns are minimized when we consider the flow through a Hele–Shaw cell, instead of a porous medium.

In the undisturbed flow, each fluid has the velocity $\mathbf{u} = U\mathbf{e}_x$, which is also the velocity of the interface. For subsequent analysis, it is convenient to use a steadily moving frame of reference in which the unperturbed interface lies always in the plane $x = 0$. Let θ be the angle between the positive x axis and the acceleration of gravity. Finally, denote properties ahead of the advancing interface by subscript 1, and those behind by subscript 2.

The macroscopic velocity of a creeping flow under these constraints is related to pressure and gravity forces in Section 11.11. This analysis gives, for the basic flow in hand,

$$U = \frac{k}{\mu_1}\left(\frac{\partial p_1}{\partial x} - \rho_1 g \cos\theta\right) = -\frac{k}{\mu_2}\left(\frac{\partial p_2}{\partial x} - \rho_2 g \cos\theta\right)$$

The permeability of the porous medium, k, is assumed to be uniform throughout, and the density and viscosity are assumed to be constants for each fluid. Thus, the quantities

$$\phi_1 = -\frac{k}{\mu_1}[p_1(x) - p_1(0) - \rho_1 g x \cos\theta] - Ux \tag{13.65}$$

and

$$\phi_2 = -\frac{k}{\mu_2}[p_2(x) - p_2(0) - \rho_2 g x \cos\theta] - Ux \tag{13.66}$$

are zero in the basic flow, and will serve as perturbation potentials for the perturbed flow.

Suppose now that the interface is given a small initial sinusoidal corrugation, which may grow or decay exponentially with time, so that on the interface

$$x = \eta(y, z, t) = \varepsilon \exp[i(my + nz) + \sigma t]$$

If σ turns out to be positive, the corrugation will grow, and the basic flow is unstable.

In the perturbed flow, ϕ_1 and ϕ_2 will exhibit small variations proportional to the amplitude ε of the interface displacement. The continuity equation for each fluid is $\nabla \cdot \mathbf{u} = 0$, and the perturbation velocity obeys an equation $\mathbf{u} = \nabla \phi$, so the perturbation potentials ϕ_1 and ϕ_2 will satisfy the Laplace equation. The solutions that approach zero far from the interface are found, by separation of variables, to be

$$\phi_1 = A_1(t) \exp \left[i(my + nz) - ax \right] \tag{13.67}$$

and

$$\phi_2 = A_2(t) \exp \left[i(my + nz) + ax \right] \tag{13.68}$$

provided that

$$a^2 = m^2 + n^2$$

One boundary condition at the interface states that the interface is a material surface, on which

$$\frac{\partial \eta}{\partial t} + v \frac{\partial \eta}{\partial y} + w \frac{\partial \eta}{\partial z} = u$$

This is linearized by dropping the terms proportional to v and w, which are of the second order in the small amplitude ε, and by applying the condition on the plane $x = 0$ rather than on the deformed interface. Using $u = \partial \phi / \partial x$ for either fluid, and requiring that the boundary condition be satisfied for both fluids, we find that

$$A_2(t) = -A_1(t) = \frac{\varepsilon \sigma}{a} \exp (\sigma t) \tag{13.69}$$

On the interface, if $a\eta \ll 1$ as assumed, the linearized relationships between the perturbation potentials and the interface displacement are

$$\phi_1 = -\phi_2 = -\frac{\sigma}{a} \eta \tag{13.70}$$

The other boundary condition at the interface requires that any pressure difference across it be accounted for by surface tension. For this purpose, the sum of the principal curvatures of the interface is approximated by $\partial^2 \eta / \partial y^2 + \partial^2 \eta / \partial z^2 = -a^2 \eta$. Since the unperturbed interface is flat, $p_1(0) = p_2(0)$. The pressure in either region can be found from the velocity potential for that region. The boundary condition is then

$$p_1(x) - p_2(x) = (\rho_1 - \rho_2)gx \cos \theta - \left(\frac{\mu_1}{k} (\phi_1 + Ux) - \frac{\mu_2}{k} (\phi_2 + Ux) \right)$$

$$= \Sigma \left(\frac{\partial^2 \eta}{\partial y^2} + \frac{\partial^2 \eta}{\partial z^2} \right) = -a^2 \Sigma \eta \tag{13.71}$$

where Σ is the surface tension. Setting $x = \eta$, and using Eq. (13.70) to eliminate ϕ_1 and ϕ_2, we get

$$-a^2\Sigma = (\rho_1 - \rho_2)g \cos\theta + (\mu_1 + \mu_2)\frac{\sigma a}{k} - (\mu_1 - \mu_2)\frac{U}{k}$$

This can be rearranged into the dimensionless expression

$$\frac{\sigma a}{U} = \frac{\mu_1 - \mu_2}{\mu_1 + \mu_2} - \left(\frac{\rho_1 - \rho_2}{\mu_1 + \mu_2}\right)\frac{gk}{U}\cos\theta$$

$$- \frac{a^2\Sigma k}{U(\mu_1 + \mu_2)} \tag{13.72}$$

From this we identify the destablizing or stabilizing influences of viscosity differences, density differences, and surface tension, as follows. We shall always define U as a positive quantity. Then we see that:

1. The displacement of a more viscous fluid by a less viscous one, $(\mu_1 - \mu_2) > 0$, tends to be *unstable*.
2. Any displacement process in which the less-dense fluid is uppermost, $(\rho_1 - \rho_2)\cos\theta > 0$, is stabilized by the gravity force.
3. Surface tension always produces a stabilizing effect, and provides a short-wave cutoff for any instability.

Notice that there is a possibility for two kinds of instability, one basically caused by a destabilizing viscosity difference, which may be aided or hindered by the gravity effect, and another basically caused by a destabilizing density difference, which may be aided or hindered by a viscosity difference.

Critical Displacement Speed

If $\sigma = 0$ in (13.72), indicating neutral stability, there is a critical displacement speed

$$U^* = [(\rho_1 - \rho_2)g\cos\theta + a^2\Sigma]\frac{k}{\mu_1 - \mu_2} \tag{13.73}$$

If the viscosity difference is destablizing, the process will be stable for $U < U^*$, unstable if $U > U^*$. If gravity is destabilizing, but the process is stabilized by a viscosity difference, the process will be further stabilized by making $U > U^*$. Hill discovered and explained both these effects,[42] because the refining of sugar involved both the downward displacement of sugar syrup by pure water, followed later by the downward displacement of the water by a fresh charge of syrup.

---

[42] Except for surface tension, because the syrup and the fresh water are slowly miscible.

13.12 THE TRANSITION TO TURBULENCE. EXPERIMENT AND THEORY

Much attention has been paid to the linearized stability theory in the preceding sections. It is now time to review what has been learned, and to combine that knowledge with insights obtained from nonlinear analyses and from laboratory or numerical experiments, to see how close we are to a satisfactory description or prediction of the transition from laminar to turbulent flow.

A review of linear stability theory. The examples of instability theory presented here are quite varied, but have certain common features. The typical product of a linear theory is the identification of important dimensionless basic-flow parameters, such as a Reynolds number, and important elementary-disturbance parameters, such as wave number, frequency, and amplification rate. A comprehensive theoretical analysis may aim to determine the curves or surfaces of constant amplification rate, in the space of the other parameters. Special emphasis is placed on the locus of zero amplification rate, the so-called neutral curve or surface. Such an analysis also provides a kinematic description of typical disturbances, usually at some points along the neutral locus, and possibly along the locus of maximum amplification rate. Since more than one eigensolution usually exists for each point in the space of given parameters, the different solutions being called different *eigenmodes*, a comprehensive theoretical study will often include the identification of a number of modes, to make sure that the one on which primary attention is focused is the most dangerous one.

With rare exceptions, an analysis of normal modes gives only approximate solutions to the partial differential equations governing the evolution of small perturbations, because variable coefficients in these equations are being treated as constants. The justification for this is the expectation that the coefficients vary slowly with position or time, compared to the rate of variation of the disturbance quantities. There is still no unanimous agreement as to the most effective way to reduce the partial differential equations to ordinary differential equations in many cases; this partially explains the large number of theoretical papers devoted to analysis of each case.

Even if the results of linear stability theory were complete and unquestionably correct, it would not be clear, without reference to a nonlinear theory or to experiment, what practical use can be made of these results. What if experiment seems to show, as it did for many years in the case of plane Poiseuille flow, that a given basic flow is unstable when linear theory would prove it to be stable? What kinds of disturbance does linear theory lead one to expect to see, when the Reynolds number exceeds the minimum critical value and disturbances with many different wave vectors and frequencies are predicted to have roughly the same rate of amplification? What happens to a single disturbance mode when its amplitude gets so large that linear theory is no longer appropriate? What range of disturbance amplitudes is "too large"? How do coexisting disturbances interact when they get large enough to do so? How do disturbances come into existence, or become *excited* in the first place? A brief consideration of experiments and of nonlinear theory is required to start the search for answers to these questions.

Experimental evidence. Modern experimental studies of instability tend to be focused rather sharply on one or more of the questions left unanswered by linear stability theory, as well as on the verification of specific predictions of that theory. The experiments require ingenuity of design, extraordinary control of conditions, often elaborate instrumentation and data-processing techniques. The pioneering experiments, such as Taylor's experiments on the vortices that bear his name,[43] and Schubauer and Skramstad's verification of the theory of TS waves,[44] rest securely in the Hall of Fame of fluid mechanics research.

The nature, convective or absolute, of the instability has a characteristic effect on the kind of experiment that is usually done. Convective instabilities are usually studied in facilities, such as low-turbulence wind tunnels, that provide a very low level of random disturbances. It is then possible to introduce artificial disturbances with precisely controlled frequencies, wave-number vectors, and phases, either singly or in interesting combinations.

The artificial disturbances can be strong enough to stand out clearly against any random background, yet weak enough to be governed, for the early part of their development, by linear theory. When the artificial stimulus is removed, the disturbances wash away. By and large, experiments of this type have established great confidence in the linear theory. Critical values of Reynolds number, rates of amplification, and shapes of eigenfunctions have been confirmed for many cases. Much has been learned about the effectiveness of various means of disturbance excitation, and about nonlinear developments.

Absolute instabilities are studied with the most minute attention to initial and boundary conditions, and even to the history of changes in boundary conditions. Artificial disturbances are usually not introduced; one waits, often with enormous patience, to see what will develop out of the random disturbances that are inevitably present. A fantastic variety of successive instabilities, and of complicated flows that are stable within narrow ranges of the controllable parameters, has been catalogued.[45] Experimental protocols

---

[43] G. I. Taylor (1922) *op. cit.*

[44] G. B. Schubauer and H. K. Skramstad, "Laminar Boundary-Layer Oscillations and Transition on a Flat Plate," NACA Report No. 909, 1947.

[45] D. E. Coles, "Transition in Circular Couette Flow," *J. Fluid Mech.* **21**:385–425 (1965). Coles reported finding as many as 26 distinct laminar flows for some combinations of parameters! See also H. A. Snyder, "Wave Number Selection at Finite Amplitude in Rotating Couette Flow," and "Change in Wave Form and Mean Flow Associated with Wavelength Variations in Rotating Couette Flow. Part I," *J. Fluid Mech.* **35**:273–298 and 337–352 (1969). (These articles are also models of fine technical writing.)

have been developed that allow one to select, by control of the history of the boundary conditions, a particular one of many flows that can be stable under the same steady boundary conditions!

Let us now take up some of the questions left unanswered by linear theory.

Why Do Some Flows Become Turbulent, When Linear Theory Predicts Stability?

Twenty years ago, the linear theory of flow instability seemed, experimentally, to be irrelevant for certain basic flows, such as plane Poiseuille flow. Such experiments as were then available exhibited turbulent flow at Reynolds numbers as low as 1000, although the linear theory revealed no unstable eigenfunctions until Re exceeded 5772. The response of theoreticians was to investigate the possibility of *subcritical instability*. That means that they looked for disturbances of some specified form that could maintain themselves or grow when Re $\leq$ Re$_c$, provided that they already had a sufficiently large amplitude. Since linear theory predicts either stability or instability without regard to amplitude, it is clear that a nonlinear theory is needed to answer the new question. A theory was constructed, and the answer was yes, subcritical instability is a possibility,[46] but the disturbances that were found theoretically showed little resemblance to those that seemed to dominate the experimental scene. A more elaborate theory was necessary, and has subsequently been devised. It will be sketched late in this section, under the subheading *Secondary Instabilities*.

Meanwhile, the design of experimental flow channels was improved sufficiently, and disturbances convected into the channel from the inlet, or radiated in as sound waves, were sufficiently reduced, that the predictions of linear theory were confirmed.[47] Nishioka and colleagues were able to maintain laminar plane Poiseuille flow at Reynolds numbers as high as 8000 by reducing the level of random disturbances to about 0.05% rms. They then introduced weak periodic disturbances by vibrating a small ribbon in the region of fully-developed flow. When the maximum amplitude of u' was less than about 1% of the maximum flow velocity, the artificial disturbances evolved in the manner expected from the linear theory of TS waves. If they corresponded to unstable eigenfunctions, they might, depending on the amplitude of excitation, grow until u' was about 2.5% of U_{max}, whereupon they become rapidly distorted, as will be described later.

For Reynolds numbers below the critical value, such as Re = 4000, the

[46] B. J. Bayly, S. A. Orszag, and T. Herbert, "Instability Mechanisms in Shear-Flow Transition," *Ann. Rev. Fluid Mech.* **20**:359–391 (1988).

[47] M. Nishioka, S. Iida, and Y. Ichikawa, "An Experimental Investigation of the Stability of Plane Poiseuille Flow," *J. Fluid Mech.* **72**:731–751 (1975).

level and frequency of the artificial excitation was varied, to define the threshold amplitude above which a Tollmien–Schlichting wave would maintain itself or grow. For the range $4000 \le \mathrm{Re} \le 6000$, these amplitudes varied between about 1% and 2%, depending rather sharply on frequency. Note how very small these numbers are, and you may understand why it took so many years before they were experimentally determined

What Kinds of Disturbances Should One Expect to See, When Linear Theory Shows that Many Kinds Might Grow?

This is really two questions in one. First we consider the one that can be be answered by linear theory: what does a flow look like, when many different growing disturbances are linearly superposed? Later we shall briefly consider the more difficult one: what are the typical consequences of competitive or cooperative nonlinear interactions between disturbances?

The outcome of linear superposition depends, in the first place, on the nature of any instability: whether it is convective or absolute.

Convective instabilities. Remember that this is the kind of instability in which disturbances propagate away from the point of origin, somewhat like the waves caused by dropping a rock in a stream. Linear theory predicts that disturbances of different wavelengths or frequencies or angular orientations will have different rates of amplification or damping. In principle, the theory permits calculation of the downstream evolution of a field of disturbances if the properties of that field are known for all time at the upstream edge of the region of interest. This presumes, of course, that no fresh disturbances are fed into the flow downstream as a result, for example, of roughness elements on the wall. The method would involve analyzing the incoming disturbances into a sum of normal modes, and then adding the elements together in the downstream region, after each had propagated and amplified or decayed in its own way.

If there is no obvious pattern in the incoming disturbances, i.e., no special phase relationships between the constituent normal modes, there will be no conspicuous pattern downstream, and all that the linear theory can predict is the evolution of statistical properties of the disturbance field. For example, it might predict that the peak of the energy spectrum[48] of u' would shift in a certain way as the point of observation is moved downstream.

If there is an obvious pattern in the incoming disturbances, much more can be said. As a result, there have been some fascinating theoretical and experimental studies. For example, Gaster and Grant studied the evolution of

---

[48] An operational definition of this term is given in Section 14.2.

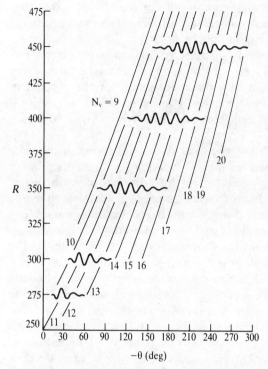

FIGURE 13.46
Spiral band of disturbances growing from a roughness element on a spinning disk. (Reproduced, by permission, from L. M. Mack, 1985.[37])

the disturbance of a Blasius boundary layer, caused by a short pressure pulse applied through a small hole in the wall;[49] Mack and Kendall studied the downstream evolution of temporally harmonic disturbances, continually injected at a fixed point in a Blasius boundary layer;[50] and Wilkinson and Malik studied the disturbance field generated by a single tiny roughness element on a rotating disk.[51] Companion theoretical studies have been made for each of these problems, with generally gratifying results.[52] It is shown, for example, that the striking spiral vortex pattern in rotating-disk flow appears in a widening spiral band that can be traced back to an origin at the roughness element. This band is approximately centered on a spiral trajectory that is everywhere tangent to the group velocity of the most rapidly amplifying normal mode. The theoretical prediction is shown in Fig. 13.46.

---

[49] M. Gaster and I. Grant, "An Experimental Investigation of the Formation and Development of a Wave Packet in a Laminar Boundary Layer," *Proc. Roy. Soc. London* **A347**:253–269 (1975).

[50] L. M. Mack and J. M. Kendall, "Wave pattern produced by a localized harmonic source in a Blasius boundary layer." Unpublished work at the Jet Propulsion Laboratory, Pasadena, CA (1983)

[51] S. P. Wilkinson and M. R. Malik, "Stability Experiments in Rotating-Disk Flow," AIAA Paper No. 831760 (1983).

[52] All theories are reviewed by L. M. Mack (1984) op. cit., Sections 7 and 12.

While these results are encouraging and provocative, it is clear that they fall short of a complete explanation of observed coherent disturbance patterns that cover, for example, the full circumference of a rotating disk, because a simple linear superposition of steady patterns originating from randomly placed roughness elements would be just a jumble of waves with every imaginable phase. The development of a dominant phase for the permanent spiral vortex pattern can only be understood as a result of a nonlinear interaction, or competition, between growing disturbances with different phases.

In the case of absolute instability, the most dangerous disturbances often do not disperse by traveling at different speeds. An arbitrary initial disturbance field, if left to evolve without any subsequent introduction of fresh disturbances from the boundaries, would seem destined to become dominated by the normal modes with the greatest amplification rate. However, these modes may be inconsistent with boundary conditions that were ignored when the normal-mode analysis was proposed. In the case of rotating cylinders, for example, the length of the cylinders does not enter into the normal-mode analysis, and yet it is obviously relevant. Experimentally, a change in the length of the cylinders, produced by displacing a movable end wall, affects the development of the disturbances, but it takes a very long time, proportional to L^2/ν, to do so. The fact that this is a diffusive time scale, rather than a convective one, suggests that a linear theory may be all that is needed to describe the evolution of the disturbed flow if the Reynolds number only slightly exceeds the critical value. One might then expect the final pattern of Taylor vortices to be descended from the most strongly growing normal mode that has the correct phase and wavelength to satisfy the end conditions. Experimental confirmation of this expectation is not, however, at hand, and a nonlinear theory may again be essential.

What Happens to an Isolated Normal Mode When its Amplitude Becomes Large?

With modern computers and algorithms, it is quite easy to study the evolution of a two-dimensional, x-periodic, temporally-amplifying disturbance on a free shear layer far past the point at which a linear theory must be abandoned. Figure 13.47 shows an example, involving the most dangerous normal mode growing on a basic flow that had an error-function velocity profile at the time at which the disturbance was introduced. The growth occurs in three phases: (1) an initial phase governed by linear theory; (2) a phase of nonlinear evolution on what is called a *convective time scale*, i.e., a time of order $k\,\Delta U$, where k is the disturbance wavenumber and ΔU is the shear across the layer; (3) a phase of slow restoration of streamwise homogeneity, on what is called a *viscous time scale*, i.e., a time of order $(k^2\nu)^{-1}$. The calculations are done with the full Navier–Stokes equations, and usually focus on the middle phase. The first phase is handled with linear stability theory, and the last phase is rarely

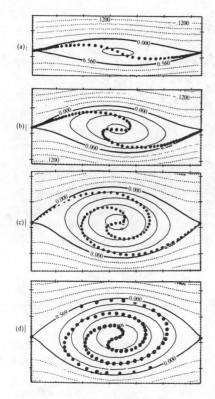

FIGURE 13.47
Computed nonlinear evolution of the most dangerous normal mode in a free shear layer. (a) $\Delta Ut/\delta = 0.5$, (b) 1.0, (c) 1.5, $d(2.0)$. - - - - streamlines; ——— the cat's eyelid; $\cdots$ material particles originally at $y = 0$. (Reproduced, from G. M. Corcos and F. S. Sherman, *J. Fluid Mech.* **139**: 29–65, 1984.)

studied extensively because other more complicated events usually dominate the corresponding experimental developments.

A simple analytical cartoon of the nonlinear rearrangement of vorticity on a convective timescale can be made.[53] In it, all the vorticity is assigned to either one of two special regions, *cores* and *braids*, as sketched in Fig. 13.48. The velocity field induced by vorticity in the cores extrudes vorticity out of the braids and into the cores, somewhat as an extensible cable would be wound onto a periodic array of spinning reels. The instantaneous rate of addition of vorticity to a core is proportional to the product of the amount of vorticity already in the core and the amount remaining in a neighboring braid. When the available vorticity is all reeled in, the main convective phase of the instability is over, and the cartoon depicts a steady state.

Numerical calculations of the nonlinear development of Taylor vortices are also comparatively straightforward, if the axial periodicity of the solution is supposed to be known, and if the values of the chosen parameters are such

[53] G. M. Corcos and F. S. Sherman, "Vorticity Concentration and the Dynamics of Unstable Free Shear Layers," *J. Fluid Mech.* **73**:241–264 (1976).

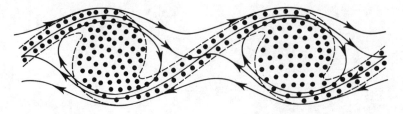

FIGURE 13.48
Cartoon of nonlinear mixing layer instability. (Reproduced, by permission, from G. M. Corcos and F. S. Sherman, 1976.[53])

that the flow approaches a steady, axisymmetric limit. An example is shown in Fig. 13.49. The nonlinear evolution of flows in which the initial disturbance is a simple axisymmetric normal mode but the final state involves one or more azimuthally propagating waves has also been successfully computed, but great care is required in the design and execution of the numerical algorithm.[54]

Systematic Observation and Description of Finite-Amplitude Disturbances

To make sense of written descriptions of the evolution and nonlinear interaction of initially infinitesimal normal-mode disturbances, you will need some terminology and some thought about possible ambiguities in the use of

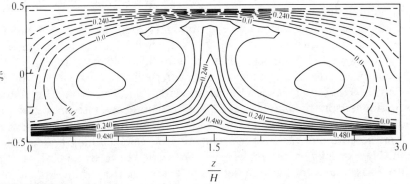

FIGURE 13.49
Supercritical axisymmetric Taylor-vortex flow, $\eta = 0.875$, $\mu = 0$, $\alpha = 2.0944$, $\text{Re} = 8 \ \text{Re}_{\text{crit}}$. Contours or constant v, which would be straight horizontal lines in the basic flow. (Figure kindly supplied by Katie Coughlin.)

[54] P. S. Marcus, "Simulation of Taylor–Couette Flow. Parts I and II," *J. Fluid Mech.* **146**:45–64 and 65–113 (1984).

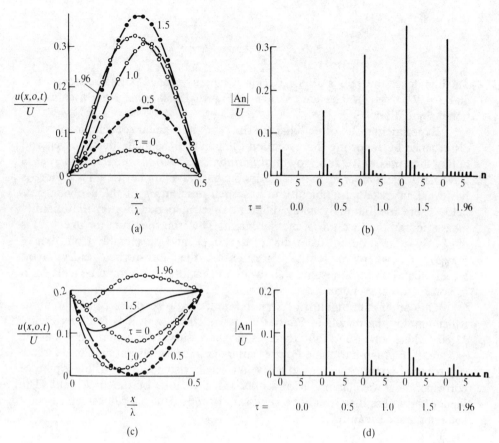

FIGURE 13.50

Different ways to characterize nonlinear transient mixing layer development: (a) evolving distribution of $v(x, 0, t)$; (b) evolving spectrum of v' at $y = 0$; (c) evolving distribution of $u(x, 0, t)$; (d) evolving spectrum of u' at $y = 0$.

standard terms. It is necessary to think about how the same phenomenon will appear if observed in different ways. Consider, for example, the computed evolution of an isolated normal mode disturbance on a free shear layer, shown in Figs. 13.47 and 13.50.

Figure 13.47 shows the evolution of a marked material surface, and of the streamline pattern. One might characterize the growth of the disturbance by plotting the maximum height of the material surface or of the cat's-eye boundary versus time, but these quantities are not easily measured. Panels (a) and (c) of Figure 13.50 show the evolution of the distributions of velocity components u and v at the middle of the shear layer. These could readily be measured, and the measurements could be converted into the spectra shown in panels (b) and (c). The spectra are exhibited as bar graphs, which show the evolution of the wave-number dependence of the coefficients in a Fourier sum such as

$$v(x, 0, t) = \sum_0^N A_n \sin(2\pi n x / \lambda),$$

which matches the values of $v(x, 0, t)$ at N equally-spaced points between $x = 0$ and $x = N\lambda/(N + 1)$. The data shown in Figure 13.50 come from a computational simulation.

Experimentalists often characterize the non-linear development of a disturbance by reporting the evolution of spectra of velocity fluctuations, and noting the appearance and growth of *harmonics of various orders*. If the initial wave has wavenumber k, the nth harmonic has wavenumber $(n + 1)k$. There is however, an essential ambiguity in any such description of the evolving flow field. There are infinitely many different spectra, depending on the quantity measured and the location of measurement. Thus the spectrum for $v(x, 0, t)$ is obviously quite different from that of $u(x, 0, t)$, and presumably from that of $v(x, \delta/2, t)$, where δ is the thickness of the undisturbed shear layer. Experimentalists soon become well aware of this, and you should be alert for it as you read their reports.

One other caution must be kept in mind, as you read descriptive accounts which employ the results of Fourier analysis. This is well illustrated by Figure 13.50, where one sees that there is no necessary connection between the existence of a high-harmonic Fourier component of significant amplitude in the spectrum of some observed quantity, and the existence of a visible ripple of corresponding wavelength in any observable feature of the flow field. The higher harmonics may simply be necessary to represent a locally steep slope of some largescale feature.

What Kinds of Important Nonlinear Interactions Occur Between Normal Modes that Grow to Finite Amplitudes?

As mentioned above, one can describe important nonlinear interactions of elementary disturbances as either *competitive* or *cooperative*. Competitive interactions must explain the appearance of a dominant wavelength and phase of the orderly finite disturbances that appear, for example, when an initially horizontal tube half-filled with each of two liquids of different density is tilted.[55] The result is shown in Fig. 13.51. Nothing has been done to seed the initial conditions with disturbances of any particular wavelength, or to establish the phase of the wave pattern. The dominant wavelength in the final pattern corresponds fairly well with the length of the most rapidly growing linear

---

[55] This classic experiment is due to S. A. Thorpe, "Experiments on the Instability of Stratified Shear Flows: Miscible Fluids," *J. Fluid Mech.* **46**:299–320 (1971).

FIGURE 13.51
Wave growth in a tilting tube. (Reproduced, by permission, from S. A. Thorpe, 1971.[55])

disturbance. Once the tube is tilted to its final angle, there is just a fixed amount of vorticity in the shear layer. The competition is, so to speak, an effort to gather this vorticity into alternative patterns that cannot long coexist. The pattern that is establish most quickly induces a strain field that tears apart or reels in weaker and incompatibly located concentrations of vorticity. Admittedly, this is a vague argument, but neither theory nor computation yet offers much more.

Another important competition is seen in the spontaneous evolution of an unstable free shear layer, such as exists at the edge of a jet. Figure 13.52 shows an example. It is obvious that the first visible disturbances soon interact in ways that leave them permanently altered. The most obvious of these interactions is called *vortex-pairing*, although it occasionally involves a clustering of more than two vortex cores at a time. The mathematical name for this is *subharmonic interaction*; a wave of wavenumber k interacts with its subharmonic of wavenumber $k/2$, and is finally swallowed up by it.

This kind of interaction, in which the perturbed flow remains two-dimensional, can occur also in boundary layers but is rarely observed, because an initially two-dimensional disturbance is likely to interact more dramatically with three-dimensional disturbances. Some of these interactions can be isolated by ingenious experiments, with results such as appear in Figs. 13.53 and 13.54. These pictures illustrate what may be called a cooperative interaction, in which two or more elementary disturbances combine in a way that greatly enhances the growth rate of some of them. This combination

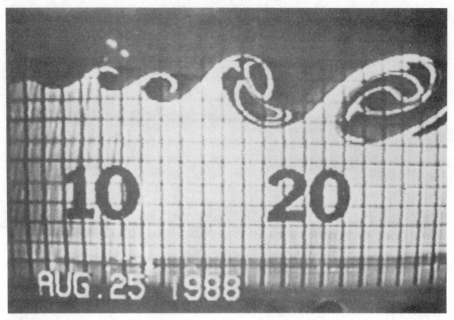

FIGURE 13.52
Vortex pairing in a free shear layer $U_{top} = 11.0$ cm/s, $U_{bot} = 2.88$ cm/s. Laser-induced fluorescence in water of lower layer. (Photo courtesy of D. I. Polinsky.)

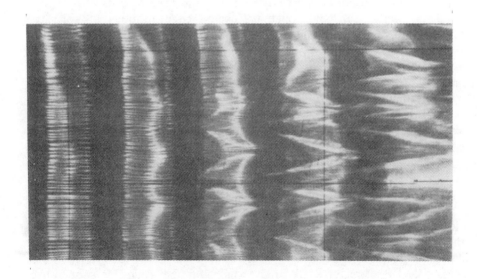

FIGURE 13.53
Secondary instabilities in boundary layers. (Reproduced, by permission, from T. Herbert, 1988.[56])

FIGURE 13.54
Secondary instabilities in boundary layers. (Adapted, by permission, from T. C. Corke "Resonant Three-Dimensional Modes in Transitioning Boundary Layers—Structure and Control," AIAA Paper 89-1001, 1989.)

quickly produces flow features that would not be seen if the disturbances were linearly superposed.

Secondary Instabilities

The kind of cooperative interaction that is easiest to recognize involves a *primary disturbance*—like a TS wave, a Taylor–Görtler vortex, or a crossflow vortex—and one or more *secondary disturbances*. The primary disturbance first grows to a finite amplitude, thereby rearranging the vorticity field in such a way as to promote explosive growth of the secondary disturbances. The interaction is cooperative, because the secondary disturbances would grow only slowly, or not at all, if the primary disturbance had not prepared the way.

The theory of secondary instability has taken giant strides in recent years, hand-in-hand with some very meticulous and ingenious experiments. There are excellent published reviews of this progress,[56] which has involved an extraordinary degree of international cooperation.[57] The brief discussion given here is drawn largely from the review by T. Herbert, and aims primarily to introduce *Floquet theory*.

---

[56] T. Herbert, "Secondary Instability of Boundary Layers," *Ann. Rev. Fluid Mech.* **20**:487–526 (1988).

[57] See, as recent examples, *Laminar–Turbulent Transition*, proceedings of the IUTAM Symposium, Novosibirsk, 1984, edited by V. V. Koslov, Springer-Verlag, 1985; also *Turbulence Management and Relaminarization*, IUTAM Symposium, Bangalore 1987, edited by H. W. Liepmann and R. Narasimha, Springer-Verlag, Berlin and New York.

Floquet theory. Herbert's analysis of secondary instability starts with the usual representation of the flow quantities as a superposition of a basic flow and a small perturbation. For example, we write $\mathbf{u}(\mathbf{r}, t) = \mathbf{u}_2(\mathbf{r}, t) + \mathbf{u}_3(\mathbf{r}, t)$. The basic flow, $\mathbf{u}_2(\mathbf{r}, t)$, is in turn represented as the superposition of the *undisturbed laminar flow*, e.g., a Blasius boundary layer, and the *primary disturbance*. Thus, $\mathbf{u}_2(\mathbf{r}, t) = \mathbf{U}(\mathbf{r}) + A\mathbf{u}_1(\mathbf{r}, t)$. The additional small perturbation, $\mathbf{u}_3(\mathbf{r}, t)$, is called the *secondary disturbance*.

We now proceed to make a linearized analysis of the evolution of secondary disturbances. We substitute $\mathbf{u} = \mathbf{u}_2 + \mathbf{u}_3$ and a corresponding decomposition of the pressure into the Navier–Stokes equations and boundary conditions, and note that $\mathbf{u}_2$ and p_2 satisfy the same equations and boundary conditions as do $\mathbf{u}$ and p. Terms quadratic in $\mathbf{u}_3$ are then dropped, leaving the linear, homogeneous equations

$$\frac{\partial \mathbf{u}_3}{\partial t} + (\mathbf{u}_2 \cdot \nabla)\mathbf{u}_3 + (\mathbf{u}_3 \cdot \nabla)\mathbf{u}_2 + \rho^{-1}\nabla p_3 - \nu \nabla^2 \mathbf{u}_3 = 0 \qquad (13.74)$$

$$\nabla \cdot \mathbf{u}_3 = 0$$

The particular phenomenon which is to be analyzed is the rapid development of three-dimensional secondary disturbances on a two-dimensional basic flow. The prototype of the primary disturbance would be a two-dimensional, streamwise-propagating TS wave. We use the symbol x^* for streamwise displacement in the usual stationary frame of reference, reserving the symbol x for later use. We assume, for the simplest development of the theory, that the undisturbed flow is strictly parallel, with velocity profile $U(y)$. The primary disturbance has scalar velocity components $u_1(x^*, y, t)$ and $v_1(x^*, y, t)$ normalized so that the maximum value of u_1 equals a specified fraction, A, of the maximum value of U. Next, we assume that $u_1(x^*, y, t)$ and $v_1(x^*, y, t)$ depend on x^* and t only as periodic functions of the quantity $x = x^* - ct$, where c is the phase velocity of the propagating primary disturbance. Mathematically,

$$\mathbf{u}_1(x^*, y, t) = \mathbf{u}_1(x, y) = \mathbf{u}_1(x + \lambda_x, y) \qquad (13.75)$$

Here, λ_x is the wavelength of the primary disturbance. Finally, we assume that A is a constant.

In summary, we have assumed a steady, parallel undisturbed flow, and a periodic traveling-wave primary disturbance of constant amplitude. Although this could be literally true in a subcritical plane Poiseuille flow, we expect that the analysis will yield useful insights even when the undisturbed flow is not quite parallel, and when the primary disturbance is slowly amplifying or decaying, and slowly changing its normalized form. This hope is based on the expectation that interesting secondary disturbances grow very rapidly compared to the rate of growth of the primary disturbance, and even more rapidly compared to the rate of streamwise change of the undisturbed flow.

We now rewrite (13.74) as an equation with independent variables x, y, and t (instead of x^*, y, and t), noting that $(\partial/\partial t)_{x^*,y} = (\partial/\partial t)_{x,y} - c(\partial/\partial x)_{y,t}$, where the subscripts identify the variables that are held constant in the partial differentiation. The result is

$$\frac{\partial \mathbf{u}_3}{\partial t} + \left((u_2 - c)\frac{\partial}{\partial x} + v_2\frac{\partial}{\partial y}\right)\mathbf{u}_3 + \left(u_3\frac{\partial}{\partial x} + v_3\frac{\partial}{\partial y}\right)\mathbf{u}_2 + \rho^{-1}\,\nabla p_3 - v\,\nabla^2 \mathbf{u}_3 = 0$$

(13.76)

$$\nabla \cdot \mathbf{u}_3 = 0 \qquad\qquad (13.77)$$

The coefficients in this set of equations are independent of t and z, so that we may set

$$\mathbf{u}_3 = \cos(\beta z)\exp(\sigma t)\,\mathbf{V}(x, y) \qquad\qquad (13.78)$$

The oscillatory behavior in z, with $\beta = 2\pi/\lambda_z$, is thought of as the result of a matched pair of obliquely propagating waves, with equal but opposite z-components of phase velocity. For the moment, σ is allowed to be complex. When (13.78) is substituted into (13.76), the result is an equation for $\mathbf{V}(x, y)$, with coefficients which are periodic functions of x.

The Floquet theory deals with ordinary differential equations with periodic coefficients. This theory suggest that $\mathbf{V}(x, y)$ can be further separated, into the product of an exponentially amplified or damped oscillation, and a function with the same periodicity as the coefficients. Thus, we can write

$$\mathbf{V}(x, y) = \exp(\gamma x)\,\mathbf{V}^*(x, y) \qquad\qquad (13.79)$$

with $\qquad \gamma = \gamma_R + i\gamma_I \qquad$ and $\qquad \mathbf{V}^*(x, y) = \mathbf{V}^*(x + \lambda_x, y)$

Because $\mathbf{V}^*(x, y)$ is periodic in x, it can be represented by a Fourier series, to wit

$$\mathbf{V}^*(x, y) = \sum_{m=-\infty}^{\infty} \mathbf{v}_m^*(y)\exp(im\alpha x) \qquad\qquad (13.80)$$

with

$$\alpha = \frac{2\pi}{\lambda_x}$$

The analysis is continued by substitution of (13.78), (13.79) and (13.80), and a corresponding representation of p_3, into (13.76) and (13.77). The primary disturbance is ordinarily weak enough so that the x-dependence of $\mathbf{u}_1$ is well approximated by $\exp(i\alpha x)$. The resulting equations can be arranged into an infinite sequence of terms, each of the form $F_m(y)\exp[(\gamma + im\alpha)x]$, which adds up to zero. Since the exponential factors are linearly independent of one another, each y-dependent factor must vanish. This produces an infinite sequence of ordinary differential equations for the $\mathbf{v}_m^*(y)$. The equations are coupled because complex exponential factors in the coefficients are multiplied by similar factors in the Fourier series. In principle, no one of the $\mathbf{v}_m^*(y)$ can be calculated without simultaneous calculation of all the rest. As a practical matter, this turns out to be an exaggeration; the Fourier series can be truncated at a very low level without appreciable loss of accuracy.[58]

The boundary conditions require that $u_3 = 0$ at $y = y_1$ and $y = y_2$. Thus, each of the $v_m^*(y)$ must vanish on these planes. This can only happen when the parameters of the secondary disturbance (β, σ, and γ) assume eigenvalues, connected to the values of the parameters of the undisturbed flow and the primary disturbance by a dispersion equation

$$F(\beta, \sigma, \gamma; \mathrm{Re}; \alpha, A) = 0 \qquad (13.81)$$

We recall that α and β are real numbers, respectively equal to the streamwise wavenumber of the primary disturbance and the transverse wavenumber of the secondary disturbance. Their values are ordinarily specified, as are values of Re and of A, the amplitude of the primary disturbance. The remaining parameters of the secondary disturbance (σ and γ) may be complex numbers, and we again face the choice of studying *temporal growth* (σ complex and γ imaginary) or *spatial growth* (γ complex and σ real). The former choice leads to somewhat simpler calculations but both have been employed.

Classification of secondary instabilities. Consider the case of temporal growth. The factor $\exp(i\gamma_I x)$ can be multiplied into the Fourier series for $V^*(x, y)$, so that the mth term now contains the harmonic factor $\exp[i(\gamma_I + m\alpha)x]$. The value of this factor, for a given value of x, is unchanged if γ_I is replaced by $\gamma_I \pm k\alpha$, k being any integer. Thus, all numerically significant values of γ_I are confined to the range $-\alpha/2 \le \gamma_I < \alpha/2$. Continuing to follow Herbert, we introduce the *subharmonic wavenumber*, $\alpha^* = \alpha/2$, and the dimensionless *tuning factor*, $\varepsilon = \gamma_I / \alpha^*$. The value of ε, like the values of α, β, Re, and A, is prescribed; it will distinguish different classes of secondary instability. The general form of u_3 is now

$$u_3 = \exp(\sigma_R t) \exp(i\sigma_I t) \sum_{m=-\infty}^{\infty} v_m^*(y) \exp[i(2m + \varepsilon)\alpha^* x] \qquad (13.82)$$

The amplification of the disturbance is given by the first exponential factor; the second factor describes the phase of the secondary disturbance relative to that of the primary disturbance. The values of σ_R and σ_I are the eigenvalues we seek.

Named modes of secondary instability are associated with special values of the tuning factor as follows.

---

[58] The ordinary differential equations, at a low level of truncation, are written out in full for Poiseuille flow, in T. Herbert, "Secondary Instability of Shear Flows," Section 7 of AGARD Report No. 709, Special Course on Stability and Transition of Laminar Flow (1984). This is a tutorial account, which supplies many helpful details of analysis and of numerical methods.

1. $\varepsilon = 0$ gives *fundamental* modes. Commonly used terms that are synonymous or intimately associated are *primary resonance, aligned modes, K-type modes, peak-valley splitting*. The terms in the Fourier series include one, for $m = 0$, that is independent of x and describes a steady spanwise-periodic modification of undisturbed flow, associated with streamwise vortices. The next longest wave component has the wavelength of the primary distur-bance; the remaining components are harmonics of this. The wave patterns for these modes are invariant under the transformation of coordinates $(x, z) \rightarrow (x + \lambda_x, z + \lambda_z)$; these patterns are aligned, in the sense shown in Fig. 13.55(a). They were first produced and studied by Klebanoff, hence the name K-type modes. The peaks and the valleys are loci of constant z, a distance $\lambda_z/2$ apart. The arrowheads of the *lambda vortices* shown in Fig. 13.55(a) form on the peaks, a distance λ_x apart.

2. $\varepsilon = 1$ gives *subharmonic* modes. Commonly used terms that are synony-mous or intimately associated are *principal parametric resonance, staggered modes, C-type modes, H-type modes*. All terms in the Fourier series include odd multiples of α^*; hence there is no x-independent component. The longest wave component has twice the wavelength, and at a fixed point half the frequency, of the primary disturbance; the remaining components,

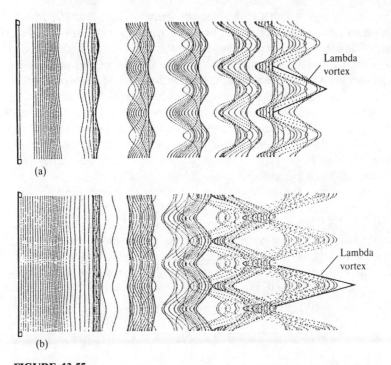

(a)

(b)

FIGURE 13.55
Classification of secondary instabilities: (a) fundamental modes; (b) subharmonic modes. (Reproduced, by permission, from T. Herbert and G. R. Santos, AIAA Paper 87-1201.)

which modify the waveform of the disturbance, have wavelengths $\frac{2}{3}\lambda_x$, $\frac{2}{5}\lambda_x$, etc.[59]

The wave patterns for these modes are invariant under the transformation of coordinates $(x, z) \rightarrow (x + 2\lambda_x, z + \lambda_z)$; thus, the arrowheads of the lambda vortices shown in Fig. 13.55(b) form a staggered array. A special case of subharmonic resonance that survives in the limit as $A \rightarrow 0$ was predicted theoretically by Craik, hence the name C-modes. The more general analysis, which includes Craik's result as a special case, was made by Herbert: hence the name H-modes.

3. $0 < |\varepsilon| < 1$ gives *detuned* modes. Associated terms are *combination modes* or *combination resonance*, names that arise because the physical disturbance consists of the real quantity produced by combining the solution with positive ε with its complex conjugate, which has negative ε. These modes are the most complicated to describe, and probably the most frequently occurring in a process of natural transition, i.e., a transition that is is not initiated by the artificial introduction of carefully matched primary and secondary disturbances.

Objectives of the theory and some typical results. The initial objective of the theory of secondary instability is to find the eigenvalues of the secondary disturbance. These are σ_R, which determines the growth rate; and σ_I, which determines the phase. As in the analysis of primary instability, it turns out that there are many different complex eigenvalues for each set of the given parameters. It often turns out that only one of these has $\sigma_R > 0$, to indicate a positive growth rate, and that the corresponding imaginary part is $\sigma_I = 0$, indicating that the growing secondary disturbances are *phase-locked* with the primary disturbance; the two disturbances move downstream with the same phase speed. Because of the coupling of the ordinary differential equations, all Fourier components of the secondary disturbances move with the same phase speed and have the the same rate of amplification or decay; in short, the secondary disturbance has a permanent shape.

The next objective is to study the dependence of growth rate on the values of the given parameters. Figures 13.56 and 13.57 show some of Herbert's results for the Blasius boundary layer. The main feature that catches the eye is the importance of the amplitude, A, of the primary disturbance. When A becomes as high as 0.01, secondary disturbances are amplified at significant rates for a wide range of values of the obliqueness β/α and of the detuning factor ε. For very small values of A, secondary instability is fairly selective; subharmonic disturbances are most dangerous, and there is a fairly

---

[59] For a good explanation of terms such as *principal parametric resonance*, see Chapter 5 of A. H. Nayfeh and D. T. Mook, *Nonlinear Oscillations*, Wiley-Interscience, 1979. This chapter also gives mathematical proofs for Floquet theory.

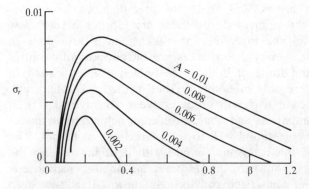

FIGURE 13.56
Subharmonic growth rate versus spanwise wavenumber and primary wave amplitude. Primary disturbance is on Branch II at Re = 606. (Reproduced, by permission, from T. Herbert, 1988.[56])

strongly preferred spanwise wavenumber. For boundary layers with uninflected basic velocity profiles, the sort of two-dimensional subharmonic interaction (vortex-pairing) that is so prominent in free shear layers does not appear. It is easy to find conditions for which the amplification rate of a secondary disturbance is very large compared to the maximum rates of growth of primary disturbances.

Experimental confirmation and importance of secondary instability. This theory of secondary instability has been applied to plane Poiseuille flow and to

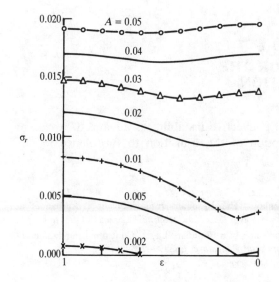

FIGURE 13.57
Secondary growth rate versus detuning factor and primary wave amplitude. Primary disturbance is on Branch II at Re = 606. (Reproduced, by permission, from T. Herbert, 1988.[56])

two-dimensional boundary layers, and analogous investigations have been made for cases in which the primary disturbances are Görtler or crossflow vortices. Meanwhile, careful and ingenious experiments have been done in which both primary and secondary disturbances are introduced with control over the frequency and amplitude of both.[60] The selective amplification of combination modes that were only randomly present in the background, and the phase-locking of these onto artificially produced weak TS waves, has been amply documented. The amplitude and phase profiles of subharmonic modes have been measured, and correspond well to the predictions of theory. The spontaneous generation of the companion mode with detuning $-\varepsilon$, as the secondary disturbance evolves downstream of a vibrating ribbon which inserts the mode with detuning $+\varepsilon$, has been convincingly shown. It is now even possible to control the spanwise wavelength of the secondary disturbance without introducing permanent obstacles that would bias the outcome in favor of the fundamental modes by introducing a steady source of streamwise vorticity.[61]

For a comprehensive review of the outcome, current as of 1988, you should really study the article by Herbert.[62] It is sufficient to say here that the key features of the theory are well confirmed by experiment and by brute-force numerical solutions of the full Navier–Stokes equations, and that the theory provides valuable guidance for those who need to plan or to interpret either physical or numerical experiments. For example, Herbert notes that the standard techniques that are used to reduce the turbulence level in a wind tunnel intended for studies of transition may produce a different mix of secondary instabilities from that which would be encountered in free flight. The theory has also been very successful in explaining why different experimenters, carrying out what seems nominally to be the same experiment with different facilities and slightly different techniques, have sometimes obtained strikingly different results.

13.13 THE RELEVANCE OF INSTABILITY THEORY FOR THE PREDICTION OF TRANSITION. COMPETING THEORIES

No matter how far the theory of convective instability is advanced, it cannot singlehandedly provide a description of the transition to turbulence in any

---

[60] Y. S. Kachanov and V. Y. Levchenko, "The Resonant Interaction of Disturbances at Laminar-Turbulent Transition in a Boundary Layer," *J. Fluid Mech.* **138**:209–248 (1984).

[61] T. C. Corke and R. A. Mangano, "Transition of a Boundary Layer: Controlled Fundamental–Subharmonic Interactions," *IUTAM Symposium Bangalore*, pp. 199–213, 1987.

[62] T. Herbert (1988) op cit., especially pp. 504–522.

given case, because it only describes what will happen to a given supply of incoming disturbances. It is almost silent on the question "How do disturbances get into a laminar flow in the first place?"

The theory does, however, point the search for an answer in definite directions. It shows that a completely irrotational flow is stable, and hence narrows the question a bit, into "How do disturbances first appear in the parts of a laminar flow that contain vorticity?" How, for example, do disturbances pass from the outside to the inside of a boundary layer? Since some of those disturbances may simply be bits of vorticity incorporated into the boundary layer as new streamlines enter it, and that vorticity must have come from some identifiable upstream source, an application of stability theory to the source region may yield useful hints.

Receptivity

At the very beginning of this chapter, reference was made to two exciting recent articles, the one by Herbert which provided much of the material for Section 13.12, and one by Morkovin (1988), which is largely devoted to the question presently in hand. This final section will serve as a brief introduction to terminology and ideas that appear prominently in Morkovin's article, and that will probably be at a focus of intensive research for years to come. The first term is *receptivity*.

Receptivity is related to instability in much the same way as forced vibration is related to free vibration. In a boundary layer, for example, the unstable disturbances such as TS waves or Görtler vortices may extend somewhat into the region of external irrotational flow, but they die away with increasing distance from the wall. It is clear that they are not being forced by some nonuniformity in that external flow; they represent a free response of the boundary layer to an accidental or deliberate rearrangement of its incoming vorticity.[63]

Of course, there may be disturbances, for example sound waves from a loudspeaker, present in the external flow, and these presumably affect the vorticity distribution in the boundary layer somewhat, wherever they impinge on it. How effectively do they produce unstable rearrangements of vorticity, so as to trigger the subsequent free growth of, say, TS waves? How effectively are these sound waves received by the boundary layer, when the measure of effective reception is rapid development of potential instabilities? On the other hand, how effective are vibrations of the wall itself, or roughness and waviness of the wall?

A great deal of research, both theoretical and experimental, has been

---

[63] Receptivity was one of the major topics of discussion in the 1984 IUTAM Symposium at Novosibirsk. See *Laminar–Turbulent Transition*, cited earlier (p. 525).

devoted to these questions. Just to get the flavor of it, let us briefly consider the receptivity of a boundary layer to sound waves. This is a nice topic, because some of the most important results are intuitively appealing and draw on some of the basic notions about unsteady boundary layers, which were presented in Chapter 12.

Sound waves with frequencies that match those of unstable TS waves can effectively stimulate the growth of the TS waves in a boundary layer, but it is not immediately clear how they do so. There is an enormous discrepancy between the wavelengths of the two kinds of waves, hence they cannot become locked in phase with one another, no matter what the direction of propagation of the sound. However, if the wall on which the boundary layer grows starts with a sharp leading edge, the wavelength of the sound becomes irrelevant. The streaming motions that accompany a sound wave are ordinarily very weak, but they can be enormously amplified in the immediate vicinity of a sharp edge. By action of the no-slip condition, the periodic flow past the edge generates a very localized periodic source of vorticity, which appears first at the wall. The nature of the forced disturbance is thus converted from that of a sound wave into that of a streamwise-periodic excess or defect of vorticity, which is convected downstream with a speed between zero and the speed outside the boundary layer. This is the stuff of which TS waves are born; hence the receptivity of the boundary layer to this kind of external disturbance is high.[64] The edge at which the sound waves are converted to vorticity waves does not have to be really sharp; but its radius of curvature should be much less than the wavelength of the sound. For a frequency of 100 Hz, which is fairly typical of TS waves in the laboratory, the acoustic wavelength for air at room temperature is about 3.5 meters, so this criterion is easy to meet.

Transition Beyond the Stage of Secondary Instability

Even when the incoming disturbances are artificially simplified, laboratory and computational experiments show that the amplification of secondary instabilities does not bring a typical flow to an identifiable end of the transition to turbulence.

To make sense of this statement, we need some agreement about terminology. This topic will be carried much further in Chapter 14; it suffices here to describe some of the criteria that occur naturally to an experimentalist. Most of these have to do with time-averaged properties of the flow.

1. *Mean streamwise velocity profile.* In the *transition zone*, which starts where the primary disturbances first grow to detectable magnitudes, and ends

---

[64] For an elaboration of this discussion, see M. V. Morkovin and S. V. Paranjape, "On Acoustic Excitation of Shear Layers," *Zeit. f. Flugwissen* **19**:328–335 (1971).

when the flow is fully turbulent, there is an obvious reshaping of the time-averaged velocity profile. The transition zone may also be characterized by spanwise nonuniformities of the time-averaged flow that are detectable in the basic laminar flow, and which gradually disappear in the fully turbulent flow. In centrifugally unstable flows, the spanwise nonuniformities may be very persistent. Both of these phenomena are illustrated in Chapter 14.

2. *Streamwise distribution of time-averaged skin friction, heat transfer, or mass transfer.* One of the oldest ways of locating transition on models or full-scale vehicles exploits the fact that time-averaged rates of exchange of mass, momentum, or heat between the fluid and the wall rise rapidly through the region where secondary disturbances become prominent, and reach local maxima somewhere in the transient zone. They then settle down into a qualitatively different dependence on x than was shown in the region upstream of transition, attaining values much higher than would exist if laminar flow had persisted. The china-clay technique of locating transition, illustrated in Fig. 13.58 exploits the change in mass-transfer rates. The clay is initially sprayed with oil of wintergreen, which gives the surface a dark appearance. Evaporation of the oil, which is hastened when the vapor is effectively convected away, as it is in the transitional and turbulent regions, makes the surface look much lighter.

3. *Evolution of frequency spectra.* Suppose that a hot-wire anemometer, or other instrument capable of resolving the temporal variations of velocity, is placed at a fixed point in the transition zone, and a long record of velocity versus time is resolved into a frequency spectrum. The appearance of the record, as displayed on an oscilloscope screen or paper chart, and the shape

FIGURE 13.58
Location of transition by the china clay technique. Flow from right to left. (Reproduced, by permission, from D. I. A. Poll, *J. Fluid Mech.* **150**: 329–356, 1985.)

of the spectrum will evolve dramatically as the point of observation is moved downstream through the transition zone. This statement applies to convective instabilities; for absolute instabilities, a somewhat similar evolution may be seen at fixed point if the Reynolds number is very slowly increased with time.

In a boundary layer, if the incoming disturbances are very weak, the hot wire signal will be almost steady where the local Reynolds number is less than critical. If artificial disturbances with a single frequency are introduced, the resulting signal will start out as a sine wave, and the spectrum will exhibit a single prominent spike at the forcing frequency. At positions farther downstream, where secondary disturbances have reached significant amplitudes, the signal may look like an amplitude-modulated sine wave, and the spectrum is likely to exhibit new, broad, peaks near the first subharmonic frequency. Examples are shown in Fig. 13.59.

If the strength of the primary disturbance is sufficient so that

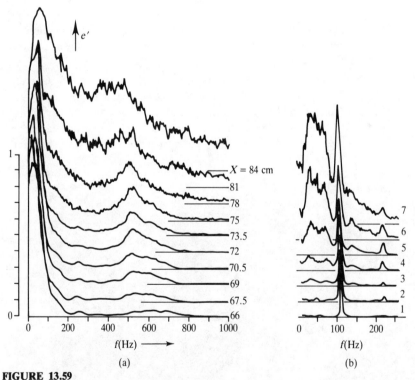

FIGURE 13.59
Evolution of u' spectra during transition. (a) Natural transition in wind tunnel with high level of low-frequency freestream disturbances. Dominant TS frequency $\approx 550\,\text{Hz}$. (Reproduced, by permission, from D. Arnal. AGARD Report 709, pp. 3–64, 1984.) (b) Transition in a very quiet environment, stimulated by ribbon vibrations at 110 Hz, dominant TS frequency. (Reproduced, by permission, from Y. S. Kachanov and V. Y. Levchenko, 1984.[60])

INSTABILITY OF VISCOUS FLOWS **537**

fundamental, or K-type secondary modes become prominent, the spectrum may by this stage begin to exhibit, at selected spanwise locations, a rapid emergence of sharp peaks at the second and higher harmonics of the primary disturbance frequency. At points slightly farther downstream, and at an appropriate depth in the boundary layer, sharp spikes begin to appear on the oscilloscope trace. In carefully controlled experiments, the appearance of the spikes may be very reproducible, with one or more spikes appearing always at the same phase of the primary disturbances. Authors then speak of the *one-spike stage*, or the *three-spike stage,* of transition.

Coincident with the appearance of spikes is a filling of gaps between prominent spectral lines throughout the range of lower frequencies. Eventually, identifiable lines are submerged in a broad, nearly continuous spectrum, which then nearly retains a constant nondimensionalized shape as the distance downstream is further increased.[65]

4. *Visual appearance of the flow.* If flow visualization studies are made of this hypothetical experiment, say by marking time lines with a smoke wire or bubble wire, these lines remain nearly straight in plan view while primary disturbances are growing. They become distorted, and reveal patterns of lambda-vortices, as a result of the sudden growth of secondary instabilities. The tips of the lambdas are then suddenly torn apart, presumably at the same place where high-frequency spikes appear in the time record. This is sometimes called the onset of a *tertiary instability.* Eventually, most visual evidence of any kind of orderly evolution of the flow disappears.[66]

In the course of this definition of terms, we have described one possible scenario for transition. It is important to realize that there are many such scenarios, and that a clearly recognizable sequence of primary and secondary disturbances is not always seen, particularly if the incoming disturbances are strong and irregular. This does not mean that the theory is wrong or inapplicable. In fact the theory itself warns us, by showing that a large variety of disturbances may have nearly equal rates of amplification, that a complicated result is to be expected.

Turbulent Spots

When disturbances are artificially inseminated in certain ways, it is possible to produce nice orderly arrays of lambda vortices, and to initiate the appearance

[65] A detailed documentation of this process is given by Yu. S. Kachanov, V. V. Kozlov, and V. Ya. Levchenko, in *Laminar-Turbulent Transition,* op cit. 57. See also Yu. S. Kachanov and Y. Ya Levchenko, "The resonant interaction of disturbances at laminar-turbulent transition in a boundary layer," *J. Fluid Mech.* **138**:209–248 (1984). For a masterly review and discussion, read Yu. S. Kachanov, "On the Resonant Nature of the Breakdown of a Laminar Boundary Layer," *J. Fluid Mech.* **184**:243–75 (1987).

[66] However, see Chapter 14 for a discussion of those coherent features that can be recognized as randomly-occurring elements of fully-turbulent flows.

of high-frequency spikes at regularly-spaced intervals of space and time. One can either do this by manipulation of primary and secondary disturbances or one can bypass this process and get more directly to a somewhat similar stage of transition by periodically injecting spatially and temporally localized disturbances at a selected array of points on the wall.[67]

In many natural circumstances involving convective instability, the events one might associate with tertiary instability of the flow near the tips of the lambda vortices seem to appear, very recognizably but randomly in space and time. From these beginnings evolves a dramatic flow feature called a *turbulent spot*.[68] It is usually described as a growing region of turbulent flow, surrounded

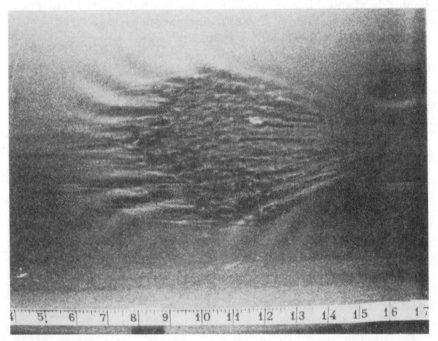

FIGURE 13.60
A turbulent spot in plane Poiseuille flow. (Reproduced, by permission, from D. R. Carlson et al., 1982.[68])

---

[67] See T. C. Corke and R. A. Mangano (1987) op. cit., and Ö. Savas and D. E. Coles, "Coherence Measurements in Synthetic Turbulent Boundary Layers," *J. Fluid Mech.* **160**:421–446 (1985).

[68] Turbulent spots were first noticed and characterized by H. W. Emmons, "The Laminar–Turbulent Transition in a Boundary Layer—Part 1," *J. Aero. Sci.* **18**:490–498 (1951). Some of the most revealing pictures are by D. R. Carlson, S. E. Widnall and M. F. Peeters, "A Flow-Visualization Study of Transition in Plane Poiseuille Flow," *J. Fluid Mech.* **121**:487–505 (1982).

by laminar flow. A particularly fine photograph of one is shown in Fig. 13.60. Spots can appear in flows that should be stable to infinitesimal disturbances, but that are subjected to rather strong[69] incoming disturbances.

After its birth, a spot grows in the streamwise and lateral directions, eventually merging with neighboring spots. According to a recently devised theoretical model,[70] the spot acts as a sort of obstacle in the way of the laminar flow, described mathematically as a traveling distribution of body force. When it moves at an optimal speed, about 0.4 times the midplane speed of a basic plane Poiseuille flow, this obstacle constitutes an effective source of wavelike perturbations that travel along with it, rather like the surface waves produced by a boat. It seems likely that these waves are strong enough to engender rapid growth of secondary disturbances and some sort of local generation of intense small-scale and high-frequency motions. It is thought that these motions intensify the momentum transfer to the wall under the flanks of the spot, and that this effectively makes the spot a larger obstacle. Thus, it engenders still more waves, which move out, break down, add more to the size of the obstacle, and so on until another spot or some other boundary is met. The outward progress of the boundary seems to be fairly independent of time, so that in plan view the spot sweeps through a wedgelike region, with apex at the point where the spot was born. In plane Poiseuille flow, the half-angle of the wedge is about 8°.

Direct Computational Simulations of Transition

A number of significantly successful computational simulations have been made, and more will undoubtedly be forthcoming. These have some special features which are worth mentioning.

1. The full Navier–Stokes equations are solved, for a three-dimensional, transient flow.
2. The solution is assumed to be periodic in x and z, so that the variation of unknowns in those directions can be represented by finite sums of sine waves.
3. The y-coordinate is transformed into a variable that runs from 0 to 1, if it does not naturally do so. This transformation is usually intended to concentrate computational effort in the region where disturbances are expected to be largest or most rapidly varying. Variation of quantities with y is usually represented by finite-difference approximations, or by sums of Chebyshev polynomials.

---

[69] Remember that a disturbance velocity as small as 1% of the maximum local basic flow velocity is strong enough to trigger rapid growth of secondary disturbances.

[70] F. Lei and S. E. Widnall, "Wave Patterns in Plane Poiseuille Flow Created by Concentrated Disturbances," Submitted to *J. Fluid Mech.*

4. Initial conditions are imposed in a way that imitates some known experimental data. Often the initial data include specified disturbances drawn from linear theory of the primary and secondary disturbances discussed above. Most of the calculations done to date have employed initial conditions that lead to the fundamental, or K-type, secondary instability.[71]

Because the computations assume periodic behavior in x, they can only be compared to experiments in which a dominant temporal periodicity is imposed, by use of something like a vibrating ribbon. The spanwise phase of oblique disturbances must also be controlled experimentally, so that the experimental records from any fixed point will be periodic functions of time. Real records will of course contain some aperiodic noise, which may be rendered insignificant by recording the signal over many cycles of the basic period, and then determining an average record for a single period. The result is called a *phase-averaged record.* Such records are obtained for different values of y, and may be composed into a snapshot of a temporal cycle, by plotting a contour map of the measured quantity in the $y-t$ plane, for fixed values of x and z. An example is shown in Fig. 13.61(a). This map is then compared to a corresponding map for one x-wavelength of the computed flow. An example is shown in Fig. 13.61(b). Some adjustment is usually allowed, to find the value of t for which the computational map best matches the experimental map for a given value of x. The comparison is usually made only for values of z at which the most dramatic flow features appear. If the two maps show the same features in the same locations, and if they evolve similarly as t of the computation and x of the experiment increase, the computation is judged to be successful.

By this standard, the calculations have quite successfully simulated the experimental flow, from the initial growth of primary instabilities until about the three-spike stage, which climaxes the development of secondary instabilities. Often the calculations are continued beyond this stage, but the corresponding experimental flow becomes strongly affected by uncontrolled disturbances, so that only a statistical comparison can be made. Even these computations are often quite impressive.

A Competing View of the Onset of Complex Motions. Chaos

The theory of hydrodynamic instability, to which the bulk of this chapter has been devoted, is about a century old. Many of the analytical and numerical

---

[71] For an example, with some details on method, see A. Wray and M. Y. Hussaini, "Numerical Experiments in Boundary-Layer Stability," *Proc. Roy. Soc. London* **A392**:373–389 (1984). For a recent overview, and relevant comments on vortex dynamics, see B. Bayly, S. A. Orszag, and Th. Herbert (1988), op. cit.

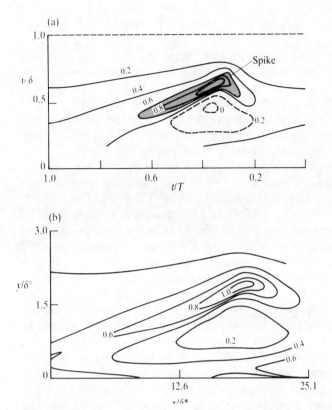

FIGURE 13.61
Maps of $\partial u / \partial y$ at the one-spike stage of transition: (a) transformed from experimental time series; (b) computed with assumption of streamwise periodicity. (Reproduced, by permission, from A. Wray and M. Y. Hussaini, 1984.[71])

tools with which its implications are revealed are quite new, but the basic idea, that simple but unstable flows may give way to somewhat more complicated ones, which may in turn become unstable and give way to still more complicated ones, has been little changed throughout the years. Lurking somewhere behind the theory is the concept of a fluid as a dynamical system possessed of many, perhaps infinitely many, *degrees of freedom*. The terminology is familiar from analysis of the motions of a rigid body, with its six degrees of freedom: three orthogonal modes of translation and three of rotation. Authors who write of the degrees of freedom of a fluid motion seldom bother to say just what they mean by the term, but in the context of this chapter they often seem to equate the excitation and amplification of a new normal-mode disturbance, when the value of a parameter such as Reynolds number increases past some threshhold, with a phenomenon like the excitation of molecular vibrations in a gas like nitrogen, when the temperature passes a certain threshold. We call the molecular vibration a new degree of freedom and may, by a rather fuzzy analogy, apply the same term to the new normal mode of fluid motion.

We have seen that raising the value of the Reynolds number past the first critical threshhold permits the excitation of many normal modes, and so we may say that it endows the fluid motion with many new degrees of freedom. If we disregard all but the most rapidly amplified, or most strongly excited normal modes, we may think of the fluid motion as having relatively few important degrees of freedom, until the onset of secondary instabilities. Then the number of important degrees of freedom increases again.

Continuing to develop this terminology, we are likely to say that any flow of really great complexity, and especially any flow that appears aperiodic and indeed random, must be a flow with a vast number of important degrees of freedom. If we are to discover the nature and contribution of each of these by an analysis at least as complex as that which yielded our identification of the important primary and secondary normal modes, the task seems hopeless. We may be able to compute realistic simulations of complex flows, and in this enterprise may speak of each computational element, such as a mesh point, a Fourier mode, or a discrete vortex blob, as a degree of freedom. Again, vast numbers of degrees of freedom seem to be necessary for a satisfying simulation of complexity, and the nature and importance of any one of them is hard to grasp.

About twenty-five years ago, it became clear that there exist some physical systems which have only a few degrees of freedom, but which are capable of astonishingly complex, aperiodic behavior. Mathematically, the possibility of complexity is inherent in the nonlinearity of the differential equations that effectively model the behavior. In many of the governing equations, the nonlinear terms are multiplied by a parameter that plays a role somewhat like that of the Reynolds number. As the value of the parameter is increased, the qualitative behavior of the system may change at certain threshhold values. The term often associated with this phenomenon is *bifurcation*, which we have encountered in a different context in the discussion of boundary-layer separation. Some bifurcations are reminiscent of, and in some cases strikingly similar to, the appearance of a new normal mode of instability. Some bifurcations lead to a type of nonperiodic behavior, which appears random under casual inspection, but is in fact deterministic and reproducible. This behavior is called *chaotic*, and the study of such systems and their mathematical models has come to be called *chaos science*.[72]

Whether these discoveries will somehow facilitate the prediction, or even the description, of the transition from laminar to turbulent flow is not yet decided. Morkovin[73] has presented arguments that would lead to a pessimistic

---

[72] See J. Gleick, op. cit. for a sense of the possible meanings of this term.

[73] M. Morkovin (1988) op. cit., particularly Appendix A.

view of the possibilities for *open systems* like boundary layers, whose convective instabilities are continually fed by disturbances that enter the system, i.e., the boundary layer, from an external source that is not affected by the nonlinear behavior of the system. A more optimistic view seems reasonable for more nearly closed systems, such as cylindrical Couette flows, whose absolute instabilities need only an initial tickling, and subsequently feed on themselves.

In any event, you may find it valuable to follow this subject as it develops in the coming years.

13.14 SUMMARY

The vorticity distribution of a laminar flow may be unstable, so that a tiny initial perturbation will amplify, interact with other perturbations, and substantially alter the future development of the flow. The initial development of infinitesimal disturbances of a sufficiently simple basic flow can be predicted with linear equations, so that an arbitrary initial disturbance can be treated as a sum of elementary disturbances. The typical elementary disturbance treated in this chapter is a *normal mode,* often recognized as a standing or travelling wave with a characteristic wavelength, orientation, speed of propagation, and rate of amplification or damping. The disturbance is assumed to vanish at the boundaries of the region of interest, and the mathematical problem becomes a search for *eigensolutions.*

Surfaces of constant amplification rate can be imagined to exist in a space whose coordinates are suitable dimensionless parameters of the disturbance, such as wavelength or frequency, and significant parameters of the basic flow, such as a Reynolds number. Special attention is paid to the neutral surface, on which the amplification rate is zero, and to the locus along which the amplification rate is maximum. A vast and rapidly growing literature is devoted to the theoretical and experimental exploration of this type of parameter space, for all the basic vorticity distributions which are simple enough to permit the necessary analysis.

Viscosity affects instability in a variety of ways. Together with the no-slip condition, it is historically responsible for the creation of potentially unstable distributions of vorticity, such as are found in jets, wakes, and boundary layers. By diffusing the concentrations of vorticity that may be convectively formed by a disturbance, viscosity may have a stabilizing effect, particularly if the Reynolds number and the wavelength of the disturbance are small. By diffusing perturbation vorticity that must be generated at a wall to keep the no-slip condition in effect, viscosity can be destabilizing. Extreme variations of viscosity can be a direct cause of instability, when a less viscous fluid displaces a more viscous one from a porous medium.

Engineers are often interested in instability as a necessary precursor to the transition from one kind of laminar flow to another, or from laminar to turbulent flow. Theoretical and computational studies focus on the nonlinear interactions between elementary disturbances that have grown large enough to affect each other's development. Clever experiments, featuring an extraordinary control over the disturbance environment to which a basic flow is submitted, are constantly being developed. Part of the motivation for these studies is the hope that instabilities may be somewhat suppressed or selectively enhanced, by manipulation of the disturbance environment. Since the disturbances that eventually affect the stability of a boundary layer or a jet may have distant origins, such as the screens of a wind tunnel or a nearby source of noise, it is important to understand how these disturbances are received into the basic flow of interest. Significant progress in all these areas of study has been made in the last two decades.

Although much is already known about the instability of laminar flows, and the contents of even this one chapter may cause you a temporary attack of intellectual indigestion, there is hardly any other aspect of fluid dynamics that attracts, at this time, more attention from researchers.

EXERCISES

13.1. Fig. 13.5 has been drawn rather naively, as though the velocity induced at point P of the perturbed sheet can be reliably estimated by just placing equally strong line vortices at A and B. A first step towards a more realistic analysis accounts for the fact that there are infinitely many points like A and B, periodically spaced along the sheet.

To draw the vector $d\mathbf{u}_A$ more accurately, you can use the streamfunction for an infinite array of equal line vortices at $kx = 0$, $\pm 2\pi$, $\pm 4\pi$, $\pm 6\pi$, and so on. This is

$$\psi = -(\Gamma/4\pi) \ln (\cosh ky - \cos kx).$$

(a) Suppose the point P has the coordinates $kx = \pi/2 + \varepsilon$, $ky = \varepsilon$. For the value $\varepsilon = 0.5$, calculate the dimensionless velocity $d\mathbf{u}_A^* = -(4\pi/k\Gamma) \, d\mathbf{u}_A$ at point P. Draw the vector, and compare it to the corresponding velocity induced by s single vortex of strength G, located at A.

(b) Do the corresponding analysis for $d\mathbf{u}_B$, by shifting the vortex array and the coordinate axes half a wavelength to the right. Note that the new coordinates of P are $kx = -\pi/2 + \varepsilon$, $ky = \varepsilon$.

(c) Draw both vectors to a common scale, and attach them to point P. Is Fig. 13.5 at least qualitatively correct?

13.2. Another way in which Fig. 13.5 is naively drawn concerns the fact that the vorticity in the deformed sheet is more strongly concentrated near B than near A, whereas the lengths of $d\mathbf{u}_A$ and $d\mathbf{u}_B$ in the figure are simply inversely proportional to the distances of P from A and B. (What does that imply?)

To estimate the ratio, S_B/S_A, where S represents the circulation per unit arclength of the sheet, you must remember that the vorticity of a material particle cannot change in an inviscid, barotropic, two-dimensional flow. When the sheet was unperturbed, the material arcs AP and PB were equally long and had equal circulations. In the deformed sheet, they retain equal circulations, but AP has become longer than PB.

Assume that the locus of the deformed sheet is given parametrically by

$$kx = kx_0 + \varepsilon \sin kx_0, \qquad ky = \varepsilon \sin kx_0,$$

where $(x_0, 0)$ is the initial position of the particle now at (x, y) neglect the curvature of the sheet near points A and B. By analyzing what has happened to the lengths of small material elements of sheet near A and B, prove that

$$S_B/S_A = [(1 + 2\varepsilon + 2\varepsilon^2)/(1 - 2\varepsilon + 2\varepsilon^2)]^{1/2}.$$

Evaluate this ratio when $\varepsilon = 0.5$, and comment on the implications of this analysis for Fig. 13.5.

13.3. Show that the roots of Eq. (13-8) are $\lambda_1 = i\alpha$, $\lambda_2 = -i\alpha$,

$$\lambda_3 = i\{\alpha^2 + \text{Re}\,(\sigma + \alpha U_\infty)\}^{1/2}, \qquad \lambda_4 = -i\{\alpha^2 + \text{Re}\,(\sigma + \alpha U_\infty)\}^{1/2}.$$

13.4. Prove that a dry atmosphere is statistically stable if the atmospheric entropy increases upward. Find an expression for the vertical temperature gradient for neutral stability, assuming a perfect gas. (Imitate the plausible but non-rigorous analysis of Section 13.2.)

13.5. Suppose that $g(y) = g_R(y) + ig_I(y)$ is a complex function, that $k = $ is a complex constant, and that ω is a real constant.
(a) Expand the equation

$$\Omega'(x, y, t) = \text{Re}\,\{g(y) \exp\,[i(kx - \omega t)]\}$$

into an equation involving only real quantities.
(b) If you have not already done so, write the right hand side of your equation in the form

$$A(t)G(y) \cos\,\{kx - \omega t + \theta(y)\},$$

and show that $G(y)$ and $\theta(y)$ can be calculated from given values of $g_R(y)$ and $g_I(y)$.
(c) Derive an explicit expression for kx, which can be used to plot contours of constant Ω' in the x-y plane, for a given value of t.

13.6. Execute the algebraic steps that lead to Eq. (13-28), and verify that the equation is correct.

13.7. Derive the eigenvalue criterion for sinuous modes of the Bickley jet, given by Eq. 13-48. Verify that the corresponding condition for varicose modes is $2E_{11\,m-1} + B_{11\,m} = 0$.

13.8. Given that $\psi'(x, y, t) = f(y) \exp\,\{i(kx - \sigma t)\}$ is the perturbation streamfunction for a particular normal-mode disturbance, show that the corresponding vorticity perturbation is given by

$$\Omega'(x, y, t) = -[f''(y) - k^2 f(y)] \exp\,\{i(kx - \sigma t)\}.$$

13.9. Given that $\psi'(x, y, t) = f(y) \exp\,\{i(kx - \sigma t)\}$ is the perturbation streamfunction for a particular normal-mode disturbance of a basic flow with velocity $U(y)$, show that the corresponding pressure perturbation is given by

$$p'(x, y, t) = i\rho\{[(\sigma - kU)f'(y) + kU'(y)f(y)] \exp\,\{i(kx - \sigma t)\}.$$

if the flow is inviscid.

13.10. (a) Verify that the no-penetration and no-slip boundary conditions, applied to the disturbances of plane Poiseuille flow, imply the following boundary conditions for the Thomas algorithm.

$$E_{111} = 0, \qquad E_{121} = 0, \qquad E_{211} = -2/B_{121}, \qquad E_{221} = 0.$$

(b) Derive the corresponding equations for the case in which $f(-1) = 0$ for zero penetration, and $s(-1) = 0$, for zero surface vorticity.

(c) Finally, verify that the conditions $f(-1) = 0$, and $s'(-1) = 0$, for zero flux of vorticity through the wall, are $E_{111} = 0$, $E_{121} = 0$, $E_{211} = 0$, $E_{221} = -2/B_{121}$.

13.11. Consider the cumulative amplification of Tollmien–Schlichting waves of a fixed frequency, growing on a self-similar Falkner–Skan type boundary layer. The basic flow characteristics are $U = ax^m$ and $\delta^{*2} = bvx^{1-m}$, in which a, b, and m are constants.

You are to use the approximate formula (13-51), and need to convert it into an equation involving the dimensionless quantities $\alpha_I = k_I \delta^*$ and $\text{Re} = U \delta^*/v$.

As a first step, show that

$$k_I \, dx = ab\alpha_I \, d\text{Re}$$

13.12. Fig. 13.34 presents contours of a constant dimensionless spatial amplification rate, $k_I v/U$, in a plot of dimensionless frequency, $\omega v/U^2$, versus Reynolds number, $U\delta^*/v$. Note that it uses logarithmic scales on both axes.

Suppose that the basic flow is a Falkner–Skan boundary layer, so that $U = ax^m$ and $\delta^{*2} = bvx^{1-m}$, in which a, b, and m are constants.

Show that a disturbance of fixed dimensional frequency ω, as it propagates downstream through this basic flow follows a trajectory that appears on this map as a straight line of slope $-4m/(1 + m)$.

13.13. Consider a Tollmien–Schlichting wave of fixed frequency, growing as it propagates downstream in a Blasius boundary layer. The dimensionless spatial amplification rate, $-\alpha_I$, depends on the Reynolds number, $U \delta^*/v$, as shown in Fig. 13.34. This dependence is fairly well approximated by the simple formula

$$\alpha_I(\omega v/U^2, \text{Re}) = \alpha_{I \max}(\omega v/U^2)(1 - \xi^2),$$

where

$$\xi = (2 \text{Re} - \text{Re} 2 - \text{Re} 1)/(\text{Re} 2 - \text{Re} 1).$$

Here Re 1 and Re 2 are the values of $U \delta^*/v$ where the line of constant $\omega v/U^2$ cuts Branch I and Branch II of the neutral stability curve, and $\alpha_{I \max}$ is the maximum value of α for that value of $\omega v/U^2$.

(a) Using the approximation $\delta^{*2} = 3vx/U$, which is quite accurate for a Blasius boundary layer, show that the corresponding result for the total amplification between Branch I and a point between the two branches is given by

$$\ln (A/A1) = -(\alpha_{I \max}/9)(\text{Re} 2 - \text{Re} 1)(3\xi - \xi^3 + 2).$$

(b) Use this equation and Fig. 13.34 to estimate $\ln (A2/A1)$ for $=$, and check your result against the precisely calculated values shown in Fig. 13.35.

13.14. The conditions for the determination of the spectra shown in Fig. 13.59(b) were defined by the values $U = 9.18$ m/s, $v = 1.5 \times 10\text{-}5$ m^2/s, $0.30m \leq x \leq 0.76m$, and $f = 111$ hz. The laminar boundary layer developed under zero pressure gradient.

Use the estimate $\delta^* = 1.72(vx/U)^{1/2}$, and calculate the range of $U\delta^*/v$ covered by the experiment. Make a copy of Fig. 13.34 and draw the locus of the experimental conditions on it.

13.15. In the reference from which Fig. 13.59(a) was drawn, it is stated that the boundary layer grew under a stream with constant speed, $U = 33$ m/s. Assume that $v = 1.5 \times 10\text{-}5$ m^2/s, and use the estimate $\delta^* = 1.72(vx/U)^{1/2}$.

Calculate $U\delta^*/v$ for the value of x that identifies each spectrum in Fig. 13.59(a). Then use Fig. 13.34 to find the corresponding dimensionless frequency for the point that lies on Branch II of the neutral curve, at that value of the Reynolds number. Using the given values of U and v, calculate the corresponding disturbance frequency in hz. Remember that $f = \omega/2\pi$.

Finally, make a copy of Fig. 13.59(a) and mark the frequency you have just calculated for each spectrum. What do you conclude from this exercise?

13.16. Derive a closed-form equation for the projection of the streamlines of Taylor-vortex flow on the r-z plane, given that the radial and axial velocity components are

$$u(r, z) = u^*(r) \cos(kz) \quad \text{and} \quad w(r, z) = -(1/k)(du^*/dr) \sin(kz).$$

{Answer: $kz = \arcsin(A/u^*(r))$, where $A = \sin(kz_0)/u^*(r_0)$, and the streamline passes through the point (r_0, z_0).}

13.17. A crude but illuminating analytic cartoon of the spontaneous redistribution of the vorticity of a shear layer into periodically spaced cores and braids, as shown in Fig. 13.48, embodies the simple ordinary differential equation

$$d\Gamma_c/dt = A\Gamma_c(\Gamma - \Gamma_c),$$

in which A is a constant, Γ_c is the circulation of a core, and $\Gamma - \Gamma_c$ is the circulation of a braid. The total circulation of the shear layer per wavelength of the periodic disturbance is Γ, a constant. The equation states that the rate at which circulation is captured by a core is proportional to the amount it already has, Γ_c, and to the amount remaining to be captured, $\Gamma - \Gamma_c$.

Noting that the differential equation is separable, find the solution that meets the initial condition $\Gamma_c = 0.55\Gamma_c$.

CHAPTER
14

TURBULENT FLOW

14.1 INTRODUCTION

INSTABILITY AND INDETERMINACY. We have seen in Chapter 13 that flows containing vorticity may be *unstable*, so that two flows with infinitesimally different distributions of vorticity at some initial time may develop very differently thereafter, even if they are subject to the same boundary conditions. One can imagine that infinitesimal differences in the boundary conditions may also perturb the distribution of vorticity, and that a flow that is unstable to such perturbations will evolve in an essentially unpredictable way.[1] In such a case the flow history is *indeterminate*, because it depends upon factors (initial and boundary perturbations) that one cannot know. Viscous flows characterized by high Reynolds numbers often exhibit this kind of unpredictable or *random behavior*. Such flows are called *turbulent*.

MEAN FLOW BEHAVIOR. We can appreciate that a theory of turbulent flow will be very different from the theory expounded and illustrated in our

---

[1] The word *infinitesimal* is applied here to a quantity that is not only mathematically small but experimentally unobservable.

548

previous chapters. However, it is believed that turbulent flows are constrained at any instant by the Navier–Stokes and continuity equations, and it is observed that the *mean* behavior of a large sample of turbulent flows, all of which are constrained by nominally identical initial and boundary conditions, is uniquely determined by these conditions.[2]

For many practical applications, one would be content to predict or to control this mean behavior. Unfortunately, the mean values of velocity and pressure do not satisfy the Navier–Stokes equations, or any other set of equations that has yet been discovered. What is known about mean velocity and pressure distributions comes mostly from direct measurement.

PRACTICAL CONSEQUENCES OF TURBULENCE. From a practical viewpoint, turbulence can be either bad or good, and the designer of flow systems may make special efforts either to prevent it or to stimulate it.

When the object of design is to promote molecular mixing between fluid bodies of disparate chemical composition, to disperse an unwanted concentration of one fluid into a large body of another, or to enhance the rate of heat or mass transfer between a bounding wall and a fluid, turbulence is good. It is also good when the object is to reduce the extent of separated flow in aerodynamic applications or in fluid machinery. The benefits in these applications arise from the large instantaneous transverse velocities[3] induced by concentrated vortex filaments, and by the large and sustained rates of deformation experienced by fluid elements near such filaments. Sometimes, the unsteadiness and large local accelerations in a turbulent flow are also beneficial, when the object is to coagulate suspended matter, such as smoke particles or water droplets.

The bad side of turbulence is a characteristically large rate of dissipation of mechanical energy, which has to be made up by pumps or other power sources.[4] This is the undesirable aspect of high rates of deformation. An enhanced rate of transfer of heat, mass, or momentum between a stream and an adjacent wall may be the last thing a designer wants, in certain circumstances. The unsteadiness of turbulent flow is also often a nuisance. High-frequency velocity fluctuations can be a source of ear-splitting noise. Lower-frequency, large-scale fluctuations can upset aircraft or excite destructive resonant oscillations in buildings.

---

[2] *Nominal* conditions are those we can measure and control by ordinary means. Thus, a wall is nominally smooth if it feels smooth to the touch or looks smooth to the unaided eye. A flow is nominally uniform and steady if no gradients or transients can be detected with high-quality, commercially available velocimeters.

[3] A component of the instantaneous velocity, normal to the mean velocity at that point.

[4] However, separation control by the judicious stimulation of turbulence in one part of the flow may reduce external power requirements. High dissipation rates in the turbulent boundary layer are offset by lowered dissipation rates in the wake.

THE ROLES OF VISCOSITY IN TURBULENT FLOW. Viscosity is both midwife and executioner in the life story of turbulence. As midwife, it delivers the vorticity from its birthplace at the wall. Thus, the flow receives something without which it cannot be unstable, and hence possibly turbulent. As executioner, it applies the torques that rub out the local concentrations of vorticity, without which the turbulent velocity fluctuations quietly die away.

Although viscous diffusion never ceases in a turbulent flow, it must compete with macroscopic deformation of fluid elements, which often tends to concentrate vorticity. It appears, from observation and from theoretical *cartoons*[5] of turbulence, that viscosity is effective in only a small part[6] of the turbulent flow field at any given instant. It is in this part of the flow that most of the conversion of mechanical energy to thermal energy occurs by viscous dissipation. Paradoxically, the time- and volume-averaged rate of viscous dissipation in turbulent flows at very high Reynolds numbers depends very weakly on the value of the viscosity coefficient. You will read more about this, after a formal discussion of ways of observing and describing turbulent flow.

This chapter focuses mostly on the description of turbulent flow. The art of description turns out to be far from trivial, because different observers often see a different significance in complex events. To the extent that turbulence is random, we require statistical descriptions. The main concepts and procedures are presented in Section 14.2.

The statistical approach led historically to Osborne Reynolds' decomposition of each flow quantity into a mean value and a fluctuation around the mean, and to his search for the implications of time-averaged conservation equations. These are presented in Section 14.3.

One cannot long contemplate statistical data without wondering what they mean, or more precisely, what sort of individual events account for the statistical trends. Believing that it is not enough to postulate kinematic scenarios whose appearance has no dynamical explanation, I present some dynamical cartoons, which partially demonstrate the possibility of certain events that might be common in turbulent flow. These appear in Section 14.4.

---

[5] "A cartoon was first an outline or pattern drawn on cardboard to guide the weaver of a rug or tapestry. It is now a caricature, a representation of something or someone which simplifies and exaggerates deliberately certain features of the object for various purposes which, at least in the mind of the cartoon's author, can be summarized as clarification". G. M. Corcos, "The Role of Cartoons in Turbulence," in *Perspectives in Fluid Mechanics* (D. Coles, ed.), Lecture Notes in Physics 320, Springer-Verlag, Berlin (1989).

[6] The rather vague word *part* is used advisedly. Recent theoretical work suggests that we should not even think of a fractional volume, but that the geometrical object that instantaneously contains the vorticity of a turbulent flow at very high Reynolds number is a *fractal of dimension about 2.5,* something between a surface and a volume.

See, for example, A. J. Chorin, "The Evolution of a Turbulent Vortex," Comm. Math. Physics *83*, 517–535 (1982).

Next, you will find a somewhat lengthy description of relatively simple turbulent flows: cleaned up enough to be replicated in different laboratories but sufficiently like practically important flows to be worth understanding. Section 14.5 briefly sets the stage for this description. Section 14.6 deals with flow along solid walls; Section 14.7 deals with flow far from solid walls.

The final section, 14.8, is devoted to description of various computational schemes that simulate individual realizations of turbulent flow, either in complete detail or with an approximate statistical treatment of motion at the smallest scales. The simulations of individual realizations provide a new source of data, restricted to cases with very simple or deliberately idealized boundary conditions, which are believed to be at least as reliable as laboratory measurements and which present unprecedented opportunities for detailed analysis.

14.2 THE DESCRIPTION OF TURBULENT FLOW

VISUAL IMPRESSIONS. An excellent first look at a variety of turbulent flows is provided in an educational film,[7] which you should see before getting deeply into this chapter. Vivid impressions of randomness, unsteadiness, and large transverse velocities are conveyed by the visible motions of boundaries between fluid masses of different color, opacity, or density. A particularly clever sequence illuminates the competition between viscous diffusion and macroscopic deformation by showing how the impression of fine-grained bumpiness of the surface of a cloud of smoke depends on Reynolds number.

Movies of this or earlier vintage are often somewhat confusing because they show motions that occur simultaneously all along a line of sight through a transparent fluid; or are frustrating because they show only the moving surface of a mass of opaque fluid. Also, the framing speed, exposure time, and depth of focus may limit the resolution of high-frequency and small-scale details. With tunable lasers to illuminate a very thin region of fluid for a very brief instant, and with fluorescent dyes or smoke to scatter that light, one can now produce snapshots of startling clarity, which have had great influence on our mental picture of the spatial variations of properties in individual realizations of a turbulent flow.[8]

We learn much by looking at turbulent flows, but logic and previous knowledge must discipline the conclusions we draw from what we see. This is particularly true when we try to describe what was happening at the time of a

---

[7] *Turbulence,* film principle, R. W. Stewart. Produced by NCFMF/EDC. Available from Brittanica Films & Video, 425 No. Michigan Avenue, Chicago, Illinois 60611.

[8] A fine collection of snapshots has been assembled by M. D. van Dyke, as part of *An Album of Fluid Motion,* Parabolic Press, Stanford, 1982.

single snapshot. For example, a view of a clearly defined spiral of dyed fluid strongly suggests the presence of a concentrated vortex, but is inconclusive because the vortex that caused the spiral may have diffused away before the picture was taken. We need to see the *motion* within the spiral to know that the vortex is still there. Even movies must be watched with care, if we would see velocity or vorticity fields. It is hard to keep one's eye on an identified bit of fluid when entire regions have been marked with smoke or dye.

STATISTICAL DESCRIPTION. Given a very large sample of experiments carried out under identical nominal constraints, one could study measured values of some flow quantity at a specified position and time, to see how they varied from experiment to experiment. Presumably this variation would be random, and statistical tools would be necessary for a quantitative description. One could calculate mean values, mean-squared deviations from the mean values, the probability of observing certain values, etc. However, such data, drawn from an *ensemble* of nominally identical experiments, is almost never available. Most available quantitative information about turbulence refers to flows that are *statistically stationary*. These are flows that started long ago, and have developed under constant nominal constraints. An example would be flow in a pipe between two reservoirs of constant level, long after the valve is opened. Statistically stationary flows can be characterized by averaging measured quantities over a sufficiently long time. Thus, for example, the *mean* pressure at point $\mathbf{r} = \mathbf{r}_0$ is given by

$$P(\mathbf{r}_0) = T^{-1} \int_{t_0}^{t_0 + T} p(\mathbf{r}_0, t) \, dt \tag{14.1}$$

which introduces two parameters: the averaging time T, and the starting time t_0. If the flow is statistically stationary, t_0 will unimportant, but T must be large enough so that further increases have no significant effect on the measured value of $P(\mathbf{r}_0)$.

The *variance* of p is

$$\langle p'^2(\mathbf{r}_0) \rangle = T^{-1} \int_{t_0}^{t_0 + T} [p(\mathbf{r}_0, t) - P(\mathbf{r}_0)]^2 \, dt \tag{14.2}$$

The square root of the variance is called the *standard deviation*. The standard deviation of the streamwise component of velocity, divided by the mean value of that component, is often cited as a local measure of the *intensity* of turbulence.

Higher powers of the pressure fluctuation, $p' = p - P$, can be averaged to give some further idea of the probability of finding a fluctuation of given size, or the data can be analyzed more directly, to yield a *probability density function*. Figure 14.1 illustrates how this is done.

We see a record of pressure versus time. The shaded horizontal strip represents a small range of pressure values. The darkened segments of the time axis represent the intervals during which p falls within the shaded band.

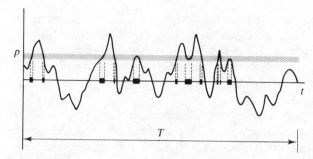

FIGURE 14.1
Construction of a probability density function.

The *probability* that p should fall within the shaded band is the fraction of the total sampling time, T, during which it does so. For the example, it is about 0.10. The probability increases to unity if the shaded band blankets the whole record. The limit of the ratio of the probability to the width of a band that shrinks to a line is called the *probability density*. The value of the probability density will depend on the value of p; this introduces the concept of the *probability density function* (PDF) for pressure fluctuations. Graphs of the PDFs for pressure and for other quantities, such as a velocity component, temperature, or concentration of a chemical species, serve well to identify a given turbulent flow, and often suggest interesting insights into the processes active in that flow. Figure 14.2 shows a modern example: a plot of probability density for the concentration of dye in a mixing layer between parallel streams of dyed and clear fluid, with concentration as one independent variable, and distance across the layer as another.

FOURIER ANALYSIS OF TIME SERIES. An experimental record such as that of Fig. 14.1, or a set of values taken from such a record at successive, equally

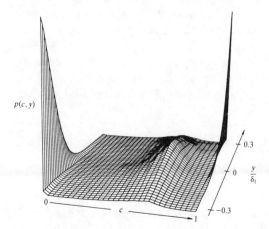

$p(c, y)$

0.3

0 $\dfrac{y}{\delta_1}$

−0.3

0

c

1

FIGURE 14.2
Probability density function as a function of position in a mixing layer. (Reproduced, by permission, from M. M. Koochesfahani and P. E. Dimotakis, *J. Fluid Mech.* **170**; 83–112, 1986.)

spaced instants, is often called a *time series*. It can be Fourier-analyzed, to reveal other characteristics of the signal that, like the probability density function, are independent of t_0 and T is the flow is statistically stationary and the record is long enough. This analysis reveals the *frequency content* of the signal, and can sometimes help one to identify the physical processes that cause the observed fluctuations.

If the experimental record arrives as a continuously varying voltage, it can be Fourier-analyzed by a variety of analog means involving electronic filters. A popular modern alternative operates digitally on the discrete time series. The procedure is particularly fast when the records contains 2^m values, m being an integer—typically about eleven.[9]

One starts by adding a final value, equal to the first value, and then fits the observed values to a sum of sine waves, the first of which has period T, the second period $T/2$, and so on. Each wave has *amplitude* and *phase*, as well as its period or frequency. The fitting is accomplished by a special linear algebraic computation called Fast Fourier Transform, which yields $N/2 + 1$ amplitudes and $N/2$ phases if N is the number of points of the original data series. The extra amplitude is the mean of the recorded values. Suppose the record is of the velocity component u. We write

$$u(t) = A_0 + \sum_{n=1}^{N/2} A_n \sin(\omega_n t + \phi_n) \tag{14.3}$$

in which $\omega_n = 2\pi n/T$.

For any *finite* value of T, the nth amplitude, A_n, and the nth phase, ϕ_n, depend not only on the frequency, ω_n, but on T and t_0. However, the distribution of the amplitude A_n, as a random function of t_0, has a variance that approaches zero as T approaches infinity.

A commercial spectrum analyzer operating on this principle allows adjustment of T, but has only a fixed maximum number of readings in each sample. The maximum useful value of T will be constrained by the maximum frequency that must be resolved, and the statistical scatter in the recorded amplitudes and phases is dealt with by repeating the entire operation until cumulative averages of A_n have settled down to an acceptable degree. More repetitions are needed as ω_n decreases.

If we accept the sum of sine waves as an adequate approximation to the original time series, we can use it to compute the integral of u^2 over the timespan t_0 to $t_0 + T$. Since T is an integral multiple of the period of each component sine wave, the result is simply

$$\int_0^T u^2 \, dt = T[A_0^2 + \tfrac{1}{2}(A_1^2 + \cdots + A_{N/2}^2)] \tag{14.4}$$

---

[9] A very helpful reference book is A. V. Smol'yakov and V. M. Tkachenko, *The Measurement of Turbulent Fluctuations*. An English translation (1983) is published by Springer-Verlag. See especially pp. 23–25 and 95–105.

Thus, for each record, A_0^2 gives an approximation to U^2. For $n > 0$, $A_n^2/2$ approximates the contribution of the sine wave of frequency ω_n to the variance of u. For a single record, a bar graph of the A_n versus the ω_n can be called the *amplitude spectrum* of the record. A corresponding plot of phase versus frequency would be a *phase spectrum*. The properties and applications of the two types of spectra are very different, to wit:

1. In a statistically stationary turbulent flow, only the *amplitude* spectrum converges to something useful as we repeat the sampling process. The cumulative average of ϕ_n is zero for all n. Thus, only the converged amplitude spectrum is cited, as is the probability density function, to characterize the flow statistically.
2. Individual records that look entirely different may have the same amplitude spectrum; the phase data are needed to distinguish them. Thus, one must refrain from constructing a mental picture of the typical record that would produce a specified amplitude spectrum.[10]

ENERGY SPECTRUM. The bar graph of converged values of $A_n^2/4$ versus ω_n can be called an *energy spectrum* of the turbulence, although it clearly represent the contribution of only one velocity component to the kinetic energy.

A typical energy spectrum from a turbulent pipeflow or boundary layer exhibits a gradual variation without sharp peaks. If peaks are present, they can usually be traced to some measurable periodicity in the nominal constraints, such as pump discharge pulsations at a blade-passing frequency.

When turbulence evolves from a steady laminar flow, the shape of the spectrum of velocity fluctuations changes in a distinctive way as the point of observation is taken farther downstream. The first observable fluctuations may occupy a relatively narrow band of frequency, determined by the selective amplification that was studied in Chapter 13.

Harmonics of this narrow band may next appear, as the first observable disturbances amplify and lose their sinusoidal waveforms. The perturbed flow may next become unstable to a different class of disturbances, possibly leading to subharmonics of the original frequency or to frequencies that are neither integral multiples nor fractions of the original. The spectrum rapidly fills in, although a residual peak may linger. Examples will be shown later, as parts of descriptions of certain classic flowfields have been intensively investigated.

CORRELATION FUNCTIONS. Frequency spectra quantify the unsteadiness of a fluctuating signal in a particular way. Another way, mathematically related

---

[10] A fascinating and vivid illustration of this point is given by L. Armi and P. Flament, "Cautionary Remarks on the Spectral Interpretation of Turbulent Flows," *J. Geophys. Res.* **90**; 11,779–11,782 (1985).

but conceptually quite different, is provided by *correlation analysis*. The simplest example of this asks the question: how well does the current value of a local flow property correlate with its value a time τ later? The answer is given by the normalized integral

$$R(\tau) = \frac{\int_{t_0}^{t_0+T} u'(t)u'(t+\tau)\, dt}{\langle u'^2 \rangle} \tag{14.5}$$

If the fluctuation of the signal is random, we expect $R(\tau)$ to approach zero as τ becomes very large. One empirical way to judge whether the sampling time, T, is long enough is to make sure that $R(\tau)$ has decayed to zero for some $\tau \ll T$. A typical measured curve of R versus τ is shown in Fig. 14.3.

It turns out that $R(\tau)$ contains exactly the same statistical information as does the energy spectrum for u. For a given record of length T, we can use the sum of sine waves to approximate the integrand in (14.5), and can easily show that

$$R(\tau) = \frac{\sum A_n^2 \cos(2\pi n\tau/T)}{\sum A_n^2} \tag{14.6}$$

Spatial correlations can also be calculated if simultaneous records are obtained at two spatially separated points, from the integral

$$R(\rho) = \frac{\int_{t_0}^{t_0+T} u'(\mathbf{r}_0, t)u'(\mathbf{r}_0+\rho, t)\, dt}{\langle u'^2 \rangle} \tag{14.7}$$

The determination of a curve of $R(\rho)$ requires a lot of measurement, since the integral has to be repeated for enough discrete values of ρ to define the curve. A whole of family of such curves could be produced by varying the direction of the line connecting the two observation points, but this is rarely done.

Finally, *space–time* correlations are sometimes produced, usually for points separated in the streamwise direction. These can suggest whether

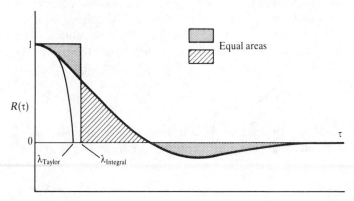

FIGURE 14.3
Correlation function for u' in a boundary layer.

approximately constant patterns of fluctuation, perhaps the signals associated with drifting vortices, move downstream at a statistically well-defined speed.

Correlations between the instantaneous, local, values of different flow properties, such as two orthogonal components of velocity, are often reported in statistical descriptions of turbulence. They play important roles in the equations that constrain the behavior of the mean flow, and you will see more of them later.

SCALES OF TURBULENCE. The areas under the curves $R(\tau)$ and $R(\rho)$ have dimensions of time and length, respectively, and can serve as intuitively attractive local time and length scales, useful for a condensed statistical description of the fluctuations. They are called the *integral scales* of turbulence. As is shown by the construction in Fig. 14.3, a correlation curve that showed perfect correlation out to the integral scale, and zero correlation for greater separations, would bound the same area as does the real curve.

Another timescale, easily conceived and measured, is the *Taylor microscale*. It is simply the standard deviation of some fluctuating quantity, divided by the standard deviation of the local time derivative of that quantity. It can be shown that it equals the value of τ at which the parabola that fits the correlation function $R(\tau)$ near $\tau = 0$ cuts the τ axis.

A final important pair of scales are the *Kolmogorov* or *dissipation* scales of length and time. They are often said to characterize the smallest-scale concentrations of vorticity that can be maintained by macroscopic deformation against the smoothing effect of viscosity. The argument used to evaluate these scales presumes that most of the viscous dissipation of mechanical energy is associated with the rates of deformation induced by these smallest eddies. The mean rate of dissipation per unit mass, called ε, can frequently be computed for a given control volume, by simply reckoning the excess of mean kinetic energy inflow over outflow. Kolmogorov argued that the scales of the motion responsible for the value of ε must be determined only by that value and by the kinematic viscosity. Since ε and ν have dimensions L^2/T^3 and L^2/T respectively, we can combine them to obtain the length and time scales

$$\lambda = \left(\frac{\nu^3}{\varepsilon}\right)^{1/4} \quad \text{and} \quad \tau = \left(\frac{\nu}{\varepsilon}\right)^{1/2} \tag{14.8}$$

From the energy accounting, which will be more fully discussed in the next section, one concludes that ε is proportional to U^3/L, where U and L are scales characteristic of the overall flow. For example, L might be the thickness of a turbulent boundary layer, while U is the velocity at the outer edge of the layer. Replacing ε by U^3/L in the formulas above, we find that $\lambda/L \sim \mathrm{Re}^{-3/4}$, and that $U\tau/L \sim \mathrm{Re}^{-1/2}$, where $\mathrm{Re} = UL/\nu$. Thus, the Kolmogoroff scales are much smaller than the overall scales when the Reynolds number is very large.

There is, however, substantial reason to question the logic that asserts that these scales characterize the smallest possible concentrations of vorticity owing to evidence that the volume-averaged rate of energy dissipation is substantially lower than typical local rates, where these are nonzero, because a large fraction of the volume contributes little to the spatial average at any given instant. This concern leads us to the concepts of the next subsection.

INTERMITTENCY AND CONDITIONED SAMPLING. Many turbulent flows involve transient intrusions of vorticity-laden fluid into a surrounding region of irrotational fluid, and a wildly meandering instantaneous boundary between the two. It is shown in previous chapters that this boundary is always somewhat diffuse, but we should not be too surprised to learn that it is very thin, at high Reynolds numbers, compared to the spatial integral scale. Snapshots hint at this, although we have been warned that they do not define vorticity distributions.

In a statistically stationary flow, a fixed point of observation may be repeatedly traversed by this boundary, so that it is sometimes in the rotational flow and sometimes in the irrotational flow. Depending on what is being measured, a continuous data record from such a point may more or less obviously switch back and forth between two very distinct kinds of trace. An example is shown in Fig. 14.4.

The record is characterized by the fraction of its length, T, during which the sensor was apparently in the rotational flow. This fraction, and the phenomenon itself, are both called *intermittency*. The term is also applied to a somewhat different phenomenon, in descriptive studies of the transition between laminar and turbulent flow in pipes and boundary layers. Then the sensor is always in a region of rotational flow, but on one side of a fairly sharp boundary the vorticity is comparatively uniformly distributed, while on the other side it is sharply concentrated in tangled filaments that induce relatively high-frequency velocity fluctuations. Whereas the boundary between irrotational and rotational flows must be very nearly a material surface if the Reynolds number is high, the boundary in the second kind of intermittent flow seems to propagate through the fluid, along the vortex lines, like a wave front.

Whatever the cause or nature of intermittency, its discovery led to a kind of statistical averaging called *conditioned sampling*. This involves averaging over only those segments of a time record that meet some defined condition, selected so as to enhance the chance of sampling a given phenomenon. Sometimes the condition for accepting parts of a continuous record of variations of one flow variable comes from simultaneous measurements of a different variable. For example, cool fluid may be made to pass over a warm wall; and velocity and temperature may then be simultaneously recorded at

FIGURE 14.4
Temperature trace, showing intermittency.

nearly the same point in the turbulent boundary layer. The velocity record might then be subdivided into sections recorded while the temperature exceeded a certain threshold value, or while the temperature was increasing, and so on.

One sometimes needs special care to avoid unintentionally conditioned sampling. For example, data from a turbulent jet might be taken with a laser Doppler velocimeter, which returns a measurable signal only when one or more light-scattering particles are in the illuminated measuring volume. If particles are added only to the fluid issuing from the nozzle, and the fluid entrained into the jet from the surroundings is free of particles, velocity data will be gathered only when jet fluid occupies the observation volume. The velocity of clear fluid entrained into the jet will completely escape detection.

14.3 THEORY OF THE MEAN FLOW. REYNOLDS-AVERAGED EQUATIONS

In 1895, Osborne Reynolds[11] manipulated the Navier–Stokes and continuity equations into a form that has served to organize the discussion of empirical studies of turbulent flows, and to launch many efforts to predict their time-averaged behavior. At the time, he was seeking a theoretical explanation for the empirical criterion $\rho D u_m / \mu \leq 2000$ that rules out turbulent flow in tubes of circular cross section. He was also trying to understand why the scales of turbulent motion, which he could estimate from visual observations, are so much larger than those of random molecular motion, which he could estimate from the kinetic theory of gases.

Reynolds represented the instantaneous value of any dependent variables of an Eulerian flow description, as a sum of a *mean* value and a *fluctuation* from the mean. Thus, $f(\mathbf{r}, t) = F(\mathbf{r}, t) + f'(\mathbf{r}, t)$, where $F(\mathbf{r}, t)$ symbolizes the mean value.

Reynolds was concerned with statistically stationary flows, in which the mean value, F, would depend on $\mathbf{r}$ alone. The averaging operation, for which we shall use the symbol $\langle \ \rangle$, was for him the *time average* defined in Section 14.2 above. You will also encounter the term *ensemble average*. That concept is based on the notion that it is possible to draw samples from an ensemble of statistically independent flows, all of which are subjected to the same nominal boundary conditions. All samples are drawn at the same time and the same position relative to the flowfield boundaries, and averaged by simply adding individual values and then dividing by the number of values. This allows for the possibility of unsteady mean flow, although temporal variations of the mean are likely to be much slower than those of individual realizations.

For either kind of averaging operation, it is understood that

---

[11] O. Reynolds, *Phil. Trans. Roy. Soc. London*, **A186**; 123–164 (1895).

$\langle f(\mathbf{r}, t)\rangle = F(\mathbf{r}, t)$, and that $\langle f'(\mathbf{r}, t)\rangle = 0$. The averaging process has properties of distribution and commutation, such that

$$\langle f_1(\mathbf{r}, t) + f_2(\mathbf{r}, t)\rangle = F_1(\mathbf{r}, t) + F_2(\mathbf{r}, t) \tag{14.9}$$

$$\langle G(\mathbf{r}, t) f(\mathbf{r}, t)\rangle = G(\mathbf{r}, t) F(\mathbf{r}, t) \tag{14.10}$$

$$\langle \operatorname{grad} f(\mathbf{r}, t)\rangle = \operatorname{grad} F(\mathbf{r}, t) \tag{14.11}$$

$$\langle \partial f(\mathbf{r}, t)/\partial t\rangle = \partial F(\mathbf{r}, t)/\partial t \tag{14.12}$$

We shall consider only flows in which ρ and μ are constant. The continuity equation, $\operatorname{div} \mathbf{u} = 0$, is linear and yields $\operatorname{div} \mathbf{U} = 0$ after averaging. From (14.8), it follows that $\operatorname{div} \mathbf{u}' = 0$. Thus both the mean velocity field and the instantaneous departures from it are solenoidal.

The momentum equation, in divergence form

$$\rho\left(\frac{\partial \mathbf{u}}{\partial t} + \operatorname{div}(\mathbf{u}; \mathbf{u})\right) + \operatorname{grad} p = \mu \operatorname{div} \operatorname{def} \mathbf{u} \tag{14.13}$$

becomes $\quad \rho\left(\dfrac{\partial \mathbf{U}}{\partial t} + \operatorname{div}(\mathbf{U}; \mathbf{U})\right) + \operatorname{grad} P = \operatorname{div}(\mu \operatorname{def} \mathbf{U} - \rho\langle \mathbf{u}'; \mathbf{u}'\rangle) \tag{14.14}$

This equation differs from the Navier–Stokes equation for the mean flow because of the extra term on the right-hand side, which involves averaged products of fluctuation velocity components. This quantity,

$$\mathsf{R} = -\rho\langle \mathbf{u}'; \mathbf{u}'\rangle \tag{14.15}$$

is called the *Reynolds stress*. To appreciate the name, imagine a control surface that moves with the local value of $\mathbf{U}$ and is oriented by a unit normal vector $\mathbf{n}$. At a given instant, mass moves across this surface, from the region into which n points, at the rate $-\rho\mathbf{n} \cdot \mathbf{u}'$ per unit surface area. This fluid carries momentum $\mathbf{u}'$ per unit mass, in the frame of reference of the moving control surface. The momentum in a control volume that is locally bounded by this surface consequently increases, just as through a stress equal to $-\rho(\mathbf{n} \cdot \mathbf{u}')\mathbf{u}'$, or to $-\mathbf{n}(\rho\mathbf{u}'; \mathbf{u}')$ were acting on the surface. This stress fluctuates, and only its averaged value, which equals $\mathbf{n}\mathsf{R}$, appears in Eq. (14.14).

INDETERMINACY AGAIN. Because R appears in (14.14), that equation plus the equation $\operatorname{div} \mathbf{U} = 0$ only constrain, but do not determine, the fields of $\mathbf{U}$ and P, even when initial and boundary conditions for these latter quantities are given. There are now visibly more unknowns than equations. This situation is familiar from Chapter 5, where the basic conservation equations for the instantaneous flow were presented. In that case a relatively simple postulate, $\tau = \mu \operatorname{def} \mathbf{u} + \lambda \operatorname{div} \mathbf{u} \mathsf{I}$, which could be supported by theory for simple gases and by experiment for a host of useful fluids, proved to be accurate in a wide range of circumstances. Unfortunately, there seems to be no comparably simple and generally useful connection between the Reynolds stresses and the mean velocity field in turbulent flow.

ENERGY ACCOUNTING. In spite of indeterminacy, one can gain occasionally useful insights by continuing this line of analysis. For example, one can scalar-multiply each term of the mean momentum equation (14.14) by the mean velocity and rearrange the result into the form

$$\frac{\rho}{2}\frac{\partial}{\partial t}(U \cdot U) + \text{div}\left[\rho U(\tfrac{1}{2}U \cdot U + gh) + U(P| - \mathsf{T} - \mathsf{R})\right] = -(\mathsf{T} + \mathsf{R}) \cdot \text{grad } U$$

(14.16)

In this, T is the time-averaged viscous-stress tensor. $\mathsf{T} \cdot \text{grad } U$ is the part of the mean viscous dissipation rate that is due to the spatial variability of U, and is intrinsically positive.

Now substitute $u = U + u'$, and $p = P + p'$ into Eq. (5.33), the instantaneous budget of mechanical energy, specialized for constant ρ and μ. Time-averaging each term of the resulting equation, and subtracting Eq. (14.16) from the result, we obtain

$$\frac{\partial e}{\partial t} = -\text{div}\left(U e + \tfrac{1}{2}\rho\langle u'(u' \cdot u')\rangle - \langle u'p'\rangle - \mu\langle u' \text{ def } u'\rangle\right) + \mathsf{R} \cdot \text{grad } U - \phi'$$

(14.17)

in which $$e = \tfrac{1}{2}\rho u' \cdot u'$$

The quantity e is called the *kinetic energy of the turbulence*; ϕ' is that part of the mean viscous dissipation rate that involves the gradients of u'. Equation (14.17) is called the *turbulent energy budget*. Approximate versions of it are very frequently used in semiempirical analyses.

PRODUCTION AND DISSIPATION OF TURBULENCE. Equation (14.17) was the main tool of Reynolds' theoretical argument. It has been arranged so that many terms appear inside the divergence operator. Using the divergence theorem, he considered the integral of e throughout a control volume into which there is no net import of e. The volume might be either an imaginary one, on the surface of which all fluctuating quantities are zero, or a short section of pipe. With the latter choice, it is assumed that the flow is statistically independent of distance downstream. The result in either case is

$$\frac{dE}{dt} = \iiint (\mathsf{R} \cdot \text{grad } U - \phi')\, dV$$

(14.18)

where $$E = \iiint e\, dV$$

Remembering that ϕ' is part of the viscous dissipation rate, which must be positive to avoid violation of the Second Law of Thermodynamics, we see that E will decrease with time unless the volume integral of $\mathsf{R} \cdot \text{grad } U$ is positive. In fact, $\mathsf{R} \cdot \text{grad } U$ is found, almost always, to be locally positive; therefore it is given the name *turbulence production*.

Reynolds then noted that both R and ϕ' depend quadratically on the intensity of turbulence, and that ϕ' alone depends on the *scale*[12] of the turbulence. He then estimate the integrals in (14.17) as follows:

$$\iiint R \cdot \text{grad } U \, dV = \frac{C_1 E U_m}{D} \qquad \iiint \phi' \, dV = \frac{C_2 \mu E}{\rho \lambda^2}$$

U_m is the time-averaged flowrate, divided by the area of the pipe; D is the pipe diameter; C_1 and C_2 are dimensionless constants. If the turbulence is to be sustained with constant E, these integrals must be equal. Then

$$\left(\frac{\lambda}{D}\right)^2 = \left(\frac{C_2}{C_1}\right) \frac{\mu}{\rho D U_m} \qquad (14.19)$$

Scaling arguments of this sort have always played a major role in theoretical discussions of turbulent flow. They may be very useful, but one must be careful not to overstate what they prove. Reynolds contented himself with the following deductions.

1. Given fixed values of U_m, D, ρ, and μ, one cannot expect to observe arbitrarily fine-scaled turbulence, because its kinetic energy would be too swiftly dissipated by viscosity.

2. Since one cannot expect λ to exceed D, turbulent flow in a tube cannot persist unless the dimensionless parameter $\rho D U_m / \mu$ exceeds a minimum value given approximately by

$$\left(\frac{\rho D U_m}{\mu}\right)_{\min} \approx \frac{C_1}{C_2}$$

The theory suggested that the critical value of the parameter should depend only on the cross-sectional shape of the tube; his famous experiment provided the critical value for a circular cross section.

To help us remember its mathematical expression, the turbulence production rate, R · grad **U**, is sometimes called the *work of deformation by the turbulence stresses*.[13] This name gives imaginary objects, to wit the body being deformed and the stresses that deform it, potentially misleading qualities of physical reality. As an antidote, the following analysis yields the same mathematical expression as a direct consequence of our accounting procedure, which require that **u**′ be measured relative to a nonuniformly moving frame of reference.

---

[12] Reynolds writes of the *periods* rather than the scales of the fluctuations, even when he refers to spatial variations. See O. Reynolds, *Phil. Trans. Roy. Soc. London*, Part III, p. 935 (1833).

[13] See, for example, J. O. Hinze, *Turbulence*, 2nd ed., p. 72, McGraw-Hill, New York (1959), and H. Tennekes and J. L. Lumley, *A First Course in Turbulence*, p. 60, MIT Press, Cambridge, MA and London (1972).

Consider a fluid particle that at the time and place of observation, has zero acceleration relative to a rigid, inertial, frame of reference. Call its velocity in that frame of reference $\mathbf{u}$. Now remeasure its velocity, relative to a local frame of reference that is moving the velocity $\mathbf{U(r)}$. In this new frame, its velocity is $\mathbf{u}'$. If $\mathbf{U}$ varies along the path of the particle, $\mathbf{u}'$ will consequently vary, at a rate equal to the product of the speed with which the particle moves through the frame of reference, u', and the negative of the rate at which $\mathbf{U}$ varies with distance along the path. Mathematically,

$$\frac{d\mathbf{u}'}{dt} = -\mathbf{u}' \text{ grad } \mathbf{U}$$

In the moving frame of reference, the kinetic energy is reckoned to be $\frac{1}{2}\rho\mathbf{u}' \cdot \mathbf{u}'$, the rate of change of which is

$$\frac{1}{2}\rho\frac{d\mathbf{u}' \cdot \mathbf{u}')}{dt} = \rho\mathbf{u}' \cdot \frac{d\mathbf{u}'}{dt} = -\rho\mathbf{u}'(\mathbf{u}' \cdot \text{ grad } \mathbf{U}) = -\rho(\mathbf{u}';\mathbf{u}') \cdot \text{ grad } \mathbf{U}$$

The average value of this rate is $\mathsf{R} \cdot \text{grad } \mathbf{u}$.

This analyis emphasizes that the Reynolds decomposition scheme tells us nothing directly about the physical mechanisms that produce turbulence. It is fundamentally illogical to speak of various interactions between the mean flow and the turbulence, as though these accounting entities were mechanical systems capable of exerting forces or storing energy. To become familiar with the mechanical events that lie behind the statistical trends and correlations, we must study individual realizations of the flow. There is some analogy to the need to study the deterministic dynamics of intermolecular collisions in order to make a quantitatively complete kinetic theory of gases.

14.4 DETERMINISTIC CARTOONS OF TURBULENCE

Because any particular realization of a turbulent flow may be susceptible, at various times, to the influences of infinitesimal disturbances, it may appear unrewarding to analyze what it would do in the absence of disturbances, or in the presence of selected disturbances. However, certain deterministic solutions of the Navier–Stokes equations, or even the Euler equations, display phenomena that seem to be closely akin to things we see in real turbulent flows. These solutions always involve highly idealized initial and boundary conditions but are substantially more complicated than the special solutions analyzed in Chapters 10, 11, and 12. Sometimes they are obtained analytically, but often involve a substantial amount of numerical solution. We shall call these *cartoons* of turbulence, and shall start with the simplest.

The Burgers Vortex

A simple laminar flow, originally analyzed by J. M. Burgers,[14] exhibits the competition between viscous diffusion of a line vortex, and a superposed irrotational straining motion that tends to concentrate the vorticity. The velocity field is axially symmetric, with components

$w = 2Az$, which *stretches* the vortex

$u = -Ar$, which competes with radially outward diffusion

$v = \dfrac{\Gamma}{2\pi r}\left[1 - \exp\left(-\dfrac{r^2}{4\delta^2}\right)\right]$, induced by the vortex

These satisfy the equations of motion, and an initial condition $\delta(0) = \delta_0$, if

$$\delta^2 = \frac{v}{A} + \left(\delta_0^2 - \frac{v}{A}\right)\exp(-At) \tag{14.20}$$

The vorticity has a single, axial component

$$\Omega = \frac{\Gamma}{\pi\delta^2}\exp\left(-\frac{r^2}{4\delta^2}\right) \tag{14.21}$$

The dissipation function has the local value

$$\Phi = 12\mu A^2 + \frac{\mu\Gamma^2}{16\pi^2\delta^4}\left(\exp(-\xi) - \frac{1 - \exp(-\xi)}{\xi}\right)^2 \tag{14.22}$$

in which
$$\xi \equiv \frac{r^2}{4\delta^2}$$

Equation (14.20) shows that $\delta \Rightarrow \sqrt{v/A}$, whatever the value of δ_0, when At becomes large. If δ initially exceeds this asymptotic value, the inward radial convection will concentrate the vorticity. In the opposite case, the outward radial diffusion dominates until δ increases to its asymptotic value.

To relate this model to our discussion of the range of scales in a turbulent motion, let us call the irrotational flow that stretches the vortex the *large-scale motion*. We assign it length and velocity scales U and D, so that its time scale $1/A$ is identified as D/U. We call the swirling motion induced by the vortex, for which the characteristic length scale is δ, the *small-scale motion*. Then, as $At \to \infty$, the ratio of length scales approaches the value

$$\frac{\delta}{D} = \sqrt{\frac{v}{UD}}$$

This reminds us of Reynolds' result for the scale ratio λ/D, and we wonder whether the δ of the vortex is somehow related to the λ of Reynolds's analysis.

Examining Eq. (14.22), we see that the local value of the dissipation

---

[14] J. M. Burgers, "A Mathematical Model Illustrating the Theory of Turbulence," *Advances in Applied Mechanics*, Vol. 1, 171–196, Academic Press, New York (1948).

function consists of a constant, contributed by the large-scale motion, plus a function of r and t, contributed by the small-scale motion. Figure 14.5 shows the second function, along with the distribution of vorticity. We see that the part of Φ associated with the vortex reaches peak values of order $\mu\Gamma^2/\delta^4$, but is negligible outside of a circular area of order δ^2. Thus, the total rate of vortex-induced dissipation, per unit length of a cylinder of infinite radius, is of order $\mu\Gamma^2/\delta^2$. When convection and diffusion are balanced, so that $\delta^2 = \nu/A$, the integrated rate of vortex-induced dissipation becomes independent of viscosity!

Finally, we investigate conditions under which the vortex-induced dissipation is very large compared to the dissipation induced by the large-scale flow. This will be true if Γ, the circulation of the vortex, is independent of viscosity and of order UD. Then the vortex-induced dissipation will dominate, by a factor proportional to the Reynolds number, UD/ν, which will be large in a turbulent flow.

Some experience with the studies initiated in Chapters 8 and 13 is necessary to make these assumptions about Γ seem reasonable. Remember, from Chapter 8, that viscosity determines the rate at which vorticity diffuses away from a wall at which is it is created as a consequence of the no-slip condition, but it does not determine the amount of vorticity so created. For example, a circular cylinder of diameter D, impulsively accelerated to speed U, adds a circulation $\Gamma = 2UD$ to the fluid above it, and a circulation $\Gamma = -2UD$ to the fluid below it. At a later time, U and D are good velocity and length scales for the large-scale motions in the wake.

Remember, also from Chapter 8, that when the Reynolds number is large the motion of vorticity that is not very close to a wall is seldom much affected by viscosity. The exceptional events may be typified by the Burgers vortex model, which exhibits a role of viscosity that is acted out in a very tiny part of the fluid. Thus, the processes that lead to spatial concentrations of vorticity, and that determine the circulation of individual concentrated vortices, are largely inviscid. This has been illustrated in Chapter 13, in the discussion of the instabilities of free shear layers.

Finally, it seems likely that individual vortices in a turbulent flow inherit their circulation from the large-scale motions. The total circulation of a large-scale concentration of vorticity being of order UD, the share given to each of a number of heirs is of this same order. If, by viscous diffusion, some get more, others will get less and the average will be unaffected.

We need now to collect our thoughts about this cartoon and its implications, taking care not to leap to logically indefensible conclusions.

1. Turbulent flows are observed to dissipate mechanical energy at a volume-integrated rate that is nearly independent of viscosity. The cartoon shows that this can happen also in a simple laminar flow, if the parameters of that flow are chosen in a particular way.

2. The smallest observable length scales in turbulent flow decrease, as a

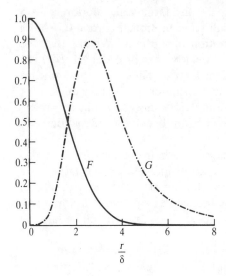

FIGURE 14.5
Vorticity and viscous dissipation in the Burgers vortex. $\Omega = \left(\dfrac{\Gamma}{\pi\delta^2}\right)F\left(\dfrac{r}{\delta}\right); \; \Phi = \left(\dfrac{\mu\Gamma^2}{16\pi^2\delta^2}\right)G\left(\dfrac{r}{\delta}\right).$

fraction of the largest observable scales, as a Reynolds number based on the U and D of large-scale motion increases. The cartoon shows explicitly that $(\delta/D)^2 \sim \nu/UD$.

3. Experimental recordings of velocity versus time at a point in a turbulent flow rarely reveal very large instantaneous velocities, while in this cartoon $v_{max}/U \sim D/\delta$, which is a very large number. Also, statistical observations of turbulence indicate that motion on the scales that account for most of the viscous dissipation contributes little to the variance of the velocity record.

A straightforward calculation can be made, to compare the kinetic energy of the small-scale motion to that of the large-scale motion, in a finite cylindrical volume coaxial with the Burgers vortex.[15] For example, if the control volume has radius $R = 10\delta$ and length $L = 10\delta$; and if $A = U/R$ and $\Gamma = 2\pi UR$, we find that the small- and large-scale motions contribute almost equally to the integral of kinetic energy. However, if the Reynolds number, $\mathrm{Re} = UR/\nu$, is very large, the small-scale motion contributes almost all the viscous dissipation within that same volume. This calculation is suggestive, but not at all conclusive. The suggestion is that instantaneous flow features somewhat like the Burgers vortex might, if they occupied a very small fraction of the total volume of turbulent fluid, accomplish most of the viscous dissipation, while the probability of observing their very localized high velocities might be small enough to explain their small contribution to measured velocity variances. The lack of conclusiveness is due to several factors, one prominent one being the lack of any dynamically plausible

---

[15] See Exercise 14.1.

scenario, in which Burgers vortices arise naturally as a consequence of transition from laminar to turbulent flow.

Vortex Sheets into Vortex Tubes: The Corcos–Lin Cartoon

By focusing on an axially symmetric situation, Burgers could make a very simple analysis. Equal simplicity can be found in the superposition of a plane potential flow and a plane shear layer, such that

$$w = Az \qquad v = -Ay \qquad \text{and} \qquad u = U \operatorname{erf}\left(\frac{y}{2\delta}\right)$$

providing that
$$\delta^2 = \frac{\nu}{A} + \left(\delta_0^2 - \frac{\nu}{A}\right) \exp(-At)$$

However, this model of a stretched vortex *sheet* does not exhibit the interesting results for integrated rate of dissipation, or for contrast of length scales, that come out of the Burgers vortex model.[16]

It turns out, however, that a vortex sheet with periodically alternating initial shear, such that

$$u(x, y, t)_{t=0} = U \sin(\beta x) \operatorname{erf}\left(\frac{y}{2\delta_0}\right)$$

may, in the presence of the plane potential-flow $V = -Ay$, $W = Az$, spontaneously evolve into a periodic array of counterrotating diffuse line vortices, each of which acts much like a Burgers vortex. The analysis is now more complicated, because both the potential flow and the flow induced by vorticity, as well as diffusion, tend to redistribute the vorticity in the x–y plane.[17] However, the vortex lines still remain straight, and the vorticity is independent of distance along the lines. The outcome depends on the values of three parameters:

1. The initial *aspect ratio* of a cell of the shear layer (see Fig. 14.6)
2. A Reynolds number based on the circulation around each cell, Γ/ν
3. A ratio that determines the importance of the flow due to vorticity compared to that due to the imposed stretching, say $\Gamma^* = \Gamma/A\lambda^2$, where $\lambda = 2\pi/\beta$.

When Γ^* is very small, the stretching motion combats diffusion of vorticity in the y-direction, but not in the direction of the sinusoidal variation. Diffusion in the latter direction weakens the vorticity and Γ dies away in proportion to $\exp(-\beta^2 \nu t)$. When Γ^* is larger, the events sketched in Fig. 14.7 occur.

---

[16] See Exercise 14.2.

[17] For details of the analysis, see S. J. Lin and G. M. Corcos, *J. Fluid Mech.* **141**; 139–178 (1984).

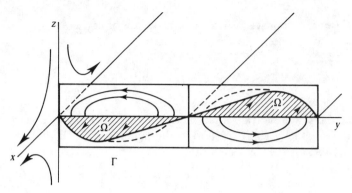

FIGURE 14.6
A stretched shear layer with alternating vorticity.

The key feature of the figure is the velocity-vector diagram, shown at points P and Q. The solid arrows represent the velocity induced by vorticity; the dashed arrows represent the velocity contributed by the added potential flow. Note that the resultant velocity has an x-component that tends to shorten the x dimension of the cross section of each vortex. By reducing the gradient of vorticity across the boundary between vortex cells, the vorticity-induced motion reduces the rate of diffusive decay of Γ. Figure 14.8 shows computed equivorticity contours at various times for the case $\Gamma^* = 0.32$ and $AR = \lambda/4\delta = 7.95$. The value of δ was set equal to $\sqrt{\nu/A}$, corresponding to an equilibrium between stretching and diffusion. This gives the Reynolds number the value $\Gamma/\nu = 16\Gamma^* AR^2 = 324$. Already when $At = 1.5$, the vortices are nearly circular in cross section. An asymptotic analysis[18] shows that if $\Gamma/\nu \to \infty$, the final structure of each vortex is just that of a Burgers vortex in an axisymmetric strain field of strength $A/2$.

Generation of New Scales of Motion: Streamwise Vortex in Shear Flow

The foregoing examples involve stretching and lateral displacement of vortex lines, and only slight temporal changes in the spatial scales of motion. The

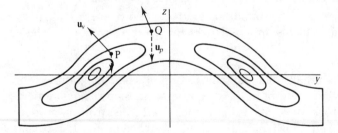

FIGURE 14.7
Concentration of vorticity into filaments because of flow due to vorticity and stretching.

[18] J. C. Neu, "The Dynamics of Stretched Vortices," *J. Fluid Mech.* **143**; 253–276 (1984).

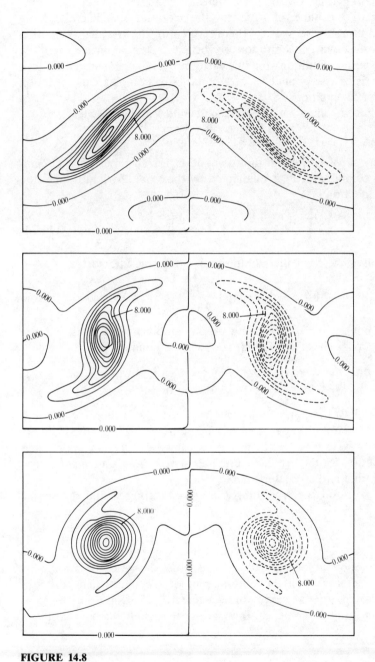

FIGURE 14.8
Computed evolution of columnar vortices from a perturbed shear layer. (Reproduced, by permission, from S. J. Lin and G. M. Corcos, 1984.[17])

following example suggests how the vortex lines of an initially linear shear flow can be wound up, like fishline on a reel, by the action of a sufficiently strong and concentrated streamwise vortex, thereby creating new scales of motion and significant convective concentration of vorticity. It is admittedly hard to imagine how the initial conditions for this cartoon could appear in nature, but they allow a surprisingly easy analysis of a remarkably complex exact solution of the Navier–Stokes equations.[19]

The velocity in this flow has the form, in cylindrical coordinates,

$$u = v(r, t)\mathbf{e}_\theta + w(r, \theta, t)\mathbf{e}_z \qquad (14.23)$$

This automatically satisfies the continuity equation for all times. The circumferential velocity component, v, is unaffected by the axial flow, and obeys the simple diffusion equation

$$\frac{\partial v}{\partial t} = v\left[\frac{\partial^2 v}{\partial r^2} + \left(\frac{1}{r}\right)\frac{\partial v}{\partial r} - \frac{v}{r^2}\right] \qquad (14.24)$$

The cartoon involves the particular solution for a diffusing line vortex

$$v = \frac{\Gamma}{2\pi r}\left[1 - \exp\left(-\frac{r^2}{4vt}\right)\right]$$

or its initial distribution, $v = \Gamma/2\pi r$. The cartoon exhibits a hydrostatic balance of pressure in the axial direction, so that w obeys the equation

$$\frac{\partial w}{\partial t} + \left(\frac{v}{r}\right)\frac{\partial w}{\partial \theta} = v\left[\frac{\partial^2 w}{\partial r^2} + \left(\frac{1}{r}\right)\frac{\partial w}{\partial r} + \left(\frac{1}{r^2}\right)\frac{\partial^2 w}{\partial \theta^2}\right] \qquad (14.25)$$

We call $w(r, \theta, 0)$ the *basic shear flow*, and call $v(r, t)$ the flow due to a *streamwise vortex*. The basic shear flow depends only on $r \sin \theta$; for simplicity we take $w(r, \theta, 0) = \Omega r \sin \theta$. Far from the streamwise vortex, w approaches this initial distribution for any $t \geq 0$. Figure 14.9 shows the assumed configuration at the initial time and at a later time.

If we set $v = 0$, and correspondingly use $v = \Gamma/2\pi r$, we find the simple result

$$w = \Omega r \sin(\theta - \eta), \qquad (14.26)$$

in which $\eta = \Gamma t/2\pi r^2$. This result is sketched in Fig. 14.10 for $\theta = \pi/2$. The circumferential component of vorticity, Ω_θ, is plotted against dimensionless distance, for this same ray, in the bottom half of the figure. The distance between zero-crossings of the curve for w decreases in proportion to r^3 as $r \to 0$. We expect to lose some of this fine-scale detail when viscosity is

---

[19] The analysis has been given by C. F. Pearson and F. H. Abernathy, *J. Fluid Mech.* **146**, 271–283 (1984). The solution has been further analyzed by G. M. Corcos (1989), *The Role of Cartoons in Turbulence, op. cit.*

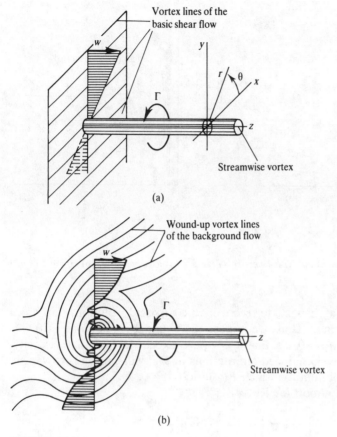

FIGURE 14.9
Effect of a streamwise vortex in a shear layer: (a) $t = 0$; (b) $t \geq 0$.

included in the analysis, the degree of loss depending on the value of the parameter Γ/v.

The analysis that includes viscosity employs a form of separation of variables. We seek a solution for w in the form

$$\frac{w}{\Omega r} = F(\eta) \sin \theta + G(\eta) \cos \theta \tag{14.27}$$

in which $\eta = \Gamma t/(2\pi r^2)$. The ordinary differential equations governing F and G are

$$\eta^2 F'' - \text{Re} \left\{ F' - \left[1 - \exp\left(-\frac{\text{Re}}{\eta} \right) \right] G \right\} = 0$$

and

$$\eta^2 G'' - \text{Re} \left\{ G' + \left[1 - \exp\left(-\frac{\text{Re}}{\eta} \right) \right] F \right\} = 0$$

In these equations, $\text{Re} = \Gamma/8\pi v$.

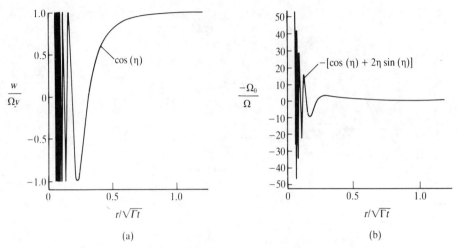

FIGURE 14.10
Vorticity distribution in wound-up shear layer, $\eta = \Gamma t/2\pi r^2$: (a) axial velocity; (b) circumferential vorticity.

When $\eta = 0$, we expect to recover the inviscid solution, in which $F = \cos \eta$ and $G = -\sin \eta$. Thus, we impose the boundary conditions $F(0) = 1$, $G(0) = 0$. If $r = 0$ and $t > 0$, so that $\eta = \infty$, we expect to find that viscosity, acting on the convectively induced alternations of Ω_θ, will cause Ω_θ to vanish. Thus, we study the solution for which $F(\infty) = G(\infty) = 0$.

An asymptotic solution for Re $\gg 1$ is[20]

$$\frac{w}{\Omega r} = \exp\left(-\frac{\eta^3}{3\text{Re}}\right) \sin(\theta - \eta)$$

with the corresponding approximation for Ω_θ,

$$\Omega_\theta = \Omega \exp\left(-\frac{\eta^3}{3\text{Re}}\right)\left[\left(\frac{1 + 2\eta^3}{\text{Re}}\right) \sin(\theta - \eta) + 2\eta \cos(\theta - \eta)\right]$$

This shows that Re must be quite large, if the distinctive oscillations of w and Ω_θ are to survive. Figure 14.11 shows $w/\Omega r$ and Ω_θ/Ω for $\theta = \pi/2$ and Re = 1000.

How is this cartoon relevant to a study of turbulence? Qualitatively, it illustrates one of many ways in which a fairly simple initial configuration of vorticity can evolve into a more complex one, with one scalar component of vorticity stretching vortex lines that are orthogonal to it. It shows how viscosity defeats, at a sufficiently small spatial scale, the tendency of convection to produce large gradients of vorticity.

---

[20] This result comes from Corcos (1989), *op. cit.*

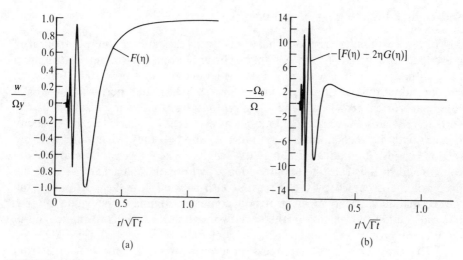

FIGURE 14.11
Smoothing effect of viscosity, Re = 1000: (a) axial velocity; (b) circumferential vorticity.

The vortex-induced distortion of the shear flow also produces concentrated regions of high viscous dissipation rate. An analysis of the dissipation rate, integrated over a constant cylindrical volume large enough to include such regions, show the following.

1. A constant value proportional to the volume, and to $\mu\Omega^2$, contributed by the basic shear flow.

2. A contribution from the swirling motion, which is proportional to $\rho\Gamma^2 L/t$ if Eq. (9.10) describes the structure of the vortex. L is the length of the vortex; the radius of the volume of integration does not appear in this contribution.

3. A constant value, again proportional to L but independent of the radius of the volume, that depends on viscosity in a complex way. As $\Gamma/v \to \infty$, this contribution, due to the distortion of the shear flow, is proportional to $\rho v^{2/3}\Gamma^{1/3}\Omega^2$.

Although implications of this sort of analysis are necessarily somewhat vague,[21] it seems that this type of stretching of vortex lines is not primarily responsible for the average degree of spatial concentration of dissipation, although it does lead to the appearance of easily recognizable new scales of motion.

---

[21] P. Saffman, Lectures on Homogeneous Turbulence, in *Topics in Nonlinear Physics,* Springer-Verlag, New York, 1968, pp. 560–561.

Self-Induced Stretching of a Vortex

All of the foregoing cartoons have involved the interaction of arbitrarily prescribed flow fields, e.g., a columnar vortex and an axisymmetric stretching fiow, without indicating what might be the source of those flows. The following cartoon, due to A. Chorin,[22] shows how small initial perturbations of a columnar vortex grow and lead to a profound spatial reorganization of the vorticity. Because the cartoon ignores any effects of viscosity, the vorticity remains within a recognizable tube, but the tube becomes twisted and its cross section varies dramatically along its length. The cartoon was originally constructed to illustrate the spontaneous growth of the integral of Ω over the fixed volume occupied by the vortex. It also showed how the portions of the vortex that have been most stretched come to dominate the value of that integral, so that the fraction of the initial volume that accounts for almost all of the integral grows even smaller as time goes on.[23]

The cartoon is better appreciated after we review some concepts from Chapter 8. Consider a *vortex filament*, which instantaneously coincides with a material body bounded by vortex lines, see Fig. 14.12. Its cross-sectional area A may vary with distance s measured along the filament. The total length of the segment is L; its volume is V. The flux of vorticity through it, Γ, is independent of s.

A differential *segment* of the filament, between two neighboring cross sections, has volume

$$dV = \int_A ds\, dA$$

The flux of vorticity is

$$\Gamma = \int_A |\Omega|\, dA$$

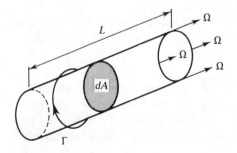

FIGURE 14.12
Material segment of a vortex tube.

[22] A. J. Chorin, Evolution of a Turbulent Vortex, op-cit 6.
[23] In fact, the geometrical object that ultimately *supports* the integral is found to be not even a volume, but something between a volume and a surface.

The contribution of dV to a volume integral of $|\Omega|$ is

$$\int_A |\Omega|\, ds\, dA = \Gamma L$$

If the filament evolves as part of an inviscid, incompressible, barotropic flow, each segment may be distorted, but the flux of vorticity through it will be unchanged. The contribution of a segment to the volume integral of $|\Omega|$ will increase if that segment is *stretched* $(dL/dt > 0)$. Eventually, the volume integral will be dominated by the contributions of the segments that have been extruded into long slender shapes.

The cartoon represents this phenomenon by the computed evolution of a filament that runs initially from the center of one side of a fixed cube to the center of the opposite side. The configuration of the filament is reproduced in an infinite set of adjoining cubes (Fig. 14.13).

The actual distortion of a filament would include changes in the size and shape of its local cross sections. The cartoon ignores these, and represents each segment of the filament by a circular cylinder, indentified for the purposes of calculation by the endpoints of its axis. The computation tracks the motion of these endpoints by an approximate application of the Biot–Savart law in the form (8.20), and substitutes two half-segments for any segment that gets too long. A key quantity associated with each segment is the number of times its length has been halved. This local measure of vortex stretching is used in various ways, to quantify the growth of the volume integrals of $|\Omega|$ and of Ω^2, and to investigate the evolution of the geometrical figure composed of the segments that make the dominant contributions to these integrals. The details of the calculation are very interesting.

The filament initially has a small kink at mid length. Figure 14.14 shows the filament at a sequence of later times.

Figure 14.15 shows the computed history of the integral of $|\Omega|$ over the

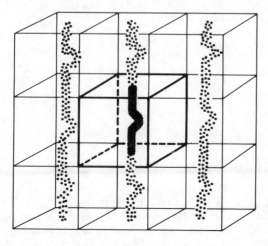

FIGURE 14.13
Periodic replications of the vortex at $t = 0$.

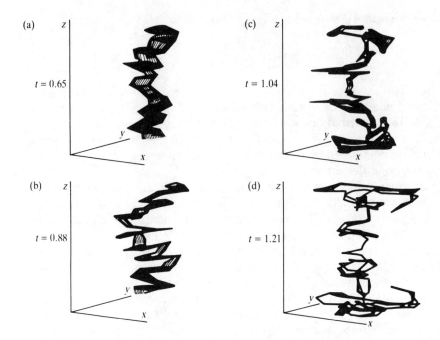

FIGURE 14.14
Stretching and distortion of a vortex filament. (Reproduced, by permission, from A. J. Chorin, 1981.[22])

volume of the basic cube. Note that it begins to grow explosively as $t^* = \Gamma t/H^2$ approaches 1, suggesting that there is no limit to the intensification of vorticity in a three-dimensional flow, except that imposed by viscosity. Another interesting feature is the tendency of the filament to fold sharply back upon itself many times in a volume. Chorin argues that this is necessary, if stretching is

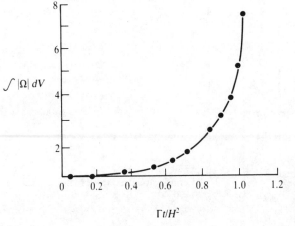

FIGURE 14.15
Growth of vorticity due to stretching. (Reproduced, by permission, from A. J. Chorin, 1981.[22])

to continue without a growing kinetic energy in the cube. The velocity induced by one leg of a folded filament nearly cancels that induced by a neighboring leg, except in the region between them, which is very small if the folding is tight.

14.5 EMPIRICAL GENERALIZATIONS FOR STATISTICALLY STATIONARY FLOWS

Introduction. In this section we shall briefly explore the vast literature that describes experimental studies of turbulent flow. Only statistically stationary flows will be described, because that will give us more than enough food for thought at this level of study. Experimental studies describe, more or less accurately, facts that in themselves may seem devoid of much intellectual interest. The challenge that we face as we read through the following recital of fact is to discover unifying themes in the data, and to identify fruitful questions that are implicitly posed by the data. The questions may suggest a search for new data, or a challenge to see what use we can make of the theoretical ideas developed in previous chapters.

Classification of data. Empirical descriptions of stationary turbulent flows, accumulated during nearly a century of research, can be grouped roughly into three categories.

1. Performance data of engineering devices that involve turbulent flow of a working fluid. This includes data on the mechanical energy losses in flows through conduits such as pipes, flumes, nozzles, valves, diffusers and all sorts of plumbing fixtures. The focus of interest is on the influence of conduit shape, Reynolds number, and surface roughness. This kind of data is well represented in elementary textbooks, and will not be reproduced here. However, one should try to understand, at least partially, why the value of the Reynolds number seems to be so unimportant in some of these flows, and why the surface roughness is so important in others, when the Reynolds number is very large.

2. Flowfield data, including measurements of mean velocity and various quantities defined in Section 14.2 that statistically characterize the velocity fluctuations. These have been accumulated mostly for easily reproduced flows, such as fully-developed pipe flow; or for those local regions of rotational flow, such as boundary layers, jets, wakes, and slipstreams, that often form recognizable components of a practically interesting flowfield. The data are organized into a more or less systematic taxonomy of turbulent flows, a knowledge of which should be useful to those who must create novel designs or predict the behavior of previously unexplored flows.

3. Observations of the instantaneous states of flows that have already been characterized statistically. It is hoped that these will reveal typical phenom-

ena that can be understood or modeled deterministically, and which lie behind the statistical correlations. Modern computerized instrumentation makes it possible to gather and analyze such data in unprecedented amounts and ways, but it is not yet clear how effectively this will enhance our understanding and predictive capabilities.

Selection and organization of material. The second and third types of result will be described here. Mean velocity profiles, which are available for the broadest class of flows, are shown and analyzed first. Samples of the sort of statistical data described in Section 14.2, which are available for a somewhat narrower subset of flows, are shown next. Studies of instantaneous flow patterns, which have been made for only a very few flows, but which are rapidly proliferating, are presented last.

Citations of literature. Experimental investigations of very comparable scope have often been made in different laboratories with different instrumentation, and frequently with different interpretations of the results. A textbook is not the place for a critical and comprehensive literature review,[24] and I have been selective in the citations of original sources. Except for occasional references to pioneering works, I have preferred to cite the most recent and easily available works, especially when they contain comprehensive lists of relevant earlier sources. When you consult this literature, do not be discouraged by its formidable appearance. You will find lots of details about experimental technique, which is very important for a critical evaluation of the findings, but which can be skimmed over in a first reading. Refer back to Section 14.2 for basic concepts behind the techniques, and to the references given there for more detailed explanations.

Interpretations of the data. Most good experimentation is guided by some preconceived scheme of interpretation of the results, or some preformulated question—to be answered by the data. Much of the work on which this section draws was part of a search for empirical correlations among statistical quantities, which might approximately resolve the closure problem associated with the Reynolds-averaged equations of motion. Thus the final result exhibited in a paper might be a profile of mixing length or eddy viscosity, calculated from measurements of mean velocity and Reynolds stress, or a

---

[24] However, critical comparison of competing or overlapping experiments is an important enterprise, well exemplified by the following two references: *Computation of Turbulent Boundary Layers, Vol. II, Compiled Data* (D. E. Coles and E. A. Hirst, eds), Thermosciences Division, Department of Mechanical Engineering, Stanford University, Stanford, CA (1968); *Complex Turbulent Flows, Vol. I, Evaluation of Data* (S. J. Kline, B. J. Cantwell and G. M. Lilley, eds), Thermosciences Division, Department of Mechanical Engineering, Stanford University, Stanford, CA (1981).

presentation of the relative magnitudes of some of the terms in Eq. (14.17), the turbulence energy budget. That guiding philosophy has been stripped away from this account of the results, not because I do not value it, but because it makes so little use of what you have tried to learn from the preceding chapters. Also, other textbooks[25] expound that philosophy with much more enthusiasm and knowledge than I can muster.

14.6 TURBULENT FLOW ALONG A WALL

Turbulent flows are full of vorticity. In barotropic flow, vorticity first enters the fluid at solid walls, where it is generated to satisfy the no-slip condition. It is then not surprising that turbulent flow is often bounded by a solid wall: by a pipe or flume, a wing or a ship's hull, the ground, or the floor of the ocean. The easiest case to investigate is fully-developed flow in a pipe. Pioneering studies[26] showed that the surface finish of the wall is important if a roughness Reynolds number exceeds a certain value. For laboratory studies, it is relatively easy to make a wall that is effectively smooth. It is next to a smooth wall that viscosity has its most obvious and direct effect on turbulent flow, so this case is presented first.

The Dual Structure of Turbulent Wall Layers

Recall, for a moment, the discussion of the transport of vorticity near a wall, which was presented in Section 8.7. Very close to a smooth, solid, stationary wall, convection is ineffective, because the velocity must be nearly zero. Any vorticity freshly generated at the wall, as required by the no-slip condition, must diffuse away from it. If the flow a little farther from the wall is laminar and nearly parallel to the wall, diffusion will continue to be the dominant mode of vorticity transport in the normal direction. For transport in the streamwise direction, convection usually dominates if the Reynolds number is large. These are the basic notions of laminar boundary-layer theory. The vorticity can be imagined to be concentrated into sheets that are nearly plane and parallel to the wall.

Suppose, however, that this spatial distribution of vorticity is unstable. The vortex sheets ripple, and the vorticity distribution in planes parallel to the wall becomes nonuniform. The new vorticity distribution induces locally intense convective motions normal to the wall. Very close to the wall, these are nearly canceled by the velocity induced by an image field of vorticity

---

[25] See, for example, H. Tennekes and J. L. Lumley, *op. cit.*, or F. M. White, *Viscous Fluid Flow*, McGraw-Hill, New York (1984).

[26] J. Nikuradse, "Strömungsgesetze in rauhen Rohren," *Forschg. Arb. Ing.-Wes.* No. 361 (1933).

behind the wall, and diffusion remains dominant. Farther from the wall, the effect of the image vorticity dies off, and the convective motions become the dominant mode of transport, even in the normal direction. The fresh vorticity brought in at the wall to keep the no-slip condition in force will now be spatially nonuniform, and may help perpetuate the postulated instability.

This world picture is of course not a real theory, but it prepares one a bit for what is seen in the data, which is the prominent *dual structure* of turbulent flows close to a smooth wall. In the quite varied cases described below, you will discover that a very thin *viscous sublayer*, found right against the wall, seems to behave in a way that is surprisingly independent of the detailed behavior of the rest of the turbulent flow, which differs considerably from case to case. This remaining *outer flow*, however it may depend on the past history of the turbulence and on local values of wall curvature and pressure gradient along the wall, is quite independent of the value of the viscosity, at least in its mean and large-scale features.

The Law of the Wall

In the viscous sublayer, and in a part of the layer above it, the profile of the tangential component of mean velocity is described by an empirically-determined relationship called the *Law of the Wall*. If there is more than one tangential velocity component, the law governs the profile of the component that is parallel to the mean tangential stress, τ_w, exerted by the fluid on the wall.[27] Let U denote that velocity component; introduce the *friction velocity* $u^* = (\tau_w/\rho)^{1/2}$; and let y denote distance normal to the wall. The Law of the Wall asserts that

$$u^+ = f(y^+) \qquad (14.28)$$

in which
$$u^+ = \frac{U}{u^*} \quad \text{and} \quad y^+ = \frac{u^* y}{\nu}$$

The function f is sketched in Fig. 14.16, first with a linear scale for y^+, then with a logarithmic scale. There is of course some scatter of measured data around this mean curve, but the measurements have been made, particularly for $y^+ < 50$, by a great variety of techniques and in dozens of different laboratories. Anyone whose measured values miss the curve by as much as 5% now starts to look seriously for sources of systematic errors of measurement.

The startling feature of the law is that it effectively represents the influence of a large number of factors, such as the past history of the flow, the curvature of the wall, and the character of any external potential flow, all by

---

[27] B. van den Berg, A. Elsenaar, J. P. F. Lindhout, and P. Wesseling, "Measurements in an Incompressible. Three-Dimensional Turbulent Boundary Layer, *J. Fluid Mech.* **70**: 127–148 (1975). See Fig. 7.

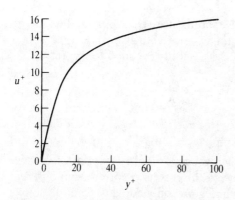

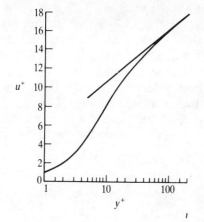

FIGURE 14.16
Law of the Wall for mean velocity.

the local value of a single parameter, u^*. These external, or historical, factors influence both the value of u^*, and the range y^+ for which the law is accurate. Let us call this range the *domain* of the law. It appears to extend from $y^+ = 0$ to at least $y^+ = 50$, unless the flow is in the process of laminar-turbulent transition or is separating from the wall. An effective fitting formula for the function f is given implicitly,[28] by

$$y^+ = u^+ + \exp\left(-\kappa C\right)\left[\exp\left(\kappa u^+\right) - 1 - \kappa u^+ - \frac{(\kappa u^+)^2}{2} - \frac{(\kappa u^+)^3}{6}\right] \quad (14.29)$$

with $\qquad\qquad\qquad \kappa = 0.41 \qquad C = 5.0$

κ is called the *von Kármán constant*, in honor of Theodore von Kármán.

While the name Law of the Wall traditionally applies to the u-profile, the Reynolds-averaged x-momentum equation supplies a connection between the

---

[28] D. B. Spalding, *J. Appl. Mech.* **28**: 455–457 (1961).

mean velocity field and the field of the Reynolds stresses. Applying this fact to a flow in which u^* serves as the significant scale of velocity, while v/u^* serves as the corresponding scale of length, one is logically inclined to expect results such as

$$\frac{\langle u'^2 \rangle}{u^{*2}} = f_1(y^+) \qquad \frac{\langle v'^2 \rangle}{u^{*2}} = f_2(y^+) \qquad \frac{\langle u'v' \rangle}{u^{*2}} = f_3(y^+) \qquad \text{etc.}$$

as corollaries of the Law of the Wall.

One does not easily find unambiguous experimental support for these corollaries. There are discrepancies between the results reported from different experiments, perhaps because the corollaries are only roughly true, or simply because the velocity fluctuations are extremely hard to measure in the region of interest.[29] Some fairly typical results are shown in Figs. 14.17 and 14.18. One of the very important things to be noticed in Fig. 14.17 is that the velocity fluctuations attain their largest values very close to the wall. For example, the standard deviation of u reaches its highest value, equal to about $2.8u^*$, at $y^+ \approx 15$. If $y^+ < 5$, the standard deviation of u equals about $0.33U$. Figure 14.18 shows a measured probability distribution function for the fluctuating values of $\partial u/\partial y$ in the region $y^+ < 5$. Note that it is quite strongly skewed.

The search for corollaries can be pressed further, with the proposal that the spectra of velocity fluctuations at points where the Law of the Wall governs

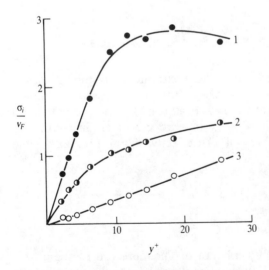

FIGURE 14.17
Standard deviations of velocity fluctuations very near a smooth wall in fully developed channel flow: ● streamwise; ◐ spanwise; ○ normal. (Reproduced, by permission, from S. S. Kutateladze et al., *Turbulent Shear Flows I*, pp. 91–103, Springer-Verlag, Berlin, Heidelberg, New York, 1979.)

[29] An excellent review is given by W. W. Willmarth and T. J. Bogar, "Survey and New Measurements of Turbulent Structure Near the Wall," *Phys. Fluids* **20**: No. 10, pt II, S9–S21 (1977). See also W. W. Willmarth, "Structure of Turbulence in Boundary Layers," *Advances in Applied Mechanics*, Vol. 15, pp. 159–254, Academic Press, New York (1975).

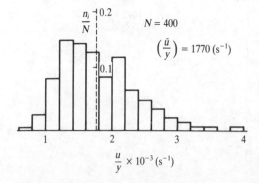

FIGURE 14.18
Probability distribution of u/y for $y^+ < 5$. Fully developed, smooth wall channel flow. (Reproduced, by permission, from S. S. Kutateladze et al., *Turbulent Shear Flows I*, pp. 91–103, Springer-Verlag, Berlin, Heidelberg, New York, 1979.)†

the profiles of mean velocity and Reynolds stresses should become independent of Reynolds number and pressure gradient if, for example, Φ_{uu}/u^{*2} is plotted versus the nondimensional wavenumber kv/u^*, for a given value of y^+. Here, Φ_{uu} is a spatial spectral density, such that

$$\langle u'^2 \rangle = \int_0^\infty \Phi_{uu} d(kv/u^*)$$

This proposal does not suggest how the shape of the curve might vary with the value of y^+, but there are interesting studies of this question, which will be mentioned later.

Figure 14.19 shows spectra of u'^2, v'^2, and w'^2 versus streamwise wavenumber, at $y^+ \approx 5.4$ and 149. These are taken from a computational experiment. Note that the spectra for the three velocity components differ prominently in the range of wavenumbers that contributes most to the standard deviation. However, they differ much less in the range of higher wavenumbers that contributes most to the viscous dissipation of kinetic energy.

THE DOMAIN OF THE LAW OF THE WALL. It was mentioned above that the range of y^+ for which the Law of the Wall collapses mean-velocity data onto a single curve is affected by factors such as Reynolds number, dimensionless pressure gradient, or wall curvature. Some examples of these effects are shown in Figs. 14.20 and 14.21.

Very recent studies of the boundary layer on a wall with strong concave curvature appear to show that, even when the vorticity distribution in the outer regions of the boundary layer is spectacularly different from that in a pipe or

---

† S. S. Kutateladze, E. M. Khabakhpasheva, V. V. Orlov, B. V. Perepelitsa and E. S. Mikhailova. "Experimental investigation of the structure of near-wall turbulence and viscous sublayer." Pp. 91–103 of "Turbulent Shear Flows I," editors F. Durst, B. E. Launder, F. W. Schmidt and J. H. Whitelaw. Springer-Verlag, Berlin, Heidelberg, New York, 1979.

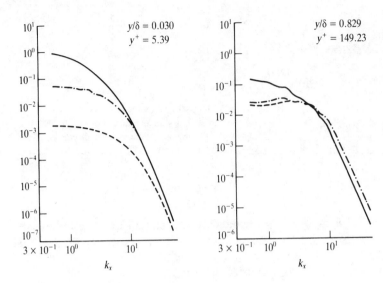

FIGURE 14.19
Spectra of velocity fluctuations in the domain of the Law of the Wall. Data from numerical simulations of fully developed channel flow. (Reproduced, by permission, from J. Kim et al., *J. Fluid Mech.* **177**: 133–167, 1987.)†

on a flat plate, the Law of the Wall and its corollaries provide good descriptions, out to about $y^+ = 50$.[30]

INSTANTANEOUS FLOW IN THE DOMAIN OF THE LAW OF THE WALL. We see that both $\partial U/\partial y$ and R_{yx} reach their maximum values in the domain of the Law of the Wall, and calculation shows that almost all the production of turbulent kinetic energy occurs there. These discoveries have provoked intense study of the instantaneous flow in this region, with the hope that a more detailed knowledge may lead to some ability to control its characteristics.

At the present, many researchers are attempting to identify, by flow visualization and conditionally sampled statistical measurements, typical kinematic features of the instantaneous flow. For this purpose, both laboratory flows and computationally simulated flows are used. When the flow is marked by visible particles, bubbles, dye or smoke, continuously emitted from a spanwise line source placed in or very close to the wall, and is viewed with a line of sign normal to the wall, one sees the visible material gathering into

---

† J. Kim, P. Moin, and R. Moser, Turbulence Statistics in Fully-Developed Channel Flow at Low Reynolds Number. *J. Fluid Mech.* **177**: 133–167, 1987.

[30] R. S. Barlow and J. P. Johnston, Structure of Turbulent Boundary Layers on a Concave Surface, *J. Fluid Mech.* **91**: 137–176 (1985).

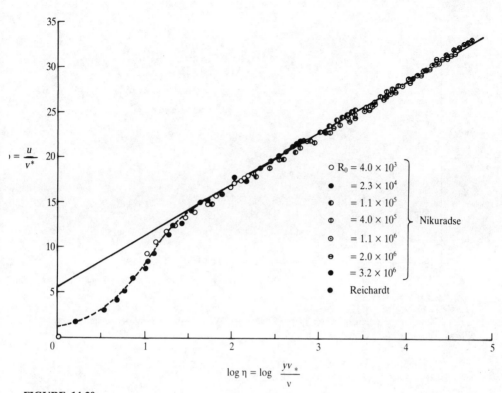

$) = \dfrac{u}{v^*}$

$\log \eta = \log \dfrac{yv_*}{v}$

FIGURE 14.20
Law of the Wall. Effect of Reynolds number in fully developed pipe flow. (Adapted, by permission, from H. Schlichting, "Boundary Layer Theory", op. cit)

coherent streamwise streaks.[31] Individual streaks appear and disappear at random: the probability of appearance being independent of spanwise position unless some nonuniformity of nominal conditions exists upstream.[32]

If this marked flow is illuminated by a thin sheet of light normal to the mean flow direction, and viewed with a line of sight parallel to the mean flow, we see that the streaks correspond to local eruptions of marked fluid. A sample of both views is shown in Fig. 14.22.

Examination of many snapshots, or of corresponding correlations between velocities measured simultaneously at points slightly separated in the spanwise direction, reveals that the average spanwise separation between streaks is about 100 times the viscous length v/u^*. These same studies reveal

---

[31] For a vivid visual impression, drawn from numerical simulation, see S. Robinson and S. J. Kline, NASA TN, 1989.

[32] Barlow and Johnston (1985), *op. cit.,* pp. 81, 82.

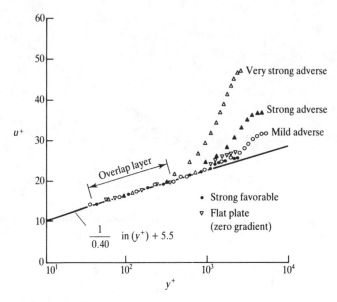

FIGURE 14.21
Effect of pressure gradient on range of the Law of the Wall. (Adapted, by permission, from F. M. White, "Viscous Fluid Flow", op. cit)

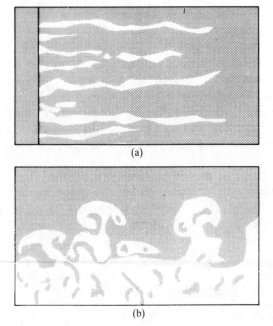

FIGURE 14.22
Instantaneous features of a turbulent wall layer: (a) plan view; (b) streamwise view.

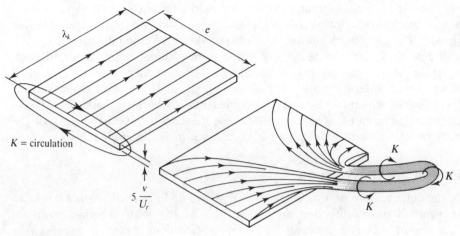

FIGURE 14.23
Evolution of a hairpin vortex. (Reproduced, by permission, from A. E. Perry and M. S. Chong, *J. Fluid Mech.* **119**: 173–218, 1982.)†

that instantaneous flow properties in the viscous sublayer vary strongly with z, so that fluctuations cannot be accurately resolved by instruments that average over a spanwise length much greater than $10v/u^*$.

Fairly often, the cross-sectional photographs show configurations of marked fluid that suggest the presence of a streamwise pair of counterrotating vortices. Figure 14.23 suggests how the vorticity distribution of such a pair may develop, although the degree of symmetry implied by this sketch is no longer thought to be typical.

It has long been suspected that such *hairpin vortices* are a basic building block of the turbulent fluctuations near the wall.[33] Recent studies of the sublayer structure in water to which a tiny amount of a *drag-reducing polymer* has been added show that the average spanwise spacing of streaks is increased, in units of v/u^*, and that the vigor of eruptions over the streaks is reduced by the presence of the polymer. It is speculated that the polymer solution specifically resists the stretching of concentrated vortex filaments, and that this may be the mechanism responsible for drag reduction.[34]

Many investigators have used hot-wire, hot-film, or laser-Doppler anem-

[33] T. Theodoresen, *Mechanism of Turbulence*, Proc. 2nd Midwestern Conf. on Fluid Mech., 1962.

[34] G. L. Donohue, W. G. Tiedermann and M. M. Reischman, "Flow Visualization in a Drag-Reducing Channel Flow," *J. Fluid Mech.* **56**: 559–576. See also D. K. Oldaker and W. G. Tiedermann, "Spatial Structure of the Viscous Sublayer in Drag-Reducing Channel Flows," *Phys. Fluids* **20** (10), Part II: S133–144 (1977).

† A. E. Perry and M. S. Chong, "On the Mechanism of Wall Turbulence," *J. Fluid Mech.* **119**: 173–218, 1982.

ometers, or have photographically tracked marked particles, hoping to provide a quantitative description of the phenomena associated with the sublayer streaks. The early work showed that the streaks move gradually away from the wall until they reach a level $y^+ \approx 15$, where they suddenly are dispersed, with evidence of violent velocity fluctuations. These sudden dispersals, named *burst*, occur intermittently at a fixed point, and are associated with a recognizable sequence of events.[35] A great effort has been made to obtain a statistical picture of this sequence, because it seems to reveal the story of the birth or intermittent rejuvenation of turbulence in wall-bounded flows.

Flow in the Outer Layers

In the outer regions of a turbulent wall flow, we find the vorticity that has been transported out of the domain of the Law of the Wall at some upstream location, or that is a remnant of a laminar-turbulent transition process that occurred still farther upstream. The time-averaged distribution of this vorticity evolves under the influence of the following factors, which are known in advance.

1. *The presence or absence of a potential flow outside the turbulent region.* Fluid containing vorticity that was generated at the wall becomes entangled with fluid from the outer flow, in a process that depends somewhat on the vorticity, if any, that is already in the outer flow. For a boundary layer on an object that moves through a quiescent atmosphere, the external flow will be irrotational. If a model of that object is placed in a wind tunnel, the external flow will have very nearly zero mean vorticity, but may have influential vorticity fluctuations: so-called *freestream turbulence*. In a fully-developed pipe or pressure-driven channel flow, the external fluid will have significant vorticity, which is on the average counterrotating relative to the vorticity from the wall of interest.

2. *The velocity and acceleration of the external potential flow, if one exists.* To understand this, we remind ourselves that the local, instantaneous, velocity is the sum of three contributions: one is due to the external potential flow, even inside the boundary layer; one induced by the vorticity in the fluid and calculated as though there were no walls; and one due to an image distribution of vorticity, which cancels the normal component of the second contribution at points on the wall. We cannot calculate the last two contributions in our head when the vorticity distribution is complicated, but we can fairly easily imagine some of the consequences of the first contribution. This flow is partially responsible for the convection of vorticity. If it is accelerated, it will deform vortex filaments in a way that

---

[35] W. W. Willmarth (1975) *op. cit.*

may initiate, reinforce, or counteract self-induced or image-induced deformations. These deformations can have a great effect on the subsequent motion of the filaments, and eventually on the mean distributions of vorticity and velocity.

3. *The curvature of the wall.* Besides being one possible cause of streamline curvature in the external potential flow, curvature of the wall modifies the third contribution to the velocity field. When the mean flow is axisymmetric, the transverse curvature of the wall leads to stretching or foreshortening of axisymmetric vortex filaments, as they move radially outward or inward. This in turn affects the stability of their shape: stretching stabilizes a circular vortex ring, while shrinking destabilizes it.

4. *The nature and location of devices that may be used upstream to accelerate transition from laminar to turbulent flow, to shorten the distance required to establish fully-developed flow in a pipe or channel, or to thicken a boundary layer.* Strips of coarse sandpaper or of artfully serrated adhesive tape; rods laid on the wall transverse to the flow; transverse rows of solid pins or of fluid jets; fences with artfully spaced rails: all have have been used for these purposes. The common objective is to generate distributions of vorticity that induce strong convective transport normal to the wall. The devices work only when the Reynolds number is high enough to permit the breakdown, through the amplification of suitable perturbations, of unstable distributions of vorticity. The hope is that nature will select a chain of instabilities that soon eradicates any memory of the specific initial configuration of vorticity, so that the mean flow and turbulence found suitably downstream from any of these devices will be in some sense independent of the particular device employed. The hope seems to be well rewarded, as far as mean profiles are concerned, but recent experiments show that at least some sets of initial disturbances are remembered for a very long time.[36]

VELOCITY-DEFECT LAWS. In spite of the potential complexity suggested by the number of factors that might influence a given outer-layer flow, some valuable generalizations have been found. We shall consider two of them, called *velocity-defect laws,* which describe, between them, a large number of interesting outer flows. These flows do not involve curved potential-flow streamlines, an important complication that will later be discussed briefly in its own right.

Prandtl's outer-flow law. While examining mean velocity profiles for fully-developed pipe flow, Prandtl noticed that the the profiles for different values of the Reynolds number, and for both smooth and rough-walled pipes, collapse

---

[36] Ö. Savas and D. Coles, "Coherence Measurements in Synthetic Turbulent Boundary Layers," *J. Fluid Mech.* **160**: 421–446 (1985).

onto a single curve when the dimensionless *velocity defect*, $\{U(r) - U(0)\}/u^*$, is plotted versus r/a. Only points from the region very near the wall, specifically for $a - r < 100v/u^*$ for smooth walls, deviated significantly from the average curve. $U(0)/u^*$ is not a very strong function of Reynolds number, but u^* seems clearly to be the best scale for the velocity defect. The resulting functional relationship:

$$U - U(0) = u^*g(\eta) \quad \text{with} \quad \eta = \frac{y}{a} \quad (14.30)$$

is called a *velocity-defect law*. Subsequent research showed that the same plotting scheme captures the effect of Reynolds number for fully-developed channel flow, and for many boundary-layer flows. For boundary layers, $U(0)$ is replaced by the speed of the external potential flow, U_e; the pipe radius is replaced by some measure of the boundary-layer thickness. The appropriate scale for velocity defect remains u^*, except in cases of strong adverse pressure gradient, which we shall discuss subsequently. A selection of velocity-defect profiles is plotted in Fig. 14.24.

Quasi-equilibrium turbulent boundary layers. A turbulent boundary layer can be artificially cultured, by control of the streamwise variation of U_e, so that the dimensionless defect function, $g(y/\delta)$, remains nearly independent of x. The necessary variation of U_e can be deduced, with the help of an appropriate momentum-integral equation.[37] A basic requirement is that $\delta \, dP/dx = -\beta\tau_w$, where β is a constant. It is then expected that the shape of the defect function will depend only on the value of β. A boundary layer for which g is strictly independent of x is called an *equilibrium* boundary layer.

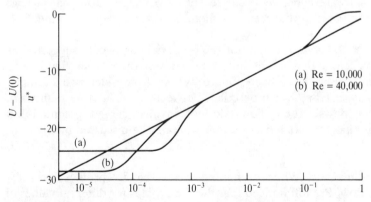

FIGURE 14.24
Velocity-defect profiles for turbulent flow in a smooth pipe.

---

[37] J. C. Rotta, *Prog. Aero. Sci.* **2**: 3 (1962).

If β varies slowly with x, and if no other factors, such as a sudden change in wall roughness, act to change the shape of the velocity profile, we may expect that g will also change slowly, and we may hope that g will be determined by the local value of β. When this is approximately true, we speak of *quasi-equilibrium*, or *moving-equilibrium* boundary layers.

The Law of the Wake. A concise fitting formula for the velocity-defect laws of pipe and channel flows, which works also for equilibrium and quasi-equilibrium boundary layers, has been presented by D. E. Coles.[38] It has the form

$$\frac{U - U_e}{u^*} = \frac{1}{\kappa}\left\{\ln\left(\frac{y}{\delta}\right) + 2\Pi\left[\sin^2\left(\frac{\pi y}{2\delta}\right) - 1\right]\right\} \tag{14.31}$$

Coles found that mild and slowly varying pressure gradients simply affect the value of Π, and he described the term involving Π as the *wake component* of the profile. The universal shape of the wake component is called the *Law of the Wake*.

Favorable pressure gradients, $dp/dx < 0$, correspond to small values of Π, so that the logarithmic term dominates the composite formula in almost the whole range of y/δ. White[39] has extracted a relationship between π and β from the data collected by Coles and Hirst, and gives the fitting formula, for quasi-equilibrium boundary layers,

$$\Pi \approx 0.8(\beta + 0.5)^{3/4}$$

Unfavorable gradients, $dp/dx > 0$, correspond to large values of Π, so that the logarithmic component dominates only for very small values of y/δ. In extreme cases, as when a boundary layer nears separation, it is hard to find any range of y within which the profile becomes straight in a semilogarithmic plot. These trends are illustrated in Fig. 14.21.

The Overlap Layer

The fitting formula for the Law of the Wall asymptotically approaches a semilogarithmic relationship as y^+ increases beyond about 100, and the data follow this trend if the Reynolds number is large enough and Π is small enough. Coles' fitting formula for the Velocity-Defect Law exhibits a semilogarithmic asymptote for $y \ll \delta$, and the data follow this asymptote under the same conditions: large Re, small Π. When these conditions are met, we

---

[38] See D. Coles, "The Law of the Wake in Turbulent Boundary Layers," *J. Fluid Mech.* **1**: 191–226 (1956); and D. E. Coles and E. A. Hirst, Proceedings, *Computation of Turbulent Boundary Layers—Volume II, Compiled Data,* pp. 1–19, 1968.

[39] In Section 6.4 of Viscous Fluid Flow, *op. cit.,* F. M. White gives a much more extensive treatment of this topic.

discover a range of y, called the *overlap domain*, within which *both* methods of presenting the data successfully suppress the effect of Reynolds number. This fact alone suffices to determine the analytic form of the velocity profile in the overlap domain.[40] The derivation starts with the observation that in the overlap layer there are, by definition, two equally successful representations of $\partial(U/u^*)/\partial y$:

$$\frac{\partial(U/u^*)}{\partial y} = \left(\frac{u^*}{v}\right)\frac{df}{dy^+} = \left(\frac{1}{\delta}\right)\frac{dg}{d\eta}$$

Multiplying through by y, we get

$$y\frac{\partial(U/u^*)}{\partial y} = y^+\frac{df}{dy^+} = \eta\frac{dg}{d\eta}$$

The second term in this equation, being a function of y^+ alone, can equal the third term, which is a function of η alone, only if both terms equal a constant, say $(1/\kappa)$. Thus we find, for the *overlap layer*,

$$f = \frac{1}{\kappa}\ln y^+ + C \qquad \text{and} \qquad g = \frac{1}{\kappa}\ln\eta + B$$

Our logic requires that κ be independent of both δ and v, and that C and B be independent of Reynolds number. In fact, the evidence reviewed above shows that C is also independent of pressure gradient, while B depends rather strongly on it. By comparing the equation for g in the overlap region with Coles equation for the Law of the Wake, we see that $B = -2\Pi/\kappa$. Since the basis for this development is a collection of data which covers only a modest range of Re, and which exhibits a significant amount of scatter, there is some disagreement about the best values to be used for κ and C, and the best way to determine values of Π, and hence B.

Strong Adverse dP/dx. The Schofield–Perry Velocity-Defect Law

When a turbulent boundary layer approaches separation, a velocity defect normalized by u^* becomes very large. It is then natural to seek a new velocity scale, say U_s, such that the quantity $(U - U_e)/U_s$ remains finite as $u^* \to 0$. We might use U_e itself, but a more effective choice was deduced by A. E. Perry and W. H. Schofield.[41] They found that a large collection of profiles for

---

[40] We follow the presentation of C. B. Millikan, *Proceedings of the Fifth International Congress on Applied Mechanics*, Cambridge, Mass., pp. 386–392, Wiley, New York, 1938.

[41] W. H. Schofield, "Equilibrium Boundary Layers in Moderate to Strong Adverse Pressure Gradients," *J. Fluid Mech.* **113**: 91–122 (1981).

equilibrium and *quasi-equilibrium boundary layers* in adverse pressure gradients could be collapsed to a single curve, which is well described, very near the wall, by the equation

$$\frac{U}{U_e} = 1 - \frac{U_s}{U_e} + 0.47\left(\frac{U_s}{U_e}\right)^{3/2}\left(\frac{y}{\delta^*}\right)^{1/2} \tag{14.32a}$$

The displacement thickness, δ^*, and the external flow velocity, U_e, are directly measurable; the ratio U_s/U_e can be obtained by plotting U/U_e versus $(y/\delta^*)^{1/2}$ and comparing the data with curves drawn from Eq. (14.32), as shown in Fig. 14.25.

A modified version of this equation, namely

$$\frac{U}{U_e} = 1 - \frac{U_s}{U_e} + 0.4\left(\frac{U_s}{U_e}\right)^{3/2}\left(\frac{y}{\delta^*}\right)^{1/2} + 0.6\left(\frac{U_s}{U_e}\right)\sin\left(\frac{\pi y U_s}{5.72\delta^* U_e}\right) \tag{14.32b}$$

will then provide a good fit to the velocity profile in the outer 98 to 99 percent of the boundary layer. The success of this procedure in collapsing a large number of measured profiles onto one curve is shown in Fig. 14.26. Many more details have been given recently by Schofield.[42] Some the most important are the following.

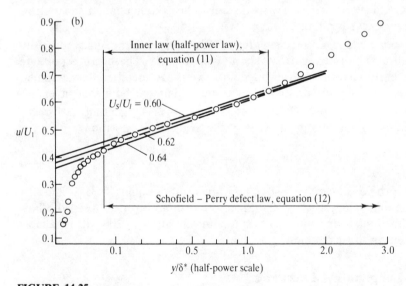

FIGURE 14.25
Schofield–Perry velocity-defect law. Determination of U_s by fitting to the half-power law. (Adapted, by permission, from W. H. Schofield (1981.[43])

---

[42] W. H. Schofield (1981) *op. cit.*

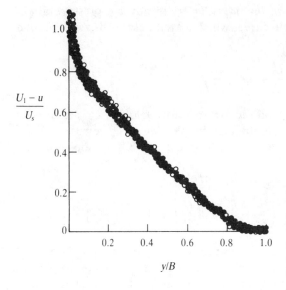

FIGURE 14.26
Variety of profiles, collapsed by the Schofield–Perry velocity-defect law. (Adapted, by permission, from W. H. Schofield, 1981.[41])

1. Equilibrium boundary layers are possible only when $U_e = a(x - x_0)^m$, with $m > -0.3$. Then $U_s = b(x - x_0)^m$ and $\delta^* = c(x - x_0)$.

2. The Schofield–Perry defect law works only when the maximum shear stress in the layer exceeds about $1.5\tau_w$.

3. For the same value of m, say $m = -0.23$, there can be many different values of U_s/U_e, ranging from about 0.45 to 1.0. These correspond to different early histories of the boundary layers. Remember that a similar result was discussed in our study of self-similar laminar boundary layers.

4. As $y \to 0$, the Schofield–Perry defect law joins tangentially with the logarithmic continuation of the Law of the Wall, at the position

$$\frac{y_c}{\delta^*} = 53\left(\frac{U_e}{U_s}\right)^3 C_F \qquad (14.33)$$

For the profile shown in Fig. 14.25, this ratio equals 0.317, and the corresponding value of y_c^+ is 203. Since the overlap layer between the Law of the Wall and an outer law starts at about $y^+ = 100$, it is almost nonexistent in this example.

Effects of Longitudinal Curvature

Turbulent wall flows are very sensitive to longitudinal curvature of the wall. Even when δ/R_T is as small as 0.01, modifications of the mean velocity profile and of the profiles of turbulent intensity and Reynolds stress are easily noticed. If a straight wall enters a curve, and eventually straightens out again, the effects of curvature are quick to appear and slow to go away. The effects, both qualitative and quantitative, depend greatly on whether the wall curves toward the fluid (concave curvature), or away from the flow (convex curvature). The general explanation, to the extent that there is one, was given by Prandtl in the 1930s, and has been seen by us in Chapter 13 in the discussion of the Taylor–Görtler instability of laminar boundary layers. If the product of the mean velocity, U, and the distance from the center of curvature, R, increases as R increases, radial excursions of vortex filaments will, on the average, be inhibited. This happens in the boundary layer on a convex wall. If UR decreases as R increases, as happens in the boundary layer on a concave wall, chance radial excursions will be enhanced.[43]

Large-scale disturbances strongly reminiscent of Görtler vortices are typically observed in the outer regions of turbulent boundary layers on concave walls. They appear more or less randomly, unless there are small stationary disturbances upstream, which may increase the probability of their appearance in specific spanwise locations. These disturbances may produce no detectable effect in the boundary layer on a flat wall, only to be amplified into easily observable magnitudes when that wall subsequently curves towards the flow. This is an interesting reminder of the long memory of a flow with great inertia. In extreme cases, the pattern of large-scale streamwise vortices becomes nearly stationary, and can be associated with spanwise variations of u^* and δ as large as $\pm 20\%$. This could be important to design or performance, but it is still not well understood or controlled. In some cases, seemingly huge disturbances appear necessary to fix the vortex pattern;[44] in others, a fixed pattern seems to appear with almost no observable provocation.[45]

A velocity-defect law is hard to define for a boundary layer with strong longitudinal curvature, because the speed of the potential flow would vary significantly over the distance occupied by the boundary layer. Some profiles taken on a concave wall are shown in Fig. 14.27, with the mean velocity normalized by the potential-flow velocity for that distance from the wall. At almost all values of y/δ, the mean streamwise velocity is substantially increased as the effect of curvature accumulates.

---

[43] Barlow and Johnston (1985) *op cit.*

[44] See P. Bradshaw, "Effect of Streamline Curvature on Turbulent Flow," AGARDograph 169 (1973) for a review which stimulated lots of research. A more recent article with a good introduction is by J. C. Gillis and J. P. Johnston, "Turbulent Boundary-layer Flow and Structure on a Convex Wall and its Redevelopment on a Flat Wall," *J. Fluid Mech.* **135**: 123–154 (1983).

[45] Fascinating examples of this are documented by A. J. Smits, S. T. B. Young, and P. Bradshaw, "The effect of Short Regions of High Surface Curvature on Turbulent Boundary Layers," *J. Fluid Mech.* **94**: 209 (1979), and by M. R. Head and I. Rechenberg, "The Preston Tube as a Means of Measuring Skin Friction," *J. Fluid Mech.* **14**: 1–17 (1962). The latter authors do not identify concave curvature as a factor in the mystery they unraveled, but the sketch of their flow circuit makes me suspect it. The article is beautifully written, and provides a classic example of the care with which empirical descriptions of turbulent flow must be constructed.

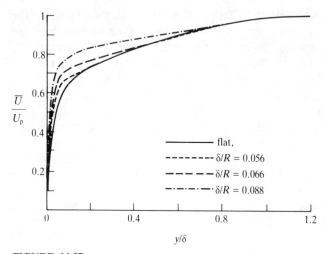

FIGURE 14.27
Velocity profiles for turbulent boundary layers on longitudinally concave walls. (Reproduced, by permission, from R. S. Barlow and J. P. Johnston, 1988.[30])

The Measurement of Turbulent Skin Friction

By now you should be getting a little nervous about empirical correlations that are based on a quantity, u^*, that cannot usually be known in advance and that seems likely to be hard to measure. For fully developed pipe flows, it is easy in principle to measure the axial pressure gradient and then compute the skin friction by a simple momentum balance. Remember, however, that a pipe flow at really high Reynolds number has lots of inertia, so that the slightest taper of the pipe, or the slightest bend or deviation from a circular cross section, may introduce significant changes of momentum in a force balance from which they are supposed to be absent. For a boundary layer, it is possible, again in principle, to determine the skin friction acting on a short streamwise segment of wall by careful measurement of the velocity profiles and the pressure, at its upstream and downstream ends. We get the wall drag, however, as a small difference between large numbers. Unavoidable random measurement errors are likely to produce an unacceptable uncertainty in the derived quantity. Systematic errors, due to slight deviations from perfect two-dimensionality of the mean flow, are also often serious.

In principle, it would seem straightforward to measure $U(y)$ in the region $y^+ < 5$, and to determine $\partial U / \partial y$ at the wall from the slope of a line through the points. This requires very special instrumentation and great care, and can hardly be recommended for routine use. The same can be said for the implantation, in the wall, of a floating-element drag balance.

Fortunately, the very existence of a Law of the Wall saves the day. This is one of the reasons for interest in a law that describes the flow in such a small domain.

The Preston tube. The vast majority of mean velocity profiles of turbulent pipe flows and boundary layers have been measured with tiny Pitot tubes. When the idea of a universal law of the wall was gaining acceptance, it occurred to J. H. Preston[46] that there should be a relationship between the skin friction and the Pitot pressure measured when the tube lies right on the wall, if the tube is small enough to lie entirely in the region described by the Law of the Wall. Since the Law of the Wall describes pipe flow and boundary-layer flow equally well, the relationship can be found empirically by tests on pipe flow, for which τ_w can be independently measured.

Preston argued that Δp, the difference between the Pitot pressure and a local static pressure, would depend only on τ_w, u^*, ρ, v, the probe diameter d, and the shape of the tip of the probe. For a family of geometrically similar probes, dimensional analysis yields the relation

$$\frac{\Delta p}{\tau_w} = f\left(u^*\frac{d}{v}\right)$$

or the more convenient form

$$\tau_w\frac{d^2}{\rho v^2} = F\left(\Delta p\frac{d^2}{\rho v^2}\right)$$

The calibration curve found by Patel[47] for round tubes shown in Fig. 14.28.

Preston tubes have been used successfully in a wide variety of applications. Just be sure that $u^*d/v < 50$ if the tube is to be used in adverse pressure gradients, or in other circumstances where there is little or no overlap layer similar to that in pipe flow. When in doubt, use tubes of different diameter, and make sure that τ_w turns out to be independent of d.

Clauser plots. The other popular way of determining u^* was presented by F. H. Clauser.[48] Let the Law of the Wall be rearranged, into

$$\frac{U}{U_e} = \frac{u^*}{U_e}f\left[\left(\frac{u^*}{U_e}\right)\left(\frac{U_e y}{v}\right)\right]$$

Plot this equation for various values of the unknown parameter u^*/U_e, or of the equivalent parameter

$$C_F = 2\left(\frac{u^*}{U_e}\right)^2$$

---

[46] J. H. Preston, "The Determination of Turbulent Skin Friction by Means of Pitot Tubes," *J. Roy. Aer. Soc.* **58**, 109 (1954). See also Head and Rechenberg (1962) *op. cit.*, who laid to rest any doubts of the validity of Preston's idea.

[47] V. C. Patel, "Calibration of the Preston Tube and Limitations on its Use in Pressure Gradients," *J. Fluid Mech.* **23**: 185–208 (1965).

[48] F. H. Clauser, "Turbulent Boundary Layers in Adverse Pressure Gradients," *J. Aero Sci.* **21**: 91–108 (1954)—see Fig. 5.

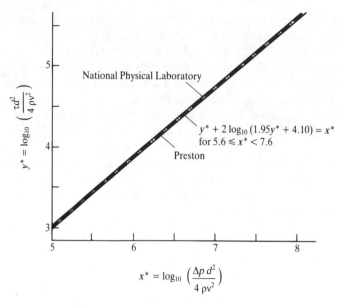

FIGURE 14.28
Calibration of the Preston tube. (Reproduced, by permission, from V. C. Patel, 1965.[47])

When an experimental velocity profile, for which the value of u^* is as yet unknown, is superposed on such a plot, the correct value of C_F can be estimated by eye. The procedure works best when the experimental profile exhibits a well-defined region of logarithmic behavior. Thus, the plots are usually made on semilogarithmic paper. An example is shown in Fig. 14.29.

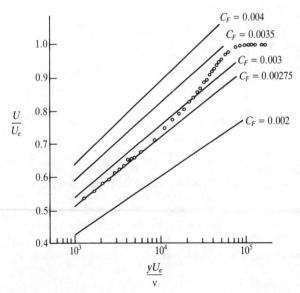

FIGURE 14.29
Clauser plot for determination of u^*. $C_F = 2(u^*/U_e)^2$.

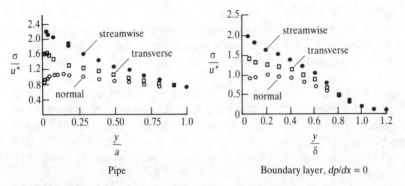

FIGURE 14.30
Rms velocity fluctuations in pipe and boundary-layer flows. (a) Pipe; (b) boundary layer, $dp/dx = 0$.

For relatively easy applications, say to two-dimensional boundary layers on flat walls, with favorable or only mildly adverse pressure gradient, the Preston-tube and Clauser-plot methods agree within a few percent, which is usually within the limits of the precision of either method. All methods of measuring u^* become more difficult when strong effects of pressure gradient, wall curvature, or three-dimensionality are present.

Statistics of Fluctuations in the Outer Regions of Turbulent Wall Flows

We cannot compactly characterize fluctuations in the outer regions of turbulent wall flows, as we can in the domain of the Law of the Wall, but we can identify some features that differ only moderately from case to case. We start by presenting two classic cases: fully-developed pipe flow and the equilibrium boundary layer with zero pressure gradient. Profiles of the standard deviations of streamwise, spanwise, and normal velocity fluctuations (u', w', and v') are shown in Fig. 14.30; profiles of the Reynolds stress component $\langle u'v' \rangle$ are shown in Fig. 14.31.

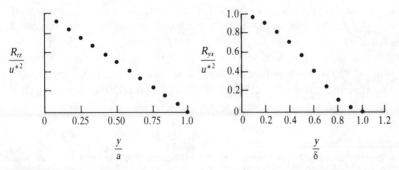

FIGURE 14.31
Reynolds stress in pipe and boundary-layer flows. (a) Pipe; (b) boundary layer, $dp/dx = 0$.

Note again the *anisotropy* of the fluctuations; most of the kinetic energy of the turbulence resides in u', the next most in w', the least in v'. When the energy is distributed over a frequency spectrum, we see that this anisotropy is most evident in the low-frequency components, and nearly absent at high frequencies.

Note also that easily-observable fluctuations are found out beyond the nominal edge of a boundary layer, where the mean velocity is uniform, but that the Reynolds stress is negligible there. Finally, note that the Reynolds stress is nearly exactly proportional to r in the pipe flow, except in a thin layer near the wall. Only in this thin layer is the mean viscous stress significant compared to the Reynolds stress.

Spectra of the velocity fluctuations in the outer region, and particularly in the overlap region, have been the object of a very interesting study by Perry, Henbest, and Chong.[49] They considered how spatial spectra might be plotted, for various ranges of wavenumber, so as to reveal features that are independent of the value of y^+. In particular, they establish the concept of three, partially overlapping wavenumber domains, with different appropriate scales of length and velocity for each.

For the lowest wavenumbers (the longest waves), the appropriate scale of length is δ, the thickness of the boundary layer, radius of a pipe, or halfheight of a plane-walled channel. The appropriate scale of velocity is u^*. Thus, one plots, for example, Φ_{uu}/u^{*2} versus $k\delta$, understanding that the integral of Φ_{uu} over the full range of $k\delta$ equals the variance of u. This is called *outer-flow scaling*.

For intermediate wavenumbers, the appropriate scale of length is y, the distance from the wall. The appropriate scale of velocity is again u^*. One plots, for example, Φ_{uu}/u^{*2} versus ky, understanding that the integral of this Φ_{uu} over the full range of ky equals the variance of u. This is called *inner-flow scaling*. Clearly, the spectral density for the inner-flow scaling equals δ/y times the spectral density of the outer-flow scaling.

For the highest wavenumbers (the shortest waves) the appropriate scale of length is λ, the Kolmogoroff length. Remember, from Eq. (14.8), that $\lambda = (v^3/\varepsilon)^{1/4}$, where v is the kinematic viscosity and ε is the local volumetric rate of energy dissipation. The appropriate scale of velocity is $(v\,\varepsilon)^{1/2}$. Thus, one plots, for example, $\Phi_{uu}/(v\,\varepsilon)$ versus $k\lambda$, understanding that the integral of this Φ_{uu} over the full range of $k\lambda$ equals the variance of u. This is called *Kolmogoroff scaling*.

Figure 14.32 shows a collection of spectra, taken in the rather extensive overlap region of fully-developed pipe flow, and plotted with the inner-flow scaling. Notice that the curves for different values of y/δ come closest together

---

[49] A. E. Perry, S. Henbest, and M. S. Chong, "A Theoretical and Experimental Study of Wall Turbulence," *J. Fluid Mech.* **165**: 163–200 (1986).

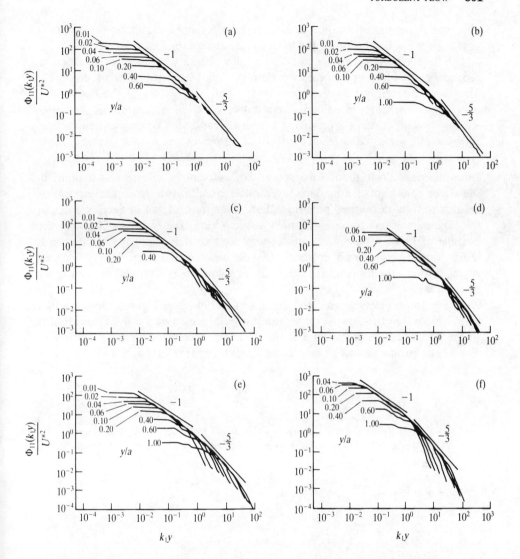

FIGURE 14.32
Streamwise wavenumber spectra in the outer region of fully developed pipe flow: (a) Re = 200,000; (b) 175,000; (c) 150,000; (d) 125,000; (e) 100,000; (f) 75,000. (Reproduced, by permission, from A. E. Perry et al., 1986.[49])

where $ky \approx 1$, and that they stay close together up to about $ky = 30$, at the highest values of the Reynolds number. As Re decreases, the value of ky at which it is necessary to shift from inner-flow scaling to Kolmogoroff scaling also decreases, showing that viscous diffusion is affecting larger and larger scales of motion. The authors of this collection successfully show that outer-flow scaling brings all the curves fairly close together for low values of k, and that Kolmogoroff scaling does the same for high values of k, providing that the spatial resolution of the velocity sensor is sufficient.

These results, even for the classic flows they represent, are harder to reproduce in different laboratories than are those for the Law of the Wall. They are apparently fairly sensitive to the detailed history of the flow. They may also be strongly modified, quantitatively, by such complicating circumstances as wall curvature or strong pressure gradient. Convex wall curvature, for example, greatly depresses the level of Reynolds stress in the outer region.

Intermittency. Continuous time records of velocity or vorticity at a point in the outer region of a boundary layer differ qualitatively from corresponding records in fully-developed pipe flow. Examples were shown in Section 14.2. In the boundary layer, the high-frequency fluctuations that we associate with the passage of small-scale regions of concentrated vorticity appear only intermittently, interspersed with episodes of relatively constant or slowly varying signal. In the pipe flow, all the fluid contains some vorticity, and intermittency is much less obvious.

The intermittency profile of the flat-plate boundary layer is shown in Fig. 14.33. Note that the flow is only continuously turbulent if $y < 0.4\delta$, and that is turbulent about 7% of the time at $y = \delta$. The profile is well fitted by a time at $y = \delta$. The profile is well fit by a complementary error function, to wit

$$\gamma\left(\frac{y}{d}\right) = 0.5 \, \text{erfc}\left[5.44\left(\frac{y}{d} - 0.82\right)\right].$$

The constants, 5.44 and 0.82, vary slightly from experiment to experiment.

An intermittent flow can be characterized by the usual time-averaged

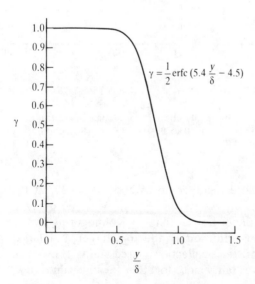

$$\gamma = \frac{1}{2}\text{erfc}\left(5.4\,\frac{y}{\delta} - 4.5\right)$$

FIGURE 14.33
Intermittency in a turbulent boundary layer with zero pressure gradient.

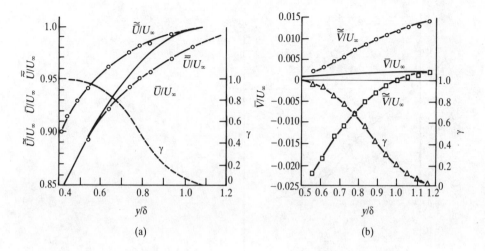

FIGURE 14.34
Zone-averaged profiles of rms streamwise velocity fluctuations. (Reproduced, by permission, from Kovasznay et al., 1970.)[50]

properties, or by separate *zone averages,* one for the intervals of rotational flow, the other for the intervals of irrotational flow. Figure 14.34 shows the zone-averaged profiles of u and v, along with the profiles of intermittency, γ, and the averages over all time, U and V.

Figure 14.35 shows the corresponding picture, for the r.m.s. fluctuations in u. The r.m.s. fluctuation of u is considerably higher in the rotational flow than in the irrotational flow.

A second kind of conditional averaging, called *point averaging,*[50] samples the continuous time records only during brief instants that correspond to the passage, over the sensor, of the interface between rotational and irrotational flow. Two such averages can be defined: one for the times when rotational flow arrives at the sensor, the other for the times when rotational flow departs. These times correspond to the passage of the fronts and the backs of the zones of rotational flow. Figure 14.36 shows that the fronts are everywhere approaching the wall; while the backs exhibit a level, at about $y = 0.7\delta$, above which they move away from the wall, below which they approach it. These features are what one would expect if the rotational zones contain, on average, concentrated spanwise vorticity.

---

[50] The data shown, and the concepts of zone and point averages, are from the pioneering work of L. S. G. Kovasznay, V. Kibens and R. F. Blackwelder, "Large-scale Motions in the Intermittent Region of a Turbulent Boundary Layer," *J. Fluid Mech.* **41:** 283–326 (1970). The Reynolds number for the tests was fairly low, but the large-scale features studied do not appear to be much different at higher Reynolds number.

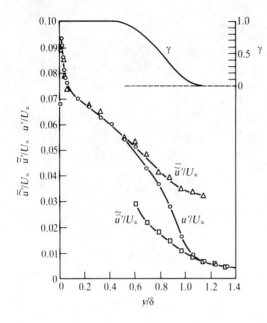

FIGURE 14.35
Zone-averaged profiles of rms stream-wise velocity fluctuations. (Reproduced, by permission, from Kovasznay et al., 1970.)[50]

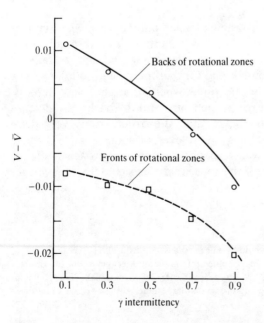

FIGURE 14.36
Point averages of normal velocity. (Data reproduced, by permission, from Fig. 9 of Kovasznay et al., 1970.)[50]

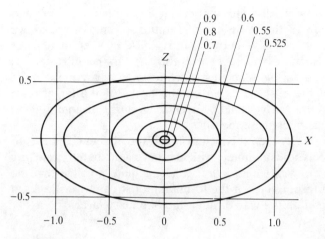

FIGURE 14.37
Probability that the flow will be simultaneously rotational at the two points: $(0, 0.8, 0)$ and $(X, 0.8, Z)$. $X = [(X - X_0)/\delta]$, $Z = [(Z - Z_0)/\delta]$, $y/\delta = 0.8$ corresponds to $\gamma = 0.5$. (Adapted, by permission from Kovasznay et al., 1970.)[50]

The rotational zones are not, however, features that extend coherently to great spanwise distances. Figure 14.37 maps the probability that the flow will be simultaneously rotational at two points in the same plane, $y = 0.8\delta$. Note that there is only a single peak, where the two points coincide. This shows that the zones of rotational flow appear, over a long-time average, at random positions in the $x-z$ plane, rather than in arrays with statistically preferred spacings. The probability, which approaches the intermittency as the separation of the two points increases, falls off to 55% at a spanwise separation of 0.4δ and a streamwise separation of 0.9δ.

Instantaneous flow fields in the outer region of a boundary layer. When the fluid in the boundary layer is marked with smoke, dye, or a small temperature difference, a different discrimination can be made between fluid masses that have moved outward from the wall, and those that have been entrained from the external irrotational flow.[51] Slightly different zone- and point-averaged values are obtained, but the general picture described above is confirmed, and interesting new details are revealed. It turns out that the backs of the rotational flow zones are the loci of particularly intense fluctuations, which

---

[51] See R. E. Falco, "Coherent Motions in the Outer Region of Turbulent Boundary Layers," *Phys. Fluids* **20** (10), Part II: S124–S132 (1977); C.-H. P. Chen and R. F. Blackwelder, Large-scale Motion in a Turbulent Boundary Layer: A Study Using Temperature Contamination," *J. Fluid Mech.* **89**: 1–32 (1978); and M. R. Head and P. Bandyopadhyay (1981), *op. cit.* The interpretation given here is that of the last authors.

appear to be associated with discrete vortices which either ride up and downstream, over the tops of the intrusions of rotational fluid, or are drawn back toward the wall, in the valleys between intrusions. Further, it appears that at relatively high Reynolds numbers the vorticity in the intrusions is organized, on average, into hairpin filaments which form an angle of about 45° to the wall. The tips of these filaments lie on the backs of the larger-scale intrusions, which in turn are fairly straight, forming an angle of about 20° to the wall. Figure 14.38 illustrates these ideas.

A discrepancy of scale between the intrusion and the filaments appears to be a strong function of Reynolds number, with the intrusions better scaled by δ, and the filaments by the wall length v/u^*. It seems quite possible that the vortex filaments that are responsible for the formation of streamwise streaks in the viscous sublayer extend all the way through the boundary layer.

Streamwise development of u^* and δ. The parameters u^* and δ, which we have used to frame these descriptions, have constant values in fully-developed flows, but depend on x (and possibly z) in boundary-layer flows. Since these variations are often of prime practical importance, persistent efforts have been made to define them. These face great difficulty when the region to be described includes a zone of transition from laminar to turbulent flow, because the location and length of that zone are very sensitive to factors, such as the intensity of freestream turbulence, that may vary significantly from one application to another. Thus, empirical formulas often describe only $d\delta/dx$ for the fully-turbulent region, assuming that this quantity will be fairly independent of the specific history of transition. The user of the formula then has to provide an initial value of δ appropriate to the specific flow of interest. Once δ is evaluated, u^* is often described by a formula involving δ, v, and U_e, other relevant parameters such as dP/dx or the wall curvature. For quasi-equilibrium boundary layers that are well described by the combination of the Law of the

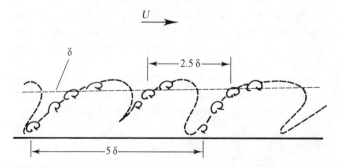

FIGURE 14.38
Falco's picture of coherent vorticity in the intermittent region of a turbulent boundary layer. (Reproduced, by permission, from R. E. Falco, *Phys. Fluids* **20**, No. 10, Part II, S124–S132, 1977.)

Wall and the Law of the Wake, this connection is expressed by the formulae

$$\frac{U_e}{u^*} = \lambda \qquad \frac{U_e \delta}{\nu} = \lambda \exp\left[\kappa(\lambda - C) - 2\Pi\right] \tag{14.33}$$

which is displayed graphically in Fig. 14.39 for a few values of Π. We use the values $\kappa = 0.41$, $C = 5.0$.

The same formula used to derive (14.33) implies the following formulae for the shape factors of the velocity profile:

$$\frac{\delta^*}{\delta} = \frac{1 + \Pi}{\kappa\lambda} \qquad \frac{\theta}{\delta} = \frac{\delta^*}{\delta} - \frac{2 + 3.2\Pi + 1.5\Pi^2}{(\kappa\lambda)^2} \tag{14.34}$$

These can be used, together with the momentum-integral equation for the mean flow

$$\frac{d\theta}{dx} = -(2\theta + \delta^*)d(\ln U_e)\,dx + \lambda^2 \tag{14.35}$$

to estimate $d\delta/dx$. Typical values of λ and $d\delta/dx$ fall in the ranges

$$0 \le 1/\lambda \le 0.05 \qquad 0.015 \le d\delta/dx \le 0.10$$

There is a large and scattered literature for this topic, the best entry to which may be found in the proceedings of conferences organized to explore the effectiveness of various semiempirical schemes for computation of boundary-

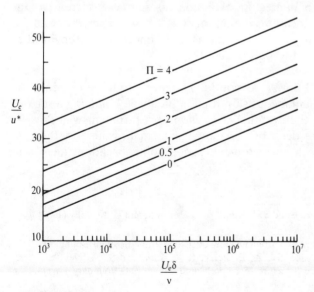

FIGURE 14.39
Relationship between parameters of quasi-equilibrium boundary layers.

layer development.[52] We only note here that boundary layers on curved walls are not well described by the Eqs. (14.34). Convex curvature tends to decrease u^* and increase δ; concave curvature does the opposite, and additionally promotes spanwise variations of u^* and δ.

Flow over rough walls. The viscous length scale, v/u^*, is very small in many practical situations, so that very small asperities on a nominally smooth wall may have significantly large *roughness Reynolds numbers*. If a wall is fairly uniformly covered with asperities of average height k, we can classify measured velocity profiles according to the value of this Reynolds number, u^*k/v. Handbooks of hydraulic data attempt to characterize all sorts of ordinary surfaces by effective values of k.

When $u^*k/v < 5$, the wall roughness has no measurable effect on the development of u^* or δ, although it must have some effect on properties in the viscous sublayer unless $u^*k/v < 1$. When $u^*k/v > 70$, viscosity seems to have no effect on u^*, δ, or the shape of the profile. In no case does the wall roughness have any measurable effect on velocity-defect profiles, and it is claimed, although the data base is as yet very limited, that the characteristic structural features of the outer region of the wall layer are much the same, whether the wall be rough or smooth.[53] If this is strictly true, it would seem to imply that viscosity plays a very subtle role, if any, in the formation and evolution of these structures.

Other factors being constant, u^* increases with increasing wall roughness; δ is not much affected. The profiles of r.m.s. fluctuations of u and v, normalized by u^*, show an interesting evolution, as u^*k/v is increased. In particular, the prominent local maximum of $\langle u'^2 \rangle^{1/2}/u^*$ near the wall is gradually erased, suggesting that viscosity is an important factor in the formation of this peak, when the wall is smooth.

Separation of turbulent wall flows. As you might expect from the evidence of large fluctuations in the instantaneous skin friction on a wall bounding a region of *attached* turbulent flow, the eventual *separation* of a turbulent wall flow is not trivially easy to define or describe.[54] At any fixed, small, value of y there is a range of x through which the probability of finding backflow increases

---

[52] See Coles and Hirst (1968), *op. cit.*, and S. J. Kline, B. J. Cantwell, and G. M. Lilly (1982), *op. cit.*

[53] A. J. Grass, "Structural Features of Turbulent Flow over Smooth and Rough Boundaries," *J. Fluid Mech.* **50**: 233–256 (1971). See also A. E. Perry, K. L. Lim, and S. M. Henbest, "An Experimental Study of Turbulence Structure in Smooth- and Rough-wall Boundary Layers," *J. Fluid Mech.* **177**: 437–4666 (1987).

[54] See, for example, R. L. Simpson, Y.-T. Chew, and B. G. Shivaprasad, "Structure of a Separating Turbulent Boundary Layer, Part I, "*J. Fluid Mech.* **113**: 23–51 (1981). Parts II and III follow in the same volume.

gradually from zero to one, as x increases. In the region where backflow predominates, spanwise meandering of marked fluid masses may be prominent. A time-averaged separation line can be defined as the locus where the time-averaged skin friction vanishes. Although a well-defined return flow may approach the separation line from downstream, this flow does not seem to exhibit any of the structural features typical of the wall layer upstream of separation.

Discussion. At the time of writing, an intense research effort is continuing, aimed at further definition, clarification, and extension of the description of turbulent wall flows. You are entitled to ask why this is thought to be so important, and whether all this study is producing any practical or intellectual benefits.

The search for generally useful statistical correlations that can be added to the mean momentum and continuity equations to yield a mathematical problem with as many equations as unknowns does not seem to be very fruitful. The correlations discovered from a study of a limited class of flows, such as equilibrium boundary layers on flat walls, can be successfully embedded in a calculation scheme that works well for interpolation between members of that class, but the scheme fails to predict phenomena that appear when the class of flows is enlarged to include, for example, the effects of wall curvature.

It is still too soon to be certain, but it seems unlikely that the structural details found in the outer regions of the simplest boundary layers, in a very limited range of Reynolds numbers, will be usefully recognizable in more complicated flows. It is also far from clear that recognition of these structures will lead to more effective means of prediction of statistical properties of interest.

It would be nice if one could truly say that recognition of the presence and key importance of stretching streamwise vortex loops in the viscous sublayer had led to the discovery of the drag-reducing properties of polymer additives in water flows, but that would be a false history. However, an active field of current research involves a direct and informed attempt to reduce turbulent drag by the insertion, into the outer regions of the boundary layer, of fixed obstacles that disrupt the natural evolution of the large-scale concentrations of vorticity. An interesting early finding is that a relatively local disruption has significant effects surprisingly far downstream. This sort of effect is very hard to predict with any sort of theory that postulates that the local structure of turbulence is very nearly determined by local values of parameters such as u^* or δ, but if understood even qualitatively it can be put to good practical use.

At the very least, the detailed studies of instantaneous flow pictures help us to recognize and rationalize certain features of the mean flow which may seem paradoxical. Figure 14.40 illustrates one such feature. It compares the vorticity profile of a turbulent boundary layer on a flat plate, at a position just

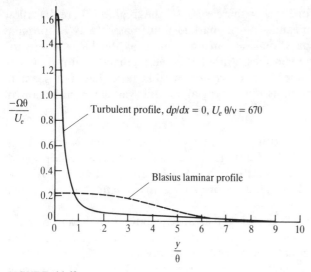

FIGURE 14.40
Vorticity profiles of boundary layers just before and after transition to turbulent flow.

downstream of a natural transition from laminar flow, with that of the laminar boundary layer just upstream of the beginning of the transition zone.

We see that the onset of turbulence has carried some vorticity much farther from the wall that it would have been found, had the laminar flow continued. This is the sort of thing one might intuitively expect. Note, however, that the turbulence has also carried some vorticity back towards the wall. The mean vorticity right at the wall, which is proportional to the skin friction, has been considerably increased by the transition to turbulent flow. Somehow, the increase in skin friction is expected, but the reconcentration of vorticity near the wall seems surprising. Turbulence is supposed to enhance rates of dispersion, and hence to reduce, rather than increase, local concentrations. What is going on?

The usual verbal description of instantaneous velocity fields in the turbulent wall layer invokes *ejections* of low-momentum fluid, and *inrushes* of high-momentum fluid, as events that dominate the time-averaged transfer of momentum across the layer. The correlation of positive v' with negative u' and vice versa, is strong during these events, and $\langle u'v' \rangle$ is negative throughout the layer. It appears that the fate of the fluctuations in spanwise vorticity, Ω', is more complicated, and that $\langle \Omega'v' \rangle$ is positive near the wall and negative farther out. Remember that $\Omega = \partial v/\partial x - \partial u/\partial y$, and hence $\langle \Omega \rangle$ is negative in the boundary layer. Thus $\Omega' > 0$ for a particle that spins more slowly than the average, around the z-axis, and $\langle \Omega'v' \rangle$ is positive if there is positive correlation between slow spin and outward movement. Thus, macroscopic mixing of Ω very near the wall does not simply transport vorticity down the gradient of its time-averaged distribution.

The secret of separation control is to keep the peak of the mean vorticity profile as close to the wall as possible. We see that in attached turbulent boundary layers, even in adverse pressure gradients, the peak stays at or very near the wall. This contrasts strongly with the situation in laminar boundary layers. We may hypothesize that the way to forestall separation of a turbulent wall layer is to identify the fluctuating concentrations of vorticity that induce a time-averaged transport of vorticity towards the wall, and then to encourage their formation. The hope is to do this subtly, rather than by brute-force methods that incur a large penalty of excess drag or energy dissipation.[55] The possibility of subtle intervention for useful purposes resides in the inherent instability of the transient vorticity distributions of turbulent flow, and the possibility of introducing disturbances that give a desired bias to the future development of the unstable flow.

Three-dimensional turbulent boundary layers. When the direction of the velocity component parallel to a wall varies with distance from the wall, the description of the velocity profile is additionally complicated. We note here only the intriguing fact that even in this case there appears to be a well-defined near-wall region, in which the direction of that component is very nearly constant, and in which the variation of speed with y is again well described by the law of the wall.[56] What can be happening in the wall region, that seems so universal and in some ways so independent of the average state of affairs in the rest of the boundary layer? One cannot yet be certain, but this kind of behavior, in which the response of part of a fluid seems oddly independent of the boundary conditions imposed upon it, is reminiscent of the phenomenon of selective amplification by an unstable system, studied in Chapter 13.

14.7 FREE TURBULENT FLOWS

INTRODUCTION. Many important turbulent flows are found far from solid walls, and hence are often called *free turbulent flows*. Such flows, if *barotropic,* have typically separated from a solid wall somewhere upstream; and their vorticity is, in an overall sense, inherited from a wall flow. Examples include *mixing layers* between uniform streams of different mean velocity, *jets*, and *wakes*. Flows with *baroclinically* generated vorticity are common in nature, and a few examples will be included in this discussion, without detailed consideration of the complex chemical and thermodynamic processes that may cause the baroclinicity. Examples include *buoyant plumes* and *clouds*.

---

[55] The vortex generators which used to decorate the wings of large passenger aircraft provides a good example of brute force methods.

[56] B. van den Berg, A. Elsenaar, J. P. F. Lindhout, and P. Wesseling, "Measurements in an Incompressible Three-Dimensional Turbulent Boundary Layer," *J. Fluid Mech.* **70**: 127–148 (1975). See Fig. 7.

The principal phenomena one needs to understand are *entrainment* and *convective enhancement of molecular mixing*. The conceptual distinction between these phenomena may not be very sharp, but may be clarified by an analogy. We shall use the term entrainment to denote the process by which a time-averaged flow of fluid is induced to cross an imaginary surface separating two regions that are distinguished by some qualitative difference in time-averaged flow properties. The difference may be between rotational and irrotational flow, or between chemically pure and chemically mixed fluid, etc.

Entrainment is somewhat analogous to immigration, a process by which foreigners are brought across a boundary, frequently in groups large enough to prevent intimate mixing with the folk of the new country, even while the groups are moving far in from the boundary. We say that a turbulent jet entrains fluid from the region into which the jet is discharged. By this process, the mass flow in what one perceives to be the jet increases with distance from the origin, as newly-entrained fluid enters and is brought up to speed.

When a body of fluid is entrained into a region of turbulent flow, it is fairly rapidly deformed, and the surface area across which it confronts fluid of contrasting properties is correspondingly enlarged. On the average, this process steepens the gradients of the contrasting properties, and accelerates the processes of molecular diffusion, by which the contrasts are eventually erased. The analogous processes in the history of immigrant groups are things like shopping and going to school or work, which increase contact with folk who came before and accelerate the diffusion of language and culture.

A third phenomenon, of great practical importance in many free turbulent flows, is generation of noise. This motivates much of the research on turbulent jets, and much of one's curiosity about instantaneous flow structures is focused on their effectiveness as sources of sound. Unfortunately, this fascinating study would carry us far beyond the scope of this text.

The organization of this section will parallel that of Section 14.6. Mean velocity profiles and the profiles of turbulence intensity and Reynolds stress will be described first, with emphasis on the concept of *self-similarity*. This concept limits the scope of the discussion, but helps to organize it and gives an opportunity for a little analysis based on integral versions of the time-averaged equations of continuity, momentum, and mean mechanical energy. You will see that viscosity seems to have little effect on what is observed at this level of detail.

The instantaneous structures present in each flow are described next, with discussion of hypotheses about the roles of these structures in the processes of entrainment and mixing. A special effort is made to illuminate the subtle effects of viscosity in the latter process.

Finally, some consideration will be given to schemes for manipulating either the large-scale or small-scale behavior of these flows, to achieve certain practical objectives.

PLANE MIXING LAYERS. When two nominally uniform and nearly parallel streams, flowing with different speeds U_1 and U_2, are brought into contact at

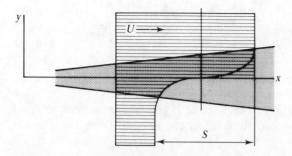

FIGURE 14.41
A free mixing layer.

the end of a splitter plate, they form a streamwise-growing turbulent shear layer, sketched in Fig. 14.41. For convenience, we introduce symbols for the mean speed, $U_m = (U_1 + U_2)/2$; and for the *shear*, $S = U_1 - U_2$.

If a Reynolds number based on S and a local thickness of the layer exceeds a few thousand, one might expect that the variation of mean velocity across the layer is independent of viscosity. Also, far downstream of the splitter plate the flow has had a chance to forget the details of its early history. Then, if the boundary conditions above and below the layer introduce no new characteristic lengths, one can postulate that the layer will occupy a wedge-like region, as depicted in Fig. 14.41. The layer just downstream of the splitter plate may not fit into this wedge, but an apparent apex of the wedge can be empirically located, and will not be too far from the end of the splitter plate. For further description, let us place the origin of a cartesian coordinate system at the apparent apex of the wedge, directing the x axis parallel to the flow in the upper, faster, stream. Assuming that the mean x-component of velocity, U, depends only on U_1, U_2, x, and y, we can use dimensional analysis to conclude that

$$U(x, y) = U_m + 0.5\, S f(\eta, R) \tag{14.36}$$

where $$\eta = \frac{y - \alpha x}{\beta x} \quad \text{and} \quad R = \frac{U_2}{U_1}$$

Measured profiles are quite well fitted[57] by a relatively simple odd function of η,

$$f = \tanh \eta (1 + \tfrac{2}{3} \operatorname{sech}^2 \eta) \tag{14.37}$$

The coefficient α specifies the slope of the locus on which $U = U_m$, while the coefficient β measures the rate of thickening of the layer. A commonly

---

[57] See M. Gaster, E. Kit, and I. Wygnanski, "Large-Scale Structures in a Forced Turbulent Mixing Layer," *Jour. Fluid Mech.* **150**: 23–39 (1985). Careful observation, and some reasoning dependent on the mean momentum equation and the boundary conditions on Reynolds stress suggest that the shape of the profile probably depends explicitly on R, and is not exactly a simple odd function around $\eta = 0$.

accepted evaluation of β is

$$\beta = (0.185 \pm 0.025)\frac{1-R}{1+R} \tag{14.38}$$

Little seems to be known about the value of α, except that it is typically negative and considerably smaller, in absolute value, than β.

For future reference, we note that mixing layers are frequently characterized by a *momentum thickness*, θ, defined by the integral

$$S^2\theta = \int_{-\infty}^{\infty} (U - U_2)(U_1 - U)\,dy \tag{14.39}$$

From the fitting formula, we calculate that $\theta = 67\beta x/270 = 0.248\beta x$.

An alternative parameter is the *vorticity thickness*, or *maximum-slope thickness*, δ_ω, defined by

$$\delta_\omega = \frac{S}{(\partial U/\partial y)_{\max}} \tag{14.40}$$

From the fitting formula, we calculate that $\delta_\omega = 2\beta x \approx 8\theta$.

ENTRAINMENT: INDUCED FLOW NORMAL TO THE LAYER. If we accept Eq. (14.36) as an adequate description of the variation of U, we can calculate the corresponding variation of V, namely

$$V(x, y) = 0.5S\{\alpha[f(\eta) - f(\infty)] + \beta[g(\eta) - g(\infty)]\} \tag{14.41}$$

in which

$$g(\eta) = \eta \tanh \eta - \ln \cosh \eta - \tfrac{1}{3}(\tanh^2 \eta - 2\eta \tanh \eta \, \text{sech}^2 \eta) \tag{14.42}$$

The functions $f(\eta)$ and $g(\eta)$ approach the asymptotic values, $f(\infty) = 1$, $g(\infty) = \ln 2 - \tfrac{1}{3}$. Since g is an even function of η, and V_1 is zero by definition, we find that $V_2 = -\alpha S$.

We can now calculate the rate at which fluid from either stream crosses the boundary of the mixing layer. We define the boundary to be the line on which $f(\eta) = \pm 0.99$, which gives $\eta = \pm 1.08$ and $g(\eta) = 0.361$, and then find

$$E_1 = \left(U\frac{dy}{dx} - V\right) \approx (1.07\beta + 0.99\alpha)U_1 \qquad \text{(for } \eta = +1.08)$$

$$E_2 = \left(V - U\frac{dy}{dx}\right) \approx (1.07\beta + 0.01\alpha)U_2 - \alpha U_1 \qquad \text{(for } \eta = -1.08)$$

We see that the entrainment may be asymmetric, and that it is likely that the greater flow rate will enter from the faster stream. The *entrainment ratio*, E_1/E_2, is a quantity of interest in further analysis of mixing and reaction; it depends only on R, since β, and presumably α, depend only on R. There is, of course, some arbitrariness in the choice of the value of η at which we set up an imaginary boundary between the mixing layer and the external streams, but this should not greatly affect the value of the entrainment ratio.

REYNOLDS STRESS PROFILES. SPECTRA. The dimensional argument that leads to Eqs. (14.36) and (14.40) leads us to expect that any component of the Reynolds stress tensor will equal S^2 times a function of η and R, and the spectral density of u' or v' will have the form

$$\phi(\omega; x, y) = \theta S F\left(\eta, R, \frac{\omega\theta}{S}\right)$$

These expectations almost define the term *fully-turbulent, self-similar mixing layer*. Typical distributions of U and $\langle u'^2 \rangle^{1/2}$ are shown in Fig. 14.42, and typical spectra in Fig. 14.43.

INTERMITTENCY. Because of the two-sided nature of the mixing layer, the concept of intermittency can be enlarged a bit. One can look for the fraction of time during which the flow appears to be rotational, getting the result shown in Fig. 14.42; or one can mark each stream chemically and study the fraction of time during which pure fluid from one stream or the other appears at any given point in the mixing layer. An elaboration of the latter concept is the three-dimensional presentation of the probability density for chemical concentration, shown in Fig. 14.3. This refers to a concentration variable that equals 1 in the faster stream and zero in the slower stream, and shows the probability of finding any specified concentration at any value of η. If the layer has reached the stage of self-similar development, the picture shown in Fig. 14.3 should be invariant to further increase of x.

SENSITIVITY TO BOUNDARY CONDTIONS. The mixing layer has been studied by many different people, using somewhat different wind or water tunnels and varied techniques of measurement and data reduction. Different measurements of nominally identical quantities have not always agreed, and it is believed that the discrepancies are not the result of uncertainties of measurement. It appears that the behavior of the mixing layer depends rather sensitively on small changes in boundary conditions, that had not initially been expected to be very important. The laminar or turbulent character of the boundary layer before it separates to form a mixing layer; the level of turbulence in the bounding streams; the acoustic environment; the spanwise width of the test section; the length of the test section; the proximity of walls parallel to the mixing layer: all seem to have some effect on the development of the layer. The reasons are still not fully understood, and the mixing layer is still a subject of active research.

MIXEDNESS. It is hard to be sure, when measuring nominally instantaneous concentrations with a particular sampling system, that measured values

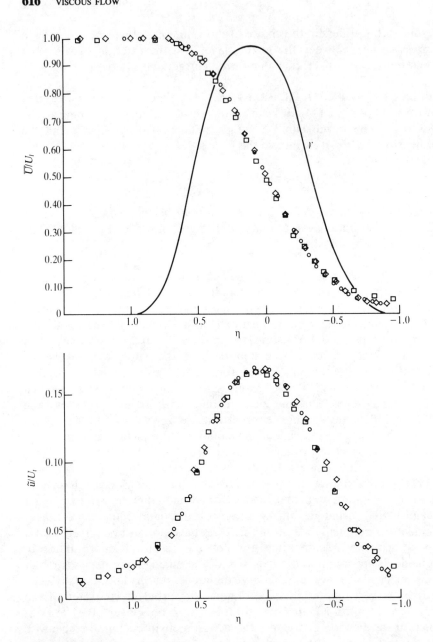

FIGURE 14.42
Self-similarity in a free mixing layer with $R = 0$. (a) Mean velocity and intermittency; (b) RMS fluctuation of streamwise velocity, $\diamond$ $\theta S/\nu = 2400$, $\square$ $\theta S/\nu = 3020$ $\bigcirc$ $\theta S/\nu = 3630$. (Adapted, by permission, from Champagne et al., 1976.)[58]

[58] F. H. Champagne, Y. H. Pao and I. J. Wygnanski, "On the two-dimensional mixing region." *J. Fluid Mech.* **74**: 209–250 (1976).

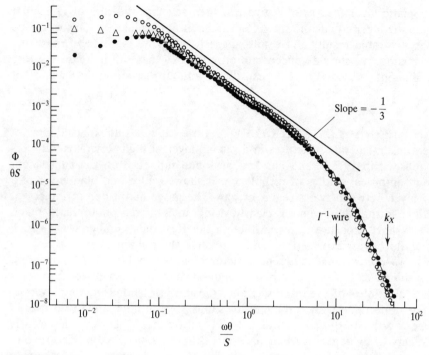

FIGURE 14.43
Frequency spectra of u'^2, v'^2, and w'^2 in a free shear layer. $R = 0$, $U/U_1 = 0.6$, $S\theta/\nu = 3630$. (Reproduced, by permission, from Champagne et al., 1976.)[58]

between 0 and 1 correspond to thorough mixing at the molecular level, rather than to the capture of an inhomogeneous sample that mixed after capture but before analysis. Recently, a more direct test for *molecular mixedness* has been employed. It depends on the appearance of a detectable product, formed by chemical reaction between trace constituents of the two streams.[59] If the chemical reaction is suitably ideal,[60] the total amount of reaction product encountered along a line parallel to the y axis will increase with x, at a rate proportional to the average rate at which molecular mixing is occurring along that line.[61] If we find that the amount of product increases at the same rate as

---

[59] R. E. Breidenthal, "A Chemically Reacting Turbulent Shear Layer," *AIAA Jour.* **17**: 310–311 (1978).

[60] The ideal is a reaction that has a constant, spatially uniform probability of producing a chemically stable, nonreactive product molecule whenever two reactant molecules collide.

[61] This terminology implies a pragmatic definition of *molecular mixing*: a process that brings marked masses of fluid so close together that molecules from one mass actually *collide* with those of the other.

the kinematic layer thickness, δ, we can then say that the rate of molecular mixing in the layer equals the rate of entrainment into the layer. This would be a very remarkable result, and would suggest that if we want to increase the rate of mixing, we should concentrate on measures that will increase the rate of entrainment, provided that we can do so without disrupting the equality of rates.

INSTANTANEOUS FLOW STRUCTURES. A dramatic and unexpected view of coherent instantaneous structures in a mixing layer at high Reynolds numbers was published in 1971,[62] and has had profound influence on the subsequent course of empirical studies of turbulence. It showed edge and plan views of a mixing layer between two streams of gas. The gases had different indices of refraction; so that the layer was clearly visible in Schlieren photographs taken with very short exposure times. Although the Reynolds number was so high that the flow was expected to show an entirely random appearance, the roughly *two-dimensional* coherent structures shown in Fig. 14.44 appeared. Similar things had been seen before, but always at much lower Reynolds numbers, in studies of the instability and transition of laminar mixing layers.

The notion that truly turbulent flows must exhibit fluctuating velocities in all three coordinate directions was not violated by these pictures; they clearly show plenty of evidence of *small-scale* three-dimensional motion. The revolutionary implication of the pictures was that the *entrainment* into the mixing layer, the corresponding *growth* of layer thickness, and the development of *Reynolds stresses,* might all be dominated by nearly two-dimensional pro-

FIGURE 14.44
Instantaneous plan and side views of a turbulent mixing layer.

[62] G. L. Brown and A. Roshko, "The Effect of Density Differences on the Turbulent Mixing Layer," *Turbulent Shear Flows. AGARD Conf. Proc.* **93**: 23-1–23-11 (1971). See also G. L. Brown and A. Roshko, "On Density Effects and Large Structure in Turbulent Mixing Layers," *J. Fluid Mech.* **64**: 775–816 (1974).

cesses. The apparently great *spanwise coherence* of the visible structures suggested also that one might learn how to manipulate them.

An interesting controversy quickly arose, concerning the precise, or the most fruitful, meaning of the term *spanwise coherence,* because no suggestion of any extraordinary degree of coherence had ever been discovered in the behavior of those spectra, spanwise spatial correlations, or space–time correlations that had been measured in the past. Indeed, studies of classical correlation maps had already led to a mental picture of the dominant structural features of this flow. This picture bore no resemblance to the new one. The gradual resolution of the controversy, documented in a dozen or more publications, is extremely interesting to anyone who makes, or intends to make, detailed measurements of turbulent flows. We cannot go into detail here, but only list a few important conclusions, noting that even these may still be somewhat tentative.

If both boundary layers on the splitter plate are laminar, and the turbulence level of the streams is low, the mixing layer goes through a formative stage, with the following characteristics.

1. The vorticity in the mixing layer coalesces into extremely coherent spanwise *rolls* or *billows*, the initial spacing of which can be estimated, by application of linear instability theory, to be about ten times the momentum thickness of the boundary layer that contributes the greater flux of vorticity. The billows move downstream at about the mean speed of the mixing layer. At this stage, the layer is easily manipulated by artificial two-dimensional disturbances of frequency equal to that at which the first spontaneously-occurring billows pass a fixed point. In the absence of such artificial disturbances, the billows move with enough dispersion in speed and y-position so that they cluster in small groups, usually just pairs, which orbit and coalesce into a single, larger billow. Fluid from the irrotational streams is drawn into the growing cores of the billows, especially during clustering events, as is shown in Fig. 14.45.

Although the scenario at any instant exhibits a very distinctive and regular pattern, there is sufficient random variation in the structure, position, and arrival frequency of the billows that pass by any fixed point of

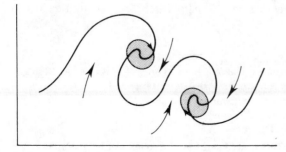

FIGURE 14.45
Rapid entrainment during the clustering of vorticity concentrations.

observation that measured velocity spectra do not reveal any preferred frequencies. The clustering events reoccur, forming larger coherent structures as the flow moves downstream, so that the mean diameter of the billows that pass a fixed point is linearly proportional to x.

2. There are significant single-point statistical differences between the mixing layer at this formative stage, and the layer that evolves sufficiently far downstream. Specifically, in the formative layer the level of the Reynolds stress R_{xy} is anomalously high, while the ratio of the rate of molecular mixing to that of entrainment is anomalously low. The rate of decay of the velocity spectrum with frequency, at the highest frequencies for which the normalized spectrum is independent of viscosity is anomalously high.[63]

3. The randomness that hides these events from the observer of frequency spectra and layer growth can be greatly reduced by introduction of high-fidelity periodic disturbances, in which case the layer can be caused to grow in a nearly stepwise way, and sharp spectral lines appear.[64]

4. The role of viscosity in all this appears to be mainly that of an ancestor, rather than that of an active participant in current events. By its influence on the thickness of the laminar boundary layers that leave the splitter plate, viscosity determines the initial spatial scales of the events that follow.

5. At a position where $x \approx 8(1 - R)\lambda/(1 + R)$, λ being the initial spacing of the billows, small-scale three-dimensional motions become clearly evident. Pictures of this are shown in Fig. 14.46. At about 1.5 times this value of x, the anomalies in Reynolds stress, mixedness, and spectral rolloff have disappeared. According to these criteria, the turbulent mixing layer achieves its self-preserving form after about *two* pairing or clustering events.[65] Another specification is that the self-preserving form appears where $x \approx (300 \text{ to } 400)(1 - R)\theta_i/(1 + R)$, where θ_i is the initial momentum thickness of the mixing layer.[66]

If the boundary layers on the splitter plate are turbulent, the formative stage of the mixing layer exhibits anomalously *low* values of R_{xy}. The visual characteristics of the flow are fuzzier, but billow formation and clustering is still discernible, and the length of the formation zone is about the same as when the boundary layers are laminar.

The controversy about the degree of *spanwise coherence* that charac-

---

[63] For the evidence to support these statements, and references to original works, see A. Roshko, "Structure of Turbulent Shear Flows—A New Look," *AIAA Jour.* **14**: 1349–1357 (1976).

[64] For a review, see C.-M. Ho and P. Huerre, "Perturbed Free Shear Layers," *Ann. Rev. Fluid Mech.* **16**: 365–424 (1984).

[65] See Ho and Huerre, *op. cit.*, pp. 394–396.

[66] F. K. Browand & T. R. Troutt, "The Turbulent Mixing Layer: Geometry of Large Vortices." *J. Fluid Mech.* **158**, 489–509 (1985).

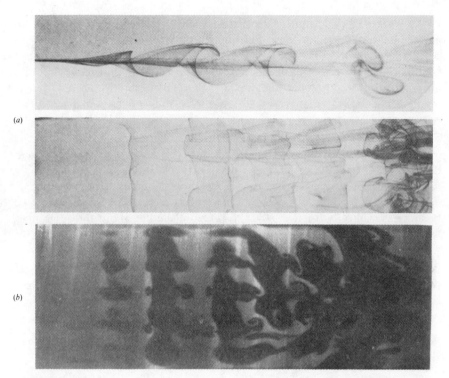

FIGURE 14.46
Pictures of the inception of three-dimensional motion in the mixing layer: (a) dye sheet produced by chemical reaction between streams; (b) fluorescence induced by sheet of laser light near $\eta = 0$, where one stream contains fluorescent dye. (Reproduced from J. C. Lasheras, J. S. Cho, and T. Maxworthy, *J. Fluid Mech.* **172**: 231–258, 1986.)

terizes the self-similar mixing layer inspired the following observations, which nearly add up to a resolution.

1. The observed correlation between simultaneous values of the velocity at the laterally displaced points (x, y, z) and $(x, y, z + \Delta z)$ depends very strongly on the value of y/δ, and on which component (u or v) of velocity is measured. Since the coherent motion is induced by vortex billows that move, on the average, along the trajectory $y = 0$, any coherent contribution to u will be maximum just above or below the billows, and will be minimum at $y = 0$. The first search for spanwise coherence involved measurements of u at $y = 0$; with 20/20 hindsight, we are no longer surprised that no remarkable coherence was found. Measurements of v at $y = 0$, or of u at $y = \pm\delta$ show much stronger spanwise correlation.[67] The correlations of v at

[67] I. Wygnanski, D. Oster, H. Fiedler, and B. Dziomba, "The Two-Dimensional Mixing Region," *Jour. Fluid Mech.* **41**: 283–325 (1970).

$y = 0$, where lots of small-scale three-dimensional motion is present, are greatly increased by low-pass filtering of the individual velocity records, before correlation. This focuses the search for *large-scale, low-frequency* coherent structures. Measurements taken in the irrotational, but still unsteady, flow just outside the mixing layer are naturally filtered, by the mutual cancelation of the far-field velocities induced by small-scale concentrations of vorticity.

2. The degree of spanwise coherence of u, measured in the external irrotational flow, decreases rapidly during the formative stage of layer development, and then increases again with increasing x. Eventually, the value of Δz for which the correlation coefficient equals 0.4 asymptotically approaches $(2.3 \pm 0.25)\delta_\omega$, which equals about 18θ. This value has as yet been measured only for $R = 1/8$, and is established at about the same place where the mixing layer begins a self-similar evolution.[68]

3. It is very helpful, in any effort to discern coherent structures in a turbulent flow, to have simultaneous observations from a large array of points, thereby getting a picture that is instantaneous, quantitative, and, to a degree, *synoptic*. For example, Fig. 14.47 shows an isometric plot of $u(z, t)$

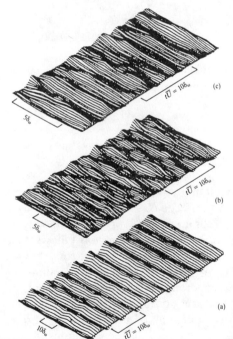

FIGURE 14.47
Relief model of distribution of $u(z, t)$ in the potential flow just above a shear layer. $R = 0.28$, $U\theta_i/\nu = 375$. (a) $x/\theta_i = 56$, (b) $x/\theta_i = 452$, (c) $x/\theta_i = 1580$. (Reproduced, by permission, from F. K. Browand and T. R. Troutt, 1980.[68])

[68] F. K. Browand and T. R. Troutt, (1985) op. cit..

at fixed x and y, constructed from data gathered by an array of 12 simultaneously recording hot-wire anemometers. This is the same array used to determine the decay of correlation with increasing Δz; in this more synoptic view, the data suggest *why* the correlation falls off as it does. There are indeed very long, coherent concentrations of vorticity, which induce the corrugated appearance of the $u(z, t)$ surface as they pass beneath the array of hot wires. They are, however, typically somewhat *skewed* to the z axis; occasionally they appear to end or merge. It is easy to imagine that this behavior is linked to the nature of the pairing or clustering of spanwise vortex filaments, which would proceed faster at spanwise locations where clustering accidentally gets a head start. If, as we believe, δ grows as a result of the clustering process; while the skewing of spanwise structures is caused by accidental disturbances of that process; we may not be surprised to find that the average length of coherent billows becomes proportional to their average streamwise spacing. This would in turn imply that the Δz for a fixed value of correlation coefficient would be proportional to θ.

MOLECULAR MIXING. A conceptually interesting model of entrainment and mixing[69] leads to some striking predictions for this sort of turbulent layer. In particular, it reminds us that even in a free turbulent flow, viscosity constrains the degree to which the deformation of fluid elements can enhance the potential for molecular mixing.

The model conceives the overall process of mixing as a sequence of three steps, or unit processes, which occur in a cross-stream slice of the mixing layer. The first is *entrainment*, which supplies pure fluid 1 at a rate $(aU_1 + bU_2)\, d\theta$ and pure fluid 2 at a rate $(aU_2 + bU_1)\, d\theta$. The second is *macroscopic deformation* of the batches of pure fluid, until the potential for molecular mixing between them and the fluid that surrounds them is as great as is permitted by the local values of S, θ, and ν. The third is *molecular diffusion*. The main assumption of the model is that the second process is much faster than either of the others. This implies that the molecular diffusion has only to transport molecules over a distance λ to achieve a molecularly uniform mixture. The mysterious length, λ, is taken to be Kolmogorov's lengthscale, which was discussed in Section 14.2. In terms of ε, the mean rate of energy dissipation per unit mass, we have $\lambda^4 = \nu^3/\varepsilon$. For the self-similar mixing layer, we can actually compute ε by making an energy balance on the cross-stream slice. The result is $\varepsilon = AS^3/\theta$.

The resulting rate of molecular mixing is thus constrained by either of two things: entrainment, proceeding on a timescale θ/S, or molecular diffusion, proceeding on a timescale λ^2/D, where D is the molecular diffusion

---

[69] J. E. Broadwell and R. E. Breidenthal, "A Simple Model of Mixing and Chemical Reaction in a Turbulent Shear Layer," *Jour. Fluid Mech.* **125**: 397–410 (1982).

coefficient. The ratio of the two timescales equals $B(D/v)(S\theta/v)^{1/2}$. We recognize $S\theta/v$ as a Reynolds number, which typically exceeds 10^4 in the self-similar mixing layer. Assuming that $B \approx 1$, we conclude that *entrainment is the slowest step unless* $D \ll v$. In gases, $D/v \approx 1$, and entrainment is the bottleneck. In liquids, however, v may exceed D by a factor of hundreds or more, so that entrainment and diffusion may both provide important constraints on the rate of the combined process. Thus, the mean profiles and probability distributions of chemical concentration in a kinematically self-similar turbulent mixing layer may depend on Reynolds number for liquids, but no such dependence is expected for gases. In particular, we should not be surprised to detect residual blobs of molecularly unmixed liquid, but not unmixed gas, in the middle of the billows.

Detailed investigations of chemical reactions in turbulent mixing layers are building up a fascinating picture of this practically very important process. We must, however, move on.

PERTURBED MIXING LAYERS. A side-effect of the intensive study of coherent structures in mixing layers has been the realization that the classic, self-similar layer may not be the type most frequently encountered in practice. Even for layers which should be past the formative stage, there appears to be a surprising scatter in rates of growth. This may be due to the proximity of walls parallel to the layer, pressure gradients in the nominally uniform streams, effects of sidewalls in planes $z = $ constant, etc.

One particularly interesting effect occurs when the mixing layer impinges on an edge or corner, or passes close by, as is sketched in Fig. 14.48. As each billow of concentrated vorticity approaches the corner, a system of image vorticity appears behind the corner, to prevent flow through the wall. The velocity induced by the image vorticity constitutes a weak but nearly periodic fluctuation on the flow as it leaves the splitter plate. The possibility of a feedback loop appears, and in some cases the randomness of the flow is greatly reduced, even to the extent that clustering of billows ceases after a certain stage, and the layer thickens very little thereafter.[70]

$U_1 \longrightarrow$

FIGURE 14.48
Immature turbulent shear layer interacting with downstream edge of a cavity.

---

[70] Ho and Huerre, *op. cit.*, pp. 394–396.

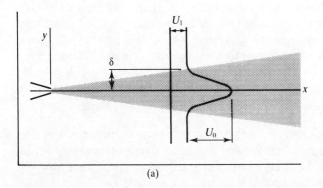

(a)

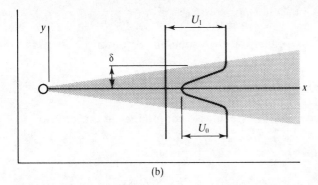

(b)

FIGURE 14.49
(a) Jet; (b) wake.

JETS AND WAKES. The simplest kinds of turbulent jets and wakes may be represented schematically by Fig. 14.49. Although many other cases are practically important, we consider only the geometrically simplest possibilities, in which the mean flow is either axisymmetric, or shows symmetry about the plane $y = 0$. We shall assume that the only bounding walls are stationary ones, which separate the inner and outer streams upstream of $x = 0$. When the inner flow is faster, we call it a *jet*; when it is slower, we call it a *wake*.

If the boundary layers on the walls are very thin compared to the initial diameter or width of the inner flow, a close view of either edge of the jet or wake may reveal a turbulent mixing layer. The layers from opposite sides grow toward each other, interact and merge. This process, particularly in the case of fast jets, is often violently unsteady and noisy, providing motivation for intensive efforts to identify and modify the instantaneous motions that are the strongest sources of sound.[71] In a round jet, issuing from a nozzle of diameter

---

[71] See S. C. Crow and F. H. Champagne, "Orderly Structure in Jet Turbulence," *Jour. Fluid Mech.* **48**: 547–591 (1971). Also see J. Laufer and T.-C. Yen, "Noise Generation by a Low-Mach-Number Jet," *Jour. Fluid Mech.* **134**: 1–31 (1983).

d, the core of potential flow, which occupies most of the nozzle exit plane, disappears at about $x/d = 4$. In a range $4 \le x/d \le 30$, the limits of which are quite approximate, there is a gradual reshaping of the profiles of mean velocity, Reynolds stresses, temperature or chemical concentration, etc. Following this, there may be a region in which these profiles, suitably scaled, become independent of x. Let us investigate the requirements for self-similarity.

Self-similar jets and wakes. Jets and wakes in general are characterized by the constant speed, U_1, of the external stream; the local range of speeds covered by the profile, $U_0(x)$; and the local width, $\delta(x)$, of the profile. The proposed self-similar form of the velocity profile is then

$$U(x, y) = U_1 + U_0(x) f'(\eta) \qquad \text{where} \qquad \eta = \frac{y}{\delta(x)}$$

From the continuity equation, we find the corresponding form for V. For *plane flow*, this is

$$V(x, y) = -f(\eta) \frac{d(U_0 \delta)}{dx} + \eta f'(\eta) U_0 \frac{d\delta}{dx}$$

The corresponding expression for the streamwise component of acceleration is

$$A_x = U_1 \left(f'(\eta) \frac{dU_0}{dx} - \eta f''(\eta) \frac{U_0}{\delta} \frac{d\delta}{dx} \right)$$
$$+ f'^2(\eta) U_0 \frac{dU_0}{dx} - f(\eta) f''(\eta) \frac{U_0}{\delta} \frac{d(U_0 \delta)}{dx}$$

This expression can represent the acceleration of a self-similar flow only if it can be reduced to the form of a function of x times a function of η. The expression contains four different functions of η, each multiplied by a different function of x. Each of these functions of x must, therefore, either vanish or be proportional to each of the remaining functions of x. The only one that can vanish without the unphysical requirement that U_0 or δ be constant, are the two that contain U_1. The case in which $U_1 = 0$ is that of a jet entering a fluid at rest. For that case, the additional requirement is that $\delta\, dU_0/dx$ be proportional to $d(U_0 \delta)/dx$, which is true if $\delta U_0^{1-\alpha}$ equals a constant. The value of α is constrained by the form of other terms in the momentum equation.

Since turbulent jets and wakes turn out to be *slender*, the x-component of the mean momentum equation can be somewhat simplified. The traditional form of the simplified equations was established by Townsend,[72] and is

$$\rho A_x + \rho \Pi_{,x} = R_{xy,y} + R_{xx,x} \tag{14.43}$$

---

[72] A. A. Townsend, *The Structure of Turbulent Shear Flow*, Cambridge University Press, London, 1956.

$$\rho\Pi - R_{yy} = \rho\Pi_\infty \qquad (14.44)$$

For the flows of interest here, it will be assumed that the outer stream is in hydrostatic equilibrium, so that $\Pi_{,x} = 0$. Then (14.43) reduces to

$$\rho A_x = R_{xy,y} + (R_{xx} - R_{yy})_{,x} \qquad (14.45)$$

Finally, we shall drop the last term of this, depending on observations that it is much smaller than the first term on the right.

We continue with the case $U_1 = 0$, assuming that the logical scaling of R_{xy} in a self-similar flow will be $R_{xy} = \rho U_0^2 g(\eta)$. Inserting this, and our expanded expression for A_x, into the truncated x-momentum equation, we have

$$\alpha f'^2(\eta) - (\alpha + \beta)f(\eta)f''(\eta) = g'(\eta) \qquad (14.46)$$

in which $\alpha = (\delta/U_0)\, dU_0/dx$ and $\beta = d\delta/dx$. For self-similarity, α and β must be constants. To see what values they may have, we integrate (14.46), using the boundary conditions $f(0) = 0$, $f'(0) = 1$, and $g(0) = 0$. The last condition comes from the realization that g, whatever its form, must be an odd function of η, given the obvious symmetry of the flow. The result is

$$g(\eta) = -(\alpha + \beta)f(\eta)f'(\eta) + (2\alpha + \beta)\int_0^\infty f'^2(\eta)\, d\eta$$

Now we apply the boundary conditions for large η, which describe the flow outside of the jet. These are $f' \to 0$, fast enough to make ff' vanish; and $g \to 0$, so that the Reynolds stress vanishes. The conclusion is that $2\alpha + \beta$ must equal zero. Since $\delta = \beta(x - x_0)$, this gives $(x - x_0)\, dU_0/dx = -U_0/2$, which implies

$$\delta = \beta(x - x_0) \qquad \text{and} \qquad U_0 = A(x - x_0)^{-1/2}$$

Thus, if $U_1 = 0$, the plane jet spreads linearly, while the peak speed decays as the inverse square root of distance from an effective origin. These equations, together with the assumed forms for the local profiles of U and R_{yx}, must all be satisfied if we are to call the jet self-similar. Experiment shows that U-profiles can be collapsed in the assumed way, even for fairly modest values of x/d, say $x/d \geq 10$, provided that U_0 and δ are treated as locally disposable parameters, but that x/d must exceed 50 or more before the R_{yx} profiles collapse and the self-consistent behavior of δ and U_0 is observed.

Effect of the external stream. When $U_1 \neq 0$, but $U_1 \ll U_0$, the foregoing analysis gives a good approximation. However, the decay of U_0 with increasing x will eventually destroy this inequality, and will lead to another interesting limiting behavior, when $U_1 \gg U_0$. The analysis of this case is somewhat less straightforward, because it is not really obvious how to scale the Reynolds stress. Should we still try $R_{yx} = \rho U_0^2 g(\eta)$, or might it make better sense to try $R_{xy} = \rho U_0 U_1 g(\eta)$? Fortunately, the two choices lead to very different conclusions about jet spreading and speed decay, so one can appeal to simple experiments for a resolution. Let us first follow the traditional path, use the

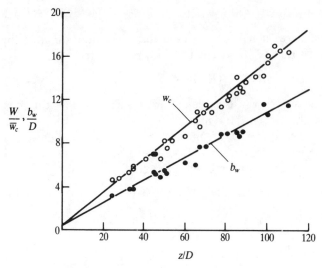

$\overline{w}_c$ = centerline velocity, b_w = radius at which $w(m, z) = e^{-1} w(0, z)$

FIGURE 14.50
Spreading of a round jet and decay of the centerline velocity. (Reproduced, by permission, from
P. N. Papanicolaou and E. J. List, 1988.)[73]

first scaling of R_{yx}, and now neglect those terms in A_x that do not contain U_1.
We derive

$$\alpha f' - \beta \eta f'' = g'$$

where now $\qquad \alpha = \left(\dfrac{\delta U_1}{U_0^2}\right) \dfrac{dU_0}{dx} \qquad$ and $\qquad \beta = \left(\dfrac{U_1}{U_0}\right) \dfrac{d\delta}{dx}$

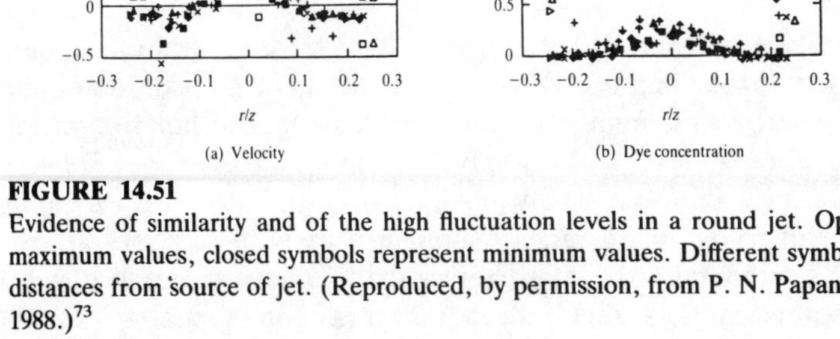

FIGURE 14.51
Evidence of similarity and of the high fluctuation levels in a round jet. Open symbols represent
maximum values, closed symbols represent minimum values. Different symbols represent different
distances from source of jet. (Reproduced, by permission, from P. N. Papanicolaou and E. J. List,
1988.)[73]

The integration and application of boundary condition leads to $\alpha + \beta = 1$, which implies that $U_0\delta$ is independent of x. If we insert $U_0\delta = A$ into the definition of β, we find

$$\delta^2 = 2\beta A \frac{(x - x_1)}{U_1} \quad \text{and} \quad U_0^2 = \frac{A U_1}{2\beta} (x - x_1)^{-1}$$

This is the same sort of speed decay as for a fast jet but a very different spreading law. We use a different subscript to identify the effective origin, which is presumably not at the same place as for the fast jet.

The analysis of the second choice, $R_{yx} = \rho U_0 U_1 g(\eta)$, follows these same lines, comes again to the conclusion that $U_0\delta = A$, but leads to linear growth of δ and inverse linear decay of U_0. Experiment favors the first scaling, and is supported in this by the argument that there would be no turbulence if there were no U_0, and therefore U_0 provides the logical scale for velocity fluctuations. This makes U_0^2 the logical scale for Reynolds stress, whether U_1 be relatively large or small.

Axisymmetric jets and wakes. For axisymmetric jets or wakes, we need to replace $R_{yx,y}$ by $y^{-1}(yR_{yx})_{,y}$. Then the analysis can be carried through as above, with the results:

$$\delta = \beta(x - x_0) \quad \text{and} \quad U_0 = A(x - x_0)^{-1} \quad \text{if } U_1 \ll U_0$$

and

$$\delta = \beta(x - x_1)^{1/3} \quad \text{and} \quad U_0 = A(x - x_1)^{-2/3} \quad \text{if } U_1 \gg U_0$$

Some sample data for round jets are shown in Figs. 14.50 and 14.51.

Since the survival of turbulence depends on the continued existence of high values of an appropriate Reynolds number, which in these flows is $U_0\delta/\nu$, we note the streamwise variation of $U_0\delta$ in each case.

In the fast plane jet	$U_0\delta$ increases, in proportion to $(x - x_0)^{1/2}$
In the slow plane jet or wake	$U_0\delta$ is constant
In the fast round jet	$U_0\delta$ is constant
In the slow round jet or wake	$U_0\delta$ decreases, in proportion to $(x - x_1)^{-1/3}$

We see that in the last case, the turbulence will eventually die out. Very little is known about this phenomenon, which may often be of only academic interest, for reasons that will now be discussed.

Effects of turbulence in the external stream. A phenomenon of great practical importance is the *dispersion* of weak jets and wakes by turbulence of the external stream. The external streams in most laboratory tests are carefully manipulated to reduce their turbulence, but real jets and wakes often appear against a background flow, such as the atmospheric wind or an ocean current, that has a significant level of turbulence. If the energetic scales of this turbulence are large compared to δ, the effect may be mostly a *meandering* of the jet or wake; if they are very small compared to δ, the effect may only be to enhance the rate of molecular mixing relative to that of entrainment. If they are about equal to δ, the external turbulence may rearrange the internally-generated concentrations of vorticity in a way that significantly modifies their mutually-induced motions, and hence has a major effect on the development of δ and U_0. The extreme case, in which the turbulence of the jet or wake is overwhelmed by that of the external flow, is often studied in the context of pollution control but has little to do with viscosity, so we drop it at this point.

Statistical properties of self-similar jets and wakes. The measurement of mean and fluctuating velocities is quite tricky in the outer regions of fast jets, where u may instantaneously be negative. With hot-wire anemometers one can resolve the direction of the velocity vector only when it is within about $\pm 30°$ of an unambiguously defined mean direction. In particular, one cannot distinguish backward flow from forward, and hence will overestimate U and underestimate u'^2.[73] This problem can be avoided by appropriate use of a laser-Doppler velocimeter, but then great care is needed to assure that the jet and the surrounding fluid are densely and equally seeded with light-scattering particles. The data presented here in Fig. 14.51a were taken with this kind of care. The dye-concentration data of Fig. 14.51b were also taken optically, by measurement of laser-induced fluorescence.

Coherent structures in jets and wakes. For the simple plane mixing layer, one can fairly readily conceive a model of coherent concentrations of vorticity that induce on each other, and in the surrounding potential flow, velocities that account for the observed motions of the billows and the observed rates of entrainment of irrotational fluid. One may naturally wonder whether this is also possible for at least the plane jets and wakes. It would be very interesting to see whether such a model would have the same or different qualitative features in the two limits, $U_1 \ll U_0$ and $U_1 \gg U_0$.

There is no doubt that the formative regions of jets and wakes exhibit coherent behavior, even at very large Reynolds numbers. Some famous examples are shown in van Dyke's *Album of Fluid Motion*. Statistical evidence of some kind of structure in self-similar fast plane jets has also been reported by a number of authors, but no satisfactory picture of the evolution of large-scale concentrations of vorticity has been proposed to explain it.[74] The

---

[73] P. N. Papanicolaou and E. J. List, "Investigations of Round Vertical Turbulent Bouyant Jets," *J. Fluid Mech.* **195**: 341–391 (1988).

[74] See R. A. Antonia, L. W. B. Browne, S. Rajagopalan, and A. J. Chambers, "On the Organized Motion of a Turbulent Plane Jet," *Jour. Fluid Mech.* **134**: 49–66 (1983) for some data and a review.

behavior of these flows in the formative region, where structure is easy to see, does not suggest a scenario that might persist in the self-similar region, somewhat disguised by small-scale irregularities. Specifically, the early behavior features a growth in the *lateral* length scale of vortex patterns, but none in the *streamwise* length scale.[75] The latter change can only result from a process, such as vortex pairing or clustering, that limits the time during which any single concentration of vorticity can be identified as a separate entity. In the simple mixing layer, we get a clear view of such processes, and we learn that they contribute a great deal to the statistical picture of velocity fluctuations. In jets and wakes, the instability that leads to vortex pairing seems to be greatly reduced when the counterrotating vortices from opposite sides of the flow are arranged in the two suitably staggered parallel rows of the famous *von Kármán vortex street*. There are a few published observations of the eventual breakdown of these relatively stable vortex streets, and of the subsequent formation of new streets with larger length scales, but no clear description of the breakdown itself has been given.

It is very difficult to perceive any coherent motions in the self-similar region of a round jet or wake, possibly because of the violent instability of vortex rings during the process of pairing,[76] and because such flows are unstable to disturbances that gather vorticity up into a *helix*, rather than a succession of separate rings.[77]

BUOYANT PLUMES AND THERMALS. To a first approximation, no external force is exerted on the fluid that occupies a control volume bounded by two planes normal to the axis of a neutrally buoyant jet. When $U_1 = 0$, no axial momentum is carried into the jet with the entrained fluid, and the jet is characterized by a constant *momentum flux, J*. Suppose now that the jet runs vertically upward into an ambient fluid of density ρ_1, but has an initial density $\rho_0 < \rho_1$. We then characterize it by a specific *buoyancy flux*, $B = g(\rho_1 - \rho)Q/\rho_0$. Here Q is the volume flux, which may vary with x as a result of entrainment, and ρ is the average density in the plume, defined so that ρQ equals the mass flux of the plume. The rate of increase of mass flux equals $\rho_1 \, dQ/dx$, since ρ_1 is the density of the fluid being entrained. If the ambient fluid is unstratified, so that ρ_1 is constant, then B will be independent of x.

As it enters the atmosphere, the buoyant jet is characterized by two given quantities, B and an initial value of the specific momentum flux, $M = J(0)/\rho_0$. We note that, whether the jet be round or plane, the quantity $\lambda = M^{3/4}/B^{1/2}$

---

[75] See Van Dyke's *Album*, pp. 56, 57.

[76] *Album*, p. 70.

[77] See A. E. Perry and T. T. Lim, "Coherent Structures in Coflowing jets and Wakes," *Jour. Fluid Mech.* **88**: 451–464 (1978), particularly Fig. 8.

serves as a reference length. The local momentum flux is now continuously increased by the buoyancy as the jet fluid rises, so that the initial momentum of the jet eventually becomes unimportant. This happens when $x \gg \lambda$. A buoyant jet with negligible initial momentum flux is called a *plume*.[78]

Consider a plume at positions far from the source, and at large Reynolds number. Let us try to discover the scaling laws for self-similar development, without specific reference to the Reynolds-averaged equations of motion. We shall use dimensional analysis, which could have been used to discover the scaling laws for self-similar jets and wakes. The basic assumption is that the gross features of the plume depend only on B and x.

For a round plume, B has the dimensions L^4/T^3; for a plane plume we use B to denote specific buoyancy flux per unit span, with dimensions $(L/T)^3$. Suppose one is interested in the local centerline velocity, U_0. We propose the relationship

$$F(U_0, B, x) = 0$$

For the plane plume, elimination of the dimension T between U_0 and B also eliminates the dimension L, so x does not belong in the proposed relationship, which becomes simply

$$U_0 = aB^{1/3}$$

By the same sort of reasoning, we discover that B does not belong in a relationship that determines the characteristic plume width δ; instead, we simply find $\delta = bx$.

For a round plume, we find the same linear law of spreading, presumably with a different value of b, say b'. The corresponding variation of U_0 now involves both B and x, and is

$$U_0 = a' \left(\frac{B}{x} \right)^{1/3}$$

The local volume flowrate, Q, is proportional to $U_0 A$, where A is the cross-sectional area of the plume. A is proportional to δ for plane plumes, and to δ^2 for round ones. Invariance of the buoyancy flux determines that the local average density deficit is proportional to $\rho_1 B/gQ$. The resultant similarity laws are thus

$$\rho_1 - \rho = \frac{cB^{2/3}}{x} \quad \text{(plane)} \qquad \text{and} \qquad \rho_1 - \rho = \frac{c'B^{2/3}}{x^{5/3}} \quad \text{(round)}$$

We can now see how the momentum flux increases, noting that dJ/dx is

---

[78] For more information, see Chapter 8 of J. S. Turner, *Bouyancy Effects in Fluids*, Cambridge University Press, London, 1973.

proportional to $(\rho_1 - \rho)gA$. This gives

$$J = dB^{2/3}x \quad \text{(plane)} \qquad \text{and} \qquad J = d'B^{2/3}x^{4/3} \quad \text{(round)}$$

These similarity laws are quite well supported by experiment, although there is considerable scatter among measured values of the proportionality constants. Very recent studies have included simultaneous optical measurements of velocity and of the concentration of a fluorescent dye added to the jet.[79] This has permitted direct measurement of the contribution of turbulent fluctuations to the vertical transport of dye and to the increase of momentum flux. The same experiment shows that entrained water, entirely free of dye, appears occasionally on the centerline of the jet or plume, no matter how large the radius of the latter becomes.

The behavior of plumes is important in environment fluid dynamics, where complicating factors such as crosswinds or density stratification of the ambient fluid are often present.[80]

Thermals. A plume or buoyant jet is maintained by some steady source of buoyancy, such as flow from a chimney. If the source of buoyancy is impulsive, like an explosion, an isolated body of buoyant fluid is generated. It is called a *thermal*, and may rise and grow in a self-similar way, which we shall treat very briefly, just to reinforce your command of dimensional analysis. We consider only a round thermal.

The thermal may be characterized by a volume V, and a mean density ρ. If the ambient density, ρ_1, is constant, we can write $dm/dt = d(\rho V)/dt = \rho_1\, dV/dt$. It follows that the total buoyant force, $F = (\rho - \rho_1)gV$, is a constant that characterizes the thermal.

As the thermal rises, the ambient fluid is accelerated and deformed. Assuming that the Reynolds number is very large, we neglect the force required to deform the surroundings, and postulate that the velocity of rise of the thermal, U_0, depends only on F, ρ_1, and the distance the thermal has risen, x.[81] A very quick dimensional analysis then shows that

$$U_0 = a\left(\frac{Fx}{\rho_1}\right)^{1/2} \qquad \text{and} \qquad \delta = bx$$

---

[79] The transition from jet to plume is beautifully documented, and a critical survey of the empirical constants is given by P. Papanicolaou and E. J. List," (1988) *op cit.*

[80] See E. J. List, "Turbulent Jets and Plumes," *Ann. Rev. Fluid Mech.* **14**: 189–212 (1982), and Turner (1973), *op. cit.*, for a discussion of these factors.

[81] We could use t, the time the puff has been rising, instead of x; but we don't need them both. Why not?

14.8 NUMERICAL SIMULATIONS OF TURBULENT FLOW

As the speed and storage capacity of digital computers increase, and as more efficient numerical algorithms are developed, the possibility of direct numerical simulation of individual realizations of a statistically simple turbulent flow increases, at least for comparatively low Reynolds numbers. Three such simulations, of fully-developed, pressure-driven flow in two-dimensional ducts with straight walls and with curved walls, and of flow in a boundary layer under an unaccelerated potential flow, have recently been made, with encouraging results.[82] Intensive efforts to see what may be learned from such simulations are being undertaken at the time of this writing.[83] They range from efforts to identify, if possible, typical instantaneous configurations of vorticity, through computations of the resonse of typical laboratory instruments exposed to the computed flow, to attempts to discover useful interrelationships of the statistical properties of the flow.

You may never need to know how to devise or to execute such a simulation, but you may need to evaluate or interpret the results of one, just as you need to interpret the results of laboratory experiments. This section calls to your attention a few key issues that must be kept in mind when you make such interpretations.

Since we have no obvious need for all the detailed information that would be available from a successful simulation of a single realization, and would probably condense it into a limited set of statistical quantities, it is obviously tempting to try to devise a numerical simulation that gives the statistics directly, without reference to individual realizations. The effort to do so raises other issues, with which you should be at least nominally acquainted.[84]

INDIVIDUAL REALIZATIONS. The special difficulties associated with numerical simulation of an individual realization of turbulent flow are due to three facts.

---

[82] P. Moin and J. Kim, "Numerical Investigation of Turbulent Channel Flow," *J. Fluid Mech.* **118**: 341–377 (1982). R. D. Moser and P. Moin, "Direct Numerical Simulation of Curved Turbulent channel Flow," *J. Fluid Mech.* **175**: 479–510 (1984). P. R. Spalart, "Direct Simulation of a Turbulent Boundary Layer up to $R_\theta = 1410$," *J. Fluid Mech.* **187**: 61–98 (1988).

[83] To get the flavor of this enterprise, see P. Moin, W. C. Reynolds, and J. Kim (eds) "Studying Turbulence Using Numerical Simulation Databases—II," Report CTR-S88, Center for Turbulence Research, NASA Ames Research Center and Stanford University, Dec. 1988.

[84] Substantial reviews of this large subject are given, for example, by R. S. Rogallo and P. Moin, "Numerical Simulations of Turbulent Flows," *Ann. Rev. Fluid Mech.* **16**: 99–138 (1984), by A. J. Chorin, *Lectures on Turbulence Theory,* Publish or Perish, Inc., Boston (1975), and by W. C. Reynolds, "Computation of Turbulent Flows", *Ann. Rev. Fluid Mech.* **8**: 183–208 (1976).

1. Physically important variations of properties occur over distances and times that are very small compared to the spatial and temporal extent of the domain of interest. The number of computational elements needed to represent these details faithfully is very large.

2. Hydrodynamic instability makes the flow extremely sensitive to minute variations of initial and boundary conditions. This makes it difficult to interpret numerical experiments designed to test the accuracy of the simulation.

3. A single realization is of little use by itself, unless it can somehow be interpreted as an ensemble of statistically independent realizations, from which statistical averages can be drawn.

Let us consider each of these difficulties in turn.

Discretization. Any numerical approximation to a solution of the Navier–Stokes equations employs a discrete description, in which the flow is represented in some way by a finite number of computational elements. These elements may be sets of values of flow quantities defined on an array of nodal points in space and time, with set rules for interpolating between the nodes; they may be sets of continuous functions of position called modes, a linear combination of which satisfies the equations of motion on a fixed array of test points or in some specified integral sense; or they may be sets of point-attached functions that are transported through space in such a way that a linear superposition of them approximates the time-evolving distribution of vorticity. These classes of computational elements correspond to the three main classes of algorithm, called (a) finite-difference or finite-element algorithms, (b) spectral or pseudo-spectral algorithms, and (c) discrete-vortex or random-vortex algorithms. For a given number of computational elements, spatial and temporal resolution can be improved by optimizing the distribution of the test points, or by the selection of particularly appropriate continuous functions. Assuming that this has been done, the amount of realistic detail revealed in the computations depends on the number of elements one can handle. The question is: how many elements are necessary for the intended purpose? How does one prove that a sufficient number was used in any given computation?

For hydrodynamically stable flows, a straightforward demonstration of convergence is possible in principle for any well-designed algorithm. We have seen examples in Chapters 10 and 12. The calculation is repeated, with ever-increasing number of computational elements, until further increases have no significant effect on the computed values of the quantities of interest. Unfortunately, this procedure is expensive, and is often terminated with results that are merely suggestive, not conclusive.

For turbulent flows, the inherent physical instability presents a new difficulty. A realistic numerical simulation will embody this instability. When the number of computational elements is increased, the computed flow field will change slightly, and the differences between new and old fields at any given instant may correspond to disturbances that would trigger an instability in the real flow. As the computations mimic the growth of these disturbances, their results diverge. As a result, a straightforward demonstration of numerical convergence of a single computed realization of turbulent flow has never been given, and it is not certain that it is possible, even in principle. In the absence of computational demonstrations of convergence, a qualitative comparison is made to physical observations, and the number and disposition of computational elements is thought to be adequate if the computation reveals realistic-looking spatial and temporal variations at least as sharp as those deduced from experimental measurements. This is usually viewed as a temporary expedient, because of the known or suspected limitations of experimental resolution, which have been mentioned in the last two sections.

Some confidence in the correctness of a computed single realization may be gained if it is possible to show that something like the probability distribution of a velocity component at a given point becomes independent of further increase of the number of computational elements in a simulation which is run for a long time with nominally constant boundary conditions. Unfortunately, this is a particularly expensive type of test, because even a perfect computational scheme will only give reliable statistical measures when the sample size is sufficiently large. At this level of prediction, one has access to a larger collection of experimental data as a standard for comparison, although great care must be taken to ensure that the data is free of instrumentation error, and that there are no systematic discrepancies between the boundary conditions imposed on the real, and the calculated, flows. When discrepancies between computation and measurement are found, one is not yet quite sure which, if either, to trust.

Boundary conditions. Frequently, the spatial domain of a computational simulation is only a portion of a larger flow field: Schemes that employ an array of nodal points usually require that values of some of the dependent variables, or of their normal derivatives, be specified at nodes that lie on the boundary of the domain, in order that corresponding values may be computed in the interior. The question is: how does one get the necessary values for boundary points that lie in the fluid?

For an inflow boundary, where fluid is entering the domain of interest, mean-flow values might be deduced from a relevant laboratory experiment. For example, one may know the mean velocity distribution in the exit plane of a pipe, from which a jet issues into the domain of interest.

However, if the pipe flow is turbulent and a single realization of the jet flow is to be simulated, one needs to specify the turbulent fluctuations in the exit of the pipe. These cannot be known experimentally at all the points where they are needed. How can one proceed?

It is conceivable that the measured velocity fluctuations in a fully-developed pipe flow could be mimicked, statistically, by drawing random values from probability distribution functions appropriate to the distance from the pipe axis. This, however, would not allow for the likely possibility that the real fluctuations in the exit of the pipe are affected by the behavior of the jet downstream. Thus, a special experiment might have to be executed, to provide the statistical distributions from which to make the random choices.

Some of these difficulties are much reduced if the inflow boundary can be located in a region of irrotational flow, or of a laminar rotational flow. This topic quickly develops into something far beyond the scope of this text, but you have been warned to be suspicious of computations of turbulent flow, for which the treatment of inflow boundary conditions is not clearly described.

At an outflow boundary, the flow is almost inevitably rotational, and often turbulent. Conditions on such a boundary cannot be independent of the behavior of the flow inside the domain of interest, and they cannot be specified arbitrarily. For example, they must allow an incompressible fluid to escape from a fixed control volume at the same rate it is allowed to enter by the specified inflow conditions. In the numerical simulations of turbulent flow that have been performed to date, that is achieved by specifying periodic boundary conditions. The flow in the outlet plane at any instant is assumed to be identical,[85] point by point, to that in the inlet plane.

Computational simulations of turbulence, and even of the early stages of a spatially amplifying instability, cannot presently utilize a very large computational domain. Thus, it cannot be expected that the outflow boundary conditions have a negligibly small effect on the flow in any part of the domain. This has been brilliantly demonstrated in a recent computational study of a spatially growing mixing layer.[86] When the computational domain was sufficiently long, the computed flow exhibited a sustained and apparently realistic instability, in spite of inflow boundary conditions that were specified to be perfectly steady. However, a mixing layer that develops in an unbounded domain is an example of a convectively unstable flow, which should remain steady if no disturbances enter from upstream. Because of this, the authors were suspicious. Careful study of the flow history, particularly that of the vorticity and pressure, disclosed a computationally induced feedback loop. Whenever a concentrated vortex passed through the downstream boundary, a pressure perturbation was generated in order to keep the postulated downstream boundary conditions in effect. This perturbation would appear simultaneously, but somewhat spread out, at the upstream boundary, where it was incompatible with the imposed condition of steady flow. This incompatibility

---

[85] In his study of turbulence in a growing boundary layer, Spalart (1988, *op. cit.*) introduces a special coordinate system and some scaling of dependent variables before applying periodic boundary conditions in the streamwise direction.

[86] J. C. Buell and P. Huerre, "Inflow/Outflow Boundary Conditions and Global Dynamics of Spatial Mixing Layers." In CTR-S88, *op. cit.*, 19–28.

was resolved by the generation of vorticity perturbations at the upstream boundary, and these perturbations were then amplified as they were convected through the unstable mixing layer. The feedback process had a characteristic period $T = L/U_m$, where L is the length of the computational domain, and U_m is the mean convective speed in the mixing layer. When this period was made to equal a sufficiently small integer multiple of the period of the most rapidly amplifying instability wave, the character of the computed flow changed radically, just as the character of a real mixing layer may be changed by an obstacle that interferes with its unconstrained development.

It is not yet known how to devise outflow boundary conditions that will avoid this kind of computational artifice, but the problem has been so well documented that a solution may soon be found.

Although an outflow boundary is not an essential feature of Lagrangian simulations that employ discrete computational elements of vorticity, one is sometimes introduced to identify a domain within which the small-scale details of the vorticity distribution are of no direct interest. Within this domain, elements are combined in some way, to reduce their number while retaining an accurate approximation to their effect in the domain where effects of all scales are of interest. In some calculations of flow in channels, vortex elements have simply been discarded after they pass an outflow boundary—a procedure that is obviously wrong in principle, but that seems to have little effect on the statistics of the flow sufficiently far upstream of the outflow boundary. Such apparently harmless, though erroneous, procedures should always be viewed skeptically until it is explained why they work.

One way to avoid, or at least to postpone, these problems is to propose that the flow repeats itself periodically, as a function of one or more spatial coordinates. This makes sense, at least statistically, in studies of pipe and channel flows, providing that wavelengths at which the flow is required to repeat itself are long compared to important physical scales of the instantaneous flow. This approach fits well with the representation of spatial distributions by sums of periodic functions, the longest period being that at which the entire flow repeats itself. The three massive calculations referred to in the introduction of this section all use this approach.

Initial conditions. If the flow is already turbulent at time zero, initial values of quantities defined at nodal points will have to be assigned in a statisticaly appropriate way, probably by drawing random numbers from appropriate distributions, with the constraint that the continuity equation be satisfied. If the simulation employs modal functions, the initial amplitude of each must be assigned; if it employs discrete vorticity elements, their initial positions and strengths must be given.

In practice, this obviously complex procedure is much simplified—because the purpose is often to gather statistical data from a calculation that has continued, with stationary boundary conditions, until the statistics become independent of the specific initial conditions. Faith that this will eventually occur is based on laboratory experiments. Whether the startup period, during which the simulation is statistically unstationary, can be shortened by a clever choice of initial conditions is as yet unproved, but it is common to initialize one calculation with the output of another.

Alternatively, one may wish to gather statistical data from the startup period, in which case the simulation must be rerun many times from slightly different initial data sets. The Lagrangian schemes, which simulate the diffusion of vorticity by random walk of discrete vorticity elements, embody an easy way to generate different realizations under identical nominal initial and boundary conditions: one simply changes the point of entry to the computer's random number string.

STATISTICAL SAMPLING. It is obviously expensive to run a simulation until (1) it represents a statistically stationary flow, and (2) the sample size, drawn from the stationary flow, is large enough to define probability distribution functions with adequate precision. If the flow is statistically independent of one or more spatial coordinates, this fact can be exploited to enlarge the sample size. You should look closely for a demonstration of both stationarity and precision before basing you own work on statistics from a simulation.

ADVANTAGES OF A COMPUTATIONALLY GENERATED DATA BASE. The foregoing warnings of potential pitfalls are important, but it is also important to appreciate some of the reasons for hope that numerical simulations will facilitate understanding of turbulent flow. One, which has a strong emotional appeal, is the fact that the computation provides both a view of local details, and an overall, synoptic, view at the same time. This is very hard to obtain experimentally. The computation can also be made to reveal values of quantities, such as vorticity, that are very hard to measure. It is felt, although far from proven, that the human mind will be able to sort out and condense this mass of new data. As a first step, much effort is being invested in a variety of three-dimensional graphical presentations, in the hope that the eye will assist the brain.

LARGE-EDDY SIMULATIONS. For many applications, the classical example being numerical weather prediction, there is no prospect of managing the number of computational elements needed to resolve all significant scales of motion. Years of research have shown that one cannot simply ignore the small-scale motions, by deliberately employing an inadequate number of computational elements. The resulting simulation of large-scale motions is likely to be very unrealistic; the specific symptoms of this depend on the details of the simulation algorithm.[87] One difficulty, which is fairly easy to imagine,

---

[87] Rogallo and Moin, *op cit.*, discuss some examples.

arises from an inadequate representation of the local rate of viscous dissipation. In a relatively crude simulation, this rate will be evaluated as μ def $\mathbf{u}_r \cdot$ def $\mathbf{u}_r$, where $\mathbf{u}_r$ is called the *resolvable velocity*. Suppose $\mathbf{u}_r$ to be a good approximation to an average of $\mathbf{u}$ over a small but finite volume. That average value might not vary much from point to point of a relatively crude computational grid, in which case the computed value of energy dissipation would be small. If $u_r h / v \gg 1$, where h is the grid interval or the wavelength of the shortest element in a spectral method, it is very possible that much of the spatial variation of $\mathbf{u}$ will be overlooked by the calculation, and that the real value of the energy dissipation will greatly exceed the calculated value. Since realistic variations of large-scale flow features, such as the mean centerline velocity of a jet, correspond to significant average rates of viscous dissipation, a theory which cannot predict the latter is very unlikely to be able to predict the former.

The alternative to ignoring the unresolved motions is an effort to discover some rule that describes their effect on the resolved motions, in terms that involve only the properties of the latter. The search for such a rule is an active one, and seems promising. One approach, designed for use with spectral methods, embodies an effective viscosity which is largest for the spectral elements with the shortest wavelengths. With even a few trustworthy fully-resolved solutions of the Navier–Stokes equations to serve as calibration standard for the simplest flows, and with ever faster and more capacious computers, it may be possible that large-eddy simulations, which resolve the fluctuations that contribute most of the turbulent energy and Reynolds stresses, and somehow assign the unresolvable viscous dissipation to the correct locations at the correct times, will provide a level of predictive power that is both practically rewarding and intellectually satisfying. This speculation, however, goes far beyond the scope of this text.

14.9 SUMMARY

Our knowledge of turbulence is based largely on experimental examination of flows that are of direct practical interest, or that are thought to embody some essential feature of a practically interesting flow. The examination usually develops a statistical picture, and supplements that with pictures of what are thought to be typical instantaneous conditions. The student of turbulent flows needs to become familiar with a variety of statistical concepts, and with a variety of measurement tools. It is important to appreciate both the intrinsic limitations of inference based on statistics, and the practical factors that limit one's ability to measure the desired statistical quantities.

Our ability to obtain enormous amounts of quantitative experimental data, and to supplement this with synoptic views that capture an unprecedented amount of detail, has been greatly enhanced in recent decades, largely because of lasers and computers. Nevertheless, the class of turbulent flows for which we have reliable and detailed experimental data is not large, consisting mostly of simple shear flows such as channel flows, boundary layers, jets, plumes and wakes. The urge to obtain finely resolved measurements of fluctuating quantities, particularly very close to walls, conflicts with the urge to obtain data at higher and higher Reynolds numbers.

A variety of useful empirical generalizations can be used to describe the simpler flows, but the scope for accurate application of these formulas is disappointingly small when a new design task needs to be faced. Although an engineer would often be satisfied to be able to predict the statistically averaged behavior of a flow, that behavior is only constrained, but not fully determined, by equations derived by averaging the equations for instantaneous flow. The hope that generally useful hypotheses might be discovered, to allow closure of the Reynolds-averaged equations, is still unfilled. On the other hand, a few highly idealized problems have been successfully analyzed numerically, using the full, unaveraged, Navier–Stokes equations, and there seems to be a promising future for large-eddy simulations. Meanwhile, modern experimental investigations of the instantaneous flow seem to have suggested some practical means of manipulating turbulent flows.

Viscosity is essential to the birth of much turbulent flow, but it then imposes important constraints on the high local rates of deformation that characterize such a flow. Its local effect on the distribution of mean-flow properties is typically very small, except near smooth solid walls, because turbulence is associated with large values of the Reynolds number.

EXERCISES

14.1. Verify Eq. (14.4), using Eq. (14.3) to evaluate the integrand.

14.2. Verify Eq. (14.6), using Eq. (14.3) to evaluate the integrand in Eq. (14.5).

14.3. (a) Show that the probability density function for the harmonically-oscillating signal, $p = P + p_0 \sin \omega t$ approaches the value $f(p) = (1/\pi)[p_0^2 - (p - P)^2]^{-1/2}$
(b) Calculate the mean and variance of $p(t)$, both by time-integration of the original signal, and by evaluation of the equations

$$\langle p \rangle = \int_{-\infty}^{\infty} pf(p)\,dp \quad \text{and} \quad \langle p'^2 \rangle = \int_{-\infty}^{\infty} (p - \langle p \rangle)^2 f(p)\,dp$$

14.4. Suppose symbols U_A and U_B are defined for flow in a mixing layer, to represent averages of velocity u over those time intervals during which the velocity sensor is immersed in fluid A and fluid B respectively. Let fT denote the time that the sensor spends in fluid A, and $(1-f)T$ the time spent in fluid B. T is the total sampling time.
   (a) Show that the conventional average of u, over all of T, is given by

$$U_C = fU_A + (1-f)U_B$$

   (b) Show that the conventional variance of u, obtained by averaging $(u - U_C)^2$ over all of T, is given by

$$\langle u_C'^2 \rangle = f\langle u_A'^2 \rangle + (1-f)\langle u_B'^2 \rangle + f(1-f)(U_A - U_B)^2$$

where $\langle u_A'^2 \rangle$ is the average of $(u - U_A)^2$ over the time when the sensor is in fluid A, and $\langle u_B'^2 \rangle$ is the average of $(u - U_B)^2$ over the time when the sensor is in fluid B.

14.5. Using the formula for the Kolmogorov length scale, $\lambda = (\nu^3/\varepsilon)^{1/4}$, and assuming that the mean rate of energy dissipation per unit mass is given by $\varepsilon = AU^3/L$, show how λ/L depends on the Reynolds number UL/ν.

Discuss the relevance of this finding to the task of detecting whether a movie sequence of a volcanic eruption shows the real thing, or a small-scale model.

14.6. Consider statistically stationary, fully-developed turbulent pipe flow of a perfect gas. Let V be the mean speed of the flow through a pipe of length L and diameter D.

Using $\Delta p = (\rho V^2/2)(L/D)f$, where f is the friction factor; $\Lambda = (\pi\gamma/2)\nu/a$, where Λ is the molecular mean free path and a is the speed of sound; and $\lambda = (\nu^3/\varepsilon)^{1/4}$, where λ is the Kolmogorov length scale and e is the average rate of energy dissipation per unit mass, derive an expression for the ratio L/λ. The expression should contain the Reynolds number VD/ν, the Mach number, V/a, the ratio of specific heats, γ, and the friction factor, f.

What quantitative conclusions do you draw from this?

14.7. Verify that Eqs. (14.20) and (14.21) satisfy the vorticity equation for axisymmetric flow,

$$\Omega_{,t} + u\Omega_{,r} + (v/r)\Omega_{,\theta} + w\Omega_{,z} - \Omega w_{,z} = (v/r)(r\Omega_{,r})_{,r} + v\Omega_{,zz}$$

14.8. Compare the volume-integrated contributions of the large-scale motion and the small-scale motion, to the total kinetic energy of the Burgers vortex velocity field. Do this for a control volume that is bounded by the cylinder $r = 10\delta$, and that extends from $z = -10\delta$ to $z = +10\delta$. Assume that $A = U/R$ and that $\Gamma = 2\pi UR$.

14.9. Duplicate the analysis of the Burgers vortex for the case of a stretched vortex sheet. The stretching velocity is $\mathbf{u}_p = -Ay\mathbf{e}_y + Az\mathbf{e}_z$; the vorticity is $\Omega = (U_0/\delta)\exp[-y^2/4\delta^2(t)]$.

In particular, show that there is again a standoff between convection and diffusion when $t \to \infty$, and evaluate the corresponding value of δ.

Also show that if $A = U_0/L$ where the values of U_0 and L are independent of viscosity, the rate of energy dissipation, D, in a cube $-L \le x \le L$, $-L \le y \le L$, $-L \le z \le L$, where $L \gg \delta$, does not become independent of ν as $\nu \to 0$.

What indeed happens to D in that limit, and why do you suppose the result differs from that for the stretched line vortex?

14.10. Show that Eq. (14.26) satisfies Eq. (14.25) if $\nu = 0$ and $v = \Gamma/2\pi r$.

14.11. Show that Eq. (14.27) represents a posible solution of Eq. (14-25) when $\nu \neq 0$ and v is the circumferential velocity of the diffusing line vortex. Derive the ordinary differential equations that must be satisfied by $F(\eta)$ and $G(\eta)$.

14.12. What effect does the three-dimensional self-induced distortion of the vortex filament in Chorin's cartoon have on the flux of vorticity through any cross section of the unit cell of the simulation? Explain your answer.

14.13. You need to convert a mean vorticity profile of a turbulent boundary layer from a plot of $\partial u^+/\partial y^+$ versus y^+ to a plot of $(\theta/U_e)\,\partial U/\partial y$ versus y/θ. Derive the conversion formulas, expressing them in terms of the friction coefficient, c_f, and the Reynolds number, $U_e\theta/\nu$.

14.14. It is sometimes hard to know what to do, and what not to do, with an empirical fitting formula. As an example, consider the Schofield–Perry velocity-defect law for quasiequilibrium boundary layers in adverse pressure gradients, as expressed by Eq. (14.32b). Can we use it to locate the edge of the boundary layer, supposing that $y = \delta$ where $\partial U/\partial Y = 0$? What result do we get from such a calculation? (Answer: $\delta/\delta^* = 3.465 U_e/U_s$.) Suppose we then check to see whether $\delta^* = \int_0^\delta (1 - U/U_e)\,dy$. How nearly is this equation satisfied, numerically?

14.15. It often takes some careful detective work to find the information you need in the technical literature. For example, read the article by Champagne, Pao & Wygnanski (1976), *J. Fluid Mech.* **74**, 209–250, and see if you can find the data that allowed their Fig. 33 to be converted into our Fig. 14.43. It is, incidentally, a fine article to read, in order to see application of the ideas of Section 14.2.

14.16. Duplicate the similarity analysis of fast plane turbulent jets for the case of slow plane jets and wakes, for which $U_0 \ll U_1$. Use the first assumption for scaling of the Reynolds stress, $R_{xy} = U_0^2 g(y/\delta)$, and confirm the results for the growth of δ and decay of U_0, given in the text.

14.17. Duplicate the similarity analysis of plane turbulent jets and wakes, for the corresponding axisymmetric flows. In particular, show that for the fast jet $\delta = \beta(x - x_0)$, while $U_0 = A/(x - x_0)$.

APPENDIX
A

MATHEMATICAL
AIDS

A. COORDINATE-FREE REPRESENTATION OF VECTORS AND TENSORS

The familiar rectangular Cartesian (x, y, z) coordinate system, used in Chapters 2, 3, 4, 5, and 8 to introduce physical concepts and conservation laws, is not always the most convenient for solution of practical problems. For general discussions of physical principles, and for the compact recording of results, a coordinate-free representation of objects such as vectors and tensors is very useful. This Appendix collects and organizes the results that are scattered throughout the book.[1]

The Algebra of Vectors and Tensors

In all that follows, a *vector* is conceived to be an object like an arrow, characterized by length and direction. Addition of vectors conforms to a parallelogram rule, shown in Fig. A1.

---

[1] A more detailed mathematical treatment, which has greatly influenced the choices made here, can be found in Chapter 2 of *Theoretical Hydrodynamics*, by L. M. Milne-Thomson, Fifth Edition, MacMillan, New York, 1968.

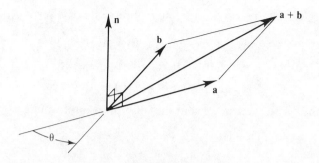

FIGURE A1

Three different results of multiplication of vectors will concern us. They are the *scalar product*

$$\mathbf{a} \cdot \mathbf{b} = ab \cos \theta \tag{A1}$$

the *vector product*

$$\mathbf{a} \times \mathbf{b} = ab \sin \theta \, \mathbf{n} \tag{A2}$$

and the *dyadic product* $\mathbf{a}; \mathbf{b}$, a second-rank tensor defined by the equation

$$\mathbf{v}(\mathbf{a}; \mathbf{b}) = (\mathbf{v} \cdot \mathbf{a})\mathbf{b} \tag{A3}$$

for any third vector $\mathbf{v}$.

A *tensor of the second rank* or second order is a homogeneous linear algebraic operator which changes one vector into another; it may change both the length and direction of the original vector. If the original and resultant vectors are called $\mathbf{v}$ and $\mathbf{u}$ respectively, and the tensor is called A, this operation may be represented symbolically by the equation

$$\mathbf{v}\mathsf{A} = \mathbf{u} \tag{A4}$$

A matter of taste or of convention is involved in the relative placement of the vector and the tensor in the product expressions in (A3) and (A4). In this book, the vector is placed in front of the tensor. The implications of this are more clearly seen after we introduce an *orthogonal triad of unit vectors* $(\boldsymbol{\alpha}, \boldsymbol{\beta}, \boldsymbol{\gamma})$ and represent an arbitrary vector as a sum of three orthogonal components, to wit

$$\mathbf{u} = u_\alpha \boldsymbol{\alpha} + u_\beta \boldsymbol{\beta} + u_\gamma \boldsymbol{\gamma} \tag{A5}$$

Because $\boldsymbol{\alpha}, \boldsymbol{\beta}$, and $\boldsymbol{\gamma}$ are orthogonal, we have

$$\boldsymbol{\alpha} \cdot \boldsymbol{\beta} = \boldsymbol{\beta} \cdot \boldsymbol{\gamma} = \boldsymbol{\gamma} \cdot \boldsymbol{\alpha} = 0 \tag{A6}$$

Once the direction of any two of the unit vectors is specified, that of the third is usually given by a *right-hand rule*, so that

$$\boldsymbol{\gamma} = \boldsymbol{\alpha} \times \boldsymbol{\beta}, \qquad \boldsymbol{\alpha} = \boldsymbol{\beta} \times \boldsymbol{\gamma}, \qquad \boldsymbol{\beta} = \boldsymbol{\gamma} \times \boldsymbol{\alpha} \tag{A7}$$

These unit vectors can be used to define components of a second-rank tensor, A, as follows:

$$A = A_{\alpha\alpha}(\alpha; \alpha) + A_{\alpha\beta}(\alpha; \beta) + A_{\alpha\gamma}(\alpha; \gamma)$$
$$+ A_{\beta\alpha}(\beta; \alpha) + A_{\beta\beta}(\beta; \beta) + A_{\beta\gamma}(\beta; \gamma)$$
$$+ A_{\gamma\alpha}(\gamma; \alpha) + A_{\gamma\beta}(\gamma; \beta) + A_{\gamma\gamma}(\gamma; \gamma) \tag{A8}$$

Using (A5) and (A8), we can expand the multiplication formula $\mathbf{v}A = \mathbf{u}$, to get

$$\mathbf{u} = (v_\alpha A_{\alpha\alpha} + v_\beta A_{\beta\alpha} + v_\gamma A_{\gamma\alpha})\boldsymbol{\alpha}$$
$$+ (v_\alpha A_{\alpha\beta} + v_\beta A_{\beta\beta} + v_\gamma A_{\gamma\beta})\boldsymbol{\beta}$$
$$+ (v_\alpha A_{\alpha\gamma} + v_\beta A_{\beta\gamma} + v_\gamma A_{\gamma\gamma})\boldsymbol{\gamma} \tag{A9}$$

Note that in each term the subscript on v is always the same as the neighboring subscript on A, while the more distant subscript on A identifies the component of the resulting vector, $\mathbf{u}$. Equation (A9) may also be written with the matrix multiplication formula

$$(v_\alpha, v_\beta, v_\gamma) = (u_\alpha, u_\beta, u_\gamma)\begin{bmatrix} A_{\alpha\alpha} A_{\alpha\beta} A_{\alpha\gamma} \\ A_{\beta\alpha} A_{\beta\beta} A_{\beta\gamma} \\ A_{\gamma\alpha} A_{\gamma\beta} A_{\gamma\gamma} \end{bmatrix}$$

The representation of a vector by a row array, rather than a column array, is consistent with the placement of $\mathbf{u}$ in front of $\mathbf{A}$ in Eq. (A4). Note that we are not narrow-minded about these conventions; in the description of the Thomas algorithm for coupled equations, in Chapter 10, we multiply a coefficient matrix times a solution vector in the other popular way, placing the vector behind the matrix, and representing it by a column, rather than a row. One convention works quite as well as the other; it is only important to know which one you have selected for a given task.

The transpose of a second-rank tensor. The tensor A is related to another tensor A*, called its *transpose*, as follows: Let u and v be any two vectors; then

$$\mathbf{v} \cdot (\mathbf{u}A^*) = \mathbf{u} \cdot (\mathbf{v}A) \tag{A10}$$

If A and A* are represented by use of the orthogonal triad of unit vectors, (α, β, γ), and we pick $\mathbf{u} = \alpha$ and $\mathbf{v} = \beta$, we easily find from (A10) that $A^*_{\alpha\beta} = A_{\beta\alpha}$. Similarly, $A^*_{\beta\gamma} = A_{\gamma\beta}$, $A^*_{\gamma\alpha} = A_{\alpha\gamma}$, $A^*_{\alpha\alpha} = A_{\alpha\alpha}$, $A^*_{\beta\beta} = A_{\beta\beta}$ and $A^*_{\gamma\gamma} = A_{\gamma\gamma}$. The matrix array for A* is simply obtained by interchanging the rows and the columns of that for A.

Symmetric and skew-symmetric tensors. Second-rank tensors can always be represented as the sum of a *symmetric tensor* and a *skew-symmetric tensor*. A symmetric tensor equals its transpose; a skew-symmetric tensor is the negative of its transpose. Thus, we can write

$$A = (1/2)(A + A^*) + (1/2)(A - A^*) \tag{A11}$$

$$\text{(symmetric)} \qquad \text{(skew-symmetric)}$$

Principal directions for a symmetric tensor. For any symmetric second-rank tensor S, there is a special orthogonal triad of projection vectors, say $(\alpha', \beta', \gamma')$, such that all the off-diagonal components of the matrix array of the tensor vanish. Thus

$$\alpha'S = S_{\alpha'\alpha'}\alpha', \qquad \beta'S = S_{\beta'\beta'}\beta', \quad \text{and} \quad \gamma'S = S_{\gamma'\gamma'}\gamma' \qquad \text{(A12)}$$

We say that this special triad of unit vectors point out the *principal axes* of the symmetric tensor. We may also say that the three vectors, $S_{\alpha'\alpha'}\alpha'$, $S_{\beta'\beta'}\beta'$, and $S_{\gamma'\gamma'}\gamma'$ constitute the *essence* of the symmetric tensor.

The unit tensor. A special symmetric tensor of great importance is the unit tensor I, which leaves a vector unchanged by multiplication. Thus, for any vector, **v**,

$$v| = v \qquad \text{(A13)}$$

Vector associated with a skew-symmetric tensor. Suppose that B is a skew symmetric tensor. For any vector **v**, we find that there is a vector **b**, such that

$$vB = -v \times b \qquad \text{(A14)}$$

In terms of scalar components, the relationship is

$$b_\alpha = B_{\beta\gamma}, \qquad b_\beta = B_{\gamma\alpha}, \qquad b_\gamma = B_{\alpha\beta}$$

One can say that the essence of a skew-symmetric tensor is its associated vector. Thus, if you find it easy to visualize a vector, but hard to visualize a second-rank tensor, you may be reassured to know that the tensor is equivalent to four vectors, the three that are the essence of its symmetric part, and the one that is the essence of its skew-symmetric part.

Multiplication of second-rank tensors. Two kinds of multiplication of second-rank tensors are important in viscous-flow theory. Let A and C be the tensors.

The *ordinary product*, P = AC, is another second-rank tensor, which acts on any vector, **v**, as follows

$$vP = v(AC) = (vA)C \qquad \text{(A15)}$$

When **v**, A, C and P are represented by (A5) and (A8), and coefficients of v_α, v_β, and v_γ are matched, we find that

$$P_{\alpha\alpha} = A_{\alpha\alpha}C_{\alpha\alpha} + A_{\alpha\beta}C_{\beta\alpha} + A_{\alpha\gamma}C_{\gamma\alpha}$$
$$P_{\alpha\beta} = A_{\alpha\alpha}C_{\alpha\beta} + A_{\alpha\beta}C_{\beta\beta} + A_{\alpha\gamma}C_{\gamma\beta} \qquad \text{(A16)}$$
$$P_{\alpha\gamma} = A_{\alpha\alpha}C_{\alpha\gamma} + A_{\alpha\beta}C_{\beta\gamma} + A_{\alpha\gamma}C_{\gamma\gamma}$$

with six similar formulas for the remaining components. These formulas follow the general rule for the multiplication of square matrices.

The *scalar product* or *inner product*, $S = A \cdot C$, may be represented as follows, using an arbitrary triad of orthogonal unit vectors

$$S = \mathsf{A} \cdot \mathsf{C} = (\alpha\mathsf{C}) \cdot (\alpha\mathsf{C}) + (\beta\mathsf{A}) \cdot (\beta\mathsf{C}) + (\gamma\mathsf{A}) \cdot (\gamma\mathsf{C})$$
$$= A_{\alpha\alpha}C_{\alpha\alpha} + A_{\alpha\beta}C_{\alpha\beta} + A_{\alpha\gamma}C_{\alpha\gamma}$$
$$+ A_{\beta\alpha}C_{\beta\alpha} + A_{\beta\beta}C_{\beta\beta} + A_{\beta\gamma}C_{\beta\gamma}$$
$$+ A_{\gamma\alpha}C_{\gamma\alpha} + A_{\gamma\beta}C_{\gamma\beta} + A_{\gamma\gamma}C_{\gamma\gamma} \tag{A17}$$

Note that if A and C are both symmetric, there will be only six independent terms in this sum. If they are not only symmetric, but share the same principal axes, the sum can be reduced to three terms by referring to the special triad $(\alpha', \beta', \gamma')$.

Note also that if A is symmetric and C is skew-symmetric, or vice versa, $\mathsf{A} \cdot \mathsf{C} = 0$.

Differential Properties of Vector and Tensor Fields

If a scalar, a vector, or a tensor is defined at every point in some spatial domain, we can speak of a *scalar field*, a *vector field*, or a *tensor field*. The following local measures of the spatial nonuniformity of such a field are important in the study of fluid mechanics.

The gradient of a scalar, s. Consider two neighboring points, P and Q, separated by an infinitesimal displacement $d\mathbf{r} = \mathbf{r}(Q) - \mathbf{r}(P)$. The gradient of s at point P, a vector that we call ∇s or grad s, may be defined indirectly, without reference to any particular set of unit vectors, by the equation

$$ds = d\mathbf{r} \cdot \nabla s \tag{A18}$$

Alternatively, it may be defined directly by the equation

$$\nabla s = \lim_{V \to 0} \left(V^{-1} \iint_A \mathbf{n} s \, dA \right) \tag{A19}$$

The integral is executed over the bounding surface, A, of a small volume, V, that encloses the point P, as shown in Fig. A2. It multiplies each element of surface area by the local value of s, and assigns to the product the direction of the local outward normal unit vector, $\mathbf{n}$. The integral will thus be a vector quantity directed away from P toward the part of the surface A where s has the largest values. In the limit as $V \to 0$, it specifies the direction in which s increases most rapidly near P, and evaluates the spatial rate of that change.

The divergence of a vector, v. The most natural definition of this quantity is analogous to the second definition of ∇s, and is written as

$$\operatorname{div} \mathbf{v} = \nabla \cdot \mathbf{v} = \lim_{V \to 0} \left(V^{-1} \iint_A \mathbf{n} \cdot \mathbf{v} \, dA \right) \tag{A20}$$

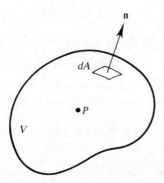

FIGURE A2

In this case, the integrand weights each infinitesimal element of boundary area with the outward normal component of the vector, $\mathbf{n} \cdot \mathbf{v}$. Thus the integral evaluates the net rate at which the vector field diverges away from point P.

The curl of a vector, v. There are again two natural definitions, one utilizing the surface integral, in the form

$$\text{curl } \mathbf{v} = \nabla \times \mathbf{v} = \lim_{V \to 0} \left(V^{-1} \iint_A \mathbf{n} \times \mathbf{v} \, dA \right) \tag{A21}$$

In this case, the integrand weights each infinitesimal element of boundary area with the tangential component of the vector, $\mathbf{n} \times \mathbf{v}$. Thus the integral evaluates the net extent to which the vector field curls around P. Curl $\mathbf{v}$ is evidently a vector, and the alternative definition evaluates an arbitrary component of that vector. Let the unit vector $\boldsymbol{\alpha}$ be normal to an infinitesimal area A, which is surrounded by a closed curve C. The α-component of curl $\mathbf{v}$ is then given by the limit of a line integral around C, to wit

$$\boldsymbol{\alpha} \cdot \text{curl } \mathbf{v} = \lim_{A \to 0} \left(A^{-1} \int_C \mathbf{v} \cdot d\mathbf{r} \right) \tag{A22}$$

Here $d\mathbf{r}$ is an element of the arclength, directed so that an observer standing on the surface A so that $\boldsymbol{\alpha}$ points up, and facing in the direction of $d\mathbf{r}$, would look left to see the surface. This formula is very convenient for calculation, as will soon be shown.

The gradient of a vector, v. Consider again two neighboring points, P and Q, separated by an infinitesimal displacement $d\mathbf{r} = \mathbf{r}(Q) - \mathbf{r}(P)$. The gradient of $\mathbf{v}$ at point P, a second-rank tensor that we call $\nabla; \mathbf{v}$ or grad $\mathbf{v}$, may be defined indirectly by the equation

$$d\mathbf{v} = d\mathbf{r}(\nabla; \mathbf{v}) = (d\mathbf{r} \cdot \nabla)\mathbf{v} \tag{A23}$$

Alternatively, it may be defined directly by the equation

$$\text{grad } \mathbf{v} \equiv \boldsymbol{\nabla}; \mathbf{v} = \lim_{V \to 0} \left(V^{-1} \iint_A \mathbf{n}; \mathbf{v} \, dA \right) \tag{A24}$$

The second definition brings out the tensor character of $\boldsymbol{\nabla}; \mathbf{v}$, but the first is generally more useful for calculation. Closely associated with grad $\mathbf{v}$ is its *transpose*

$$(\text{grad } \mathbf{v})^* \equiv (\boldsymbol{\nabla}; \mathbf{v})^* = \lim_{V \to 0} \left(V^{-1} \iint_A \mathbf{v}; \mathbf{n} \, dA \right) \tag{A25}$$

and its decomposition into symmetric and skew-symmetric parts:

$$\text{def } \mathbf{v} \equiv (1/2)[\text{grad } \mathbf{v} + (\text{grad } \mathbf{v})^*] \tag{A26}$$

$$\text{rot } \mathbf{v} \equiv (1/2)[\text{grad } \mathbf{v} - (\text{grad } \mathbf{v})^*] \tag{A27}$$

The divergence of a second-rank tensor, T. For applications to fluid mechanics, the natural definition involves the surface integral

$$\text{div } \mathsf{T} = \boldsymbol{\nabla} \cdot \mathsf{T} = \lim_{V \to 0} \left(V^{-1} \iint_A \mathbf{n} \mathsf{T} \, dA \right) \tag{A28}$$

Since $\mathbf{n}\mathsf{T}$ is a vector, the integral evaluates an average value of that vector over the surface A.

A collection of useful identities. Differential operations must often be performed on a product of scalars and/or vectors, or performed repeatedly. Here are some useful results, in which s is a scalar, $\mathbf{a}$ and $\mathbf{b}$ are vectors, A is a second-rank tensor with transpose A*, B is a skew-symmetric tensor, and I is the unit tensor.

$$\text{grad } (\mathbf{a} \cdot \mathbf{b}) = \mathbf{b}(\text{grad } \mathbf{a})^* + \mathbf{a}(\text{grad } \mathbf{b})^*$$

$$\text{div } (\mathbf{b}\mathsf{A}) = \mathbf{b} \cdot \text{div } \mathsf{A}^* + \text{grad } \mathbf{b} \cdot \mathsf{A}^*$$

$$\text{curl } (\mathbf{a}s) = -\mathbf{a} \times \text{grad } s + s \text{ curl } \mathbf{a}$$

$$\text{div } (\mathbf{b}; \mathbf{a}) = \mathbf{b} \text{ grad } \mathbf{a} + \mathbf{a} \text{ div } \mathbf{b}$$

$$\text{curl } (\mathbf{a} \times \mathbf{b}) = \text{div } [(\mathbf{b}; \mathbf{a}) - (\mathbf{a}; \mathbf{b})] = \mathbf{b} \text{ grad } \mathbf{a} + \mathbf{a} \text{ div } \mathbf{b} + \mathbf{a} \text{ grad } \mathbf{b} - \mathbf{b} \text{ div } \mathbf{a}$$

$$\text{div } (s\mathsf{I}) = \text{grad } s$$

$$\mathsf{I} \cdot \text{grad } \mathbf{a} = \text{div } \mathbf{a}$$

$$\mathbf{a} \text{ grad } \mathbf{a} \equiv (\mathbf{a} \cdot \boldsymbol{\nabla})\mathbf{a} = (1/2) \text{ grad } (\mathbf{a} \cdot \mathbf{a}) + (\text{curl } \mathbf{a}) \times \mathbf{a}$$

$$\text{div } (\text{def } \mathbf{a}) = 2 \text{ grad } (\text{div } \mathbf{a}) - \text{curl } (\text{curl } \mathbf{a})$$

$$\nabla^2 s \equiv \text{div } (\text{grad } s)$$

$$\nabla^2 \mathbf{a} \equiv \text{div } (\text{grad } \mathbf{a}) = \text{grad } (\text{div } \mathbf{a}) - \text{curl } (\text{curl } \mathbf{a})$$

The last two lines define and evaluate the *Laplacian* operator, applied to a scalar or a vector.

B. REPRESENTATION OF DIFFERENTIAL QUANTITIES IN ORTHOGONAL COORDINATE SYSTEMS

For the analysis of any particular flow problem, points in the spatial domain will be located by reference to some convenient system of coordinates. The choice of a coordinate system usually depends on the configuration of the solid boundaries, or of the streamlines of some known part of the flow, such as the flow outside a boundary layer. Usually the coordinates will be orthogonal, so that three surfaces, on each of which one coordinate is constant, meet at right angles at any given point. Let the coordinates be called α, β, and γ. At the point of intersection there will be an orthogonal triad of unit vectors, (α, β, γ), with α pointing along the intersection of surfaces of constant β and γ, in the direction in which α increases, and so on.

In general, the coordinates will not themselves be distances; the local distance between coordinate surfaces $\alpha = \alpha_1$ and $\alpha = \alpha_1 + d\alpha$ is given by the formula $dr_\alpha = h_\alpha\, d\alpha$. This introduces the *metric factor*, h_α. A general infinitesimal displacement will be represented as

$$d\mathbf{r} = (h_\alpha\, d\alpha)\alpha + (h_\beta\, d\beta)\beta + (h_\gamma\, d\gamma)\gamma \qquad (A29)$$

Variability of the unit vectors when the coordinate surfaces are curved. In general, the unit vectors (α, β, γ), and the metric coefficients $(h_\alpha, h_\beta, h_\gamma)$, all vary from point to point. Consider, Fig. A3, an infinitesimal patch of the surface $\gamma = $ constant. For the moment, we imagine the patch to be plane, and seen in true size. You can see that β at point B differs from β at point A, because it has rotated through a small angle $d\theta$. In fact, $(\partial\beta/\partial\beta)\, d\beta = -d\theta\alpha$. Looking at the other small triangle that involves $d\theta$, you can also see that $(h_\alpha\, d\alpha)\, d\theta = (\partial(h_\beta\, d\beta)/\partial\alpha)\, d\alpha$. Eliminating $d\theta$ between these equations, we find that $\partial\beta/\partial\beta = -(\alpha/h_\alpha)(\partial h_\beta/\partial\alpha)$. However, this is not the whole story for the general case, because then the path from A to B may also be curved in the surface $\alpha = $ constant, which we see on edge. Looking down the normal to that surface, we see a picture just like the one considered above, and deduce that

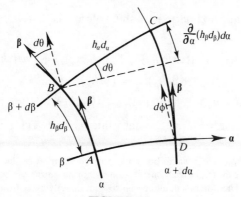

FIGURE A3

$\partial\boldsymbol{\beta}/\partial\beta = -(\boldsymbol{\gamma}/h_\gamma)(\partial h_\beta/\partial\gamma)$. In the general case, both the surface of constant α and that of constant γ are curved, and the partial effects add. Thus,

$$\partial\boldsymbol{\beta}/\partial\beta = -(\boldsymbol{\alpha}/h_\alpha)(\partial h_\beta/\partial\alpha) - (\boldsymbol{\gamma}/h_\gamma)(\partial h_\beta/\partial\gamma) \tag{A30}$$

and two corresponding equations obtained by cyclic permutations of α, β, and γ.

We see also that $\boldsymbol{\beta}$ at point D differs from $\boldsymbol{\beta}$ at point A, because it has rotated through a small angle $d\phi$. A further analysis of the sketch produces the results

$$\partial\boldsymbol{\beta}/\partial\alpha = (\boldsymbol{\alpha}/h_\beta)(\partial h_\alpha/\partial\beta) \quad \text{and} \quad \partial\boldsymbol{\alpha}/\partial\beta = (\boldsymbol{\beta}/h_\alpha)(\partial h_\beta/\partial\alpha) \tag{A31}$$

This time, these results are complete; views of the other coordinate surfaces only produce the companion relationships

$$\partial\boldsymbol{\gamma}/\partial\beta = (\boldsymbol{\beta}/h_\gamma)(\partial h_\beta/\partial\gamma), \qquad \partial\boldsymbol{\beta}/\partial\gamma = (\boldsymbol{\gamma}/h_\beta)(\partial h_\gamma/\partial\beta),$$

$$\partial\boldsymbol{\alpha}/\partial\gamma = (\boldsymbol{\gamma}/h_\alpha)(\partial h_\gamma/\partial\alpha), \qquad \partial\boldsymbol{\gamma}/\partial\alpha = (\boldsymbol{\alpha}/h_\gamma)(\partial h_\alpha/\partial\gamma).$$

These results affirm that the change in any of the unit vectors that results from a displacement in a surface normal to that vector is parallel to the displacement.[2]

The gradient operator. Let the small volume involved in Eq. (A19) be bounded by the coordinate surfaces $\alpha = \alpha_1$, $\alpha = \alpha_1 + d\alpha$, $\beta = \beta_1$, $\beta = \beta_1 + d\beta$, $\gamma = \gamma_1$, $\gamma = \gamma_1 + d\gamma$. The integral in (A19) is, to the first order of the infinitesimal quantities $d\alpha$, $d\beta$, and $d\gamma$, equal to

$$\iint = [s\boldsymbol{\alpha}h_\beta \, d\beta h_\gamma \, d\gamma]_{\alpha1+d\alpha} - [s\boldsymbol{\alpha}h_\beta \, d\beta h_\gamma \, d\gamma]_{\alpha1}$$

$$+ [s\boldsymbol{\beta}h_\gamma \, d\gamma h_\alpha \, d\alpha]_{\beta1+d\beta} - [s\boldsymbol{\beta}h_\gamma \, d\gamma h_\alpha \, d\alpha]_{\beta1}$$

$$+ [s\boldsymbol{\gamma}h_\alpha \, d\alpha h_\beta \, d\beta]_{\gamma1+d\gamma} - [s\boldsymbol{\gamma}h_\alpha \, d\alpha h_\beta \, d\beta]_{\gamma1}$$

Using the first term of a Taylor series to approximate each line, we get

$$d\alpha(s\boldsymbol{\alpha}h_\beta d\beta h_\gamma \, d\gamma)_{,\alpha} + d\beta(s\boldsymbol{\beta}h_\gamma \, d\gamma h_\alpha \, d\alpha)_{,\beta} + d\gamma(s\boldsymbol{\gamma}h_\alpha \, d\alpha h_\beta d\beta)_{,\gamma}$$

The corresponding approximation to the volume is $V = h_\alpha h_\beta h_\gamma \, d\alpha \, d\beta \, d\gamma$, so we get, initially,

$$\nabla s = (h_\alpha h_\beta h_\gamma)^{-1}[(\boldsymbol{\alpha}sh_\beta h_\gamma)_{,\alpha} + (\boldsymbol{\beta}sh_\gamma h_\alpha)_{,\beta} + (\boldsymbol{\gamma}sh_\alpha h_\beta)_{,\gamma}]$$

$$= [(\boldsymbol{\alpha}/h_\alpha)s_{,\alpha} + (\boldsymbol{\beta}/h_\beta)s_{,\beta} + (\boldsymbol{\gamma}/h_\gamma)s_{,\gamma}]$$

$$+ s(h_\alpha h_\beta h_\gamma)^{-1}[(\boldsymbol{\alpha}h_\beta h_\gamma)_{,\alpha} + (\boldsymbol{\beta}h_\gamma h_\alpha)_{,\beta} + (\boldsymbol{\gamma}h_\alpha h_\beta)_{,\gamma}]$$

However, the cofactor of s in the last line is just $\nabla(1)$, and the gradient

---

[2] To see this, insert toothpicks in a small but noticeably curved part of the surface of a piece of fruit or vegetable, taking care to get each one accurately normal to the surface. Then look straight down one of toothpicks. The others will all appear to lean directly away from that one.

of any constant is zero. This leaves

$$\mathbf{V}s = (\boldsymbol{\alpha}/h_\alpha)s_{,\alpha} + (\boldsymbol{\beta}/h_\beta)s_{,\beta} + (\boldsymbol{\gamma}/h_\gamma)s_{,\gamma} \tag{A32}$$

Inserting this result and Eq. (A29) into (A18), the indirect definition of $\mathbf{V}s$, we obtain

$$d\mathbf{r} \cdot \mathbf{V}s = s_{,\alpha}\, d\alpha + s_{,\beta}\, d\beta + s_{,\gamma}\, d\gamma = ds$$

verifying that the Chain Rule is satisfied.

This last analysis suggests that the differential operator, $\mathbf{V}$, can usefully be represented as

$$\mathbf{V} = (\boldsymbol{\alpha}/h_\alpha)\, \partial/\partial\alpha + (\boldsymbol{\beta}/h_\beta)\, \partial/\partial\beta + (\boldsymbol{\gamma}/h_\gamma)\, \partial/\partial\gamma \tag{A33}$$

In many textbooks, this equation is taken as the definition of the gradient operator, and our definitions are derived as consequences.

Divergence of a vector. Evaluating the integral in (A20) with the same choice of bounding surfaces, and noting that $\mathbf{n} \cdot \mathbf{v} = v_\alpha$ on the surface $\alpha = \alpha_1 + d\alpha$, while $\mathbf{n} \cdot \mathbf{v} = -v_\alpha$ on the surface $\alpha = \alpha_1$ and so on, we find that

$$\mathbf{V} \cdot \mathbf{v} = (h_\alpha h_\beta h_\gamma)^{-1}[(v_\alpha h_\beta h_\gamma)_{,\alpha} + (v_\beta h_\gamma h_\alpha)_{,\beta} + (v_\gamma h_\alpha h_\beta)_{,\gamma}] \tag{A34}$$

Curl of a vector. It is easiest to calculate one component at a time, using (A22). To get the α-component, let the curve C lie in the surface $\alpha = \text{constant}$, and be composed of the four segments shown in Fig. A.4. Note that on the segment $AB\ d\mathbf{r} = h_\beta\, d\beta\boldsymbol{\beta}$, and that $\mathbf{v} \cdot d\mathbf{r} = (v_\beta h_\beta\, d\beta)_{\gamma=\gamma 1}$. The line integral is, to the first order in infinitesimal quantities,

$$\int = (v_\beta h_\beta\, d\beta)_{\gamma=\gamma 1} + (v_\gamma h_\gamma\, d\gamma)_{\beta=\beta 1+d\beta} - (v_\beta h_\beta\, d\beta)_{\gamma=\gamma 1+d\gamma} - (v_\gamma h_\gamma\, d\gamma)_{\beta=\beta 1}$$

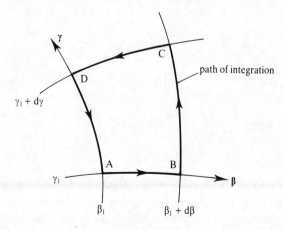

FIGURE A4

The area A is $h_\beta h_\gamma \, d\beta \, d\gamma$, and the Taylor-series expansion leads to the final result

$$\boldsymbol{\alpha} \cdot (\boldsymbol{\nabla} \times \mathbf{v}) = (h_\beta h_\gamma)^{-1}[(v_\gamma h_\gamma)_{,\beta} - (v_\beta h_\beta)_{,\gamma}]$$

The results for all three components can conveniently be collected in the formula

$$\boldsymbol{\nabla} \times \mathbf{v} = (h_\alpha h_\beta h_\gamma)^{-1} \begin{vmatrix} h_\alpha \boldsymbol{\alpha} & h_\beta \boldsymbol{\beta} & h_\gamma \boldsymbol{\gamma} \\ \partial/\partial\alpha & \partial/\partial\beta & \partial/\partial\gamma \\ h_\alpha v_\alpha & h_\beta v_\beta & h_\gamma v_\gamma \end{vmatrix} \tag{A35}$$

Gradient of a vector. If the direct definition, (A24), is analyzed as were (A19) and (A20), the preliminary result is

$$\boldsymbol{\nabla}; \mathbf{v} = (h_\alpha h_\beta h_\gamma)^{-1}[(\boldsymbol{\alpha}; \mathbf{v} h_\beta h_\gamma)_{,\alpha} + (\boldsymbol{\beta}; \mathbf{v} h_\gamma h_\alpha)_{,\beta} + (\boldsymbol{\gamma}; \mathbf{v} h_\alpha h_\beta)_{,\gamma}]$$

Again this can be simplified by expanding the derivatives of products and noting that $\boldsymbol{\nabla}(1) = 0$. The result is

$$\boldsymbol{\nabla}; \mathbf{v} = (\boldsymbol{\alpha}/h_\alpha); \partial\mathbf{v}/\partial\alpha + (\boldsymbol{\beta}/h_\beta); \partial\mathbf{v}/\partial\beta + (\boldsymbol{\gamma}/h_\gamma); \partial\mathbf{v}/\partial\gamma.$$

When differentiating $\mathbf{v}$ with respect to α, we must remember to differentiate the unit vectors as well as the scalar components. Thus

$$\partial\mathbf{v}/\partial\alpha = \boldsymbol{\alpha} \, \partial v_\alpha/\partial\alpha + v_\alpha \, \partial\boldsymbol{\alpha}/\partial\alpha + \boldsymbol{\beta} \, \partial v_\beta/\partial\alpha + v_\beta \, \partial\boldsymbol{\beta}/\partial\alpha + \boldsymbol{\gamma} \, \partial v_\gamma/\partial\alpha + v_\gamma \, \partial\boldsymbol{\gamma}/\partial\alpha$$

$$= \boldsymbol{\alpha}[\partial v_\alpha/\partial\alpha + (v_\beta/h_\beta) \, \partial h_\alpha/\partial\beta + (v_\gamma/h_\gamma) \, \partial h_\alpha/\partial\gamma] + \boldsymbol{\beta}[\ ] + \boldsymbol{\gamma}[\ ]$$

where the terms in [ ] after $\boldsymbol{\beta}$ and $\boldsymbol{\gamma}$ follow by cyclic permutation of indices. When all these results are pulled together, the final result is

$$\boldsymbol{\nabla}; \mathbf{v} = \operatorname{grad} \mathbf{v} = \frac{1}{h_\alpha}\left(\frac{\partial v_\alpha}{\partial\alpha} + \frac{v_\beta}{h_\beta}\frac{\partial h_\alpha}{\partial\beta} + \frac{v_\gamma}{h_\gamma}\frac{\partial h_\alpha}{\partial\gamma}\right)(\boldsymbol{\alpha}; \boldsymbol{\alpha})$$

$$+ \frac{1}{h_\alpha}\left(\frac{\partial v_\beta}{\partial\alpha} - \frac{v_\alpha}{h_\beta}\frac{\partial h_\alpha}{\partial\beta}\right)(\boldsymbol{\alpha}; \boldsymbol{\beta}) + \frac{1}{h_\alpha}\left(\frac{\partial v_\gamma}{\partial\alpha} - \frac{v_\alpha}{h_\gamma}\frac{\partial h_\alpha}{\partial\gamma}\right)(\boldsymbol{\alpha}; \boldsymbol{\gamma})$$

$$+ \frac{1}{h_\beta}\left(\frac{\partial v_\alpha}{\partial\beta} - \frac{v_\beta}{h_\alpha}\frac{\partial h_\beta}{\partial\alpha}\right)(\boldsymbol{\beta}; \boldsymbol{\alpha}) + \frac{1}{h_\beta}\left(\frac{\partial v_\beta}{\partial\beta} + \frac{v_\gamma}{h_\gamma}\frac{\partial h_\beta}{\partial\gamma} + \frac{v_\alpha}{h_\alpha}\frac{\partial h_\beta}{\partial\alpha}\right)(\boldsymbol{\beta}; \boldsymbol{\beta})$$

$$+ \frac{1}{h_\beta}\left(\frac{\partial v_\gamma}{\partial\beta} - \frac{v_\beta}{h_\gamma}\frac{\partial h_\beta}{\partial\gamma}\right)(\boldsymbol{\beta}; \boldsymbol{\gamma}) + \frac{1}{h_\gamma}\left(\frac{\partial v_\alpha}{\partial\gamma} - \frac{v_\gamma}{h_\alpha}\frac{\partial h_\gamma}{\partial\alpha}\right)(\boldsymbol{\gamma}; \boldsymbol{\alpha})$$

$$+ \frac{1}{h_\gamma}\left(\frac{\partial v_\beta}{\partial\gamma} - \frac{v_\gamma}{h_\beta}\frac{\partial h_\gamma}{\partial\beta}\right)(\boldsymbol{\gamma}; \boldsymbol{\beta}) + \frac{1}{h_\gamma}\left(\frac{\partial v_\gamma}{\partial\gamma} + \frac{v_\alpha}{h_\alpha}\frac{\partial h_\gamma}{\partial\alpha} + \frac{v_\beta}{h_\beta}\frac{\partial h_\gamma}{\partial\beta}\right)(\boldsymbol{\gamma}; \boldsymbol{\gamma})$$

$$\tag{A36}$$

Divergence of a tensor. Again the direct definition yields the computationally most useful form. Note that the surface $\alpha = \alpha_1 + d\alpha$ contributes $\boldsymbol{\alpha} A h_\beta h_\gamma \, d\beta \, d\gamma$ to the surface integral; then use (A8) and remember that $\boldsymbol{\alpha}(\boldsymbol{\alpha}; \boldsymbol{\alpha}) = \boldsymbol{\alpha}$, $\boldsymbol{\alpha}(\boldsymbol{\alpha}; \boldsymbol{\beta}) = \boldsymbol{\beta}$, and $\boldsymbol{\alpha}(\boldsymbol{\alpha}; \boldsymbol{\gamma}) = \boldsymbol{\gamma}$, while $\boldsymbol{\alpha}$ times any dyadic in which the first member is β or γ equals zero. The final result is

$$\nabla \cdot A = \text{div } A = \frac{\alpha}{h_\alpha h_\beta h_\gamma} \left\{ \frac{\partial}{\partial \alpha} (h_\beta h_\gamma A_{\alpha\alpha}) + \frac{\partial}{\partial \beta} (h_\gamma h_\alpha A_{\beta\alpha}) + \frac{\partial}{\partial \gamma} (h_\alpha h_\beta A_{\gamma\alpha}) \right.$$

$$\left. + h_\gamma \left(A_{\alpha\beta} \frac{\partial h_\alpha}{\partial \beta} - A_{\beta\beta} \frac{\partial h_\beta}{\partial \alpha} \right) + h_\beta \left(A_{\alpha\gamma} \frac{\partial h_\alpha}{\partial \gamma} - A_{\gamma\gamma} \frac{\partial h_\gamma}{\partial \alpha} \right) \right\}$$

$$+ \beta\{ \quad \} \gamma\{ \quad \} \tag{A37}$$

where the terms in β and γ come from those in α by cyclic permutation of indices.

Convective rate of change. This is given by the scalar operator $(\mathbf{u} \cdot \nabla)$, where $\mathbf{u}$ is the fluid velocity. Its representation is simply

$$(\mathbf{u} \cdot \nabla) = (u_\alpha/h_\alpha) \, \partial/\partial\alpha + (u_\beta/h_\beta) \, \partial/\partial\beta + (u_\gamma/h_\gamma) \, \partial/\partial\gamma \tag{A38}$$

Convective acceleration. Knowing how to evaluate $\partial\mathbf{u}/\partial\alpha$, etc, we can quickly get the result

$$\mathbf{u} \, \text{grad } \mathbf{u} = \alpha \left\{ \frac{u_\alpha}{h_\alpha} \left(\frac{\partial u_\alpha}{\partial \alpha} + \frac{u_\beta}{h_\beta} \frac{\partial h_\alpha}{\partial \beta} + \frac{u_\gamma}{h_\gamma} \frac{\partial h_\alpha}{\partial \gamma} \right) + \frac{u_\beta}{h_\beta} \left(\frac{\partial u_\alpha}{\partial \beta} - \frac{u_\beta}{h_\alpha} \frac{\partial h_\beta}{\partial \alpha} \right) \right.$$

$$\left. + \frac{u_\gamma}{h_\gamma} \left(\frac{\partial u_\alpha}{\partial \gamma} - \frac{u_\gamma}{h_\alpha} \frac{\partial h_\gamma}{\partial \alpha} \right) \right\} + \beta\{ \quad \} + \gamma\{ \quad \} \tag{A39}$$

Calculation of the metric coefficients. Most of the commonly used coordinate systems are related to a simple rectangular Cartesian (x, y, z) by transformation equations

$$x = x(\alpha, \beta, \gamma), \qquad y = y(\alpha, \beta, \gamma), \qquad z = z(\alpha, \beta, \gamma)$$

An infinitesimal displacement can be represented in either coordinate system; the square of its length is thus

$$dr^2 = dx^2 + dy^2 + dz^2 = h_\alpha^2 \, d\alpha^2 + h_\beta^2 \, d\beta^2 + h_\gamma^2 \, d\gamma^2$$

From the Chain Rule, $dx = x,_\alpha \, d\alpha + x,_\beta \, d\beta + x,_\gamma \, d\gamma$, and so on. If $d\beta$ and $d\gamma$ are zero, we get

$$dx^2 + dy^2 + dz^2 = (x,_\alpha^2 + y,_\alpha^2 + z,_\alpha^2) \, d\alpha^2 = h_\alpha^2 \, d\alpha^2$$

From this, and the corresponding special results if $d\gamma$ and $d\alpha$ are zero, of if $d\alpha$ and $d\beta$ are zero, it follows that

$$h_\alpha^2 = x,_\alpha^2 + y,_\alpha^2 + z,_\alpha^2$$
$$h_\beta^2 = x,_\beta^2 + y,_\beta^2 + z,_\beta^2 \tag{A40}$$
$$h_\gamma^2 = x,_\gamma^2 + y,_\gamma^2 + z,_\gamma^2$$

For example, in spherical polar (R, θ, ϕ) coordinates, $x = R \sin \theta \cos \phi$,

$y = R \sin \theta \sin \phi$, $z = R \cos \theta$, so $h_\phi^2 = (-R \sin \theta \sin \phi)^2 + (R \sin \theta \cos \phi)^2 + (0)^2 = R^2 \sin^2 \theta$.

Metric coefficients for popular coordinate systems

Rectangular Cartesian:

$$h_x = 1, \qquad h_y = 1, \qquad h_z = 1 \tag{A41}$$

Cylindrical polar, (r, θ, z): $x = r \cos \theta$, $y = r \sin \theta$, $z = z$;

$$h_r = 1, \qquad h_\theta = r, \qquad h_z = 1 \tag{A42}$$

Spherical polar, (R, θ, ϕ): $x = R \sin \theta \cos \phi$, $y = R \sin \theta \sin \phi$, $z = R \cos \theta$;

$$h_R = 1, \qquad h_\theta = R, \qquad h_\phi = R \sin \theta \tag{A43}$$

Plane elliptical, (Φ, Ψ, z): $x/L = \sinh \Phi \cos \Psi$, $y/L = \cosh \Phi \sin \Psi$, $z = z$.

$$h_\Phi = h_\Psi = L(\cosh^2 \Phi - \sin^2 \Psi)^{1/2}, \qquad h_z = 1 \tag{A44}$$

Plane parabolic, (ξ, η, z): $x/L = \xi^2 - \eta^2 + 1$, $y/L = 2\xi\eta$, $z = z$;

$$h_\xi = h_\eta = 2L(\xi^2 + \eta^2)^{1/2}, \qquad h_z = 1 \tag{A45}$$

C. EQUATIONS OF MOTION IN RECTANGULAR, CYLINDRICAL, AND SPHERICAL COORDINATES

With the formulas presented in A and B above, you should be able to work out the form of the equations of motion in any of the popular coordinate systems. Some of the most frequently used results are given here for easy reference. They all embody the assumption that ρ and μ are constants.

The representation of the viscous force per unit mass is worked out from the formula $\mathbf{f}_v = -v \operatorname{curl} \mathbf{\Omega}$, which is convenient in many idealized analyses in which there is only one scalar component of the vorticity. The formulas in other textbooks can be found from these by use of the fact that $\operatorname{div} \mathbf{u} = 0$. For the general orthogonal coordinate system, this representation is

$$\mathbf{f}_v = -v \operatorname{curl} \mathbf{\Omega} = -v(h_\alpha h_\beta h_\gamma)^{-1} \begin{vmatrix} h_\alpha \boldsymbol{\alpha} & h_\beta \boldsymbol{\beta} & h_\gamma \boldsymbol{\gamma} \\ \partial/\partial\alpha & \partial/\partial\beta & \partial/\partial\gamma \\ h_\alpha \Omega_\alpha & h_\beta \Omega_\beta & h_\gamma \Omega_\gamma \end{vmatrix}$$

Rectangular Cartesian Coordinates

Symbols for velocity components:

$$\mathbf{u} = u(x, y, z, t)\mathbf{e}_x + v(x, y, z, t)\mathbf{e}_y + w(x, y, z, t)\mathbf{e}_z$$

Vorticity:

$$\mathbf{\Omega} = \xi(x, y, z, t)\mathbf{e}_x + \eta(x, y, z, t)\mathbf{e}_y + \zeta(x, y, z, t)\mathbf{e}_z$$

where
$$\xi = w,_y - v,_z \qquad \eta = u,_z - w,_x \qquad \zeta = v,_x - u,_y \tag{A46}$$

Continuity equation:
$$\operatorname{div} \mathbf{u} = u,_x + v,_y + w,_z = 0 \tag{A47}$$

x-momentum:
$$u,_t + uu,_x + vu,_y + wu,_z + \Pi,_x = v(\eta,_z - \zeta,_y) \tag{A48}$$

y-momentum:
$$v,_t + uv,_x + vv,_y + wv,_z + \Pi,_y = v(\zeta,_x - \xi,_z) \tag{A49}$$

z-momentum:
$$w,_t + uw,_x + vw,_y + ww,_z + \Pi,_z = v(\xi,_y - \eta,_x) \tag{A50}$$

Viscous stresses:
$$\tau_{xx} = 2\mu u,_x \qquad \tau_{yy} = 2\mu v,_y \qquad \tau_{zz} = 2\mu w,_z$$
$$\tau_{xy} = \tau_{yx} = \mu(u,_y + v,_x) \qquad \tau_{yz} = \tau_{zy} = \mu(v,_z + w,_y)$$
$$\tau_{zx} = \tau_{xz} = \mu(w,_x + u,_z) \tag{A51}$$

Cylindrical Polar Coordinates

Symbols for velocity components:
$$\mathbf{u} = u(r,\ \theta,\ z,\ t)\mathbf{e}_r + v(r,\ \theta,\ z,\ t)\mathbf{e}_\theta + w(r,\ \theta,\ z,\ t)\mathbf{e}_z$$

Vorticity:
$$\mathbf{\Omega} = \xi(r,\ \theta,\ z,\ t)\mathbf{e}_r + \eta(r,\ \theta,\ z,\ t)\mathbf{e}_\theta + \zeta(r,\ \theta,\ z,\ t)\mathbf{e}_z$$

where
$$\xi = r^{-1}w,_\theta - v,_z \qquad \eta = u,_z - w,_r \qquad \zeta = r^{-1}[(rv),_r - u,_\theta] \tag{A52}$$

Continuity equation:
$$\operatorname{div} \mathbf{u} = r^{-1}[(ru),_r + v,_\theta] + w,_z = 0 \tag{A53}$$

r-momentum:
$$u,_t + uu,_r + r^{-1}vu,_\theta + wu,_z - r^{-1}v^2 + \Pi,_r = v(\eta,_z - r^{-1}\zeta,_\theta) \tag{A54}$$

θ-momentum:
$$v,_t + uv,_r + r^{-1}vv,_\theta + wv,_z + r^{-1}uv + r^{-1}\Pi,_\theta = v(\zeta,_r - \xi,_z) \tag{A55}$$

z-momentum:
$$w,_t + uw,_r + r^{-1}vw,_\theta + ww,_z + \Pi,_z = vr^{-1}[\xi,_\theta - (r\eta),_r] \tag{A56}$$

Viscous stresses:
$$\tau_{rr} = 2\mu u,_r \qquad \tau_{\theta\theta} = 2\mu r^{-1}(v,_\theta + u) \qquad \tau_{zz} = 2\mu w,_z$$
$$\tau_{r\theta} = \tau_{\theta r} = \mu[r^{-1}w,_\theta + v,_z] \qquad \tau_{\theta z} = \tau_{z\theta} = \mu[r(v/r),_r + r^{-1}u,_\theta]$$
$$\tau_{zr} = \tau_{rz} = \mu[w,_r + u,_z] \tag{A57}$$

Spherical Polar Coordinates

Symbols for velocity components:

$$\mathbf{u} = u(R, \theta, \phi, t)\mathbf{e}_R + v(R, \theta, \phi, t)\mathbf{e}_\theta + w(R, \theta, \phi, t)\mathbf{e}_\phi$$

Vorticity:

$$\mathbf{\Omega} = \xi(R, \theta, \phi, t)\mathbf{e}_R + \eta(R, \theta, \phi, t)\mathbf{e}_\theta + \zeta(R, \theta, \phi, t)\mathbf{e}_\phi$$

where

$$\xi = (R \sin \theta)^{-1}[(w \sin \theta)_{,\theta} - v_{,\phi}] \qquad \eta = (R \sin \theta)^{-1}u_{,\phi} - R^{-1}(Rw)_{,R}$$

$$\zeta = R^{-1}[(Rv)_{,R} - u_{,\theta}] \tag{A58}$$

Continuity equation:

$$\text{div } \mathbf{u} = R^{-2}(R^2 u)_{,R} + (R \sin \theta)^{-1}[(v \sin \theta)_{,\theta} + w_{,\phi}] = 0 \tag{A59}$$

R-momentum:

$$u_{,t} + uu_{,R} + R^{-1}vu_{,\theta} + (R \sin \theta)^{-1}wu_{,\phi} - R^{-1}(v^2 + w^2) - \Pi_{,R}$$
$$= v(R^2 \sin \theta)^{-1}[(R\eta)_{,\phi} - (R\zeta \sin \theta)_{,\theta}] \tag{A60}$$

θ-momentum:

$$v_{,t} + uv_{,R} + R^{-1}vv_{,\theta} + (R \sin \theta)^{-1}wv_{,\phi} + R^{-1}uv - (R \tan \theta)^{-1}w^2 + R^{-1}\Pi_{,\theta}$$
$$= v(R \sin \theta)^{-1}[(R\zeta \sin \theta)_{,R} - \xi_{,\phi}] \tag{A61}$$

ϕ-momentum:

$$w_{,t} + uw_{,R} + R^{-1}vw_{,\theta} + (R \sin \theta)^{-1}ww_{,\phi} + R^{-1}uw - (R \tan \theta)^{-1}vw$$
$$+ (R \sin \theta)^{-1}\Pi_{,\phi} = vR^{-1}[\xi_{,\theta} - (R\eta)_{,R}] \tag{A62}$$

Viscous stresses

$$\tau_{RR} = 2\mu u_{,R} \qquad \tau_{\theta\theta} = 2\mu R^{-1}(v_{,\theta} + u)$$

$$\tau_{\phi\phi} = 2\mu(R \sin \theta)^{-1}(w_{,\phi} + u \sin \theta + v \cos \theta)$$

$$\tau_{R\theta} = \tau_{\theta R} = \mu[R(v/R)_{,R} + R^{-1}u_{,\theta}]$$

$$\tau_{\theta\phi} = \tau_{\phi\theta} = \mu R^{-1}[\sin \theta(w \csc \theta)_{,\theta} + (\sin \theta)^{-1}v_{,\phi}]$$

$$\tau_{\phi R} = \tau_{R\phi} = \mu[(R \sin \theta)^{-1}u_{,\phi} + R(w/R)_{,R}] \tag{A63}$$

B1 SHEAR

```
C*****************************************************************************
C  THIS PROGRAM SOLVES EQUATIONS (7-26) AND (7-27) ITERATIVELY, FOR
C  GLYCEROL, TO SHOW THE EFFECT OF VISCOUS DISSIPATION ON VISCOMETRY.
C*****************************************************************************

      PROGRAM SHEAR
      IMPLICIT DOUBLE PRECISION (A-H,O-Z)

      DIMENSION V(101),T(101),ETA(101),TEMP(101)

C*****************************************************************************
C  SPECIFY VALUE OF PARAMETER EPS, AND SET UP FIRST-GUESS PROFILE OF T.
C*****************************************************************************

      NN = 101
      RN = NN-1
      DETA = 1.D0/RN

      DO 1000 I = 1,41
      EPS = .005D0*(I-1)**2

      DO 100 N = 1,NN
      ETA(N) = (N-1)*DETA
      T(N) = .5D0*EPS*(1.D0 - ETA(N)**2)
100   CONTINUE
```

```
C**********************************************************************
C   START ITERATION LOOP. SET CONVERGENCE TOLERANCE AND MAXIMUM NUMBER OF
C   ITERATIONS.  SET UNDERRELAXATION FACTOR, ALF.
C**********************************************************************

            ITMAX = 30
            TOL = .000001D0
            ITER = 1
            ALF = .5D0
999     CONTINUE

C**********************************************************************
C   INTEGRATE EQUATION (7-26), USING TRAPEZOIDAL RULE.
C**********************************************************************

            V(1) = 0.D0
            DV1 = DEXP(2.31D4*T(1)/(1.D0 + T(1))/293.D0)
            DV1 = DV1/(1.D0+ T(1))**52.4D0
         DO 200 N = 2,NN
            DV = DEXP(2.31D4*T(N)/(1.D0 + T(N))/293.D0)
            DV = DV/(1.D0+ T(N))**52.4D0
            V(N) = V(N-1) + .5D0*DETA*(DV + DV1)
            DV1 = DV
200     CONTINUE
            TAU = 1.D0/V(NN)

C**********************************************************************
C   INTEGRATE EQUATION (7-27), USING TRAPEZOIDAL RULE.
C**********************************************************************

            TINT = 0.D0
            DT1 = V(NN)
         DO 300 N = 1,NN-1
            L = NN-N
            DT = V(L)
            TINT = TINT + .5D0*(DT + DT1)*DETA
            DT1 = DT
            TEMP(L) = EPS*TINT/V(NN)**2
            T(L) = ALF*TEMP(L) + (1.D0 - ALF)*T(L)
300     CONTINUE

C           WRITE(6,3)ITER, T(1), TAU

C**********************************************************************
C   ADVANCE ITERATION COUNTER AND TEST FOR CONVERGENCE.
C**********************************************************************

        ITER = ITER + 1
            ERR = T(1) - TOLD
            TOLD = T(1)
            IF(DABS(ERR).GT.TOL.AND.ITER.LT.ITMAX) GO TO 999

C**********************************************************************
C   WRITE OUT CONVERGED TEMPERATURE AND VELOCITY PROFILES.
C**********************************************************************

            GO TO 500
            WRITE(6,1) EPS,NN
         DO 400 N = 1,NN
            V(N) = V(N)/V(NN)
            WRITE (6,2) ETA(N), V(N), T(N)
400     CONTINUE
500         WRITE(6,4) ITER, EPS, TAU, T(1)
1000    CONTINUE
1       FORMAT(5X,'EPS =',F10.5,5X,'NN =',I3,/,5X,'ETA',7X,'V',9X,'T',/)
2       FORMAT(3F10.5)
3       FORMAT(5X,'ITER =',I3,5X,'T(1) =',F10.5,5X,'TAU =',F10.5,/)
4       FORMAT(5X,I3,3F10.5)
        STOP
        END
```

B2   HAMEL

```
C******************************************************************************
C     THIS PROGRAM DEMONSTRATES THE THOMAS ALGORITHM, WITH ITERATION, TO
C     SOLVE A NONLINEAR DIFFERENTIAL EQUATION.   THE EQUATION IS
C          U'' + 2*(1 - U**2),   WITH BOUNDARY CONDITIONS U(0) = 0, AND
C     U'(NN) + 2 U(NN) = 2.   THIS IS DONE TWICE, FOR H = 0.1 AND H= 0.05,
C     THEN RICHARDSON EXTRAPOLATION IS APPLIED.
C******************************************************************************

      PROGRAM HAMEL
      IMPLICIT DOUBLE PRECISION (A-H,O-Z)

      DIMENSION ETA(81),U(81),E(81),F(81),U1(81)

         WRITE (6,1)
         ETAMAX = 4.D0
         BETA = (DLOG(1.D0+DSQRT(2.D0/3.D0))-DLOG(1.D0-DSQRT(2.D0/3.D0))
     1   )/2.D0

         DO 1000 I = 1,2

C******************************************************************************
C     SPECIFY VALUES THAT DON'T DEPEND ON ETA.
C******************************************************************************

         NN = 40*I+1
         NM = NN-1
         RN = NM
         H = ETAMAX/RN
         HI = 1.D0/H
         HI2 = HI*HI
         C = HI2
         A = HI2

C******************************************************************************
C     SET UP ETA(N) AND FIRST-GUESS VELOCITY PROFILE.
C******************************************************************************

      DO 10 N = 1,NN
         ETA(N) = (N-1)*H
         G = ETA(N)/ETAMAX
         U(N) = G*(2.D0-G)
10       CONTINUE

C******************************************************************************
C     SET UP BOUNDARY CONDITIONS.
C******************************************************************************

         ALFN = 1.D0
         E(1) = 0.D0
         F(1) = 0.D0

C******************************************************************************
C     SET ITERATION COUNTER.
C******************************************************************************

         ITER = 1

C******************************************************************************
C     TRIANGÚLARIZE COEFFICIENT MATRIX.
C******************************************************************************

2000  DO 200 N = 2,NN
         NM1 = N-1
         B = -4.D0*U(N) - 2.D0*HI2
         D = -2.D0*(1.D0 + U(N)*U(N))
         DENO= A*E(NM1) + B
```

```
          E(N) = -C/DENO
          F(N) = (D-A*F(NM1))/DENO
200   CONTINUE

C*********************************************************************
C     IMPOSE UPPER BOUNDARY CONDITION.
C*********************************************************************

          BETAN = 2.D0*ETA(NN)
          GAMN = BETAN
          AN = - ALFN*(A+C)/(2.D0*H*C)
          BN = BETAN - ALFN*B/(2.D0*H*C)
          DN = GAMN - ALFN*D/(2.D0*H*C)
          U(NN) =(DN - AN*F(NM))/(BN + AN*E(NM))

C*********************************************************************
C     BACK-SUBSTITUTE TO FIND U(N)

          DO 300 N = 1,NM
          K = NN-N
          U(K) = E(K)*U(K+1) + F(K)
300   CONTINUE

C*********************************************************************
C     CALCULATE SKIN FRICTION.
C*********************************************************************

          B = - 2.D0*HI
          D = - 2.D0
          SF = ((A+C)*U(2) +B*U(1) - D)/(2.D0*H*A)
          WRITE(6,2) ITER, SF

C*********************************************************************
C     ADVANCE AND CHECK ITERATION COUNTER.
C*********************************************************************

          ITER = ITER + 1
          IF(ITER.LT.7) GO TO 2000

C*********************************************************************
C     AFTER ITERATIONS HAVE CONVERGED, PREPARE FOR RICHARDSON EXTRAP.
C*********************************************************************

          IF(I.EQ.1) SF1 = SF
          IF(I.EQ.2) GO TO 1000
          DO 400 N = 1,NN
          U1(N) = U(N)
400   CONTINUE
1000  CONTINUE

C*********************************************************************
C     EXECUTE RICHARDSON EXTRAPOLATION AND CALCULATE EXACT ANSWER.
C*********************************************************************

          WRITE(6,3)
          NN = (NN-1)/2 + 1
          DO 500 N = 1,NN
          L = 2*N-1
          UE = (4.D0*U(L)-U1(N))/3.D0
          ANS = 3.D0*(DTANH(ETA(L)+BETA))**2 - 2.D0
          WRITE(6,4) ETA(L),U1(N),U(L),UE,ANS
500   CONTINUE

          SFE = (4.D0*SF - SF1)/3.D0
          EXACT = DSQRT(8.D0/3.D0)
          WRITE(6,5) SF1, SF, SFE, EXACT
```

B3 EKMAN

```
1       FORMAT(10X,'JEFFERY-HAMEL FLOW',/)
2       FORMAT(10X,'ITER =',.2,10X,'SF =',F10.5)
3       FORMAT(10X,'ETA',7X,'T(.1)',6X,'U(.05)',5X,'UE',7X,'ANS',/)
4       FORMAT(5X,4F10.5)
5       FORMAT(/,5X,'SF(.1) =',F10.5,5X,'SF(.05) =',F10.5,5X,'SFE =',
       1    F10.5,5X,'SFEX =',F10.5)

        STOP
        END




C*********************************************************************
C    THIS PROGRAM DEMONSTRATES THE THOMAS ALGORITHM FOR LINEAR COUPLED
C    DIFFERENTIAL EQUATIONS. THE PHYSICAL PROBLEM IS THE STEADY EKMAN
C    LAYER.
C*********************************************************************

        PROGRAM EKMAN
        IMPLICIT DOUBLE PRECISION (A-H,O-Z)
C*********************************************************************
C    ASSIGN STORAGE ARRAYS FOR INDEPENDENT AND DEPENDENT VARIABLES.
C*********************************************************************

        DIMENSION Z(121),U(121),V(121),S(121),T(121)
        DIMENSION E(2,2,121), F(2,121), G(2,2),GI(2,2), DD(2)
        DIMENSION A(2,2),B(2,2),C(2,2),D(2)

C*********************************************************************
C    SPECIFY NUMBER OF MESH POINTS.
C*********************************************************************

        DO 1000 M = 1,2
            NN = 60*M+1
            NM = NN-1
            RN = NM
            ZMAX = 12.D0
            H = ZMAX/RN
            HI = 1.D0/H
            HI2 = HI*HI
            WRITE(6,1) ZMAX,H

C*********************************************************************
C    FILL IN COEFFICIENT MATRICES.
C*********************************************************************

            A(1,1) = HI2
            A(1,2) = 0.D0
            A(2,1) = 0.D0
            A(2,2) = HI2
            B(1,1) = -2.D0*HI2
            B(1,2) = 2.D0
            B(2,1) = -2.D0
            B(2,2) = -2.D0*HI2
            C(1,1) = HI2
            C(1,2) = 0.D0
            C(2,1) = 0.D0
            C(2,2) = HI2
            D(1) = 0.D0
            D(2) = 0.D0
```

```
C*********************************************************************
C    SET BOUNDARY CONDITIONS AT Z = 0.
C*********************************************************************

          E(1,1,1) = 0.D0
          E(1,2,1) = 0.D0
          E(2,1,1) = 0.D0
          E(2,2,1) = 0.D0
          F(1,1)  = 1.D0
          F(2,1)  = 0.D0


C*********************************************************************
C    TRIANGULARIZE. FORM THE MATRIX G.
C*********************************************************************

      DO 200 N = 2,NM
          NM1 = N-1

          DO 10  I = 1,2
          DO 11 J = 1,2
            G(I,J) = 0.D0
            DO 12 K = 1,2
            G(I,J) = G(I,J) + A(I,K)*E(K,J,NM1)
12        CONTINUE
            G(I,J) = G(I,J) + B(I,J)
11        CONTINUE
10        CONTINUE


C*********************************************************************
C    INVERT THE MATRIX G, BY USE OF KRAMER'S RULE.
C*********************************************************************

          DEL = G(1,1)*G(2,2)-G(1,2)*G(2,1)
          GI(1,1) = G(2,2)/DEL
          GI(1,2) = - G(1,2)/DEL
          GI(2,1) = - G(2,1)/DEL
          GI(2,2) = G(1,1)/DEL


C*********************************************************************
C    CALCULATE THE E(I,J,N) AND F(I,N)
C*********************************************************************

          E(1,1,N) = - GI(1,1)*HI2
          E(1,2,N) = - GI(1,2)*HI2
          E(2,1,N) = - GI(2,1)*HI2
          E(2,2,N) = - GI(2,2)*HI2
          DD(1) = D(1)-A(1,1)*F(1,NM1)-A(1,2)*F(2,NM1)
          DD(2) = D(2)-A(2,1)*F(1,NM1)-A(2,2)*F(2,NM1)
          F(1,N) = GI(1,1)*DD(1) + GI(1,2)*DD(2)
          F(2,N) = GI(2,1)*DD(1) + GI(2,2)*DD(2)
200   CONTINUE


C*********************************************************************
C    SPECIFY BOUNDARY CONDITIONS AT ZMAX.
C*********************************************************************

          U(NN) = 0.D0
          V(NN) = 0.D0


C*********************************************************************
C    BACK-SUBSTITUTE.
C*********************************************************************

      DO 300 N = 1,NM
          L = NN-N
          U(L) = E(1,1,L)*U(L+1)+E(1,2,L)*V(L+1)+F(1,L)
          V(L) = E(2,1,L)*U(L+1)+E(2,2,L)*V(L+1)+F(2,L)
300   CONTINUE
```

```
C*****************************************************************************
C    CALCULATE VORTICITY COMPONENTS.
C*****************************************************************************

        S(1) = HI*(U(2)-1.D0)
        T(1) = HI*V(2)-H
     DO 400 N = 2,NM
        S(N) = 0.5D0*HI*(U(N+1)-U(N-1))
        T(N) = 0.5D0*HI*(V(N+1)-V(N-1))
400    CONTINUE

C*****************************************************************************
C    PRINT OUTPUT.
C*****************************************************************************

        WRITE(6,2)
     DO 500 N = 1,NN
        WRITE(6,3) Z(N),U(N),V(N),S(N),T(N)
500    CONTINUE
1000   CONTINUE

1      FORMAT(10X,'THE EKMAN LAYER.  ZMAX =',F7.3,10X,'H=',F5.3,/)
2      FORMAT (10X,'Z',10X,'U',10X,'V',10X,'S',10X,'T',/)
3      FORMAT( 5X,5F10.5)

        STOP
        END
```

B4 STAGPT

```
C***********************************************************************
C     THIS PROGRAM DEMONSTRATES THE USE OF THE THOMAS ALGORITHM FOR
C     NON-LINEAR DIFFERENTIAL EQUATIONS.  THE PROBLEM IS THAT OF THREE-
C     DIMENSIONAL STAGNATION-POINT FLOW.  SEE CHAPTER 10, SECTION 2.
C***********************************************************************

      PROGRAM STAGPT
      IMPLICIT DOUBLE PRECISION (A-H,O-Z)

C***********************************************************************
C     ASSIGN STORAGE ARRAYS FOR INDEPENDENT AND DEPENDENT VARIABLES.
C***********************************************************************

      DIMENSION  Z(481), U(481), V(481), W(481), R(481), S(481), P(481)
      DIMENSION  U1(481),V1(481),W1(481),R1(481),S1(481)
      DIMENSION  E(3,2,481),F(3,481)
      DIMENSION  A(3,3),B(3,3),D(3)
      DIMENSION  G(3,3)

C     ****************************************************************
C     SPECIFY NUMBER OF MESH POINTS
C     ****************************************************************

      NN = 481
      NM = NN - 1
      RN = NM
        WRITE (6,1)
        WRITE (6,2)

C***********************************************************************
C     SET UP Z(N) AND FIRST-GUESS PROFILES.
C***********************************************************************

      ZMAX = 12.D0
      H = ZMAX/RN

      DO 100 N = 1,NN
      Z(N) = (N-1)*H
      GG = Z(N)/ZMAX
      R(N) = 2.D0*(1.D0-GG)/ZMAX
      S(N) = R(N)
      U(N) = GG*(2.D0-GG)
      V(N) = U(N)
      W(N) = GG*GG*(1.D0-GG/3.D0)*ZMAX
100   CONTINUE

C***********************************************************************
C     SPECIFY VALUES OF PHYSICAL CONSTANT. THIS IS THE "C" OF THE TEXT.
C     LIMIT NUMBER OF ITERATIONS.  LOCATE OUTER BOUNDARY.
C***********************************************************************

      DO 1000 M = 1,41
      C = 1.D0 - (M-1)*0.05D0
      WRITE(6,2) C
      IF (M.EQ.1) ITM = 7
      IF (M.GT.1) ITM = 3
      IF (M.GE.40) ITM = 8
      ZMAX = 12.D0
      H = ZMAX/RN
      HI = 1.D0/H
      HI2 = HI*HI

C***********************************************************************
C     FILL IN NON-ZERO CONSTANTS IN COEFFICIENT MATRICES.
C***********************************************************************
```

```
         A(1,2) = 0.D0
         A(1,3) = 0.D0
         A(2,1) = 0.D0
         A(2,3) = 0.D0
         B(1,2) = 0.D0
         B(2,1) = 0.D0
         A(3,1) = 0.5D0
         A(3,2) = 0.5D0*C
         A(3,3) = HI
         B(3,1) = 0.5D0
         B(3,2) = 0.5D0*C
         B(3,3) = - HI

C***********************************************************************
C     TRIANGULARIZE THE COEFFICIENT MATRIX. START BY SETTING WALL
C     BOUNDARY CONDITIONS, IN TERMS OF E(I,J,1) AND F(I,1).
C***********************************************************************

         DO 10 I = 1,3
         DO 11 J = 1,2
         E(I,J,1) = 0.D0
11       CONTINUE
         F(I,1) = 0.D0
10       CONTINUE

C***********************************************************************
C     START ITERATION LOOP.
C***********************************************************************

         ITER = 1

C***********************************************************************
C     TRIANGULARIZE.
C***********************************************************************

2000     DO 200 N = 2,NN
         NM1 = N-1
         A(1,1) = HI2 - 0.5D0*HI*W(N)
         A(2,2) = A(1,1)
         B(1,1) = -2.D0*(HI2 + U(N))
         B(1,3) = R(N)
         B(2,2) = -2,D0*(HI2 + V(N)*C)
         B(2,3) = S(N)
         CC = HI2 + 0.5D0*HI*W(N)
         D(1) = W(N)*R(N) - (U(N)*U(N) + 1.D0)
         D(2) = W(N)*S(N) - C*(V(N)*V(N) + 1.D0)

C***********************************************************************
C     FORM THE MATRIX G.
C***********************************************************************

         DO 30 I = 1,3
         DO 31 J = 1,2
         G(I,J) = 0.D0
         DO 32 K = 1,3
         G(I,J) = G(I,J) + A(I,K)*E(K,J,NM1)
32       CONTINUE
         G(I,J)=G(I,J) + B(I,J)
31       CONTINUE
30       CONTINUE
         G(1,3) = B(1,3)
         G(2,3) = B(2,3)
         G(3,3) = B(3,3)
         DD1 = D(1) - A(1,1)*F(1,NM1)
         DD2 = D(2) - A(2,2)*F(2,NM1)
         DD3 = D(3) - A(3,1)*F(1,NM1) - A(3,2)*F(2,NM1) - A(3,3)*
```

```
      1   F(3,NM1)

C*********************************************************************
C     SOLVE FOR E(I,J,N) AND F(I,N) BY KRAMER'S RULE.
C*********************************************************************

      H1 = G(2,2)*G(3,3)-G(2,3)*G(3,2)
      H2 = G(2,3)*G(3,1)-G(2,1)*G(3,3)
      H3 = G(2,1)*G(3,2)-G(2,2)*G(3,1)
      H4 = G(3,2)*G(1,3)-G(1,2)*G(3,3)
      H5 = G(1,1)*G(3,3)-G(1,3)*G(3,1)
      H6 = G(1,2)*G(3,1)-G(3,2)*G(1,1)
      H7 = G(1,2)*G(2,3)-G(2,2)*G(1,3)
      H8 = G(2,1)*G(1,3)-G(2,3)*G(1,1)
      H9 = G(1,1)*G(2,2)-G(1,2)*G(2,1)
      DENO = G(1,1)*H1+G(1,2)*H2+G(1,3)*H3

      E(1,1,N) = -CC*H1/DENO
      E(2,1,N) = -CC*H2/DENO
      E(3,1,N) = -CC*H3/DENO
      E(1,2,N) = -CC*H4/DENO
      E(2,2,N) = -CC*H5/DENO
      E(3,2,N) = -CC*H6/DENO
      F(1,N) = (DD1*H1+DD2*H4+DD3*H7)/DENO
      F(2,N) = (DD1*H2+DD2*H5+DD3*H8)/DENO
      F(3,N) = (DD1*H3+DD2*H6+DD3*H9)/DENO
200   CONTINUE

C*********************************************************************
C     CALCULATE THE OUTER BOUNDARY CONDITIONS FOR U AND V.
C*********************************************************************

      IF(M.EQ.1.AND.ITER.EQ.1) GO TO 201
      IF(M.EQ.41) W(NN) = DSQRT(8.D0)
      ALF1N = 1.D0
      ALF2N = 1.D0
      BET1N = W(NN) + (3.D0+C)/W(NN)
      BET2N = W(NN) + (1.D0+3.D0*C)/W(NN)
      IF(M.EQ.41) BET1N = .5D0*(W(NN)+DSQRT(W(NN)**2+8.D0))
      IF(M.EQ.41) BET2N = .5D0* W(NN)
      GAM1N = BET1N
      GAM2N = BET2N
      E1 = E(1,1,NM) - 2.D0*H*BET1N/ALF1N
      E2 = E(2,2,NM) - 2.D0*H*BET2N/ALF2N
      F1 = F(1,NM)+2.D0*H*GAM1N/ALF1N
      F2 = F(2,NM)+2.D0*H*GAM2N/ALF2N
      AU = 1.D0 - E(1,1,NN)*E1 - E(1,2,NN)*E(2,1,NM)
      AV = -E(1,1,NN)*E(1,2,NM) - E(1,2,NN)*E2
      BU = -E(2,1,NN)*E1 - E(2,2,NN)*E(2,1,NM)
      BV = 1.D0 - E(2,1,NN)*E(1,2,NM) - E(2,2,NN)*E2
      AA = E(1,1,NN)*F1 + E(1,2,NN)*F2 + F(1,NN)
      BB = E(2,1,NN)*F1 + E(2,2,NN)*F2 + F(2,NN)
      DENO = AU*BV - AV*BU
      U(NN) = (AA*BV - BB*AV)/DENO
      V(NN) = (BB*AU - AA*BU)/DENO
      GO TO 202
201   U(NN) = 1.D0
      V(NN) = 1.D0
202   W(NN) = (E(3,1,NN)*E1 + E(3,2,NN)*E(2,1,NM))*U(NN)
     1        + (E(3,1,NN)*E(1,2,NM) + E(3,2,NN)*E2)*V(NN)
     2        + E(3,1,NN)*F1 + E(3,2,NN)*F2 + F(3,NN)

C*********************************************************************
C     BACK-SUBSTITUTE TO FIND U(N), V(N) AND W(N).
C*********************************************************************
```

```
      DO 300 N = 1,NM
      K = NN-N
      U(K) = E(1,1,K)*U(K+1)+E(1,2,K)*V(K+1)+F(1,K)
      V(K) = E(2,1,K)*U(K+1)+E(2,2,K)*V(K+1)+F(2,K)
      W(K) = E(3,1,K)*U(K+1)+E(3,2,K)*V(K+1)+F(3,K)
300   CONTINUE

C********************************************************************
C     UPDATE R(N) AND S(N) FOR ITERATIONS.
C********************************************************************

      DO 400 N = 2,NM
      R(N) = 0.5D0*HI*(U(N+1)-U(N-1))
      S(N) = 0.5D0*HI*(V(N+1)-V(N-1))
400   CONTINUE

      R(1) = HI*U(2)+0.5D0*H
      S(1) = HI*V(2)+0.5D0*H*C
      DEL1 = (1.D0+C)*Z(NN) - W(NN)

C********************************************************************
C     WRITE R(1), S(1), AND DEL1 FOR EACH ITERATION.
C********************************************************************

      WRITE(6,3) ITER, R(1), S(1), DEL1

C********************************************************************
C     ADVANCE AND CHECK ITERATION COUNTER.
C********************************************************************

      ITER = ITER+1
      IF (ITER.LE.ITM) GO TO 2000

C********************************************************************
C     CALCULATE PRESSURE.
C********************************************************************

      DO 500 N = 2,NN
      P(N) = -.5D0*W(N)**2 - U(N) - C*V(N)
500   CONTINUE

C********************************************************************
C     PRINT HEADINGS AND SELECTED VELOCITY PROFILES.
C********************************************************************

      K = (M+3)/4
      KK = (M+2)/4
      IF (K.EQ.KK) GO TO 650
550   WRITE (7,2) C
      WRITE (7,4)
      DO 600 N = 1,NN,4
      Z(N) = (N-1)*H
      WRITE (7,5) Z(N),U(N),V(N),W(N),R(N),S(N),P(N)
600   CONTINUE
C********************************************************************
C  PROJECT PROFILES FOR NEXT ITERATION.
C********************************************************************

650   CONTINUE
      DO 700 N = 1,NN
      TEMPU = U(N)
      TEMPV = V(N)
      TEMPW = W(N)
      TEMPR = R(N)
      TEMPS = S(N)
      IF(M.EQ.1) GO TO 750
```

B5 SPNCYL

```
        U(N) = 2.D0*U(N) - U1(N)
        V(N) = 2.D0*V(N) - V1(N)
        W(N) = 2.D0*W(N) - W1(N)
        R(N) = 2.D0*R(N) - R1(N)
        S(N) = 2.D0*S(N) - S1(N)
750     U1(N) = TEMPU
        V1(N) = TEMPV
        W1(N) = TEMPW
        R1(N) = TEMPR
        S1(N) = TEMPS
700     CONTINUE

1000    CONTINUE

1       FORMAT(10X, 'THREE-DIMENSIONAL STAGNATION-POINT FLOW.',/)
2       FORMAT(10X, 'C = ', F 6.3,/)
3       FORMAT( 5X, 'ITER =', I2, 5X, 'R(1) =',F9.5,5X,'S(1)=',F9.5,
      1 5X,'DEL1=',F9.5)
4       FORMAT (/,5X,'Z',9X,'U',9X,'V',9X,'W',9X,'R',9X,'S',9X,'P',/)
5       FORMAT (1X,7F10.5)

        STOP
        END




C*********************************************************************
C    THIS PROGRAM DEMONSTRATES AN IMPLICIT MARCHING SCHEME FOR LINEAR
C    PARABOLIC PARTIAL DIFFERENTIAL EQUATIONS. THE PHYSICAL FLOW IS THAT
C    OUTSIDE AN IMPULSIVELY SPUN-UP CYLINDER. SEE SECTION 10.3.
C*********************************************************************

        PROGRAM SPNCYL
        IMPLICIT DOUBLE PRECISION (A-H,O-Z)

C*********************************************************************
C    ASSIGN STORAGE FOR ETA, G, GM1, R/A, VORTICITY, AND VELOCITY.
C*********************************************************************

        DIMENSION ETA(201),G(201),GM1(201),RA(201),VOR(201),V(201)
        DIMENSION E(201),F(201)

C*********************************************************************
C    SPECIFY NUMBER OF MESH POINTS IN ETA.
C*********************************************************************

        NN = 81
        NM = NN-1
        RN = NM
        ETAMAX =8.D0
        H = ETAMAX/RN
        HI = 1.D0/H
        HI2 = HI*HI

C*********************************************************************
C    SET UP ETA(N), AND A PHONY INITIAL PROFILE FOR GM1(N).  THIS
C    FUNCTION WILL BE MULTIPLIED BY ZERO WHEN T = 0.
C*********************************************************************

        DO 100 N = 1,NN
        ETA(N) = (N-1)*H
        GM1(N) = 1.D0
100     CONTINUE
```

```
C**********************************************************************
C      SET THE NUMBER OF TIME STEPS, AND START THE MARCHING LOOP.
C**********************************************************************

         MN = 320
         MM = MN+1
         RM = MN
         WRITE(6,1) NN, MN
      DO 1000 M = 1,MN
         T1 = (M-1)/RM
         TAU = T1**2/(1.D0+T1**2)

C**********************************************************************
C      SPECIFY WALL BOUNDARY CONDITION IN TERMS OF E(1) AND F(1).
C**********************************************************************

         E(1) = 0.D0
         F(1) = 1.D0

C**********************************************************************
C      TRIANGULARIZE.
C**********************************************************************

      DO 200 N = 2,NM
         NM1 = N-1
         IF(M.EQ.1) GO TO 20
         P = 0.5D0*ETA(N) - T1/(1.D0+T1*ETA(N))
         GO TO 21

20       P = 0.5D0*ETA(N)
21       A = HI2 - 0.5D0*HI*P
         B = -2.D0*HI2 - .5D0*(M-1)
         C = HI2 + 0.5D0*HI*P
         D = -R*GM1(N)
         DENO = A*E(NM1) + B
         E(N) = -C/DENO
         F(N) = (D-A*F(NM1))/DENO
200      CONTINUE

C**********************************************************************
C      SET OUTER BOUNDARY CONDITION.
C**********************************************************************

         G(NN) = 0.D0

C**********************************************************************
C      BACK-SUBSTITUTE.
C**********************************************************************

      DO 300 N = 1,NM
         K = NN-N
         G(K) = E(K)*G(K+1) + F(K)
300      CONTINUE

C**********************************************************************
C      UPDATE GM1(N)
C**********************************************************************

      DO 400 N = 1,NN
         GM1(N) = G(N)
400      CONTINUE
```

```
C**********************************************************************
C     CALCULATE R/A AND VORTICITY, EXCEPT AT THE INITIAL AND FINAL TIMES.
C**********************************************************************

          IF(M.EQ.1.OR.M.EQ.MM) GO TO 900
      DO 500 N = 2,NM
          RA(N) = 1.D0 + T1*ETA(N)
          VOR(N) = 0.5D0*HI*(G(N+1)-G(N-1))/(T1*RA(N))
500   CONTINUE
          VOR(1) = HI*(G(2)-G(1))/((1.D0+0.5D0*H*T1)*T1)
          RA(1) = 1.D0
          RA(NN) = 1.D0 + T1*ETAMAX

          LL = (M-2)/32
          LLL = (M-1)/32
          IF(LL.EQ.LLL) GO TO 550
          T   = T1*T1
          WRITE(6,2) T,TAU,VOR(1)
550   CONTINUE

C**********************************************************************
C     PRINT PROFILES.
C**********************************************************************

          LL = (M-2)/64
          LLL = (M-1)/64
          IF(LL.EQ.LLL) GO TO 900
          WRITE (6,3)
      DO 600 N = 1,NN,2
          V(N) = G(N)/RA(N)
          WRITE (6,4) ETA(N),RA(N),G(N),V(N),VOR(N)
600   CONTINUE
900   CONTINUE
          IF(M.GT.1.AND.M.LT.MM) GO TO 1000

          WRITE (6,5)
      DO 700 N = 1,NN,2
          WRITE(6,6) ETA(N), G(N)
700   CONTINUE
1000  CONTINUE

1     FORMAT(/,10X,'NN =',I3,10X,'MN =',I3,/)
2     FORMAT(5X,'TAU = ',F10.5,5X,'NU*T/A**2 =',F10.5,5X,'VOR(1) =',
     1    F10.5,/)
3     FORMAT(10X,'ETA',7X,'R/A',8X,'G',9X,'V',7X,'VOR',/)
4     FORMAT(5X,5F10.5)
5     FORMAT(10X,'ETA',10X,'G',/)
6     FORMAT(5X, 2F10.5)
      STOP
      END
```

B6 SPNDSK

```
C**********************************************************************
C     THIS PROGRAM DEMONSTRATES AN IMPLICIT MARCHING SCHEME FOR COUPLED,
C     NON-LINEAR PARABOLIC PARTIAL DIFFERENTIAL EQUATIONS. THE FLOW IS
C     THAT ABOVE AN IMPULSIVELY SPUN-UP DISK. SEE SECTION 10.3.
C**********************************************************************

      PROGRAM SPNDSK
      IMPLICIT DOUBLE PRECISION (A-H,O-Z)

C**********************************************************************
C     ASSIGN STORAGE FOR Z, U, V, W, UM1, VM1, R ,AND S. BECAUSE F,G,
C     AND H ARE USED ELSEWHERE IN THE PROGRAM, WE USE THE FOLLOWING
C     SYMBOLS:  U FOR F, V FOR G, W FOR H, Z FOR ZETA. X IS THE INTEGRAL
C     OF W, AND P IS PRESSURE - WALL PRESSURE.
C**********************************************************************

      DIMENSION Z(121),U(121),V(121),UM1(121),VM1(121),W(121)
      DIMENSION R(121),S(121),X(121),XM1(121),P(121)
      DIMENSION E(3,2,121),F(3,121),A(3,3),B(3,3),D(3),DD(3)
      DIMENSION G(3,3)

C**********************************************************************
C     SPECIFY NUMBER OF MESH POINTS IN ETA.
C**********************************************************************

      NN = 121
      NM = NN-1
      RN = NM
      ZMAX = 6.D0
      H = ZMAX/RN
      HI = 1.D0/H
      HI2 = HI*HI

C**********************************************************************
C     SET UP ETA(N), AND PHONY INITIAL PROFILES FOR UM1(N) AND VM1(N).
C     THESE FUNCTIONS WILL BE MULTIPLIED BY ZERO WHEN T = 0.
C**********************************************************************

      DO 100 N = 1,NN
         Z(N) = (N-1)*H
         V(N) = 1.D0 - DERF(Z(N))
         UM1(N) = 1.D0
         VM1(N) = 1.D0
         XM1(N) = 1.D0
100   CONTINUE

C**********************************************************************
C     SET CONSTANT COEFFICIENTS.
C**********************************************************************

      PI = 4.D0*DATAN(1.D0)
      A(1,2) = 0.D0
      A(1,3) = 0.D0
      A(2,1) = 0.D0
      A(2,3) = 0.D0
      A(3,2) = 0.D0
      B(3,2) = 0.D0
      A(3,1) = 1.D0
      A(3,3) = HI
      B(3,1) = 1.D0
      B(3,3) = - HI
C**********************************************************************
C     SET THE NUMBER OF TIME STEPS, AND START THE MARCHING LOOP.
C**********************************************************************
```

```
          MM = MN+1
          RM = MN
          WRITE(6,1) NN, MN
          DO 1000 M = 1,MM
          TAU = (M-1)/RM
          WT = 0.D0
          T = 0.D0
          IF(M.EQ.1.OR.M.EQ.MM) GO TO 990
          WT = TAU/(1.D0-TAU)
          T = 1.D0 - DEXP(-WT)
990       IF(M.EQ.MM) T = 1.D0
          PSI = T*(1.D0-TAU)**2*RM
          WRITE(6,2) TAU, WT,T

C********************************************************************
C     SPECIFY WALL BOUNDARY CONDITION IN TERMS OF E(1) AND F(1).
C********************************************************************

          DO 10 I = 1,3
          DO 11 J = 1,2
          E(I,J,1) = 0.D0
11        CONTINUE
10        CONTINUE
          F(1,1) = 0.D0
          F(2,1) = 1.D0
          F(3,1) = 0.D0

C********************************************************************
C     PREPARE TO ITERATE IF T GT. ZERO
C********************************************************************

          ITER = 1

C********************************************************************
C     TRIANGULARIZE.
C********************************************************************

2000  DO 200 N = 2,NN
          NM1 = N-1
          P1 = 0.25D0*HI*((1.D0-T)*Z(N)+2.D0*T**2*W(N))
          A(1,1) = HI2 - P1
          A(2,2) = A(1,1)
          B(1,1) = -2.D0*HI2 - (1.D0-T) - 2.D0*T**2*U(N) - PSI
          B(1,2) = 2.D0*V(N)
          B(1,3) = T**2*R(N)
          B(2,1) = -2.D0*T*V(N)
          B(2,2) = -2.D0*HI2 - 2.D0*T*U(N) - PSI
          B(2,3) = T**2*S(N)
          C = HI2 + P1
          D(1) = -PSI*UM1(N) + V(N)**2 - T**2*(U(N)**2 - R(N)*W(N))
          D(2) = -PSI*VM1(N) - 2.D0*T*U(N)*V(N) + T**2*S(N)*W(N)
          D(3) = 0.D0

C********************************************************************
C     FORM THE MATRIX G.
C********************************************************************

          DO 20 I = 1,3
          DO 21 J = 1,2
          G(I,J) = 0.D0
          DO 22 K = 1,3
          G(I,J) = G(I,J) + A(I,K)*E(K,J,NM1)
22        CONTINUE
          G(I,J)=G(I,J) + B(I,J)
21        CONTINUE
20        CONTINUE
```

```
         G(1,3) = B(1,3)
         G(2,3) = B(2,3)
         G(3,3) = B(3,3)

         DD1 = D(1) - A(1,1)*F(1,NM1)
         DD2 = D(2) - A(2,2)*F(2,NM1)
         DD3 = D(3) - A(3,1)*F(1,NM1) - A(3,3)*F(3,NM1)

C*********************************************************************
C      SOLVE FOR E(I,J,N) AND F(I,N) BY KRAMER'S RULE.
C*********************************************************************

         H1 = G(2,2)*G(3,3)-G(2,3)*G(3,2)
         H2 = G(2,3)*G(3,1)-G(2,1)*G(3,3)
         H3 = G(2,1)*G(3,2)-G(2,2)*G(3,1)
         H4 = G(3,2)*G(1,3)-G(1,2)*G(3,3)
         H5 = G(1,1)*G(3,3)-G(1,3)*G(3,1)
         H6 = G(1,2)*G(3,1)-G(3,2)*G(1,1)
         H7 = G(1,2)*G(2,3)-G(2,2)*G(1,3)
         H8 = G(2,1)*G(1,3)-G(2,3)*G(1,1)
         H9 = G(1,1)*G(2,2)-G(1,2)*G(2,1)
         DENO = G(1,1)*H1+G(1,2)*H2+G(1,3)*H3

         E(1,1,N) = -C*H1/DENO
         E(2,1,N) = -C*H2/DENO
         E(3,1,N) = -C*H3/DENO
         E(1,2,N) = -C*H4/DENO
         E(2,2,N) = -C*H5/DENO
         E(3,2,N) = -C*H6/DENO
         F(1,N) = (DD1*H1+DD2*H4+DD3*H7)/DENO
         F(2,N) = (DD1*H2+DD2*H5+DD3*H8)/DENO
         F(3,N) = (DD1*H3+DD2*H6+DD3*H9)/DENO
200   CONTINUE

C*********************************************************************
C      SET OUTER BOUNDARY CONDITION.
C*********************************************************************

         IF (M.EQ.1) GO TO 201
         ALF1N = 1.D0
         ALF2N = 1.D0
         BET1N = W(NN)
         BET2N = W(NN)
         GAM1N = 0.D0
         GAM2N = 0.D0
         E1 = E(1,1,NM) - 2.D0*H*BET1N/ALF1N
         E2 = E(2,2,NM) - 2.D0*H*BET2N/ALF2N
         F1 = F(1,NM)+2.D0*H*GAM1N/ALF1N
         F2 = F(2,NM)+2.D0*H*GAM2N/ALF2N
         AU = 1.D0 - E(1,1,NN)*E1 - E(1,2,NN)*E(2,1,NM)
         AV = -E(1,1,NN)*E(1,2,NM) - E(1,2,NN)*E2
         BU = -E(2,1,NN)*E1 - E(2,2,NN)*E(2,1,NM)
         BV = 1.D0 - E(2,1,NN)*E(1,2,NM) - E(2,2,NN)*E2
         AA = E(1,1,NN)*F1 + E(1,2,NN)*F2 + F(1,NN)
         BB = E(2,1,NN)*F1 + E(2,2,NN)*F2 + F(2,NN)
         DENO = AU*BV - AV*BU
           U(NN) = (AA*BV - BB*AV)/DENO
           V(NN) = (BB*AU - AA*BU)/DENO
           W(NN) = (E(3,1,NN)*E1 + E(3,2,NN)*E(2,1,NM))*U(NN)
     1           + (E(3,1,NN)*E(1,2,NM) + E(3,2,NN)*E2)*V(NN)
     2           + E(3,1,NN)*F1 + E(3,2,NN)*F2 + F(3,NN)

         GO TO 202
201      U(NN) = 0.D0
         V(NN) = 0.D0
202   CONTINUE
```

```
C*********************************************************************
C     BACK-SUBSTITUTE.
C*********************************************************************

      DO 300 N = 1,NM
         K = NN-N
         U(K)  = E(1,1,K)*U(K+1)+E(1,2,K)*V(K+1)+F(1,K)
         V(K)  = E(2,1,K)*U(K+1)+E(2,2,K)*V(K+1)+F(2,K)
         W(K)  = E(3,1,K)*U(K+1)+E(3,2,K)*V(K+1)+F(3,K)
300   CONTINUE

         IF (M.EQ.1) W(NN) = W(NM)+ H*(U(NM)+U(NN))

C*********************************************************************
C     UPDATE R(N) AND S(N) FOR ITERATIONS.
C*********************************************************************

      DO 400 N = 2,NM
         R(N) = 0.5D0*HI*(U(N+1)-U(N-1))
         S(N) = 0.5D0*HI*(V(N+1)-V(N-1))
400   CONTINUE

         R(1) = HI*U(2)+0.5D0*H
         S(1) = HI*(V(2)-V(1))
         WRITE(6,3) TAU, ITER, R(1), S(1)
         ITER = ITER+1
         IF (ITER.LE.2) GO TO 2000

C*********************************************************************
C     CALCULATE X AND P.
C*********************************************************************

         X(1) = 0.D0
      DO 500 N = 2,NN
         X(N) = X(N-1) + 0.5D0*H*(W(N)+W(N-1))
         P(N) = PSI*(X(N)-XM1(N)) + 2.D0*(1.D0-T)*X(N) - 2.D0*U(N)
     1      -0.5D0*(T**2*W(N)**2 + (1.D0-T)*Z(N)*W(N))
500   CONTINUE

C*********************************************************************
C     UPDATE UM1(N),VM1(N),AND XM1(N), PREPARING FOR NEXT TIME STEP.
C*********************************************************************

      DO 600 N = 1,NN
         UM1(N) = U(N)
         VM1(N) = V(N)
         XM1(N) = X(N)
600   CONTINUE

C*********************************************************************
C     PRINT PROFILES.
C*********************************************************************

         IF(M.EQ.1.OR.M.EQ.MM) GO TO 650
         LL = (M+7)/16
         LLL = (M+6)/16
         IF(LL.EQ.LLL) GO TO 1000
650      WRITE (6,4)
      DO 700 N = 1,NN,2
         IF (N.EQ.1) GO TO 701
         IF (N.EQ.NN) GO TO 702
         THETA =(180.D0/PI)*DATAN (V(N)/U(N))
         GO TO 702
701      THETA = 90.D0
702      WRITE (6,5) Z(N),U(N),V(N),W(N),P(N),R(N),S(N),THETA
```

```
700    CONTINUE
1000   CONTINUE

1      FORMAT(/,10X,'NN =',I3,10X,'MN =',I3,/)
2      FORMAT(/,10X,'TAU = ',F10.5,10X,'WT =',F10.5,10X,'T =',F10.5,/)
3      FORMAT(5X,'TAU =',F10.5,5X,'ITER =',I2,5X,'R(1) =',F10.5,5X,
       1    'S(1) =',F10.5)
4      FORMAT(/,'Z',8X,'U',8X,'V',8X,'W',8X,'P',8X,'R',8X,'S',
       1    3X,'THETA',/)
5      FORMAT(8F9.5)

       STOP
       END
```

B7 DIRECOU

```
C*********************************************************************
C THIS PROGRAM APPROXIMATES COUETTE FLOW IN A RECTANGULAR CHANNEL BY
C CENTERED FINITE DIFFERENCES, AND SOLVES THE RESULTING ALGEBRAIC
C EQUATIONS DIRECTLY, BY USE OF FINITE FOURIER TRANSFORMS. SEE EQUATIONS
C (10-83) TO (10-87) OF TEXT.
C*********************************************************************

      PROGRAM DIRECOU
      IMPLICIT DOUBLE PRECISION (A-H,O-Z)

C*********************************************************************
C  ASSIGN STORAGE FOR REAL AND TRANSFORMED VELOCITY,FOR FLOWRATE PER
C  UNIT SPAN, AND FOR THE THOMAS ALGORITHM.
C*********************************************************************

      DIMENSION U(201,201),UT(201,201),QI(201),E(201)
      INTEGER S

C*********************************************************************
C  EVALUATE CONSTANTS.
C*********************************************************************

      H = 1.D0
      W = 5.D0
      AR = W/H
      JJ = 9
      II = 9
      JM = JJ-1
      RJ = JM
      IM = II-1
      RI = IM
      HY = H/RJ
      HZ = W/RI
      RA = HY/HZ
      PI = 2.D0*DASIN(1.D0)

C*********************************************************************
C  CALCULATE UT(S,JJ), THE TRANSFORMED VELOCITY AT THE MOVING WALL.
C*********************************************************************

      WRITE(6,1)
      DO 100 S = 1,IM
         UT(S,JJ) = 0.D0
         DO 50 I = 2,IM
            UT(S,JJ) = UT(S,JJ) + (2.D0/RI)*DSIN(PI*S*(I-1)/RI)
50       CONTINUE
         WRITE(6,2) S,UT(S,JJ)
100   CONTINUE

C*********************************************************************
C  USE THOMAS ALGORITHM TO CALCULATE UT(S,J).  FIRST APPLY NO-SLIP
C  CONDITION UT(S,1) = 0, WHICH IMPLIES E(1) = 0.
C*********************************************************************

      E(1) = 0.D0
      DO 200 S = 1,IM
         B = 2.D0*(RA**2*(DCOS(S*PI/RI)-1.D0) - 1.D0)
         DO 150 J = 2,JM
            E(J) = -1.D0/(B + E(J-1))
150      CONTINUE
         DO 175 J = 1,JM
            K = JJ-J
            UT(S,K) = E(K)*UT(S,K+1)
175      CONTINUE
200   CONTINUE
```

```
C*********************************************************************
C   SYNTHESIZE THE REAL VELOCITY.
C*********************************************************************

         DO 300 I = 2,IM
         DO 275 J = 1,JJ
            U(I,J) = 0.D0
         DO 250 S = 1,IM
            U(I,J) = U(I,J) + UT(S,J)*DSIN(PI*(I-1)*S/RI)
250      CONTINUE
275      CONTINUE
300      CONTINUE


C*********************************************************************
C   CALCULATE FLOW RATE AND DRAG ON MOVING WALL.  USE TRAPEZOIDAL RULE
C   FOR FLOW RATE. WRITE SELECTED VELOCITY PROIFILES.
C*********************************************************************

           FLOW = 0.D0
           DRAG = 0.D0
           DRAGB = 0.D0
           DRAGS = 0.D0
         DO 400 I = 2,IM
           QI(I) = HY*.5D0*(U(I,1) + U(I,JJ))
         DO 350 J = 2,JM
           QI(I) = QI(I) + HY*U(I,J)
350      CONTINUE
           FLOW = FLOW + HZ*QI(I)
           TAU = (U(I,JJ) - U(I,JM))/HY
           DRAG = DRAG + HZ*TAU
           TAUB = (U(I,2)-U(I,1))/HY
           DRAGB = DRAGB + HZ*TAUB
400      CONTINUE

         DO 450 J = 2,JM
           TAUS = (U(2,J)-U(1,J))/HZ
           DRAGS = DRAGS + 2.D0*TAUS*HY
450      CONTINUE

           WRITE(6,3) AR,JJ,M,FLOW,DRAG,DRAGB,DRAGS

           WRITE(6,4)
           JMID = JM/2+1
         DO 500 I =1,II
           Z = (I-1)*HZ
           Y = (JJ-2)*HY
           WRITE(6,5) Y,Z,U(I,JM)
500      CONTINUE

           WRITE(6,4)
           IMID = MN/2+1
         DO 600 J = 1,JJ
           Z = HZ
           Y = (J-1)*HY
           WRITE(6,5) Y,Z,U(2,J)
600      CONTINUE

1        FORMAT(5X,'S',9X,'UT(S,JJ)',/)
2        FORMAT(I3,F10.5)
3        FORMAT(/,5X,'AR =',F10.5,3X,'JJ =',I3,3X,'M =',I3,/,1X,'FLOW =',
     1          F10.5,1X,'DRAG =',F10.5,1X,'DRAGB =',F10.5,1X,'DRAGS =',
     2          F10.5)
4        FORMAT(5X,'Y',9X,'Z',9X,'U',/)
5        FORMAT(3F10.5)
         STOP
         END
```

B8 ADICOU

```
C*******************************************************************
C THIS PROGRAM APPROXIMATES COUETTE FLOW IN A RECTANGULAR CHANNEL BY
C CENTERED FINITE DIFFERENCES, AND SOLVES THE RESULTING ALGEBRAIC
C EQUATIONS ITERATIVELY, BY THE ADI METHOD.        SEE EQUATIONS
C (10-88) AND (10-89) OF TEXT.
C*******************************************************************

      PROGRAM ADICOU
      IMPLICIT DOUBLE PRECISION (A-H,O-Z)

C*******************************************************************
C  ASSIGN STORAGE FOR REAL AND TRANSFORMED VELOCITY,FOR FLOWRATE PER
C  UNIT SPAN, AND FOR THE THOMAS ALGORITHM. UH IS THE HALF-STEP VELOCITY
C*******************************************************************

      DIMENSION U(201,201),UH(201,201),UP(201,201),QI(201),E(201),F(201)

C*******************************************************************
C EVALUATE CONSTANTS.
C*******************************************************************

      H = 1.D0
      W = 5.D0
      AR = W/H
      JJ = 65
      II = 65
      JM = JJ-1
      RJ = JM
      IM = II-1
      RI = IM
      IMID = IM/2+1
      JMID = JM/2+1
      HY = H/RJ
      HZ = W/RI
      ZY =(HZ/HY)**2
      YZ =(HY/HZ)**2
      ZT = .5D0
      YT = ZT*YZ

C*******************************************************************
C SET UP INITIAL FLOW. LET IT BE THE FLOW WITHOUT SIDE WALLS.
C*******************************************************************

      WRITE(6,1)
      WRITE(6,2) AR,II,JJ
      DO 100 I = 2,IM
      DO 50 J = 1,JJ
         U(I,J) = (J-1)*HY
         UP(I,J) = U(I,J)
50    CONTINUE
100   CONTINUE

C*******************************************************************
C SET ITERATION COUNTER AND LIMIT MAXIMUM NUMBER OF ITERATIONS.
C ESTABLISH CONVERGENCE CRITERION.
C*******************************************************************

      ITER = 1.D0
      ITMAX = 60
      EPS = .00001D0
      FLOW1 = 0.D0
      DRAGB1 = 0.D0

C*******************************************************************
C TRY UNDERRELAXATION TO IMPROVE CONVERGENCE.
C*******************************************************************
```

```
            ALF = .90D0
999     DO 10 I= 1,II
        DO 9 J = 1,JJ
            U(I,J) = ALF*U(I,J)+(1.D0-ALF)*UP(I,J)
            UP(I,J)= U(I,J)
9       CONTINUE
10      CONTINUE

C**********************************************************************
C   IMPOSE NO-SLIP CONDITION AT SIDEWALLS, AND ACCOUNT FOR DIFFUSION IN
C   THE Z-DIRECTION.
C**********************************************************************

            E(1) = 0.D0
            F(1) = 0.D0
            B = -2.D0*(1.D0+ZT)
        DO 200 J = 2,JM
        DO 150 I = 2,IM
            D = 2.D0*(ZY-ZT)*U(I,J) - ZY*(U(I,J+1) + U(I,J-1))
            DENO = B + E(I-1)
            E(I) = - 1.D0/DENO
            F(I) = (D - F(I-1))/DENO
150     CONTINUE
            UH(II,J) = 0.D0
        DO 175 I = 1,IM
            K = II - I
            UH(K,J) = E(K)*UH(K+1,J) +F(K)
175     CONTINUE
200     CONTINUE

C**********************************************************************
C   ACCOUNT FOR DIFFUSION IN Y-DIRECTION.   THE BOUNDARY CONDITION AT
C   Y = 0 IMPLIES THAT E(1) AND F(1) ARE ZERO.
C**********************************************************************

            E(1) = 0.D0
            F(1) = 0.D0
            B = -2.D0*(1.D0 + YT)
        DO 300 I = 2,IM
        DO 250 J = 2,JM
            D = 2.D0*(YZ-YT)*UH(I,J) - YZ*(UH(I+1,J) + UH(I-1,J))
            DENO = B + E(J-1)
            E(J) = - 1.D0/DENO
            F(J) = (D - F(J-1))/DENO
250     CONTINUE
            U(I,JJ) = 1.D0
        DO 275 J = 1,JM
            K = JJ - J
            U(I,K) = E(K)*U(I,K+1) +F(K)
275     CONTINUE
300     CONTINUE

C**********************************************************************
C   CALCULATE FLOW RATE AND DRAG ON MOVING WALL.   USE TRAPEZOIDAL RULE
C   FOR FLOW RATE. WRITE SELECTED VELOCITY PROFILES.
C**********************************************************************

            FLOW = 0.D0
            DRAG = RA
            DRAGB = 0.D0
            DRAGS = 0.D0
        DO 400 I = 2,IM
            QI(I) = HY*.5D0*(U(I,1) + U(I,JJ))
        DO 350 J = 2,JM
            QI(I) = QI(I) + HY*U(I,J)
```

```
350     CONTINUE
            FLOW = FLOW + HZ*QI(I)
            TAU = (U(I,JJ) - U(I,JM))/HY
            DRAG = DRAG + HZ*TAU
            TAUB = (U(I,2)-U(I,1))/HY
            DRAGB = DRAGB + HZ*TAUB
400     CONTINUE

        DO 450 J = 2,JM
            TAUS = (U(2,J)-U(1,J))/HZ
            DRAGS = DRAGS + 2.D0*TAUS*HY
450     CONTINUE

            WRITE(6,3) ITER,FLOW,DRAG,DRAGB,DRAGS
            ITER= ITER + 1
            DIFF1 = DRAGB - DRAGB1
            DIFF2 = FLOW - FLOW1
            DRAGB1 = DRAGB
            FLOW1 = FLOW
            IF(DABS(DIFF1).GT.EPS.AND.ITER.LT.ITMAX) GO TO 999
            IF(DABS(DIFF2).GT.EPS.AND.ITER.LT.ITMAX) GO TO 999

            WRITE(6,4)
        DO 500 I =1,II
            Z = (I-1)*HZ
            Y = (JJ-2)*HY
            WRITE(6,5) Y,Z,U(I,JM)
500     CONTINUE

            WRITE(6,4)
        DO 600 J = 1,JJ
            Z = HZ
            Y = (J-1)*HY
            WRITE(6,5) Y,Z,U(2,J)
600     CONTINUE

1       FORMAT(5X,'COUETTE FLOW BY ADI METHOD.',/)
2       FORMAT(/,5X,'AR =',F10.5,3X,'II =',I3,3X,'JJ =',I3,/)
3       FORMAT(1X,'ITER=', I3,1X,'FLOW =',F10.5,1X,'DRAG =',F10.5,1X,
        1 'DRAGB =',F10.5,1X,'DRAGS = ',F10.5)
4       FORMAT(5X,'Y',9X,'Z',9X,'U',/)
5       FORMAT(3F10.5)
        STOP
        END
```

B9 NEWBL

```
C************************************************************************
C THIS PROGRAM INTEGRATES THE LAMINAR BOUNDARY-LAYER EQUATIONS FOR
C PLANE, STEADY, INCOMPRESSIBLE FLOW.  IT EMPLOYS A TRANSFORMATION
C OF VARIABLES THAT IS BASED ON THWAITES ANALYSIS OF THE MOMENTUM-
C INTEGRAL EQUATION, AND EMBODIES A TWO-TERM BLASIUS SERIES TO START
C THE CALCULATION.  THREE-POINT TRAILING DIFFERENCES ARE SUBSEQUENTLY
C USED FOR SECOND-ORDER ACCURACY IN XI.  THE FLOW IS THAT UNDER A LINE
C VORTEX.
C************************************************************************

      PROGRAM NEWBL
      IMPLICIT DOUBLE PRECISION (A-H,O-Z)
      DIMENSION ETA(121),U0(121),F0(121),S0(121),YOT(121),VOR(121)
      COMMON/INTEG/MT,ITER,NM,NN,IBL,IFL
      COMMON/REAL/DEL,GAM,PSI,HI,HI2,A21,A22,B21,B22,SUM,SF,DEL1,H,T
      COMMON/REAL1/ETAMAX,DN
      COMMON/ALG/E11(121),E21(121),F1(121),F2(121),FMN2(121)
      COMMON/SOLN/UM(121),FM(121),S(121),UMN1(121),FMN1(121),UMN2(121)
      DOUBLE PRECISION MOM, MTW

C************************************************************************
C  PRINT HEADING WHICH IDENTIFY THE PROBLEM.
C************************************************************************

      WRITE(6,1)

C************************************************************************
C  ASSIGN VALUES TO CONSTANTS.
C************************************************************************

         PI = 4.D0*DATAN(1.D0)
         HPI = 0.5D0*PI
         ETAMAX = 3.6D0
         NN = 121
         NM = NN-1
         RNM = NM
         DN = ETAMAX/RNM
         HI = 1.D0/DN
         HI2 = HI*HI
         A21 = .5D0
         A22 = HI
         B21 = .5D0
         B22 = -HI

C************************************************************************
C    SET UP ETA(N) AND FIRST-GUESS INITIAL PROFILES OF UM, S, AND FM.
C    THESE WILL BE REFINED BY ITERATION.
C************************************************************************

      DO 10 N=1,NN
         ETA(N) = (N-1)*DN
         G = ETA(N)/ETAMAX
         UM(N) = G*(2.D0-G)
         S(N) = 2.D0*(1.D0-G)/ETAMAX
         FM(N) = G*G*(1.D0-G/3.D0)*ETAMAX
         UMN1(N) = UM(N)
         FMN1(N) = FM(N)
         UMN2(N) = UM(N)
         FMN2(N) = FM(N)
10    CONTINUE

C************************************************************************
C    CALCULATE U0 AND F0 OF STARTING PROFILE.
C************************************************************************

         GAM = 1.D0
```

```
          DEL = 4.D0
          PSI = 0.D0
          E11(1) = 0.D0
          E21(1) = 0.D0
          F1(1) = 0.D0
          F2(1) = 0.D0
          ITER = 1
          MT = 1
          CALL THOMAS
          WRITE (7,3)
C      DO 30 I = 1,NN
C         WRITE(7,4) ETA(I),FM(I),UM(I),S(I)
30     CONTINUE
       DO 31 N = 1,NN
          U0(N) = UM(N)
          F0(N) = FM(N)
          S0(N) = S(N)
31     CONTINUE
          SF0 = SF
          DEL0 = DEL1
          SUM0 = SUM
C         WRITE (6,5)
C         WRITE(6,2)XI,V,GAM,DEL,DYDETA,SF,DEL1,SUM,TAU,DISPL,MOM,H,T


C ***********************************************************************
C      CALCULATE U1 AND F1 IN THE BLASIUS-SERIES REPRESENTATION.
C ***********************************************************************
          PSI = 4.D0
          E11(1) = 0.D0
          E21(1) = 0.D0
          F1(1) = 0.D0
          F2(1) = 0.D0
          ITER = 10
          MT = 1
          IBL = 1
          CALL THOMAS
          WRITE (7,3)
C      DO 32 I = 1,NN
C         WRITE(7,4) ETA(I),FM(I),UM(I),S(I)
32     CONTINUE
          WRITE(6,5)
          WRITE(6,2)XI,V,GAM,DEL,DYDETA,SF,DEL1,SUM,TAU,DISPL,MOM,H,T

C ***********************************************************************
C      COMBINE U0 AND U1, ETC, TO GET INITIAL PROFILES.
C ***********************************************************************
       XI = -6.D0
          DXI = .05D0
          Z = 1.D0/XI**2
          ZZ = 1.D0/(XI+DXI)**2
          ZZZ = 1.D0/(XI+2.D0*DXI)**2
       G2 = -2.D0*XI**3*(1.D0+Z*(21.D0/11.D0+Z*(120.D0-8.D0*Z)/143.D0))
          V = 1.D0/(1.D0 + XI*XI)
          DVDXI = -2.D0*XI*V*V
          DYDETA = DSQRT(G2)
          DEL = G2*DVDXI
          GAM = 9.D0 - 2.D0*DEL
          SUM = 0.D0
       DO 33 N = 2,NN
          UMN2(N) = U0(N)+Z*UM(N)
          FMN2(N) = F0(N)+Z*FM(N)
          UMN1(N) = U0(N) + ZZ*UM(N)
```

```
          FMN1(N) = F0(N) + ZZ*FM(N)
          UM(N) = U0(N) + ZZZ*UM(N)
          FM(N) = F0(N) + ZZZ*FM(N)
          S(N) = S0(N)+ZZZ*S(N)
          SUM = SUM+0.5D0*DN*(UM(N)+UM(N-1))*(1.D0-0.5D0*(UM(N)+UM(N-1)))
33        CONTINUE
          DEL1 = ETA(NM)-FM(NM)
          SF = UM(2)*HI + .50D0*DN*DEL
          UM(1) = 0.D0
          FM(1) = 0.D0
          S(1) = SF
          T = SF*SUM
          H = DEL1/SUM
          DISPL = DYDETA*DEL1
          MOM = DYDETA*SUM
          TAU = V*SF/DYDETA
C         WRITE(6,5)
C         WRITE(6,2)XI,V,GAM,DEL,DYDETA,SF,DEL1,SUM,TAU,DISPL,MOM,H,T

C*************************************************************************
C    INITIALIZE XI.  SET THE INITIAL SIZE OF DXI.
C*************************************************************************

          XI = - 5.95D0
          DXI = 0.05D0

C*************************************************************************
C    START THE MARCHING LOOP. SPECIFY DXI FOR VARIOUS RANGES OF XI.
C    CALCULATE QUANTITIES THAT DEPEND ONLY ON XI.
C*************************************************************************

          IBL = 2
          M = 1
1000      M = M+1
          MT = M
          IF(M.GT.80) DXI = 0.025D0
          IF(M.EQ.81) IFL = 1
          IF(M.GT.120) DXI = 0.0125D0
          IF(M.EQ.121) IFL = 1
          IF(M.GT.160) DXI = 0.00625D0
          IF(M.EQ.161) IFL = 1
          IF(M.GT.240) DXI = 0.003125D0
          IF(M.EQ.241) IFL = 1
          XI = XI+DXI
          V = 1.D0/(1.D0 + XI*XI)
          DVDXI = -2.D0*XI*V*V
          IF (XI. EQ. 0.D0) GO TO 43
          Z = 1.D0/XI**2
      G2 = -2.D0*XI**3*(1.D0+Z*(21.D0/11.D0+Z*(120.D0-8.D0*Z)/143.D0))
          IF(XI.LT.-7.D0) GO TO 44
43        G2 = .046875D0*(XI*(1.D0+XI*XI)**2*(279.D0+XI**2*(511.D0+
     1    XI**2*(385.D0+105.D0*XI**2)))+ 105.D0*(1.D0+XI*XI)**6*(
     2    DATAN(XI)+HPI))
44        DYDETA = DSQRT(G2)
          DEL = G2*DVDXI
          GAM = 9.D0 - 2.D0*DEL
          PSI = G2*V
          PSI = PSI/DXI

C*************************************************************************
C    SET WALL BOUNDARY CONDITIONS.
C*************************************************************************

          E11(1) = 0.D0
          E21(1) = 0.D0
          F1(1) = 0.D0
          F2(1) = 0.D0
```

```
C**********************************************************************
C      SET ITERATION COUNTER.  CALCULATE PROFILES AT NEW VALUE OF XI.
C**********************************************************************

         ITER = 1
         CALL THOMAS

C**********************************************************************
C      UNWIND TRANSFORMATION.
C**********************************************************************

         DISPL = DYDETA*DEL1
         MOM = DYDETA*SUM
         TAU = V*SF/DYDETA

C**********************************************************************
C   PRINT THESE VALUES AT EVERY LTH VALUE OF M. MM MUST EQUAL
C   I*L+1, TO GET A PRINT AT M=MM. I IS AN INTEGER.
C**********************************************************************

         L = 1
         LL = (M+L-1)/L
         LLL = (M+L-2)/L
         IF(LL.EQ.LLL) GO TO 55
50       WRITE(6,2)XI,V,GAM,DEL,DYDETA,SF,DEL1,SUM,TAU,DISPL,MOM,H,T
55    CONTINUE

C**********************************************************************
C   PRINT NORMALIZED VORTICITY PROFILES.
C**********************************************************************

         IF(M.EQ.240) GO TO 60
         IF(M.EQ.296) GO TO 60
         GO TO 99
60       WRITE(7,3)
      DO 66 I = 1,NN
         YOT(I) = ETA(I)/SUM
         VOR(I) = S(I)*SUM
         WRITE(7,4) ETA(I),YOT(I),UM(I),VOR(I)
66    CONTINUE
         WRITE(6,5)
99    IF(SF.LT.0.D0) GO TO 101
      IF (M.LT.400) GO TO 1000

C**********************************************************************
C      FORMAT WRITEOUTS.
C**********************************************************************

1     FORMAT(10X,'BOUNDARY LAYER UNDER A LINE VORTEX',//)
2     FORMAT(13F9.5)
3     FORMAT(//,4X,'ETA',7X,'Y/T',9X,'U',9X,'VOR',/)
4     FORMAT(3X,F5.3,3F10.5)
5     FORMAT(//,4X,'XI',9X,'V',6X,'GAM',6X,'DEL',5X,'DYDETA',5X,'SF',
     1 6X,'DEL1',6X,'SUM',6X,'TAU',5X,'DISPL',5X,'MOM',6X,'H',7X,'T',//)

101   CONTINUE
      STOP
      END

      SUBROUTINE THOMAS
      IMPLICIT DOUBLE PRECISION (A-H,O-Z)
      COMMON/INTEG/MT,ITER,NM,NN,IBL,IFL
      COMMON/REAL/DEL,GAM,PSI,HI,HI2,A21,A22,B21,B22,SUM,SF,DEL1,H,T
      COMMON/REAL1/ETAMAX,DN

      COMMON/ALG/E11(121),E21(121),F1(121),F2(121),FMN2(121)
      COMMON/SOLN/UM(121),FM(121),S(121),UMN1(121),FMN1(121),UMN2(121)
```

```
C ****************************************************************************
C     START THOMAS ALGORITHM.   TRIANGULARIZE.
C ****************************************************************************

1000    DO 20 N=2,NM
           IF(IFL.EQ.1) GO TO 10
           A = 1.5D0
           B = -2.D0
           C = .5D0
           GO TO 15
10         A = 4.D0/3.D0
           B = -1.5D0
           C = 1.D0/6.D0
15         X = GAM*FM(N) + PSI*(A*FM(N)+B*FMN1(N)+C*FMN2(N))
           Y = -2.D0*DEL*UM(N)-PSI*(2.D0*A*UM(N)+B*UMN1(N)+C*UMN2(N))
           A11 = HI2-.5D0*HI*X
           C11 = HI2+.5D0*HI*X
           B11 = -2.D0*HI2+Y
           B12 = (GAM + A*PSI)*S(N)
           D1 = GAM*FM(N)*S(N) - DEL*(1.D0+UM(N)**2)+ PSI*A*(S(N)*FM(N)-
      1    UM(N)**2)
           IF(IBL.EQ.1) D1 = (4.D0/11.D0)*(1.D0-UM(N)**2)-(8.D0/11.D0)*
      1    FM(N)*S(N)
           G11 = A11*E11(N-1)+B11
           G12 = B12
           G21 = A21*E11(N-1)+A22*E21(N-1)+B21
           G22 = B22
           DENO = G11*G22-G12*G21
           GI11 = G22/DENO
           GI12 = -G12/DENO
           GI21 = -G21/DENO
           GI22 = G11/DENO
           E11(N) = -GI11*C11
           E21(N) = -GI21*C11
           F1(N) = GI11*(D1-A11*F1(N-1))-GI12*(A21*F1(N-1)+A22*F2(N-1))
           F2(N) = GI21*(D1-A11*F1(N-1))-GI22*(A21*F1(N-1)+A22*F2(N-1))
20      CONTINUE

C****************************************************************************
C    SET OUTER BOUNDARY CONDITION.
C****************************************************************************

           UM(NN) =   1.D0
           IF(IBL.EQ.1) UM(NN) = 0.D0

C ****************************************************************************
C   BACK-SUBSTITUTE TO FIND UM AND FM.
C ****************************************************************************

        DO 30 N=1,NM
           K = NN-N
           UM(K) = E11(K)*UM(K+1) + F1(K)
           FM(K) = E21(K)*UM(K+1) + F2(K)
30      CONTINUE
           FM(NN) = FM(NM)+ .5D0*DN*(UM(NN)+UM(NM))

C****************************************************************************
C      UPDATE S(N).
C****************************************************************************

        DO 40 N=2,NM
           S(N) = .5D0*HI*(UM(N+1)-UM(N-1))
40      CONTINUE
```

```
C  *********************************************************************
C     ADVANCE AND CHECK ITERATION COUNTER.
C  *********************************************************************
         ITER = ITER+1
         IF(MT.EQ.1.AND.ITER.LT.10) GO TO 1000
         IF(MT.GT.1.AND.ITER.LT.3) GO TO 1000

C**********************************************************************
C  CALCULATE AND PRINT B.L. THICKNESSES, SHAPE FACTORS, AND SKIN
C  FRICTION.  UPDATE UMN1 AND FMN1.
C**********************************************************************
         SUM = 0.D0
      DO 50 N=2,NM
         SUM = SUM+0.5D0*DN*(UM(N)+UM(N-1))*(1.D0-0.5D0*(UM(N)+UM(N-1)))
         UMN2(N) = UMN1(N)
         FMN2(N) = FMN1(N)
         UMN1(N) = UM(N)
         FMN1(N) = FM(N)
50    CONTINUE
         SF = UM(2)*HI + .50D0*DN*DEL
         S(1) = SF
         DEL1 = ETAMAX - DN - FM(NM)
         H = DEL1/SUM
         T = SF*SUM
      IFL = 0
      RETURN
      END
```

B10 ENTRY

```
C*************************************************************************
C    THIS PROGRAM SOLVES A BOUNDARY-LAYER VERSION OF THE VORTICITY
C    EQUATION FOR THE ENTRY LENGTH OF A PLANE CHANNEL, IN THE REGION
C    BEFORE MERGING OF THE BOUNDARY LAYERS.
C*************************************************************************

      PROGRAM ENTRY
      IMPLICIT DOUBLE PRECISION (A-H,O-Z)

C*************************************************************************
C    ASSIGN STORAGE ARRAYS FOR INDEPENDENT AND DEPENDENT VARIABLES.
C*************************************************************************

      COMMON/SOLN/ Z(801),U(801),P(801),S(801),R(801),PM2(801),SM2(801)
      COMMON/INTEG/M,MP,NN,NM,IM,ITER
      COMMON/SOL2/ PM1(801),SM1(801),DPDXI(801),DSDXI(801)
      COMMON/ALG/ E(2,2,801), F(2,801), G(2,2),GI(2,2), DD(2)
      COMMON/REAL/ A(2,2),B(2,2),C(2,2),D(2),XI,DXI,XI0,HI,HI2,ZMAX,H

C*************************************************************************
C    SPECIFY NUMBER OF MESH POINTS.
C*************************************************************************

          NN = 321
          NM = NN-1
          RN = NM
          ZMAX = 8.D0
          H = ZMAX/RN
          HI = 1.D0/H
          HI2 = HI*HI
          MM = 101
          MN = MM-1
          MP = MM+1
          MF = 105
          RM = MN
          DXI = 1.D0/(ZMAX*RM)
          WRITE(6,1) ZMAX,H,DXI
          RM1 = 0.D0
          PINT = 0.D0

C*************************************************************************
C    FILL IN CONSTANTS IN COEFFICIENT MATRICES.
C*************************************************************************

          A(2,1) = HI2
          A(2,2) = 0.D0
          B(2,1) = -2.D0*HI2
          B(2,2) = 1.D0
          C(2,1) = HI2
          C(2,2) = 0.D0
          D(2) = 0.D0

C*************************************************************************
C    SET BOUNDARY CONDITIONS AT Z = 0.
C*************************************************************************

          E(1,1,1) = 0.D0
          E(1,2,1) = 0.D0
          E(2,1,1) =  2.D0*HI2
          E(2,2,1) = 0.D0
          F(1,1) = 0.D0
          F(2,1) = 0.D0

C*************************************************************************
C    SET UP FIRST-GUESS PROFILES AT XI = 0.
C*************************************************************************
```

```
      DO 150 N = 1,NN
         Z(N) = H*(N-1)
         X = Z(N)/ZMAX
         P(N) = ZMAX*(1.5D0*X**2 - X**3 + .25D0*X**4)
         U(N) = 3.D0*X - 3.D0*X**2 + X**3
         S(N) = (3.D0 - 6.D0*X + 3.D0*X**2)/ZMAX
         R(N) = 6.D0*(X-1.D0)/ZMAX**2
150   CONTINUE

C***************************** PART ONE *****************************
C    START THE MARCHING LOOP.
C******************************************************************

      WRITE (7,5)
      PINT = 0.D0
      DO 1000 M = 1,MM
         IM = 1
         XI = (M-1)*DXI
         ITER = 1
         CALL THOMAS

C******************************************************************
C    UNWIND TRANSFORMATIONS AND CALCULATE PRESSURE
C******************************************************************

      IF(M.EQ.1) GO TO 97
      TW = S(1)/XI
      IF(M.EQ.2) TW = (SM2(1)+DXI*S(1))/XI
      FCL = R(NN)/XI**2
      GO TO 98
97    CONTINUE
      FCL = 0.D0
98    PINT = PINT + .5D0*DXI*(XI*FCL + (XI-DXI)*FCL1)
      CP = U(NN)**2 - 1.D0 + PINT
      FCL1 = FCL

C******************************************************************
C    PRINT OUTPUT.
C******************************************************************

      WRITE(7,2) XI,U(NN),TW,FCL,CP

      L = (M+18)/20
      LL = (M+19)/20
      IF(L.EQ.LL) GO TO 900

499      WRITE(6,3)
      DO 500 N = 1,NN,4
         WRITE(6,4)  Z(N),P(N),U(N),S(N),R(N)
500   CONTINUE
900   CONTINUE

C******************************************************************
C    UPDATE PM1 AND SM1.  PROJECT FIRST GUESSES FOR NEXT STATION.
C******************************************************************

      IF(M.LT.3) GO TO 1000
      DO 475 N = 1,NN
         PM2(N) = PM1(N)
         SM2(N) = SM1(N)
         PM1(N) = P(N)
         SM1(N) = S(N)
         P(N) = 2.D0*PM1(N) - PM2(N)
         S(N) = 2.D0*SM1(N) - SM2(N)
475   CONTINUE
```

```
1000   CONTINUE

C*********************** PART TWO **********************************
C    START THE SECOND MARCHING LOOP.
C*****************************************************************

       XI0 = XI
     DO 2000 M = MP,MF
       IM = 2
       XI = XI + DXI
       ITER = 1
       CALL THOMAS

C*****************************************************************
C    UPDATE PM1 AND SM1, PM2 AND SM2.
C*****************************************************************

     DO 600 N = 2,NM
       PM2(N) = PM1(N)
       SM2(N) = SM1(N)
       PM1(N) = P(N)
       SM1(N) = S(N)
600    CONTINUE

C*****************************************************************
C    UNWIND TRANSFORMATIONS AND CALCULATE PRESSURE
C*****************************************************************

       TW = S(1)/XI0
       FCL = R(NN)/XI0**2
       PINT = PINT + .5D0*DXI*(XI*FCL + (XI-DXI)*FCL1)
       CP = U(NN)**2 - 1.D0 + PINT
       FCL1 = FCL

C*****************************************************************
C    PRINT OUTPUT.
C*****************************************************************

       WRITE(7,2) XI,U(NN),TW,FCL,CP

       L = (M+38)/40
       LL = (M+39)/40
       IF(L.EQ.LL) GO TO 2000

       WRITE(6,3)
     DO 510 N = 1,NN,4
       WRITE(6,4) Z(N),P(N),U(N),S(N),R(N)
510    CONTINUE

2000   CONTINUE

1      FORMAT(10X,'THE ENTRY LENGTH. ZMAX =',F7.3,10X,'H=',F5.3,
     1   10X,'DXI =',F5.3,/)
2      FORMAT( 5X,5F10.5)
3      FORMAT (10X,'Z',10X,'P',10X,'U',10X,'S',10X,'R',/)
4      FORMAT ( 5X,5F10.5)
5      FORMAT(10X,'XI',8X,'UCL',7X,'TW',8X,'FCL',7X,'CP',/)
       STOP
       END

       SUBROUTINE THOMAS
       IMPLICIT DOUBLE PRECISION (A-H,O-Z)
       COMMON/SOLN/ Z(801),U(801),P(801),S(801),R(801),PM2(801),SM2(801)
       COMMON/INTEG/M,MP,NN,NM,IM,ITER
       COMMON/SOL2/ PM1(801),SM1(801),DPDXI(801),DSDXI(801)
```

```
        COMMON/ALG/ E(2,2,801), F(2,801), G(2,2),GI(2,2), DD(2)
        COMMON/REAL/ A(2,2),B(2,2),C(2,2),D(2),XI,DXI,XI0,HI,HI2,ZMAX,H
999     DO 200 N = 2,NM
        IF(IM.EQ.2) GO TO 100
        A(1,1) = -.5D0*(S(N) - XI*DSDXI(N))*HI
        IF(M.EQ.2) A(1,1) = -.5D0*S(N)*HI
        B(1,1) = (1.D0 + 1.5D0*XI/DXI)*R(N)
        IF(M.EQ.2) B(1,1) = 2.D0*R(N)
        C(1,1) = - A(1,1)
        Y = .5D0*(P(N) + XI*DPDXI(N))*HI
        IF(M.EQ.2) Y = 0.5D0*P(N)*HI
        A(1,2) = HI2 - Y
        B(1,2) = -2.D0*HI2 + (1.D0 - 1.5D0*XI/DXI)*U(N)
        IF(M.EQ.2) B(1,2) = - 2.D0*HI2
        C(1,2) = HI2 + Y
        D(1) = U(N)*S(N) + P(N)*R(N)  - (1.5D0*XI/DXI)*
     1         (U(N)*S(N) - R(N)*P(N))
        IF(M.EQ.2) D(1) = 0.D0
        GO TO 150
100     A(1,1) =  .5D0*XI0**2*DSDXI(N)*HI/XI
        B(1,1) =  1.5D0*XI0**2*R(N)/(XI*DXI)
        IF(M.EQ.MP) B(1,1) = B(1,1)/1.5D0
        C(1,1) = - A(1,1)
        Y = .5D0*XI0**2*DPDXI(N)*HI/XI
        A(1,2) = HI2 - Y
        B(1,2) = -2.D0*HI2 - 1.5D0*XI0**2*U(N)/(XI*DXI)
        IF(M.EQ.MP) B(1,2) = -2.D0*HI2 - XI0**2*U(N)/(XI*DXI)
        C(1,2) = HI2 + Y
        D(1) = - 1.5D0*XI0**2*(U(N)*S(N) - R(N)*P(N))/(XI*DXI)
        IF(M.EQ.MP) D(1) = D(1)/1.5D0

C***************************************************************************
C   TRIANGULARIZE. FORM THE MATRIX G.
C***************************************************************************

150     NM1 = N-1
        DO 10  I = 1,2
        DO 11 J = 1,2
          G(I,J) = 0.D0
        DO 12 K = 1,2
          G(I,J) = G(I,J) + A(I,K)*E(K,J,NM1)
12      CONTINUE
          G(I,J) = G(I,J) + B(I,J)
11      CONTINUE
10      CONTINUE

C***************************************************************************
C   INVERT THE MATRIX G, USING KRAMER'S RULE.
C***************************************************************************

        DEL = G(1,1)*G(2,2)-G(1,2)*G(2,1)
        GI(1,1) = G(2,2)/DEL
        GI(1,2) = - G(1,2)/DEL
        GI(2,1) = - G(2,1)/DEL
        GI(2,2) = G(1,1)/DEL

C***************************************************************************
C   CALCULATE THE E(I,J,N) AND F(I,N)
C***************************************************************************

        E(1,1,N) = - (GI(1,1)*C(1,1) + GI(1,2)*C(2,1))
        E(1,2,N) = - (GI(1,1)*C(1,2) + GI(1,2)*C(2,2))
        E(2,1,N) = - (GI(2,1)*C(1,1) + GI(2,2)*C(2,1))
        E(2,2,N) = - (GI(2,1)*C(1,2) + GI(2,2)*C(2,2))
        DD(1) = D(1)-A(1,1)*F(1,NM1)-A(1,2)*F(2,NM1)
        DD(2) = D(2)-A(2,1)*F(1,NM1)-A(2,2)*F(2,NM1)
        F(1,N) = GI(1,1)*DD(1) + GI(1,2)*DD(2)
        F(2,N) = GI(2,1)*DD(1) + GI(2,2)*DD(2)
200     CONTINUE
```

```
C********************************************************************
C     SPECIFY BOUNDARY CONDITIONS AT ZMAX.
C********************************************************************

          S(NN) = 0.D0
          IF(IM.EQ.2) GO TO 224
          AA = XI
          BB = HI*(1.D0 - ZMAX*XI)
          IF(M.EQ.2) GO TO 222
          P(NN) = (1.D0+BB*F(1,NM))/(AA+BB*(1.D0-E(1,1,NM)))
          GO TO 225
222       DEL1 = ZMAX - P(NN)
          P(NN) = (H*DEL1 + F(1,NM))/(1.D0 - E(1,1,NM))
          GO TO 225
224       P(NN) = ZMAX
225       CONTINUE


C********************************************************************
C     BACK-SUBSTITUTE.
C********************************************************************

      DO 300 N = 1,NM
          L = NN-N
          P(L) = E(1,1,L)*P(L+1)+E(1,2,L)*S(L+1)+F(1,L)
          S(L) = E(2,1,L)*P(L+1)+E(2,2,L)*S(L+1)+F(2,L)
          IF(M.EQ.1) PM2(N) = P(N)
          IF(M.EQ.1) SM2(N) = S(N)
          IF(M.EQ.2) PM1(N) = PM2(N) + DXI*P(N)
          IF(M.EQ.2) SM1(N) = SM2(N) + DXI*S(N)
300       CONTINUE

C********************************************************************
C     CALCULATE VELOCITY AND VISCOUS FORCE.
C********************************************************************

          U(1) = 0.D0
          R(1) = HI*(S(2)-S(1))
          IF(M.EQ.2) R(1), = HI*(SM2(2)-SM2(1)) + DXI*HI*(S(2) - S(1))
          IF(M.EQ.2) S(1) = DXI*S(1) + SM2(1)
      DO 400 N = 2,NM
          U(N) = 0.5D0*HI*(P(N+1) - P(N-1))
          R(N) = 0.5D0*HI*(S(N+1) - S(N-1))
          IF(M.EQ.2) U(N) = 0.5D0*HI*(PM2(N+1)-PM2(N-1)) + DXI*U(N)
          IF(M.EQ.2) R(N) = 0.5D0*HI*(SM2(N+1)-SM2(N-1)) + DXI*R(N)
400       CONTINUE
          U(NN) = HI*(P(NN) - P(NM))
          IF(M.EQ.2) U(NN) = DEL1*DXI + 1.D0
          R(NN) = - HI*S(NM)
          IF(M.EQ.2) R(NN) = DXI*HI*(S(NN)-S(NM))

C********************************************************************
C     CALCULATE DPDXI AND DSDXI.
C********************************************************************

          IF(M.EQ.1) GO TO 460
      DO 450 N = 2,NN
          IF(M.EQ.MP) GO TO 455
          DPDXI(N) = .5D0*(3.D0*P(N) - 4.D0*PM1(N) + PM2(N))/DXI
          DSDXI(N) = .5D0*(3.D0*S(N) - 4.D0*SM1(N) + SM2(N))/DXI
          GO TO 450
455       DPDXI(N) = (P(N) - PM1(N))/DXI
          DSDXI(N) = (S(N) - SM1(N))/DXI
450       CONTINUE
460       CONTINUE
```

```
C*****************************************************************************
C     CHECK AND ADVANCE ITERATION COUNTER.
C*****************************************************************************
      ITER = ITER + 1
      IF(M.EQ.1.AND.ITER.LT.7) GO TO 999
      IF(M.EQ.MP.AND.ITER.LT.5) GO TO 999
      IF(M.GT.2.AND.ITER.LT.3) GO TO 999
      RETURN
      END
```

B11   BLCTRL

```
C************************************************************************
C THIS PROGRAM INTEGRATES THE LAMINAR BOUNDARY-LAYER EQUATIONS FOR
C PLANE, STEADY, INCOMPRESSIBLE FLOW.  IT EMPLOYS AN ITERATIVE STRATEGY
C TO MAINTAIN A CONSTANT DISPLACEMENT THICKNESS BY CONTROLLED
C APPLICATION OF SUCTION.
C************************************************************************

      PROGRAM BLCTRL
      IMPLICIT DOUBLE PRECISION (A-H,O-Z)
      DIMENSION ETA(161)
      COMMON/INTEG/MT,ITER,NM,NN,IBL
      COMMON/REAL/DEL,GAM,PSI,HI,HI2,A21,A22,B21,B22,SUM,SF,DEL1,H,T
      COMMON/REAL1/ETAMAX,DN,SIG,DYDETA,TARG,DERSIG
      COMMON/ALG/E11(161),E21(161),F1(161),F2(161)
      COMMON/SOLN/UM(161),FM(161),S(161),UMN1(161),FMN1(161)
      COMMON/BL/FM0(161),UM0(161),S0(161),FM1(161),UM1(161),S1(161)
      DOUBLE PRECISION MOM, MTW

C************************************************************************
C  PRINT HEADING WHICH IDENTIFY THE PROBLEM.
C************************************************************************

      WRITE(6,1)

C************************************************************************
C  ASSIGN VALUES TO CONSTANTS.
C************************************************************************

      PI = 4.D0*DATAN(1.D0)
      HPI = 0.5D0*PI
      ETAMAX = 8.D0
      NN = 161
      NM = NN-1
      RNM = NM
      DN = ETAMAX/RNM
      HI = 1.D0/DN
      HI2 = HI*HI
      A21 = .5D0
      A22 = HI
      B21 = .5D0
      B22 = -HI

C************************************************************************
C   SET UP ETA(N) AND FIRST-GUESS INITIAL PROFILES OF UM, S, AND FM.
C   THESE WILL BE REFINED BY ITERATION.
C************************************************************************

      DO 10 N=1,NN
      ETA(N) = (N-1)*DN
      G = ETA(N)/ETAMAX
      UM(N) = G*(2.D0-G)
      S(N) = 2.D0*(1.D0-G)/ETAMAX
      FM(N) = G*G*(1.D0-G/3.D0)*ETAMAX
10    CONTINUE

C************************************************************************
C   CALCULATE U0 AND F0 OF STARTING PROFILE.
C************************************************************************

      SIG = 0.D0
      E11(1) = 0.D0
      E21(1) = 0.D0
      F1(1) = 0.D0
      F2(1) = SIG
      ITER = 1
      MT = 1
```

```
          CALL INIT
          V = 0.D0
          G2 = 0.5D0
          DYDETA = DSQRT(G2)
          DISPL = DYDETA*DEL1
          MOM = DYDETA*SUM
          TAU = V*SF/DYDETA
          WRITE (6,5)
          WRITE(6,2)XI,SIG,GAM,DEL,DYDETA,SF,DEL1,SUM,TAU,DISPL,MOM,H,T
          WRITE (6,3)
          II = NM/3
       DO 30 I = 1,II
          J=I+II
          K=J+II
          WRITE(6,4) ETA(I),FM(I),UM(I),S(I),ETA(J),FM(J),UM(J),S(J),
     1    ETA(K),FM(K),UM(K),S(K)
30        CONTINUE

C*************************************************************************
C   SET TARGET VALUE FOR DEL1.
C*************************************************************************

          TARG = DEL1

C*************************************************************************
C   USE BLASIUS SERIES TO DETERMINE THE INITIAL VARIATION OF SIG. THIS
C   AMOUNTS TO EVALUATING SIG1 IN THE EXPRESSION SIG = SIG1* XI**2.
C*************************************************************************

          GAM = 4.D0
          DEL = 3.D0
          SIG = 0.D0
          F2(1) = SIG
          CALL BLAS
          DEL10 = FM(NN) - SIG
          WRITE (6,8) SIG, DEL10,SF
          SIG = 1.D0
          F2(1) = SIG
          CALL BLAS
          DEL1 = FM(NN) - SIG
          WRITE (6,8) SIG, DEL1,SF
          ERR = DEL1
          DERSIG = DEL1 - DEL10
          TEMP = DERSIG
          SIG = SIG - ERR/DERSIG
          F2(1) = SIG
          CALL BLAS
          DEL1 = FM(NN) - SIG
          WRITE (6,8) SIG, DEL1,SF
          SIGM1 = SIG
       DO 25 N = 1,NN
          UM1(N) = UM(N)
          FM1(N) = FM(N)
          S1(N) = S(N)
25        CONTINUE

          GAM = 6.D0
          DEL = 5.D0
          IBL = 2
          SIG = 0.D0
          F2(1) = SIG
          CALL BLAS
          DEL10 = FM(NN) - SIG
          SIG = 1.D0
          F2(1) = SIG
          CALL BLAS
```

```
              DEL1 = FM(NN) - SIG
              ERR = DEL1
              DERSIG = DEL1 - DEL10
              SIG = SIG - ERR/DERSIG
              F2(1) = SIG
           CALL BLAS
              DEL1 = FM(NN) - SIG
              WRITE (6,8) SIG, DEL1
              DERSIG = TEMP

C**********************************************************************
C    INITIALIZE XI.  SET THE INITIAL SIZE OF DXI.  USE THE BLASIUS
C    SERIES TO SET UP THE INITIAL PROFILES.
C**********************************************************************

              XX =  (PI/20.D0)**2
              DXX = XX
              SUM = 0.D0
           DO 37 N = 1,NN
              UM(N) = UM0(N) + UM1(N)*XX + UM(N)*XX**2
              FM(N) = FM0(N) + FM1(N)*XX + FM(N)*XX**2
              S(N)  = S0(N)  + S1(N)*XX + S(N)*XX**2
              UMN1(N) = UM(N)
              FMN1(N) = FM(N)
              IF(N.EQ.1) GO TO 37
              SUM = SUM+0.5D0*DN*(UM(N)+UM(N-1))*(1.D0-0.5D0*(UM(N)+UM(N-1)))
37         CONTINUE
              SIG = SIGM1*XX + SIG*XX**2
              VW = SIG*3.D0*DSQRT(2.D0)
              DEL1 = ETAMAX - FM(NN) + SIG
              XI = DSQRT(XX)
              V = 2.D0*DSIN(XI)
              DVDXI = 2.D0*DCOS(XI)
              G2 = 0.5D0
              DYDETA = DSQRT(G2)
              DISPL = DYDETA*DEL1
              MOM = DYDETA*SUM
              DEL = G2*DVDXI
              GAM = DEL
              PSI = 2.D0*G2*V*XI
              PSI = PSI/DXX
              AA = GAM*FM(1) + PSI*FM(1)
              SF = (HI*UM(2)+.5D0*DN*DEL)/(1.D0-.5D0*AA*DN)
              TAU = V*SF/DYDETA
              DDRAGM = 0.D0
              DDRAG = TAU*V
              DXI = XI
              DRAG = 0.5D0*(DDRAG+DDRAGM)*DXX
              DDRAGM = DDRAG
              WRITE (6,5)
              WRITE(6,2)XX,SIG,GAM,DEL,DRAG,SF,DEL1,SUM,TAU,DISPL,MOM,H,T
              SIGM = 0.D0
              VMN1 = 0.D0

C**********************************************************************
C    START THE MARCHING LOOP. FIRST CALCULATE QUANTITIES
C    THAT DEPEND ONLY ON XI.
C**********************************************************************

           DO 1000 M = 2,399
              XX = XX + DXX
              XI = DSQRT(XX)
              G2 = .5D0
              DYDETA = DSQRT(G2)
              V = 2.D0*DSIN(XI)
              DVDXI = 2.D0*DCOS(XI)
```

```
          DEL = G2*DVDXI
          GAM = DEL
          PSI = G2*V*2.D0*XI
          PSI = PSI/DXX
          TEMP = SIG
          SIG =  2.D0*SIG - SIGM
          SIGM = TEMP

C**********************************************************************
C    SET WALL BOUNDARY CONDITIONS.
C**********************************************************************

          E11(1) = 0.D0
          E21(1) = 0.D0
          F1(1) = 0.D0

C**********************************************************************
C      SET ITERATION COUNTER.  CALCULATE PROFILES AT NEW VALUE OF XI.
C**********************************************************************

          ITER = 1
          CALL THOMAS

C**********************************************************************
C      UNWIND TRANSFORMATION.
C**********************************************************************

          DISPL = DYDETA*DEL1
          MOM = DYDETA*SUM
          TAU = V*SF/DYDETA
          DDRAG = TAU*V
          DXI = XI - DSQRT(XX-DXX)
          DRAG = DRAG + 0.5D0*(DDRAG+DDRAGM)*DXI
          DDRAGM = DDRAG
          VW = DYDETA*(FM(1)*V - FMN1(1)*VMN1)/DXI

C**********************************************************************
C   PRINT THESE VALUES AT EVERY LTH VALUE OF M. MM MUST EQUAL
C   I*L+1, TO GET A PRINT AT M=MM.  I IS AN INTEGER.
C**********************************************************************

          L = 1
          LL = (M+L-1)/L
          LLL = (M+L-2)/L
          IF(LL.EQ.LLL) GO TO 55
50        WRITE(6,2)XI,SIG,VW,DEL,DRAG,SF,DEL1,SUM,TAU,DISPL,MOM,H,T
55        CONTINUE

C**********************************************************************
C   PRINT SELECTED PROFILES.
C**********************************************************************

          IF(M.EQ.100) GO TO 60
          IF(M.EQ.300) GO TO 60
          IF(M.EQ.480) GO TO 60
          IF(SF.LT.0.D0) GO TO 60
          GO TO 99
60        WRITE(6,3)
          II = NM/3
        DO 66 I = 1,II
          J=I+II
          K=J+II
          WRITE(6,4) ETA(I),FM(I),UM(I),S(I),ETA(J),FM(J),UM(J),S(J),
     1    ETA(K),FM(K),UM(K),S(K)
66        CONTINUE
          WRITE(6,5)
```

```
99      IF(SF.LT.0.D0) GO TO 101
C*****************************************************************************
C   PROJECT FM* FOR NEXT STEP.
C*****************************************************************************

        DO 1300 N = 1,NN
        TEMP = FM(N)
        IF(M.EQ.101) GO TO 1290
        IF(M.EQ.201) GO TO 1290
        IF(M.EQ.301) GO TO 1290
        FM(N) = 2.D0*FM(N) - FMN1(N)
        GO TO 1295
1290    FM(N) = 1.5D0*FM(N) - 0.5D0*FMN1(N)
1295    FMN1(N) = TEMP
        UMN1 (N) = UM(N)
1300 CONTINUE
        VMN1 = V
1000    CONTINUE


C*****************************************************************************
C       FORMAT WRITEOUTS.
C*****************************************************************************

1       FORMAT(15X,'BOUNDARY LAYER ON A CYLINDER WITH CONTROLLED SUCTION'
        1 ,//)
2       FORMAT(13F9.5)
3       FORMAT(//,4X,'ETA',7X,'F',9X,'U',9X,'S',7X,'ETA',7X,
        1 'F',9X,'U',9X,'S',7X,'ETA',7X,'F',9X,'U',9X,'S',/)
4       FORMAT(3(3X,F5.3,3F10.5))
5       FORMAT(//,4X,'XI',9X,'SIG',6X,'VW',6X,'DEL',5X,'DRAG',5X,'SF',
        1 6X,'DEL1',6X,'SUM',6X,'TAU',5X,'DISPL',5X,'MOM',6X,'H',7X,'T',//)
8       FORMAT(10X,'SIGMA1 =',F10.5,5X,'DEL1 =',F10.5,5X,'SF =',F10.5,/)
101     CONTINUE
        STOP
        END

        SUBROUTINE THOMAS
        IMPLICIT DOUBLE PRECISION (A-H,O-Z)
        COMMON/INTEG/MT,ITER,NM,NN,IBL
        COMMON/REAL/DEL,GAM,PSI,HI,HI2,A21,A22,B21,B22,SUM,SF,DEL1,H,T
        COMMON/REAL1/ETAMAX,DN,SIG,DYDETA,TARG,DERSIG
        COMMON/ALG/E11(161),E21(161),F1(161),F2(161)
        COMMON/SOLN/UM(161),FM(161),S(161),UMN1(161),FMN1(161)
        COMMON/BL/FM0(161),UM0(161),S0(161),FM1(161),UM1(161),S1(161)

C *****************************************************************************
C   START THOMAS ALGORITHM.  TRIANGULARIZE.
C *****************************************************************************

1000    F2(1) = SIG
        ITIN = 1
999     DO 20 N=2,NM
        X = (GAM+PSI)*FM(N) - PSI*FMN1(N)
        Y = -2.D0*(DEL+PSI)*UM(N) + PSI*UMN1(N)
        Z = (GAM+PSI)*S(N)
        A11 = HI2-.5D0*HI*X
        C11 = HI2+.5D0*HI*X
        B11 = -2.D0*HI2+Y
        B12 = Z
        D1 = Z*FM(N)-DEL-(DEL+PSI)*UM(N)*UM(N)
        G11 = A11*E11(N-1)+B11
        G12 = B12
        G21 = A21*E11(N-1)+A22*E21(N-1)+B21
        G22 = B22
        DENO = G11*G22-G12*G21
        GI11 = G22/DENO
```

```
          GI12 = -G12/DENO
          GI21 = -G21/DENO
          GI22 = G11/DENO
          E11(N) = -GI11*C11
          E21(N) = -GI21*C11
          F1(N) = GI11*(D1-A11*F1(N-1))-GI12*(A21*F1(N-1)+A22*F2(N-1))
          F2(N) = GI21*(D1-A11*F1(N-1))-GI22*(A21*F1(N-1)+A22*F2(N-1))
 20       CONTINUE

C*********************************************************************
C   SET OUTER BOUNDARY CONDITION.
C*********************************************************************

          UM(NN) =   1.D0

C *******************************************************************
C   BACK-SUBSTITUTE TO FIND UM AND FM.
C *******************************************************************

      DO 30 N=1,NM
          K = NN-N
          UM(K) = E11(K)*UM(K+1)  + F1(K)
          FM(K) = E21(K)*UM(K+1)  + F2(K)
 30       CONTINUE
          FM(NN) = FM(NM)+ .5D0*DN*(UM(NN)+UM(NM))

C*********************************************************************
C       UPDATE S(N).
C*********************************************************************

      DO 40 N=2,NM
          S(N) = .5D0*HI*(UM(N+1)-UM(N-1))
 40       CONTINUE

C *******************************************************************
C   ADVANCE AND CHECK ITERATION COUNTER.
C *******************************************************************

          ITIN = ITIN + 1
          IF(ITIN.LT.4) GO TO 999
          DEL1 = ETAMAX - FM(NN) + SIG
          IF (ITER.GT.1) GO TO 42
          ERROLD = DEL1 - TARG
          SIGOLD = SIG
          SIG = SIGOLD - ERROLD /DERSIG
          ITER = 2
          GO TO 1000
 42       CONTINUE
          ITER = ITER+1
          ERR = DEL1 - TARG
          DERSIG = (ERR-ERROLD)/(SIG-SIGOLD)
          SIGOLD = SIG
          ERROLD = ERR
          SIG = SIGOLD - ERROLD /DERSIG
          AA = GAM*FM(1) + PSI*(FM(1)-FMN1(1))
          SF = (HI*UM(2)+.5D0*DN*DEL)/(1.D0-.5D0*AA*DN)
          IF(DABS(ERR).GT.1.D-5.AND.ITER.LE.9)GO TO 1000

C*********************************************************************
C CALCULATE AND PRINT B.L. THICKNESSES, SHAPE FACTORS, AND SKIN
C FRICTION.
C*********************************************************************

          SUM = 0.D0
      DO 50 N=1,NM
          IF(N.EQ.1) GO TO 50
```

```
          SUM = SUM+0.5D0*DN*(UM(N)+UM(N-1))*(1.D0-0.5D0*(UM(N)+UM(N-1)))
50        CONTINUE
          S(1) = SF
          H = DEL1/SUM
          T = SF*SUM
6         FORMAT(5X,'ITER =',I3,5X,'SIG=',F10.5,5X,'ERR=',F10.5,5X,'SF=',
     1    F10.5)
          RETURN
          END

          SUBROUTINE BLAS
          IMPLICIT DOUBLE PRECISION (A-H,O-Z)
          COMMON/INTEG/MT,ITER,NM,NN,IBL
          COMMON/REAL/DEL,GAM,PSI,HI,HI2,A21,A22,B21,B22,SUM,SF,DEL1,H,T
          COMMON/REAL1/ETAMAX,DN,SIG,DYDETA,TARG,DERSIG
          COMMON/ALG/E11(161),E21(161),F1(161),F2(161)
          COMMON/SOLN/UM(161),FM(161),S(161),UMN1(161),FMN1(161)
          COMMON/BL/FM0(161),UM0(161),S0(161),FM1(161),UM1(161),S1(161)

C ********************************************************************
C    START THOMAS ALGORITHM.    TRIANGULARIZE.
C ********************************************************************

1000      DO 20 N=2,NM
          A11 = HI2-.5D0*HI*FM0(N)
          C11 = HI2+.5D0*HI*FM0(N)
          B11 = -2.D0*HI2 - GAM*UM0(N)
          B12 = DEL*S0(N)
          D1 = .5D0*(FM0(N)*S0(N) + 1.D0 - UM0(N)**2)
          IF(IBL.EQ.2) D1 = -(FM0(N)*S0(N)+1.D0-UM0(N)**2)/24.D0 - 4.D0*
     1    UM0(N)*UM1(N)/3.D0 + 5.D0*S0(N)*FM1(N)/6.D0 + .5D0*S1(N)*FM0(N)
     2    - 3.D0*(FM1(N)*S1(N)-UM1(N)**2)
          G11 = A11*E11(N-1)+B11
          G12 = B12
          G21 = A21*E11(N-1)+A22*E21(N-1)+B21
          G22 = B22
          DENO = G11*G22-G12*G21
          GI11 = G22/DENO
          GI12 = -G12/DENO
          GI21 = -G21/DENO
          GI22 = G11/DENO
          E11(N) = -GI11*C11
          E21(N) = -GI21*C11
          F1(N) = GI11*(D1-A11*F1(N-1))-GI12*(A21*F1(N-1)+A22*F2(N-1))
          F2(N) = GI21*(D1-A11*F1(N-1))-GI22*(A21*F1(N-1)+A22*F2(N-1))
20        CONTINUE

C ********************************************************************
C    SET OUTER BOUNDARY CONDITION.
C ********************************************************************

          UM(NN) =    0.D0

C ********************************************************************
C    BACK-SUBSTITUTE TO FIND UM AND FM.
C ********************************************************************

          DO 30 N=1,NM
          K = NN-N
          UM(K) = E11(K)*UM(K+1) + F1(K)
          FM(K) = E21(K)*UM(K+1) + F2(K)
30        CONTINUE
          FM(NN) = FM(NM) + .5D0*DN*(UM(NN)+UM(NM))

C ********************************************************************
C        UPDATE S(N).
```

```
C*********************************************************************
      DO 40 N=2,NM
         S(N) = .5D0*HI*(UM(N+1)-UM(N-1))
40    CONTINUE
         SF = UM(2)*HI + .5D0*DN*(3.D0*S0(1)*SIG - .5D0)
         S(1) = SF

         DEL1 = FM(NM) - SIG
      RETURN
      END

      SUBROUTINE INIT
      IMPLICIT DOUBLE PRECISION (A-H,O-Z)
      COMMON/INTEG/MT,ITER,NM,NN,IBL
      COMMON/REAL/DEL,GAM,PSI,HI,HI2,A21,A22,B21,B22,SUM,SF,DEL1,H,T
      COMMON/REAL1/ETAMAX,DN,SIG,DYDETA,TARG,DERSIG
      COMMON/ALG/E11(161),E21(161),F1(161),F2(161)
      COMMON/SOLN/UM(161),FM(161),S(161),UMN1(161),FMN1(161)
      COMMON/BL/FM0(161),UM0(161),S0(161),FM1(161),UM1(161),S1(161)

C ********************************************************************
C   START THOMAS ALGORITHM.   TRIANGULARIZE.
C ****************************************************************

1000  DO 20 N=2,NM
         A11 = HI2-.5D0*HI*FM(N)
         C11 = HI2+.5D0*HI*FM(N)
         B11 = -2.D0*HI2 - 2.D0*UM(N)
         B12 = S(N)
         D1 =  FM(N)*S(N) - 1.D0 - UM(N)**2
         G11 = A11*E11(N-1)+B11
         G12 = B12
         G21 = A21*E11(N-1)+A22*E21(N-1)+B21
         G22 = B22
         DENO = G11*G22-G12*G21
         GI11 = G22/DENO
         GI12 = -G12/DENO
         GI21 = -G21/DENO
         GI22 = G11/DENO
         E11(N) = -GI11*C11
         E21(N) = -GI21*C11
         F1(N) = GI11*(D1-A11*F1(N-1))-GI12*(A21*F1(N-1)+A22*F2(N-1))
         F2(N) = GI21*(D1-A11*F1(N-1))-GI22*(A21*F1(N-1)+A22*F2(N-1))
20    CONTINUE

C*********************************************************************
C   SET OUTER BOUNDARY CONDITION.
C*********************************************************************

         UM(NN) =   1.D0

C *******************************************************************
C   BACK-SUBSTITUTE TO FIND UM AND FM.
C *******************************************************************

      DO 30 N=1,NM
         K = NN-N
         UM(K) = E11(K)*UM(K+1) + F1(K)
         FM(K) = E21(K)*UM(K+1) + F2(K)
30    CONTINUE
         FM(NN) = FM(NM) + .5D0*DN*(UM(NN)+UM(NM))

C*********************************************************************
C      UPDATE S(N).
C*********************************************************************
```

```
      DO 40 N=2,NM
         S(N) = .5D0*HI*(UM(N+1)-UM(N-1))
40    CONTINUE
         ITER = ITER + 1
         IF(ITER.LT.10) GO TO 1000
         SF = UM(2)*HI + .5D0*DN
         S(1) = SF
         DEL1 = ETAMAX - FM(NN) + SIG
      DO 35 N = 1,NN
         UM0(N) = UM(N)
         FM0(N) = FM(N)
         S0(N) = S(N)
         IF(N.EQ.1) GO TO 35
         SUM = SUM+0.5D0*DN*(UM(N)+UM(N-1))*(1.D0-0.5D0*(UM(N)+UM(N-1)))
35    CONTINUE
         T = SF*SUM
         H = DEL1/SUM
      RETURN
      END
```

B12 TRANBL

```
C**********************************************************************
C THIS PROGRAM INTEGRATES THE LAMINAR BOUNDARY-LAYER EQUATIONS FOR
C PLANE, IMPULSIVELY-STARTED INCOMPRESSIBLE FLOW, WHEN THE PROPAGATION
C OF INFORMATION FROM THE LEADING EDGE IS IMPORTANT.
C**********************************************************************

      PROGRAM TRANBL
      IMPLICIT DOUBLE PRECISION (A-H,O-Z)
      COMMON/INTEG/M,ITER,NM,NN,IVAR
      COMMON/REAL/DEL,GAM,PSI,HI,HI2,A21,A22,B21,B22,SUM,SF,DEL1,H,T
      COMMON/REAL1/ETAMAX,DN,ALFA,EPS,BETA,THETA,COM1,COM2
      COMMON/ALG/E11(161),E21(161),F1(161),F2(161),ETA(161)
      COMMON/SOLN/UM(161),FM(161),S(161),UL(161),FL(161),UP(60,161),
     1    FP(60,161)
      DIMENSION U0(161),F0(161),UI(161),FI(161)

C**********************************************************************
C  PRINT HEADING WHICH IDENTIFY THE PROBLEM.
C**********************************************************************

      WRITE(6,1)

C**********************************************************************
C  ASSIGN VALUES TO CONSTANTS.
C**********************************************************************

      PI = 4.D0*DATAN(1.D0)
      RPI = 1.D0/DSQRT(PI)
      ETAMAX = 8.D0
      NN = 161
      NM = NN-1
      RNM = NM
      DN = ETAMAX/RNM
      HI = 1.D0/DN
      HI2 = HI*HI
      A21 = .5D0
      A22 = HI
      B21 = .5D0
      B22 = -HI

C**********************************************************************
C  PICK VALUES OF THE EXPONENT, M.
C**********************************************************************

      DO 1000 MM = 1,2
      EM = 1.D0 - .5D0*MM
      ALF = DSQRT(PI)*(EM+1.D0)/4.D0
      WRITE(6,2) EM

C****************** SECTION ONE ****************************************
C    SET UP ETA(N) AND INITIAL PROFILES OF UM, S, AND FM.
C  MARCH FORWARD UNTIL FIRST SIGNAL ARRIVES FROM LEADING EDGE.
C**********************************************************************

      DO 10 N=1,NN
      ETA(N) = (N-1)*DN
      UM(N) = DERF(ETA(N)/2.D0)
      S(N) = RPI/DEXP(ETA(N)**2/4.D0)
      FM(N) = ETA(N)*UM(N)+ 2.D0*(S(N)-RPI)
      UL(N) = UM(N)
      FL(N) = FM(N)
10    CONTINUE
      SUM = 0.5D0*DN*UM(2)*(1.D0-UM(2))
      DO 11 N = 3,NN
      SUM = SUM+0.5D0*DN*(UM(N)*(1.D0-UM(N))+UM(N-1)*(1.D0-UM(N-1)))
11    CONTINUE
```

```
         SF = S(1)
         DEL1 = ETA(NN) - FM(NN)
         T = SF*SUM
         H = DEL1/SUM
         L = 0
         CHI = 0.D0
         EPS = .5D0
         GAM = 0.D0
         DEL = 0.D0
         ALFA = 0.D0
         WRITE (6,3)
         WRITE(6,4)L,CHI,GAM,DEL,EPS,ALFA,SF,DEL1,SUM,H,T


C*********************************************************************
C    INITIALIZE CHI.  SET  DCHI.
C*********************************************************************

         CHIMAX = 1.D0/(1.D0-EM)
         DCHI = 0.01D0*CHIMAX

C*********************************************************************
C   MARCH FORWARD IN CHI UNTIL FIRST SIGNAL ARRIVES FROM X = 0.
C*********************************************************************

         IVAR = 1
      DO 2000 L = 1,100
         CHI = L*DCHI
         ALX = ALF*CHI
         IF (ALX.GT.12.D0) GO TO 301
         XX = DERF(ALX)
         EPS = .5D0/DEXP(ALX**2)
         GO TO 302
301   CONTINUE
         XX = 1.D0
         EPS = 0.D0
302   CONTINUE
         DEL = (2.D0*EM/(EM+1.D0))*XX
         GAM = XX + (EM-1.D0)*CHI*EPS
         ALFA = (2.D0/(EM+1.D0))*XX
         PSI = (EM-1.D0)*CHI*ALFA
         ALFA = ALFA/DCHI
         PSI = PSI/DCHI

C*********************************************************************
C   SET WALL BOUNDARY CONDITIONS.
C*********************************************************************

         E11(1) = 0.D0
         E21(1) = 0.D0
         F1(1) = 0.D0
         F2(1) = 0.D0

C*********************************************************************
C    SET ITERATION COUNTER.  CALCULATE PROFILES AT NEW VALUE OF XI.
C*********************************************************************

         ITER = 1
         CALL THOMAS

      DO 25 N = 1,NN
         UL(N) = UM(N)
         FL(N) = FM(N)
25    CONTINUE

2000  CONTINUE
```

```
C*****************************************************************************
C   PRINT FINAL PROFILE.
C*****************************************************************************

        ALFA = ALFA*DCHI
        WRITE(6,4)L,CHI,GAM,DEL,EPS,ALFA,SF,DEL1,SUM,H,T
        WRITE(6,5)
        II = NM/3
     DO 66 I = 1,II
        J=I+II
        K=J+II
        WRITE(6,6) ETA(I),FM(I),UM(I),S(I),ETA(J),FM(J),UM(J),S(J),
     1  ETA(K),FM(K),UM(K),S(K)
66      CONTINUE

C*****************************************************************************
C   NAME THIS PROFILE.
C*****************************************************************************

     DO 30 N = 1,NN
        UI(N) = UM(N)
        FI(N) = FM(N)
30      CONTINUE

C********************* SECTION TWO ****************************************
C   SWITCH TO THE SECOND SET OF VARIABLES.  START BY SETTING UP AND
C   STORING LEADING-EDGE PROFILES, TO BE CALLED U0(N) AND F0(N).
C*****************************************************************************

        GAM = 1.D0
        DEL = 2.D0*EM/(1.D0+EM)

C*****************************************************************************
C    SET WALL BOUNDARY CONDITIONS.
C*****************************************************************************

        E11(1) = 0.D0
        E21(1) = 0.D0
        F1(1) = 0.D0
        F2(1) = 0.D0

C*****************************************************************************
C    SET ITERATION COUNTER.  CALCULATE.
C*****************************************************************************

        ITER = 1
        CALL FINAL
     DO 20 N = 1,NN
        U0(N) = UM(N)
        F0(N) = FM(N)
20      CONTINUE
        WRITE(6,3)
        WRITE(6,4)L,CHI,GAM,DEL,EPS,ALFA,SF,DEL1,SUM,H,T

C*****************************************************************************
C   PREPARE TO MARCH FROM LEFT TO RIGHT AT CONSTANT TAU. START BY
C   PRE-COMPUTING VALUES OF XI AND DXI.
C*****************************************************************************

        LL = 60
        IVAR = 2
        TAUSTR = 1.D0/(1.D0-EM)
        RL = LL
        DTAU = TAUSTR/RL
        DXI = 1.D0/RL
```

```
C*********************************************************************
C    MARCH, FIRST INSERTING LEADING-EDGE VALUES AS UL AND FL.
C*********************************************************************

      DO 3000 L = 2,LL
      DO 40 N = 1,NN
         UL(N) = U0(N)
         FL(N) = F0(N)
40    CONTINUE

         TAU = L*DTAU
         IF(L.EQ.LL) WRITE(6,7) L,TAU
         IF(L.EQ.LL) WRITE(6,8)
         LM = L-1
      DO 4000 M = 1,LM

C*********************************************************************
C    CALCULATE COEFFICIENTS THAT DEPEND ONLY UPON TAU AND XI.
C*********************************************************************

         XI = M*DXI
         CHI = TAU/XI
         ALX = ALF*CHI
         IF (ALX.GT.12.D0) GO TO 303
         XX = DERF(ALX)
         EPS = .5D0/DEXP(ALX**2)
         GO TO 304
303   CONTINUE
         XX = 1.D0
         EPS = 0.D0
304   CONTINUE
         DEL = (2.D0*EM/(EM+1.D0))*XX
         GAM = XX + (EM-1.D0)*CHI*EPS
         PSI = (2.D0/(EM+1.D0))*XI*XX
         BETA = PSI
         THETA = DEL*TAU
         PSI = PSI/DXI
         BETA = BETA/DTAU
         THETA = THETA/DTAU
         COM1 = GAM + PSI + THETA
         COM2 = DEL + PSI + THETA

C*********************************************************************
C    SET WALL BOUNDARY CONDITIONS.
C*********************************************************************

         E11(1) = 0.D0
         E21(1) = 0.D0
         F1(1) = 0.D0
         F2(1) = 0.D0


C*********************************************************************
C    ASSIGN U1 AND F1 TO THE ROLES OF UP AND FP, IF M = LM.
C*********************************************************************

      IF(M.NE.LM) GO TO 170
      DO 50 N = 1,NN
         UP(M,N) = UI(N)
         FP(M,N) = FI(N)
50    CONTINUE
170   CONTINUE

C*********************************************************************
C    SET ITERATION COUNTER.  CALCULATE PROFILES AT NEW POINT (XI,TAU).
```

```
C******************************************************************
          ITER = 1
          CALL THOMAS

C******************************************************************
C    PRINT SUMMARY VALUES .
C******************************************************************

          IF(L.LT.LL) GO TO 159
          DIS2 = (2.D0/(EM+1.D0))*XX*DEL1**2
          WRITE(6,9)M,XI,CHI,SF,DEL1,SUM,H,T,DIS2
159       CONTINUE

C******************************************************************
C    PRINT SELECTED PROFILES.
C******************************************************************

          IF(SF.LT.0.D0) GO TO 160
          GO TO 99
160       WRITE(6,3)
          II = NM/3
       DO 67 I = 1,II
          J=I+II
          K=J+II
          WRITE(6,4) ETA(I),FM(I),UM(I),S(I),ETA(J),FM(J),UM(J),S(J),
     1    ETA(K),FM(K),UM(K),S(K)
67        CONTINUE
          WRITE(6,5)
99        IF(SF.LT.0.D0) GO TO 4000

C******************************************************************
C  ASSIGN THESE VALUES TO PROPER STORAGE LOCATIONS.
C******************************************************************

       DO 60 N = 1,NN
          UP(M,N) = UM(N)
          UL(N)  = UM(N)
          FP(M,N) = FM(N)
          FL(N)  = FM(N)
60        CONTINUE
4000      CONTINUE
3000      CONTINUE
1000      CONTINUE

C******************************************************************
C    FORMAT WRITEOUTS.
C******************************************************************

1         FORMAT(5X,'IMPULSIVELY-STARTED BOUNDARY LAYER ON A WEDGE',//)
2         FORMAT (/,10X,'EM =', F 10.5)
3         FORMAT(//,2X,'L',5X,'CHI',6X,'GAM',6X,'DEL',5X,'EPS',5X,'ALF',
     1    6X,'SF',6X,'DEL1',6X,'SUM',6X,'H',7X,'T',//)
4         FORMAT(I3,10F9.5)
5         FORMAT(//,4X,'ETA',7X,'F',9X,'U',9X,'S',7X,'ETA',7X,
     1    'F',9X,'U',9X,'S',7X,'ETA',7X,'F',9X,'U',9X,'S',/)
6         FORMAT(3(3X,F5.3,3F10.5))
7         FORMAT(/,10X,'L =',I3,5X,'TAU =',F10.5,/)
8         FORMAT(//,2X,'M',5X,'XI',7X,'CHI',6X,'SF',6X,'DEL1',5X,'SUM',5X,
     1     'H',7X,'T',6X,'DIS2',//)
9         FORMAT(I3,8F9.5)
          STOP
          END

          SUBROUTINE THOMAS
          IMPLICIT DOUBLE PRECISION (A-H,O-Z)
```

```
        COMMON/INTEG/M,ITER,NM,NN,IVAR
        COMMON/REAL/DEL,GAM,PSI,HI,HI2,A21,A22,B21,B22,SUM,SF,DEL1,H,T
        COMMON/REAL1/ETAMAX,DN,ALFA,EPS,BETA,THETA,COM1,COM2
        COMMON/ALG/E11(161),E21(161),F1(161),F2(161),ETA(161)
        COMMON/SOLN/UM(161),FM(161),S(161),UL(161),FL(161),UP(60,161),
     1  FP(60,161)

C ************************************************************************
C   START THOMAS ALGORITHM.   TRIANGULARIZE.
C ********************************************************************

1000    DO 20 N=2,NM
            IF(IVAR.EQ.2)GO TO 100
            X = (GAM+PSI)*FM(N) - PSI*FL(N) + EPS*ETA(N)
            Y = -2.D0*(DEL+PSI)*UM(N) + PSI*UL(N) - ALFA
            Z = (GAM+PSI)*S(N)
            D1 = Z*FM(N) - DEL - (DEL+PSI)*UM(N)*UM(N) - ALFA*UL(N)
            GO TO 101
100     CONTINUE
            X = COM1*FM(N) - PSI*FL(N) - THETA*FP(M,N) + EPS*ETA(N)
            Y = -2.D0*COM2*UM(N) + PSI*UL(N) + THETA*UP(M,N) - BETA
            Z = COM1*S(N)
            D1 = Z*FM(N) - DEL - COM2*UM(N)*UM(N) - BETA*UP(M,N)
101     CONTINUE
            A11 = HI2-.5D0*HI*X
            C11 = HI2+.5D0*HI*X
            B11 = -2.D0*HI2+Y
            B12 = Z
            G11 = A11*E11(N-1)+B11
            G12 = B12
            G21 = A21*E11(N-1)+A22*E21(N-1)+B21
            G22 = B22
            DENO = G11*G22-G12*G21
            GI11 = G22/DENO
            GI12 = -G12/DENO
            GI21 = -G21/DENO
            GI22 = G11/DENO
            E11(N) = -GI11*C11
            E21(N) = -GI21*C11
            F1(N) = GI11*(D1-A11*F1(N-1))-GI12*(A21*F1(N-1)+A22*F2(N-1))
            F2(N) = GI21*(D1-A11*F1(N-1))-GI22*(A21*F1(N-1)+A22*F2(N-1))
20      CONTINUE

C***********************************************************************
C   SET OUTER BOUNDARY CONDITION.
C***********************************************************************

            UM(NN) =   1.D0

C ***********************************************************************
C   BACK-SUBSTITUTE TO FIND UM AND FM.
C ***********************************************************************

        DO 30 N=1,NM
            K = NN-N
            UM(K) = E11(K)*UM(K+1) + F1(K)
            FM(K) = E21(K)*UM(K+1) + F2(K)
30      CONTINUE
            FM(NN) = FM(NM)+ .5D0*DN*(UM(NN)+UM(NM))

C***********************************************************************
C       UPDATE S(N).
C***********************************************************************

        DO 40 N=2,NM
            S(N) = .5D0*HI*(UM(N+1)-UM(N-1))

40      CONTINUE
```

```
C  ****************************************************************************
C     ADVANCE AND CHECK ITERATION COUNTER.
C  ****************************************************************************

         ITER = ITER+1
         IF(ITER.LT.3) GO TO 1000

C********************************************************************************
C  CALCULATE  B.L. THICKNESSES, SHAPE FACTORS, AND SKIN FRICTION.
C********************************************************************************

         SUM = 0.5D0*DN*UM(2)*(1.D0-UM(2))
      DO 50 N=3,NN
         SUM = SUM+0.5D0*DN*(UM(N)*(1.D0-UM(N))+UM(N-1)*(1.D0-UM(N-1)))
50    CONTINUE
         SF = UM(2)*HI + .50D0*DN*DEL
         S(1) = SF
         DEL1 = ETAMAX - FM(NN)
         H = DEL1/SUM
         T = SF*SUM
      RETURN
      END

      SUBROUTINE FINAL
      IMPLICIT DOUBLE PRECISION (A-H,O-Z)
      COMMON/INTEG/M,ITER,NM,NN,IVAR
      COMMON/REAL/DEL,GAM,PSI,HI,HI2,A21,A22,B21,B22,SUM,SF,DEL1,H,T
      COMMON/REAL1/ETAMAX,DN,ALFA,EPS,BETA,THETA,COM1,COM2
      COMMON/ALG/E11(161),E21(161),F1(161),F2(161),ETA(161)
      COMMON/SOLN/UM(161),FM(161),S(161),UL(161),FL(161),UP(60,161),
     1     FP(60,161)

C  ****************************************************************************
C    START THOMAS ALGORITHM.   TRIANGULARIZE.
C  ****************************************************************************

1000     DO 20 N=2,NM
         X = GAM*FM(N)
         Y = -2.D0*DEL*UM(N)
         Z = GAM*S(N) ,
         D1 = Z*FM(N) - DEL*(1.D0+ UM(N)*UM(N))
         A11 = HI2-.5D0*HI*X
         C11 = HI2+.5D0*HI*X
         B11 = -2.D0*HI2+Y
         B12 = Z
         G11 = A11*E11(N-1)+B11
         G12 = B12
         G21 = A21*E11(N-1)+A22*E21(N-1)+B21
         G22 = B22
         DENO = G11*G22-G12*G21
         GI11 = G22/DENO
         GI12 = -G12/DENO
         GI21 = -G21/DENO
         GI22 = G11/DENO
         E11(N) = -GI11*C11
         E21(N) = -GI21*C11
         F1(N) = GI11*(D1-A11*F1(N-1))-GI12*(A21*F1(N-1)+A22*F2(N-1))
         F2(N) = GI21*(D1-A11*F1(N-1))-GI22*(A21*F1(N-1)+A22*F2(N-1))
20    CONTINUE
```

```
C******************************************************************************
C    SET OUTER BOUNDARY CONDITION.
C******************************************************************************

          UM(NN) =    1.D0

C ******************************************************************************
C    BACK-SUBSTITUTE TO FIND UM AND FM.
C ******************************************************************************

        DO 30 N=1,NM
          K = NN-N
          UM(K)  = E11(K)*UM(K+1)  + F1(K)
          FM(K)  = E21(K)*UM(K+1)  + F2(K)
30       CONTINUE
          FM(NN) = FM(NM)+ .5D0*DN*(UM(NN)+UM(NM))

C******************************************************************************
C     UPDATE S(N).
C******************************************************************************

        DO 40 N=2,NM
          S(N) = .5D0*HI*(UM(N+1)-UM(N-1))
40       CONTINUE

C ******************************************************************************
C    ADVANCE AND CHECK ITERATION COUNTER.
C ******************************************************************************

          ITER = ITER+1
          IF(ITER.LT.9) GO TO 1000

C******************************************************************************
C CALCULATE   B.L. THICKNESSES, SHAPE FACTORS, AND SKIN FRICTION.
C******************************************************************************

          SUM = 0.5D0*DN*UM(2)*(1.D0-UM(2))
        DO 50 N=3,NN
          SUM = SUM+0.5D0*DN*(UM(N)*(1.D0-UM(N))+UM(N-1)*(1.D0-UM(N-1)))
50       CONTINUE
          SF = UM(2)*HI + .50D0*DN*DEL
          S(1) = SF
          DEL1 = ETAMAX - DN - FM(NM)
          H = DEL1/SUM
          T = SF*SUM
        RETURN
        END
```

B13 SYMMBL

```
C*************************************************************************
C THIS PROGRAM INTEGRATES THE LAMINAR BOUNDARY-LAYER EQUATIONS FOR
C STEADY INCOMPRESSIBLE FLOW, IN THE SYMMETRY PLANE OF
C A HYPERBOLIC NOZZLE.
C*************************************************************************

      PROGRAM SYMMBL
      IMPLICIT DOUBLE PRECISION (A-H,O-Z)
      COMMON/INTEG/NM,NN,ITMAX,IFL
      COMMON/REAL/DELX1,DELY2,GAM1,PSI,DEL1
      COMMON/REAL1/ETAMAX,DN,HI,HI2,DXI,WINF
      COMMON/ALG/A(3,3),B(3,3),D(3),E(3,2,401),F(3,401)
      COMMON/SOLN/UM(401),VM(401),R(401),S(401),W(401),UM1(401),VM1(401)
     1   ,Z(401),UM2(401),VM2(401)
      DIMENSION U0(401),W0(401),R0(401)

C*************************************************************************
C  PRINT HEADING WHICH IDENTIFY THE PROBLEM.
C*************************************************************************

      WRITE(6,1)

C*************************************************************************
C  ASSIGN VALUES TO CONSTANTS.
C*************************************************************************

      PI = 4.D0*DATAN(1.D0)
      RPI = 1.D0/DSQRT(PI)
      ETAMAX = 20.D0
      NN = 401
      NM = NN-1
      RNM = NM
      DN = ETAMAX/RNM
      HI = 1.D0/DN
      HI2 = HI*HI
      A(1,2) = 0.D0
      A(1,3) = 0.D0
      A(2,1) = 0.D0
      A(2,3) = 0.D0
      B(1,2) = 0.D0

C*************************************************************************
C   PICK INITIAL VALUE OF XI.   CALCULATE DXI.
C*************************************************************************

      XI = -4.D0
      MM = 140
      RM = MM
      DXI = .05D0

C*************************************************************************
C    SET UP ZETA(N) AND FIRST-GUESS PROFILES OF U, V, W, R AND S.
C*************************************************************************

      DO 10 N=1,NN
        Z(N)  = (N-1)*DN
        UM(N) = DERF(Z(N)/2.D0)
        R(N)  = RPI/DEXP(Z(N)**2/4.D0)
        W(N)  = Z(N)*UM(N)+ 2.D0*(R(N)-RPI)
        VM(N) = 0.D0
        S(N)  = 0.D0
10    CONTINUE

C*************************************************************************
C    ESTABLISH UPSTREAM PROFILES.
C    SET WALL BOUNDARY CONDITIONS.
```

```
C************************************************************************

      DO 20 I = 1,3
      DO 21 J = 1,2
         E(I,J,1) = 0.D0
21    CONTINUE
         F(I,1) = 0.D0
20    CONTINUE

         ITMAX = 7
         PSI = 0.D0
         GAM1 = -1.D0
         DELX1 = 1.D0
         DELY2 = 0.D0
         CALL THOMAS

      DO 25 N = 1,NM
         U0(N) = UM(N)
         W0(N) = W(N)
         R0(N) = R(N)
25    CONTINUE

C************************************************************************
C PRINT FIRST TERM IN BLASIUS SERIES.
C************************************************************************

         WRITE (7,4)

      DO 44 N = 1,NN,2
         WRITE (7,5) Z(N),UM(N),VM(N),W(N),R(N),S(N)
44    CONTINUE

C************************************************************************
C  CALCULATE SECOND TERM IN BLASIUS SERIES.
C************************************************************************

         CALL BLAS

C************************************************************************
C PRINT SECOND TERM IN BLASIUS SERIES.
C************************************************************************

         WRITE (7,4)

      DO 45 N = 1,NN,2
         WRITE (7,5) Z(N),UM(N),VM(N),W(N),R(N),S(N)
45    CONTINUE

C************************************************************************
C      COMPOSE STARTING PROFILE.
C************************************************************************

      DO 30 N = 1,NM
         UM2(N) = U0(N) + UM(N)/XI**2
         VM2(N) = VM(N)/XI**3
         UM1(N) = U0(N) + UM(N)/(XI+DXI)**2
         VM1(N) = VM(N)/(XI+DXI)**3
         UM(N) = U0(N) + UM(N)/(XI+2.D0*DXI)**2
         VM(N) = VM(N)/(XI+2.D0*DXI)**3
         W(N)  = W0(N) + W(N)/(XI+2.D0*DXI)**2
         R(N)  = R0(N) + R(N)/(XI+2.D0*DXI)**2
         S(N)  =  S(N)/(XI+2.D0*DXI)**3
30    CONTINUE

C************************************************************************
C PRINT STARTING PROFILE.
```

```
C**********************************************************************

          WRITE (7,4)

       DO 40 N = 1,NN,2
          WRITE (7,5) Z(N),UM(N),VM(N),W(N),R(N),S(N)
40     CONTINUE

C**********************************************************************
C    MARCH DOWNSTREAM, UNTIL SEPARATION IS INDICATED.
C**********************************************************************

          XI = -3.95D0
          WRITE(6,3)
       DO 2000 M = 2,MM
          IF(M.GT.60) DXI = 0.025D0
          IF(M.EQ.61) IFL = 1
          IF(M.GT.100) DXI = 0.0125D0
          IF(M.EQ.101) IFL = 1
          IF(M.GT.110) DXI = 0.00625D0
          IF(M.EQ.111) IFL = 1
          IF(M.GT.120) DXI = 0.003125D0
          IF(M.EQ.121) IFL = 1
          XI = XI + DXI
          SQXI = DSQRT (1.D0 + XI**2)
          GAM1 = XI/SQXI
          DELX1 = - GAM1
C         DELY2 =  1.D0/SQXI**3
          DELY2 =  0.D0
          PSI = SQXI

C**********************************************************************
C  CALCULATE PROFILES AT NEW VALUE OF XI. STORE SKIN FRICTION.
C**********************************************************************

          ITMAX = 3
          CALL THOMAS
          WRITE (6,6) XI,R(1),S(1),W(NM)
          IF(R(1).LT.0.D0) GO TO 650
          IF(M.EQ.MM) GO TO 650

C**********************************************************************
C    UPDATE UM1,VM1,UM2,VM2 IN PREPARATION FOR NEXT STEP.
C**********************************************************************

       DO 50 N = 1,NM
          UM2(N) = UM1(N)
          VM2(N) = VM1(N)
          UM1(N) = UM(N)
          VM1(N) = VM(N)
50     CONTINUE
2000   CONTINUE

C**********************************************************************
C    PRINT PROFILES.
C**********************************************************************

650       WRITE (7,4)
       DO 70 N = 1,NN
          WRITE (7,5) Z(N),UM(N),VM(N),W(N),R(N),S(N)
70     CONTINUE

C**********************************************************************
C    FORMAT WRITEOUTS.
C**********************************************************************
```

```
1        FORMAT(5X,'STEADY 3-D BOUNDARY LAYER ON SIDEWALL OF NOZZLE.',//)
3        FORMAT(10X,'XI',8X,'R(1)',7X,'S(1)',7X,'W(NM)',/)
4        FORMAT(/,5X,'Z',9X,'U',9X,'V',9X,'W',9X,'R',9X,'S',/)
5        FORMAT(6F10.5)
6        FORMAT(5X,4F10.5)
         STOP
         END

         SUBROUTINE THOMAS
         IMPLICIT DOUBLE PRECISION (A-H,O-Z)
         COMMON/INTEG/NM,NN,ITMAX,IFL
         COMMON/REAL/DELX1,DELY2,GAM1,PSI,DEL1
         COMMON/REAL1/ETAMAX,DN,HI,HI2,DXI,WINF
         COMMON/ALG/A(3,3),B(3,3),D(3),E(3,2,401),F(3,401)
         COMMON/SOLN/UM(401),VM(401),R(401),S(401),W(401),UM1(401),VM1(401)
     1      ,Z(401),UM2(401),VM2(401)
         DIMENSION G(3,3)

C******************************************************************************
C        TRIANGULARIZE.
C******************************************************************************

         IF(IFL.EQ.1) GO TO 10
         AAA = 1.5D0/DXI
         BBB = -2.D0/DXI
         CCC = .5D0/DXI
         GO TO 15
10       AAA = 4.D0/(3.D0*DXI)
         BBB = -1.5D0/DXI
         CCC = 1.D0/(6.D0*DXI)
15          ITHOM = 0
1000   CONTINUE
       DO 200 N = 2,NM
         NM1 = N-1
         P1 =   GAM1*UM(N)*Z(N)+W(N)
         A(1,1) = HI2 - .5D0*P1*HI
         A(2,2) = A(1,1)
         A(3,1) = -.5D0*HI*GAM1*(Z(N)+Z(NM1))
         A(3,2) = -.5D0*PSI
         A(3,3) = - HI
         B(1,1) = -2.D0*HI2 + GAM1*Z(N)*R(N) - 2.D0*DELX1*UM(N)
         B(1,3) = R(N)
         B(2,1) = GAM1*Z(N)*S(N) - 2.D0*DELY2*UM(N)
         B(2,2) = -2.D0*HI2 - 2.D0*PSI*VM(N)
         B(2,3) = S(N)
         B(3,1) = - A(3,1)
         B(3,2) = A(3,2)
         B(3,3) = - A(3,3)
         CC = HI2 + .5D0*HI*P1
         D(1) = P1*R(N) - DELX1*(1.D0+UM(N)**2)
         D(2) = P1*S(N) - DELY2*(1.D0+UM(N)**2) - PSI*VM(N)**2
         D(3) = 0.D0

         A(3,1) = A(3,1) - .5D0*PSI*AAA
         B(1,1) = B(1,1) - PSI*(2.D0*AAA*UM(N)+BBB*UM1(N)+CCC*UM2(N))
         B(2,1) = B(2,1) - PSI*(AAA*VM(N)+BBB*VM1(N)+CCC*VM2(N))
         B(2,2) = B(2,2) - PSI*AAA*UM(N)
         B(3,1)= B(3,1) - .5D0*PSI*AAA
         D(1) = D(1) - PSI*AAA*UM(N)**2
         D(2) = D(2) - PSI*AAA*UM(N)*VM(N)
         D(3) = D(3) + .5D0*PSI*(BBB*(UM1(N)+UM1(NM1))+CCC*(UM2(N)
     1   +UM2(NM1)))

C******************************************************************************
C     FORM THE MATRIX G.
C******************************************************************************
```

```
          DO 20 I = 1,3
          DO 21 J = 1,2
             G(I,J) = 0.D0
          DO 22 K = 1,3
             G(I,J) = G(I,J) + A(I,K)*E(K,J,NM1)
22        CONTINUE
             G(I,J)=G(I,J) + B(I,J)
21        CONTINUE
20        CONTINUE
             G(1,3) = B(1,3)
             G(2,3) = B(2,3)
             G(3,3) = B(3,3)

             DD1 = D(1) - A(1,1)*F(1,NM1)
             DD2 = D(2) - A(2,2)*F(2,NM1)
             DD3 = D(3) - A(3,1)*F(1,NM1) - A(3,2)*F(2,NM1)
        1     - A(3,3)*F(3,NM1)

C*******************************************************************
C     SOLVE FOR E(I,J,N) AND F(I,N) BY KRAMER'S RULE.
C*******************************************************************

          H1 = G(2,2)*G(3,3)-G(2,3)*G(3,2)
          H2 = G(2,3)*G(3,1)-G(2,1)*G(3,3)
          H3 = G(2,1)*G(3,2)-G(2,2)*G(3,1)
          H4 = G(3,2)*G(1,3)-G(1,2)*G(3,3)
          H5 = G(1,1)*G(3,3)-G(1,3)*G(3,1)
          H6 = G(1,2)*G(3,1)-G(3,2)*G(1,1)
          H7 = G(1,2)*G(2,3)-G(2,2)*G(1,3)
          H8 = G(2,1)*G(1,3)-G(2,3)*G(1,1)
          H9 = G(1,1)*G(2,2)-G(1,2)*G(2,1)
          DENO = G(1,1)*H1+G(1,2)*H2+G(1,3)*H3

             E(1,1,N) = -CC*H1/DENO
             E(2,1,N) = -CC*H2/DENO
             E(3,1,N) = -CC*H3/DENO
             E(1,2,N) = -CC*H4/DENO
             E(2,2,N) = -CC*H5/DENO
             E(3,2,N) = -CC*H6/DENO
             F(1,N) = (DD1*H1+DD2*H4+DD3*H7)/DENO
             F(2,N) = (DD1*H2+DD2*H5+DD3*H8)/DENO
             F(3,N) = (DD1*H3+DD2*H6+DD3*H9)/DENO
200       CONTINUE

C*******************************************************************
C     SET OUTER BOUNDARY CONDITION.
C*******************************************************************

          XX = 2.D0*DN/(Z(NM)-W(NM))

          UM(NN) = (XX + F(1,NM))/(1.D0 + XX - E(1,1,NM))
          VM(NN) = 0.D0

C*******************************************************************
C     BACK-SUBSTITUTE.
C*******************************************************************

          DO 300 N = 1,NM
          K = NN-N
          UM(K) = E(1,1,K)*UM(K+1)+E(1,2,K)*VM(K+1)+F(1,K)
          VM(K) = E(2,1,K)*UM(K+1)+E(2,2,K)*VM(K+1)+F(2,K)
          W(K)  = E(3,1,K)*UM(K+1)+E(3,2,K)*VM(K+1)+F(3,K)
300       CONTINUE

C*******************************************************************
```

```
C      UPDATE R(N) AND S(N) FOR ITERATIONS.
C***********************************************************************

       DO 400 N = 2,NM
          R(N) = 0.5D0*HI*(UM(N+1)-UM(N-1))
          S(N) = 0.5D0*HI*(VM(N+1)-VM(N-1))
400    CONTINUE

          ITHOM = ITHOM + 1
          IF(ITHOM.LT.ITMAX) GO TO 1000

          R(1) = HI*UM(2)+0.5D0*DN*DELX1
          S(1) = HI*VM(2)+0.5D0*DN*DELY2
       IFL = 0

       RETURN
       END

       SUBROUTINE BLAS
       IMPLICIT DOUBLE PRECISION (A-H,O-Z)
       COMMON/INTEG/NM,NN,ITMAX,IFL
       COMMON/REAL1/ETAMAX,DN,HI,HI2,DXI,WINF
       COMMON/ALG/A(3,3),B(3,3),D(3),E(3,2,401),F(3,401)
       COMMON/SOLN/UM(401),VM(401),R(401),S(401),W(401),UM1(401),VM1(401)
      1   ,Z(401),UM2(401),VM2(401)
       DIMENSION G(3,3)

C***********************************************************************
C      TRIANGULARIZE.
C***********************************************************************

       DO 200 N = 2,NM
          NM1 = N-1
          P1 = - UM(N)*Z(N)+W(N)
          A(1,1) = HI2 - .5D0*HI*P1
          A(2,2) = A(1,1)
          A(3,1) = -.5D0*HI*(Z(N)+Z(NM1)) + 1.D0
          A(3,2) = -.5D0
          A(3,3) =  HI
          B(1,1) = -2.D0*HI2 - Z(N)*R(N) - 4.D0*UM(N)
          B(1,3) =  R(N)
          B(2,1) = 0.D0
          B(2,2) = -2.D0*HI2 - 3.D0*UM(N)
          B(2,3) = 0.D0
          B(3,1) = .5D0*HI*(Z(N)+Z(NM1)) + 1.D0
          B(3,2) = A(3,2)
          B(3,3) = - A(3,3)
          CC = HI2 + .5D0*HI*P1
          D(1) = -.5D0*(Z(N)*UM(N)*R(N) - (1.D0-UM(N)**2))
C         D(2) = 1.D0-UM(N)**2
          D(2) = 0.D0
          D(3) =  .25D0*HI*(Z(N)+Z(NM1))*(UM(N)-UM(NM1))

C***********************************************************************
C      FORM THE MATRIX G.
C***********************************************************************

       DO 20 I = 1,3
       DO 21 J = 1,2
          G(I,J) = 0.D0
       DO 22 K = 1,3
          G(I,J) = G(I,J) + A(I,K)*E(K,J,NM1)
22     CONTINUE
          G(I,J)=G(I,J) + B(I,J)
21     CONTINUE
```

```
20      CONTINUE
          G(1,3) = B(1,3)
          G(2,3) = B(2,3)
          G(3,3) = B(3,3)

          DD1 = D(1) - A(1,1)*F(1,NM1)
          DD2 = D(2) - A(2,2)*F(2,NM1)
          DD3 = D(3) - A(3,1)*F(1,NM1) - A(3,2)*F(2,NM1)
     1        - A(3,3)*F(3,NM1)

C*********************************************************************
C     SOLVE FOR E(I,J,N) AND F(I,N) BY KRAMER'S RULE.
C*********************************************************************

          H1 = G(2,2)*G(3,3)-G(2,3)*G(3,2)
          H2 = G(2,3)*G(3,1)-G(2,1)*G(3,3)
          H3 = G(2,1)*G(3,2)-G(2,2)*G(3,1)
          H4 = G(3,2)*G(1,3)-G(1,2)*G(3,3)
          H5 = G(1,1)*G(3,3)-G(1,3)*G(3,1)
          H6 = G(1,2)*G(3,1)-G(3,2)*G(1,1)
          H7 = G(1,2)*G(2,3)-G(2,2)*G(1,3)
          H8 = G(2,1)*G(1,3)-G(2,3)*G(1,1)
          H9 = G(1,1)*G(2,2)-G(1,2)*G(2,1)
          DENO = G(1,1)*H1+G(1,2)*H2+G(1,3)*H3

          E(1,1,N) = -CC*H1/DENO
          E(2,1,N) = -CC*H2/DENO
          E(3,1,N) = -CC*H3/DENO
          E(1,2,N) = -CC*H4/DENO
          E(2,2,N) = -CC*H5/DENO
          E(3,2,N) = -CC*H6/DENO
          F(1,N) = (DD1*H1+DD2*H4+DD3*H7)/DENO
          F(2,N) = (DD1*H2+DD2*H5+DD3*H8)/DENO
          F(3,N) = (DD1*H3+DD2*H6+DD3*H9)/DENO
200     CONTINUE

C*********************************************************************
C     SET OUTER BOUNDARY CONDITION.
C*********************************************************************

          UM(NN) = 0.D0
          VM(NN) = 0.D0

C*********************************************************************
C     BACK-SUBSTITUTE.
C*********************************************************************

          DO 300 N = 1,NM
          K = NN-N
          UM(K) = E(1,1,K)*UM(K+1)+E(1,2,K)*VM(K+1)+F(1,K)
          VM(K) = E(2,1,K)*UM(K+1)+E(2,2,K)*VM(K+1)+F(2,K)
          W(K)  = E(3,1,K)*UM(K+1)+E(3,2,K)*VM(K+1)+F(3,K)
300     CONTINUE

C*********************************************************************
C     UPDATE R(N) AND S(N) FOR ITERATIONS.
C*********************************************************************

          DO 400 N = 2,NM
          R(N) = 0.5D0*HI*(UM(N+1)-UM(N-1))
          S(N) = 0.5D0*HI*(VM(N+1)-VM(N-1))
400     CONTINUE
          R(1) = HI*UM(2)-.25D0*DN*DELX1
          S(1) = HI*VM(2)+0.5D0*DN*DELY2

        RETURN
        END
```

B14   HOTDOG

```
C***********************************************************************
C     THIS PROGRAM DEMONSTRATES UPWIND DIFFERENCING, FOR THE TRANSIENT,
C     SYMMETRY-LINE FLOW BETWEEN NODAL AND SADDLE STAGNATION POINTS. THE
C     INVISCID VELOCITY IS U = AX,   V = BY*(1-Y/L).
C***********************************************************************

      PROGRAM HOTDOG
      IMPLICIT DOUBLE PRECISION (A-H,O-Z)

C***********************************************************************
C     ASSIGN STORAGE FOR Z, U, V, W, UP, VP, UPP, VPP, R ,AND S.
C     U(MM,NN) ALLOWS FOR MM XI-STATIONS, NN Z-STATIONS.
C***********************************************************************

      COMMON/ALG/Z(161),U(85,161),V(85,161),W(85,161),UP(85,161),
     1 VP(85,161),UPP(85,161),VPP(85,161),R(85,161),S(85,161),
     2 UT(85,161),VT(85,161),WT(85,161),UE(2,161),VE(2,161),WE(2,161),
     3 UEP(2,161),VEP(2,161),UEPP(2,161),VEPP(2,161), ETA(85)
      COMMON/SOLN/E(3,2,161),F(3,161),A(3,3),B(3,3),D(3)
      COMMON/COEF/GAM3,DELX1,DELY2,EPS,PSI,THET,DETA,DTAU,H,HI,HI2,
     1 CP,T,VI
      COMMON/INTEG/M,L,NN,NM,MM,IFL,IE,MSTOP,IML
      DIMENSION DEL1(85)

C***********************************************************************
C     SPECIFY NUMBER OF MESH POINTS IN Z AND IN XI. SET THE TIME STEP.
C     SET THE NUMBER OF ITERATIONS PER TIME STEP.  SET THE VALUE OF C.
C***********************************************************************

      NN = 71
      NM = NN-1
      RN = NM
      ZMAX = 7.D0
      H = ZMAX/RN
      HI = 1.D0/H
      HI2 = HI*HI
      MM = 59
      MN = MM-5
      MH = MN/2 + 3
      ITMAX = 2
      MSTOP = MM-2
      CP = .8D0

C***********************************************************************
C     SET UP Z(N), INITIAL PROFILES FOR UP(M,N) AND VP(M,N), AND PHONY
C     INITIAL PROFILES FOR UPP(M,N) AND VPP(M,N).
C***********************************************************************
      PI = 4.D0*DATAN(1.D0)
      RPI = DSQRT(PI)
      DETA = .025D0
      ETA(1) = - 2.D0*DETA
      ETA(2) = - DETA
      DO 10 M =1,MM
      IF(M.GT.29) DETA = 0.0125D0
C     IF(M.GT.33) DETA = .00625D0
      IF(M.GE.3) ETA(M) = ETA(M-1) + DETA
      DO 20 N = 1,NN
      Z(N) = (N-1)*H
      U(M,N) = DERF(.5D0*Z(N))
      V(M,N) = U(M,N)
      UP(M,N) = U(M,N)
      VP(M,N) = U(M,N)
      UPP(M,N) = U(M,N)
      VPP(M,N) = U(M,N)
      R(M,N) = DEXP(-Z(N)**2/4.D0)/RPI
```

```
               S(M,N)  =  R(M,N)
               W(M,N)  =  1.D0+CP*(1.D0-2.D0*ETA(M))
               W(M,N)  =  W(M,N)*(Z(N)*U(M,N)-2.D0*(1.D0/RPI-R(M,N)))
  20       CONTINUE
  10       CONTINUE

C***********************************************************************
C  SET UP INITIAL PROFILES OF UE,VE,WE.
C***********************************************************************

           DO 15 M = 1,2
           DO 25 N = 1,NN
               UEP(M,N) = 0.D0
               VEP(M,N) = 0.D0
  25       CONTINUE
  15       CONTINUE

C***********************************************************************
C      SET CONSTANT COEFFICIENTS, AND CONSTANT PARTS OF OTHERS.
C***********************************************************************

               A(1,2)  =  0.D0
               A(1,3)  =  0.D0
               A(2,1)  =  0.D0
               B(2,1)  =  0.D0
               A(2,3)  =  0.D0
               A(3,1)  =  .5D0
               B(3,1)  =  A(3,1)
               A(3,3)  =   HI
               B(3,3)  =  - A(3,3)

C***********************************************************************
C      SPECIFY WALL BOUNDARY CONDITION IN TERMS OF E(1) AND F(1).
C***********************************************************************

           DO 30 I = 1,3
           DO 31 J = 1,2
               E(I,J,1) = 0.D0
  31       CONTINUE
               F(I,1) = 0.D0
  30       CONTINUE

C***********************************************************************
C  SET THE NUMBER OF TIME STEPS, AND START THE MARCHING LOOP OVER TAU.
C***********************************************************************

               NT = 190
               WRITE(6,1) NN, MM, CP
C              WRITE (6,2)
               IFL = 0
               DTAU = .10D0
               DAT = .1D0
           DO 1000 L = 1,NT
               GO TO 1001
               IF(L.GT.30) DTAU = .0100 D0
               IF(L.EQ.31) IFL = 1
               IF(L.GT.40) DTAU = .0050 D0
               IF(L.EQ.41) IFL = 1
               IF(L.GT.60) DTAU = .0025 D0
               IF(L.EQ.61) IFL = 1
               IF(L.GT.80) DTAU = .00125D0
               IF(L.EQ.81) IFL = 1
               IF(L.GT.120) DTAU = .000625 D0
               IF(L.EQ.121) IFL = 1
  1001         AT = L*DAT
               T = 1.D0 - DEXP(-AT)

               TAU = AT/(AT+1.D0)
           DO 1500 K = 1,ITMAX
```

```
C*************************************************************************
C   CALCULATE U, V, AND W, AND THEIR ETA-DERIVATIVES AT END POINTS.
C*************************************************************************


        DO 1750 M = 3,MSTOP,MN
           IE = 0
1725       GAM3 = T
           DELX1 = T
           DELY2 = CP*T*(1.D0-2.D0*ETA(M))
           VI = CP*ETA(M)*(1.D0-ETA(M))
           PSI = 0.D0
        IF(IE.EQ.1) PSI = DELY2
           EPS = .5D0*(1.D0-T)
           THET = T
           G2 = T
           DZDZET = DSQRT(G2)
        CALL THOMAS
        IF(IE.EQ.1) GO TO 33
        DO 32 N = 1,NN
           U(M,N) = UT(M,N)
           V(M,N) = VT(M,N)
           W(M,N) = WT(M,N)
           IF(N.EQ.1.OR.N.EQ.NN) GO TO 32
           R(M,N) = 0.5D0*HI*(UT(M,N+1)-UT(M,N-1))
           S(M,N) = 0.5D0*HI*(VT(M,N+1)-VT(M,N-1))
32      CONTINUE
           R(3,1) = HI*U(3,2)+0.5D0*H*T
           S(3,1) = HI*V(3,2)+0.5D0*H*CP*T
           R(MSTOP,1) = HI*U(MSTOP,2)+0.5D0*H*T
           S(MSTOP,1) = HI*V(MSTOP,2)-0.5D0*H*CP*T
           IE = 1
           GO TO 1725
33      DO 34 N = 1,NN
           UE(1,N) = UT(3,N)
           VE(1,N) = VT(3,N)
           UE(2,N) = UT(MSTOP,N)
           VE(2,N) = VT(MSTOP,N)
34      CONTINUE
35      CONTINUE
           RE1 = HI*UE(1,2)+.5D0*H*T
           SE1 = HI*VE(1,2)+.5D0*H*CP*T
           RE2 = HI*UE(2,2)+.5D0*H*T
           SE2 = HI*VE(2,2)-.5D0*H*CP*T
1750    CONTINUE
           IE = 0

C*************************************************************************
C    START THE MARCHING LOOP OVER ETA.
C*************************************************************************

           DETA = 0.025D0
           IML = 0
        DO 2000 M = 4,MSTOP-1
           IF(M.GT.29) DETA = 0.0125D0
           IF(M.EQ.30) IML = 1
C          IF(M.GT.33) DETA = .00625D0
C          IF(M.EQ.34) IML = 1

C*************************************************************************
C    EVALUATE COEFFICIENTS THAT DEPEND ONLY ON ETA AND TAU.
C*************************************************************************
```

```
                  GAM3 = T
                  DELX1 = T
                  DELY2 = CP*T*(1.D0-2.D0*ETA(M))
                  VI = CP*ETA(M)*(1.D0-ETA(M))
                  PSI = VI*T
                  EPS = .5D0*(1.D0-T)
                  THET =  T
                  G2 = T
                  DZDZET = DSQRT(G2)

C*****************************************************************************
C     INVOKE THE THOMAS ALGORITHM.
C*****************************************************************************

                  CALL THOMAS
                  DEL1(M) = (1.D0+CP*(1.D0-2.D0*ETA(M)))*Z(NM) - WT(M,NM)

2000  CONTINUE
                  IF(K.LT.ITMAX) GO TO 2001
                  WRITE(7,7)AT,TAU,DEL1(13),DEL1(23),DEL1(37),DEL1(41),DEL1(49)
2001  CONTINUE
C*****************************************************************************
C     UPDATE U, V, W, R, AND S, PREPARING FOR NEXT ITERATION.
C*****************************************************************************

                  DO 50 M = 4,MSTOP-1
                  DELX1 = T
                  DELY2 = CP*T*(1.D0-2.D0*ETA(M))
                  DO 40 N = 1,NN
                  U(M,N) = UT(M,N)
                  V(M,N) = VT(M,N)
                  W(M,N) = WT(M,N)
                  IF(N.EQ.1.OR.N.EQ.NN) GO TO 40
                  R(M,N) = 0.5D0*HI*(UT(M,N+1)-UT(M,N-1))
                  S(M,N) = 0.5D0*HI*(VT(M,N+1)-VT(M,N-1))
40    CONTINUE
                  R(M,1) = HI*U(M,2)+0.5D0*H*DELX1
                  S(M,1) = HI*V(M,2)+0.5D0*H*DELY2
50    CONTINUE
1500  CONTINUE


C*****************************************************************************
C     UPDATE UP,VP,UPP,VPP, PREPARING FOR NEXT TIME STEP.
C*****************************************************************************

                  DO 70 M = 3,MSTOP
                  DO 60 N = 1,NN
                  UPP(M,N) = UP(M,N)
                  VPP(M,N) = VP(M,N)
                  UP(M,N) = U(M,N)
                  VP(M,N) = V(M,N)
60    CONTINUE
                  IF(L.EQ.10) GO TO 65
                  IF(L.EQ.40) GO TO 65
                  IF(L.EQ.90) GO TO 65
                  IF(L.EQ.123) GO TO 65
                  IF(L.EQ.190) GO TO 65
C                  IF(L.EQ.120) GO TO 65
C                  IF(L.EQ.200) GO TO 65
                  GO TO 70
65                DEL1(M) = (1.D0+CP*(1.D0-2.D0*ETA(M)))*Z(NM) - W(M,NM)
                  WRITE(6,3) TAU, ETA(M), R(M,1), S(M,1), DEL1(M)
70    CONTINUE
                  DO 80 M = 1,2
                  DO 75 N = 1,NN
                  UEPP(M,N) = UEP(M,N)
```

```
              VEPP (M,N) = VEP (M,N)
              UEP (M,N) = UE (M,N)
              VEP (M,N) = VE (M,N)
75         CONTINUE
80         CONTINUE

           IF(L.EQ.10) GO TO 85
           IF(L.EQ.40) GO TO 85
           IF(L.EQ.90) GO TO 85
           IF(L.EQ.123) GO TO 85
           IF(L.EQ.190) GO TO 85
C          IF(L.EQ.120) GO TO 85
C          IF(L.EQ.200) GO TO 85
           GO TO 1000
85         WRITE(6,6) TAU, RE1, SE1, RE2, SE2
1000    CONTINUE

C*************************************************************************
C     WRITE PROFILES.
C*************************************************************************

           M = 37
           WRITE (7,4) TAU,M
           R(M,NN) = 0.D0
           S(M,NN) = 0.D0
        DO 700 N = 1,NN
           WRITE (7,5) Z(N),U(M,N),V(M,N),W(M,N),R(M,N),S(M,N)
700     CONTINUE


1       FORMAT(/,10X,'NN =',I3,10X,'MN =',I3,10X,'C =',F10.5,/)
2       FORMAT(/,5X,'TAU',7X,'ETA',8X,'R(1)',7X,'S(1)',7X,'DEL1',/)
3       FORMAT(5F10.5)
4       FORMAT(/,5X,'TAU =',F10.5,5X,'M = ',I3,/,5X,'Z',7X,
       1 'U',9X,'V',9X,'W',9X,'R',9X,'S',/)
5       FORMAT(6F10.5)
6       FORMAT(5F10.5)
7       FORMAT(7F10.5)
        STOP
        END


        SUBROUTINE THOMAS
        IMPLICIT DOUBLE PRECISION (A-H,O-Z)
        COMMON/ALG/Z(161),U(85,161),V(85,161),W(85,161),UP(85,161),
       1  VP(85,161),UPP(85,161),VPP(85,161),R(85,161),S(85,161),
       2  UT(85,161),VT(85,161),WT(85,161),UE(2,161),VE(2,161),WE(2,161),
       3  UEP(2,161),VEP(2,161),UEPP(2,161),VEPP(2,161),ETA(85)
        COMMON/SOLN/E(3,2,161),F(3,161),A(3,3),B(3,3),D(3)
        COMMON/COEF/GAM3,DELX1,DELY2,EPS,PSI,THET,DETA,DTAU,H,HI,HI2,
       1  CP,T,VI
        COMMON/INTEG/M,L,NN,NM,MM,IFL,IE,MSTOP,IML
        DIMENSION G(3,3)

C*************************************************************************
C     SET VALUES Oi AA, ETC. ACCORDING TO VALUE OF L.
C*************************************************************************

           IF(IML.EQ.0) GO TO 10
           AAA = (4.D0/3.D0)/DETA
           BBB = - 1.5D0/DETA
           CCC = (1.D0/6.D0)/DETA
           GO TO 11
10         AAA = 1.5D0/DETA
           BBB = -2.D0/DETA
           CCC = .5D0/DETA
11         IF(L.EQ.1)GO TO 2
```

```
          IF(IFL.EQ.1)GO TO 1
          AA = 1.5D0/DTAU
          BB = - 2.D0/DTAU
          CC = .5D0/DTAU
          GO TO 3
1         AA = (4.D0/3.D0)/DTAU
          BB = - 1.5D0/DTAU
          CC = (1.D0/6.D0)/DTAU
          GO TO 3
2         AA = 1.D0/DTAU
          BB = - 1.D0/DTAU
          CC = 0.D0
3     CONTINUE

C**********************************************************************
C     TRIANGULARIZE.
C**********************************************************************

      DO 200 N = 2,NM
          NM1 = N-1

C**********************************************************************
C SET UP ALL QUANTITIES THAT ARE INDEPENDENT OF THE SIGN OF U.
C**********************************************************************

          X = GAM3*W(M,N) + EPS*Z(N)
          A(1,1) = HI2 -.5D0*HI*X
          A(2,2) = A(1,1)
          A(3,2) = .5D0*CP*(1.D0-2.D0*ETA(M))
          B(3,2) = A(3,2)
          B(1,2) = 0.D0
          B(1,3) = GAM3*R(M,N)
          B(2,3) = GAM3*S(M,N)
          CT = HI2 + .5D0*HI*X
          B(1,1) = -2.D0*HI2 - 2.D0*DELX1*U(M,N) - AA*THET
          B(2,2) = -2.D0*HI2 - 2.D0*DELY2*V(M,N) - AA*THET
      IF(IE.EQ.1) GO TO 4
          D(1) = GAM3*W(M,N)*R(M,N)-DELX1*(1.D0+U(M,N)**2)
     1    + THET*(BB*UP(M,N) + CC*UPP(M,N))
          D(2) = GAM3*W(M,N)*S(M,N)-DELY2*(1.D0+V(M,N)**2)
     1    + THET*(BB*VP(M,N) + CC*VPP(M,N))
          D(3) = 0.D0
      GO TO 5

C**********************************************************************
C     SPECIAL COEFFICIENTS FOR ENDPOINT DERIVATIVES.
C**********************************************************************

4         A(3,2) = 2.D0*A(3,2)
          B(3,2) = A(3,2)
          B(1,1) = B(1,1) - PSI*V(M,N)
          B(2,2) = B(2,2) - PSI*V(M,N)
          IF(M.EQ.3) K = 1
          IF(M.EQ.MSTOP) K = 2
          D(1) = THET*(BB*UEP(K,N) + CC*UEPP(K,N))
          D(2) = THET*(BB*VEP(K,N) + CC*VEPP(K,N)) + 2.D0*CP*T*
     1    (1.D0 - V(M,N)**2)
          D(3) = CP*(V(M,N) + V(M,NM1))
      GO TO 9

C**********************************************************************
C     ESTABLISH UPWIND-DIFFERENCING FORMULAS.
C**********************************************************************

5     CONTINUE
      IF(V(M,N).LT.0.D0) GO TO 7
```

```
      IF(M.EQ.4) GO TO 6
          B(1,1) = B(1,1) - PSI*AAA*V(M,N)
          B(2,2) = B(2,2) - PSI*(2.D0*AAA*V(M,N) + BBB*V(M-1,N)
     1            + CCC*V(M-2,N))
          B(1,2) = - PSI*(AAA*U(M,N) + BBB*U(M-1,N) + CCC*U(M-2,N))
          A(3,2) = A(3,2) + .5D0*AAA*VI
          B(3,2) = A(3,2)
          D(1) = D(1) - PSI*AAA*U(M,N)*V(M,N)
          D(2) = D(2) - PSI*AAA*V(M,N)**2
          D(3) = -.5D0*VI*(BBB*(V(M-1,N)+V(M-1,NM1))
     1            + CCC*(V(M-2,N)+ V(M-2,NM1)))
          GO TO 9

6         B(1,1) = B(1,1) + PSI*BBB*V(M,N)
          B(2,2) = B(2,2) + PSI*(BBB*(2.D0*V(M,N) - V(M-1,N))+ VE(1,N))
          B(1,2) = PSI*(BBB*(U(M,N) - U(M-1,N)) + UE(1,N))
          A(3,2) = A(3,2) - .5D0*BBB*VI
          B(3,2) = A(3,2)
          D(1) = D(1) + PSI*BBB*U(M,N)*V(M,N)
          D(2) = D(2) + PSI*BBB*V(M,N)**2
          D(3) = -.5D0*VI*(BBB*(V(M-1,N)+V(M-1,NM1))
     1            - (VE(1,N) + VE(1,NM1)))
          GO TO 9

7     IF(M.EQ.MSTOP-1) GO TO 8
          B(1,1) = B(1,1) + PSI*AAA*V(M,N)
          B(2,2) = B(2,2) + PSI*(2.D0*AAA*V(M,N) + BBB*V(M+1,N)
     1            + CCC*V(M+2,N))
          B(1,2) = + PSI*(AAA*U(M,N) + BBB*U(M+1,N) + CCC*U(M+2,N))
          A(3,2) = A(3,2) - .5D0*AAA*VI
          B(3,2) = A(3,2)
          D(1) = D(1) + PSI*AAA*U(M,N)*V(M,N)
          D(2) = D(2) + PSI*AAA*V(M,N)**2
          D(3) = +.5D0*VI*(BBB*(V(M+1,N)+V(M+1,NM1))
     1            + CCC*(V(M+2,N)+ V(M+2,NM1)))
          GO TO 9

8         B(1,1) = B(1,1) - PSI*BBB*V(M,N)
          B(2,2) = B(2,2) + PSI*(BBB*(V(M+1,N)- 2.D0*V(M,N))+ VE(2,N))
          B(1,2) = PSI*(BBB*(U(M+1,N) - U(M,N)) + UE(2,N))
          A(3,2) = A(3,2) + .5D0*BBB*VI
          B(3,2) = A(3,2)
          D(1) = D(1)'- PSI*BBB*U(M,N)*V(M,N)
          D(2) = D(2) - PSI*BBB*V(M,N)**2
          D(3) = .5D0*VI*(BBB*(V(M+1,N)+V(M+1,NM1))
     1            + (VE(2,N) + VE(2,NM1)))
9     CONTINUE

C*********************************************************************
C     FORM THE MATRIX G.
C*********************************************************************

      DO 20 I = 1,3
      DO 21 J = 1,2
          G(I,J) = 0.D0
      DO 22 K = 1,3
          G(I,J) = G(I,J) + A(I,K)*E(K,J,NM1)
22    CONTINUE
          G(I,J)=G(I,J) + B(I,J)
21    CONTINUE
20    CONTINUE
          G(1,3) = B(1,3)
          G(2,3) = B(2,3)
          G(3,3) = B(3,3)

          DD1 = D(1) - A(1,1)*F(1,NM1)
```

```
          DD2 = D(2) - A(2,2)*F(2,NM1)
          DD3 = D(3) - A(3,1)*F(1,NM1) - A(3,2)*F(2,NM1)
     1        - A(3,3)*F(3,NM1)
C***************************************************************************
C      SOLVE FOR E(I,J,N) AND F(I,N) BY KRAMER'S RULE.
C***************************************************************************

          H1 = G(2,2)*G(3,3)-G(2,3)*G(3,2)
          H2 = G(2,3)*G(3,1)-G(2,1)*G(3,3)
          H3 = G(2,1)*G(3,2)-G(2,2)*G(3,1)
          H4 = G(3,2)*G(1,3)-G(1,2)*G(3,3)
          H5 = G(1,1)*G(3,3)-G(1,3)*G(3,1)
          H6 = G(1,2)*G(3,1)-G(3,2)*G(1,1)
          H7 = G(1,2)*G(2,3)-G(2,2)*G(1,3)
          H8 = G(2,1)*G(1,3)-G(2,3)*G(1,1)
          H9 = G(1,1)*G(2,2)-G(1,2)*G(2,1)
          DENO = G(1,1)*H1+G(1,2)*H2+G(1,3)*H3

          E(1,1,N) = -CT*H1/DENO
          E(2,1,N) = -CT*H2/DENO
          E(3,1,N) = -CT*H3/DENO
          E(1,2,N) = -CT*H4/DENO
          E(2,2,N) = -CT*H5/DENO
          E(3,2,N) = -CT*H6/DENO
          F(1,N) = (DD1*H1+DD2*H4+DD3*H7)/DENO
          F(2,N) = (DD1*H2+DD2*H5+DD3*H8)/DENO
          F(3,N) = (DD1*H3+DD2*H6+DD3*H9)/DENO
200       CONTINUE

C***************************************************************************
C      SET OUTER BOUNDARY CONDITION.
C***************************************************************************

          UT(M,NN) = 1.D0
          VT(M,NN) = 1.D0
          IF(IE.EQ.1) UT(M,NN) = 0.D0
          IF(IE.EQ.1) VT(M,NN) = 0.D0

C***************************************************************************
C      BACK-SUBSTITUTE.
C***************************************************************************

          DO 300 N = 1,NM
          K = NN-N
          UT(M,K) = E(1,1,K)*UT(M,K+1)+E(1,2,K)*VT(M,K+1)+F(1,K)
          VT(M,K) = E(2,1,K)*UT(M,K+1)+E(2,2,K)*VT(M,K+1)+F(2,K)
          WT(M,K) = E(3,1,K)*UT(M,K+1)+E(3,2,K)*VT(M,K+1)+F(3,K)
300       CONTINUE
          WT(M,NN) = 1.D0
          IFL = 0
          IML = 0
      RETURN
      END
```

B15 EIGEN

```
C*********************************************************************
C  THIS PROGRAM SEARCHES FOR EIGENVALUES OF THE RAYLEIGH EQUATION BY
C  USE OF THE THOMAS ALGORITHM.  THE BASIC FLOW IS THE ERROR-FUNCTION
C  SHEAR LAYER.
C*********************************************************************

      PROGRAM EIGEN
      IMPLICIT DOUBLE PRECISION (A-H,O-Z)
      DIMENSION ETA(2561),U(2561),UPP(2561)
      COMPLEX*16 B,C,CNEW,COLD,CORR,TARG,E(2561),F(2561),VOR(2561)

C*********************************************************************
C  ASSIGN VALUES TO ETAMAX, ETAMIN.  EVALUATE GENERAL CONSTANTS.
C*********************************************************************

         ETAMAX = 4.D0
         ETAMIN = -4.D0
         NN = 641
         NE = 199
         NM = NN-1
         NMM = NM/2 + 1
         RN = NM
         DN = (ETAMAX-ETAMIN)/RN
         PI = 2.D0*DASIN(1.D0)
         SPI = DSQRT(PI)

C*********************************************************************
C   PICK WAVENUMBER. SET E(1),TARGET,AND ERROR TOLERANCE.
C*********************************************************************

         ITMAX = 12
         EPS = .00000001D0
         GAM = .85D0

C*********************************************************************
C  CALCULATE BASIC-FLOW QUANTITIES
C*********************************************************************

      DO 100 N = 1,NMM
         ETA(N) = (N-1)*DN + ETAMIN
         U(N) = DERF(.5D0*SPI*ETA(N))
         UPP(N) =  -.5D0*PI*ETA(N)/DEXP(.25D0*PI*ETA(N)**2)
C        WRITE(6,4)ETA(N),U(N),UPP(N)
100   CONTINUE

C*********************************************************************
C  START ITERATION LOOP WHICH SEEKS AN EIGENVALUE BY THE SECANT METHOD.
C  ENTER FIRST GUESS OF C.
C*********************************************************************

C     DO 20  M = 1,18
         ALF = .50D0
         WRITE (6,1) ALF
         BETI = 1.D0/DEXP (ALF*DN)
         E(1) = DCMPLX(BETI,0.D0)
         ITER = 1
         C = DCMPLX(0.0D0,.93D0-ALF)

11    DO 200 N = 2,NMM
         B = -2.D0 - DN**2*(ALF**2 + UPP(N)/(U(N)-C))
         E(N) = -1.D0/(B+E(N-1))
200   CONTINUE

         ERR = REAL(2.D0*E(NMM-1)+ B)
         WRITE(6,2) ITER, C, ERR

         IF (ITER.GT.1) GO TO 15
         COLD = C
         ERROLD = ERR
```

```
C**********************************************************************
C  ENTER SECOND GUESS OF C.
C**********************************************************************

          C = .95D0*C
          ITER = 2
          GO TO 11

C**********************************************************************
C  CONTINUE ITERATIONS TO MEET CONVERGENCE TEST, OR TO MAXIMUM LIMIT.
C**********************************************************************

15        IF(DABS(ERR).LT.EPS) GO TO 20
          CORR = - ERR*(C - COLD)/(ERR - ERROLD)
          CNEW = C + GAM*CORR
          COLD = C
          ERROLD = ERR
          C = CNEW
          ITER = ITER + 1
          IF (ITER.EQ.ITMAX)GO TO 20
          GO TO 11
20        CONTINUE

C**********************************************************************
C    SPECIFY F(NMM) AND CALCULATE EIGENFUNCTIONS.
C**********************************************************************

          F(NMM) = DCMPLX(1.D0,0.D0)

      DO 300 N = 1,NMM
          L = NMM-N
          LP = L+1
          F(L) = E(L)*F(LP)
300       CONTINUE

C**********************************************************************
C  PREPARE AND PRINT EIGENFUNCTIONS FOR STREAMFUNCTION AND VORTICITY,
C  BOTH BY REAL AND IMAGINARY PARTS, AND BY AMPLITUDE AND PHASE ANGLE.
C**********************************************************************

          WRITE(6,6)
      DO 400 N = 2,NMM-1
          VOR(N) = ALF**2* F(N) - (F(N+1)-2.D0*F(N)+F(N-1))/DN**2
          WRITE(6,5)ETA(N),F(N),VOR(N)
400       CONTINUE

          WRITE(6,7)
      DO 500 N = 2,NMM-1
          ABF = CDABS (F(N))
          PHF = DATAN(IMAG(F(N))/REAL(F(N)))
          ABV = CDABS(VOR(N))
          PHV = DATAN(IMAG(VOR(N))/REAL(VOR(N)))
          WRITE (6,5) ETA(N), ABF,PHF,ABV,PHV
500       CONTINUE

C**********************************************************************
C  FORMAT PRINTOUTS.
C**********************************************************************

1         FORMAT (/,5X,'EIGENVALUE SEARCH FOR ALPHA =', F10.5,5X,/)
2         FORMAT (2X,'ITER =',I3,2X,'C =',2F10.5,2X,'ERR =',2E13.5)
3         FORMAT (2X,I4,2X,2E13.5)
4         FORMAT (6X,3F10.5)
5         FORMAT(5(3X,F10.5))
6         FORMAT (/,7X,'ETA',7X,'FR',8X,'FI',8X,'VORR',6X,'VORI',/)
7         FORMAT(/,7X,'ETA',7X,'ABF',7X,'PHF',7X,'ABV',7X,'PHV',/)
          STOP
          END
```

B16 TAYLOR

```
C*********************************************************************
C     THIS PROGRAM SOLVES THE EIGENVALUE PROBLEM FOR THE ONSET OF TAYLOR
C     VORTICES IN THE NARROW-GAP, CO-ROTATING APPROXIMATION.
C*********************************************************************

      PROGRAM TAYLOR
      IMPLICIT DOUBLE PRECISION (A-H,O-Z)

C*********************************************************************
C     ASSIGN STORAGE FOR V,R,S, AND E.
C*********************************************************************

      DIMENSION ETA(201),S(201),V(201),U(201),E(3,3,201),W(201)
      DIMENSION B(3,3),F(3,3),G(3,3)

C*********************************************************************
C     SPECIFY NUMBER OF MESH POINTS IN ETA.
C*********************************************************************

          NN = 201
          NM = NN-1
          RN = NM
          ZMAX = .5D0
          H = ZMAX/RN
          HI = 1.D0/H
          HI2 = HI*HI

C*********************************************************************
C     SET UP ETA(N).
C*********************************************************************

      DO 100 N = 1,NN
          ETA(N) = -ZMAX + (N-1)*H
100   CONTINUE

C*********************************************************************
C     SET CONSTANT COEFFICIENTS.
C*********************************************************************

          B(1,3) = 0.D0
          B(2,1) = 0.D0
          B(3,1) = - H**2

C*********************************************************************
C     SET ITERATION PARAMETERS.
C*********************************************************************

          ITMAX = 10
          EPS = .000000001D0
          GAM = 1.0D0

C*********************************************************************
C     PICK VALUE OF WAVENUMBER.
C*********************************************************************

          ALF = 3.117D0
          WRITE (6,1) ALF, H
          B(1,1) = - (2.D0+(ALF*H)**2)
          B(2,2) = B(1,1)
          B(3,3) = B(1,1)

C*********************************************************************
C     SPECIFY WALL BOUNDARY CONDITION IN TERMS OF E(1) AND F(1).
C*********************************************************************

      DO 10 I = 1,3
```

```
        DO 11 J = 1,2
          E(I,J,1) = 0.D0
11      CONTINUE
10      CONTINUE
          E(1,3,1) = 2.D0*HI2

C**********************************************************************
C    MAKE FIRST GUESS AT EIGENVALUE OF REYNOLDS NUMBER.
C**********************************************************************

          RE = 30.D0
          ITER = 1

C**********************************************************************
C    TRIANGULARIZE.
C**********************************************************************

2000  DO 200 N = 2,NN
          B(1,2) = - 2.D0*(ALF*H)**2*RE
          B(2,3) =  H**2*RE
          NM1 = N-1

C**********************************************************************
C    FORM THE MATRIX G.
C**********************************************************************

        DO 20 I = 1,3
        DO 21 J = 1,3
          G(I,J) = B(I,J)+ E(I,J,NM1)
21      CONTINUE
20      CONTINUE


C**********************************************************************
C    INVERT G BY USE OF KRAMER'S RULE.
C**********************************************************************

        H1 = G(2,2)*G(3,3)-G(2,3)*G(3,2)
        H2 = G(2,3)*G(3,1)-G(2,1)*G(3,3)
        H3 = G(2,1)*G(3,2)-G(2,2)*G(3,1)
        H4 = G(3,2)*G(1,3)-G(1,2)*G(3,3)
        H5 = G(1,1)*G(3,3)-G(1,3)*G(3,1)
        H6 = G(1,2)*G(3,1)-G(3,2)*G(1,1)
        H7 = G(1,2)*G(2,3)-G(2,2)*G(1,3)
        H8 = G(2,1)*G(1,3)-G(2,3)*G(1,1)
        H9 = G(1,1)*G(2,2)-G(1,2)*G(2,1)
        DENO = G(1,1)*H1+G(1,2)*H2+G(1,3)*H3

        E(1,1,N) = - H1/DENO
        E(2,1,N) = - H2/DENO
        E(3,1,N) = - H3/DENO
        E(1,2,N) = - H4/DENO
        E(2,2,N) = - H5/DENO
        E(3,2,N) = - H6/DENO
        E(1,3,N) = - H7/DENO
        E(2,3,N) = - H8/DENO
        E(3,3,N) = - H9/DENO

200     CONTINUE

C**********************************************************************
C    TEST FOR EIGENVALUE.
C**********************************************************************

        DO 25 I = 1,3
        DO 24 J = 1,3
```

```
          F(I,J) = B(I,J) + 2.D0*E(I,J,NM)
24    CONTINUE
25    CONTINUE
          ERR = F(1,1)*(F(2,2)*F(3,3)-F(2,3)*F(3,2)) + F(1,2)*(F(2,3)*
     1    F(3,1)-F(2,1)*F(3,3))+F(1,3)*(F(2,1)*F(3,2)-F(2,2)*F(3,1))
          ERR = ERR*HI2
          WRITE(6,2) ITER, RE, ERR
          IF (ITER.GT.1) GO TO 15
          REOLD = RE
          ERROLD = ERR

C*************************************************************************
C ENTER SECOND GUESS OF RE.
C*************************************************************************

          RE = .95D0*RE
          ITER = 2
          GO TO 2000

C*************************************************************************
C CONTINUE ITERATIONS TO MEET CONVERGENCE TEST, OR TO MAXIMUM LIMIT.
C*************************************************************************

15        IF(DABS(ERR).LT.EPS) GO TO 40
          CORR = - ERR*(RE - REOLD)/(ERR - ERROLD)
          RENEW = RE + GAM*CORR
          REOLD = RE
          ERROLD = ERR
          RE = RENEW
          ITER = ITER + 1
          IF (ITER.EQ.ITMAX)GO TO 40
          GO TO 2000
40    CONTINUE

C*************************************************************************
C     SET AMPLITUDE OF V. CALCULATE AMPLITUDES OF R AND S.
C*************************************************************************

          V(NN) = 1.D0
          DENO = F(1,1)*F(2,3) - F(1,3)*F(2,1)
          S(NN) = -(F(1,2)*F(2,3)-F(1,3)*F(2,2))/DENO
          U(NN) = -(F(1,1)*F(2,2)-F(1,2)*F(2,1))/DENO

C*************************************************************************
C     BACK-SUBSTITUTE.
C*************************************************************************

      DO 300 N = 1,NM
          K = NN-N
          S(K) = E(1,1,K)*S(K+1)+E(1,2,K)*V(K+1)+E(1,3,K)*U(K+1)
          V(K) = E(2,1,K)*S(K+1)+E(2,2,K)*V(K+1)+E(2,3,K)*U(K+1)
          U(K) = E(3,1,K)*S(K+1)+E(3,2,K)*V(K+1)+E(3,3,K)*U(K+1)
300   CONTINUE


C*************************************************************************
C     PRINT PROFILES.
C*************************************************************************

650       WRITE (6,4)
      DO 700 N = 1,NN,2
          W(N) = (U(N+1)-U(N-1))/(2.D0*ALF*H)
          W(1) = 0.D0
          W(NN) = 0.D0
          WRITE (6,5) ETA(N),W(N),V(N),U(N)
700   CONTINUE
```

```
41      CONTINUE
1000    CONTINUE

1       FORMAT(/,10X,'ALF =',F10.5,10X,'H =',F10.5,/)
2       FORMAT(/,10X,'ITER = ',I4,5X,'RE =',E13.8,5X,'ERR =',E13.5)
4       FORMAT(/,5X,'X',9X,'W',9X,'V',9X,'U',/)
5       FORMAT(4F10.5)
6       FORMAT (2X,F10.5,3(3X,E15.8))
        STOP
        END
```

AUTHOR INDEX

SUBJECT INDEX